IB DIPLOMA PROGRAMME

Mathematical Studies

Course Companion

Stephen Bedding
Jane Forrest
Paula Waldman de Tokman
Beryl Fussey
Mal Coad

OXFORD
UNIVERSITY PRESS

IN COOPERATION WITH ib

OXFORD
UNIVERSITY PRESS

Great Clarendon Street, Oxford OX2 6DP

Oxford University Press is a department of the University of Oxford.
It furthers the University's objective of excellence in research, scholarship,
and education by publishing worldwide in

Oxford New York

Auckland Cape Town Dar es Salaam Hong Kong Karachi
Kuala Lumpur Madrid Melbourne Mexico City Nairobi
New Delhi Shanghai Taipei Toronto

With offices in

Argentina Austria Brazil Chile Czech Republic France Greece
Guatemala Hungary Italy Japan Poland Portugal Singapore
South Korea Switzerland Thailand Turkey Ukraine Vietnam

First published 2007

British Library Cataloguing in Publication Data

Data available

ISBN-978-0199151219

10 9 8 7 6 5

Printed in China by Printplus

Acknowledgements
The publisher would like to thank the following for their kind permission to reproduce copyright material:

Cover image © D. Hurst/Alamy;

We are grateful for permission to reprint the following

p 25:cartoon, copyright © 1997 Randy Glasbergen, reprinted with permission of the author, www. glasbergen. com; p 82: FoxTrot cartoon, copyright © 2001 Bill Amend, reprinted with permission of Universal Press Syndicate. All rights reserved.; p 187: Close to Home cartoon, copyright © 1997 John McPherson, reprinted with permission of Universal Press Syndicate. All rights reserved.; p 296: cartoon, copyright © 2003 Randy Glasbergen, reprinted with permission of the author, www. glasbergen.com; p 307: cartoon, copyright © 2002 Randy Glasbergen, reprinted with permission of the author, www. glasbergen.com; p 349: Calvin and Hobbes cartoon, copyright © 1994 Bill Watterson, distributed by Universal Press Syndicate, and reprinted with permission. All rights reserved; p 401: cartoon, copyright © Harly Schwadron, reprinted with permission of www.CartoonStock.com

P8: Oxford University Press; P 61: Corbis UK Ltd; p 71: Wikimedia/Oxford University Press; p 75: Oxford University Press: p 83: Science Photo Library: p 90: Oxford University Press; p 92: Oxford University Press; p 124: Silvia Salmi/Bettmann/Corbis UK Ltd; p 126: Oxford University Press; p 132: Oxford University Press; p 135l: Corbis UK Ltd; p 135r: Oxford University Press; p 149: Oxford University Press; p 150: Yevgeny Khaldei/Corbis UK Ltd; p 204: Wikimedia/Oxford University Press; p 230: Science Photo Library; p 267t: Danita Delimont/Alamy; p 267m: Oxford University Press; p 272: Kadu Niemeyer/Arcaid/Alamy; p 284: Oxford University Press (both); p 290: Oxford University Press; p 309: Mediacolor's/Alamy; p 317: Science Photo Library; p 325: Oxford University Press; p 349t: Wikimedia/Oxford University Press; p 349b: Bettmann/Corbis UK Ltd; p 382: Torsten Blackwood/AFP/Getty Images.

Artwork is by Barking Dog, Stefan Chabluk and Tim Kahane.

Biographies

Beryl Fussey has been teaching mathematics for over thirty years. She teaches at Dartford Grammar School, UK and is Deputy Chief Examiner for IB Diploma Programme mathematical studies SL.

Jane Forrest studied mathematics at Edinburgh University, Scotland. She is now deputy head of Rotterdam International Secondary School (Netherlands). Jane was deputy chief examiner for IB Diploma Programme mathematical studies SL for five years.

Mal Coad teaches mathematics at Pembroke School in South Australia. He has been an assistant examiner/moderator for IB Diploma Programme mathematical studies SL since 1995, team leader from 1998 and is currently deputy chief examiner for the subject.

Paula Waldman de Tokman has been a mathematics professor at Buenos Aires University (UBA), Argentina and currently teaches at St. Andrew's Scots School. She has been deputy chief examiner for IB Diploma Programme mathematical studies SL since 2004.

Stephen Bedding studied at King's College, London and has taught at the University of Technology in Lae, Papua New Guinea and at La Trobe University in Melbourne. He was chief examiner for IB Diploma Programme mathematical studies SL from 2000 to 2006.

Boyd Roberts has been engaged with the International Baccalaureate for thirty years. He was head of the Amman Baccalaureate School and of St Clare's, Oxford and is consultant for International Mindedness for the Course Companion series.

Manjula Solomon has taught in India, Iran, the US and Indonesia and has held a variety of roles associated with the IBO, most recently as Deputy Chief Assessor for TOK. She is consultant for Theory of Knowledge for the Course Companion series.

Special Thanks
We extend special thanks to our colleague and fellow deputy chief examiner Alison Ryan and to Sarah Jones, one of the subject area managers for group 5 subjects.

These people have earned our highest respect for their competence and professionalism, but they have also become good and trusted friends. We value their friendship highly.

Indeed we have quickly come to value all our personal interactions with the many dedicated staff, examiners and teachers associated with the International Baccalaureate.

Dedication
We dedicate this work to our families and friends in recognition of their constant support and encouragement. They have had to be very tolerant both during the writing of this book and in coping with our many journeys and extended periods spent away at IBCA offices and at teacher workshops.

Course Companion definition

The IB Diploma Programme Course Companions are resource materials designed to provide students with extra support through their two-year course of study. These books will help students gain an understanding of what is expected from the study of an IB Diploma Programme subject.

The Course Companions reflect the philosophy and approach of the IB Diploma Programme and present content in a way that illustrates the purpose and aims of the IB. They encourage a deep understanding of each subject by making connections to wider issues and providing opportunities for critical thinking.

These Course Companions, therefore, may or may not contain all of the curriculum content required in each IB Diploma Programme subject, and so are not designed to be complete and prescriptive textbooks. Each book will try to ensure that areas of curriculum that are unique to the IB or to a new course revision are thoroughly covered. These books mirror the IB philosophy of viewing the curriculum in terms of a whole-course approach; the use of a wide range of resources; international-mindedness; the IB learner profile and the IB Diploma Programme core requirements: theory of knowledge; the extended essay; and creativity, action, service (CAS).

In addition, the Course Companions provide advice and guidance on the specific course assessment requirements and also on academic honesty protocol.

The Course Companions are not designed to be:

- study/revision guides or a one-stop solution for students to pass the subjects
- prescriptive or essential subject textbooks.

IB mission statement

The International Baccalaureate aims to develop inquiring, knowledgeable and caring young people who help to create a better and more peaceful world through intercultural understanding and respect.

To this end the IB works with schools, governments and international organizations to develop challenging programmes of international education and rigorous assessment.

These programmes encourage students across the world to become active, compassionate and lifelong learners who understand that other people, with their differences, can also be right.

The IB learner profile

The International Baccalaureate aims to develop internationally minded people who, recognizing their common humanity and shared guardianship of the planet, help to create a better and more peaceful world. IB learners strive to be:

Inquirers They develop their natural curiosity. They acquire the skills necessary to conduct inquiry and research and show independence in learning. They actively enjoy learning and this love of learning will be sustained throughout their lives.

Knowledgeable They explore concepts, ideas and issues that have local and global significance. In so doing, they acquire in-depth knowledge and develop understanding across a broad and balanced range of disciplines.

Thinkers They exercise initiative in applying thinking skills critically and creatively to recognize and approach complex problems, and make reasoned, ethical decisions.

Communicators They understand and express ideas and information confidently and creatively in more than one language and in a variety of modes of communication. They work effectively and willingly in collaboration with others.

Principled They act with integrity and honesty, with a strong sense of fairness, justice and respect for the dignity of the individual, groups and communities. They take responsibility for their own actions and the consequences that accompany them.

Open-minded They understand and appreciate their own cultures and personal histories, and are open to the perspectives, values and traditions of other individuals and communities. They are accustomed to seeking and evaluating a range of points of view, and are willing to grow from the experience.

Caring They show empathy, compassion and respect towards the needs and feelings of others. They have a personal commitment to service, and act to make a positive difference to the lives of others and to the environment.

Risk-takers They approach unfamiliar situations and uncertainty with courage and forethought, and have the independence of spirit to explore new roles, ideas and strategies. They are brave and articulate in defending their beliefs.

Balanced They understand the importance of intellectual, physical and emotional balance to achieve personal well-being for themselves and others.

Reflective They give thoughtful consideration to their own learning and experience. They are able to assess and understand their strengths and limitations in order to support their learning and personal development.

A note on academic honesty

It is of vital importance to acknowledge and appropriately credit the owners of information when that information is used in your work. After all, owners of ideas (intellectual property) have property rights. To have an authentic piece of work, it must be based on your individual and original ideas with the work of others fully acknowledged. Therefore, all assignments, written or oral, completed for assessment must use your own language and expression. Where sources are used or referred to, whether in the form of direct quotation or paraphrase, such sources must be appropriately acknowledged.

How do I acknowledge the work of others?

The way that you acknowledge that you have used the ideas of other people is through the use of footnotes and bibliographies.

Footnotes (placed at the bottom of a page) or endnotes (placed at the end of a document) are to be provided when you quote or paraphrase from another document, or closely summarize the information provided in another document. You do not need to provide a footnote for information that is part of a "body of knowledge". That is, definitions do not need to be footnoted as they are part of the assumed knowledge.

Bibliographies should include a formal list of the resources that you used in your work. 'Formal' means that you should use one of the several accepted forms of presentation. This usually involves separating the resources that you use into different categories (e.g. books, magazines, newspaper articles, Internet-based resources, CDs, and works of art) and providing full information as to how a reader or viewer of your work can find the same information. A bibliography is compulsory in the Extended Essay.

What constitutes malpractice?

Malpractice is behaviour that results in, or may result in, you or any student gaining an unfair advantage in one or more assessment component. Malpractice includes plagiarism and collusion.

Plagiarism is defined as the representation of the ideas or work of another person as your own. The following are some of the ways to avoid plagiarism:

- words and ideas of another person to support one's arguments must be acknowledged
- passages that are quoted verbatim must be enclosed within quotation marks and acknowledged
- CD-ROMs, email messages, web sites on the Internet, and any other electronic media must be treated in the same way as books and journals
- the sources of all photographs, maps, illustrations, computer programs, data, graphs, audio-visual, and similar material must be acknowledged if they are not your own work
- works of art, whether music, film, dance, theatre arts, or visual arts and where the creative use of a part of a work takes place, must be acknowledged.

Collusion is defined as supporting malpractice by another student. This includes:

- allowing your work to be copied or submitted for assessment by another student
- duplicating work for different assessment components and/or diploma requirements.

Other forms of malpractice include any action that gives you an unfair advantage or affects the results of another student. Examples include, taking unauthorized material into an examination room, misconduct during an examination, and falsifying a CAS record.

Contents

Introduction

IB Diploma Programme Mathematical Studies is a standard level course in mathematics designed for candidates who do not expect to have a serious need for mathematics in their future careers. For these candidates (whose strengths often lie in other directions), it is not an easy option. It is however, constructed to give insight into areas of mathematics which might impact on their lives whether they intend this or not.

This book is intended as a companion to students as they embark upon their study of the International Baccalaureate Diploma Programme in mathematics. This book will help clarify what is and is not included in the syllabus and act as a source of questions and examples written in the IB style. The Companion has been designed to facilitate the learning process in a number of ways:

- As a student enjoying the privilege of taking the IB Diploma Programme, you are strongly encouraged to live up to all the ideals of the IB learner profile. However, the study of mathematics is likely to be particularly stimulating in certain areas of the profile. The unique combination of mathematics, TOK and course advice in this book along with the emphasis on international examples encourages the reader to develop an enquiring mind, to think creatively and critically, to learn to communicate in the language of mathematics, to be persistent in grappling with difficult problems, to gain an understanding of how mathematics has helped to shape our societies, and contributed to their history and cultural variety. All of this encourages students to consider the broad picture of mathematics in order to appreciate how it may be used to help guide personal and commercial undertakings in an ethical and peaceful manner.
- There are many worked examples provided. Often these examples are similar in style to questions that the candidates will encounter in examinations. There is also a healthy supply of exercises for the candidate to attempt and a selection of actual past examination questions from both papers. Official IB notation is used throughout the Companion.
- The book contains up to date examination tips and advice coming directly from members of the current, or recent past, senior examining team. This advice reflects the accumulated wisdom acquired from many years of setting, marking and standardizing IB Diploma Programme examinations.
- Extensive advice for approaching the internal assessment project is presented, including discussion of the marking criteria and things to be considered when choosing a topic.
- The book integrates material on Theory of Knowledge, a core element of the IB Diploma Programme model, through the use of *TOK boxes*. These help students to understand that aspects of TOK exist in, and apply to, mathematics but also that TOK ideas which have their origin in mathematics actually permeate all areas of life and learning.
- Topic 1 is devoted entirely to guidance in the use of the graphic display calculator (GDC). The two most commonly used brands, TI and Casio are covered in parallel. Screen dumps have been used within the text of the other topics to alert the reader to the possibility of GDC use at that point and examples from within each topic are worked through with the GDC when appropriate. Readers should appreciate that when they see a screen dump embedded in a piece of text, they are not immediately expected to be able to produce that screen. Rather, the picture is a flag which says *'turn to Topic 1 and investigate the GDC procedures further'*. Note also that in the interest of equity, both TI and Casio screen dumps appear in the text. Casio has been used throughout Topic 4.
- The international nature of mathematics is made evident by the use of examples originating from different countries and cultures. Short histories of mathematical pioneers are included, along with pictures of these people.

BF, JF, MEC, PWT and SPB.
February, 2007.

TOPIC 1

Introduction to the graphic display calculator

The aim of this topic is to introduce the numerical, graphical and listing facilities of the graphic display calculator (GDC).

The topic is divided into sections. Each section corresponds to the topic in the Course Companion with matching numbering. For example, Section 1.4 discusses the use of the GDC for Topic 4, Functions. The reader should study Topic 1 in parallel with the later topics rather than attempting to read the whole topic independently.

The topic addresses the most commonly needed facilities (in mathematical studies) on the Texas Instruments (TI) and Casio GDC's.

It is important to note that many downloadable applications are not permitted in IB Diploma Programme examinations. Before your examinations, your calculator will be reset to default condition. It is your responsibility to make sure that you restore important settings such as DEGREE mode.

Use of the downloadable applications POLYSMLT, XPOLYNOM and SIMULT for the Texas instruments GDC is permitted by the IBO. Equivalent facilities are built in to the Casio.

The TI 84+ is being used to write the left-hand columns of this topic. Most of what is written here is the same for the TI 83 and 83+.

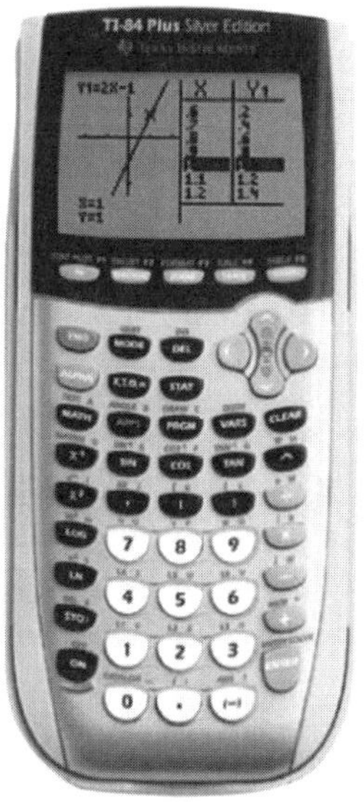

The Casio 9850 and 9860 are being used to write the right-hand columns of this topic. All of what is written here is the same or very similar for both models.

1.1 Basic usage of the GDC including graph drawing and lists

1.1.1 Arithmetic calculations; use of the GDC to graph a variety of functions

TI	Casio
When you turn on the calculator you see a blank workspace used for performing calculations.	When you turn on the calculator you see a menu. Select RUN·MAT mode as the workspace for performing calculations.
The four basic arithmetic operations are on the right of the keyboard. Use the x^2 button for a square and x^{-1} for a reciprocal.	The four basic arithmetic operations are on the right of the keyboard. Use the x^2 button for a square and SHIFT) for a reciprocal.
Other powers are entered using ^.	Other powers are entered using ^.
The three common trigonometric functions have their own buttons SIN, COS, TAN.	The three common trigonometric functions have their own buttons sin, cos, tan.
To enter the blue commands *above* each button, press the 2nd button. For example the square root operation √ is obtained as 2nd x^2. The inverse trigonometric functions and powers of 10 and the special numbers π and *e* are also obtained in this way.	To enter the yellow commands *above* each button, press the SHIFT button. For example the square root operation √ is obtained as SHIFT x^2. The inverse trigonometric functions and powers of 10 and the special numbers π and *e* are also obtained in this way.
Text can be entered using the green ALPHA button to obtain the green alternative entries above each button to the right.	Text can be entered using the ALPHA button to obtain the red alternative entries above each button to the right.
Just type in the expression you want to evaluate then press ENTER.	Just type in the expression you want to evaluate then press EXE.
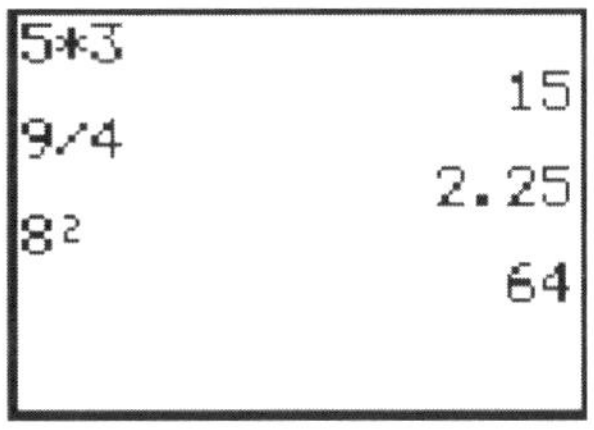 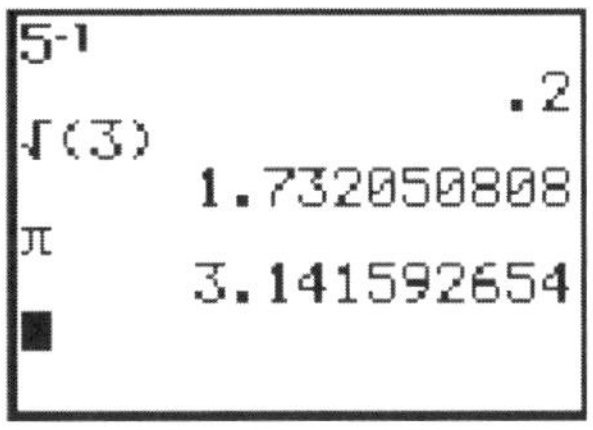	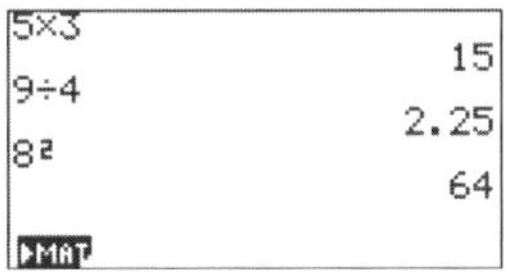 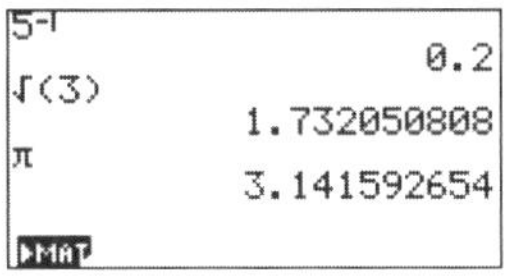
• Multiplication appears on the screen as * • There are two minus symbols, (-) for a negative number and − for subtraction.	• There are two minus symbols (-) for a negative number and − for subtraction.

Press [MODE]. 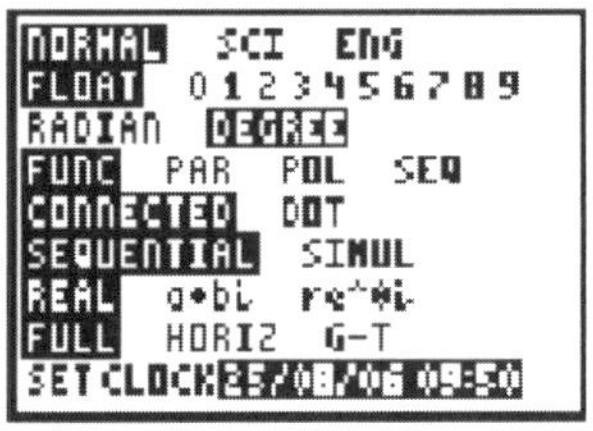	Press SHIFT MENU 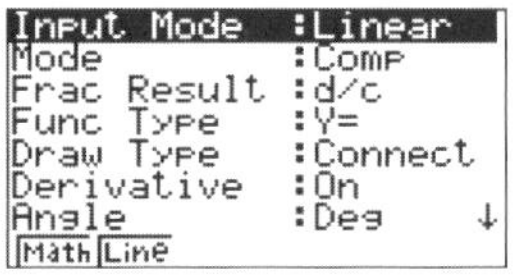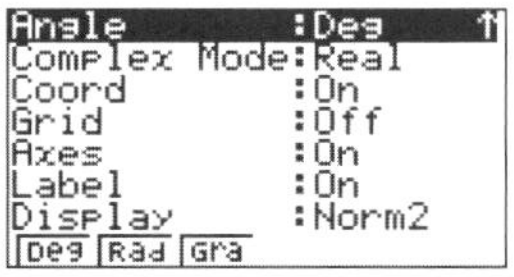
Most of the time, the settings in white against a black highlight you see here are the ones you should choose and keep. In particular, make absolutely sure that the **DEGREE** mode is highlighted. If it is not, then use [▼] [▶] to highlight it and press [ENTER]. To quit the mode screen (or any other screen) use [2nd] [MODE]. If your GDC gets changed to default settings (before an exam for instance), this will switch it back to **RADIAN** mode. Radians are not used in mathematical studies and you will make errors if you forget to switch to degrees.	Most of the time, the settings you see here are the ones you should choose and keep. In particular, make absolutely sure that the **DEGREE** mode is highlighted. If it is not, then highlight it and press Trace F1 . To quit the mode screen (or any other screen) use EXIT .
To complete the introductions we mention two more useful facilities. When an answer has appeared on the screen, the GDC remembers it. You can call it back as a new input in a later line using [2nd] [−] (**ANS**). For example, you are using Pythagoras' rule but you forgot to enter the $\sqrt{\ }$ symbol. No problem. Enter [√] [2nd] [(-)] [)] 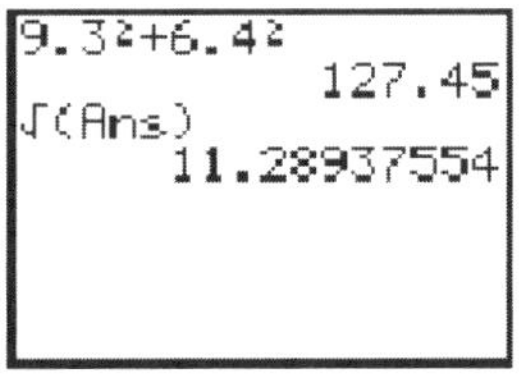	To complete the introductions we mention two more useful facilities. When an answer has appeared on the screen, the GDC remembers it. You can call it back as a new input in a later line using SHIFT [−] (**ANS**). For example, you are using Pythagoras' rule but you forgot to enter the $\sqrt{\ }$ symbol. No problem. Enter SHIFT [x^2] [ANS]. 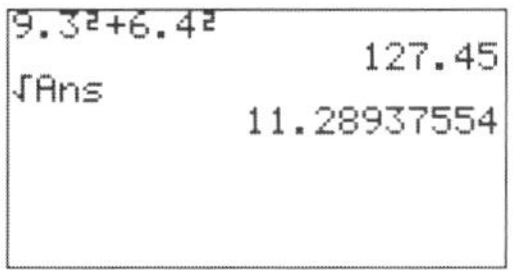
You can recall a whole previous entry using [2nd] [ENTER] (**ENTRY**). For example, you have entered a cosine rule but then realise you entered the angle as 60 instead of 30. No problem. Use the button sequence above then backtrack with [◀] and change the 6 to a 3.	You can correct a previous entry using the arrow keypad ➞. For example, you have entered a cosine rule but then realise you entered the angle as 60 instead of 30. No problem. Use the arrow keys to position the cursor then overwrite the 6 with a 3.

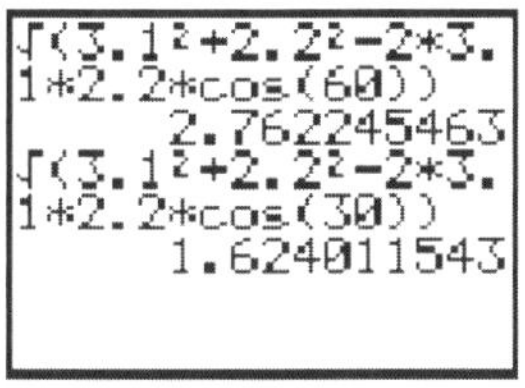

To insert rather than overwrite a symbol use [2nd] [DEL] **(INS)**.

To insert rather than overwrite a symbol use SHIFT DEL **(INS)**.

Example 1.1.1a

a Calculate the area of a circle of radius 3 cm.
b Find sin 30°.
c Convert 0.1625 to a fraction in lowest terms.

TI solution

a, b The entries are clear.
c Type **0.1625** [MATH] [ENTER] [ENTER].
(The first [ENTER] chooses **1:▶ Frac**)

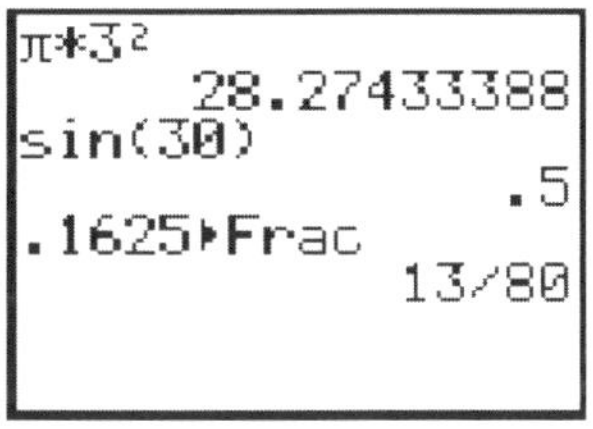

Casio solution

a, b The entries are clear.
c Type 0.1625 EXE then press F↔D for fraction.

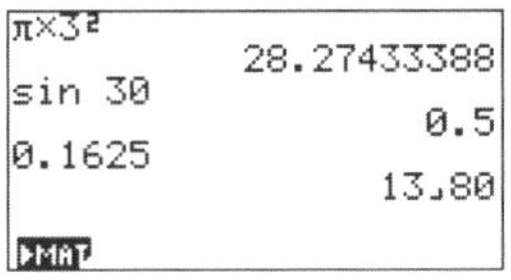

Answers: a 28.3 cm^2 (3 s.f.) b 0.5 c $\frac{13}{80}$

Drawing a graph

Example 1.1.1b (from *4.1.4b*)
Draw the graph of $g(x) = x^2$.

TI solution

Press [Y=] [CLEAR] to clear previous equations.
Type in [X,T,Θ,n] [x^2] [ENTER]

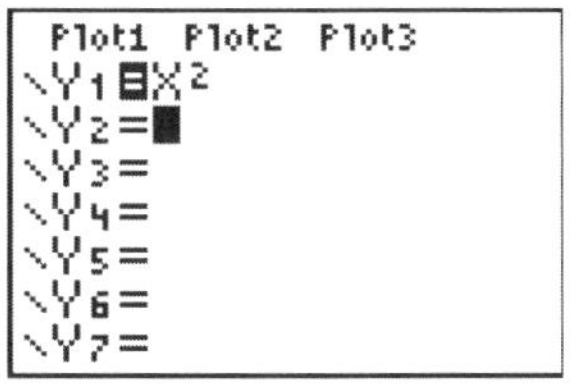

Press [ZOOM] **6** (for standard size).

Casio solution

Select GRAPH from the **MENU**.
Input X,θ,T x^2 EXE

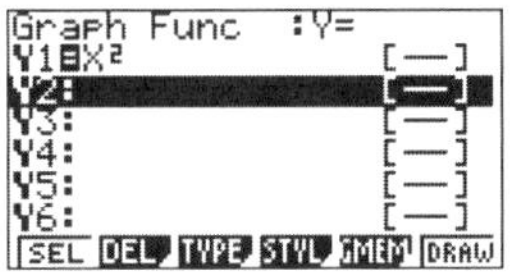

Select **VIEW WINDOW** using SHIFT F3 (V-Window)
then standard screen F3 (V-Window)

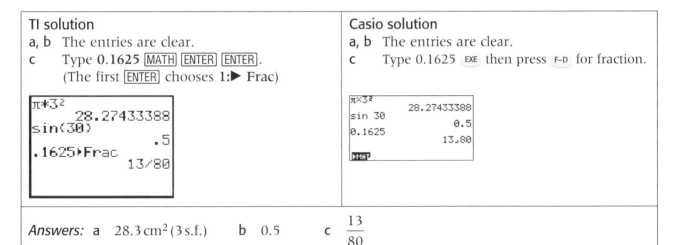

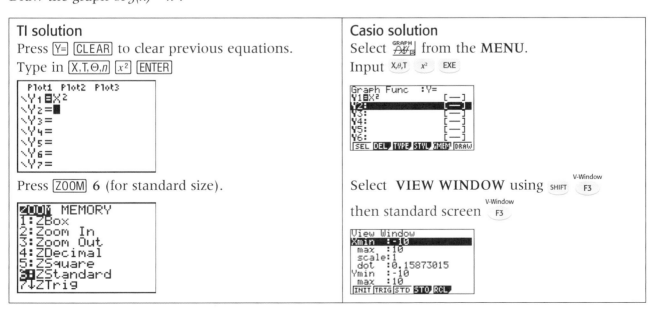

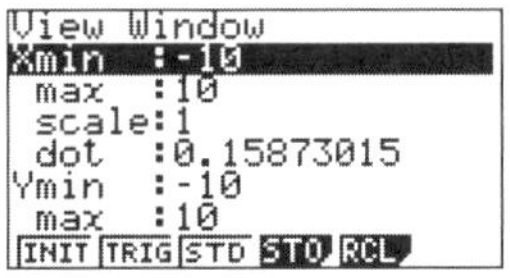

Press [GRAPH]

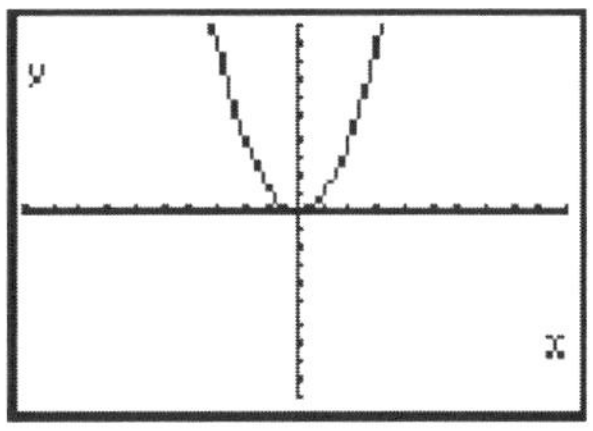

The inclusion of labels, axes and so on is adjusted using [2nd][ZOOM] (**FORMAT**).

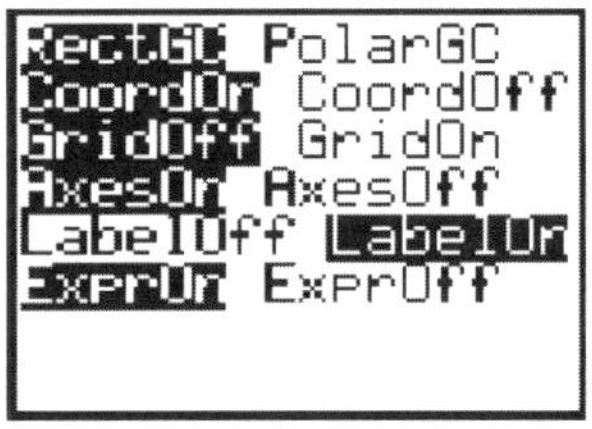

EXE Select **DRAW** G–1 F6

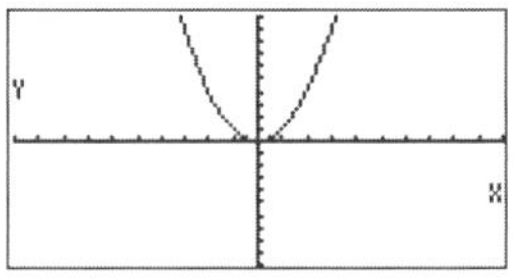

The inclusion of labels, axes and so on is adjusted using SHIFT MENU (**SET UP**).

Locating a graph: Sometimes a graph does not appear on the screen. The graph can be found using [ZOOM] **0:ZoomFit**.

For example, draw the graph of $x^2 + 15$. Press [ZOOM] **6**.

Note that the graph does not appear.

Press [ZOOM] **0:ZoomFit** to fit the graph to the screen. (NOTE: the Zoom 0 setting is the tenth entry off the bottom of the screen, use [▼] to find it.)

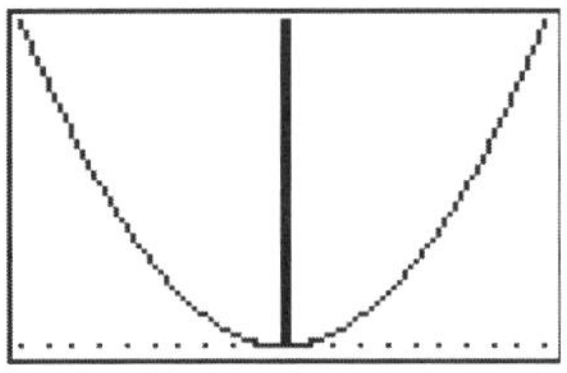

Locating a graph: Sometimes a graph does not appear on the screen. The graph can be found using Zoom F2 G-Solv F5 (Zoom auto).

For example, draw the graph of $x^2 + 15$. Use a standard view window, **STD**. EXE G–1 F6

Note that the graph does not appear and the screen is empty.

Press Zoom F2 G-Solv F5 to fit the graph to the screen.

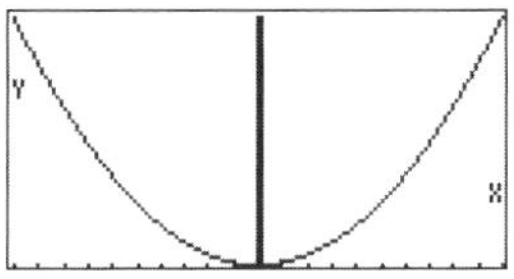

Finding values from a graph

Example 1.1.1c

Given $g(x) = x^2$ (example above), **a** find $g(2.5)$ **b** find x when $g(x) = 7$.

TI solution

a Press [2nd] [TRACE] (that is, **CALC**), **1:Value**, [ENTER]

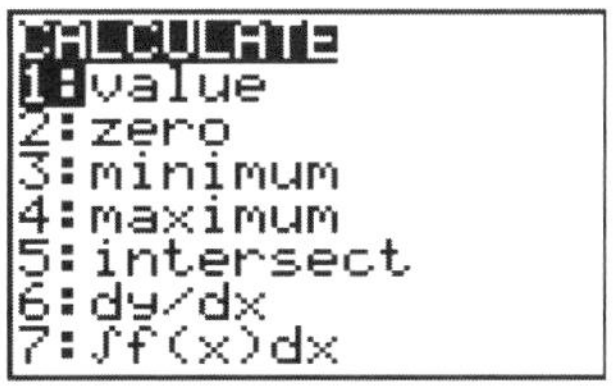

Type in 2.5 (the value of x) [ENTER]

Casio solution

Select **G_SOLV** by SHIFT G-Solv F5 then G–1 F6 to access **Y-CAL** and **X-CAL**.

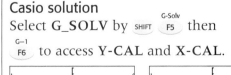

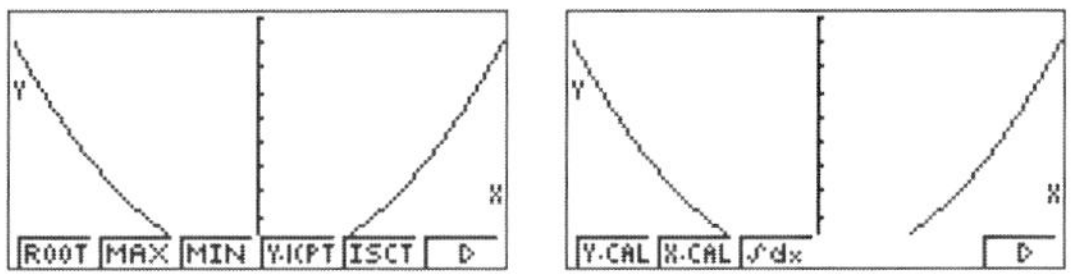

a Select **Y-CAL** using Trace F1 then input **2.5** EXE.

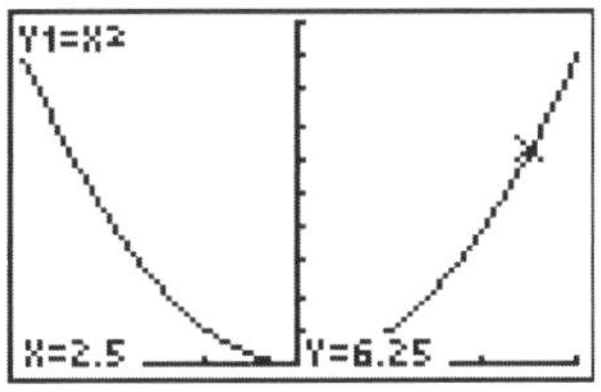

The value of y (= $g(2.5)$) is 6.25.

b This is done by finding the intersection with the line $y = 7$.

[Y=] then type 7 in the Y_2 = position.

[GRAPH], [2nd] [TRACE] (**CALC**), **5:intersect**, [ENTER], [ENTER], [ENTER].

(The last three [ENTER]s tell the GDC to choose the curves suggested on the screen then make its own guess about the intersection.)

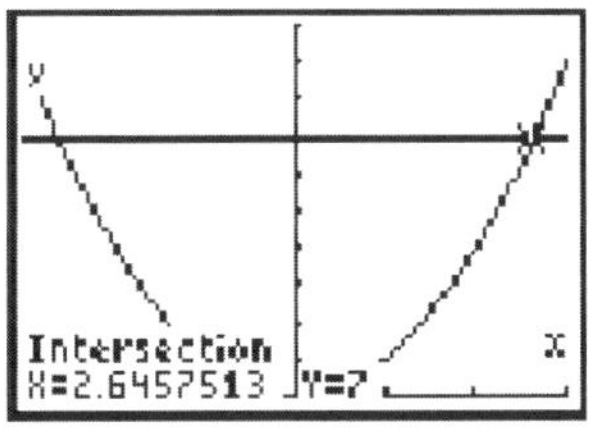

b

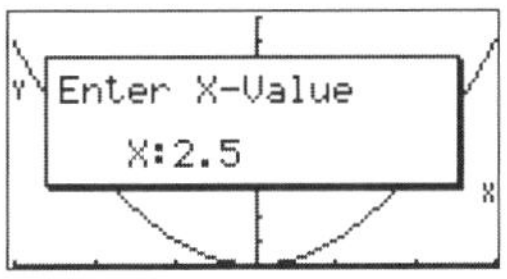

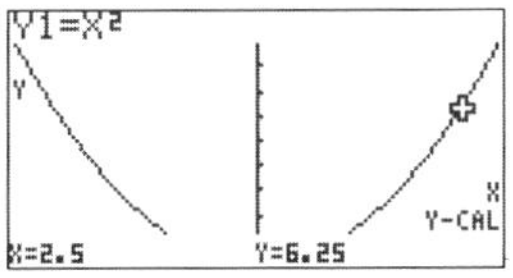

exit to **G_SOLV** then select **X-CAL** using Zoom F2. Input **7**, EXE.

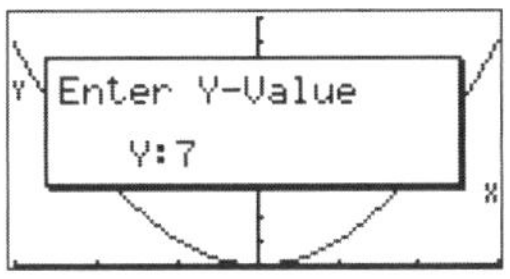

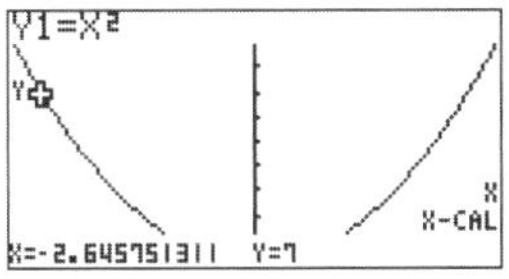

Answers: a $g(2.5) = 6.25$ b $x = 2.65$ (3 s.f.)

TABLE can also be used to find individual values for a function.

Example 1.1.1d (from *4.2.1a*)

Given $f(x) = 2x + 3$ find a $f(25)$ b $f(38.6)$

TI solution

Press [Y=] [CLEAR] Type in 2 [X,T,Θ,n] + 3.

Make sure you choose a window that includes the x-value you are interested in.

a Use [2nd], [TRACE] (**CALC**), **1:value** and type in 25. This is the quickest way.

Alternatively press [2nd] [GRAPH] (**TABLE**).

You can read off values of y for whole number values of x. Just scroll down to 25 using [▼].

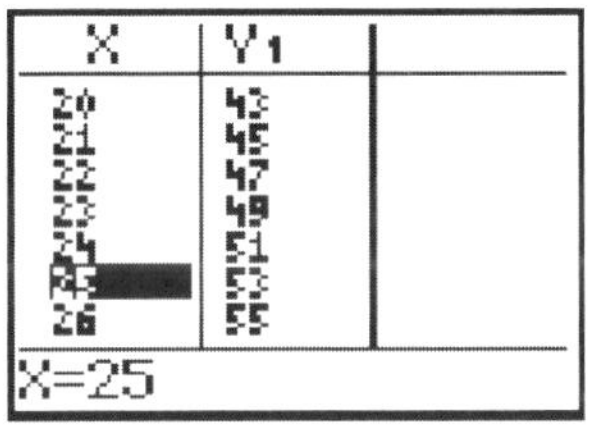

We see that $f(25) = 53$.

Casio solution

Select TABLE from the MENU.

Input $2x + 3$.

Select **SET** (or **RANGE**)*.

G-Solv F5 Input suitable values

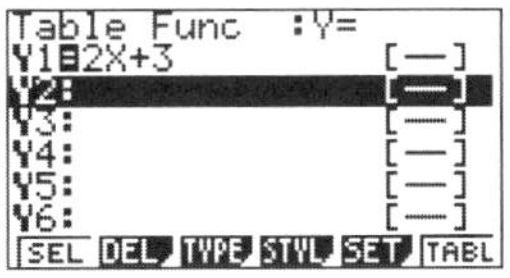

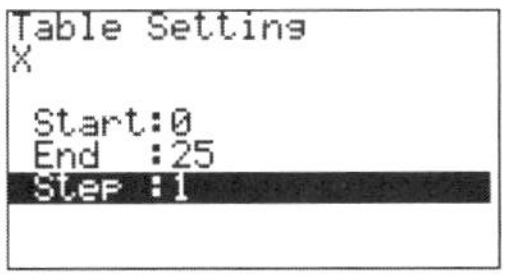

*Note that this is model dependent.

EXE Select **TABLE** G↔T F6

a Scroll down the table to 25.

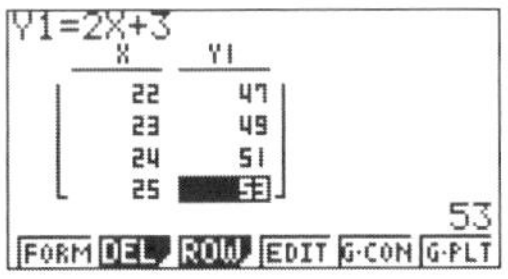

We see that $f(25) = 53$.

b We can make the table to start at other values for x or to take different sized steps.

Press [2nd] [WINDOW] (**TBLSET**).

Type 38.6 in **TblStart**.

Now use [2nd] [GRAPH] (**TABLE**) again.

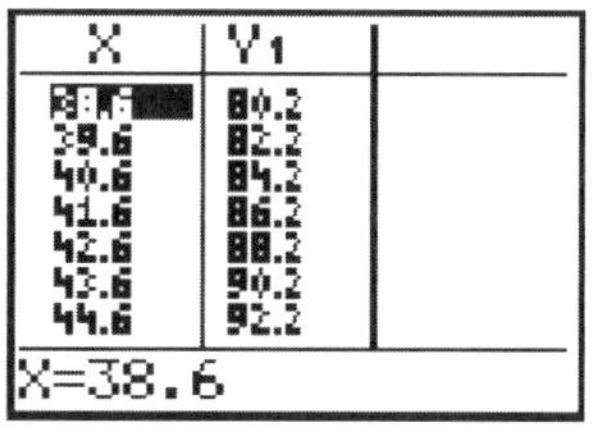

b We do not need to change the table to find other values. Any x-value can be input directly to the screen. Note that the cursor needs to be over an x-value to input.

Type 38.6 (EXE).

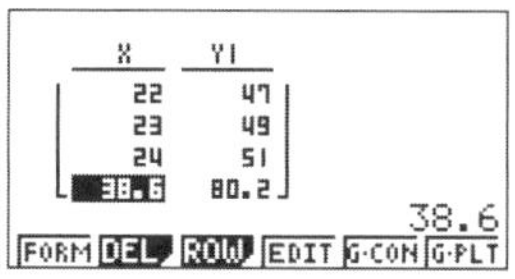

You can now identify the answer as 80.2.

Answers: a $f(25) = 53$ b $f(38.6) = 80.2$

1.1.2 Appropriate choice of 'window'; use of 'zoom' and 'trace' (or equivalent) to locate points to a given accuracy

Example 1.1.2

Investigate the intersection of the curves $f(x) = \dfrac{1}{x^2+1}$ and $g(x) = \sqrt{x}$.

Use values of x from –3 to 3 and y from 0 to 2.

Find a solution of the equation $\dfrac{1}{x^2+1} - \sqrt{x} = 0$.

Taken from *M06P1q15*

TI solution

[Y=] then input *exactly* as on the screen below. Note the brackets.

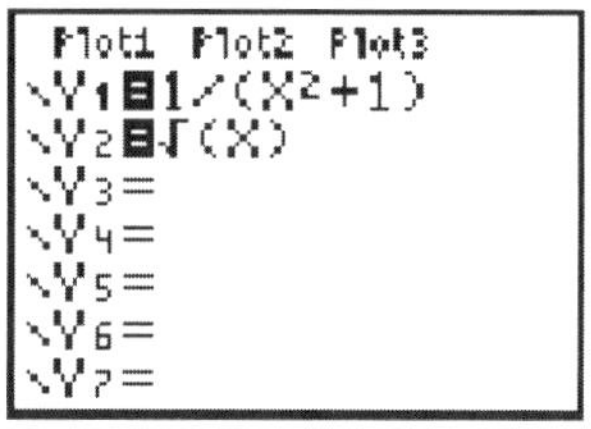

Set the window size with [WINDOW].

Casio solution

Input the functions as below.

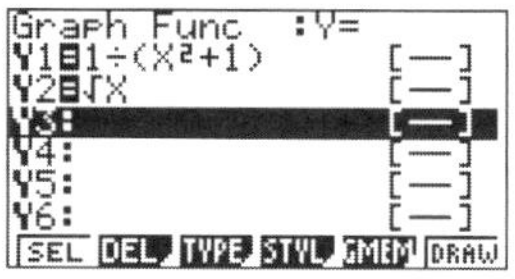

Set the view window (SHIFT) (F3) (V-Window).

Take **Xmin** = –3, **Xmax** =3, **Ymin** = 0 and **Ymax** = 2.

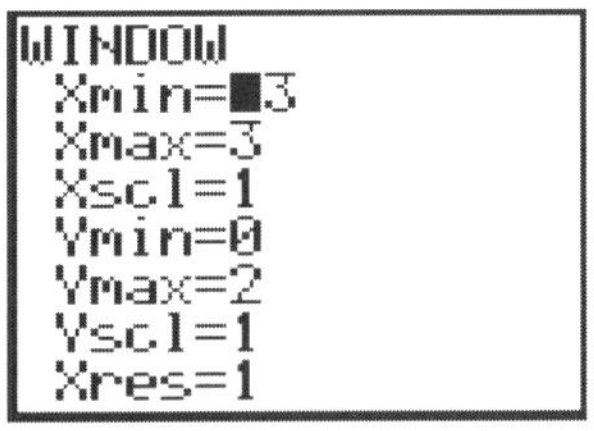

[GRAPH] the curves.

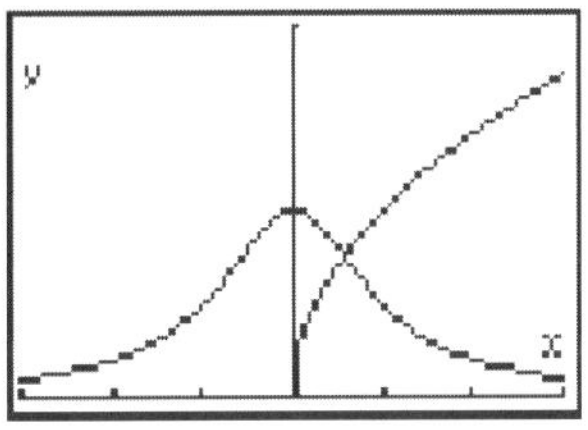

Press [TRACE] then use the arrows [◄][►] to move the flashing cursor onto the intersection.

Now [ZOOM], **2:Zoom In,** [ENTER].

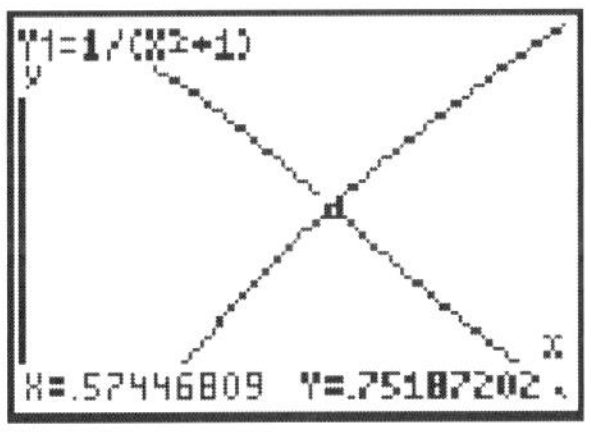

You can read off the approximate coordinates of the intersection at the bottom of the screen (0.574, 0.752).

Repeat this procedure as often as necessary.

Re-position the cursor when it strays from the intersection.

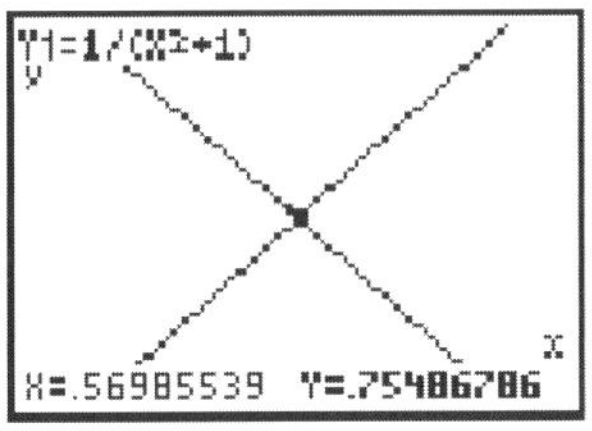

The intersection now can be read as (0.570, 0.755) (3 s.f.).

In fact a more accurate approximation is $x = 0.56\,984\,029$, $y = 0.75\,487\,767$.

Take **Xmin** = –3, **Xmax** =3, **Ymin** = 0 and **Ymax** = 2.

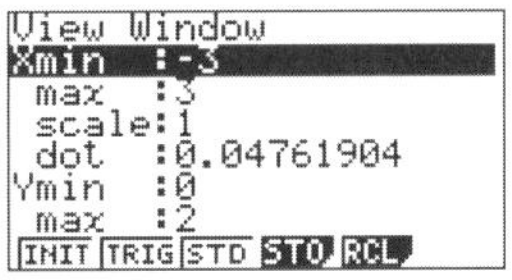

EXE then G–1 F6 to draw the graphs.

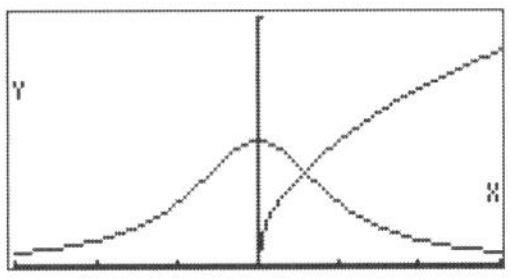

Select G-Solv F5 (**GSOLV**).

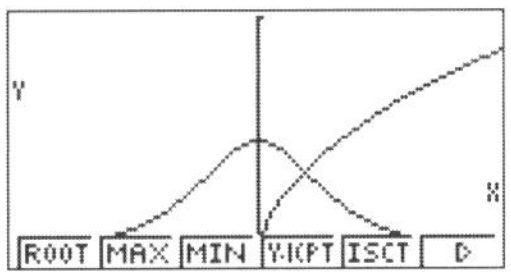

then G-Solv F5 (**ISCT**).

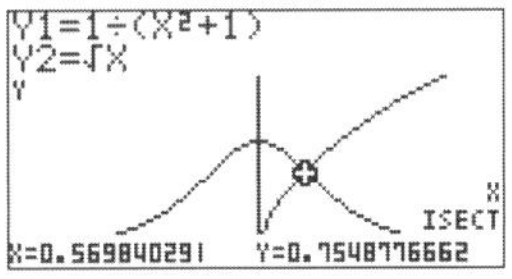

The intersection now can be read as (0.570, 0.755) (3 s.f.).

In fact a more accurate approximation is $x = 0.56\,984\,029$, $y = 0.75\,487\,767$.

Answer: (0.570, 0.755) (3s.f.)

1.1.3 Explanations of commonly used buttons

TI

By now you have probably already used some of the following buttons:

[CLEAR] Clears the screen. Sometimes it's needed repeatedly to finish the job.

[DEL] Deletes the entry covered by the flashing cursor.

[2nd][DEL] **(INS)** allows you to insert symbols without overwriting.

[MATH] is useful. Experiment with the facilities here.

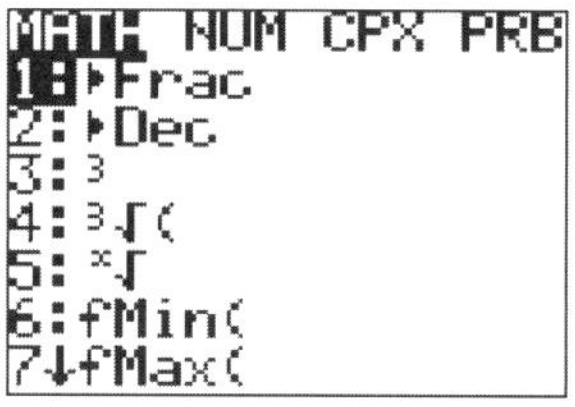

Notice there are eighth and ninth choices at the bottom of the list and off the screen (use [▼]). One of these is the **SOLVER**. It is very useful.

Casio

By now you have probably already used some of the following buttons:

To clear the screen press (AC/ON).

Use (DEL) to delete an entry.

Use the arrow pad (REPLY) to place the cursor where you wish to insert symbols without overwriting.

Use buttons (F1 Trace) to (F6 G–1) to select the options at the bottom of the screen.

In many cases, pressing (F6 G–1) offers more choices.

Example 1.1.3a

Convert 0.3162 to a fraction in its lowest terms.

TI solution

Enter 0.3162. Press [MATH], Choose **1:FRAC**, [ENTER].

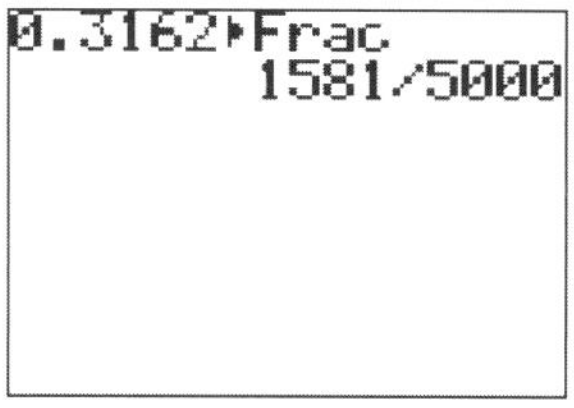

Casio solution

Enter 0.3162. Press (EXE) (F–D).

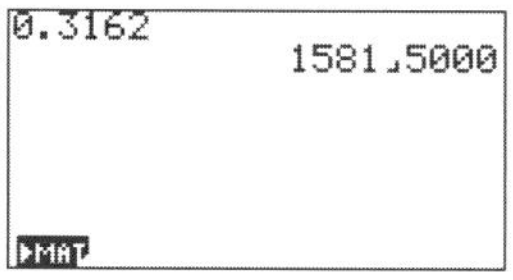

Answer: $\dfrac{1581}{5000}$

TI

Pressing [2nd] [x^{-1}] brings up the **MATRIX** facility. Here you can edit matrices (tables of values of various sizes).

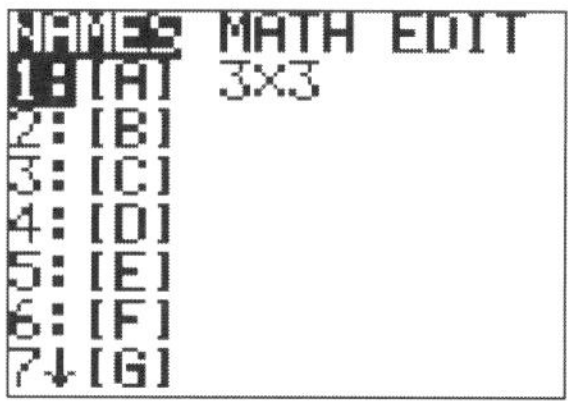

Casio

Select **MAT** to bring up the **MATRIX** facility. Here you can edit matrices (tables of values of various sizes).

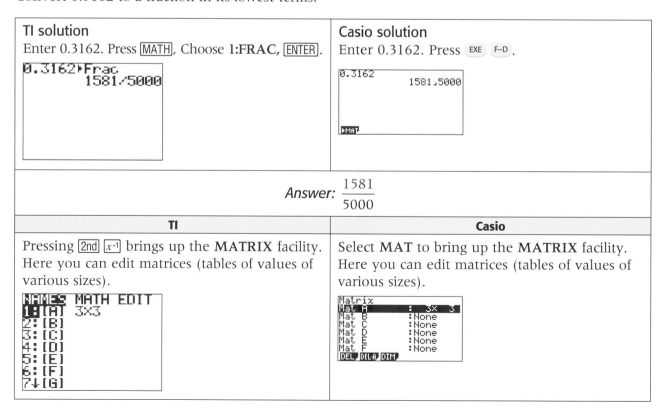

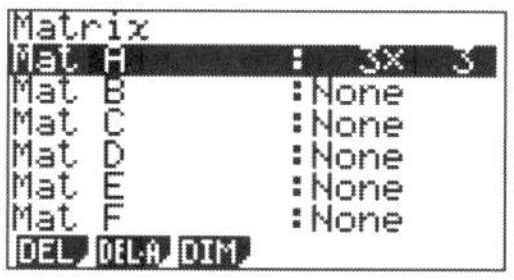

Matrices are used to perform a χ^2 test on the GDC.	Matrices are used to perform a χ^2 test on the GDC.
The [VARS] button allows you to read or select values of any of the variables you have stored during calculations. This can be useful, for example, to input values already calculated in a statistics process.	The VARS button allows you to read or select values of any of the variables you have stored during calculations. This can be useful, for example, to input values already calculated in a statistics process.
[STO▸] allows you to store values or lists from the screen.	→ allows you to store values or lists from the screen.
For example, four numbers have been stored in lists **L1** and **L2**. (See **1.1.4** for lists).	For example, four numbers have been stored in lists **List 1** and **List 2**. (See **1.1.4** for lists).
L1 is {1,2,3,4}, **L2** is {2,4,6,8}.	**List 1** is {1,2,3,4}, **List 2** is {2,4,6,8}.
	Select Run from the menu: OPTN F1 (Trace) F1 (Trace) 1 × F1 (Trace) 2 EXE EXIT
[L1] [×] [L2] multiplies corresponding entries in the lists then [STO▸] [L3] stores the new numbers in list 3. 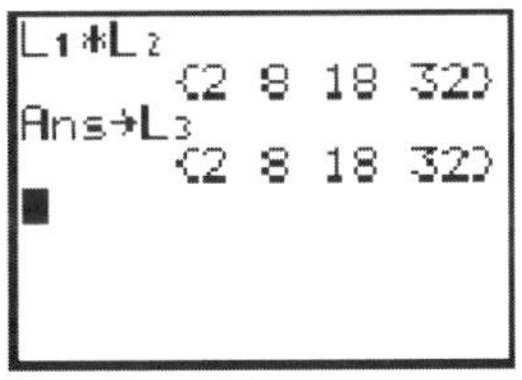	multiplies corresponding numbers in the lists 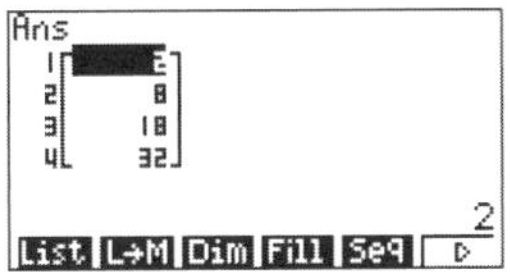 F1 (Trace) (List) SHIFT (−) (Ans) → F1 (Trace) (List) (−) stores the answer in List 3. 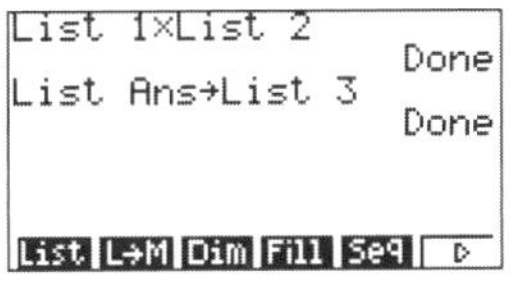
Finally press the button marked [APPS]. What you see here will depend on the applications on your GDC. Many are not permitted in IBO examinations.	The Casio has many inbuilt applications. They are all allowed in IBO examinations.
One that is permitted is the **1:FINANCE** application. More on this in Section **1.8**. 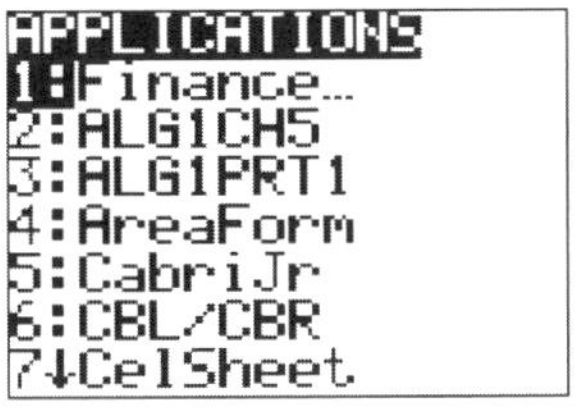	For example, **FINANCE**. Select **TVM** from the **MENU**. 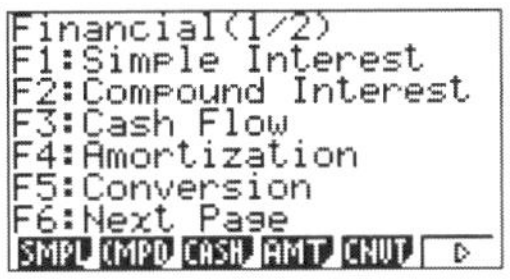

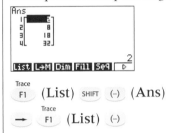

Example 1.1.3b Using the Solver

Use the solver facility to find an angle x satisfying

$$\frac{\sin x}{5.3} = \frac{\sin 28.4}{2.7}$$

TI solution

Press [MATH] **0:Solver**

Position cursor on **0=**

The solver requires an equation with 0 on the left. We rearrange our equation and type it in as $0 = \frac{\sin x}{5.3} - \frac{\sin 28.4}{2.7}$ and [ENTER] it.

Now with cursor positioned on **X=** simply press [ALPHA] [ENTER].

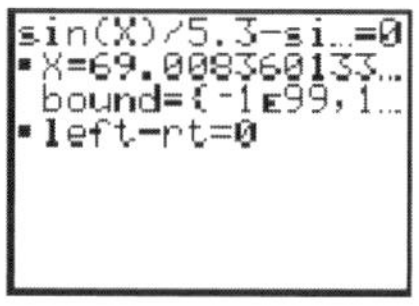

You don't usually have to worry about the *bound*. If you think there might be more than one solution, then guess a number close to the one you want and enter it in X= before using [ALPHA] [ENTER].

Casio solution

Enter the equation as written, using SHIFT • for the "=" sign.

$\frac{\sin x}{5.3} = \frac{\sin 28.4}{2.7}$ EXE.

Now simply press G–1 F6 **(SOLV)**.

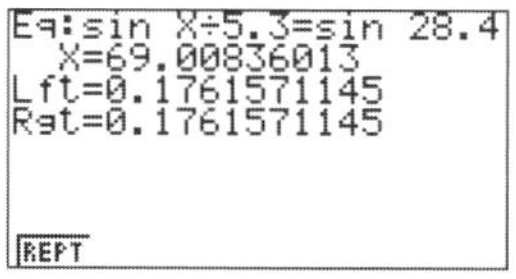

The answer is 69.0°. The cosine rule can be solved in a similar manner.

Did you get this wrong first time? Perhaps your calculator was not set in degrees.

Only a small part of the facilities available on your GDC have been covered here. You are encouraged to experiment and see what other things you can find out.

1.1.4 Entering data in lists

Lists are an extremely important feature of all GDC's. You will need them often to do statistical calculations. They can also be used to evaluate truth tables in mathematical logic.

TI

To create a list press [STAT] then choose **1:EDIT**.

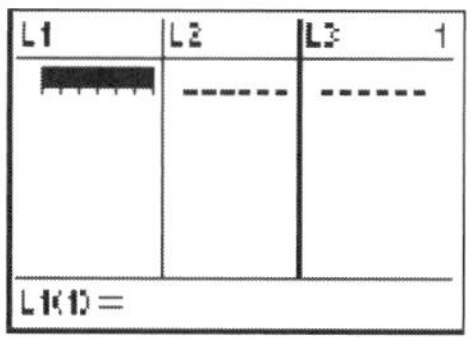

There are six lists called [L1]-[L6] already set up. Numbers can be added to a highlighted list in the usual way using [ENTER]. Move between lists with the arrow keys.

To remove all entries from a list set the cursor on the list name at the top and press [CLEAR]. **(Be careful not to use [DEL] as this removes the list entirely from the menu and it can be hard to get back.)**

Casio

Select **STAT** from the **MENU**.

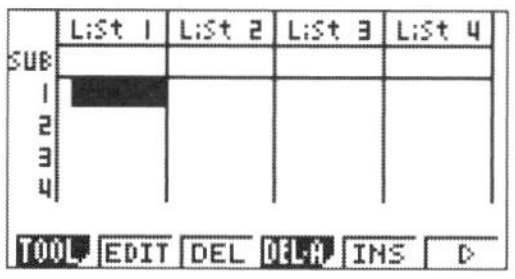

There are six lists already set up. Numbers can be added to a highlighted list in the usual way using EXE. Move between lists with the arrow keys. To remove all entries from a list set the cursor on the list and select **DEL-A** from the choices at the bottom of the screen G–1 F6 then Sketch F4.

You can create your own list names in text using the [ALPHA] button. Arithmetic can be performed on lists as long as they have the same length.	You can create your own list names in text using the ALPHA button. Arithmetic can be performed on lists as long as they have the same length.

Example 1.1.4

Enter two lists of numbers of equal length. Multiply the entries from each list in corresponding positions and put the answer in a third list.

TI solution

Take for example 1, 2, 3, 4, 5 in [L1] and 2, 4, 6, 8, 10 in [L2].

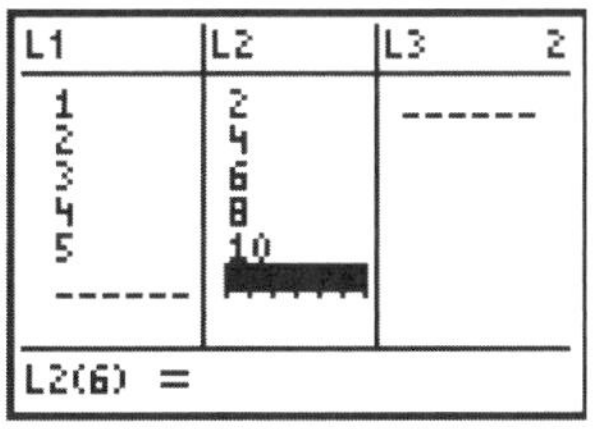

Move the cursor to the title bar of [L3].

Type [L1] [×] [L2].

(You can find the six list names above the numbers 1–6 using [2nd].)

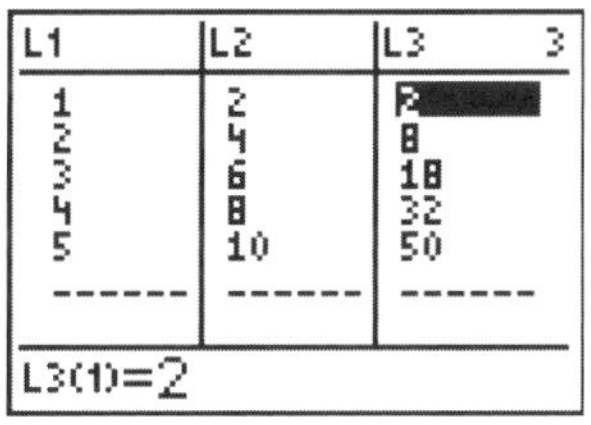

The entries in [L3] are now the products of the entries on the corresponding row in [L1] and [L2].

To quit the list menu press [2nd] [MODE].

Once back on the workspace you can still manipulate lists.

For example [L1][+][L3] [STO▸] [L4] adds the entries of [L1] and [L3] and puts them in [L4].

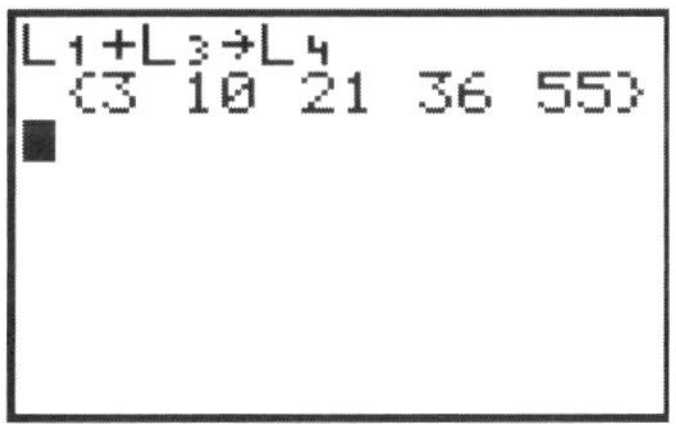

Casio solution

Take for example 1, 2, 3, 4, 5 in **List 1** and 2, 4, 6, 8, 10 in **List 2**.

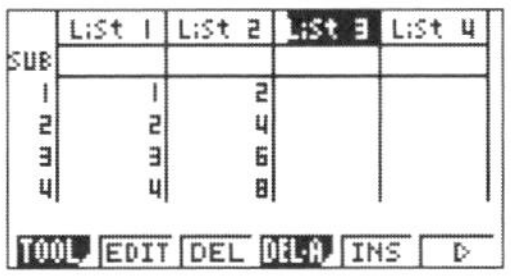

Move the cursor to the title bar of **List 3**.

Select OPTN then **LIST** F1 (Trace) and F1 (Trace) again to access the specific list you want.

Type **1**, then **x**, then OPTN F1 (Trace) **2** EXE

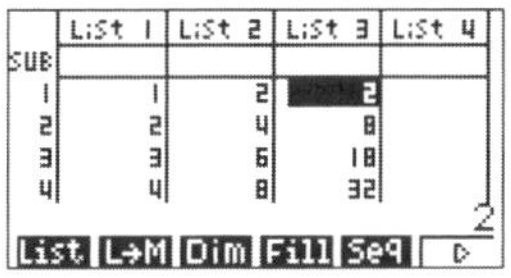

The entries in **List 3** are now the products of the entries on the corresponding row in **List 1** and **List 2**.

Once back in **RUN MODE** you can still manipulate lists (use OPTN as above).

For example **List 1** + **List 2** adds the entries of **List 1** and **List 2** and shows them on the Screen.

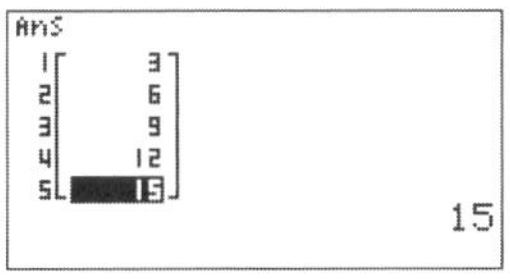

1.2 Number and algebra

1.2.1, 1.2.4 Sets of numbers; SI units

There are no GDC issues to discuss in these subsections.

1.2.2 Rounding

Example 1.2.2 (from *2.2.2a*)

Write down 2.065 correct to 1 d.p.

TI solution

Enter [MODE], select **FLOAT** then **1**.

[2nd] [MODE] (**QUIT**)

Input 2.065, [ENTER].

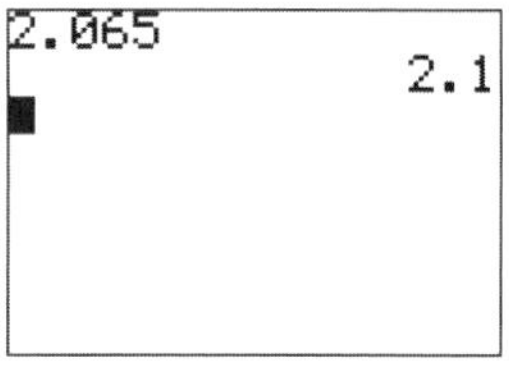

Remember to change the [MODE] back to **NORMAL**.

Casio solution

Select RUN·MAT **from the** SET UP MENU.

Enter SHIFT, MENU.

Scroll down the screen to reach **Display**.

Select **Fix** Trace F1 then **1**.

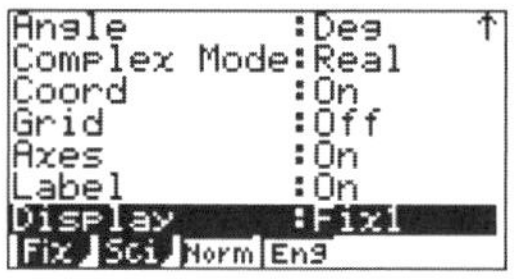

EXE, input 2.065, EXE

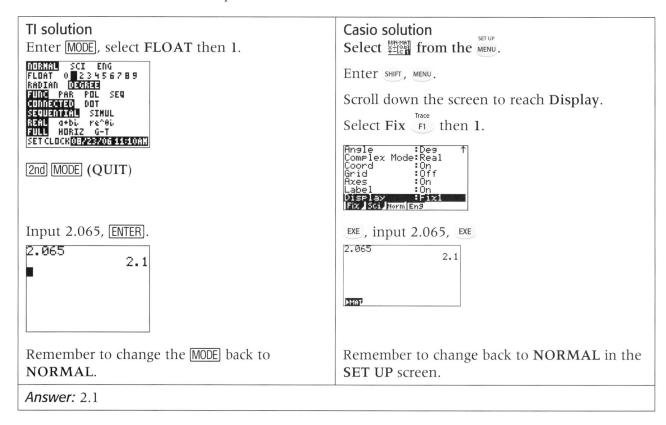

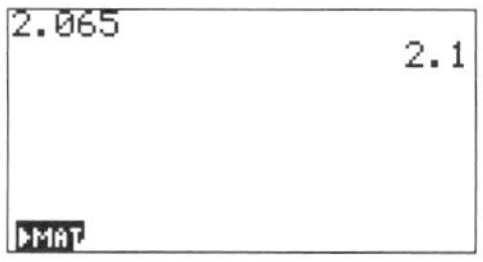

Remember to change back to **NORMAL** in the **SET UP** screen.

Answer: 2.1

1.2.3 Standard form

Example 1.2.3 (from *2.3.1c*)

These numbers are given in full. Write them in standard form.

a 546 000 000 000 b 0.003 242

TI solution

Enter [MODE] select **SCI** using [▶], [ENTER] [2nd] [MODE] to quit.

Input 546 000 000 000 [ENTER].

Input 0.003 242 [ENTER].

Casio solution

SHIFT MENU, scroll down the screen to **Display**.

Select Zoom F2 (**SCI**) Sketch F4 (**3**), EXE

Input 546 000 000 000 EXE.

Input 0.003 242 EXE.

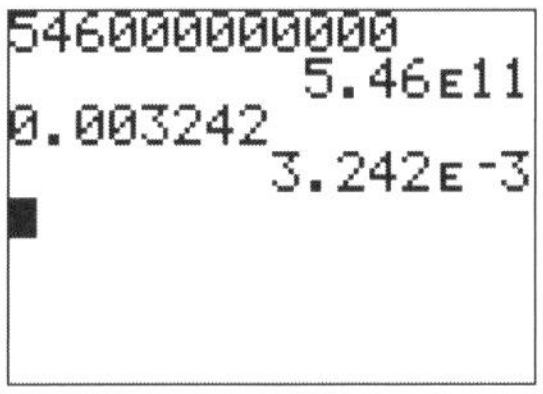

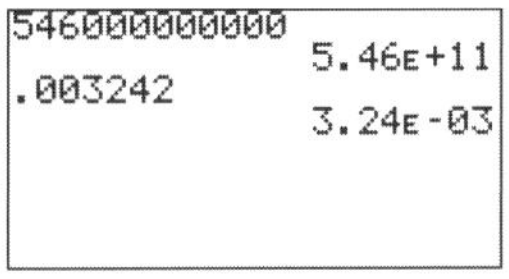

Note that we have chosen 3. This means that the answer will be given correct to 3 significant figures.

You must remember that E11 and E -3 stand for 10^{11} and 10^{-3} respectively.
Your answer **must** be written in correct mathematical notation, (see below).

Remember to change the [MODE] back to **NORMAL**.

Remember to go back to the **SET UP** screen and return to V-Window F3 **(NORMAL)**.

Answer: **a** 5.46×10^{11} **b** 3.24×10^{-3} (3 s.f.) or 3.242×10^{-3} (exact)

1.2.5 and 1.2.6 Arithmetic and geometric sequences

Example 1.2.5a (from *2.5.2d*).

List the first four terms of the sequence $a_n = \frac{1}{2}(n+1)$

TI solution

USING LIST

[2nd] [STAT] **(LIST) OPS** select **5**.

Input $\left(\frac{1}{2}(x+1), x, 1, 4\right)$ [ENTER].

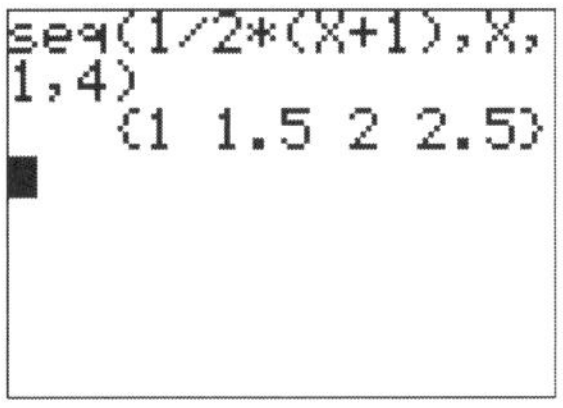

Alternatively using TABLE [Y=] and input [1] × [2] [(] [X,T,Θ,n] Press [÷] [+] [1] [)] [ENTER].

The screen should look like that below.

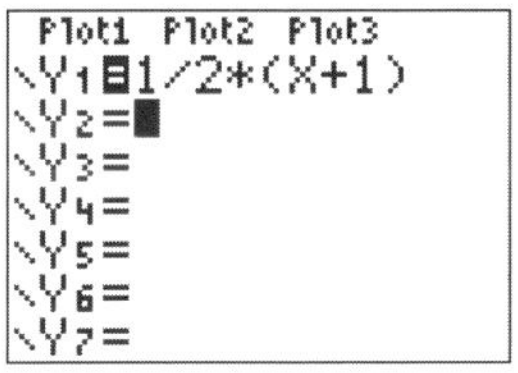

Casio solution

RECURSION MODE

Select RECUR from the MENU.

V-Window F3 **(TYPE)** select Trace F1 (a_n).

Input $\frac{1}{2}(n+1)$ as follows. Note that the variable n is obtained from the bottom of the screen using Sketch F4. Hence type

1 ÷ 2 × (Sketch F4 + 1) EXE

to obtain

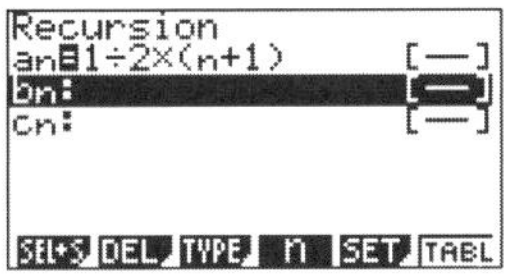

Now G-Solv F5 **(SET)***

Input **1**, EXE **4**, EXE EXE

G-Solv F5 **(TABL)**.

Using TABLE

Now press [2nd] [WINDOW] (**TBLSET**).

Input **1**, [ENTER], **1** then [2nd] [GRAPH] (**TABLE**).

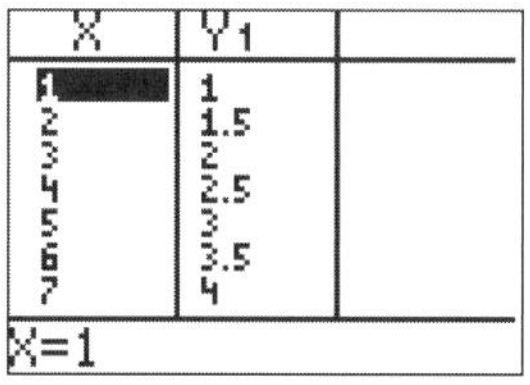

The required values appear in the second column titled under $\mathbf{Y_1}$. To exit from this screen press [2nd] [MODE] as always.

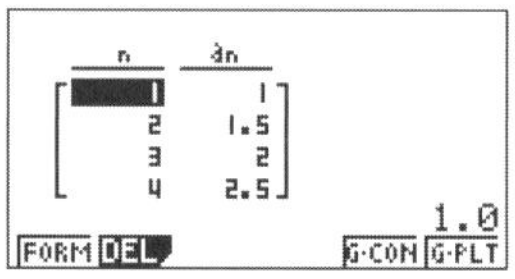

The required values appear in the second column.

*Model dependent, might be (RANG).

Alternatively using TABLE MODE

Select TABLE from the MENU.

Input $\frac{1}{2}(x+1)$ EXE.

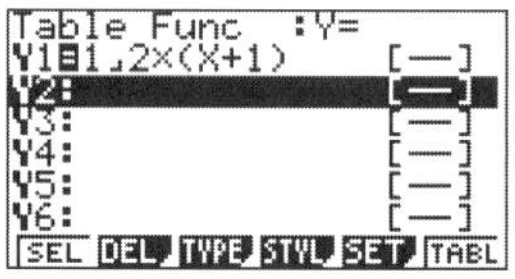

Then G-Solv F5 (**SET**). Input the Range.

G–1 F6 (**TABL**).

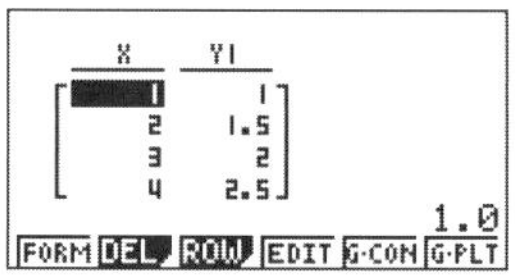

The required values appear in the second column under $\mathbf{Y_1}$ as in this example we have entered the general term in $\mathbf{Y_1}$.

Answer: $a_1 = 1;\ a_2 = 1.5;\ a_3 = 2;\ a_4 = 2.5$

The **TABLE** method is also useful when you need to find the first term greater or smaller than a given value or to compare two sequences. It is also useful when you need to find the number of terms in a finite sequence.

Example 1.2.5b (from *2.5.5b*)

Find the sum of the first 25 multiples of 7.

TI solution

[2nd] [STAT] (**LIST**) then **MATH 5**

[2nd] [STAT] (**LIST**) then **OPS 5**.

Now input **7** [X,T,Θ,*n*], [X,T,Θ,*n*], **1, 25))** (including all the commas and the end brackets).

Casio solution

Select RECUR from the MENU.

SHIFT MENU and turn **Σ Display** on.

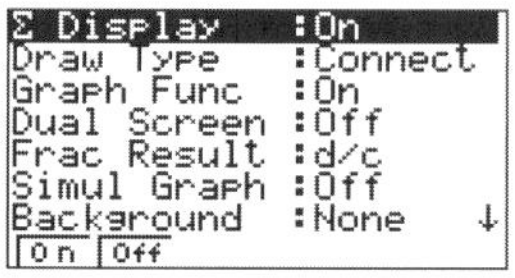

[ENTER]

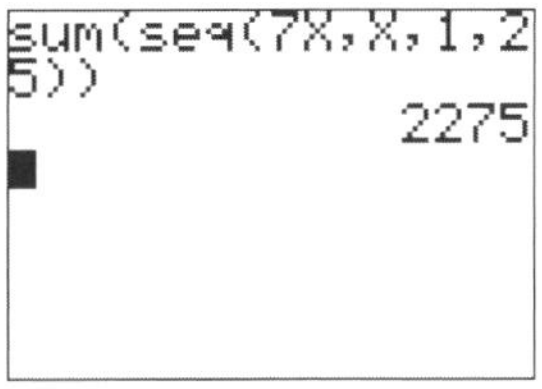

V-Window F3 then Trace F1 (a_n).

Input 7 Trace F1 EXE G-Solv F5 1, EXE 25

EXE EXE G−1 F6

Note that Trace F1 here is model dependent—it might be Sketch F4.

Scroll down the screen.

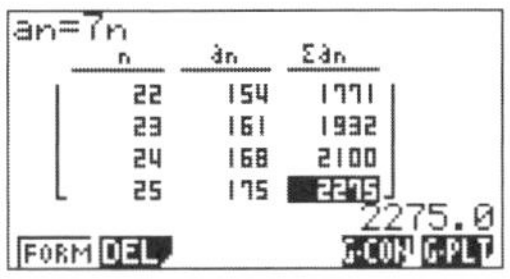

Note that in the last column you have the cumulative sums, that is, this column contains $S_1, S_2, \ldots, S_{25}$.

Answer: $S_{25} = \frac{25}{2}(2 \times 7 + 24 \times 7) = 2275$

Your GDC will sum *any* finite sequence in an analogous manner. Simply enter the formula for the terms in the input space. In Exercise 1, question 2b on page 54, you can practise with a geometric sequence.

1.2.7 Simultaneous and quadratic equations

Example 1.2.7a (from *2.7.1b*)

Solve this pair of linear equations: $2x + y = 7, \quad 5x - 2y = 4$

TI solution

POLYSMLT

The form of each of the equations is $ax + by = c$

Press [APPS] and select **PolySmlt,** [ENTER].

(If you have a long list of Applications you can save time scrolling by pressing [ALPHA] [8] which jumps to letter **P**.)

Select **2:SimultEqnSolver**.

We have two equations in two unknowns so enter this information.

Casio solution

EQUATION

The form of each of the equations is $ax + by = c$

Select EQUA from the MENU and then select Trace F1 **(Simultaneous)**.

Select **Number of unknowns 2** with Trace F1

Input the coefficients and constants

Select Trace F1 (**SOLV**) to see the solutions.

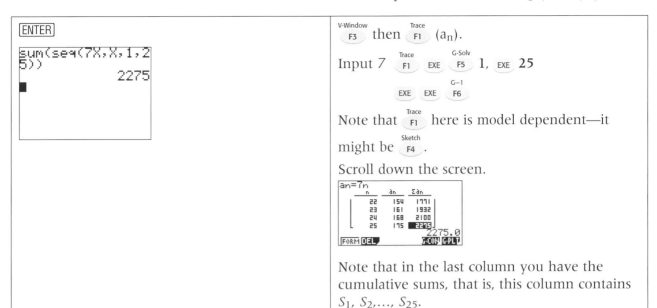

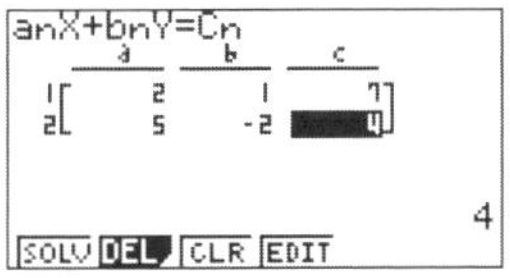

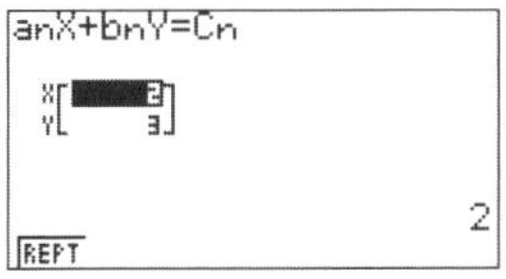

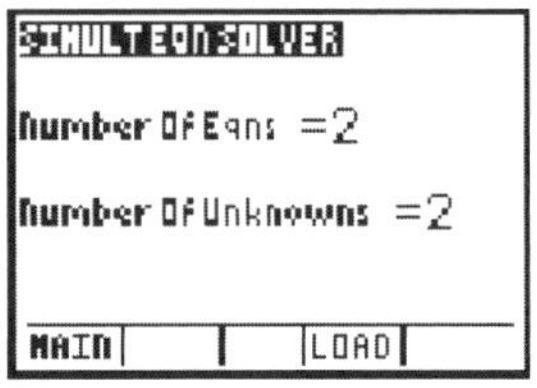

[ENTER].

Now type and [ENTER] the coefficients from the equations we are solving.

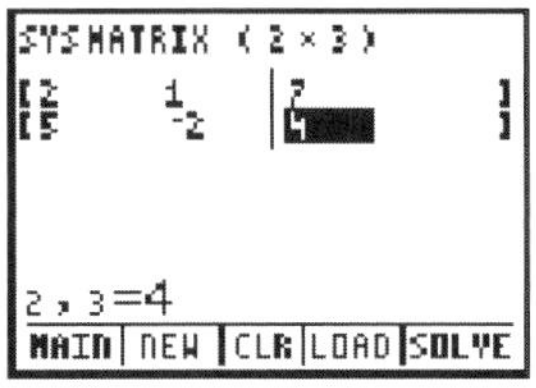

Press [F5] **(SOLVE)** to see the solution:

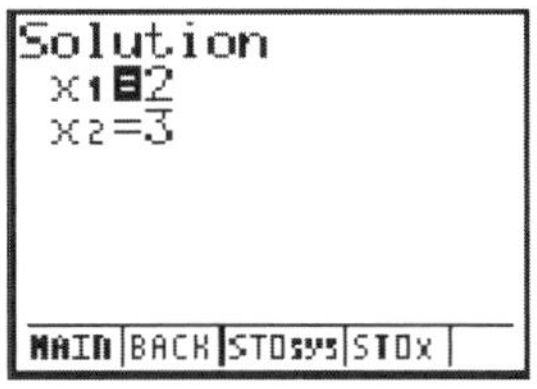

You can exit **PolySmlt** from the **MAIN MENU** using **6:QuitPolySmlt**.

The equations can also be solved using the graph intersection method described in **1.1**.

They must first be rewritten as $y = -2x + 7$ and $y = 2.5x - 2$.

GRAPH

The form of each of the equations is $y = mx + c$

The equations must be rewritten as $y = -2x + 7$ and $y = 2.5x - 2$.

Input both equations. Use G-Solv F5 **(GSOLV)** then G-Solv F5 **(ISCT)** to find the point of intersection.

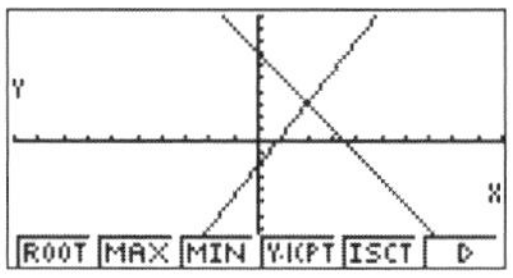

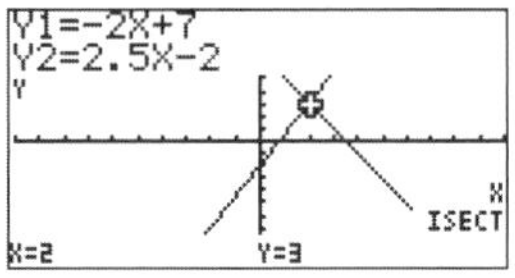

Answer: $x = 2$, $y = 3$

Example 1.2.7b (from *2.7.3g*)
Solve: $2x^2 - 7x - 4 = 0$ using the GDC.

TI solution

The form of the quadratic equation is $ax^2 + bx + c = 0$

If you are still in **SimultEqnSolver** then press [F1] (**MAIN**), otherwise open **PolySmlt**. This time choose **1:PolyRootFinder**
For a quadratic equation enter degree 2.
Now enter the coefficients of the equation:

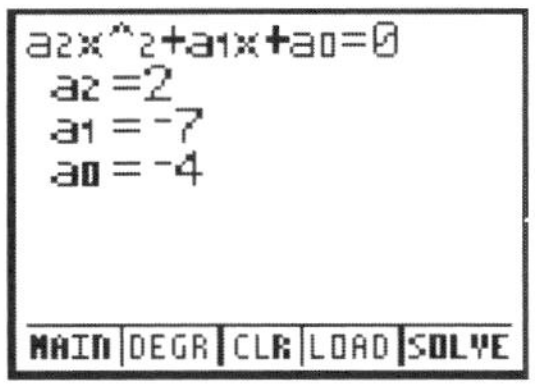

Casio solution
EQUATION

The form of the quadratic equation is $ax^2 + bx + c = 0$

Select **Polynomial** Zoom F2.

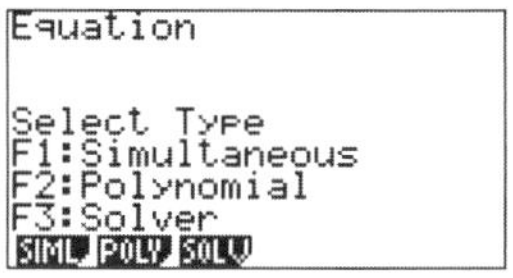

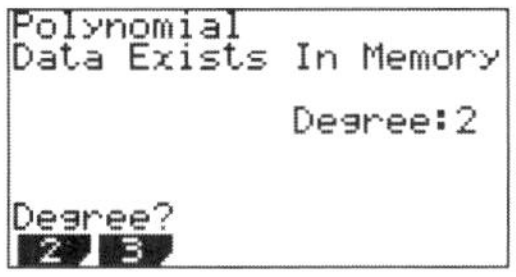

Select **Degree 2** Trace F1.

Input the coefficients and constant.

Press [F5](**SOLVE**). 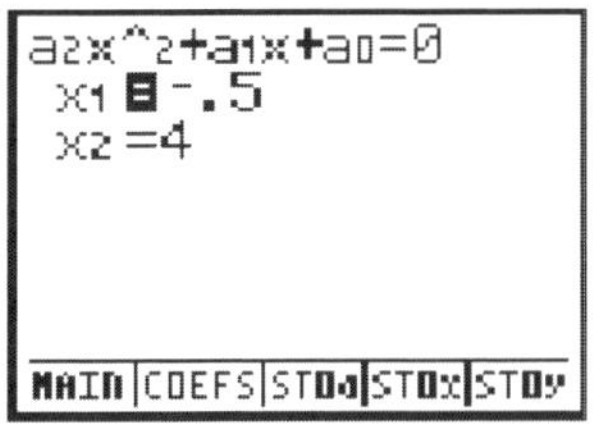 	Select Trace F1 (**Solv**). 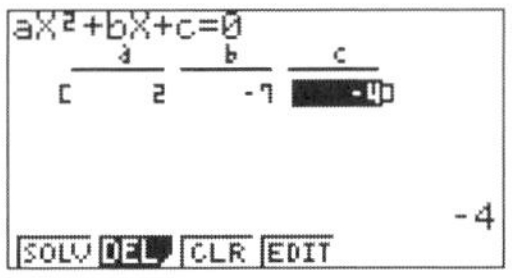 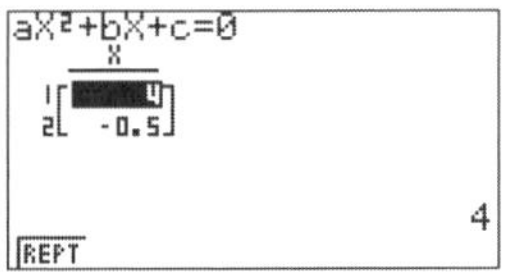 **GRAPH** GRAPH Input the function $Y_1 = 2x^2 - 7x - 4$.
The roots can also be found using the graph. Draw the graph of the quadratic then use [2nd] [TRACE] ([CALC]). Choose **2:zero**. Move the cursor to the left of a root, [ENTER] then to the right, [ENTER], [ENTER]. 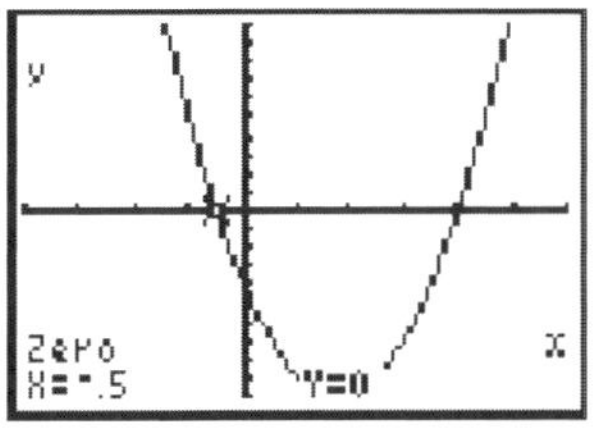 Repeat for the second root.	Draw the graph of Y_1 and then use G-Solv F5 (**GSOLV**) followed by Trace F1 (**ROOT**) to find the first x-intercept. Use the right arrow to find the second value. 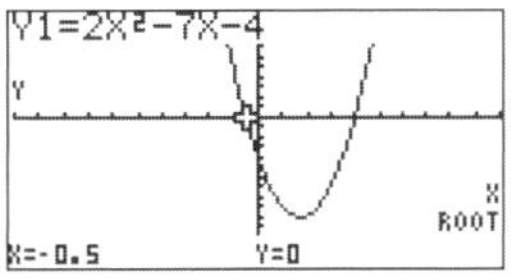 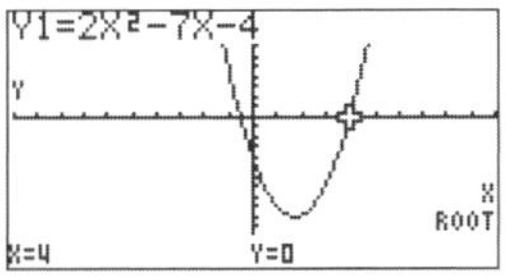
Answer: $x = -0.5$ and 4.	

1.3 Sets, logic and probability

1.3.1 to 1.3.3 GDC for set theory

At present, graphic display calculators do not incorporate commands for set operations such as union, intersection and so on. Lists of numbers can be thought of as sets and the GDC uses set brackets for these, but there is little that one can do with them in terms of set theory.

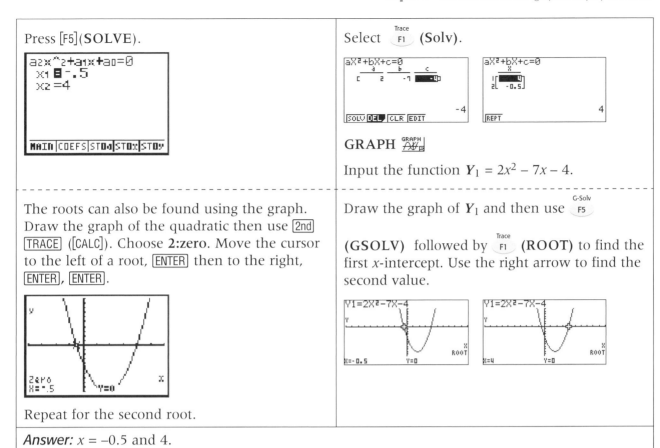

"There aren't any icons to click. It's a chalk board."

1.3.4 to 1.3.7 GDC for logic

The result of a logic operation can be evaluated on the GDC by using the list facility to enter truth values from the columns of a truth table.

The value T is represented by a 1 in the list and F is represented by 0. Simply enter these values in a list in the usual way.

Using 1 for True and 0 for False, enter values for propositions p and q (and r if needed) in your list facility in the usual way.

Example 1.3.6a

Complete the truth table for $p \vee \neg q$.

TI solution

Enter truth values for p and q in any convenient lists; here we use [L1] and [L2].

Move the cursor to the title bar of [L3] and enter $\neg q$ as follows:

Enter [TEST] (Use [2nd] [MATH]).

[TEST] - gives equal, not equal, less than etc signs.

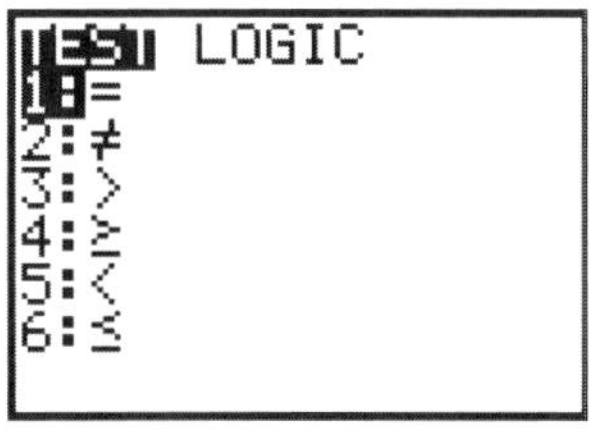

Highlight **LOGIC** to get the **and** (∧), **or** (∨), **xor**(⊻) and **not** (¬) functions

(**xor** stands for *exclusive or* also known as disjunctive union).

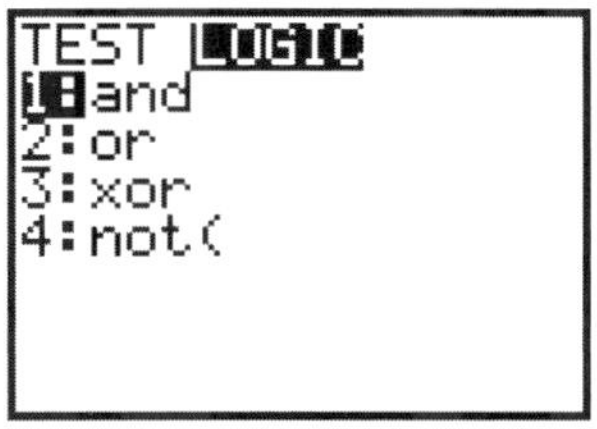

Select **not(** then [L2].

Your screen should look like:

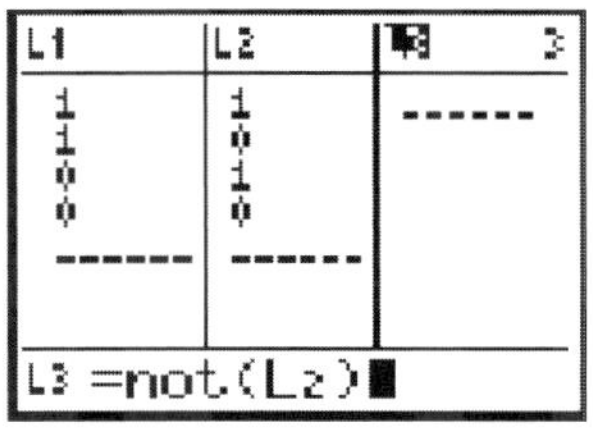

[ENTER] now puts the values of $\neg q$ in [L3].

Repeat the process to put ([L1] or [L3]) into [L4]. This is the solution of the problem.

Casio solution

Enter truth values for p and q in any convenient lists; here we use **LIST 1** and **LIST 2**.

Set cursor on **LIST heading** where you want the next column. (**LIST 3**)

Select [OPTN]

Select [F6] (G–1) **LOGIC**,*

Choices **And**, (∧), **Or**, (∨) and **Not**, (¬) are available.

Choose [F3] (V-Window) **(Not)**.

The **LOGIC** button is model dependent, it might be [F3] (V-Window) or [F5] (G-Solv).

Select [OPTN] [F1] (Trace) [F1] (Trace) Choose the **LIST**

so input **2**, [EXE].

Your screen should look like this with **Not**(q) in **LIST 3**.

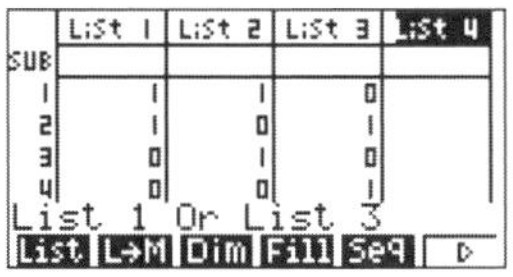

To put **LIST 1 Or LIST 3** into **LIST 4**.

Set cursor on **LIST 4 heading**.

Select [OPTN] [F1] (Trace) [F1] (Trace) Input **1**

Select [OPTN] [F6] (G–1).

Select **LOGIC** Choose **Or**

Select [OPTN] [F1] (Trace) [F1] (Trace) Input **3**.

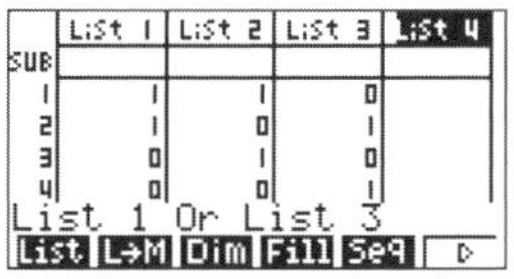

[EXE]

The final screen looks like this.

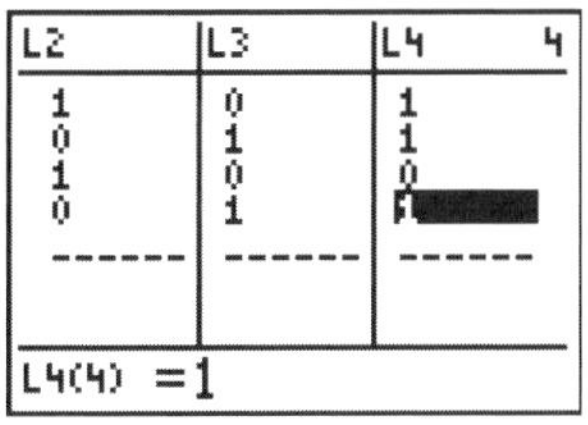

The final screen looks like this and the solution of the problem is now in **LIST 4**.

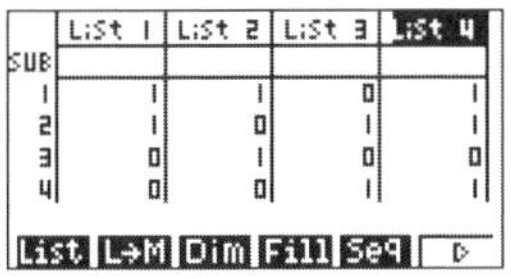

If you decide to work in the workspace instead of the list menu, use the menus described above to enter

[L1] or **not** [L2] (TI) or **LIST 3 Or Not(LIST 4)** (Casio).

In either case you can send the answer to a new list using the → facility of the calculator.

The intermediate 'not' step can be evaluated separately first if you wish.

- Careful use of brackets is important when entering a complicated expression.
- Candidates should show **all** intermediate steps as working even although the final answer can be obtained from the GDC.
- If you have learned the logic symbols it is usually quicker to work out truth tables by hand rather than by GDC. The GDC gives a useful double-check though.

1.3.8 to 1.3.10 GDC for probability

The only use you will make of the GDC for probability calculations is likely to be for basic arithmetic operations. Remember that many of the probabilities you calculate will be exact fractions and can safely be left in that form. If you must approximate them with decimals, don't forget to use three significant figures.

Both the TI and the Casio have facilities that you might find useful if your internal assessment project requires probability.

These include

- A random number generator **RAND** on the TI, **RAN#** on the Casio
- The factorial operation, denoted by $n!$ where for example $6! = 6 \times 5 \times 4 \times 3 \times 2 \times 1$
- The ${}_n\mathbf{P}_r$ and ${}_n\mathbf{C}_r$ operations where ${}_nP_r = \dfrac{n!}{(n-r)!}$ and ${}_nC_r = \dfrac{n!}{n!\,(n-r)!}$

  On the Casio these operations are available using OPTN F3 (V-Window) **(PRB)**.

  On the TI use MATH then choose **PRB**.

1.4 Functions

Topic 4 is one of the two topics for which you will use your GDC extensively. (The other is Topic 6.) It is necessary to list some introductory material for these two topics.

TI	Casio
The following settings are appropriate for mathematical studies.	

Remember to turn off statistical plots before sketching graphs. [2nd][Y=][STAT PLOT] **Ref 4: Plots off** [ENTER].

TI

MODE

Press [MODE].

NORMAL SCI ENG— These are the ways in which numbers can be viewed on the screen. Use **NORMAL** most of the time. **SCI** stands for scientific notation. It can be useful in mathematical studies.

FLOAT — you can select the number of decimal places or leave it to the GDC.

RADIAN DEGREE — always keep in **DEGREE** mode.
FUNC PAR POL SEQ — always keep in **FUNC** mode
CONNECTED DOT — draws graph as continuous function or as dots.

SEQUENTIAL SIMUL — draws graphs one by one or simultaneously.
REAL $a + bi$ — always keep in **REAL**.
FULL HORIZ G-T — draws a full graph, splits the display horizontally or shows graph and table of values on one screen.
Variable is entered using the [X,T,Θ,n] key.
To delete a function place the cursor over the entry to be deleted and select [DEL].

Casio

GRAPH

SHIFT MENU (GRAPH SETUP)

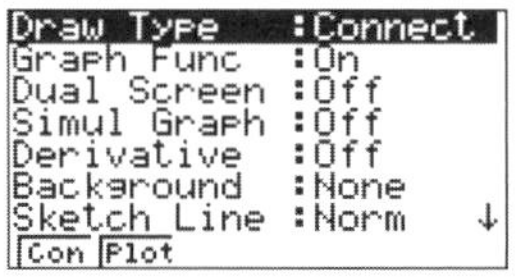

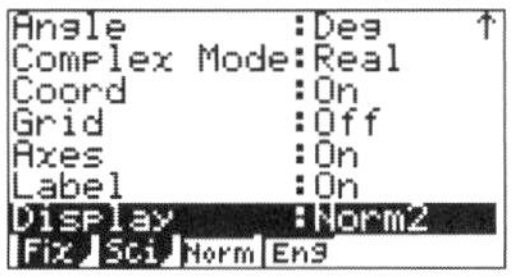

Draw Type — draws graph as a continuous function or as dots

Graph Func — on. Type - always **Y =**

Dual Screen — Graph and Table on same screen

Simul Graph — draws graphs one by one or simultaneously

Derivative — on for analysis of functions

Angle — always in degree mode.
Complex Mode — always real numbers.
Coords — coordinates on
Grid — vertical gridlines can be useful with trace function.

Axes — on.
Label — on to show x and y labels for axes.

Enter the function using the X,θ,T key for the variable.

To delete a function select DEL

then F1 (**YES**).

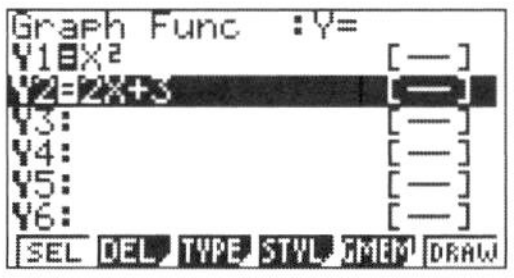

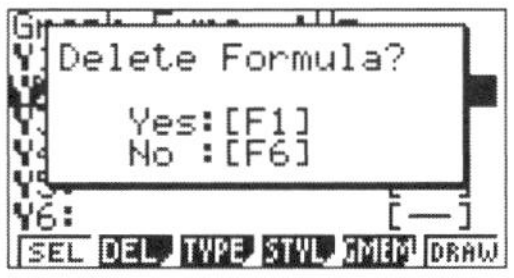

1.4.1 Domain and range

The domain is often given for a function. The range can be found using TABLE, by considering the maximum and minimum values of the function for the given domain.

Example1.4.1 (from *4.1.4b*)
Draw the graph of $g(x) = x^2$, with domain $-3 \leq x \leq 3$.
Identify the range for this domain.

TI solution

Select [Y=] and enter input x^2 as explained in Section 1.1.

Select [2nd] [WINDOW]

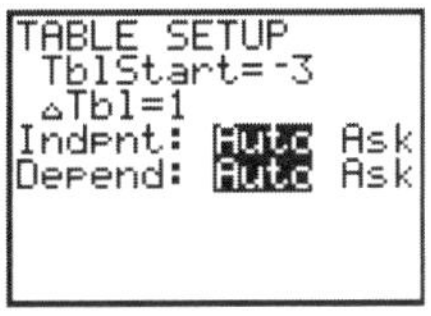

Input starting value (–3) and size of the step.

Select [2nd] [GRAPH] ([TABLE])

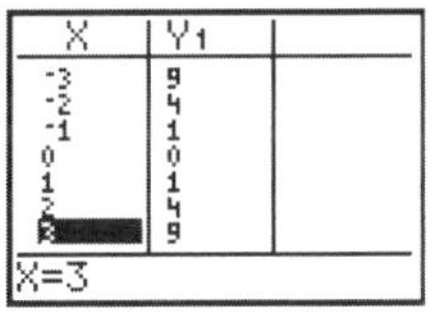

Casio solution

Select TABLE from the MENU and enter input x^2 as explained in Section 1.1.

Select G-Solv F5 (**SET**) or (**RANG**). Input the domain.

[–3. 3]. EXIT Select G–1 F6 (**TABLE**).

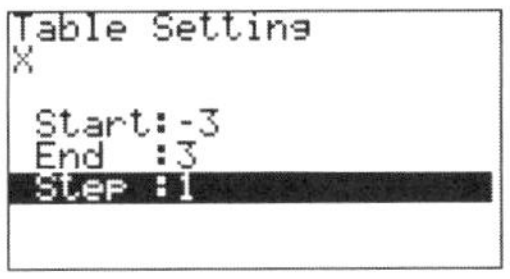

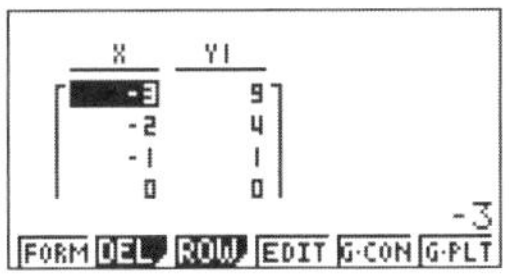

Answer: The maximum value is 9. The minimum is 0.

Select [WINDOW] and enter values from –3 to 3 for the x and 0 to 10 for the y-values.
Scales of 1 would be suitable here.

Then select [GRAPH].

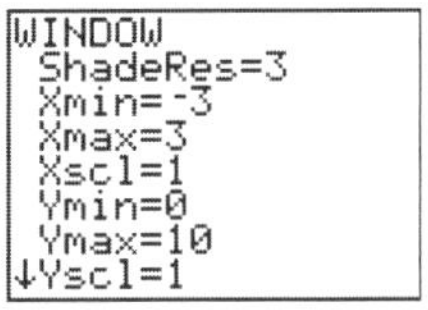

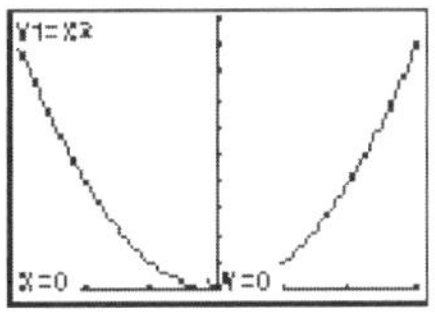

Now select GRAPH from the MENU.

Select SHIFT V-Window F3 (**VIEW WINDOW**).

Enter [–3, 3] for domain, [0, 10] for range.

Scales of 1 would be suitable here.

EXIT Select G–1 F6 (**DRAW**).

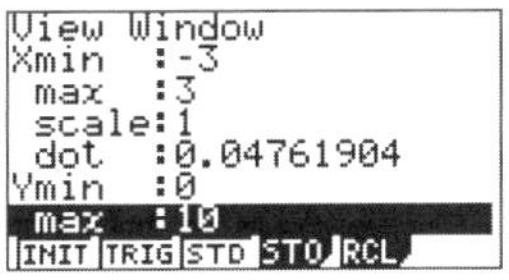

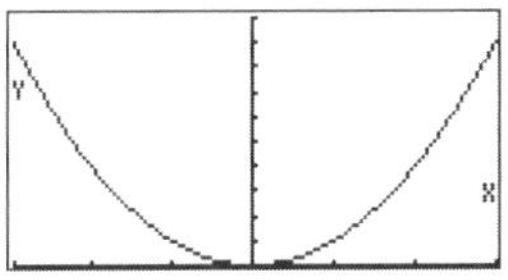

1.4.2 Linear functions

Finding points of intersection

Example 1.4.2a (from *4.2.6*)
Where do the graphs of $f(x) = x + 4$ and $g(x) = 3x + 2$ meet?

TI solution

Select [Y=].

Input [X,T,Θ,n] + 4 in Y_1 and 3 [X,T,Θ,n] + 2 in Y_2.

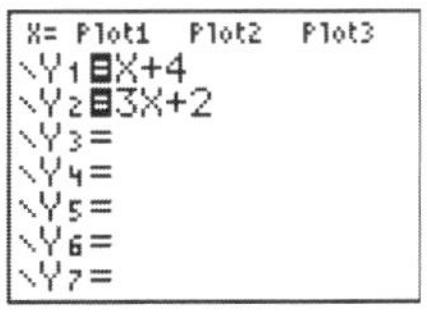

Casio solution

Select GRAPH from the MENU.

Input $Y_1 = x + 4$ and $Y_2 = 3x + 2$.

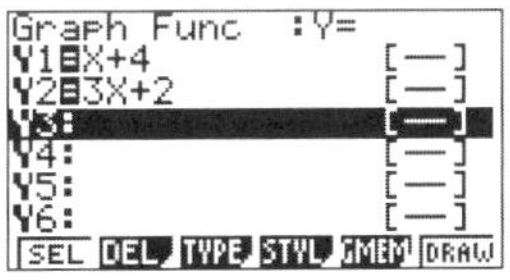

Select a standard window by pressing [ZOOM] then **6:ZStandard**.

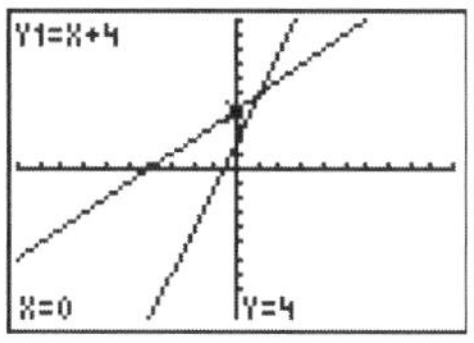

Select [CALC] ([2nd] [TRACE]).

Then select **5:intersect**.

Press [ENTER] three times to find the point of intersection.

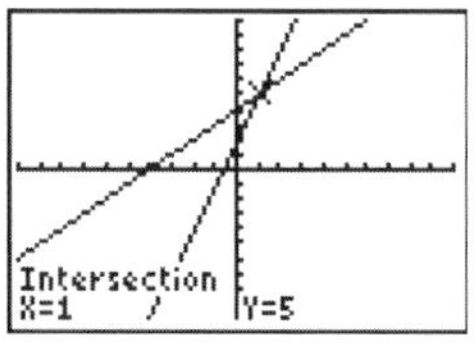

Select SHIFT F3 (V-Window) **(VIEW WINDOW)**.

Select standard screen, F3 (V-Window) **(STD)**.

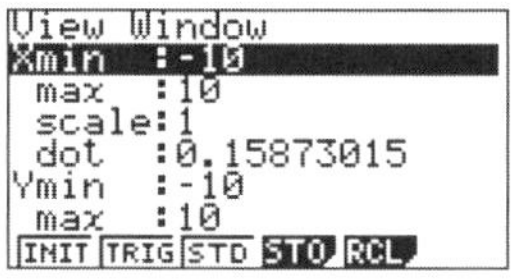

EXE Select F6 (G↔T) **(DRAW)**.

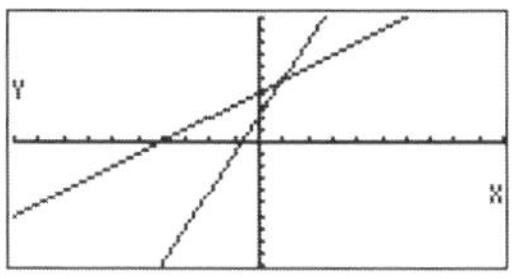

To find the point of intersection, select F5 (G-Solv) **(G-SOLV)** then F5 (G-Solv) **(ISCT)**.

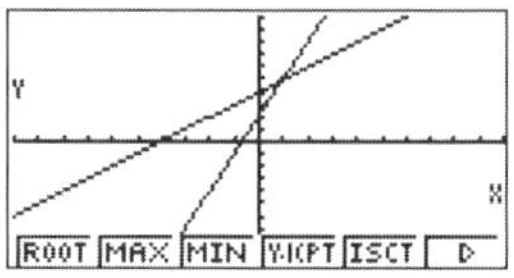

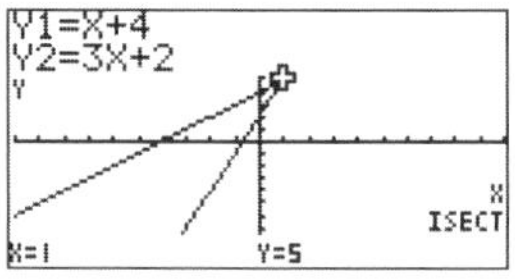

Answer: The lines meet at (1, 5).

Finding a linear function from given coordinates

Example 1.4.2b (from *4.2.3a*)

Find the linear function $f(x)$ described by the following table of values.

x	−2	−1	0	1	2
$f(x)$	−7	−4	−1	2	5

TI solution

Select MENU **EDIT** and enter the x-values in L1 and $f(x)$ values in L2.

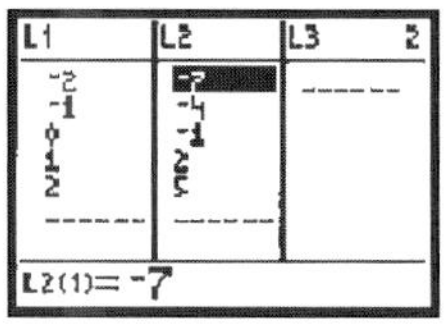

Now select [ENTER] **CALC** then **4:LineReg(ax + b)** and type **L1, L2** [ENTER].

Casio solution

Select STAT from the MENU.

Input the x-coordinates to list 1 and the y-coordinates to list 2.

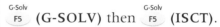

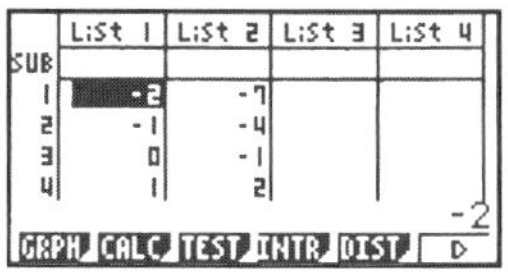

Select F2 (Zoom) **(CALC)** then F6 (G↔T) **(SET)**.

(The 2Var lists should be 1 and 2 with a frequency of 1.)

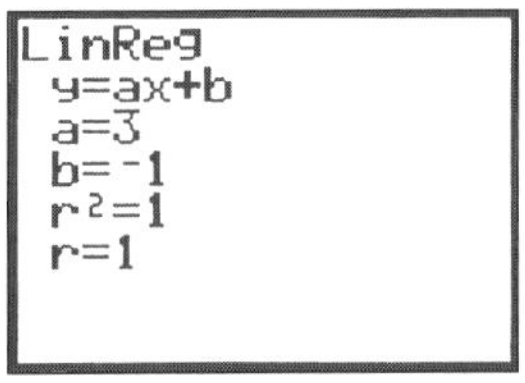

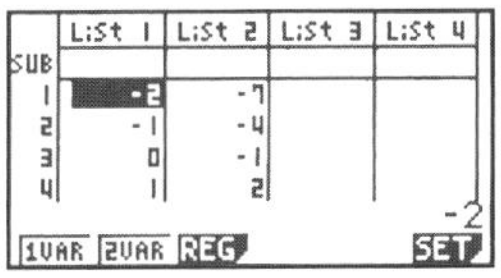

EXIT Select V-Window F3 (REG) then Trace F1 (x).

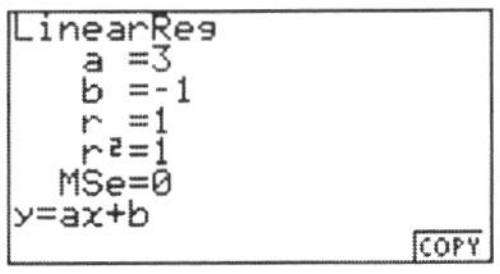

For linear functions only two coordinate pairs need to be entered. Equations of quadratic, cubic and quartic functions can also be found by regression using a minimum of 3, 4 or 5 coordinate pairs respectively. You can also fit functions like sines and exponentials this way. (Could be useful in your project!) The variable r should be shown as having a value 1. This confirms that the data fits the equation exactly.

Answer: The equation of the function is $y = 3x - 1$.

1.4.3 Graphs of quadratic functions

Vertex and axis of symmetry

Example 1.4.3a (from *4.3.3a*)

Draw the graph of $f(x) = 2x^2 - 4x + 5, \quad -2 \leq x \leq 2$.

a Find the coordinates of the vertex.

b Write down the equation of the axis of symmetry.

TI solution

Select [Y=].

Input $2x^2 - 4x + 5$.

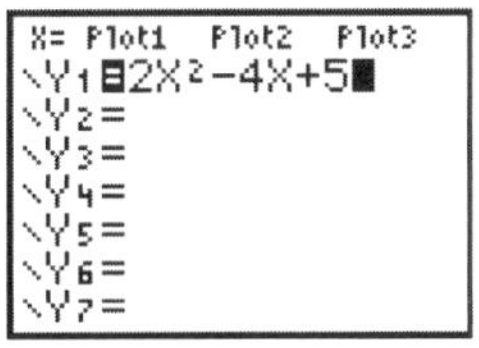

Select [TBLSET] Input the domain [–2, 2].

Select [TABLE].

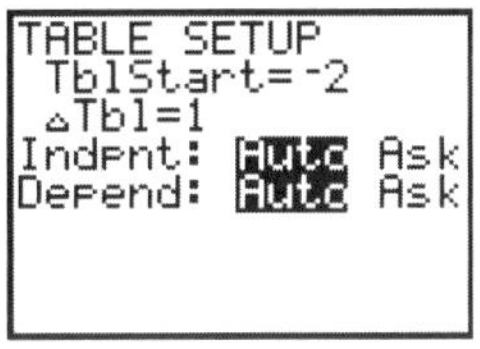

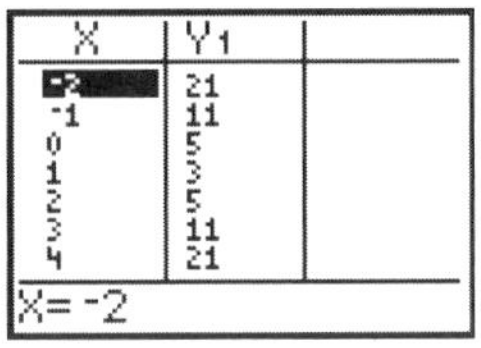

Casio solution

Select TABLE from the MENU.

Input $2x^2 - 4x + 5$.

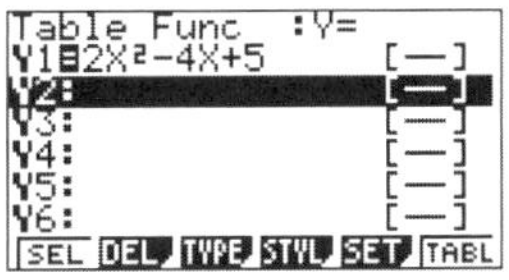

Select G-Solv F5 (SET).

Input the domain. [–2, 2].

EXIT Select G–1 F6 (TABLE)

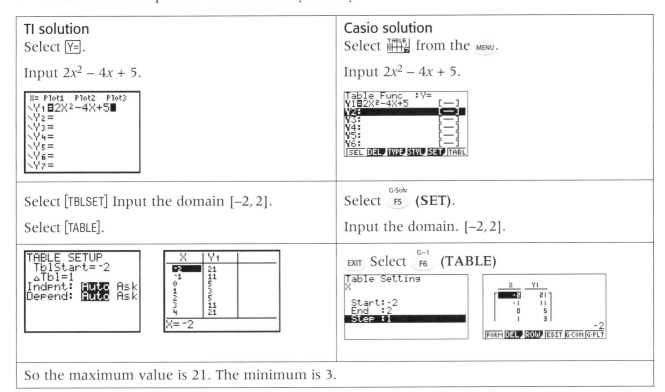

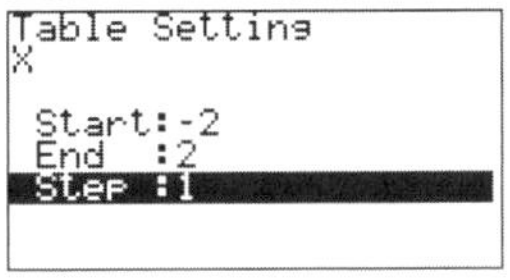

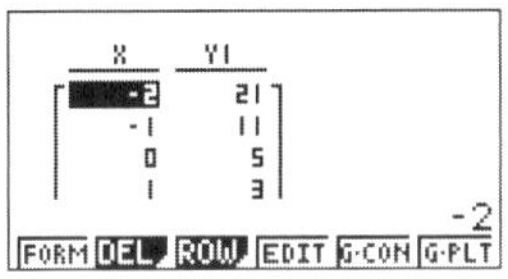

So the maximum value is 21. The minimum is 3.

Now select [WINDOW] and enter suitable values for x and y then select [GRAPH].

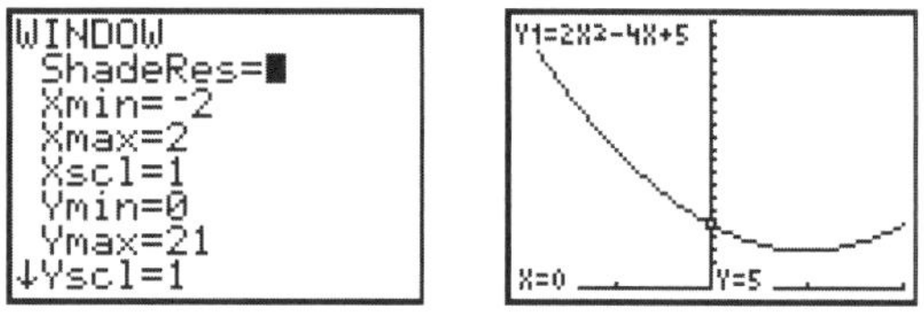

Select [2nd] [TRACE], ([CALC]) and **3:minimum**.

Place the cursor to the left of the minimum point and press [ENTER] then move it to the right and press [ENTER] then press [ENTER] one more time to get the coordinates.

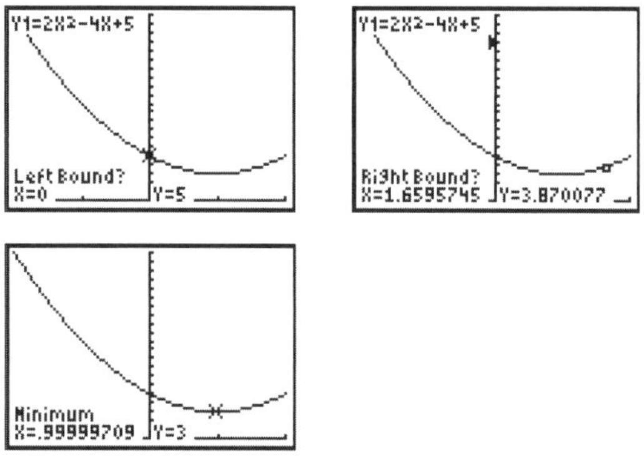

Now select GRAPH from the MENU.

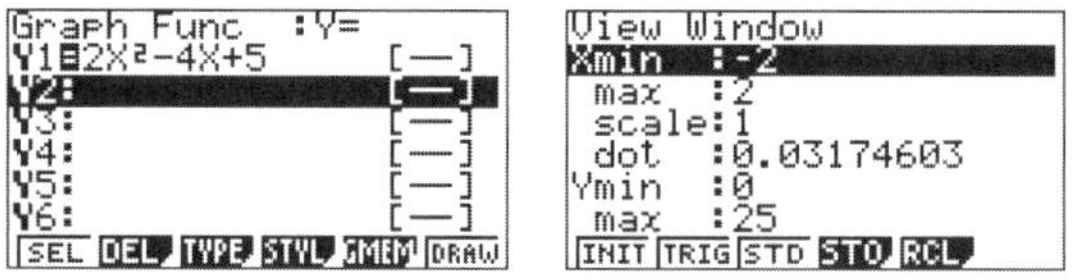

Select SHIFT F3 (V-Window) **(VIEW WINDOW)**

Input appropriate values for the domain and range EXIT.

Select F6 (G↔T) **(DRAW)**, F5 (G-Solv) **(G-SOLV)**

then F3 (V-Window) **(MIN)**.

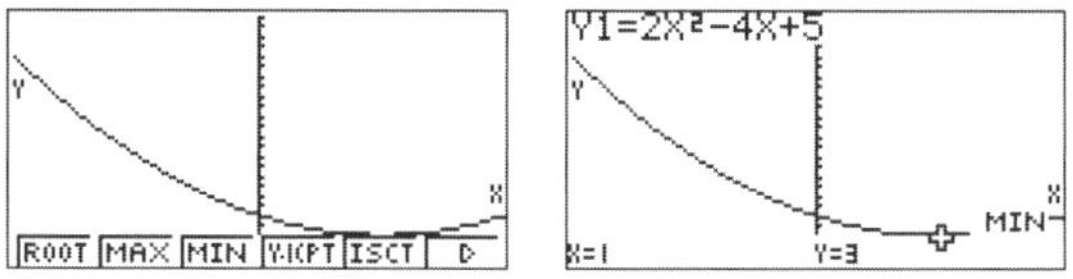

The minimum is at (1,3).

Answers: a The coordinates of the vertex are (1, 3).
b The axis of symmetry is at $x = 1$.

Graphs of quadratic functions: axis intercepts

Example 1.4.3b (from *4.3.4c*)

Find the axis intercepts for $f(x) = x^2 + 2x - 3$.

TI solution

Draw the graph of $y = x^2 + 2x - 3$ using a standard ([ZOOM] **6:ZStandard)** window.

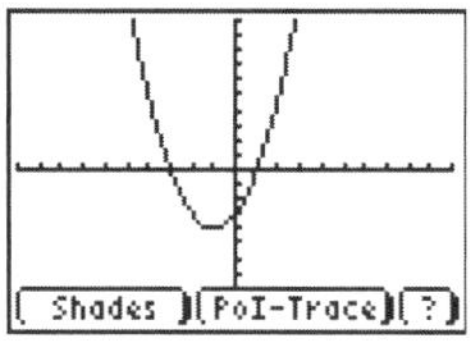

Select [CALC], **2:zero**

Place cursor to the right of the zero and press [ENTER], then to the left and press [ENTER], then press [ENTER] to get the zero.

Casio solution

Draw the graph of $y = x^2 + 2x - 3$ using a standard

SHIFT F3 (V-Window) F3 (V-Window) VIEW WINDOW.

Select F5 (G-Solv) (G-SOLV), F1 (Trace) (ROOT)

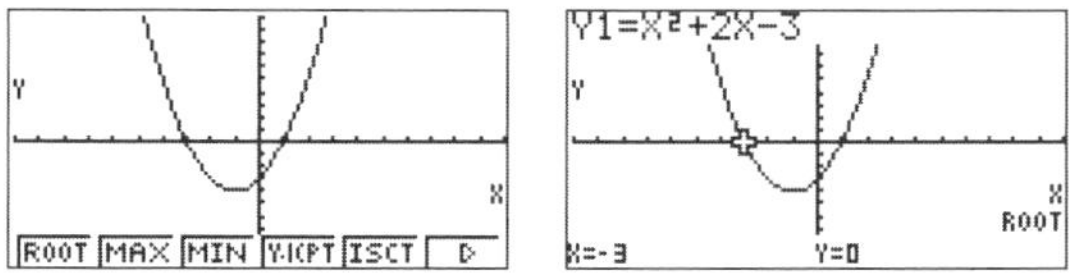

one x-intercept is –3.

Repeat this procedure to find the second zero.

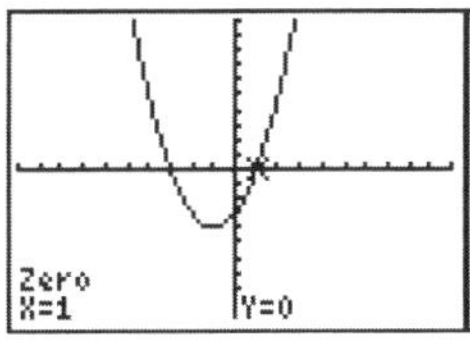

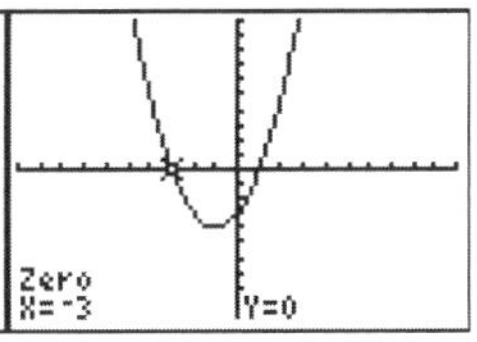

To find the y-intercept, press [CALC], **1:value, 0,** [ENTER].

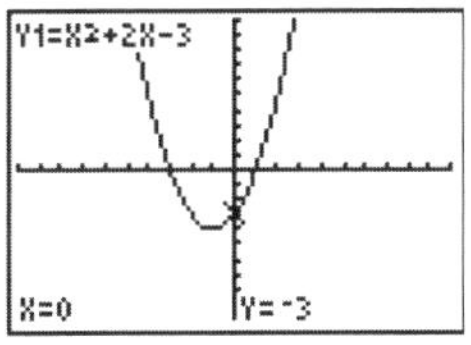

Now tap the right arrow pad (REPLAY) to find the other intercept.

Select G-Solv F5 **(G-SOLV),** Trace F1 **(Y-ICPT)** to find the y-intercept.

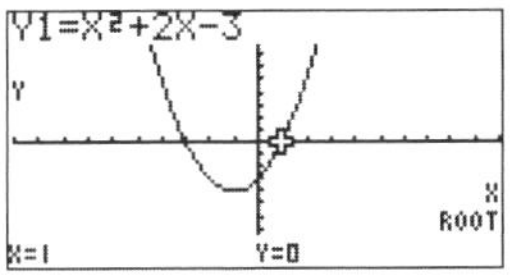

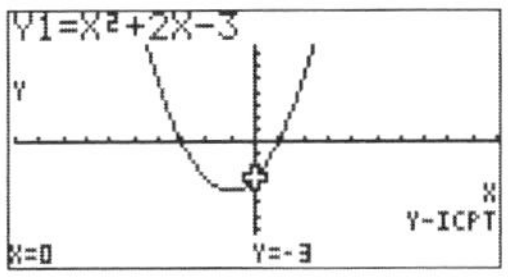

Answer: The x-intercepts are –3 and 1, the y-intercept is –3.

1.4.4 Graphs of exponential functions

Example 1.4.4 (from *4.4.3c*)

Consider the functions $y = 2^{-x}$ and $y = 2^{x}$.

Draw the graphs for $-2 \le x \le 2$, $0 \le y \le 5$.

Observe the nature and properties of the graphs.

TI solution

Select [Y=].

Input $2^{(-x)}$ into $\mathbf{Y_1}$ and 2^{x} into $\mathbf{Y_2}$.

Note the use of a bracket for the negative exponent.

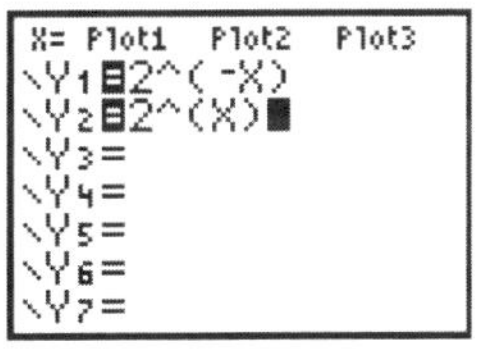

Select [WINDOW] and enter suitable values.

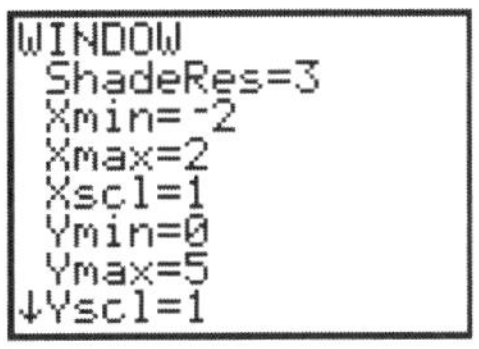

Press [GRAPH]

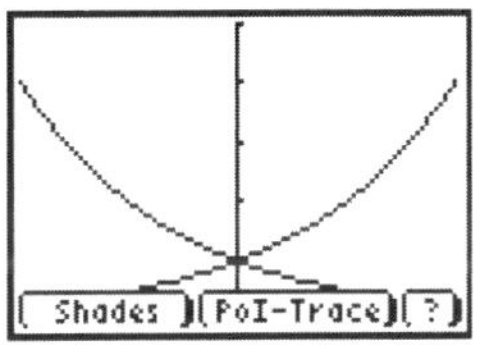

Casio solution

Select GRAPH from the MENU.

Input $\mathbf{Y_1} = 2^{(-x)}$ and $\mathbf{Y_2} = 2^{x}$.

Note the use a bracket for the negative exponent.

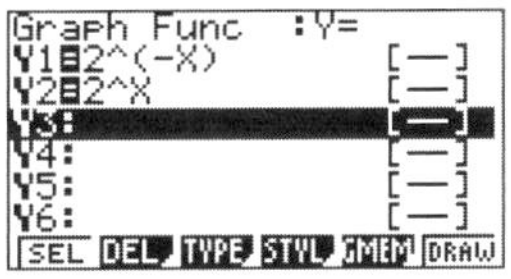

Select SHIFT V-Window F3 **(VIEW WINDOW).** Input given values for the domain and range EXIT. Select G↔T F6 **(DRAW).**

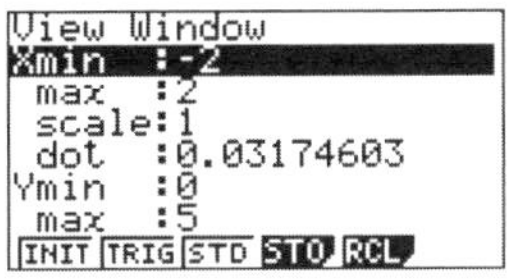

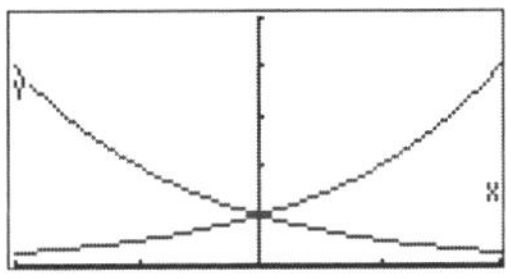

Answer: Both graphs cut the y-axis at (0,1), are asymptotic to the x-axis, and have the same shape. The graph of $y = 2^{-x}$ is the reflection of the graph of $y = 2^{x}$ in the y-axis.

1.4.5 Graphs of trigonometric functions

Example 1.4.5a (from *4.5.5a*)

Determine the amplitude of the function $f(x) = 2\sin 2x - 1$, $0° \leq x \leq 360°$.

TI solution

Select [Y=].

Input 2sin(2x) – 1 in $\mathbf{Y}_1$.

Select [WINDOW] and put in suitable values.

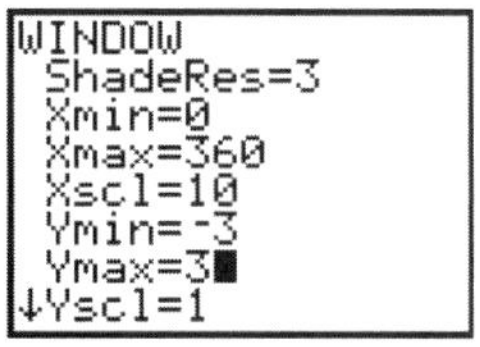

Press [GRAPH].

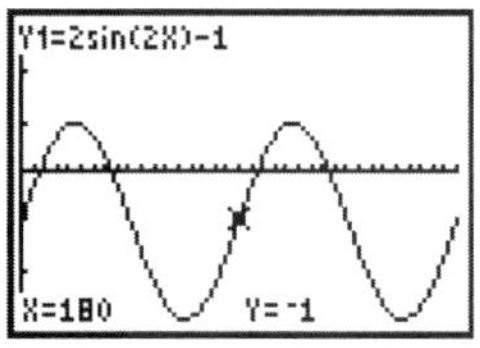

Use [CALC] to find the maximum and minimum values for the function.

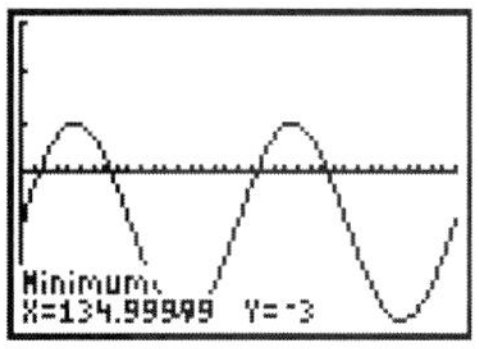

Casio solution

Select GRAPH from the MENU.

Input 2sin(2x) – 1.

Select SHIFT F3 (V-Window) **(VIEW WINDOW)**

then select F2 (Zoom) **(TRIG)**.

Input the given values for the domain.

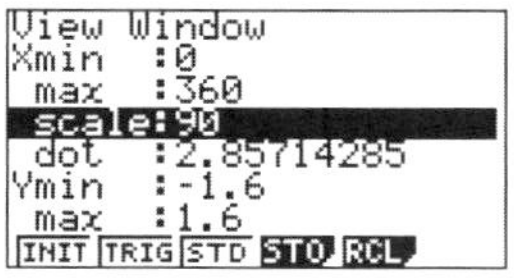

EXIT Select F6 (G↔T) **(DRAW)**.

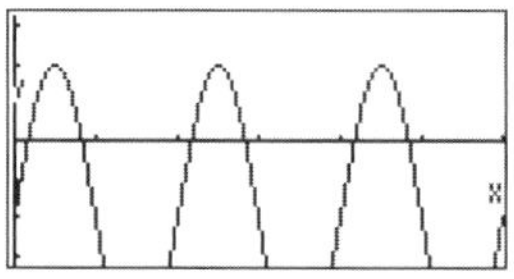

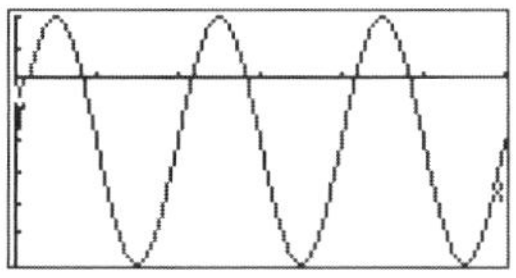

Select F2 (Zoom) F5 (G-Solv) **(ZOOM AUTO)** to fit the graph to the screen.

Select F5 (G-Solv) **(G-SOLV)** then F3 (V-Window) **(MIN)** and F5 (G-Solv) then F2 (Zoom) **(MAX)** to find the minimum and maximum values for the function.

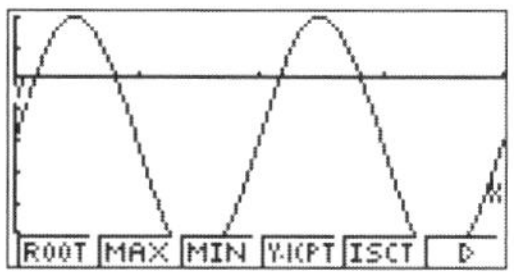

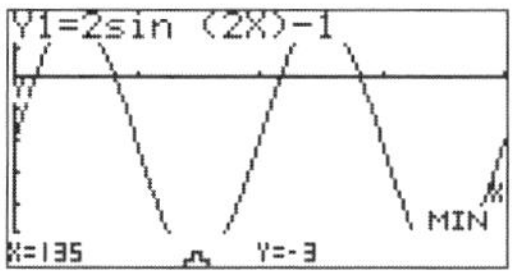

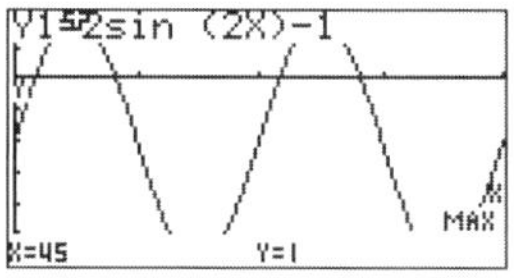

Answer: Amplitude $= \dfrac{-3+1}{2} = 1$

Solving trigonometric equations: using intersecting lines

Example 1.4.5b (from *4.5.8c*)

Solve for x given $2\cos x - 1 = -2$, $-360° \le x \le 0°$.

TI solution

Select [Y=].

Input $2\cos(x) - 1$ in Y_1 and -2 in Y_2.

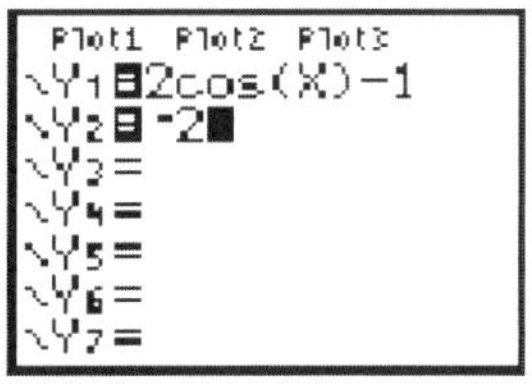

Select [WINDOW]

Input the given values for the domain.

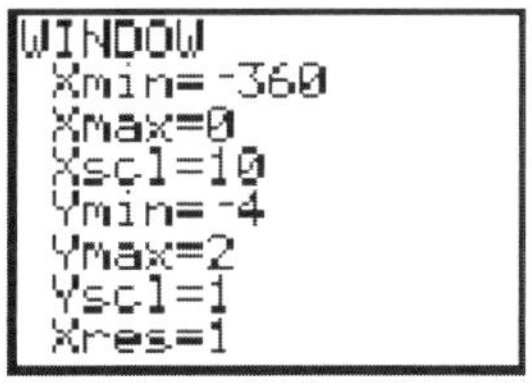

Press [GRAPH].

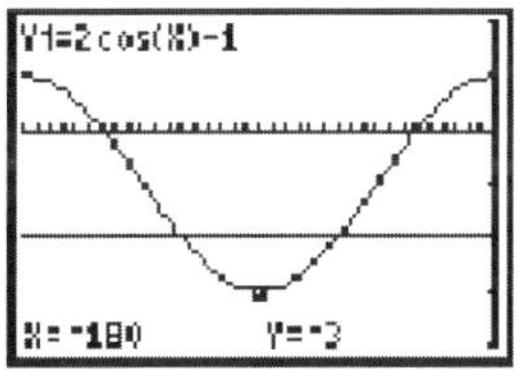

Select [CALC], **5:intersection.**

Place cursor close to first point of intersection and press [ENTER] three times. Repeat the process for the other points of intersection.

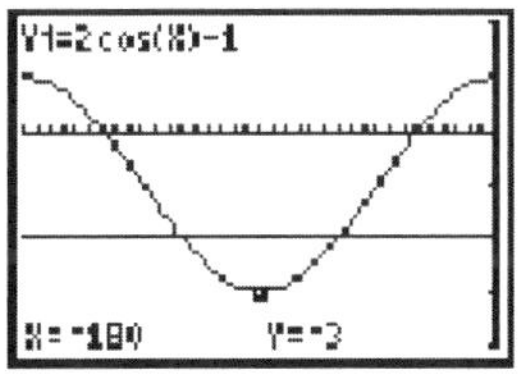

Casio solution

Select GRAPH from the MENU.

Input **Y1 :** $2\cos(x) - 1$ and **Y2 :** -2.

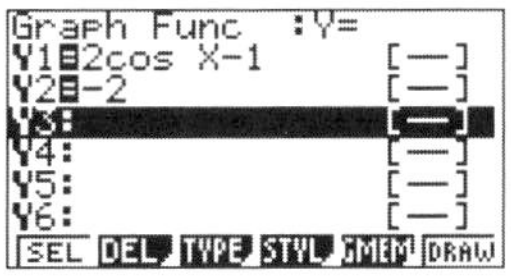

Select **VIEW WINDOW** SHIFT F3 (V-Window)

then select **TRIG** F2 (Zoom) Input the given values for the domain.

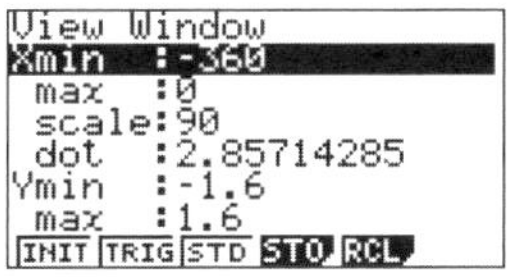

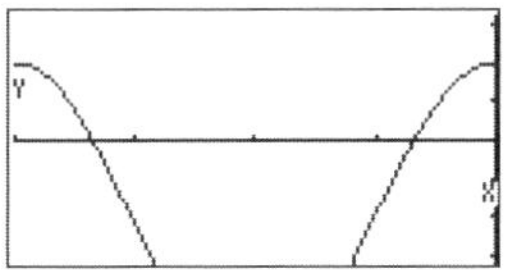

EXIT Select **DRAW** F6 (G↔T).

Use **ZOOM AUTO** F2 (Zoom) F5 (G-Solv) to fit the graph to the screen.

Select **G-SOLV** F5 (G-Solv) then **ISCT** F5 (G-Solv) to find where the lines intersect.

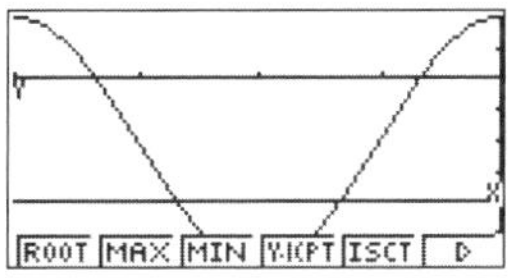

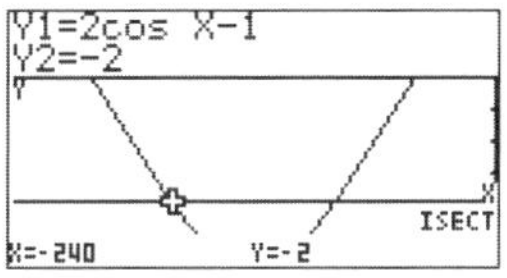

Tap the right arrow (REPLAY) to find the other intercept.

Answers: $x = -240°$ and $-120°$

1.4.6 Accurate graph drawing

There is nothing special to add under this title. The GDC can assist you in planning your graphs, deciding on scales and positions of axes. It can be used to check your graphs after they are drawn.

1.4.7 Unfamiliar functions

The GDC is a very valuable tool when you need to plot an unfamiliar function.

Finding asymptotes: use of ZOOM AUTO and VIEW WINDOW

Example 1.4.7a (from *4.7.5c*)

Identify the asymptotes of $y = \dfrac{x+1}{x^2 - 2x - 8}$.

TI solution

The graph of $y = \dfrac{x+1}{x^2 - 2x - 8}$ is drawn below using a standard view window.

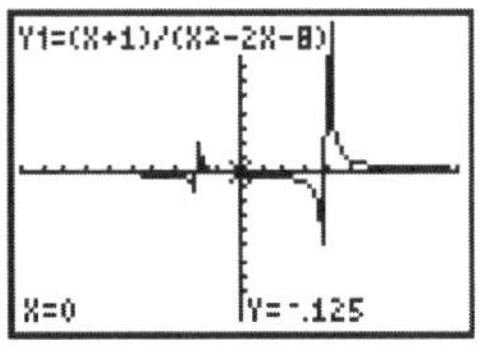

The asymptotes cannot be seen clearly.

Try **ZoomFit**.

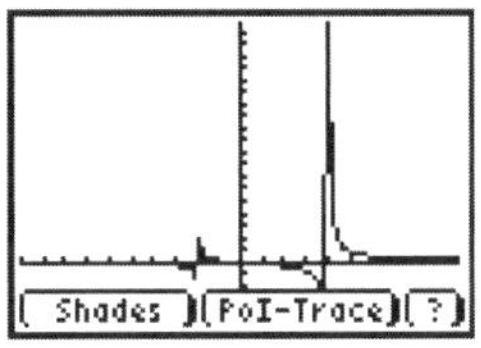

The asymptotes are still not clear.

Select [WINDOW].

Input values to narrow the domain and range.

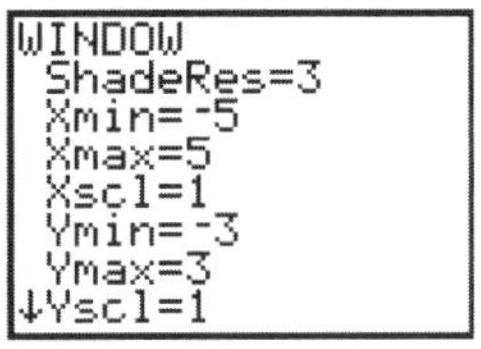

The asymptotes are clear.

Use [TRACE] to identify the asymptotes.

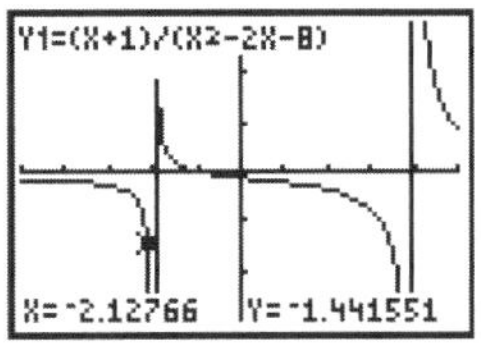

Casio solution

The graph of $\dfrac{x+1}{x^2 - 2x - 8}$ is drawn below using a standard view window.

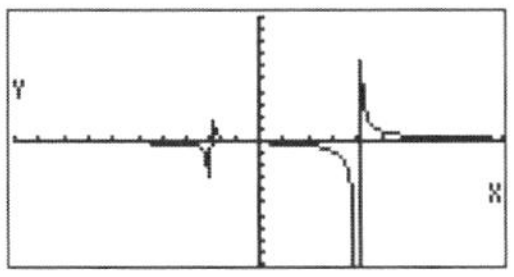

The asymptotes cannot be seen clearly.

Try Zoom F2 G-Solv F5 **(ZOOM AUTO)**

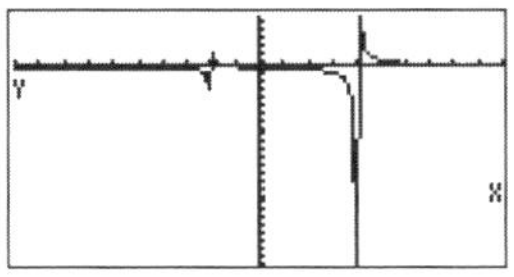

The asymptotes are still not clear.

Select SHIFT V-Window F3 **(VIEW WINDOW).**

Input values to narrow the domain and range.

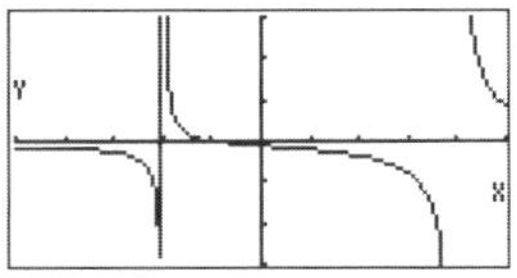

The asymptotes are clear.

Use SHIFT Trace F1 **(TRACE)** to identify the asymptotes.

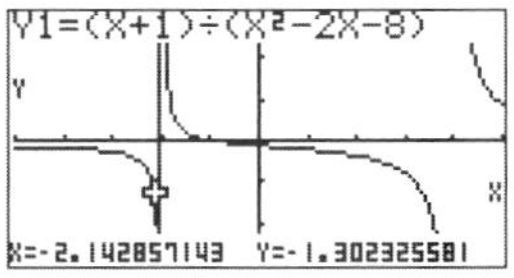

Answer: Horizontal asymptote at $y = 0$, vertical asymptotes at $x = -2$ and $x = 4$.

Finding vertical asymptotes using the TABLE facility

Example1.4.7b (from *4.7.5c*)

Identify the vertical asymptotes of $y = \dfrac{x+1}{x^2 - 2x - 8}$.

TI solution

Input $\dfrac{x+1}{x^2-2x-8}$ in $\mathbf{Y_1}$.

Select [TABLE].

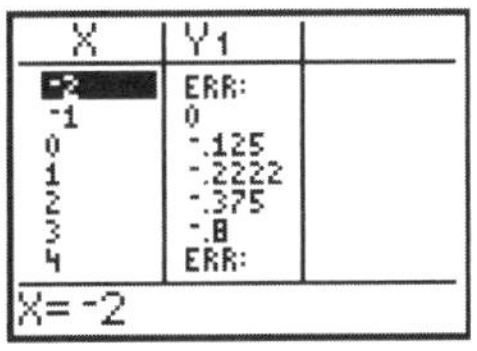

The asymptotes appear as **ERROR** in the lists.

Casio solution

Select TABLE from the MENU.

Input $\dfrac{x+1}{x^2-2x-8}$.

Select G–1 F6 **(TABL)**.

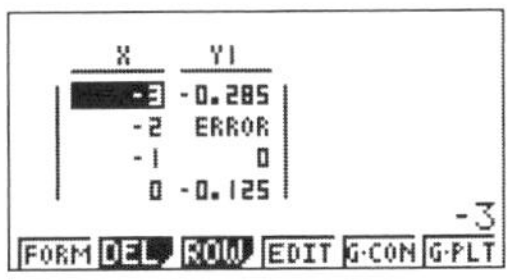

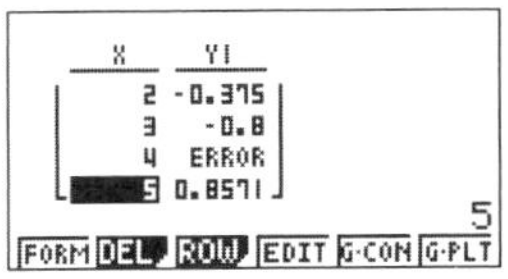

The asymptotes appear as **ERROR** in the lists.

Answer: $x = -2$ and $x = 4$.

1.4.8 Combinations of unfamiliar functions

Solving more complicated equations: using intersecting lines

Example 1.4.8 (from *4.8*)

Solve for x: $x - 2 = \dfrac{1}{x}$.

TI solution

Select [Y=].

Input $x - 2$ in $\mathbf{Y_1}$ and $\dfrac{1}{x}$ in $\mathbf{Y_2}$.

Draw the graphs using a standard view window.

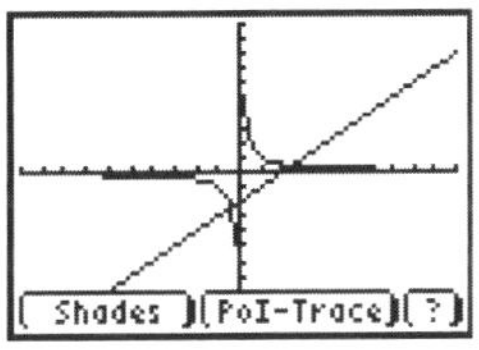

Select [CALC], **5:intersection**.

Place cursor close to intersection point and press [ENTER] three times. Repeat process to find the other point of intersection.

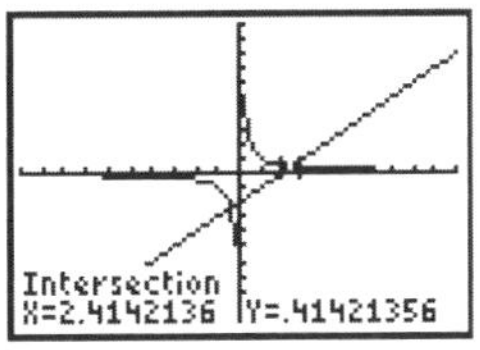

Casio solution

Select STAT from the MENU.

Input $\mathbf{Y_1} = x - 2$ and $\mathbf{Y_2} = \dfrac{1}{x}$.

Draw the graphs using a standard view window.

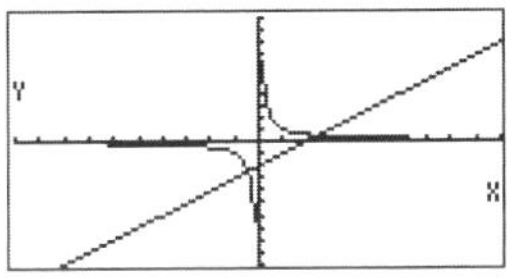

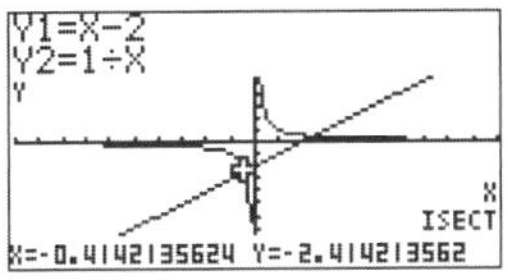

Select G-Solv F5 **(G-SOLV)** then G-Solv F5 **(ISCT)** to find where the lines intersect.

Tap the right hand arrow REPLY to find other points of intersection.

Answer: $x = -0.414$ (3 s.f.) and $x = 2.41$ (3 s.f.).

1.5 Geometry and trigonometry

Use of the GDC in this topic is limited to basic numerical calculations involving trigonometric functions and geometric formulae. Candidates are **strongly reminded** that they must ensure that their GDC is set in **DEGREE** [MODE].

1.6 Statistics

1.6.1 Statistical functions of the GDC

Here are some general descriptions of the facilities you will use to study statistics on the GDC.

TI

Press [2nd] [Y=] [STAT PLOT] [ENTER], **Plot1 ON** [ENTER] scroll down to the type of graph that you want to display – fill in the lists that your data is in – set your window – press [GRAPH].

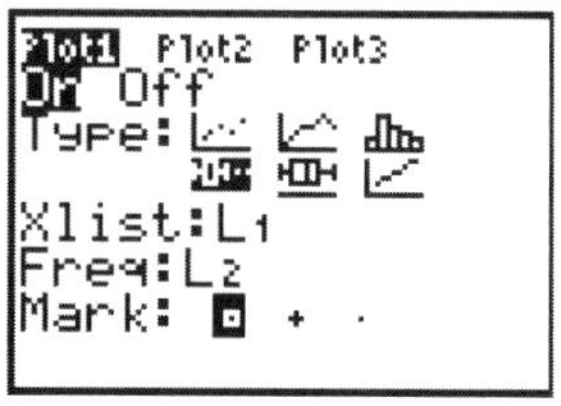

Scatter plot

Frequency polygon

Frequency histogram

Boxplot with outliers separate

Boxplot with outliers included

If a non-statistical graph appears as well as the statistical one you want, then you will need to choose **PLOTSOFF** [ENTER] first. Then turn on the individual statistical plot required.

It is also a good idea to turn off the individual plots within the **STATPLOT** menu when you have finished with them.

EDIT – allows you to enter data in lists, sort the lists in ascending (**SortA:**) or descending (**SortD:**)order and clear the lists. This feature was discussed in Section 1.1.6.

CALC – allows you to calculate various statistics.

Casio

Select STAT from the MENU.

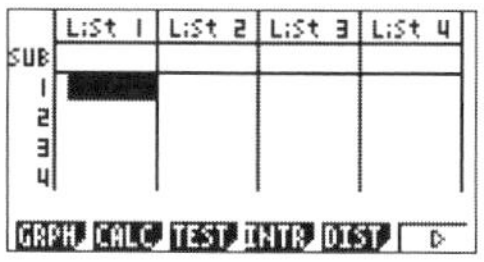

F1 (Trace) **(GRPH)** – allows you to draw various graphs from the lists of data including scatter diagrams, histograms, box plots.

F2 (Zoom) **(CALC)** – allows you to calculate various statistics.

F3 (V-Window) **(TEST)** – allows you to perform various tests for example, χ^2 test.

F5 (G-Solv) **(DIST)** – a number of statistical distributions including normal and chi-squared and *t*-distribution. Can be useful for projects.

Press F6 (G↔T) for more selections.

F1 (Trace) **(TOOL)** – sorting lists in ascending or descending order.

F3 (V-Window) **(DEL)** – deletes individual values.

F4 (Sketch) **(DEL-A)** – deletes all entries in a list.

You will use **1:1-Var Stats** **2:2-Var Stats** and **4:LinReg(ax+b)** often. You might also find the other regression commands useful in your project. 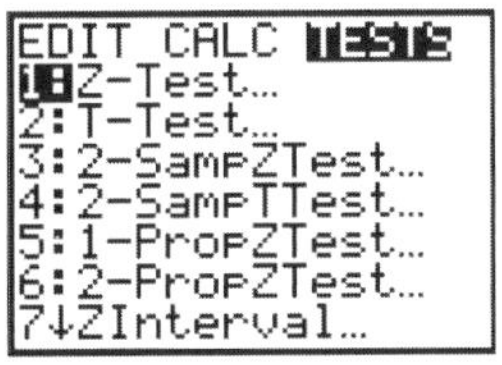 **TESTS** – allows you to perform various tests. You will use this for performing $\chi\chi^2$ tests but it might also be useful in a project for doing a *t*-test.	

1.6.2 Frequency polygons

Example 1.6.2 (from *6.2.2*)

The table shows the time, in minutes to the nearest three minutes, of 100 telephone calls made from the same cell phone in one month. Draw a frequency polygon to represent this information.

Time (*t* minutes)	Frequency
3	42
6	23
9	8
12	5
15	6
18	4
21	4
24	2
27	6

TI solution	Casio solution
Enter the data into **L1** and **L2**. 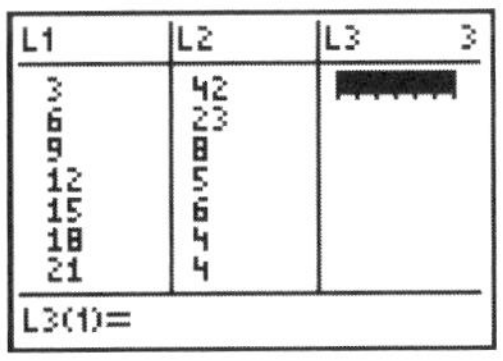 	Select STAT from the MENU. 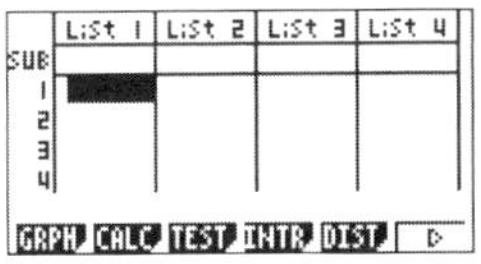
2nd MODE (QUIT) 2nd Y= **(STAT PLOT)**, ENTER, Plot1 On, ENTER. 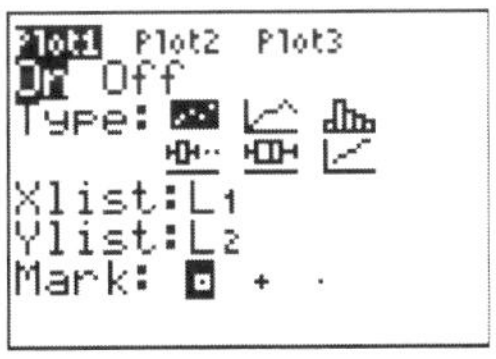 	Enter the data into **List 1** and **List 2**. 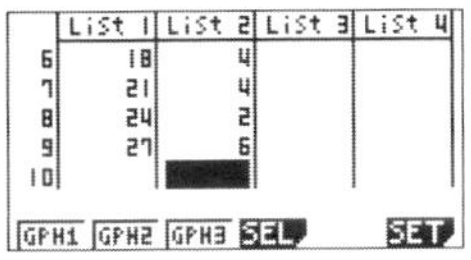 Select F1 (Trace) (GRPH). Select F6 (G↔T) (SET).

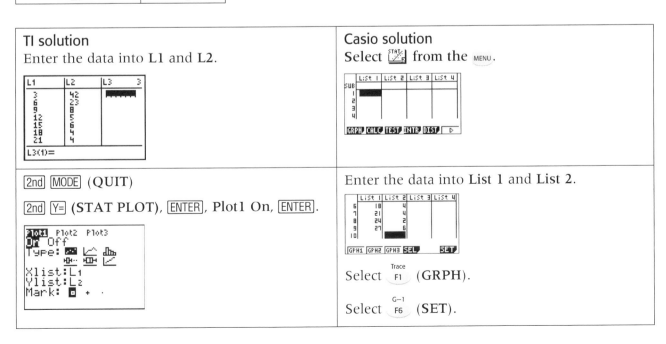

Scroll down to **Type:** – select the second graph ⌐⌐, ENTER.

```
Plot1 Plot2 Plot3
On Off
Type: ⌐⌐  ▩  ▥
      ▥  ▥  ⌐⌐
Xlist:L1
Ylist:L2
Mark: ■ + ·
```

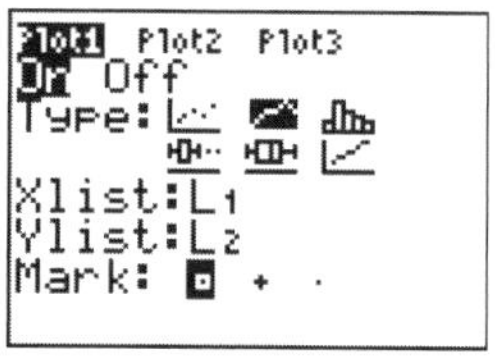

Enter **L1** in **XList** and **L2** in **YList**.

Choose a suitable window or use ZOOM 9: **ZoomStat**.

```
WINDOW
 Xmin=0
 Xmax=30
 Xscl=1
 Ymin=0
 Ymax=50
 Yscl=1
 Xres=1
```

Press GRAPH.

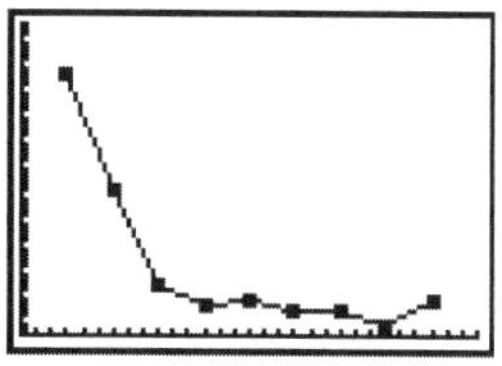

Graph Type – press F2 (Zoom) (*x y*)

F1 (Trace) (List 1) F2 (Zoom) (List 2) F1 (Trace) (1).

EXIT Select **GPH1** F1 (Trace).

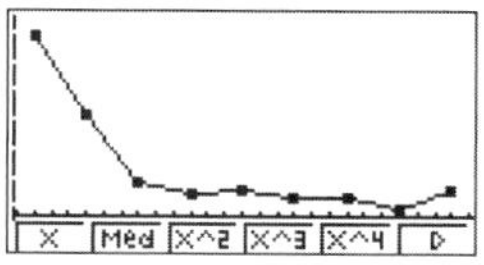

1.6.3 Frequency histograms

Example 1.6.3 (from *6.3.1a*)

A survey was carried out in a shopping mall to find out how old people were when they passed their driving test. 150 people between the ages of 15 and 92 were interviewed and the results are shown in the table below.

Plot a histogram for this data.

Age (*x*)	Frequency
$15 \leq x < 25$	85
$25 \leq x < 35$	33
$35 \leq x < 45$	14
$45 \leq x < 55$	8
$55 \leq x < 65$	3
$65 \leq x < 75$	4
$75 \leq x < 85$	2
$85 \leq x < 95$	1

TI solution

Enter the data into **L1** and **L2**.

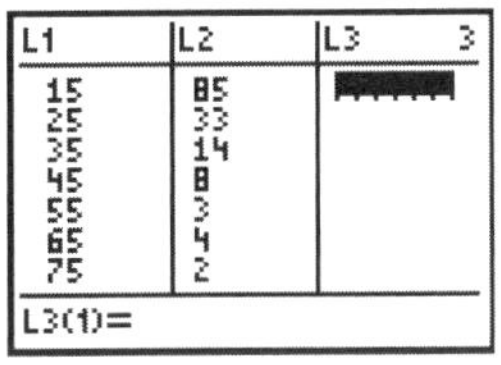

[2nd][MODE] (**QUIT**).

[WINDOW]– make sure that the *x*-scale is the size of the bars (width 10) or use **ZoomStat** as before.

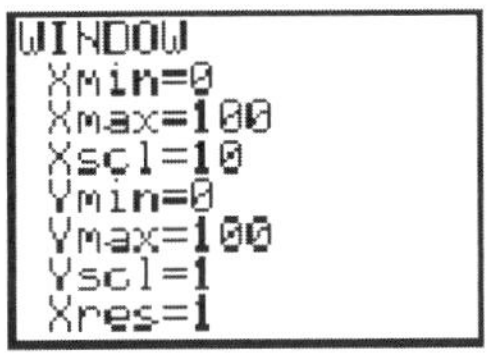

[2nd] [Y=](**STAT PLOT**), [ENTER], **Plot1 On**, [ENTER].

Scroll down to **Type:** – select the third graph, [ENTER].

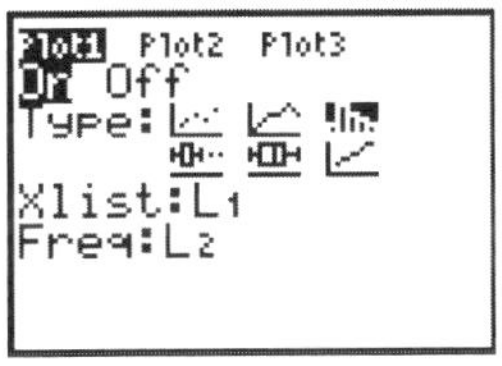

Enter **L1** in **XList** and **L2** in **Freq**.

Press [GRAPH].

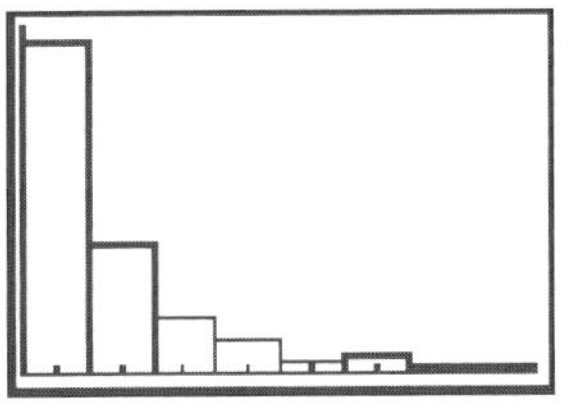

Casio solution

Select STAT from the MENU.

Enter the data into List 1 and List 2

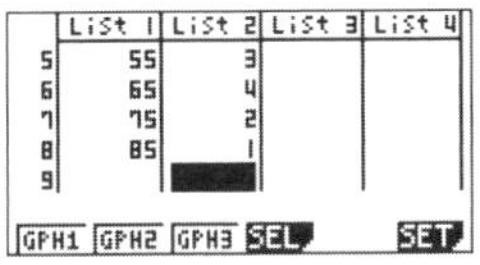

Select F1 (Trace) (**GRPH**).

Select F6 (G–1) (**SET**).

Graph Type – press F6 (G–1)

and choose F1 (Trace) (**Hist**).

XList is List 1 F1 (Trace).

Frequency is List 2 F3 (V-Window).

EXIT F1 (Trace)

Input 15 for **Start** and 10 for **Pitch** (bar width).

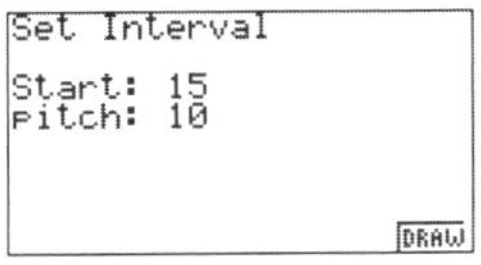

F6 (G–1) (**Draw**)

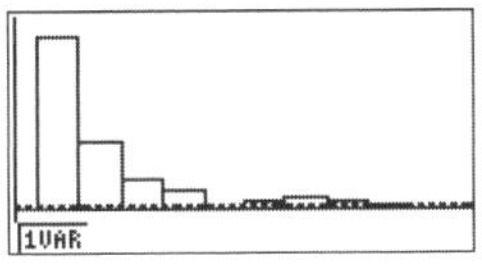

1.6.4 Box plots

Example 1.6.4 (from *6.4.4b*)

The temperature each day, in degrees Celsius, at 12 noon in Tokyo in the month of July was 31°, 28°, 30°, 27°, 46°, 32°, 31°, 28°, 30°, 27°, 30°, 31°, 30°, 30°, 28°, 29°, 32°, 27°, 29°, 30°, 31°, 30°, 32°, 27°, 29°, 30°, 31°, 30°, 28°, 31°, 31°.

a Represent this information in a box and whisker plot.

b Are there any outliers?

TI solution

a [STAT] – EDIT – [ENTER].

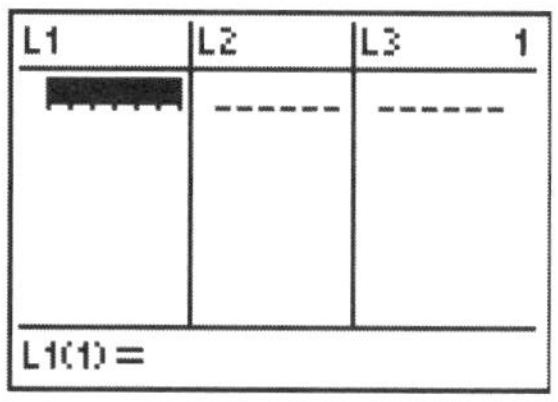

Make a frequency table and enter into [L1] and [L2].

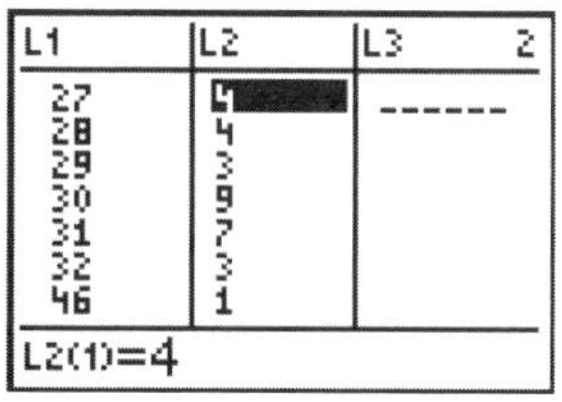

[STAT PLOT]– [ENTER]–

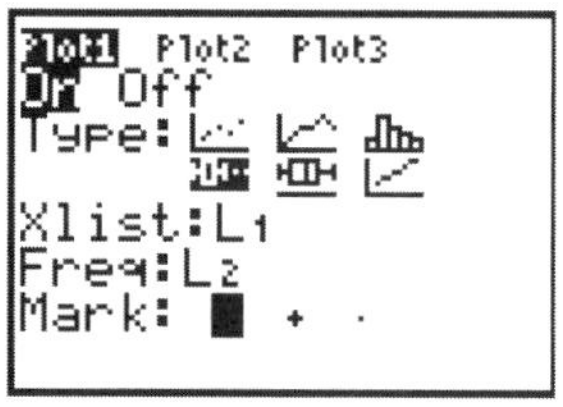

Set your window – [GRAPH].

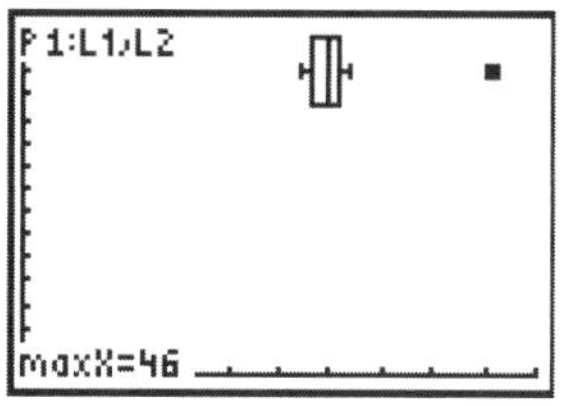

If you press the [TRACE] button then you can scroll to get the five-figure summary and any outliers.

Casio solution

a Select STAT from MENU.

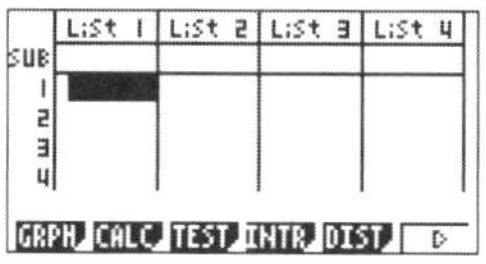

Make a frequency table and enter into **List 1** and **List 2**.

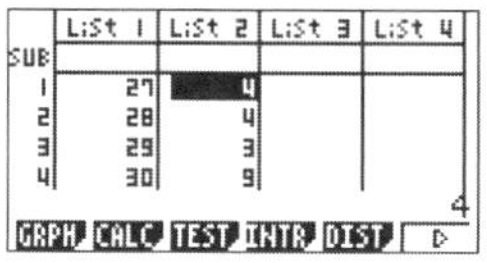

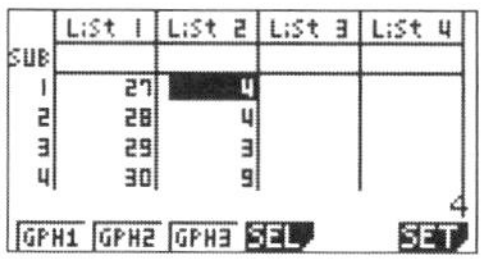

Select F1 **(GRPH)**.

Select F6 **(SET)** **Graph Type** –

press F6

and choose F2 **(Box)**.

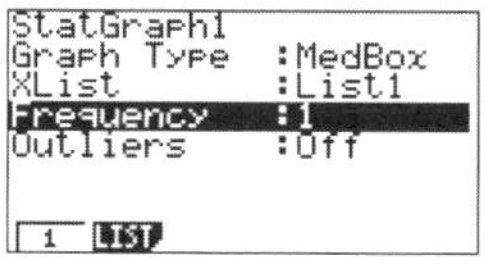

The frequency is **List 2** – Select **LIST** F2.

Input **2.** EXE Outliers **ON** F1.

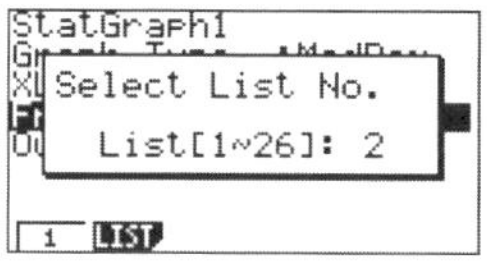

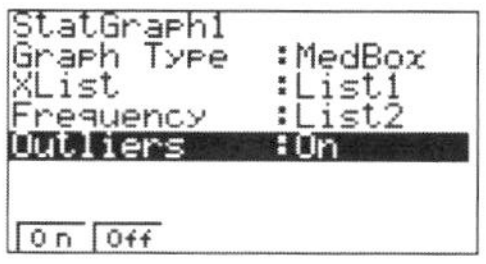

EXIT Select **GPH1** F1.

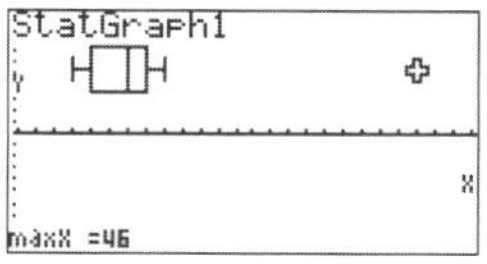

If you press **L** F1 **(Trace)** then you can scroll to get the five-figure summary and any outliers.

The box plot can also be plotted using the raw data without first creating a frequency table.

Answer: b 46 is an outlier.

1.6.6 Measures of dispersion

Example 1.6.6 (from *6.6.3a*)

Find the mean and standard deviation of the following numbers:

3 4 4 7 10 11 12 12 13 15

Each value appears with frequency 1 so this is a **1-VAR** problem.

TI solution

Enter the numbers into [L1].

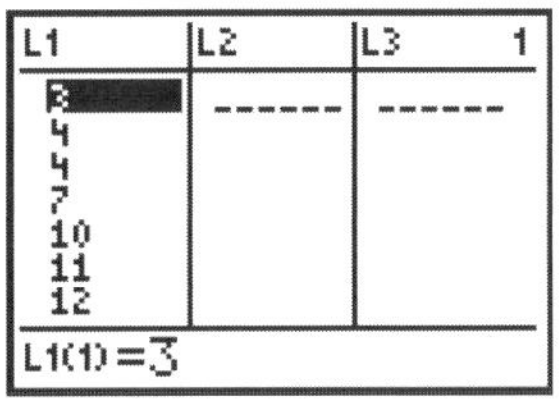

[STAT] – [CALC]– **1-VAR Stats** – [L1] ([2nd] 1) – [ENTER]

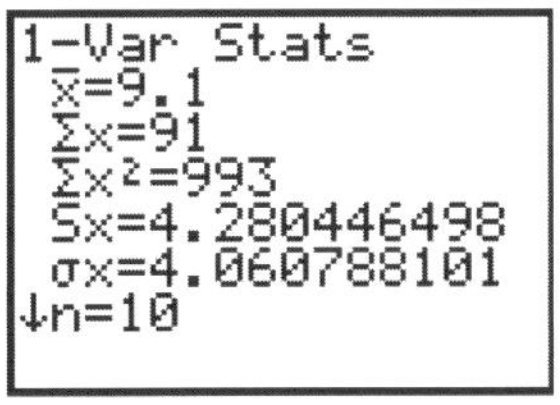

When you have data in **L1** and frequencies in **L2** then you must type **L1, L2** to get the correct answers.

In mathematical studies we select the smaller value of the standard deviation: *σx*.

Casio solution

Select STAT from MENU.

Enter the numbers into **List 1**.

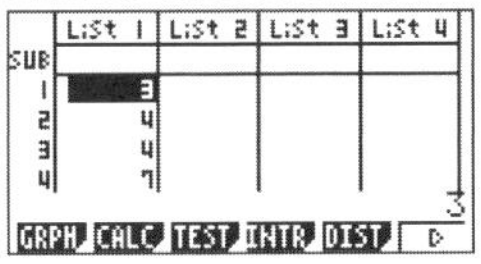

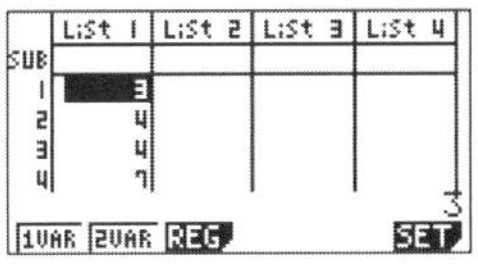

Select **CALC** (Zoom F2). Select **SET** (G–1 F6).

The data is in **List 1**. Each entry occurs once so the frequency is 1.

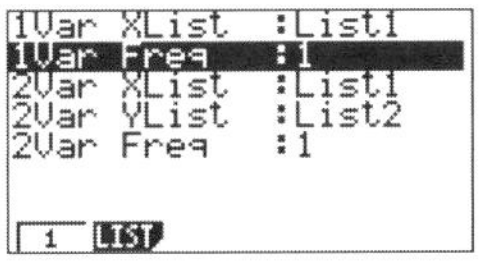

EXIT Select **1VAR** (Trace F1).

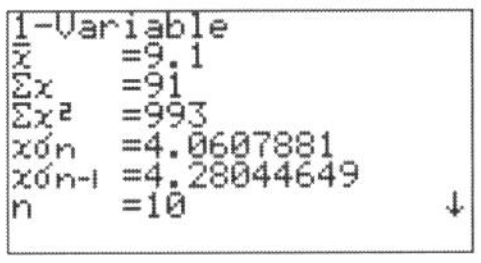

Select *σx* for the standard deviation.

Answer: Mean is 9.1, standard deviation is 4.06.

1.6.7 and 1.6.8 Scatter diagrams and regression

Example 1.6.7a (from *6.7.1a*)

It is thought that the number of observed specimens of jellyfish in an atoll depends on the temperature. A record of the temperature and the number of jellyfish was kept over a two-week period.

Plot these points on a scatter diagram and comment on the type of correlation.

Temperature (°C)	20	21	22	19	24	26	31	27	24	21	21	20	21	19
Number of jellyfish	135	138	150	135	162	201	263	221	168	155	149	152	148	124

TI solution

Enter the data into [L1] and [L2].

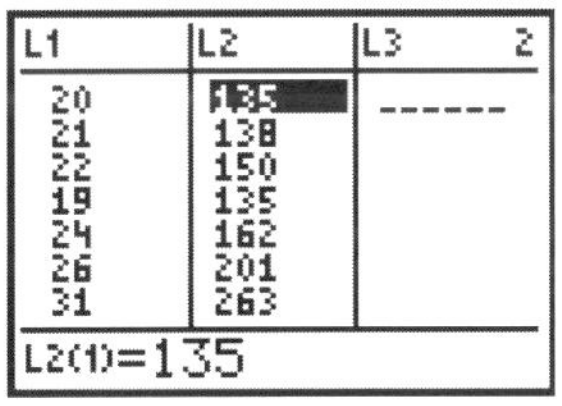

Casio solution

Select STAT from MENU.

Enter the data into **List 1** and **List 2**.

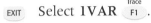

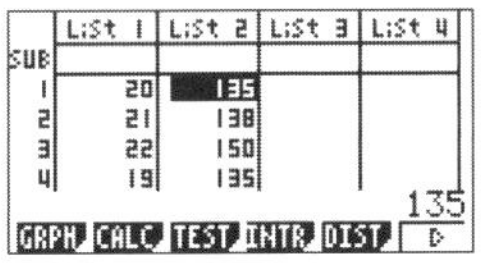

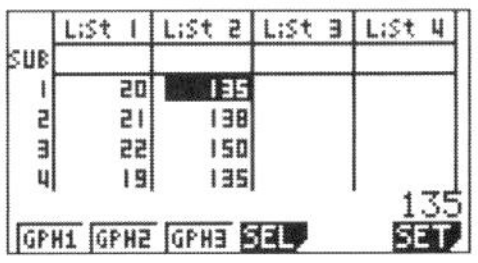

[STAT PLOT] – **Plot On**

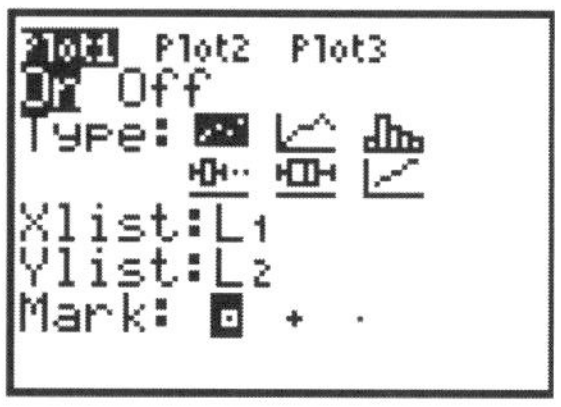

Set appropriate [WINDOW] then [GRAPH] or use [ZOOM] 9: **ZoomStat**.

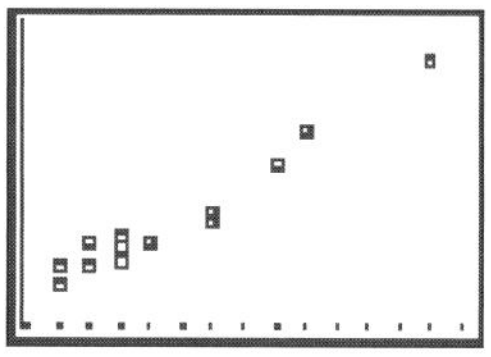

Select Trace F1 **(GPH1)**, select G–1 F6 **(SET)**.

Graph type – **SCAT** Frequency is 1

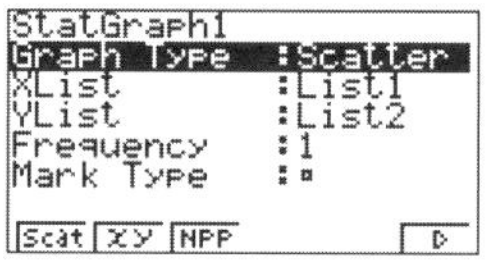

EXIT Select Trace F1 **(GPH1)**

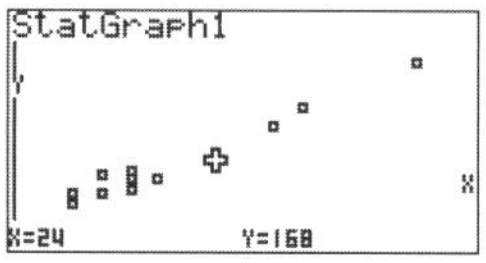

The graph shows strong linear correlation. This can be confirmed by calculating the regression coefficient r (see Example *1.6.7b*).

Example 1.6.7b (from *6.7.5a*)

The data given below for the Dutch football first division show the position of the team and the number of goals scored. Given that $s_{xy} = -74.4$, calculate the correlation coefficient, r, and comment on this value.

Position	1	2	3	4	5	6	7	8	9	10	11	12	13	14	15	16	17	18
Goals	69	62	58	49	44	43	43	41	39	38	35	33	28	27	27	24	20	8

TI solution

Enter the data into **L1** and **L2**.

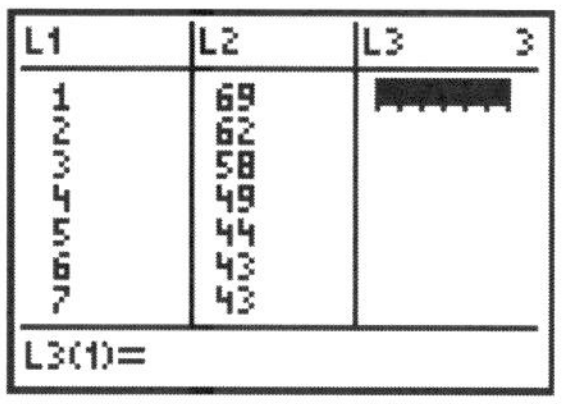

STAT – [CALC]– **4** – [L1] , [L2]– [ENTER]

The correlation coefficient r is calculated.

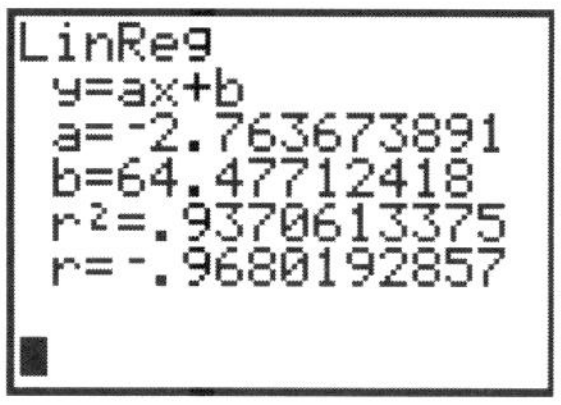

You are also given the equation of the regression line of y on x.

Casio solution

Select STAT from MENU.

Enter the data into **List 1** and **List 2**.

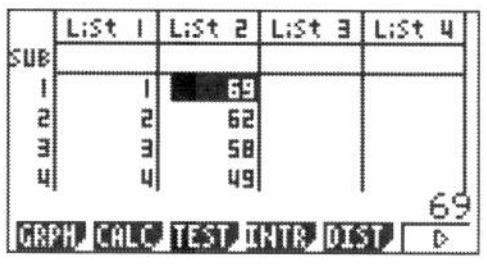

Select Zoom F2 **(CALC)**. Select G–1 F6 **(SET)**.

2VAR Xlist is Trace F1 **List 1**

2VAR Ylist is Zoom F2 **List 2 I-Var Freq** is set to 1.

EXIT

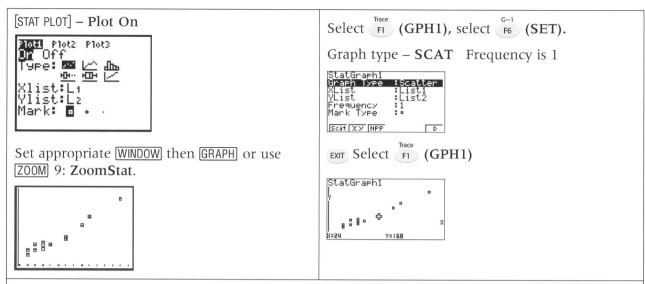

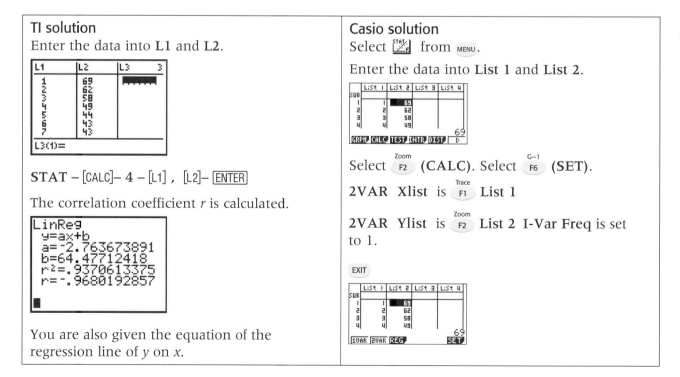

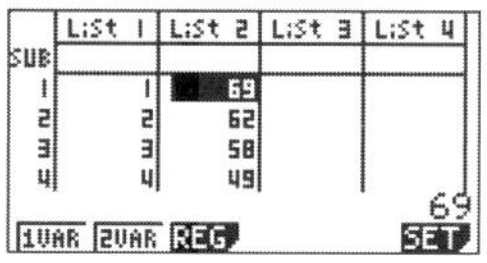

If you do not see all of the above items you need to turn on 'Diagnostic' as follows:
2nd 0 (that's zero) (**CATALOG**).
Scroll down to DiagnosticOn ENTER ENTER.
(Scrolling can be speeded up by pressing x^{-1} which here stands for letter 'D'.)

If you want to save the equation of the regression line (say in Y_1) then type STAT –
[CALC]– 4:**LinReg(ax+b)** – ENTER– [L1] , [L2], VARS
– **Y-VARS** – ENTER– ENTER

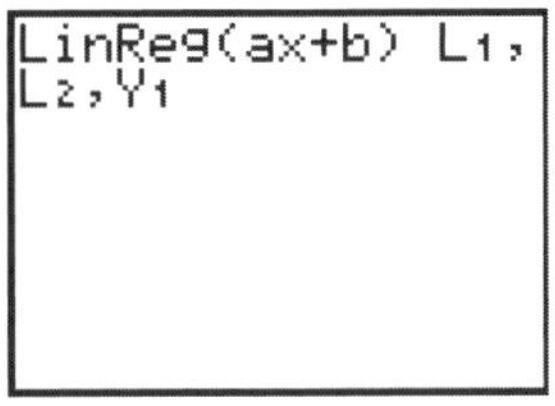

ENTER

```
LinReg
 y=ax+b
 a=-2.763673891
 b=64.47712418
 r²=.9370613375
 r=-.9680192857
```

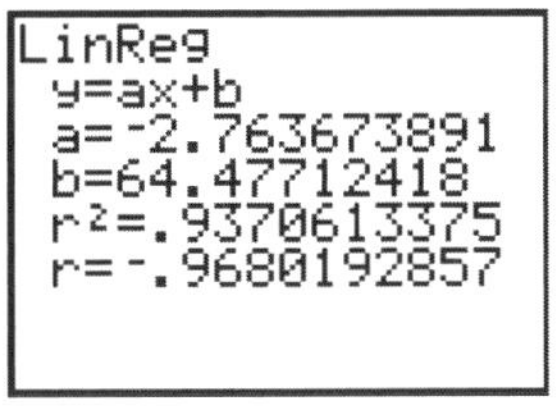

You can look in **Y1**, find the equation and graph it if you need to.

```
X= Plot1  Plot2  Plot3
\Y1■-2.763673890
6089X+64.4771241
83007
\Y2=
\Y3=
\Y4=
(=)(<)(≤)(>)(≥)
```

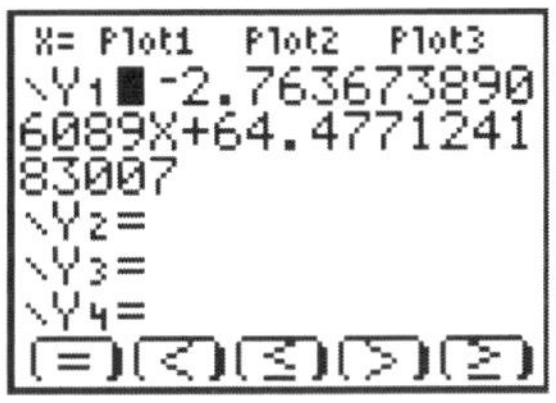

Press GRAPH to plot the scatter graph with the regression line.

```
P1:L1,L2

X=1 ........ Y=69
```

If you use screen dumps in your project then remember to put in the scales.

To find the equation of the regression line you can also use STAT – **TESTS** – **E:**
LinRegTTest – ENTER.

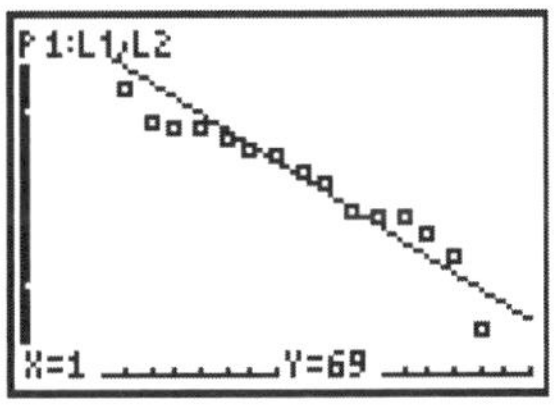

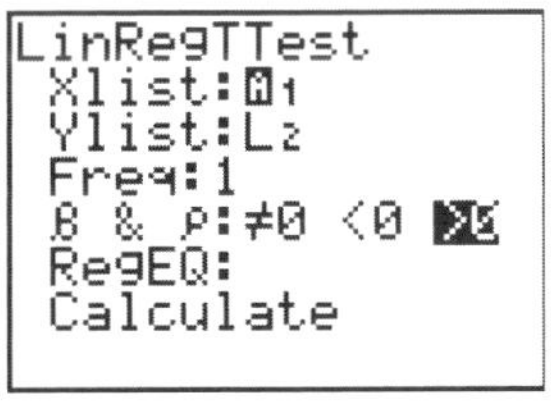

```
LinRegTTest
 Xlist:■1
 Ylist:L2
 Freq:1
 β & ρ:≠0 <0 >0
 RegEQ:
 Calculate
```

Select F3 (**REG**). [V-Window]

Select F1 (**X**). [Trace]

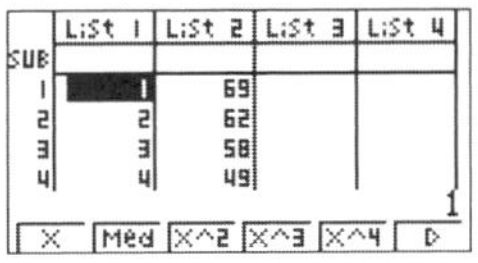

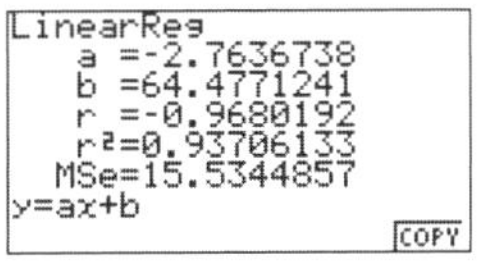

You are also given the equation of the regression line of *y* on *x*.

If you want to save the equation of the regression line (say in Y_1) then select
COPY F6 EXE . [G–1]

The equation of the regression line has been copied to **GRPH**. You can draw a graph of the regression line.

EXIT EXIT EXIT

Select F1 (**GRPH**) then select F1 (**GPH1**). [Trace] [Trace]

Select F2 (**DefG**). [Zoom]

Select F6 (**DRAW**). [G–1]

If you use screen dumps in your project then remember to put in the scales.

The equation of the regression line can also be found using F1 (**GRPH**). [Trace]

Select F6 (**SET**) Graph type F6 F6 [G–1] [G–1] [G–1]
choose **X**.

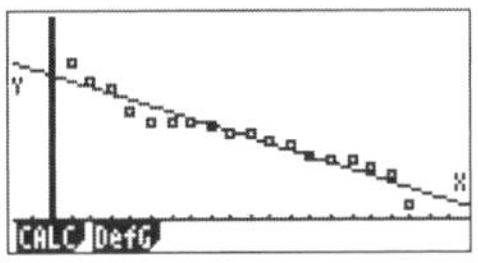

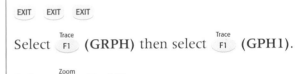

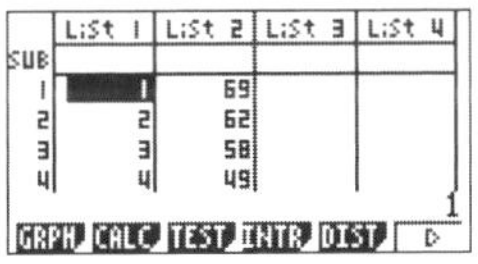

Put [L1] as the **Xlist** and [L2] as the **Ylist** then scroll down to **CALCULATE** then [ENTER].

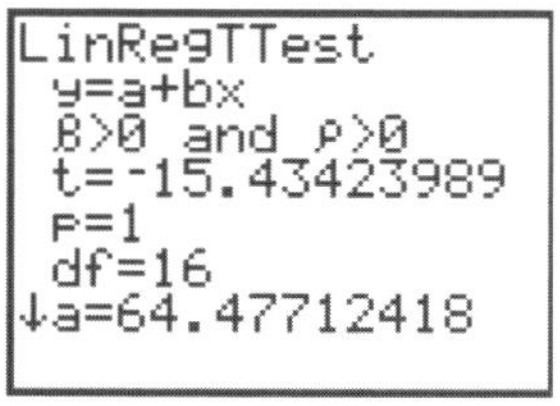

You need to scroll down to find more values.

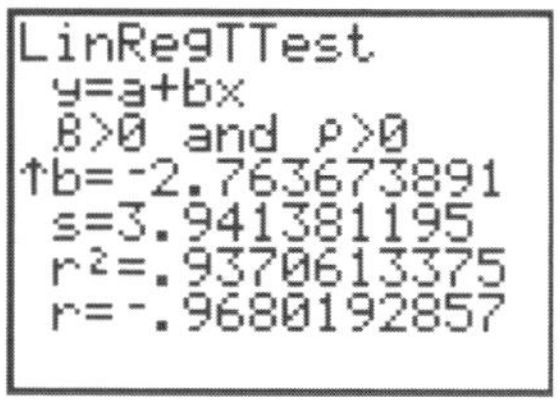

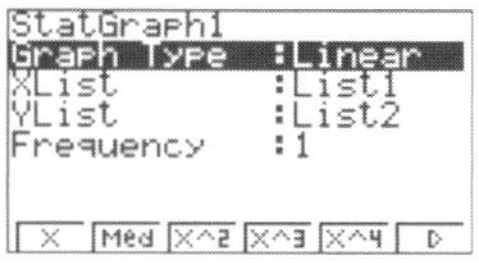

EXIT then select F1 (Trace) (GPH1).

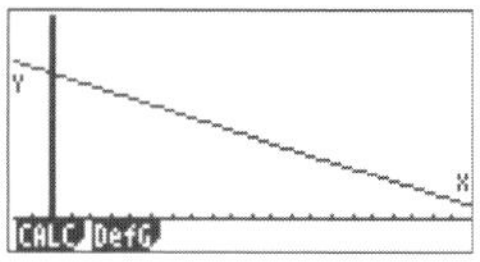

Select F1 (Trace) (CALC).

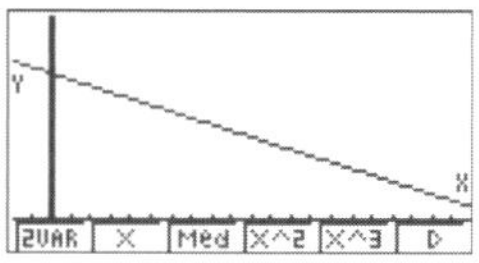

Select F2 (Zoom) (X).

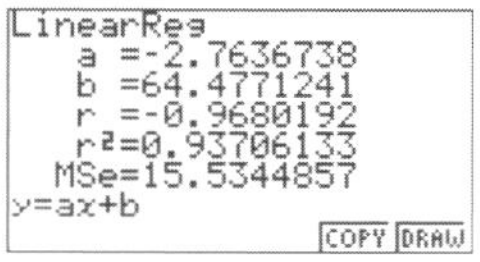

The answer is $r = -0.968$ (3 s.f.). This is very strong negative correlation. Clearly the teams with the highest goal scores reach positions in the league with the lowest numbers (that is, 1st, 2nd, 3rd, etc.).

1.6.9 The χ^2 test for independence

Example 1.6.9 (from *6.9a*)

A survey was conducted to find out which type of flower males and females preferred.

Eighty people were interviewed outside a florist's shop and the results are shown below.

	Rose	**Carnation**	**Lily**	**Freesia**	**Totals**
Male	16	10	5	8	39
Female	19	6	4	12	41
Totals	35	16	9	20	80

Using the χ^2 test, at the 5% significance level, determine whether the favourite flower is independent of gender.

a State the null hypothesis and the alternative hypothesis.

b Show that the expected frequency for female and rose is approximately 17.9.

c Write down the number of degrees of freedom.

d Write down the χ^2_{calc} value for these data.

e Comment on your result.

TI solution

You must enter your *observed* data into a 2 × 4 matrix.

[MATRIX] – **EDIT** – [ENTER]

[2nd][x^{-1}] Entering the matrix menu might be different on older models.

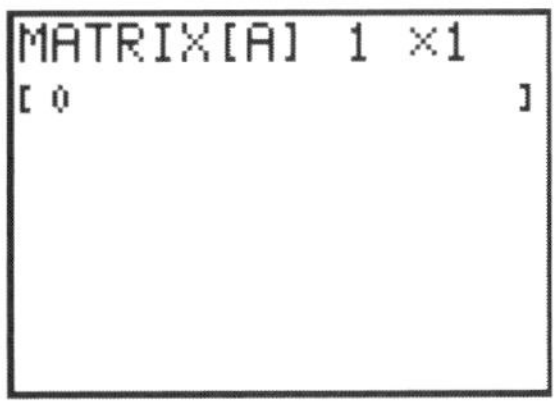

You need to change the size of the matrix. Type **2** – [ENTER] – **4** – [ENTER] and then enter the observed data. Remember to press [ENTER] after each number.

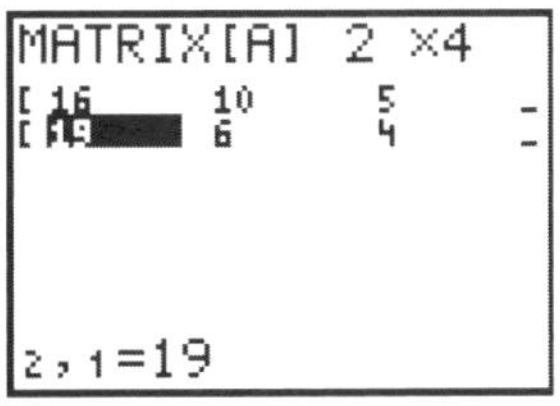

[STAT] – **TESTS** – **C: X²-TEST** – [ENTER].

Observed data are in **[A]** and let expected data be in **[B]** – you can enter the **[A]** and **[B]** by going to the **MATRIX** menu (**NAMES**), scrolling to the correct matrix and pressing [ENTER].

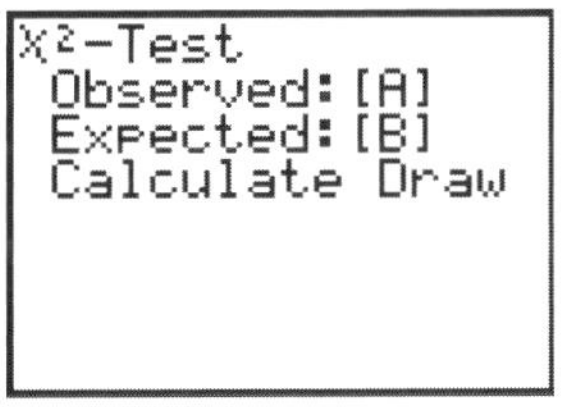

Then move your cursor down to **CALCULATE** and press [ENTER].

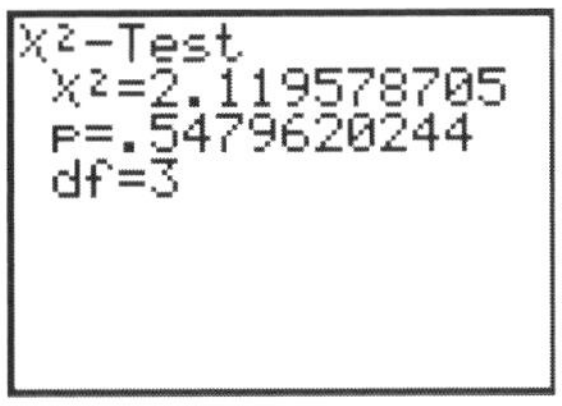

Casio solution

You must enter your *observed* data into a 2 × 4 matrix.

Select MENU then **RUN-MAT**.

Select F1 (Trace) **(MAT)**.

Input **2** EXE. Input **4** EXE EXE.

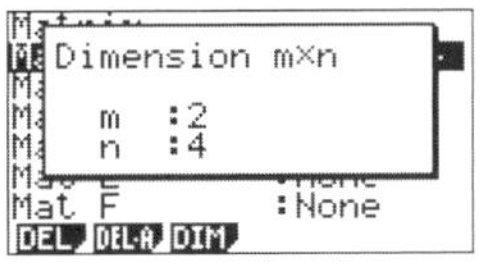

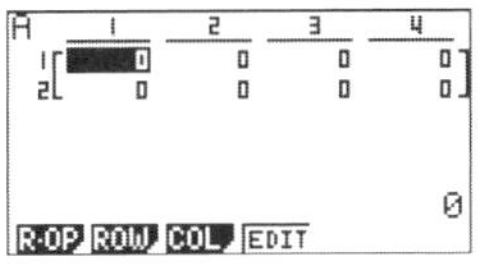

Input the observed data. Remember to press EXE after each number.

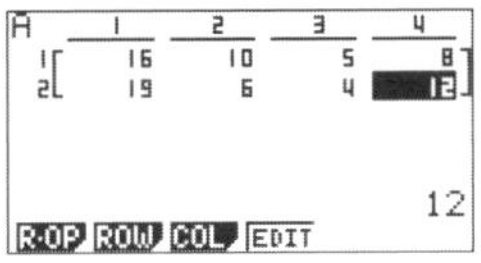

Select MENU then **STAT**.

Select F3 (V-Window) **(TEST)**. Select F3 (V-Window) **(CHI)**.

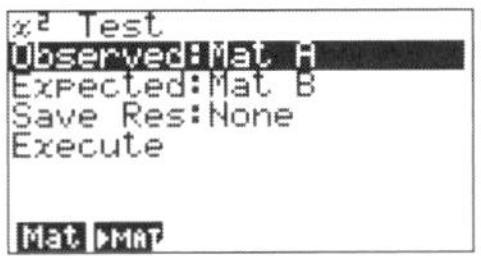

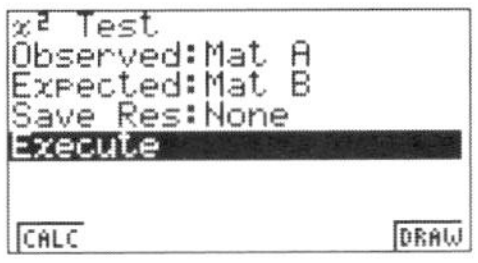

Observed data are in **Mat A** and the expected data will be found in **Mat B** (or MatANS).

Move your cursor down to **Execute** and press F1 (Trace) **(CALC)**.

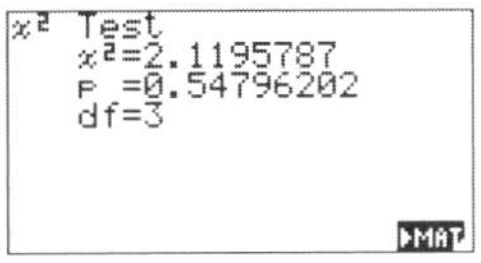

Here you have the χ^2-value, the p-value and the number of degrees of freedom.

Answer:

a Preference for flower type is independent of gender.

b Expected values can be seen in the matrix B.

c degrees of freedom = 3

d $\chi^2 = 2.12$ and $p = 0.548$

e From the information booklet, the critical value is 7.815. Since 2.12 < 7.815 or alternatively using the p-value, 0.548 > 0.05, so we do not reject the null hypothesis and hence can conclude that favourite flower is independent of gender.

1.7 Calculus

Note to TI users: Remember to turn off statistical plots before sketching graphs. [2nd][Y=] (**STAT PLOT**) then **4:PlotsOFF** [ENTER].

1.7.1–1.7.3 Gradients and tangents

Example 1.7.1

Use your GDC to write down the gradient of the curve whose equation is $y = x^3 - 2x + \frac{3}{x}$ at the point where $x = 2$.

TI solution

Press [Y=] [CLEAR].

Input $x^3 - 2x + \frac{3}{x}$.

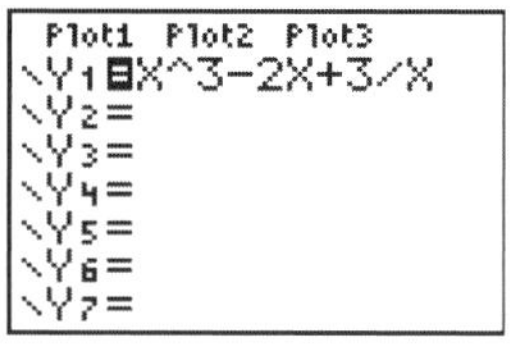

Press [2nd] [TRACE] **6** (dy/dx).

Type in **2** (for $x = 2$), [ENTER].

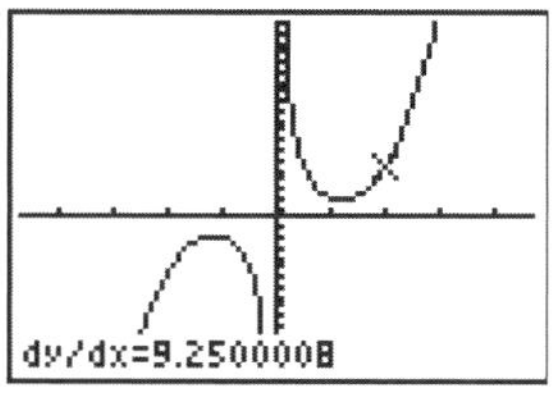

Casio solution

Select SHIFT MENU (**SET UP**).

Turn **Derivative On** F1 (Trace).

Select TABLE from MENU (SET UP).

Input $x^3 - 2x + \frac{3}{x}$.

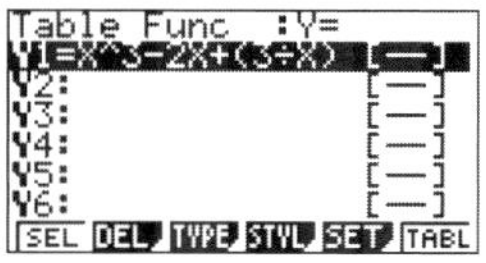

Select F5 (G-Solv) (**SET**).

Be sure that the value $x = 2$ is included in your range. EXIT

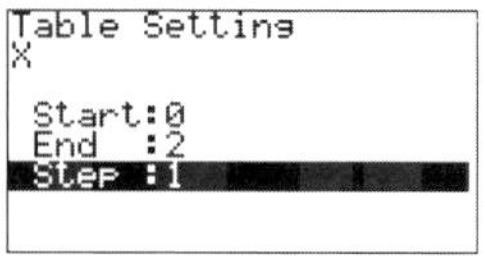

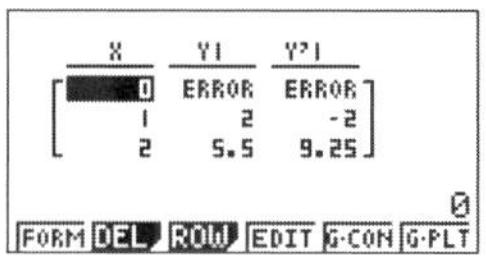

Note that there is a third column as we have set the derivative **On**.

Answer: The gradient is 9.25.

Example 1.7.2

Sketch the curve whose equation is given by $y = \frac{x^4}{4} + 3x^2 - x - 5$, write down the equation of the tangent at the point where $x = 1$ and draw this tangent in the same sketch.

TI solution

Press [Y=][CLEAR].

Input $\frac{x^4}{4}+3x^2-x-5$.

Press [ZOOM] **6** for the standard curve.

If it looks OK press [2nd] [PRGM] (that is, [DRAW]) **5** (**Tangent**), [ENTER] **1** [ENTER] (because we are asked for the tangent where $x = 1$)

[ENTER]

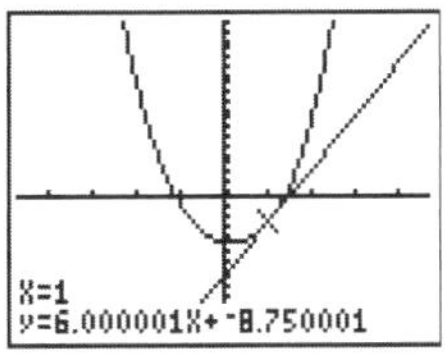

Casio solution

Select TABLE from MENU.

Input $\frac{x^4}{4}+3x^2-x-5$.

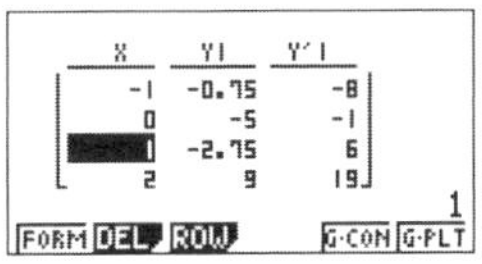

The gradient when $x = 1$ is 6.

The equation is $y=6x+c$ and the tangent passes through the point (1, –2.75) so,

$-2.75 = 6 \times 1 + c \quad \Rightarrow \quad c = -8.75$

Answer: The equation of the tangent is

$y = 6x - 8.75$.

Since the question said 'write down' then this is sufficient. However, if the question had said 'find' the equation of the tangent, then full working must be shown — see later examples.

1.7.4, 1.7.5 Increasing and decreasing functions; maximum and minimum points

Example 1.7.5 (from *7.5.2b*)

Find turning points for the function $y=5000x-300x^2+4x^3$.

TI solution

Enter the function as **Y1** and plot the graph. Here a window of [0,50] was used for x and [–25 000, 25 000] for y.

To find the maximum use [2nd] [TRACE] ([CALC]) **4: maximum**.

Use [▶], [◀] as often as necessary to position the cursor to then left of the maximum then [ENTER], then to the right and [ENTER] again. Now press [ENTER] a third time.

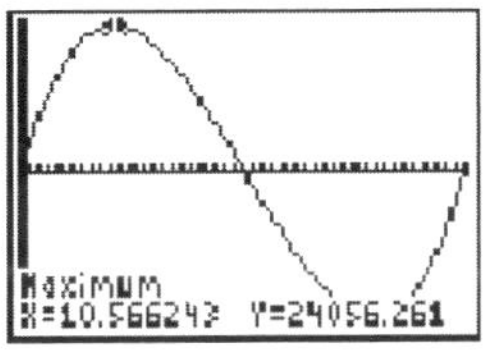

Repeat for the minimum using **3:minimum** on the [CALC] menu.

Casio solution

Select GRAPH from MENU.

Enter the function as **Y1** and plot the graph. Here a window of [0,50] was used for x and [–25 000, 25 000] for y.

To find the maximum

select F5 (G-Solv) **(GSOLV)** F2 (Zoom) **(MAX)**.

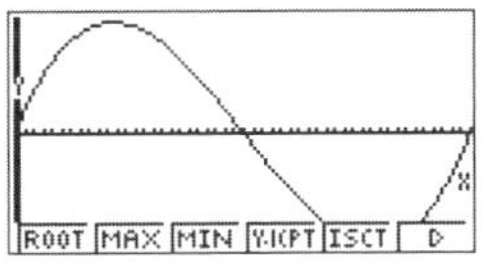

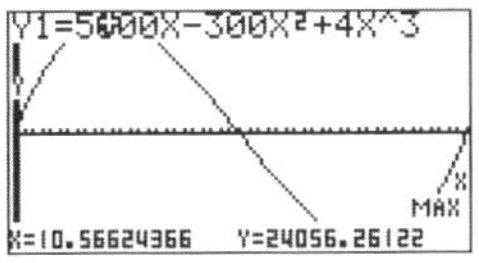

For the minimum use F5 (G-Solv) F3 (V-Window) **(MIN)**.

Answers: Maximum point at (10.6, 24 100), Minimum point at (39.4, –24 100) (all 3 s.f.).

1.8 Financial mathematics

1.8.1, 1.8.2 Currency conversions; simple interest

The material in these subsections does not have any particular need for the GDC apart from the execution of standard arithmetic procedures. The Casio Finance application does have a simple interest calculator but it is just as easy to do the calculation by other means. If you wish to use this application, then the technique is similar to that explained in Section 1.8.3 for compound interest (only simpler).

1.8.3 Compound interest

After entering values in the compound interest formula, this can be rearranged to solve for the unknown variable. After rearrangement, the GDC is used to evaluate the expression on the right-hand side. If the unknown variable is the time period then you can use a guess and check method or use the log function (not required) or use the SOLVER facility on the GDC. Alternatively, and probably easiest once you know how, use the finance application.

All examples are taken from Topic 8.3 with corresponding numbering.

IMPORTANT NOTE: In Paper 2 examinations it is not good enough to find an answer on your GDC then simply write down the answer with no explanation. The safest course of action is to record the appropriate compound interest formula with substituted values and to state which variable is being calculated. Use IB recommended notation, not the variable names in the GDC. Give answers initially to greater than 2 d.p. accuracy then apply whatever rounding is required in the problem, or if not required, then use 2 d.p. or 3 s.f. This is also sound advice for Paper 1, even though correct answers in the answer box receive full marks there.

TI	Casio
Open the **FINANCE** application on your GDC. (See Section 1.1.3)	
Choose 1:**TVM Solver**. The variables here stand for: **N:** number of time periods **I:** interest rate (**note: for depreciation, the value of I can be negative**) **PV:** present or principal value (**we call it C in mathematical studies**) **PMT:** extra payments to the account per year	Select **Compound Interest** [Zoom F2]. The variables here stand for: **n:** number of time periods **I%:** interest rate (**note: for depreciation, the value of I can be negative**) **PV:** present or principal value (**we call it C in mathematical studies**). **PMT:** payment

FV: final or forward value **P/Y:** number of interest payments to the account per year **C/Y:** compounding events per year **PMT: END BEGIN** gives a choice of when to apply interest in the period. In the final screen the quantity which has been calculated has a small black square ■ to its left. In all following examples make sure that **END** is highlighted at the bottom of the screen unless otherwise instructed.	**FV:** final or forward value **P/Y:** payments per year **C/Y:** compounding events per year **END/BEGIN** found in **SETUP** SHIFT MENU (SET UP). To calculate a particular quantity, select the appropriate key, F1 (Trace) to F5 (G-Solv). In all following examples make sure that **END** appears at the top of the screen unless otherwise instructed.

The PV is entered relative to the investor. Money invested is regarded as an outgoing so it is entered with a minus sign. FV is positive because it is regarded as being paid to the investor at the end of the investment. When we look at loans, money borrowed is incoming so PV is entered as positive and the PMT will be negative. If there are no extra payments made, set P/Y = 1.

Example 1.8.3a (from *8.3.1a*)
Lars invests 10 000 Swedish krona (SEK) at a rate of 5.1% compounding yearly. Calculate the amount in Lars's account after four years and find how much of that amount is interest. Give answers correct to 2 d.p.

TI solution	Casio solution
Enter **N** = 4, **I** = 5.1, **PV** = –10 000, **PMT** = 0, **P/Y** = 1, **C/Y** = 1. Set cursor on the unknown variable **FV**. Press ALPHA, ENTER 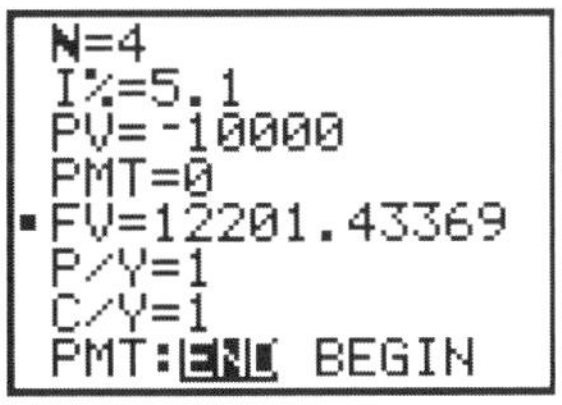	Enter **n** = 4, **I** = 5.1, **PV** = –10 000, **PMT** = 0, **P/Y** = 1, **C/Y** = 1 Select **FV** F5 (G-Solv) 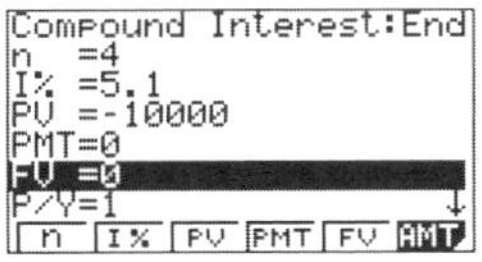 
Answers: FV=12 201.43 SEK. Interest is **FV** – 10 000 = 2201.43.	

Example 1.8.3b (from *8.3.1b*)
Giovanni's bank manager told him that if he invests 3000 EUR now, compounding yearly, it will be worth 4600 EUR in 5 years' time.

What is the rate of interest p.a?

TI solution

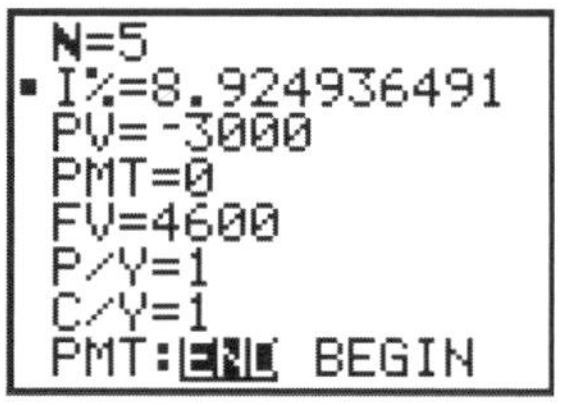

Casio solution

Select I% Zoom F2

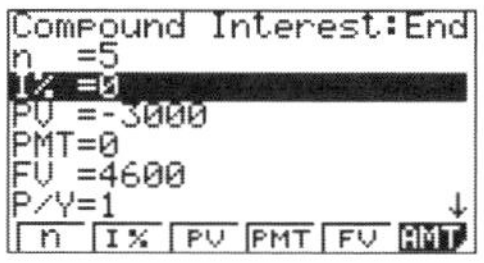

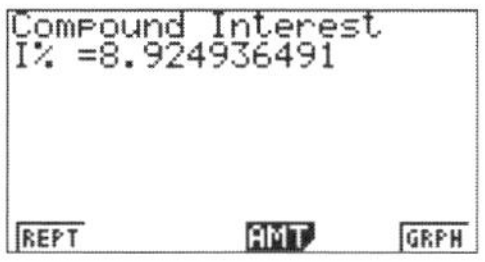

Answer: The rate is 8.92%.

Example 1.8.3c (from *8.3.1c*)
Marina is saving to buy a small boat that costs 35 000 USD. She has 28 000 USD in an account that pays 5.34% interest compounding yearly. How long must Marina wait before she can buy the boat?

This example is important because the GDC is the only method available for finding the time period, other than trial and error.

TI solution

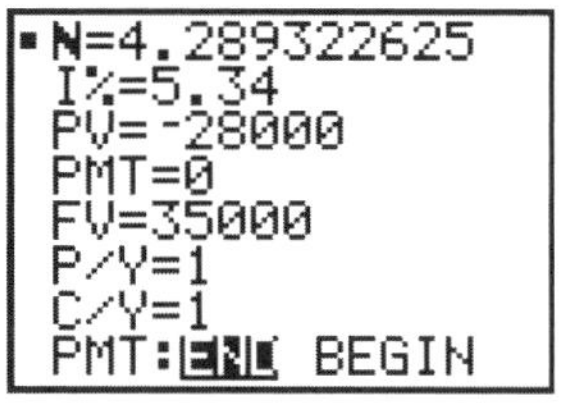

N = 4.29 years but we round up to make sure the final amount has been added.

The answer is 5 years.

If you prefer to solve this problem with the GDC solver it goes as follows:

Refer to the first line of the solution in Topic 8, Example 8.3.1c.

$$\left(1+\frac{5.34}{100}\right)^n=\frac{5}{4}$$

Enter the solver facility.

In **eqn:0=** type (1+.0534)[^][X,T,Θ,*n*] –5/4 [ENTER] then enter a guess of X = 4.

[ALPHA][ENTER].

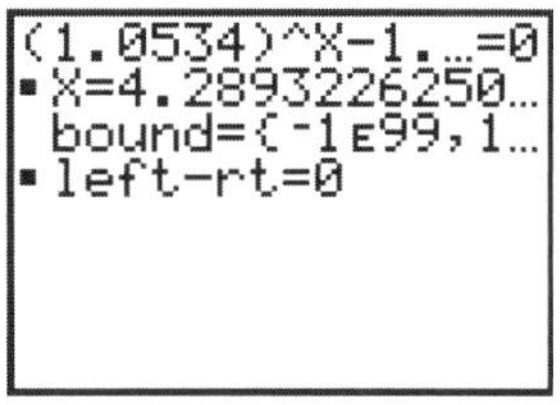

Casio solution

Select Trace F1

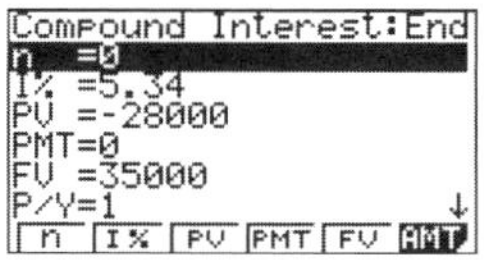

n = 4.29 years but we round up to make sure the final amount has been added.

The answer is 5 years.

If you prefer to solve this problem with the GDC solver it goes as follows:

Refer to the first line of the solution in Topic 8, Example 8.3.1c.

$$\left(1+\frac{5.34}{100}\right)^n=\frac{5}{4}$$

Select EQUA from the **MENU**.

Select **Solver** V-Window F3.

Input

(1 + 5.34 ÷ 100) ∧ X,θ,T SHIFT • 5 ÷ 4

EXE then enter a guess of **X** = 4. EXE

Select **SOLV** G–1 F6.

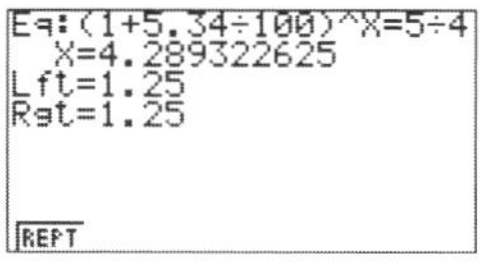

Example 1.8.3d (from *8.3.1d*)

a The bank in Grabiton is advertising a nominal yearly rate of 5% with compounding applied quarterly. State the number of compounding periods for a 3-year investment and find the actual interest rate applied after **each** time period.

b Fleur invests 500 GBP in this bank for three years. Calculate the total capital in the account after this time.

c Suppose Fleur invests this money at a rate of 5% p.a. compounding only once a year. How much less interest would she receive?

TI solution

a There are $3 \times 4 = 12$ periods. The rate for one time period is $\frac{5}{4}\% = 1.25\%$.

b

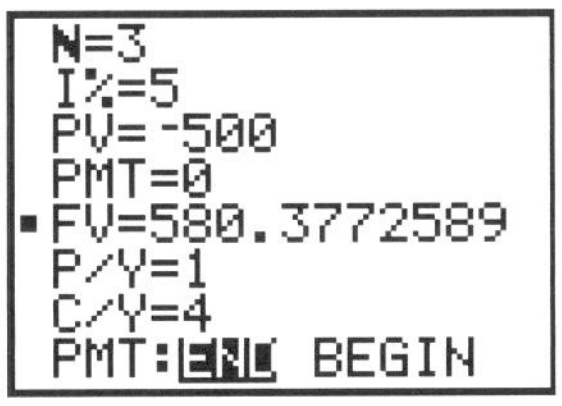

The total is 580.38 GBP.

The same result is obtained if we enter N = 12 and **P/Y** = 4.

c

Casio solution

a There are $3 \times 4 = 12$ periods. The rate for one time period is $\frac{5}{4}\% = 1.25\%$.

b Note that **n** is in periods

Select **FV** G-Solv F5.

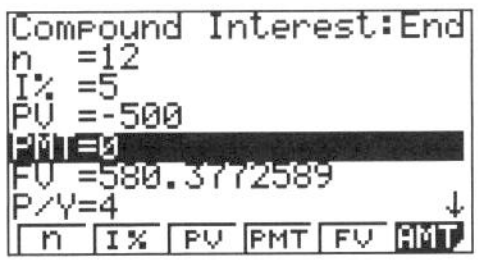

The total is 580.38 GBP.

c Select **FV** G-Solv F5

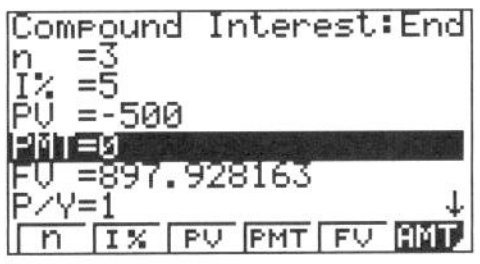

Answer: The total is now 578.81 and the difference is 1.57 GBP.

1.8.4 Construction of tables

The GDC can be used to fill in repayments and other information in loan tables.
This simply amounts to familiar use of the finance application.

Example 1.8.4

Use your graphic display calculator to fill in the missing values x, y, z, w and v in the adjacent loan table. The interest rates given are nominal per annum rates but are to compound weekly. The repayment is calculated on a loan of $1000 and is to be made on a weekly basis. The repayment time in the left-hand column is in years. The values for x and y should be whole numbers.

Time in years	4%	6%	y%
1	19.6253	z	20.2268
2	10.0088	10.2094	v
x	4.2451	w	4.8911

TI solution

Use the **FINANCE** application. Calculate x first.

Weekly repayments and compounding mean you must enter 52 for **P/Y** and **C/Y**.

Enter **I, PV** and **PMT** as given.

Calculate **N**.

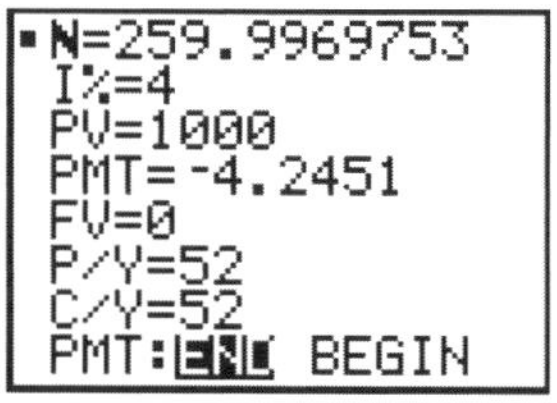

Round **N** to 260. This is the number of weeks, which translates to 5 years.

$x = 5$. Use this value of x to calculate w.

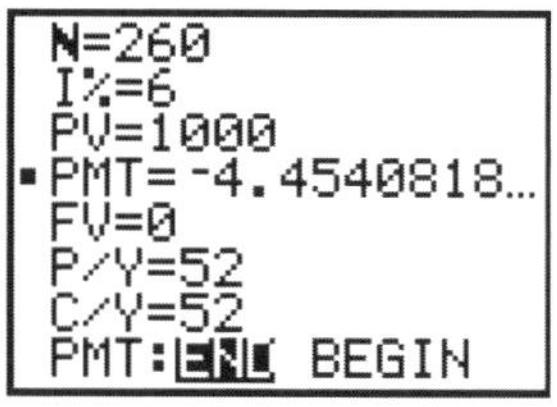

$w = 4.4541$.

Casio solution

Use TVM. Calculate x first.

Weekly repayments and compounding mean you must enter 52 for **P/Y** and **C/Y**.

Enter **I, PV** and **PMT** as given.

Calculate **n**.

Select **n** (Trace F1)

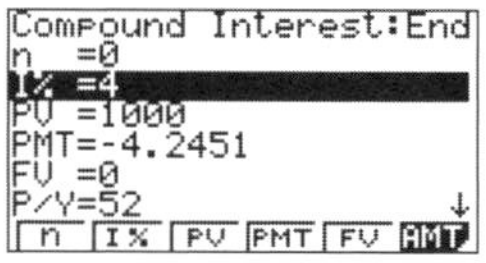

Round **n** to 260. This is the number of weeks, which translates to 5 years.

$x = 5$. Use this value of x to calculate w.

Select **PMT** (Sketch F4)

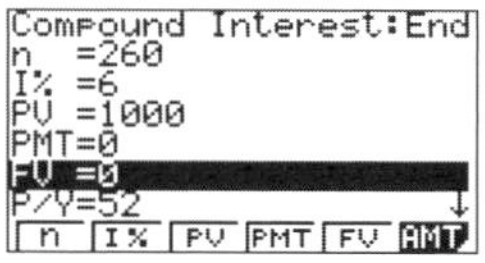

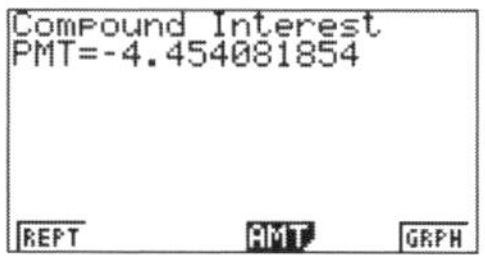

$w = 4.4541$.

Continue in this way until all unknown values are found.
(The full table appears in the answers for Exercise 8.4.)

Exercise 1

In all the following questions, find the answer using your GDC. Methods used in question number 1 are things you might need to do for a question from Topic 1 and so on.

1 Evaluate

a $\sqrt{8.34^2 - 2.16^2}$

b $\dfrac{122.75 - 121.18}{122.75} \times 100\%$ (answer as a percentage)

c $\sqrt{14.1^2 + 7.3^2 - 2 \times 14.1 \times 7.3 \cos(13°)}$

d $\dfrac{2^{\sqrt{3}}(\sin 16° - \tan 28°)}{\sqrt[3]{5}} \times \tan^{-1} 1.2$

2 Use your GDC to calculate

a the 17th, 18th and 19th terms

b the sum of the first 122 terms of the sequences whose first three terms are:

i 2.20 3.55, 4.90…

ii −8.6000, 8.1700, −7.7615…

3 Use your GDC to work out the truth table entries for the proposition $(p \Rightarrow q) \wedge (r \Rightarrow - q) \wedge (p \wedge r)$. What kind of proposition is this?

4 a Sketch the curve $y = x^4 - 2x^3 + 2x - 7$ showing the coordinates of any axis crossings.

b Solve the equation $\cos x = \frac{2^{\frac{x}{90}}}{3}$ for all x (in degrees) in the interval [–90, 90]

i using a graphical method

ii using the solver facility.

c Sketch the curve $y = \frac{4x-1}{x+1}$, identifying any zeros and other axis crossings, horizontal and vertical asymptotes.

5 a Evaluate the angle θ resulting from a sine rule application with $\frac{\sin\theta}{6} = \frac{\sin 65}{8.5}$.

b Evaluate the angle θ resulting from a cosine rule application with $\cos\theta = \frac{11.8^2 + 7.7^2 - 12.3^2}{2 \times 11.8 \times 7.7}$.

c A sphere has the same volume as a box, whose dimensions are 3.5 cm × 4.1 cm × 7.2 cm. Find the radius of the sphere.

6 a A survey was carried out in a supermarket to find out how much people spent on shopping for food each week. Two hundred people were questioned and the results are shown in the table below. Plot a histogram for these data.

Spending (x) in AUD	Frequency
$50 \le x < 100$	11
$100 \le x < 150$	21
$150 \le x < 200$	48
$200 \le x < 250$	53
$250 \le x < 300$	36
$300 \le x < 350$	24
$350 \le x < 400$	7

b The table below shows the number of days absent in the year for students at a certain school, rounded to the nearest four days. Draw a frequency polygon to represent this information.

Days in the year absent	0	4	8	12	16	20	24	28	32	36	40	44	48	52
Frequency	37	65	46	50	31	32	25	22	11	12	4	6	3	1

c For the data in part **b** draw a box plot, find the mean and the median number of days absent and comment on outliers.

d Anton wrote a project to prove that there is a correlation between the number of rainy days in the month and the number of mosquitoes around his house. He caught mosquitoes on a sticky pad designed to attract them. He recorded data every month during a period of two years.

Rainy days *R*	3	7	15	22	23	18	12	9	5	4	1	0
Number of mosquitoes *M*	3	5	12	25	30	21	15	8	6	2	0	0
Rainy days *R*	2	7	18	23	26	17	14	10	7	4	3	1
Number of mosquitoes *M*	2	4	16	22	22	19	17	6	4	4	1	0

Calculate the correlation coefficient. Plot a scatter plot and if justified, plot the line of regression of M on R over the scatter plot. (For $M \leq 5$ collect data together in a single entry.)

e Jakob thought that favourite food might depend on nationality so he gave out a questionnaire in his school. He collected the following data.

	Pizza	**Moussaka**	**Curry**	**Fried Rice**	**Hamburger**	**Totals**
Italian	8	6	6	5	8	33
Greek	10	9	8	8	9	44
Indian	10	9	12	11	12	54
Chinese	9	5	8	13	8	43
American	12	11	6	9	14	52
Totals	49	40	40	46	51	226

Using the χ^2 test, at the 5% significance level, determine whether the favourite food is independent of nationality.

i State the null hypothesis and the alternative hypothesis.
ii Write down the number of degrees of freedom.
iii Write down the χ^2_{calc} value and the p-value for these data.
iv Comment on your result.

7 a For the function $y = \frac{1}{\sqrt{x}}$ Find the value of $\frac{dy}{dx}$ at $x = 2$.

b Find the coordinates of all maxima and minima of the function $y = (x^2 - 5x + 1) \times 3^{-x^2}$

c Draw the graph of $y = x \sin x$ for $x \in [-900, 900]$. Decide on an appropriate range for the y-window to fit the graph in. Find the coordinates of the first maximum and first minimum to the right of the y-axis.

8 a Evaluate $1500\left(1 - \frac{3.8}{12 \times 100}\right)^7$ using the finance application on the GDC.

b Find the number of years required for \$0.10 to grow larger than \$100 when invested at a rate of 9% nominal, compounding monthly.

TOPIC 2

Number and algebra

2.1 The sets of numbers ℕ, ℤ, ℚ and ℝ

2.1.1 The set of natural numbers, ℕ

We say, "March has 31 days" or "There are 15 students in my mathematics class." We count with the numbers 1, 2, 3, 4, ...

We also use these numbers for ordering. We say "This is the first year of my mathematical studies course" or "Alison came second in the race".

If we include the number 0 with 1, 2, 3, 4, ... then we make up a special collection of numbers named the set of **natural numbers**.

The set of all natural numbers is $\mathbb{N} = \{0, 1, 2, 3, 4, 5, \ldots\}$

In Topic 3 all these ideas will be developed and further examples given.

We can represent ℕ on a **number line** by setting an origin and a unit length.

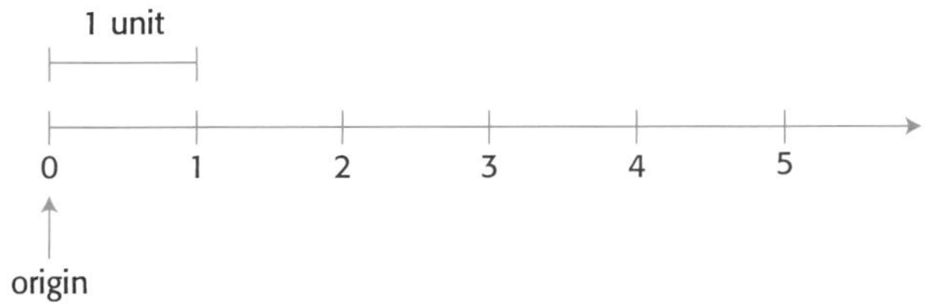

The arrow at the right end of the number line indicates the direction in which the numbers increase.

- The natural numbers, "0, 1, 2, 3, 4, ..." have been enclosed within a pair of brackets "{"and "}". We use the brackets to enclose the **elements** of a set. For instance, 3 is an element of the set ℕ but 1.5 is not an element of ℕ. We write this symbolically as "$3 \in \mathbb{N}$" and "$1.5 \notin \mathbb{N}$".
- The dots "..." are used to indicate that there are more elements in the set and that the pattern continues.

Example 2.1.1

Which of these expressions can be used to show that the difference between two natural numbers is *not always* a natural number?

a $8 - 5$ **b** $15 - 0$ **c** $6 - 10$ **d** $123 - 120$

Solution

Expression **c** is equal to -4 which is negative and so it is not a natural number.

We conclude from this example that the difference between two natural numbers is *not always* a natural number.

However, the sum of two natural numbers and the product of two natural numbers is *always* a natural number.
For example: $3 + 2 = 5$, $4 \times 3 = 12$, ...

2.1.2 The set of integers, \mathbb{Z}

We have seen that the difference between two natural numbers is not always a natural number. For instance, if I have \$1200 in the bank and make a withdrawal of \$1600 from my account then I would owe the bank \$400. This situation is represented by the number -400.

The set of all **integers** is $\mathbb{Z} = \{\ldots, -3, -2, -1, 0, 1, 2, 3, \ldots\}$

\mathbb{Z} is an extension of \mathbb{N}. This means that any natural number is also *always* an integer.

We represent \mathbb{Z} on a number line like this.

Example 2.1.2a

Which of these expressions can be used to show that the **quotient** of two integers is *not always* an integer?

a $-10 \div (-2)$ **b** $-15 \div 10$ **c** $0 \div (-4)$ **d** $16 \div 4$

Solution
Expression **b** is equal to -1.5 which is not an integer.

We conclude from this example that the quotient of two integers is *not always* an integer.

However, the sum, product and difference of two integers is *always* an integer.
For example,

$$-5 + -3 = -8, \quad -2 \times 6 = -12, \quad 9 - (-3) = 12, \ldots$$

Example 2.1.2b

a Which of these equations have a solution for x in \mathbb{Z}?

 i $\;-x = 4$ **ii** $\;x \times x = 9$ **iii** $\;11 = -5x$

b Solve for x those equations that have a solution in \mathbb{Z}.

Solution

i	**ii**	**iii**
$-x = 4$ $x = -4$ therefore $x \in \mathbb{Z}$ (x is an element of \mathbb{Z}). **a** This equation has a solution in \mathbb{Z}. **b** $x = -4$	$x \times x = 9$ then $x = 3$ or $x = -3$ therefore this equation has two possible values for x and both are elements of \mathbb{Z}. **a** This equation has solutions in \mathbb{Z}. **b** $x = 3$ or $x = -3$	$11 = -5x$ then $-\dfrac{11}{5} = x$ therefore the solution of this equation is *not* an integer. **a** This equation has no solution in \mathbb{Z}.

\mathbb{Z} stands for Zahlen (German for "numbers").

Integers are also known as whole numbers.
- An integer is **positive** if it is greater than zero.
- An integer is **negative** if it is less than zero.
- Zero is defined as neither negative nor positive.
- The distance from a to 0 is the same as the distance from $-a$ to 0.

Remember that $-a = -1 \times a$

Quotient means ratio.

2.1.3 The set of rational numbers, $\mathbb{Q}$

In Figure 1, ABCD is a square with area 1 unit^2 and P, Q, R and S are the midpoints of the sides. O is the point of intersection of PR and QS.

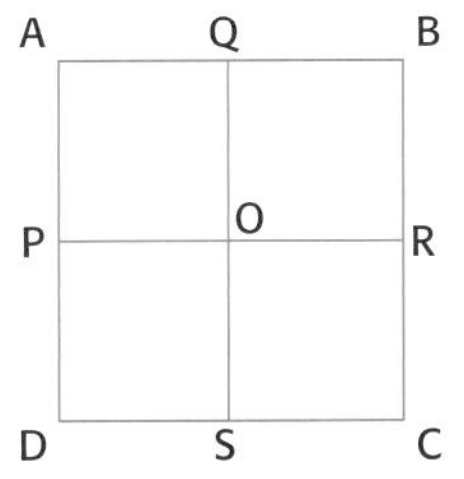

Figure 1

The area of ABRP is $\frac{1}{2}$. The area of DPOS is $\frac{1}{4}$.

Both $\frac{1}{2}$ and $\frac{1}{4}$ are rational numbers.

The set of rational numbers is

$$\mathbb{Q} = \left\{ \frac{p}{q} \text{ where } p \text{ and } q \text{ are integers and } q \neq 0 \right\}$$

$\mathbb{Q}$ stands for "Quotient".

This means that a number is rational if and only if[1] it can be written as a quotient of two integers where the denominator is different from zero.

Examples of rational numbers are -2, 0.1, $0.\dot{5}$ and $6.00\dot{4}\dot{5}$:

- -2 is a rational number as we can write $-2 = -\frac{2}{1}$, where both -2 and 1 are integers.
- 0.1 is a rational number as we can write $0.1 = \frac{1}{10}$, where both 1 and 10 are integers.
- $0.\dot{5} = 0.555\ldots$ is a rational number as we can write $0.\dot{5} = \frac{5}{9}$, where both 5 and 9 are integers.
- $6.00\dot{4}\dot{5} = 6.00454\ldots$ is a rational number as we can write $6.00\dot{4}\dot{5} = \frac{1321}{220}$, where 1321 and 220 are integers.

The decimal expansion of a rational number will either have a finite number of decimal places or will recur. In the latter case we say that the numbers have a **period**. The period is the digit or group of digits that is repeated after the decimal point. For instance in $0.\dot{5}$ the period is 5 and in $6.00\dot{4}\dot{5}$ the period is 45.

$6.00\dot{4}\dot{5}$ can be written $6.00\widehat{45}$. There are also other notations.

The set of rational numbers is an extension of the set of integers.

We represent the rational numbers on the number line like this.

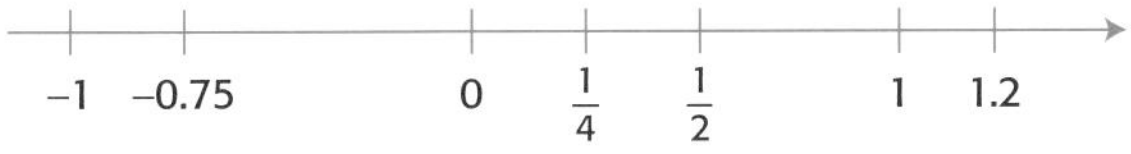

Example 2.1.3a

Show that $0.\dot{4}$ is a rational number.

Solution

Let $n = 0.\dot{4}$ then $n = 0.44444\ldots$

$$10n = 4.4444\ldots$$

$$\therefore 10n - n = 4$$

$$\therefore 9n = 4$$

$$\therefore n = \frac{4}{9} \text{ with } 4 \in \mathbb{Z} \text{ and } 9 \in \mathbb{Z}.$$

> ***Theory of knowledge***
>
> How logical is mathematics? Consider the following:
>
> $0.1111\ldots = \frac{1}{9}$
>
> $0.2222\ldots = \frac{2}{9}$
>
> ...
>
> $0.8888\ldots = \frac{8}{9}$
>
> $0.9999\ldots = \frac{9}{9} = 1$
>
> Why can we say that $0.9999\ldots = 1$?

$\therefore$ is mathematical notation for "therefore".

[1] You will find an explanation on how to use the expression "if and only if" in Topic 3.5.1.

Example 2.1.3b

a Find a rational number that lies on the number line between $\frac{1}{2}$ and $\frac{3}{4}$.

b Find another rational number that lies on the number line between $\frac{1}{2}$ and $\frac{3}{4}$.

c How many rationals are there on the number line that lie between $\frac{1}{2}$ and $\frac{3}{4}$?

Solution

a One such number could be the **arithmetic mean** of these two numbers $\left(\frac{1}{2}+\frac{3}{4}\right)\div 2=\frac{5}{8}$

b Another number could be $\left(\frac{1}{2}+\frac{5}{8}\right)\div 2=\frac{9}{16}$.

c Halfway between $\frac{1}{2}$ and $\frac{9}{16}$ as shown in this diagram:

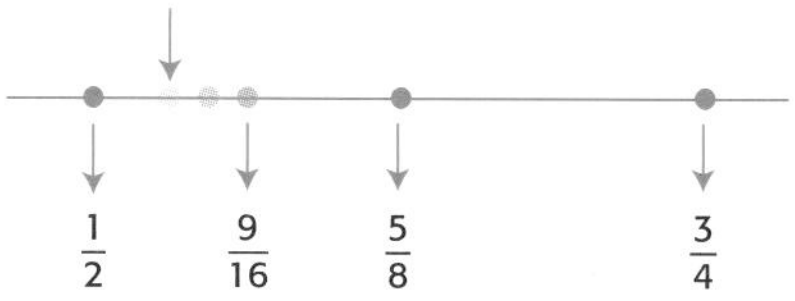

is another rational number.

This process is endless; we can always find a rational number between any two other rational numbers.

There is an infinite number of rationals between $\frac{1}{2}$ and $\frac{3}{4}$.

$\frac{5}{8}$ is halfway between $\frac{1}{2}$ and $\frac{3}{4}$.

$\frac{9}{16}$ is halfway between $\frac{5}{8}$ and $\frac{1}{2}$.

2.1.4 The set of real numbers, $\mathbb{R}$

Consider the side of a square 1 unit long. What is the length of its diagonal?

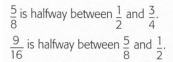

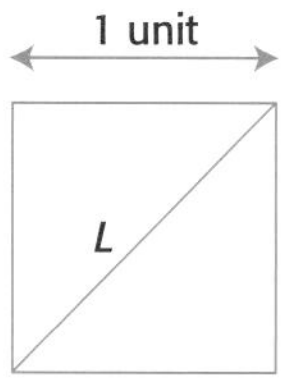

We can solve this problem using Pythagoras' theorem.

Let the length of the diagonal be L then

$$L=\sqrt{1^2+1^2}$$

$$\therefore L=\sqrt{2}$$

Is L a rational number? What is the decimal expansion of $\sqrt{2}$? Does it recur?

The number $\sqrt{2}$ is not a rational number as it cannot be written as a quotient of two integers, therefore its decimal expansion does not recur nor terminate.

$\sqrt{2}=1.414\ 213\ 562\ldots$

Theory of knowledge

Does the proof that $\sqrt{2}\neq a\div b$ convince everyone, now and forever, that $\sqrt{2}$ is not rational?

There is a set of numbers that are not rational numbers that can be represented on the number line. This is the set of **irrational numbers**. The decimal expansion of these numbers does not recur or terminate.

Examples of irrational numbers are $\sqrt{2}$, π, $\sqrt[3]{11}$, $-\frac{\sqrt{3}}{4}$

A **real number** is any number that can be placed on the number line.

The set of rational numbers together with the set of irrational numbers form the set of **real numbers, $\mathbb{R}$,** and completes the number line.

Theory of knowledge

What do we mean when we say one number is bigger than another? How does this relate to the number line? Do numbers have to be represented on a line to have an order? Could we order (and hence count) integer coordinate pairs marked on a piece of graph paper?
(Think spiral!)

Georg Cantor (1845–1918)

Georg Cantor was born in Saint Petersburg, Russia. In 1856 the family moved to Germany and he continued his education in German schools, earning his doctorate from the University of Berlin in 1867.

Cantor is best known as the creator of modern set theory. He recognized that sets with an infinite number of elements can have different sizes, distinguished between countable and uncountable sets and proved that the set of all rational numbers $\mathbb{Q}$ is **countable** while the set of all real numbers $\mathbb{R}$ is **uncountable** and hence strictly bigger. Cantor's ideas faced significant resistance. Today, the majority of mathematicians accept Cantor's work on set theory and recognize it as a *paradigm shift* of major importance.

Exercise 2.1

1 Copy and complete these statements to make them true by using *always, sometimes* or *never*.

- **a** The sum of two naturals numbers is .... a natural number.
- **b** The difference of two natural numbers is .... a natural number.
- **c** If a and b are two natural numbers different from zero and b is greater than a then $\frac{a}{b}$ is .... a natural number.

2 Find an example to demonstrate that the statement "the quotient of two natural numbers is a natural number" is false.

3 a Copy and complete this table.

a	-3	4	$-\frac{1}{2}$	$-\sqrt{2}$	0.8		x	
$-a$						6		t

b Complete these statements using > or <.

- **i** If $a > 0$ then $-a \ldots 0$.
- **ii** If $a < 0$ then $-a \ldots 0$.
- **iii** If $-a < 0$ then $a \ldots 0$.
- **iv** If $-a > 0$ then $a \ldots 0$.

4 a Which of these equations have a solution for x in $\mathbb{Z}$?

b Solve for x those equations that have a solution in $\mathbb{Z}$.

i $x + x = +6$ ii $x + \frac{1}{2}x = 3$

iii $8x = +2$ iv $x \times x = 100$

5 State whether these statements are true or false. If they are false give an example to justify your answer.

a The difference of two negative integers is a negative integer.

b If a and b are two integers and $b+0$, then $\frac{a}{b}$ is sometimes an integer.

6 Explain why we write $q \neq 0$ in the definition of rational numbers.

7 Write down a rational number whose decimal expansion

a recurs

b is finite

c has a period that starts in the third digit after the decimal point.

8 Show that these numbers are rational.

a -2.15 b 4 c $1.\dot{8}$

9 Use Figure 1 on page 59 to find the area of

a triangle BCD b triangle SRC

c 3 times triangle SRC d 5 times AQOP.

10 a Find three rational numbers between $\frac{11}{12}$ and $\frac{12}{13}$.

b How many rational numbers can be found between $\frac{11}{12}$ and $\frac{12}{13}$?

11 Given the numbers $1+\sqrt{3}$, $\frac{\pi}{2}$, $\frac{16}{4}$, $-\frac{2}{3}$, 0

a state whether they are rational or not

b sort them into descending order.

12 Copy and complete these tables using ✓ if the number is an element of the set.

Table A

Number / Set	$1.03\dot{2}$	$\frac{30}{6}$	$-\frac{10}{5}$	$\frac{\sqrt{3}}{4}$
$\mathbb{N}$				
$\mathbb{Z}$				
$\mathbb{Q}$				
$\mathbb{R}$				

Table B

Number \ Set	−2.1	π	0	$-\frac{5}{10}$
$\mathbb{N}$				
$\mathbb{Z}$				
$\mathbb{Q}$				
$\mathbb{R}$				

13 State whether these measurements are rational or not.

a The perimeter of a rectangle whose length and width are 6 cm and 1.8 cm respectively.

b The area of a circle whose radius is 4 cm.

c The length of the hypotenuse of a right-angled triangle whose other sides are 6 cm and 8 cm.

14 State whether these statements are true or not.

a All real numbers are rational numbers.

b There are rational numbers that are integers.

c $\sqrt{\frac{4}{9}}$ is a rational number.

d π cannot be written as the quotient of two integers.

2.2 Approximation, rounding and estimation

2.2.1 Rounding

Rounding a number is the process of approximating this number to a given degree of accuracy.

We round a number when we don't need the exact value of that number, we need only an estimate of it.

For example, "About 12 000 people went to the football match last Sunday" or "Loss of habitats and biological diversity continues, with more than 10 000 species considered under threat."

The following rules apply when rounding numbers correct to some place value to the *left* of the decimal point.

- If the digit after the one that is being rounded is *less than 5* then keep the rounded digit unchanged and change all the remaining digits to the right of this to 0.
- If the digit after the one that is being rounded is *5 or more* then add 1 to the rounded digit and change all remaining digits to the right of this to 0.

Example 2.2.1a

The GNI (gross national income) per capita in Algeria is $2730. Round this figure correct to the nearest hundred dollars.

Solution
There are two ways to do this.

a Using the number line
If we place 2730 in the number line we can clearly see that it is between two multiples of 100, namely 2700 and 2800, but 2700 is closer to 2730 so we choose 2700.

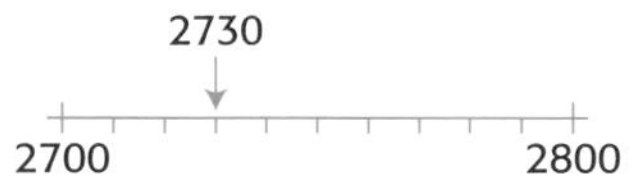

b Using the rules
As we want to round 2730 correct to the nearest 100 we have to look at the 7 and the first digit after the 7, which is 3. Since *3 is less than 5* keep the 7 hundreds and change the figures that follow to zeros.

Answer: \$2730 = \$2700 to the nearest \$100.

Example 2.2.1b
The surface area of the Dominican Republic is 48 734 km^2. Round this figure correct to the nearest 1000 km^2.

Solution
a 48 734 is between 48 000 and 49 000, and it is closer to 49 000.

b If we are rounding to the nearest thousand then we have to look at the 8 and notice that the number to the right of that is 7. As *7 is greater than 5* we increase the 8 thousands by 1 and change the figures that follow to zeros.

Answer: 48 734 km^2 = 49 000 km^2 correct to the nearest 1000 km.

It seems that "the number line method" can be easier than "using the rule". In the following case "using the rule" is more helpful.

Example 2.2.1c
In 2006 the number of IB World Schools teaching the Diploma Programme was 1195. Round this number correct to the nearest 10.

Solution
a The multiples of 10 immediately each side of 1195 are 1190 and 1200. 1195 is exactly in the *middle*.
In this situation we need the rule – see part b.

b As we are rounding to the nearest 10 we have to look at the 9 and the next figure which is 5. As *5 is greater than or equal to 5,* we add 1 to the 9. Since the 9 has now become 10 we change the 1 in the hundreds column to a 2 and change the figures that follow to zeros.

Answer: 1195 = 1200 correct to the nearest 10.

2.2.2 Decimal places

In order to round a decimal number correct to a given number of decimal places (d.p.) these rules apply:

- If the digit after the one that is being rounded is *less than 5* then keep the rounded digit unchanged and delete all the following digits.
- If the digit after the one that is being rounded is *5 or more* then add 1 to the rounded digit and delete all the following digits.

Example 2.2.2a

It costs \$2.065 per child to immunize a community against measles. Write down this figure correct to 2 d.p.

Solution

a 2.065 is between 2.06 and 2.07 and exactly in the *middle* so *we round up* to 2.07.

b Since 6 is in the second decimal place we look at the next digit, a 5, which is *5 or more,* so we increase the hundredths by 1 and delete the other digits.

Answer: \$2.065 = \$2.07 correct to 2 d.p.

Example 2.2.2b

Write down 2.065 correct to 1 d.p.

Solution

a If we place 2.065 on the number line we see that it is between 2.0 and 2.1, but that it is closer to 2.1.

b Since the first decimal place contains a 0, we look at the next digit, a 6. This is *5 or more,* so we increase 0 by 1 and delete the digits that follow.

Answer: 2.065 = 2.1 correct to 1 d.p.

- Rounding correct to 1 d.p. is equivalent to rounding to the nearest tenth.
- Rounding correct to 2 d.p. is equivalent to rounding to the nearest hundredth and so on.

Example 2.2.2c

Use your GDC to find $\dfrac{\sqrt[3]{2.001}}{5\times 1.2^4}$ correct to 3 d.p.

Solution

You can change the mode setting for decimals to *3 decimal places.*

Answer: 0.122

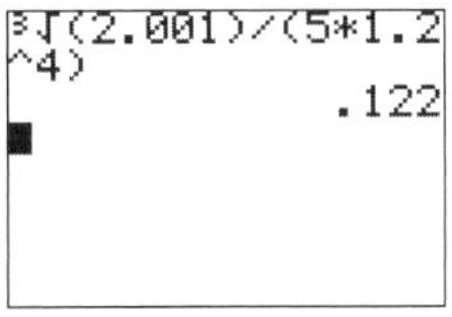

2.2.3 Significant figures

The number of significant figures in a result is the number of figures that are known with some degree of reliability.

The rules regarding significant figures are these.

1 All non-zero digits are significant.

1.235 g has 4 significant figures
1.6 g has 2 significant figures
3000 km has 1 significant figure.

2 Zeros between non-zero digits are significant.

2005 km has 4 significant figures
1.02 ml has 3 significant figures.

3 Zeros to the left of the first non-zero digit are not significant; such zeros merely indicate the position of the decimal point.

0.05 cm has only 1 significant figure
0.063 g has 2 significant figures.

4 Zeros placed after other digits but to the right of the decimal point *are* significant.

0.200 ml has 3 significant figures.

Here what the zeros are actually telling us is that we believe in the accuracy of these figures. If we just write this number to one significant figure as 0.2 we are saying that it might be for example, 0.208 or 0.213 to a higher level of accuracy, but we do not know for sure.

5 For non-decimal numbers, zeros after the last non-zero digit are not significant.

345 700 m has 4 significant figures.

Rules for rounding to a given number of significant figures

If we need to round a number correct to *n* significant figures (s.f.) then these rules apply.

- If the $(n + 1)th$ figure is *less than 5* then keep the *nth* figure unchanged,
- If the $(n + 1)th$ figure is *5 or more* then add 1 to the *nth* figure,
- In both cases all the figures to the right of figure *n* should be deleted if they are to the right of the decimal point and should be replaced by zeros if they are to the left of the decimal point.

Example 2.2.3a

a To enclose the irrigated area used for cultivation in a farm 2306.51 m of wire is needed. Write this length correct to 1 s.f.

Solution

The first significant figure is 2 so look at the next figure, 3, which is *less than* 5, therefore leave the 2 and replace the 306 by zeros.

Answer: 2306.51 m = 2000 m correct to 1s.f.

b The Museo del Prado exposition held in Tokyo in July 2006 had 501 932 visitors.
Write down this number correct to 3 s.f.

Solution

The third significant figure is 1, so we look at the next figure, 9, which is *more than* 5, so we round the 1 up to 2 and replace the 932 by three zeros.

Answer: 501 932 visitors = 502 000 visitors correct to 3 s.f.

c In a certain school it costs $0.00873 to print out one page.
Write this figure correct to 1 s.f.

Solution
The first significant figure is 8 so look at the next figure, 7, which is *more than* 5, so round 8 *up* to 9 and delete the figures that follow.

Answer: \$0.00 873 = \$0.009 correct to 1 s.f.

Example 2.2.3b

a Use your GDC to find $\dfrac{2.51^3}{\sqrt{2}+0.3}$. Give your full calculator display.

b Give your answer to part **a** correct to 4 s.f.

Solution

a

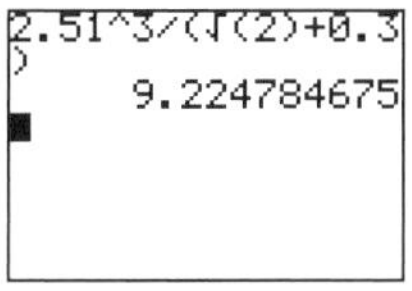

b 9.225

Use of brackets is essential when entering the expression in the GDC.

Answer: 9.224 784 675

When doing multi-step calculations, keep *at least* one more significant figure in intermediate results than needed in your final answer. For instance, if a final answer requires 3 significant figures, then carry at least 4 significant figures in calculations. The general rule in mathematical studies is

"Unless otherwise stated in the question answers must be given exactly or correct to 3 significant figures."

Example 2.2.3c

Find the area of this right-angled triangle.

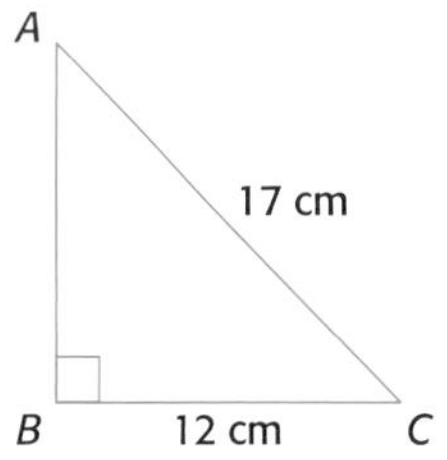

Solution
We need first to find the length of side *AB*.

$$AB = \sqrt{17^2 - 12^2} = \sqrt{145} = 12.041\ldots$$

To find the area we can either work with exact values or use the calculator display for $\sqrt{145}$ (that is 12.041…) and round 12.041 to *at least 4* s.f. as the answer must be given exactly or to 3 s.f. so

$$\text{Area} = \frac{AB \times BC}{2} = \frac{\sqrt{145} \times 12}{2} = 6\sqrt{145}\ \text{cm}^2 \rightarrow \text{exact answer}$$

or

$$\text{Area} = \frac{AB \times BC}{2} = \frac{12.04 \times 12}{2} = 72.2\ \text{cm}^2 \rightarrow \text{answer given}$$

correct to 3 s.f.

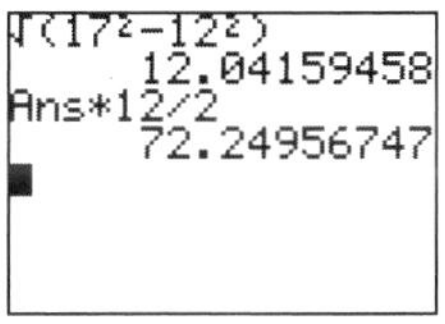

2.2.4 Estimation

An **estimate** of a quantity is an approximation that is usually used to check the reasonableness of an answer. When we estimate a quantity we try to gain an idea of the size of that quantity. This is done by rounding the numbers involved in the calculations to make them more straightforward.

Example 2.2.4a

Estimate the area of a rectangular piece of land which has a length of 126.8 m and a width of 48.6 m.

Solution

$$126.8 \times 48.6$$
$$\downarrow \quad \downarrow$$
$$100 \times 50 = 5000$$

An estimate of the area is 5000 m^2.

If we accidentally enter 26.8×48.6 on the calculator and find the answer 1302.48, this differs so much from the estimate it should tell us that we have made a mistake and ought to look again.

Example 2.2.4b

Ivo made a journey of 286.30 km on his bicycle. He did it in 8 stages of equal length. Estimate the length of each stage.

Solution

$$\frac{286.30}{8} \begin{matrix}\rightarrow\\\rightarrow\end{matrix} \frac{300}{10} = 30$$

Each stage was about 30 km long.

2.2.5 Percentage errors

When we estimate or measure a quantity, it is important to consider the difference between any of these values and the actual value of the quantity. This difference is called the **error**.

Errors arise because no measurement can be absolutely accurate or exact. Measuring instruments and human limitations cause measurements to deviate from the "exact values".

Error does not necessarily mean **mistake.**

The **error** in a measurement is the difference between the value found by measurement or estimation and the exact value of the quantity.
Error $= v_A - v_E$ where v_E represents the *exact value* and v_A represents the *approximated value.*

Example 2.2.5a

a Paula estimates that the weight of a herring is 250 g. The actual weight of the herring is 246 g. Find the error that Paula made in her estimation.

Solution

Error = 250 − 246 = 4 g

b Zhou estimates that the weight of four herrings is 1000 g. The actual weight of the herrings is 996 g. Find the error in Zhou's estimation.

Solution

Error = 1000 − 996 = 4 g

In both **a** and **b** the error is the same, 4 g, however, the second estimate is more accurate than the first one as

$\frac{4}{246} \times 100\% = 1.63\%$ correct to 3 s.f. but $\frac{4}{996} \times 100\% = 0.402\%$ correct to 3 s.f.

We see from these calculations that in Example **a** the error of 4 g represents 1.63% of our total while in Example **b** the error of 4 g represents 0.402% of our total. These percentages help us to have a better idea of the accuracy of the estimation.

In these examples we have calculated **percentage errors**.

$$\text{Percentage error} = \frac{v_A - v_E}{v_E} \times 100\%$$

Where v_E represents the *"exact value"* and v_A represents the *"approximated value"* or *"estimated value"*.

Example 2.2.5b

The size of angle *A* is 24.6°. Isabella measured *A* with her protractor and found that its size was 24°. Find the percentage error made by Isabella when measuring *A*.

Solution

$$\text{Percentage error} = \frac{24 - 24.6}{24.6} \times 100\%$$

$$= \frac{-0.6}{24.6} \times 100\% = -2.44\% \text{ correct to 3 s.f.}$$

Example 2.2.5c

Use Example 2.2.3c to find

a the area of triangle ABC if *AB* is rounded *prematurely* correct to 3 s.f.

b the percentage error made.

Solution

a $AB = 12.041... = 12.0$ to 3 s.f.

$\text{Area} = \dfrac{12.0 \times 12}{2} = 72\,\text{cm}^2$

b 72 cm^2 → this is the approximate value

$6\sqrt{145}\text{ cm}^2$ → this is the exact value

$\text{Percentage error} = \dfrac{72 - 6\sqrt{145}}{6\sqrt{145}} \times 100\% = 0.345\%$ to 3 s.f.

This is not a large percentage error. However, sometimes such premature rounding can make a big difference to an answer. This tends to happen when you round prematurely before taking differences of quantities, particularly if such differences later appear in a denominator.

> *Theory of knowledge*
>
> How important is it to be exact? Discuss this with regard to different scenarios in mathematics and other areas of knowledge such as science, medicine, architecture etc.

Exercise 2.2

1 Round correct to the nearest 10.

a 235 b 1009 c 1 d 745.26

2 Round correct to the nearest hundred.

a 481 b 1309 c 18 152.91 d 355

3 Round correct to the nearest 1000.

a 2149 b 109 642 c 1500.10 d 19 901

4 Write down these numbers correct to the number of decimal places stated.

a 12.053 i 1 d.p. ii 2 d.p.
b 0.009 24 i 3 d.p. ii 4 d.p.
c 5.9908 i 3 d.p. ii 1 d.p.

5 Use the GDC to find the value of

a $2.30 + 15.1 \times 2.65$ i correct to 1 d.p. ii correct to 2 d.p.

b $\dfrac{2.455 - \frac{1}{3}}{12.05 + 0.021^2}$ i correct to 3 d.p. ii correct to 6 d.p.

6 Write these numbers correct to the number of significant figures stated.

a 108.25 i 2 s.f. ii 4 s.f.
b 36 001 i 1 s.f. ii 3 s.f.
c 0.03 046 i 3 s.f. ii 1 s.f.

7 Use your GDC to do these calculations and then write your answers correct to 3 s.f.

a $12.018 \div 2.334$ b $\dfrac{31.2 \times 6.07}{20.1 \times 1.18}$ c 2.17^6

d $\sqrt{2\pi}$

8 Write these measurements correct to the accuracy required.

a 12.18 to 1d.p. b 268 901 to 3 s.f. c 0.024 to 1 s.f.
d 931 to the nearest hundred e 0.6426 to the nearest tenth
f $\dfrac{2}{7}$ to 4 d.p.

9 Rewrite these quantities to a reasonable degree of accuracy. In the news it was reported that

a The actress Diana Mountain is 2.134 m tall.
b Yesterday 79 854 people attended the football match Barcelona-Real Madrid in the Santiago Bernabeu stadium.
c The current temperature in Zurich is 1.42 °C.

10 The weight W is 3.5 kg correct to 2 s.f. Write down

a three possible values for W less than 3.5
b three possible values for W greater than 3.5.

11 Akio estimated that the wall was 2.50 m high when in fact it was 2.38 m.

a Find the error in his estimate.
b Find the percentage error Akio made in his estimate.

12 Look at the map of Brazil. Read the scale carefully and then comment on the statement "Brasilia is 546 miles away from Rio de Janeiro."

13 Let $a = 2.13$ and $b = 51.2$.

a Calculate the exact value of $4a + b^2$.
b Write your answer to part **a** correct to
  i 1 decimal place
  ii 2 significant figures.

Antonio estimates that the value of $4a + b^2$ is 2500.

c Find the percentage error made by Antonio in this estimation.

14 The measurements of the length and width of a rectangular room are 4.75 m and 3.42 m respectively.

a Calculate the area of the room.
b Write down the length and width of the room rounded to the nearest whole number.
c Find the percentage error made if the area is calculated using the width and the length rounded to the nearest whole number.

15 Estimate

a $2.97^2 \times 18.3$
b the perimeter of a square of side 18.3 cm
c the average speed of a car that travelled 2136 km in 18.5 hours.

16 The wheel of a bike has an outer radius of 28 cm.

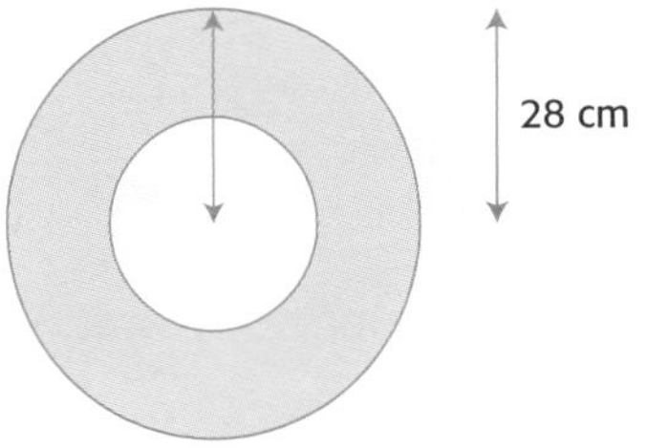

a Calculate how far the bicycle has travelled after one complete turn of the wheel. Give your answer in metres.

Julia used π correct to 2 s.f. to answer part **a**.

b Find the percentage error she made.

2.3 Use of scientific notation

2.3.1 Standard form

"The volume of the Earth is considered to be 1.08×10^{12} km^3."

"The faintest sound that the typical human ear can detect has an intensity of 1×10^{-12} W/m^2."

"The speed of light in a vacuum is defined to be approximately 1.86×10^5 miles per second."

These examples show different numbers written in **standard form**. Standard form is also known as **scientific notation** as this is the form mostly used by scientists to write very large numbers or very small numbers in terms of powers of ten.

When numbers are written in standard form

- it is easier to perform different operations with them
- it is easier to compare them.

A number is written in standard form if it is in the form $\boldsymbol{a} \times \mathbf{10}^{\boldsymbol{k}}$ where $1 \le a < 10$ and $k \in \mathbb{Z}$.

Example 2.3.1a

In the table, the numbers in the first column are written in the form $a\times10^k$ where $1\le a<10$ and $k\in\mathbb{Z}$. Complete the table by writing down the values of a and k.

Number in standard form	a	k
a 1.08×10^{12}		
b 1×10^{-12}		
c 1.86×10^{8}		

Solution

a $a = 1.08$, $k = 12$, b $a = 1$, $k = 12$, c $a = 1.86$, $k = 8$.

Example 2.3.1b

Some of these numbers are *not* in scientific notation. Which ones? Explain your answers.

- 9.4×10^{-3}
- $2.30 \times 10^{\frac{1}{2}}$
- 2.54
- 12.5×10^{4}
- 6.89×10^{-19}
- 10^{-6}

Solution

12.5×10^4 is not given in scientific notation as $12.5 \ge 10$.

$2.30 \times 10^{\frac{1}{2}}$ is not given in scientific notation as $\frac{1}{2} \notin \mathbb{Z}$.

Since $2.54 = 2.54 \times 10^0$, it is written in standard form. This is a very special case!

$10^{-6} = 1\times10^{-6}$, so it is also written in standard form.

Example 2.3.1c

These numbers are given in full. Write them in standard form.

a 546 000 000 000   b 0.003 242

Solution

Let's see how it works.

a 1 The first significant figure is 5 so $546\,000\,000\,000 = 5.46 \times 10^k$ and $a = 5.46$

2 5.46 000 000 000

We need to move the decimal point in a 11 places to the right to get 546 000 000 000 therefore $k = 11$ and $546\,000\,000\,000 = 5.46 \times 10^{11}$

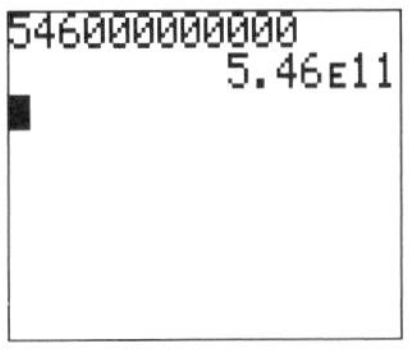

- Moving the decimal point 11 places to the right is equivalent to multiplying by 10^{11}.

b 1 The first significant figure is 3 so $0.003242 = 3.242 \times 10^k$ and $a = 3.242$

2 0 003.242

We need to move the decimal point in a three places to the left to get 0.003 242 therefore $k = -3$ and $0.003\,242 = 3.24 \times 10^{-3}$ correct to 3 s.f.

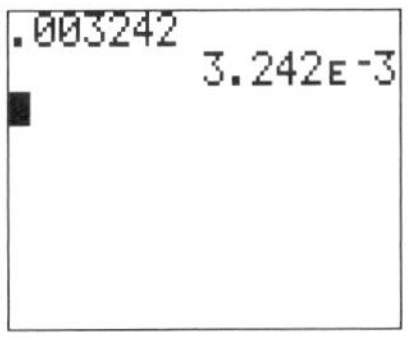

- Moving the decimal point three places to the left is equivalent to multiplying by 10^{-3}.

The steps needed to write a number in standard form are:

1 Write down a: write all the significant figures of the number and place the decimal point immediately after the first significant figure.

2 Find k.

Example 2.3.1d

a Evaluate $\dfrac{\sqrt{81}+3}{5^3+3}$. Write your answer in full.

b Write your answer to part **a** correct to 3 s.f.

c Write your answer to part **b** in the form $a \times 10^k$ where $1 \le a < 10$ and $k \in \mathbb{Z}$.

Solution

a

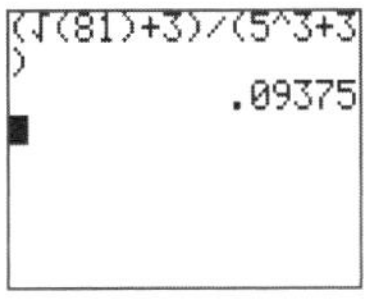

Answer: 0.09375

b 0.0938

c Set your GDC in Scientific notation

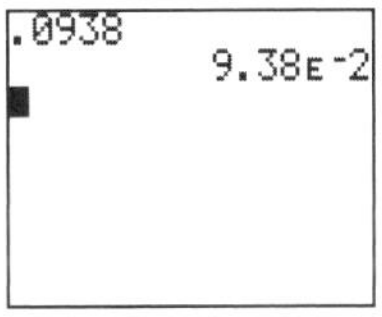

Answer: 9.38×10^{-2}

2.3.2 Operations with numbers expressed in the form $a \times 10^k$ where $1 \leqslant a < 10$ and $k \in \mathbb{Z}$

Example 2.3.2a

Let $x = 3.24 \times 10^4$ and $y = 8.32 \times 10^6$.

a Evaluate $x + 2y$.

b Write your answer to part **a** correct to 3 s.f.

c Give your answer to part **b** in the form $a \times 10^k$ where $1 \leq a < 10$ and $k \in \mathbb{Z}$.

Solution

a 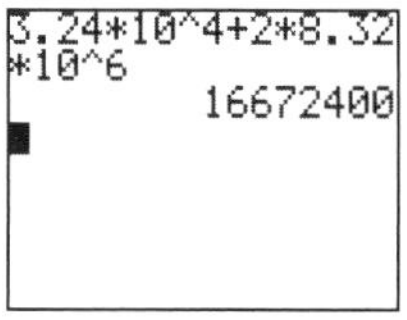

Answer: 16 672 400

b 16 700 000

c 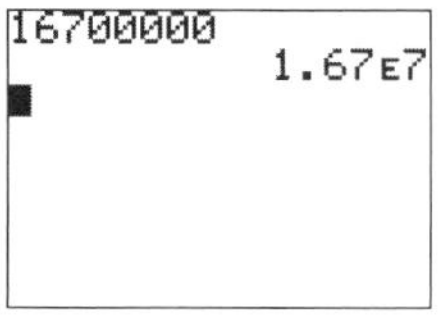

Answer: 1.67×10^7

Example 2.3.2b

Let $x = 8 \times 10^7$ and $y = 3.2 \times 10^{-5}$. Evaluate $\frac{x}{y}$ and give the answer in scientific notation.

Solution

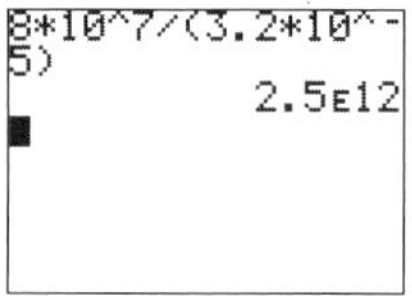

Answer: 2.5×10^{12}

Exercise 2.3

1 Find the value of k in each of these.

a $6800 = 6.8 \times 10^k$

b $0.00508 = 5.08 \times 10^k$

c $9.36 \times 10^k = 936$ million

d $167.3 = 1.673 \times 10^k$

2 Write these numbers in standard form.

a 351 million b 0.000 000 000 291 c 493 000

d 32.8×10^5 e $\frac{12}{1000}$

3 Which of these numbers is *not* written in scientific notation? Explain your answers.

a 3.50 b 11×10^{-9} c 1.01×10^{12}

d 0.32×10^4 e $6 \times 10^{2.3}$

4 Write these numbers in order of magnitude starting with the smallest.
1.8×10^4; 8.94×10^2; 612×10^{-3}; 0.032×10^2

Hint: first write them in standard form.

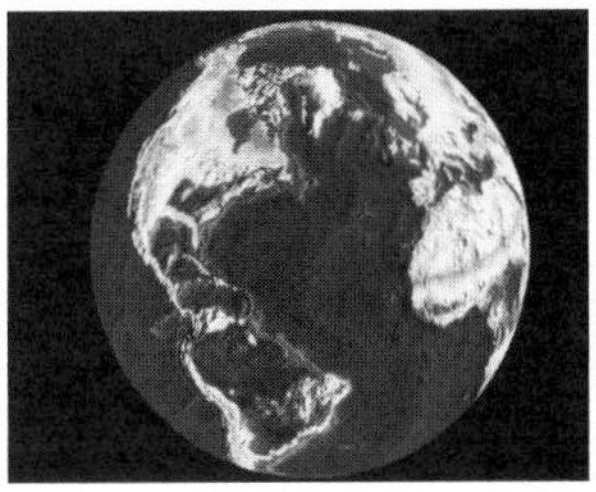

5 Write these numbers in full.

a The volume of the Earth is considered to be 1.08×10^{12} km^3.

b The faintest sound that the typical human ear can detect has an intensity of 1×10^{-12} W/m^2.

c The speed of light in a vacuum is approximately 1.86×10^5 miles per second.

d The average distance of Neptune from the Sun is 2.74×10^9 miles.

6 Let $a = 2.65 \times 10^4$ and $b = 4.10 \times 10^7$. Use your GDC to evaluate these expressions. Give your answers in standard form.

a $a \times b$ b $\frac{a}{b}$ c $a + b$ d $a - b$ e $\frac{b}{2a}$

7 The universe is made up of many galaxies. There are about 1.25×10^{12} galaxies in the universe and about 10^{11} stars in our galaxy. Estimate the number of stars that there are in the universe if we suppose that all galaxies have the same number of stars.

8 The average distances from Earth and Saturn to the Sun are 1.50×10^8 km and 1.43×10^9 km respectively. Calculate what the distance between them would be if they were lined up at these distances from the Sun. Give your answer in scientific notation. (Assume the Sun is in between Earth and Saturn.)

9 Let $x = 12.5$ and $y = 60.9$.

a Write in full the number $x^2 + 4y$.

b Write down the answer to part **a** correct to 3 s.f.

c Write down the answer to part **b** in scientific notation.

2.4 SI and other basic systems of units

2.4.1 The SI

The SI is the international abbreviation for the international system of units (*Système International d'Unités* in French) that was adopted in 1960 by the 11th General Conference on Weights and Measures (CGPM).

The CGPM conferences are attended by representatives of all the industrial countries and international scientific and engineering organizations.

The SI is a practical system of units for international use and is founded on seven SI **base units**. The definition of each unit is absolutely independent of the others. These units are:

- the **metre** (m) for distance
- the **kilogram** (kg) for mass
- the **second** (s) for time
- the **ampere** (A) for electric current
- the **kelvin** (K) for temperature
- the **mole** (mol) for amount of substance
- the **candela** (cd) for intensity of light.

The SI has other units, called **derived units,** that are defined algebraically in terms of the seven base units or other derived units. They are products or quotients of the base units.

Examples of these units are:

- the **square metre** (m^2) for area
- the **cubic metre** (m^3) for volume
- the **metre per second** (m/s or ms^{-1}) for speed or velocity
- the **ampere per square metre** (A/m^2) for current density.

There are some units that are not part of the SI. Some of these units are accepted (some temporarily) for use with the SI as they are essential and widely used. Examples of these are given in this table along with their equivalence with SI units.

Table 2.4.1

Name	Symbol	Value in SI units
litre	L, 1	$1\,L = 1\,dm^3 = 10^{-3}\,m^3$
minute (time)	min	1 min = 60 s
hour (time)	h	1 h = 60 min = 3600 s
day (time)	d	1 d = 24 h = 86 400 s
metric ton*	t	$1\,t = 10^3\,kg$
are	a	$1\,a = 10^2\,m^2$
hectare†	ha	$1\,ha = 10^4\,m^2$

*This unit is called "tonne" in many countries.

†Unit used to express agrarian areas, temporarily accepted to be used with SI as well as "the are".

Example 2.4.1

a The acceleration of an object has units measured in metres per second squared. Write down the symbol used for acceleration.

Solution

ms^{-2} or $m{\cdot}s^{-2}$ or m/s^2

b The SI unit for **force** is the newton. It is abbreviated N. The force F needed to produce an acceleration a on an object with mass m is given by

$F = m \times a$ (newtons)

This equation is known as **Newton's second law**.

i Write down the correct combination of SI units (m, kg, s) for force.

ii Calculate the force needed to cause an acceleration of $20\,m/s^2$ on a soccer ball that has a mass of 0.450 kg.

Solution

i $kg \cdot m \cdot s^{-2}$

ii $F = m \times a$

$F = 0.450\,kg \times 20\,m/s^2$

$F = 9\,kg \times m/s^2$

$F = 9\,N$

In SI the following conventions hold.

1 The letter "s" is always used to indicate seconds and never to indicate a plural, so ms means "milliseconds" and not "metres".

2 Capitals are not used to name the unit, even if the unit is named after a person, though in that case the abbreviation is capitalised (e.g. 1 newton = 1 N). There is just one exception, which is that L can be used for litre, though it is not a person's name.

3 Expressions involving units never have a dot at the end unless it is a normal full stop (period).

2.4.2 Decimal multiples and submultiples of SI units (SI prefixes)

Table 2.4.2 gives some prefixes that allow us to avoid writing very small or very large values.

Table 2.4.2

Factor	10^3	10^2	10^{-1}	10^{-1}	10^{-2}	10^{-3}
Name	kilo*	hecto	deka	deci	centi	milli
Symbol	k	h	da	d	c	m

* The **kilo**gram is the only SI unit with a prefix as part of its name and symbol. In the case of the gram the prefix names are used with the unit name "gram" and the prefix symbols are used with the unit symbol "g".

Example 2.4.2a

1 km = 10^3 m where 'km' reads 'kilometre'
1 ms = 10^{-3} s where 'ms' reads 'millisecond'
1 hl = 10^2 l where 'hl' reads 'hectolitre' and
1 kg = 10^3 g

Example 2.4.2b

Convert 1230 hl to l. Write your answer in standard form.

Solution

1 hl = 10^2 l therefore $1230 \text{ hl} = 1230 \times 10^2 \text{ l} = 1.23 \times 10^5 \text{ l}$

Example 2.4.2c

Convert 0.006 491 millimetres to metres.

a Give your answer correct to two significant figures.

b Give your answer to part **a** in standard form.

Solution

a 1 mm = 10^{-3} m

$0.006\,491 \text{ mm} = 0.006491 \times 10^{-3} \text{ m} = 0.000\,006\,491 \text{ m}$

$= 0.000\,0065 \text{ m}$ correct to 2 s.f.

An alternative answer is 0.0065×10^{-3} m correct to 2 s.f. as you are not asked to give the answer in standard form.

b 6.5×10^{-6} m

Example 2.4.2d

Convert 5 129 000 mg to kg. Give your answer correct to the nearest kg.

Solution

1 kg = 10^3 g $\Rightarrow$ 10^{-3} kg = 1 g
1 mg = 10^{-3} g = $10^{-3} \times 10^{-3}$ kg = 10^{-6} kg
5 129 000 mg = 5 129 000 × 10^{-6} kg = 5.129 kg = 5 kg correct to the nearest kg

Example 2.4.2e

Show that 1 km^2 = 10^6 m^2.

Solution

We know that 1 km = 10^3 m therefore 1 km^2 = $(10^3 \text{ m})^2$ = 10^6 m^2.

Example 2.4.2f

Convert 2 days 9 h 12 min to seconds.

Solution

$$\left.\begin{array}{l} 1\text{ d} = 86\,400\text{ s} \Rightarrow 2\text{ d} = 172\,800\text{ s} \\ 1\text{ h} = 3600\text{ s} \Rightarrow 9\text{ h} = 32\,400\text{ s} \\ 1\text{ min} = 60\text{ s} \Rightarrow 12\text{ min} = 720\text{ s} \end{array}\right\} \quad 172\,800 + 32\,400 + 720 = 205\,920\text{ s}$$

2.4.3 Speed

Speed is a measure of how fast an object is moving. As an object (for example a car) moves, it may change its speed. For instance, an average trip to the town 60 km away takes 1 h. Along the road sometimes we have to slow down, at other times we increase speed and sometimes we stop (then speed = zero). However, if our speed were constant over the 60 km for the whole hour, then it would be 60 km/h. This is an **average speed**. The formula to calculate average speed is

$$\text{Average speed} = \frac{\text{distance travelled}}{\text{time taken}}$$

Example 2.4.3

A car travelled 1200 km.

a Calculate the average speed if the trip took 12 h.

b Calculate the time taken if the average speed of the car was 110 km/h. Give your answer correct to the nearest hour.

Solution

a Average speed $= \frac{1200}{12}$ km/h $= 100$ km/h

b Let t be the time taken by the car. Then 110 km/h $= \frac{1200}{t}$ km/h

$t = \frac{1200}{110}$ h $= 11$ h correct to the nearest hour.

2.4.4 Temperature scales

There are three temperature scales that are used today. The two most widely used are the **Celsius** (°C) and the **Fahrenheit** (°F) scales. The Celsius temperature scale is used in most countries of the world other than the United States, where the Fahrenheit scale is used. The third scale is the Kelvin scale and the **kelvin** (K) is the only official SI basic unit of temperature, used mainly by scientists. The degree Celsius is regarded as a derived unit in SI. It is identical in magnitude to the kelvin but the zero value for the Celsius scale is set at a different temperature (the freezing point of water) to that of the zero for the Kelvin scale, which is absolute zero (−273 °C).

2.4.5 Conversion between the three temperature scales Fahrenheit, Celsius and Kelvin (see Further applications of linear functions, page 180)

In the table the freezing and boiling points of water for the three scales are given.

Table 2.4.5

Scale	Freezing point of water	Boiling point of water
Fahrenheit (°F)	32	212
Celsius (°C)	0	100
Kelvin (K)	273.15	373.15

- To calculate temperatures in °F when they are given in °C this formula is used.

$t_F = \frac{9}{5}t_C + 32$ where t_F represents the temperature in °F and t_C represents the temperature in °C

- To calculate temperatures in °C when they are given in K this formula is used.

$t_C = t_K - 273.15$ where t_K represents the temperature in K and t_C represents the temperature in °C

Example 2.4.5a

Check that the first formula works for the values given in the table.

When $t_C = 0$, $t_F = \frac{9}{5} \times 0 + 32 = 32$ When $t_C = 100$, $t_F = \frac{9}{5} \times 100 + 32 = 212$

Example 2.4.5b

Convert

a −10 °C to °F b 41 °F to °C c 280 K to °C.

Give your answer to the nearest degree.

Solution

a $t_F = \frac{9}{5}(-10) + 32 = 14\,°F$

b $41 = \frac{9}{5}t_C + 32$

$t_C = (41 - 32) \times \frac{5}{9}$ $t_C = 5\,°C$

c $t_C = 280 - 273.15$

$t_C = 6.85\,°C = 7\,°C$ to the nearest degree

Example 2.4.5c

Convert 280 K to °F.

Solution

To convert from K to °F we first convert K to °C and then the temperature in °C into °F.

K ⟶ °C ⟶ °F

$280 - 273.15$ $\quad$ $\frac{9}{5} \times 6.85 + 32$

280 ⟶ 6.85 ⟶ 44.33

Therefore 280 K = 44.3 °F

Exercise 2.4

1 Convert each of these to the unit stated.

a 304.25 m to km b 601 hm to m c 2300 mm to dam

2 Convert each of these to the unit stated.

a 23 hl to l b 340.5 l to dl c 12×10^3 ml to dal
d 2 m^3 to l

3 Convert each of these to the unit stated.

a 245.6 kg to dg b 32 000 mg to g c 10 000 kg to t

4 Convert each of these to the unit stated.
Give your answers in standard form.

a 2591 km^2 to m^2 b 18 900 ha to m^2 c 6255 m^2 to km^2

5 Convert each of these to the unit stated.
Give your answers in scientific notation.

a 6 d 12.5 h to s b 64 800 h to days c 268 s to ms

6 A skein of wool is 15 m long and has to be cut into pieces 12.4 cm in length.

a Calculate the maximum number of pieces that can be cut from this skein.

b Calculate the length of the last remaining piece of wool. Give your answer in cm.

7 A rectangular field is 560 m long and 125 m wide.

a Calculate the perimeter of the field. Give your answer in metres.

b Calculate the area of the field. Give your answer in
i m^2 ii hectares.

8 The distance from Madrid to London is 1725 m. An aeroplane flies at an average speed of 750 km/h.

a Calculate how long it takes for this aeroplane to get from one city to the other. Give the time in hours and minutes.

b If the plane leaves Madrid at 10:15 am, find the time at which it lands in London.

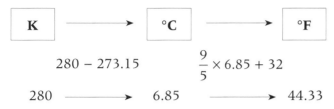

9 Convert

a 15 km/h to m/s b 1200 m s^{-1} to km h^{-1}.

10 On a journey of 240 km, a car travels the first 60 km at an average speed of 120 km/h. It then takes 1 h 12 min to travel the rest of the journey.

a Calculate how long it takes for the car to travel the first 60 km.

b Calculate the average speed of the car for the second part of the journey.

c Calculate the average speed of the car for the entire journey.

11 a Show that the formula $t_C = (t_F - 32) \times \frac{5}{9}$, where t_F represents the temperature in °F and t_C represents the temperature in °C, returns the temperature in °C when it is given in °F.

Hint: rearrange the first formula that is given in Section 2.4.5.

b The temperature of combustion for paper is 451 °F . This is why the title of the book by Ray Bradbury is *Fahrenheit 451*. Convert 451°F to °C.

12 Copy and complete this table. The temperatures were registered in January 2006. Give your answers to the nearest degree.

	Buenos Aires	**Paris**	**Melbourne**	**Los Angeles**	**Casablanca**	**Ottawa**	**New Delhi**
°F				55		32	
°C	28	2	19		14		12

13 a Given that t_K represents the temperature in K and t_C represents the temperature in °C, show that the formula $t_K = t_C + 273.15$ converts °C into K.

b Convert 60 °F to K. Give your answer to the nearest kelvin.

Hint: use a diagram like the one given in Example 2.4.5c.

14 The 2005 production of the Tinto winery is stored in 400 barrels of wine each one containing 100 hl. Calculate how many litres of wine were produced in 2005. Give your answer in standard form.

15 a A *light-year* is a unit of distance. It is the distance that light can travel in one year. Light moves at a velocity of about 300 000 km/s. Show that one light-year is 9.46×10^{12} km.

b The *parsec* is another unit of distance used in astronomy. The parsec (symbol pc) is 3.26 light-years (correct to 3 s.f.).

i Convert 1 pc to km. Write your answer in standard form.

ii Our galaxy (the Milky Way) is about 100 000 light-years in diameter. Convert this distance to parsecs. Write your answer in standard form.

iii The distance from Earth to the centre of the Milky Way is 8.6 kpc (kiloparsec). Write this length in light-years.

Note that neither the *light-year* nor the *parsec* is an SI unit.

2.5 Arithmetic sequences and series and applications

2.5.1 Number sequences

Example 2.5.1a

Look at this sequence of models and consider the number of dots needed to build each of them

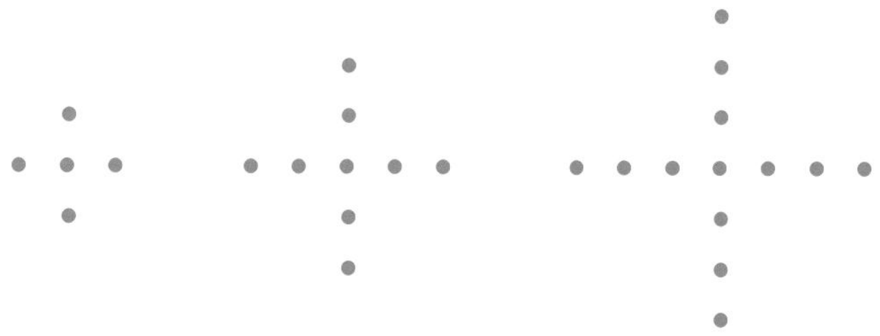

Model 1 5 dots | Model 2 9 dots | Model 3 13 dots

How many dots are needed to draw model 4? And to draw model 5? Now suppose we want to know how many dots are needed to draw model 12. Would it be necessary to draw all the previous models to find out? If we draw model 4 we will see that it is made by 17 dots.

In the *list* 5, 9, 13, 17, … , the first place is occupied by the number of dots needed to draw model 1, the second place by the numbers of dots needed to draw model 2 and so on. This list of numbers is arranged in a definite order. It is a **number sequence**. We see that in this number sequence there is a *pattern* as each number is 4 more than the previous number.

Example 2.5.1b

Let's consider the decimal expansion of π. Using the GDC it can be seen that $\pi = 3.141\,592\,654\ldots$

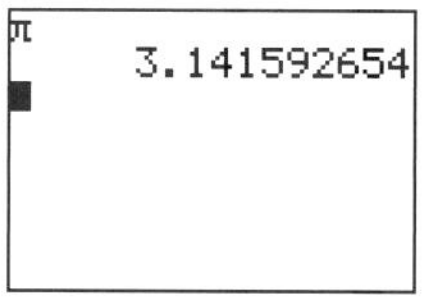

We say that the list 1, 4, 1, 5, 9, 2, 6, 5, 4,… is also a number sequence as the numbers are also arranged in a definite order. However in this list a pattern can not be found as π is not a rational number.

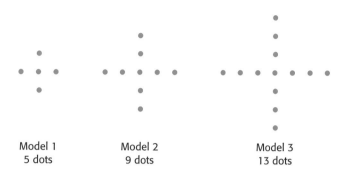

A **number sequence** is a list of numbers arranged in a definite order. The list can be finite or infinite.
The elements of a sequence are called its **terms.**

In this course we study only those number sequences from which patterns can be found.

Example 2.5.1c

For each of the three number sequences

a 4, 2, 0, −2, . . .
b 1, 4, 9, 16, 25, . . .
c 100, 50, 25, 12.5, . . .
 i describe the pattern
 ii write down the next two terms in the sequence.

Solution

a i This sequence starts at 4 and each term is 2 less than the previous term.
 ii −4 and −6.

b i This is the sequence of *square* numbers starting from 1.
 ii 36 and 49.
c i This sequence starts at 100 and each term can be found by multiplying the previous term by 0.5.
 ii 6.25 and 3.125.

Example 2.5.1d

Find the next two terms of these number sequences.

a 1, 8, 27, 64 . . .
b 0.4, 0.2, 0, −0.2, . . .
c 0, 1, 1, 2, 3, 5, . . .

Solution

a This is the sequence of *cube* numbers. We can also write it as 1^3, 2^3, 3^3, 4^3 and the next two terms are 5^3 and 6^3.

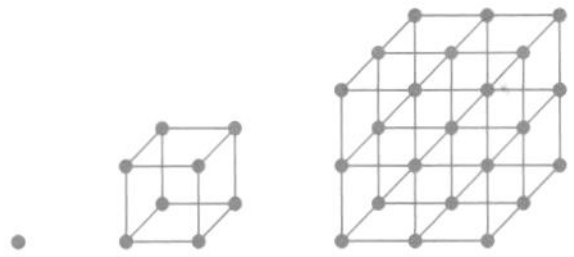

b Each term is 0.2 less than the previous term, therefore the next two terms are −0.4 and −0.6.
c We obtain each term by adding the preceding two terms. This rule is valid as from the third term. The next two terms are 8 and 13. This sequence is known as the *Fibonacci sequence*.

Fibonacci (1170–1250)

Fibonacci's full name was Leonardo Fibonacci, but he was known as Leonardo of Pisa (Leonardo Pisano) since he was born in Pisa (Italy), the city with the famous Leaning Tower, in about 1175 AD. Pisa was an important commercial town in its day and had links with many Mediterranean ports. Leonardo's father was a kind of customs officer in a North African town, so Leonardo grew up with a North African education under the Moors. He was in contact with many merchants and learned of their systems of doing arithmetic. He soon realized the many advantages of the "Hindu-Arabic" system over all the others. He was one of the first people to introduce this number system into Europe – the positional system we use today – based on ten digits with its decimal point and a symbol for zero.

In his book, *Liber Abbaci* (meaning *Book of the Abacus* or *Book of Calculating*), he persuaded many European mathematicians of his day to use this "new" system. The book describes the rules we all now learn at elementary school for adding numbers, subtracting, multiplying and dividing, together with many problems to illustrate the methods.

2.5.2 The general term of a number sequence

So far we have seen that a number sequence may be defined by

- listing the first terms and using "..." after them to indicate that the sequence follows the same pattern
- describing the number sequence in words.

Another way in which a number sequence can be defined is by writing its **general term**.

The terms of a sequence are designated in the following way:
$u_1, u_2, u_3, u_4, \dots, u_n, \dots$.
This means that u_1 represents the first term, u_2 represents the second term, etc.
The ***nth term*** of the sequence is represented by u_n and is called the general term.

Example 2.5.2a

For the sequence 1, 4, 9, 16... we write

$u_1 = 1$, $u_2 = 4$, $u_3 = 9$, $u_4 = 16, \dots$ and the general term is $u_n = n^2$

where n is a positive integer.

Example 2.5.2b

Given the sequence 100, 50, 25, 2.5,... we write

$u_1 = 100$, $u_2 = 50$, $u_3 = 25$, $u_4 = 12.5, \dots$ and the general term is $u_n = 100 \times 0.5^{n-1}$

where n is a positive integer.

The **general term** is a **formula** which is expressed in terms of ***n***, a positive integer. This formula allows us to work out any term of the sequence.

Example 2.5.2c

Consider the sequence 5, 9, 13, 17,....

a Show that the formula $u_n = 5+(n-1)\times 4$ works when $n = 1$, $n = 2$ and $n = 3$.

b Use the formula $u_n = 5+(n-1)\times 4$ to find u_{12}.

c Are these terms members of the sequence?
i 307 ii 401

Solution

a We have to substitute $n = 1$, $n = 2$ and $n = 3$ into the formula of the general term and check that we obtain 5, 9 and 13 respectively.

$u_1 = 5 + (1-1) \times 4 = 5$ $u_2 = 5 + (2-1) \times 4 = 9$

$u_3 = 5 + (3-1) \times 4 = 13$

b To find u_{12} we have to substitute n by 12 into the formula. Therefore $u_{12} = 5 + (12-1) \times 4 = 49$

You can find the value of u_{12} without finding the previous term (that is, u_{11}). This is the advantage of having the general term of a number sequence – you can find any term of the sequence without knowing or finding the previous term.

c i Let $u_n = 307$

$\therefore 5 + (n-1)4 = 307$

$\therefore (n-1)4 = 302$

$\therefore n - 1 = 75.5$

$\therefore n = 76.5$

n is not a positive integer so 307 is not a member of this sequence.

ii Let $u_n = 401$

$\therefore 5 + (n-1)4 = 401$

$\therefore (n-1)4 = 396$

$\therefore n - 1 = 99$

$\therefore n = 100$

n is a positive integer so 401 is a member of this sequence. It is the 100th term.

The general term of a sequence can be represented by u_n, a_n, b_n, and so on.

Example 2.5.2d

List the first four terms of the sequence

a $a_n = \frac{1}{2}(n+1)$ b $b_n = 2^n - 1$

Solution

a 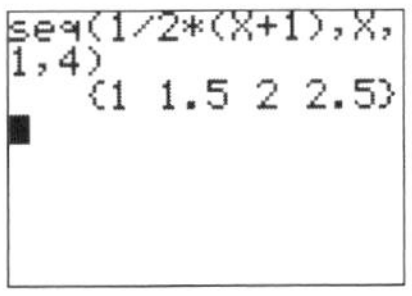

b 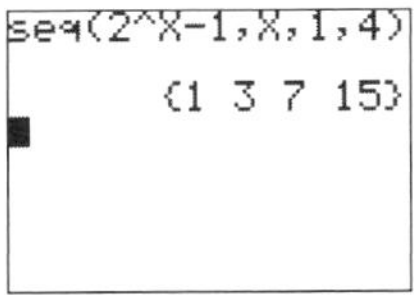

Answer: $a_1 = 1; a_2 = 1.5; a_3 = 2; a_4 = 2.5$ Answer: $b_1 = 1; b_2 = 3; b_3 = 7; b_4 = 15$

2.5.3 Arithmetic sequences

The **arithmetic sequences** are a special type of number sequence.

A number sequence is arithmetic if there is a **constant** difference between each term and the previous one. This constant is called the **common difference.** Given the arithmetic sequence $u_1, u_2, u_3, u_4, \ldots$ the common difference is $d = u_2 - u_1 = u_3 - u_2 = u_4 - u_3 = \ldots$

Example 2.5.3a

Consider the number sequence 5, 9, 13, 17,...

This sequence is arithmetic as the difference between each term and its preceding term is always the same number.

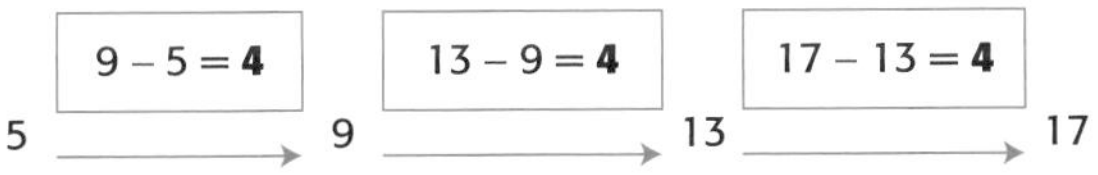

This is an arithmetic sequence with **common difference** 4.

Example 2.5.3b

Consider the number sequence 4, 2, 0, −2,....

This sequence is also arithmetic.

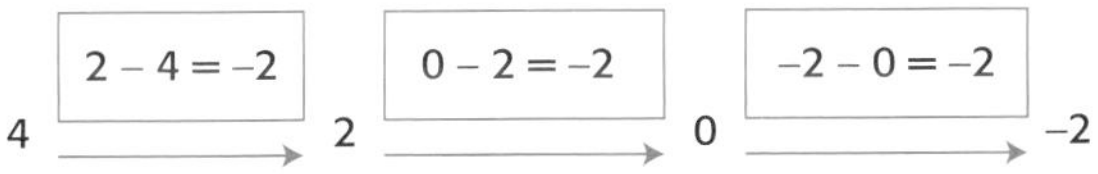

There is a **common difference**: the difference between each term and the preceding term is always –2.

The common difference may be a positive or a negative number.

Example 2.5.3c

Consider the sequence 1, 4, 9, 16,.... Show that this sequence is *not* arithmetic.

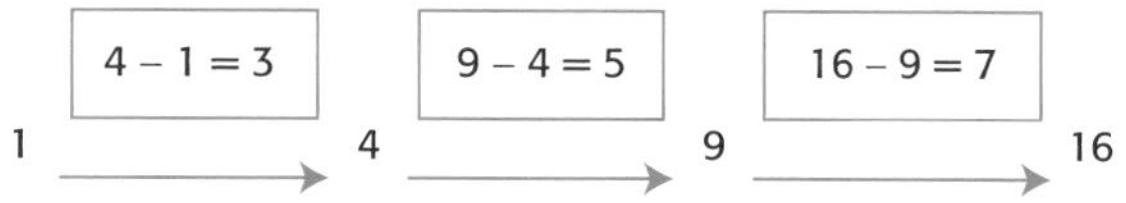

The difference between consecutive terms is *not* constant ($3 \neq 5$ and $5 \neq 7$). There is no common difference and therefore this is not an arithmetic sequence.

Example 2.5.3d

The sequence 5, x, 12,... is arithmetic.

a Find x, the second term of the sequence.

b Find the common difference, d.

Solution

a If the sequence is arithmetic and the common difference is d then

$d = x - 5$ and $d = 12 - x$ therefore $x - 5 = 12 - x$

$$2x = 17$$

$$x = 8.5$$

The second term of the sequence is 8.5.

b To find d we can substitute $x = 8.5$ in either $d = x - 5$ or $d = 12 - x$.

Therefore $d = 12 - 8.5 = 3$.

The common difference, $d = 3.5$.

2.5.4 The general term of an arithmetic sequence

Let's call the first term of an arithmetic sequence u_1 and the common difference d.

Then we can generate the sequence as shown here.

u_1

$u_2 = u_1 + 1d$

$u_3 = u_1 + d + d = u_1 + 2d$

$u_4 = u_1 + d + d + d = u_1 + 3d$

$u_5 = u_1 + d + d + d + d = u_1 + 4d$

and in general

$u_n = u_1 + (n - 1)\, d$

The **general term** (or ***n*th** term) of an **arithmetic sequence** with first term u_1 and common difference d is $u_n = u_1 + (n - 1)d$ where n is a positive integer.

Example 2.5.4a

The first term of an arithmetic sequence is 6 and its common difference is $\frac{3}{4}$.

a Find the second and third term of this sequence.

b Write down an expression for the *n*th term.

c Is 25.5 a term of this sequence?

Solution

a $u_1 = 6$

$u_2 = 6 + \frac{3}{4} = \frac{27}{4}$

$u_3 = \frac{27}{4} + \frac{3}{4} = \frac{30}{4}$

b We substitute 6 for u_1 and $\frac{3}{4}$ for d in the general term formula therefore $u_n = 6 + (n - 1) \times \frac{3}{4}$

c Let $u_n = 25.5$

$6 + (n - 1) \times \frac{3}{4}$

$= 25.5$

$(n - 1) \times \frac{3}{4}$

$= 19.5$

$n - 1 = 26$

$n = 27$

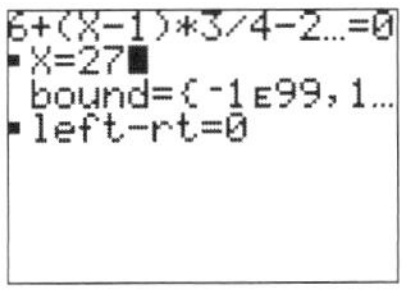

As n is a positive integer we can conclude that 25.5 is a term of this sequence. Furthermore 25.5 is the 27th term of this sequence.

Example 2.5.4b

When "The International Company" opened up it had 300 employees. The General Director decided to increase the number of employees by 15 at the beginning of each year.

a How many employees will "The International Company" have during the tenth year?

b In which year will the company first have more than 500 employees?

Solution

a We can write

$u_n = 300 + (n - 1) \times 15$ where u_n represents the number of employees in the company during the *n*th year.

$u_{10} = 300 + (10 - 1) \times 15 = 435$

There will be 435 employees during the tenth year.

b We need to find the n for which $u_n > 500$

$300 + (n - 1) \times 15 > 500$

$\therefore (n - 1) \times 15 > 200$

$\therefore n > 14.\overset{\frown}{3}$

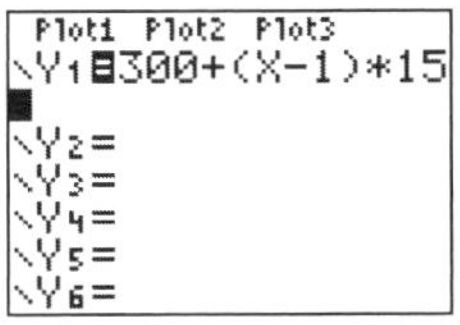

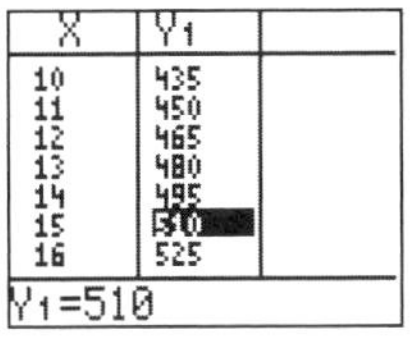

Therefore at the beginning of year 15 there will first be more than 500 employees in the company.

Example 2.5.4c

The second term of an arithmetic sequence is 18 and the fifth term is 12.

a Find the common difference, d.

b Find the first term, u_1.

c Find the 100th term.

Solution

Two methods to solve **a** and **b** are shown below.

Method 1

a As $u_2 = 18$ then $18 = u_1 + (2-1) \times d$
As $u_5 = 12$ then $12 = u_1 + (5-1) \times d$
These equations can be simplified to

Here we have a pair of simultaneous equations where the unknowns are u_1 and d.

$$\begin{cases} 18 = u_1 + d \\ 12 = u_1 + 4d \end{cases}$$

We solve this pair of simultaneous equations using the GDC so $d = -2$

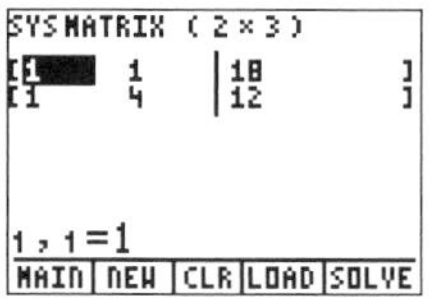

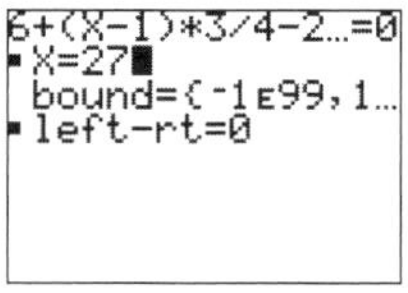

b $u_1 = 20$

Method 2

a You can make a diagram like this one.

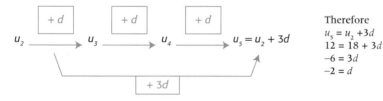

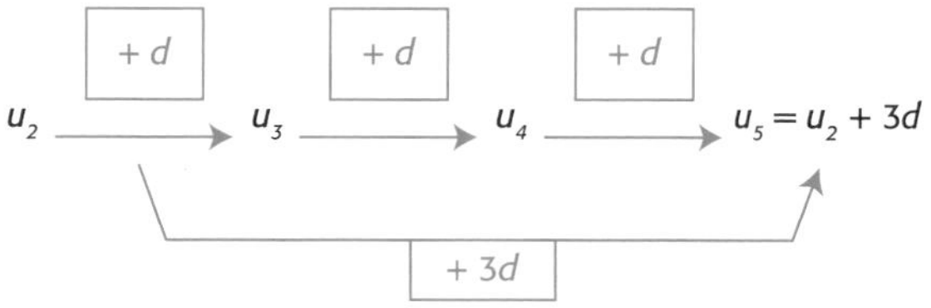

Therefore
$u_5 = u_2 + 3d$
$12 = 18 + 3d$
$-6 = 3d$
$-2 = d$

b We know that $d = u_2 - u_1$

$$-2 = 18 - u_1$$

$$u_1 = 18 + 2 = 20$$

Regardless of the method used in **a** and **b**.

c $u_{100} = 20 + (100 - 1) \times -2$

$u_{100} = -178$

Example 2.5.4d

Consider the finite arithmetic sequence –3, –1, 1, ..., 81.

a Find the common difference, d.

b Find the number of terms in this sequence.

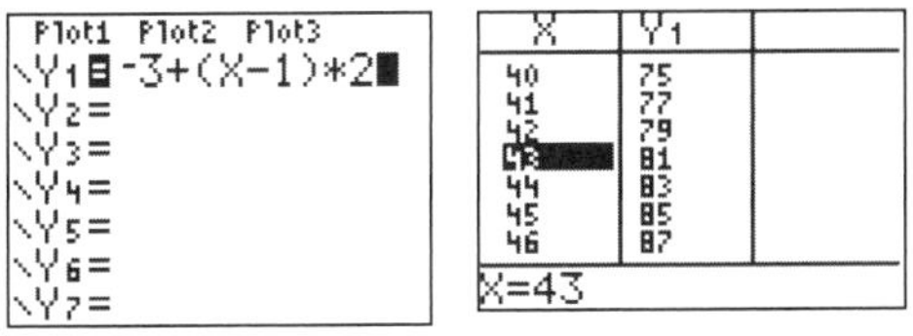

Solution

a The common difference is the difference between any term and the previous one so $d = u_2 - u_1 = -1 - (-3) = 2$

b The number of terms is a number n such that $u_n = 81$.

$$\begin{aligned} -3 + (n-1) \times 2 &= 81 \\ (n-1) \times 2 &= 84 \\ n - 1 &= 42 \\ n &= 43 \end{aligned}$$

Hence there are 43 terms in this sequence.

2.5.5 Arithmetic series; sum of the first *n* terms of an arithmetic sequence

Let's see what Gauss's idea was.

He needed to calculate 1 + 2 + 3 + 4 + … + 97 + 98 + 99 + 100.

Let's call this sum S.

$$\begin{array}{rl} S = & 1 + 2 + 3 + 4 + \ldots + 97 + 98 + 99 + 100 \\ +S = & 100 + 99 + 98 + 97 + \ldots + 4 + 3 + 2 + 1 \\ \hline 2S & \underbrace{101 + 101 + 101 + 101 + \ldots + 101 + 101 + 101 + 101}_{\text{This is } 100 \times 101} \end{array}$$

Therefore $2S = 100 \times 101$ so $S = \dfrac{100 \times 101}{2} = 50 \times 101 = 5050$

The sum 1 + 2 + 3 + 4 + … + 97 + 98 + 99 + 100 is an example of an **arithmetic series**.

These are also examples of arithmetic series.

a 3 + 6 + 9 + 12 + … + 303
as 3, 6, 9, 12, … is an arithmetic sequence with common difference $d = 3$.

b 14 + 13.5 + 13 + 12.5 + … + 4.5
as 14, 13.5, 13, 12.5, … is an arithmetic sequence with common difference $d = -0.5$.

Let's call the sum of the first n terms of an arithmetic sequence, S_n. This means that

$S_1 = u_1$

$S_2 = u_1 + u_2$

$S_3 = u_1 + u_2 + u_3$

and

$S_n = u_1 + u_2 + u_3 + \ldots. + u_n$

This sum, S_n, of the first n terms of the arithmetic sequence $u_1, u_2, u_3, \ldots$ can be calculated using the formula $S_n = \dfrac{n}{2}(u_1 + u_n)$.

Did you know that…?

Carl Friedrich Gauss (1777–1855) was a German mathematician and astronomer. At the age of seven he started elementary school, and his potential was noticed almost immediately. His teacher and his assistant were amazed when Gauss summed the integers from 1 to 100 instantly by spotting that the sum was 50 pairs of numbers, each pair summing to 101.

An **arithmetic series** is the **sum** of the terms of an arithmetic sequence.

If we substitute the expression $u_1 + (n-1)d$ for u_n, the formula can be rewritten as

$$S_n = \frac{n}{2}(u_1 + u_1 + (n-1)d) = S_n$$
$$= \frac{n}{2}(2u_1 + (n-1)d)$$

Example 2.5.5a

Calculate the sum of all the integers from 1 to 100 by using the formula for the sum of the first n terms of an arithmetic sequence.

Solution

$1 + 2 + 3 + 4 + \ldots + 97 + 98 + 99 + 100$ is an arithmetic series as $1, 2, 3, \ldots, 100$ is an arithmetic sequence with common difference $d = 1$. There are 100 terms therefore $n = 100$, $u_1 = 1$, $u_{100} = 100$ and substituting into the formula we have

$$S_{100} = \frac{100 \times (1 + 100)}{2} = \frac{100 \times 101}{2} = 5050$$

That is exactly the calculation that Gauss did!

Example 2.5.5b

Find the sum of the first 25 multiples of 7.

Solution

The first 25 multiples of 7 are 7, 14, 21, 28, …., 175, which clearly make up an arithmetic sequence with common difference 7.

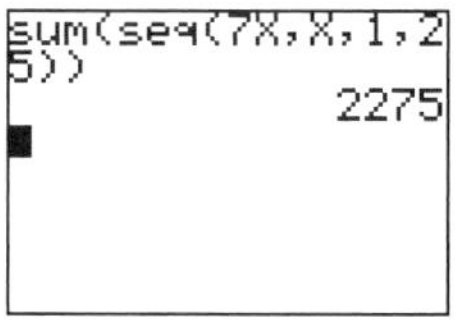

$n = 25$, $u_1 = 7$ and $u_{25} = 175$. Substituting into the formula of S_n

$$S_{25} = \frac{25}{2}(7 + 175) = 2275$$

Therefore the first 25 multiples of 7 add up to 2275.

The general term can be simply written as $u_n = 7n$.

Example 2.5.5c

The "Find Seats Theatre" has the following number of seats in each row starting from the first row: 20, 21, 22, 23, … and the last row has 40 seats. Find the total seating capacity in the theatre.

Solution

20, 21, 22, 23,….,40 is an arithmetic sequence with common difference $d = 1$.

We need to calculate $20 + 21 + 22 + 23 + \ldots + 40$.

$u_1 = 20$ and $u_n = 40$ but we still need to find n, the number of rows in the theatre.

As $u_n = 40$ then

$$40 = 20 + (n - 1) \times 1 \therefore n = 21 \text{ and } S_{21} = \frac{21}{2}(20 + 40) = 630$$

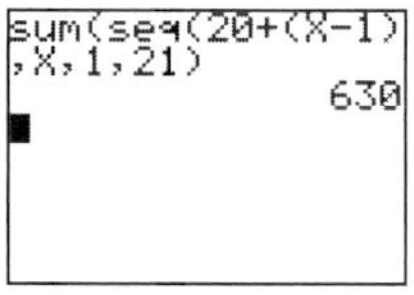

Therefore the theatre has 630 seats.

Exercise 2.5

1 Find the next two terms of these numbers sequences.

a 9, 6, 3, 0, … b $\frac{1}{2}, \frac{2}{3}, \frac{3}{4}, \frac{4}{5}, \ldots$ c 6, −6, 6, −6, …

d 4, 5, 9, 14, 23, …

2 a Which of these number sequences are arithmetic?

b Write down the common difference where appropriate.

i 2, 6, 10, 13, 17 ii $5, \frac{23}{4}, \frac{13}{2}, \frac{29}{4}, \ldots$ iii $1, -1, -3, -5, \ldots$

3 Consider this sequence of models.

Model 1 Model 2 Model 3 Model 4

a Write down the number of dots needed to draw model 5 and to draw model 6.

b Consider the arithmetic sequence 3, 5, 7, 9, …

i Write down its common difference.

ii Hence find the number of dots needed to draw model 20.

4 The nth term of a sequence is $a_n = 3n(n + 1)$ where n is a positive integer.

a Write down the values of a_1, a_2 and a_3.

b Show that this sequence is *not* arithmetic.

5 Consider the arithmetic sequence 10, 7, 4, 1, …

a Write down the common difference, d.

b Write down the next two terms of the sequence.

c Find the 30th term.

d Is −44 a member of the sequence? If so which term is it?

6 An arithmetic sequence is defined by $b_n = \frac{1 + 3n}{2}$ where n is a positive integer.

a Find b_1 and b_2.

b Write down the common difference.

c Find b_{21}.

d Find the first term of the sequence that is greater than 45.

e Use your graphic display calculator to find the sum of the first 90 terms of b_n.

7 a Find the common difference d and the first term u_1 of an arithmetic sequence given that $u_3 = 12$ and $u_7 = 20$.

b Find the common difference d and the first term u_1 of an arithmetic sequence given that $u_6 = -4$ and $u_{10} = -12$.

8 Calculate the sum of the first 200 positive integers.

9 Calculate

a 4 + 9 + 14 + 19 + … to 15 terms

b $3 + \frac{7}{3} + \frac{5}{3} + 1 + \ldots$ to 20 terms

c 1 + 0.9 + 0.8 + … to 18 terms.

10 Calculate the sum of the first 30 multiples of 4.

11 Consider the arithmetic sequence 9, 11, 13, …, 51.

a Write down the common difference.
b Calculate the number of terms of this sequence.
c Hence or otherwise find the sum 9 + 11 + 13 + … + 51.

12 The cost of renting a motorbike is \$30 for the first hour and \$8 for each additional hour.

a Write down the cost of renting the motorbike for
i 2 hours ii 3 hours.
b Calculate the cost of renting the motorbike for 12 hours.

The motorbike can be rented daily at a special rate of \$160.

c Calculate the *least* number of hours that the motorbike must be rented so that the special rate is cheaper than paying per hour.

13 The sequence 3, a, 10, b … is arithmetic.

a Find the value of a.
b Write down the common difference.
c Find b.
d Find the 11th term.

14 Three consecutive terms of an arithmetic sequence are $3k + 1$, $3 + 4k$ and $7k - 1$.

a Find k.
b Write down the values of these three terms.

15 In the arithmetic sequence $u_1, u_2, u_3, \ldots$ it is known that $S_1 = 5$ and $S_2 = 12$.

a Write down
i u_1 ii u_2.
b Find the common difference, d.
c Find S_4.

16 The perimeter of a hexagon is 48 cm and its shortest side is 3 cm. It is known that the sides of this hexagon are in arithmetic sequence. Let the length of the first side be a_1, the length of the sixth side be a_6 and the perimeter P.

a Write down an equation involving a_1, a_6 and P.
b Hence or otherwise find a_6.

17 In the following statements u_n is an arithmetic sequence with common difference d.
Decide whether or not these statements are true.

a $d = u_7 - u_6$ b $u_{12} = u_3 + 9d$ c $u_6 = 2\ \ u_3$
d $S_{25} = S_{24} + u_{24}$ e $S_{10} = 5(u_1 + u_{10})$

2.6 Geometric sequences and series and applications

2.6.1 Geometric sequences

Geometric sequences are another special type of number sequence.

A sequence is **geometric** if the quotient between any term of the sequence and its previous term is constant.
This constant is called **common ratio** of the sequence.
Given the geometric sequence $u_1, u_2, u_3, \ldots$ the common ratio is

$$r = \frac{u_2}{u_1} = \frac{u_3}{u_2} = \frac{u_4}{u_3} = \ldots$$

Example 2.6.1a

Here are some examples of geometric sequences.

a $1, 2, 4, 8, 16, \ldots$ As $\frac{2}{1} = \frac{4}{2} = \frac{8}{4} = \ldots = 2$ the sequence is geometric with common ratio $r = 2$.

b $100, -50, 25, -12.5, \ldots$ As $\frac{-50}{100} = \frac{25}{-50} = \frac{-12.5}{25} = \ldots = -0.5$ the sequence is geometric with common ratio $r = -0.5$.

c $3, 1, \frac{1}{3}, \frac{1}{9}, \ldots$ As $\frac{1}{3} = \frac{\frac{1}{3}}{1} = \frac{\frac{1}{9}}{\frac{1}{3}} = \ldots = \frac{1}{3}$ the sequence is geometric with common ratio $r = \frac{1}{3}$.

The common ratio can be either positive or negative.

Example 2.6.1b

In a geometric sequence $u_1 = 3$ and the common ratio is -2.
Find u_2 and u_3.

Solution

As $\frac{u_2}{u_1} = -2 \therefore u_2 = -2 \times 3 \therefore u_2 = -6$

As $\frac{u_3}{u_2} = -2 \therefore u_3 = -2(-6) \therefore u_3 = 12$

2.6.2 The general term of a geometric sequence

Let the terms of the sequence be $u_1, u_2, u_3, \ldots$ and the common ratio r then

$u_2 = u_1 \times r$
$u_3 = u_2 \times r = u_1 r^2$
$u_4 = u_3 \times r = u_1 \times r^3$

and in general

$u_n = u_1 \times r^{n-1}$

Therefore the **general term** (or the *n*th term) of a **geometric sequence** with first term u_1 and common ratio r is $u_n = u_1 \times r^{n-1}$.

Example 2.6.2a

Given the geometric sequence 100, 25, 6.25,...

a find the common ratio, r

b write down the fourth term

c find the general term, u_n

d hence, find u_8, giving your answer in standard form.

Solution

a $r = \dfrac{u_2}{u_1} = \dfrac{25}{100} = 0.25$

b $u_4 = u_3 \times r = 6.25(0.25) = 1.5625$

c $u_n = u_1 \times r^{n-1} \therefore u_n = 100 \times 0.25^{n-1}$

d $u_8 = 100 \times 0.25^{8-1} = 0.00610$ (3 s.f.) and written in standard form is 6.10×10^{-3} (3 s.f.).

Example 2.6.2b

In a geometric sequence $a_3 = 0.5$ and $a_7 = 40.5$. It is known that the common ratio r is positive.

a Find r.

b Find a_1.

Solution

a **Method 1**

Write a pair of simultaneous equations where the unknowns are a_1 and r.

$$a_3 = a_1 \times r^2 \therefore 0.5 = a_1 \times r^2 \therefore \frac{0.5}{r^2} = a_1$$

$$a_7 = a_1 \times r^6 \therefore 40.5 = a_1 \times r^6 \therefore \frac{40.5}{r^6} = a_1$$

Equating both expressions for a_1 we get

$$\frac{0.5}{r^2} = \frac{40.5}{r^6}$$

$$0.5r^6 = 40.5r^2$$

$$r^4 = 81$$

$$r = 3 \text{ as } r > 0$$

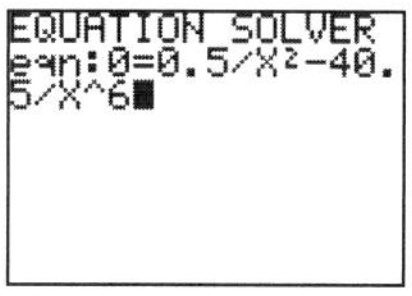

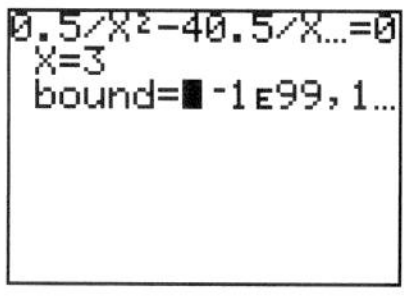

Method 2

Make a diagram like this one to see that $a_7 = a_3 \times r^4$.

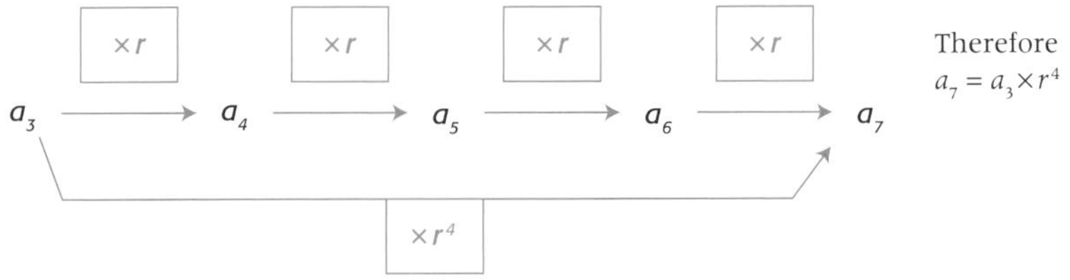

Substituting

$40.5 = 0.5 \times r^4 \therefore 81 = r^4 \therefore r = 3$ as $r > 0$

b Find a_1 by substituting the value of r into any of the two equations used for Method 1

$$a_1 = \frac{0.5}{3^2} = \frac{1}{18} = 0.0556 \ \text{(3 s.f.)}$$

Example 2.6.2c

Find the number of terms in the geometric sequence 0.3, 0.9, . . . , 53 144.1.

Solution

$a_1 = 0.3$ and $r = \frac{0.9}{0.3} = 3$. The last term, $u_n = 53\,144.1$. Find n such that $u_n = 0.3 \times 3^{n-1} = 53\,144.1$

Solve this equation by making a list or by using the solver or using any alternative method.

Therefore the number of terms in this sequence is 12.

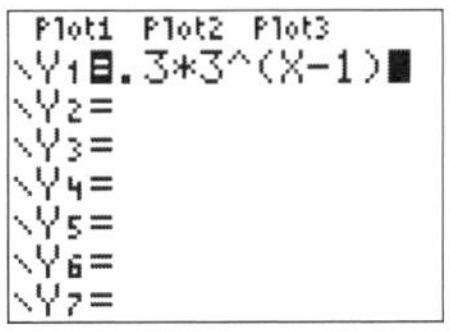

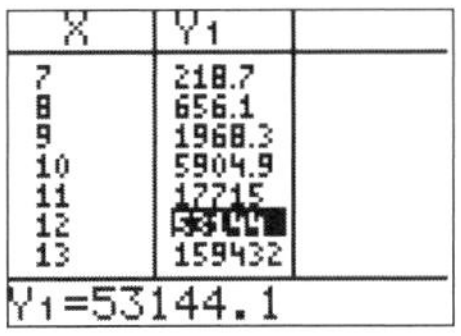

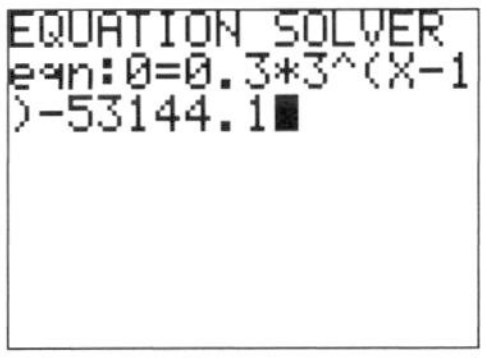

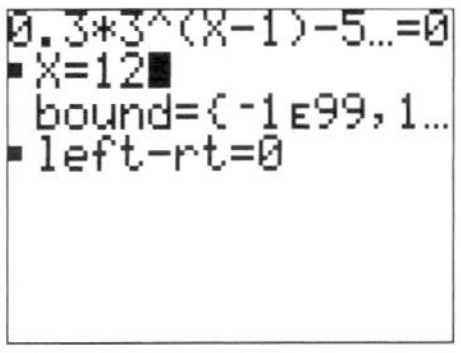

Example 2.6.2d

Given the sequences $a_n = 8 + 2n$ and $b_n = \left(\frac{3}{2}\right)^n$ where n is a positive integer:

a List the first three terms of a_n.

b List the first three terms of b_n.

c Find the first value of n for which $a_n < b_n$.

Solution

a and **b**

Answer to part **a**: $a_1 = 10, a_2 = 12, a_3 = 14$

Answer to part **b**: $a_1 = 1.5, a_2 = 2.25, a_3 = 3.375$

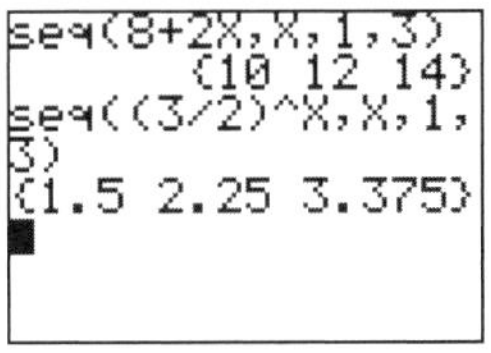

c We need to find the first n for which $a_n < b_n$.

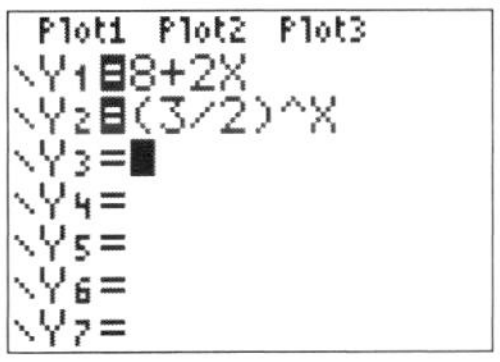

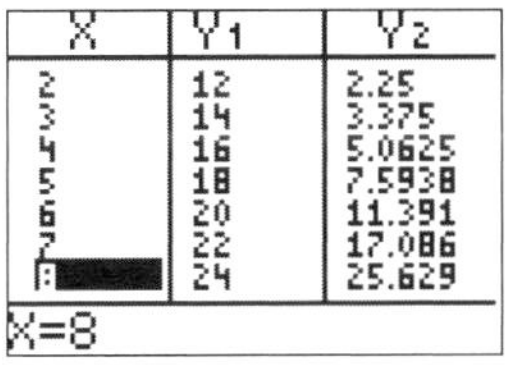

The first n for which $a_n < b_n$ is $n = 8$.

2.6.3 Applications of geometric sequences

Example 2.6.3a

The value of a car when it was first bought was \$45 000. The car loses 20% of its value each year.

a Calculate the value of the car after one, two and three years.
b Calculate the value of the car six years after it was first bought.
c Find how long it will take for the value of the car to fall below \$9000.

Solution

a If the car loses 20% of its value each year this means that each year it will be worth 80% of its previous year's value. Therefore to find the value of the car each new year we have to multiply by 0.80 each time.

Initial value of the car (\$)	**45 000**
Value of the car after one year (\$)	$45\,000 \times 0.80 = 36\,000$
Value of the car after two years (\$)	$45\,000 \times 0.80 \times 0.80 =$ $45\,000 \times 0.80^2 = 28\,800$
Value of the car after three years (\$)	$45\,000 \times 0.80^2 \times 0.80 =$ $45\,000 \times 0.80^3 = 23\,040$

45 000, 36 000, 28 800, 23 040, … is a geometric sequence with common ratio 0.80 and first term 45 000. In this context the first term of the sequence represents the initial value of the car. We can write the formula for this geometric sequence as $v_n = 45\,000 \times 0.8^n$ where v_n represents the value of the car after n years have passed and $n = 0, 1, 2, \ldots$

We are letting n be 0 as $n = 0$ means that *no* years have passed. The exponent of the common ratio is not $n - 1$, but n and therefore it coincides with the number of years that have passed. The formula $v_n = 45000 \times 0.8^{n-1}$ for $n = 1, 2, 3, \ldots$ is also correct though the value of n in this formula does not coincide with the number of years that have passed. It would represent one year more.

Let's check that the formula $v_n = 45\,000 \times 0.8^n$ for $n \in \mathbb{N}$ works.

n	0	1	2	3
$\mathbf{v_n}$	$45\,000 \times 0.8^0$ $= \$45\,000$	$45\,000 \times 0.8^1$ $= \$36\,000$	$45\,000 \times 0.8^2$ $= \$28\,800$	$45\,000 \times 0.8^3$ $= \$23\,040$

b Now we are ready to find the value of the car 6 years after it was first bought. Using this formula:

$v_6 = 45\,000 \times 0.80^6 = \$\ 11\,796.48$

c We need to find n such that $v_n = 45\,000 \quad 0.8^n < 9000$

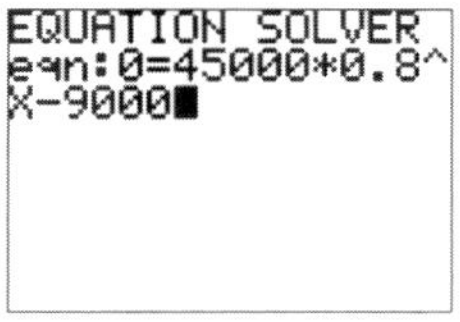

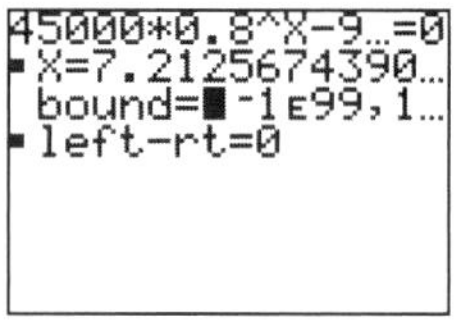

So it will take 7.21 years for the value of the car to fall below \$9000.

Example 2.6.3b

The population of a small town at the start of the year 2000 was 3500 and it is estimated that it increases by 3% each year.

Calculate the estimated population of this town

a at the start of 2001 and 2002

b at the start of 2006.

Give your answers to the nearest hundred.

Solution

a The population increases by 3% each year, therefore, to calculate the population of each subsequent year, we have to multiply by 1.03%. In this situation we can take as the initial value the population of the town at the start of 2000.

Year	Population
2000	3500 (initial population)
2001 (after 1 year)	$3500 \times 1.03 = 3605$
2002 (after 2 years)	$3500 \times 1.03 \times 1.03 =$ $3500 \times 1.03^2 = 3713.15$

The predicted population at the start of 2001 and 2002 is 3600 and 3700 to the nearest hundred respectively.

b As we see, 3500, 3605, 3713.15, ... is a geometric sequence with common ratio 1.03. But we can write the general term of this sequence as

$$p_n = 3500 \times 1.03^n \text{ for } n = 0, 1, 2, \ldots$$

Notice that the exponent of the common ratio coincides with the number of years that have passed, therefore if we need to find the predicted population for 2006 we have to substitute n by 6.

Therefore $p_6 = 3500 \times 1.03^6 = 4200$ to the nearest hundred.

The predicted population at the start of 2006 is 4200 correct to the nearest hundred.

2.6.4 Geometric series; sum of the first *n* terms of a geometric sequence

A **geometric series** is the **sum** of the terms of a geometric sequence.

For example

a $1+2+4+8+16+\ldots+2^{63}$ is a geometric series as 1, 2, 4, 8, ... is a geometric sequence.

b $2-1+0.5-0.25+\ldots+2\left(-\frac{1}{2}\right)^{25}$ is a geometric series as $2, -1, 0.5, -0.25, \ldots$ is a geometric sequence.

Did you know that...?

There is a popular legend in which a man named Sissa ibn Dahir invented the game of chess for an Indian king. This king liked the game so much that he had chessboards placed in all the Hindu temples. The king wished to reward Sissa so told him to make a wish. The wise man replied, "Then I wish that one grain of wheat shall be put on the first square of the chessboard, two on the second, and that the number of grains shall be doubled until the last square is reached; whatever the quantity this might be, I desire to receive it." When the king first heard Sissa's wish he found it very simple but after a while the king realized that all the wheat in the world would not suffice so he commended Sissa for formulating such a wish and pronounced it even more clever than his invention of chess.

Example 2.6.4a

Calculate S, the number of grains of wheat needed to accomplish Sissa's wish.

Solution

First we draw a table to see the number of grains needed to put in each square.

Square	1	2	3	4	5	n	64
Number of grains of wheat	1	2	2^2	2^3	2^4	2^{n-1}	2^{63}

Remember that a chess board has 64 squares.

The number of grains of wheat for squares 1, 2, 3, … , 64 is a geometric sequence with common ratio 2 so

$$S = 1 + 2 + 2^2 + 2^3 + 2^4 + \ldots + 2^{62} + 2^{63}$$

$$2S = 2(1 + 2 + 2^2 + 2^3 + 2^4 + \ldots + 2^{62} + 2^{63})$$ Multiplying both sides by **2**

$$2S = \underbrace{2 + 2^2 + 2^3 + 2^4 + 2^5 + \ldots + 2^{63}}_{\text{This is } S-1} + 2^{64}$$ Distributing

where 1 represents the first term of the sequence.

Therefore

$$2S = S - 1 + 2^{64}$$
$$2S - S = -1 + 2^{64}$$
$$S = 2^{64} - 1$$

It is clear that the calculation $2^{64} - 1$ is much simpler than $1 + 2 + 2^2 + 2^3 + 2^4 + \ldots + 2^{62} + 2^{63}$

Therefore to accomplish Sissa's wish the king would have needed $2^{64} - 1$ grains of wheat. In scientific notation, this is 1.84×10^{19}, a very large number!

Using this method it can be proved that

The sum, S_n, of the n terms of the geometric sequence $u_1, u_2, u_3, \ldots$ with common ratio r is given by $S_n = \dfrac{u_1(r^n - 1)}{r - 1}$ where $r \neq 1$.

$S_n = \dfrac{u_1(r^n - 1)}{r - 1} = \dfrac{u_1(1 - r^n)}{1 - r}$ – these expressions are equivalent.

Notice that $r \neq 1$ as this formula has no meaning when $r = 1$.

If $r = 1$ the sequence would be $u_1, u_1 \times 1, u_1 \times 1^2, u_1 \times 1^3, \ldots, u_1 \times 1^{n-1}$ that is $u_1, u_1, u_1, \ldots, u_1$ and this is *not* a geometric sequence.

However the sum of its first n terms can be calculated using $S_n = n \times u_1$.

Example 2.6.4b

Use the formula for the sum of the n terms of a geometric sequence to calculate

$3 + 9 + 27 + 81 + \ldots$ to ten terms.

Solution
To use the S_n formula we need n, u_1 and r.
$n = 10$, $u_1 = 3$ so we still need to find r.

$$r = \frac{9}{3} = \frac{27}{9} = \ldots = 3 \text{ then } S_{10} = \frac{3(3^{10}-1)}{3-1} = \frac{3(3^{10}-1)}{2} = 88572$$

Example 2.6.4c

Use the graphic display calculator to find $2 - 1 + 0.5 - 0.25 + \ldots$ to 12 terms.

Solution

$n = 12$, $u_1 = 2$ and $r = \frac{-1}{2} = \frac{0.5}{-1} = \ldots = -0.5$ then $u_n = 2\times\left(-\frac{1}{2}\right)^{n-1}$

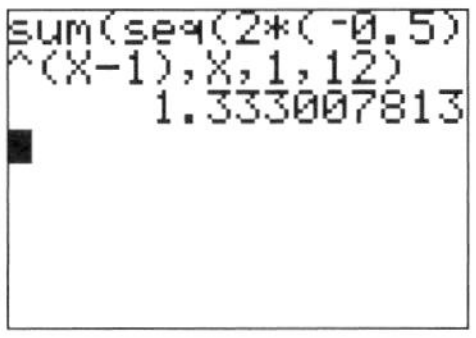

$S_{12} = 1.33$ (3 s.f.)

Example 2.6.4d

Find the sum of the geometric series $1 + 2 + 4 + \ldots + 1024$.

Solution
To use the formula $S_n = \frac{u_1(r^n - 1)}{r-1}$ we have $u_1 = 1$ but still need to find r and n.

1 Find r.

$$r = \frac{2}{1} = \frac{4}{2} = \ldots = 2$$

2 Find n by using the formula of the *nth* term of a geometric sequence.

$1024 = 1\times 2^{n-1} \therefore n = 11$

3 Substitute $u_1 = 1$, $r = 2$ and $n = 11$ in the formula of S_n.

$$S_{11} = \frac{1(2^{11}-1)}{2-1} = 2047$$

Example 2.6.4e

A company is offering Abid a job with an initial salary of $28 000 and a 4% raise each year after that. This 4% raise continues every year.

a Find what Abid's salary will be after five years.

b Calculate the amount of money that Abid will have earned after 15 years in his career.

Solution

a Each year the salary increases by 4%, therefore to find the new salary we multiply the previous salary by 1.04%. This is a geometric sequence with first term $28 000 and common ratio 1.04.

Year	Initial salary	After 1 year	After 2 years	After 3 years	After 5 years
Salary	\$28000	$28000 \times 1.04 =$ \$29120	$28000 \times 1.04^2 =$ \$30284.80	$28000 \times 1.04^3 =$ \$31496.19	$28000 \times 1.04^5 =$ \$34 066.28

After 5 years Abid's salary will be \$34066.28.

b We need to calculate 28000 + 29120 + 30284.80 + … up to the 15th term. This is S_{15}.

$$S_{15} = \frac{28000(1.04^{15} - 1)}{1.04 - 1} = \$560\,660.45$$

After a 15-year career Abid will have earned \$560660.45.

Theory of knowledge

Zeno's paradox

Zeno, an ancient Greek possibly too smart for his own good, developed a paradox. It postulated that, in a race between the hare and the tortoise, if distance is infinitely divisible and the tortoise is given a head start the hare can never beat the tortoise.

The advanced fable example

Assuming everyone is familiar with the story of the tortoise and the hare, Zeno's paradox shows how the hare never would have stood a chance had the tortoise been given a head start.

Once upon a time, Hare decided to race the other members of the animal kingdom. "You're too fast!" they all said, "We don't want to race you. We don't stand a chance." Only one animal was willing to race, Tortoise. You see, he had read about Zeno's Paradox and knew that all he needed was a head start.

"You just want a head start? Fine!" snapped Hare. "I'm 10 times faster than you are, so you can start 10 metres ahead of me."

Tortoise started the race. After he was 10 metres along, Hare started and caught up fast. When Hare got to the 10 metre marker Tortoise was at only 11 metres, and when Hare got there Tortoise was at 11.1 metres, and when Hare got there Tortoise was at 11.11 metres. As hard as he ran, Hare couldn't pass the Tortoise! By the end of the race, Hare was just a tiny bit behind. Of course, the "Fastest Animal" trophy couldn't be awarded because Tortoise was affected by Zeno's Paradox too. Neither one of them finished the race.

Why does Zeno's Paradox not apply to the real world?
Do you know any more paradoxes?

Exercise 2.6

1 Decide which of the following sequences are geometric. For those that are geometric write down their common ratio.

a 2, 4, 6, 8, …   b 2, 4, 8, 16, 32, …
c 3, 0.6, 0.12, 0.024, …   d 8, 4, 2, 1, 0.25

2 In the table, four sequences are given in the first column. One of these sequences is arithmetic, one is geometric and the other two are neither arithmetic nor geometric.

a Copy and complete the table with a ✔ in the appropriate column.

Sequence	Arithmetic	Geometric	Neither geometric nor arithmetic
10, 9.7, 9.4, 9.1,…			
$1, \frac{1}{2}, \frac{1}{3}, \frac{1}{4}, \ldots$			
12, 1.2, 0.12, 0.012			
1, 3, 4, 7, 11, …			

b Write down the common difference for the arithmetic sequence.
c Write down the common ratio for the geometric sequence.

3 Find the common ratios of these geometric sequences.

a $3, -1, \frac{1}{3}, -\frac{1}{9}, \ldots$
b $1, 3, 9, 27, \ldots$
c $-5, 25, -125, 625, \ldots$
d $\pi, \pi^2, \pi^3, \pi^4, \ldots$

4 Find the stated term for each of these geometric sequences.

a 100, 10, 1, 0.1, … the 10th term
b $-6, -12, -24, -48, \ldots$ the 12th term
c 32, 24, 18, 13.5, … the 9th term
d $ax, ax^2, ax^3, ax^4, \ldots$ the 10th term with $a \neq 0, x \neq 0$.

5 Calculate the number of terms in each of these geometric sequences.

a $1, \frac{3}{2}, \frac{9}{4}, \ldots, \frac{729}{64}$
b $2, 20, 200, \ldots, 2 \times 10^{11}$
c $\frac{100}{3}, \frac{80}{3}, \frac{64}{3}, \ldots, \frac{4096}{375}$

6 Calculate

a $1 + 3 + 3^2 + \ldots + 3^{10}$
b $2 - 1 + 0.5 - 0.25 + \ldots + 2\left(-\frac{1}{2}\right)^{24} + 2\left(-\frac{1}{2}\right)^{25}$

7 Calculate

a $1 + 1.2 + 1.44 + 1.728 + \ldots$ to 10 terms
b $3 + \frac{9}{4} + \frac{27}{16} + \frac{81}{64} + \ldots$ to 12 terms

8 Consider the geometric sequence $1, -2, 4, -8, \ldots, 65\,536$. Find

a the common ratio
b the number of terms
c the sum of its terms.

9 The first three terms of a geometric sequence are $2, x$ and 18. It is known that the common ratio is a positive number.

a Write down an equation involving the first term, the common ratio, r, and the third term.
b Hence or otherwise find
  i r
  ii the second term, x
  iii the tenth term.

10 In a geometric sequence the second term is 15 and the fifth term is 1.875.

a Find the common ratio, r.

b Hence find the first term of the sequence.

c Calculate the first term of the sequence that is less than 0.5.

11 The zoom function in an art software package is programmed to enlarge an image by 20%. In a photo the height of a tree is 3.2 cm.

a Calculate the height of the tree in a picture after the zoom function has been applied three times. Give your answer to the nearest mm.

b The art designer needs the height of the tree to be more than 10 cm. Calculate the least number of times he has to apply the zoom function if he starts from the original photo.

12 A rubber ball rebounds to 90% of the height from which it has dropped. The first time it is dropped from a height of 8 m onto ground level.

a Copy and complete the table with the heights that the ball reaches after n bounces.

n (number of bounces)	0	1	2	3	6
Height after n bounces (in metres)	8				

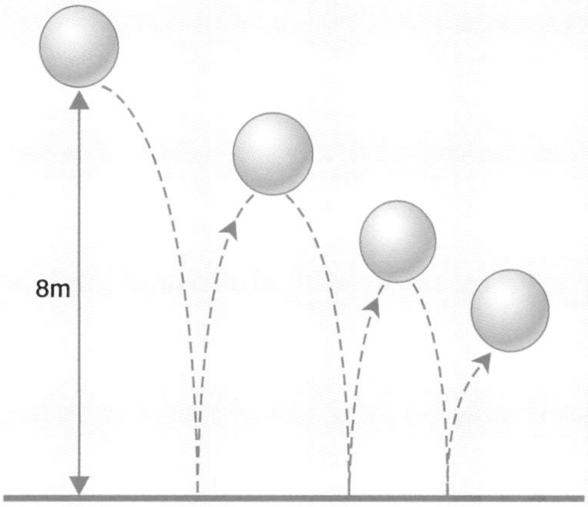

b Calculate the number of times the ball has to rebound for it to reach a height less than 2 m.

13 At 11 am, Dmitrieva and Emiko were told a secret. After half an hour each of them had told this secret to three different friends who did not know it. After another half an hour each of these friends told the secret to another three and so on.

a Calculate the number of people that will be told the secret between 12:30 and 1 pm.

b Calculate the total number of people that will be told the secret by 1 pm.

c At a certain time in the afternoon the secret had already been told to 2186 people. Calculate what that time was.

14 Each year a machine loses 15% of its value. One year after it was bought the value of the machine was $5100.

a Calculate the initial value of the machine.

b Calculate the value of the machine four years after it was bought.

c The owner of the machine says that he will keep the machine while its value is greater than $2000. Calculate the number of years that the owner will keep the machine.

15 In the following statements u_n is a geometric sequence with common ratio r. Decide whether or not each statement is true.

a $u_8 = u_3 \times r^5$ b $\frac{u_{12}}{u_6} = u_2$

c $S_{15} = S_{13} + u_{14} + u_{15}$ d $r^2 = \frac{u_9}{u_7}$

2.7 Solution of linear and quadratic equations using the GDC

The solution of a pair of linear equations in two variables can be found using the graphic display calculator (GDC) or by analytical methods. Finding the solution is also known as "solving simultaneous equations". It means that we must find values of *each variable* that will solve both equations. For lines with different gradients the solution is their point of intersection. This point is the one and only point which is on both of the lines representing the equations.

A linear equation in two variables is the equation of a line.

2.7.1 Solving simultaneous equations using the GDC

Method 1 By graphing both lines

Example 2.7.1a

Solve this pair of equations simultaneously for x and y.

$y = 7 - x$

$y = x - 2$

Solution

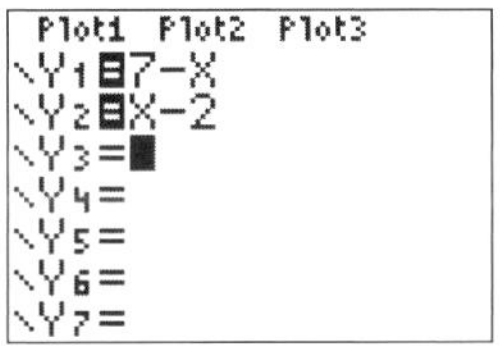

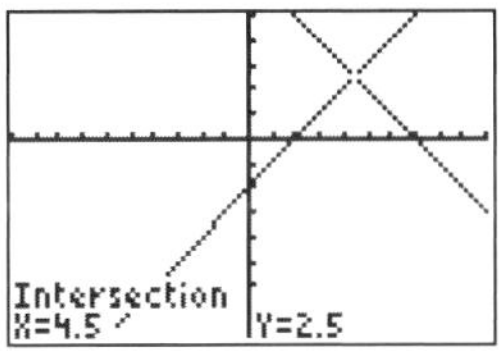

The solution is $x = 4.5$, $y = 2.5$, which is the ordered pair (4.5, 2.5).

Example 2.7.1b

Solve this pair of linear equations.

$$2x + y = 7$$
$$5x - 2y = 4$$

Solution

The equations must first be rearranged to make y the subject.

$2x + y = 7 \rightarrow\ y = 7 - 2x$ and $5x - 2y = 4 \rightarrow y = \dfrac{5}{2}x - 2$

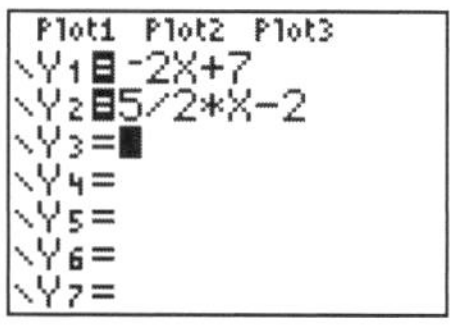

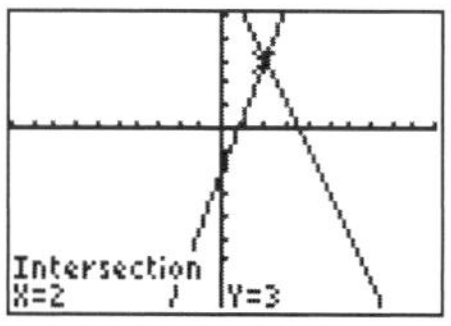

The solution is $x = 2,\ y = 3$, or $(2, 3)$.

Method 2 Using the simultaneous equation solution function

The equations need to be written in general form, that is, $ax + by = c$. The coefficients a and b, and the constant c are entered in to the GDC.

Solution *(Example 2.7.1a)*

Rearrange equations: $y = 7 - x \ \rightarrow\ \ x + y = 7$

$y = x - 2\ \rightarrow\ -x + y = -2$

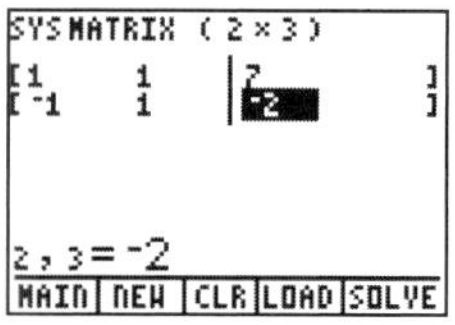

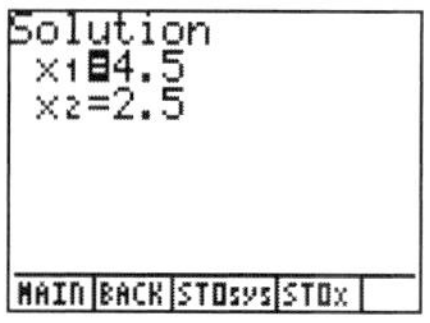

The solution is $x = 4.5$, $y = 2.5$.

Solution *(Example 2.7.1b)*

The equations are in general form. $2x + y = 7$

$5x - 2y = 4$

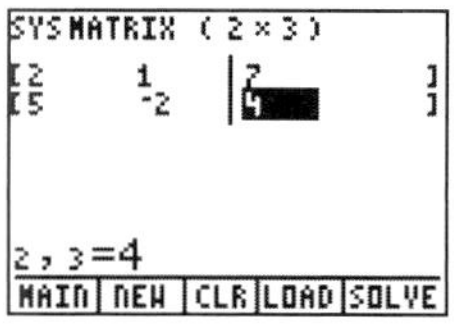

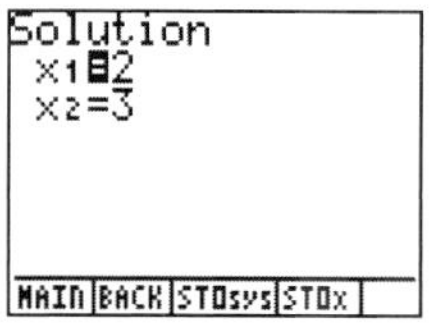

The solution is $x = 2,\ y = 3$.

2.7.2 Analytical (algebraic) methods

Method 1 The method of substitution

Solve one of the equations for *one* variable in terms of the other. Then, substitute into the other equation. The result will be an equation in one variable which can be solved. The other variable can then be found by substitution into either equation.

Example 2.7.2a *(Example 2.7.1a)*

Solve simultaneously $y = 7 - x$

$y = x - 2$

Solution

The equations are already solved for one variable (y) in terms of the other (x). Substituting for y in the second equation gives

$$7 - x = x - 2$$

Therefore $9 = 2x$ and hence $x = 4.5$.
Now, substitute for x in the first equation.

$$y = 7 - 4.5 = 2.5$$

The solution is $x = 4.5$, $y = 2.5$.

Example 2.7.2b *(Example 2.7.1b)*

Solve this pair of linear equations.

$$2x + y = 7$$
$$5x - 2y = 4$$

Solution

Rearrange the first equation to make y the subject.

$$y = -2x + 7$$

Substituting for y in the second equation gives

$$5x - 2(-2x + 7) = 4$$
$$5x + 4x - 14 = 4$$
$$9x = 18 \quad \text{so} \quad x = 2$$

Substituting for x in the first equation gives

$$y = -2(2) + 7 = 3$$

The solution is $x = 2$, $y = 3$.

Method 2 The method of elimination

If the coefficients of one variable in both equations are identical or numerically equal and of opposite sign, the equations can be subtracted or added respectively to eliminate that variable. The result will be an equation in one variable which can be solved. The other variable can then be found by substitution into either equation.

Example 2.7.2c *(Example 2.7.1a)*

Solve simultaneously, $y = 7 - x$

$y = x - 2$

Solution

Rearrange to line up the variables.

$$y = -x + 7$$
$$y = x - 2$$

Alternatively
the x-variable coefficients have opposite signs.
Add the equations to eliminate the x's.
$2y = 5 \rightarrow y = 2.5$
Substitute for y in the second equation
$2.5 = x - 2 \rightarrow x = 4.5$

The y-variable coefficients are identical.
Subtract the equations to eliminate the y's.

$0 = -2x + 9 \rightarrow \quad x = 4.5$

Substitute for x in the first equation.

$y = -4.5 + 7 \rightarrow \quad y = 2.5$

The solution is $x = 4.5$, $y = 2.5$.

Example 2.7.2d *(Example 2.7.1b)*
Solve this pair of linear equations.

$$2x + y = 7$$
$$5x - 2y = 4$$

Solution
Multiply the first equation by 2 to get opposite coefficients of the y-variable.

$4x + 2y = 14$
$5x - 2y = 4$

In this case, add the two equations to eliminate the y-variable.

$9x = 18 \therefore x = 2$

Substitute for x in the first equation.

$2(2) + y = 7 \therefore y = 3$

The solution is $x = 2$, $y = 3$.

A further example *(Example 5.2.6c, page 250)*
Find the intersection of the lines $2x + y = -1$ and $6x - 2y + 8 = 0$.

Solution
Method 1 GDC
Rearrange both equations into the form $y = mx + c$ and proceed by considering the lines as the graph of the linear functions $f(x)$ and $g(x)$. Graph these functions and find the point of intersection.

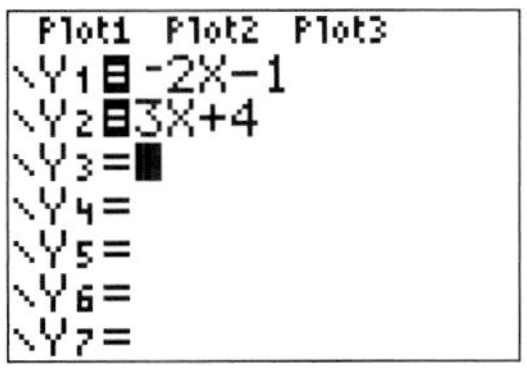

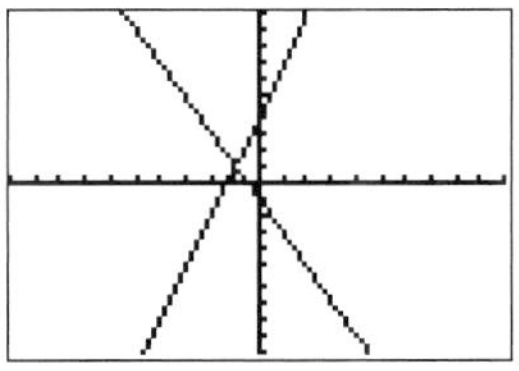

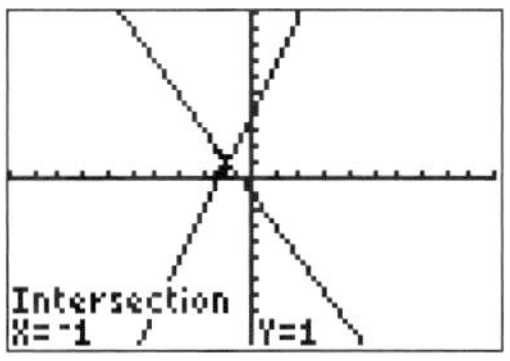

The lines intersect at the point (−1, 1).

Method 2 Substitution
Solve the pair of simultaneous equations

$2x + y = -1$

$6x - 2y = -8$

Solution
Rearrange the first equation

$y = -2x - 1$

Substitute for y in the second equation

$$6x - 2(-2x - 1) = -8$$
$$6x + 4x + 2 = -8$$
$$10x = -10$$
$$x = -1$$

Substitute for y in the first equation

$$y = -2(-1) - 1$$
$$y = 1$$

The solution is $(-1, 1)$.

Method 3 Elimination
Solve the pair of simultaneous equations

$$2x + y = -1$$
$$6x - 2y = -8$$

Solution
Multiply the first equation by 2 to get opposite coefficients of the y-variables.

$$4x + 2y = -2$$
$$6x - 2y = -8$$

Add the equations to eliminate the y-variables.

$$10x = -10$$
$$x = -1$$

Substitute for y in the first equation

$$2(-1) + y = -1$$
$$y = 1$$

The solution is $(-1, 1)$.

2.7.3 Solving quadratic equations

Solving quadratic equations means finding values of x such that $ax^2 + bx + c = 0$. The solution(s) can be found algebraically by factorizing or by using the quadratic formula.
Solutions can also be found using the graphic display calculator.

Knowledge of the quadratic formula is not required for examinations.

Method 1 Factorizing
Example 2.7.3a
Solve: $x^2 + 7x + 12 = 0$

Solution
The factors of 12 are: {1, 12}, {4, 3} and {6, 2}. Of these, 4 and 3 sum to 7.

The quadratic equation can be written, in its factor form, as $(x + 4)(x + 3) = 0$

Using the **null factor law**, either $(x + 4) = 0$ or $(x + 3) = 0$

Therefore $x = -4$ or $x = -3$.

That is, both $x = -4$ and $x = -3$ are solutions. This can be checked by substituting these values for x into the original expression, $x^2 + 7x + 12$.

$(-4)^2 + 7(-4) + 12 = 16 - 28 + 12 = 0$

$(-3)^2 + 7(-3) + 12 = 9 - 21 + 12 = 0$

Example 2.7.3b

Solve: $x^2 + 3x = 18$

Solution

Rearrange the equation to equate to zero: $x^2 + 3x - 18 = 0$

Factorize: $(x + 6)(x - 3) = 0$

The solutions are $x = -6$ or $x = 3$.

Check: $(-6)^2 + 3(-6) - 18 = 36 - 18 - 18 = 0$

$(3)^2 + 3(3) - 18 = 9 + 9 - 18 = 0$

Example 2.7.3c

Solve: $2x^2 - 7x - 4 = 0$

Solution

Factorizing the equation gives $(2x + 1)(x - 4) = 0$.

The solutions are $x = 4$ or $x = -\frac{1}{2}$.

Check: $2(4)^2 - 7(4) - 4 = 32 - 28 - 4 = 0$

$$2\left(-\frac{1}{2}\right)^2 - 7 - \frac{1}{2} - 4 = \frac{1}{2} + 3\frac{1}{2} - 4 = 0$$

Example 2.7.3d

Solve: $x^2 - 4 = 0$

Solution

Factorizing the equation gives $(x + 2)(x - 2) = 0$.

The solutions are $x = 2$ or $x = -2$.

Alternative method: $x^2 - 4 = 0 \rightarrow x^2 = 4 \rightarrow x = \pm 2$

Check: $(2)^2 - 4 = 0$ and $(-2)^2 - 4 = 0$

Method 2 Using the quadratic formula

The solutions of $ax^2 + bx + c = 0$ can be found using the formula

$$x = \frac{-b \pm \sqrt{b^2 - 4ac}}{2a}$$

Example 2.7.3e

Use the quadratic formula to solve: $x^2 + 7x + 12 = 0$

Solution

In this equation $a = 1$, $b = 7$ and $c = 12$.

Substitute for a, b and c in the formula:

$$x = \frac{-7 \pm \sqrt{7^2 - 4 \times 1 \times 12}}{2} = \frac{-7 \pm \sqrt{1}}{2} \text{ so } x = \frac{-7+1}{2} \text{ or } x = \frac{-7-1}{2}$$

The solutions are $x = -3$ or $x = -4$.

Example 2.7.3f

Use the quadratic formula to solve: $x^2 - 4 = 0$.

Solution

$$x = \frac{0 \pm \sqrt{0^2 - 4 \times 1 \times -4}}{2} = \frac{\pm\sqrt{16}}{2} = \frac{\pm 4}{2}$$

The solutions are $x = 2$ or $x = -2$

Method 3 Using the GDC

Example 2.7.3g

Solve $2x^2 - 7x - 4 = 0$ using the GDC.

i By graphing the equation

The solutions are the x-intercepts (roots) of the function $f(x) = 2x^2 - 7x - 4$.

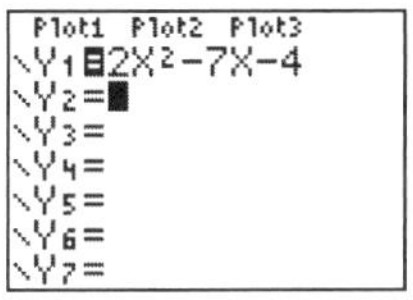

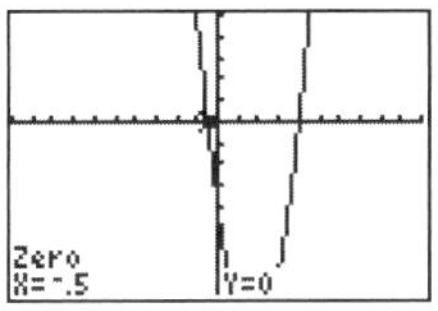

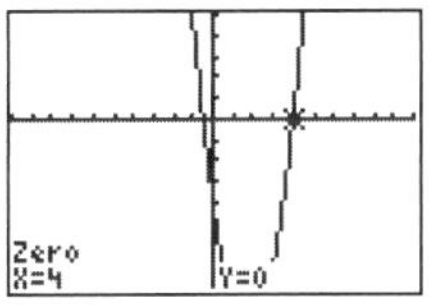

The solutions are $x = -0.5$ or $x = 4$.

ii By using the analytical function

The coefficients of x^2 and x, and the constant term are entered.

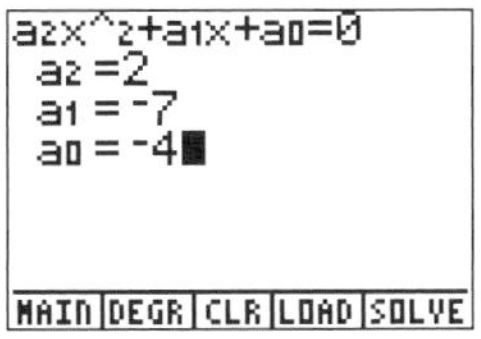

The solutions are $x = -0.5$ or $x = 4$.

Exercise 2.7

1 Solve these pairs of linear equations using the method of substitution.

a $y = 2x - 1$, $y = x + 1$ b $x + y = 11$, $x - y = 3$ c $3y = 2x + 1$, $y = x + 5$ d $y = x + 5$, $2x + y = 12$

2 Solve these pairs of linear equations using the method of elimination.

a $x + y = 10$, $x - y = 4$ b $2x - 6y = -6$, $3x + 2y = 2$ c $5x + 3y = 34$, $7x + 3y = 44$ d $x + 3y = 345$, $2x = y$

3 Solve for x and y using the graph mode of your GDC.

a $y = 4 - x$, $y = x$ b $2y = 3x + 12$, $y = 5 - 2x$ c $x + \frac{y}{3} = 5$, $y - 4x = 1$ d $\frac{x}{2} = 4 + 3y$, $3y + 8x = -2$

4 Solve all the simultaneous equations in questions 1 and 2 using the graph mode of your GDC.

5 Factorize these expressions.

a $x^2 + 5x + 4$ b $a^2 + 9a + 18$ c $x^2 + 12x + 35$
d $x^2 - 5x + 4$ e $b^2 - 9b + 18$ f $x^2 - 12x + 35$
g $x^2 + 5x - 6$ h $c^2 + 9c - 22$ i $x^2 + 12x - 13$
j $x^2 - 5x - 6$ k $d^2 - 9d - 10$ l $x^2 - 12x - 13$

6 Factorize these expressions.

a $2x^2 + 5x + 2$ b $2a^2 + 3a - 2$ c $2x^2 - 6x + 4$
d $3x^2 - 5x - 2$ e $3b^2 - 10b + 8$ f $3y^2 - y - 30$
g $3 - 2x - 5x^2$ h $c^2 + 9c - 22$ i $5x^2 + 9x + 4$

7 Solve for x, using the quadratic formula.

a $x^2 + 5x + 4 = 0$ b $a^2 + 9a + 18 = 0$ c $x^2 + 12x + 35 = 0$
d $x^2 + 5x - 6 = 0$ e $y^2 + 9y - 22 = 0$ f $x^2 + 12x - 13 = 0$
g $30 = x^2 + x$ h $2p^2 + 3p = 2$ i $3x^2 = 5x + 2$

8 Solve for x, using a graphic display calculator.

a $3x^2 = 48$ b $2a^2 + 18a + 36 = 0$ c $5x^2 + 60x + 105 = 0$
d $6x^2 - x - 2 = 0$ e $5y^2 - 10x + 10 = 0$ f $0.1x^2 + 0.3x - 1 = 0$
g $62x^2 - 126x - 263 = 0$ h $2p^2 + 3p = 2$ i $17x^2 = -28x + 64$

Past examination questions for topic 2

Paper 1

1 A problem has an **exact** answer of $x = 0.1265$.

a Write down the **exact** value of x in the form $a \times 10^k$ where k is an integer and $1 \leq a < 10$.

b State the value of x given correct to **two** significant figures.

c Calculate the percentage error if x is given correct to **two** significant figures. *M06q1*

2 Jacques can buy six CD's and three video cassettes for $163.17 or he can buy nine CD's and two video cassettes for $200.53.

a Express this information using two equations relating the price of CD's and the price of video cassettes.

b Find the price of one video cassette.

c If Jacques has $180 to spend find the exact amount of change he will receive if he buys nine CD's. *M06q11*

3 a The first term of an arithmetic sequence is −16 and the eleventh term is 39. Calculate the value of the common difference.

b The third term of a geometric sequence is 12 and the fifth term is $\frac{16}{3}$. All the terms in the sequence are positive. Calculate the value of the common ratio. *N05q13*

4 a Convert 0.001 673 litres to millilitres (ml). Give your answer to the nearest ml.

The SI unit for energy is Joule. An object with mass m travelling at speed v has energy given by $\frac{1}{2}mv^2$ (joules).

b Calculate the energy of a comet of mass 351 223 kg travelling at speed 176.334 m/s. Give your answer correct to 6 significant figures.

In the SI system of units, distance is measured in metres (m), mass in kilograms (kg) and time in seconds (s). The momentum of an object is given by the mass of the object multiplied by its speed.

c Write down the correct combination of SI units (m, kg, s) for momentum. *M05q4*

5 A swimming pool is to be built in the shape of a letter L. The shape is formed from two squares with side dimensions x and $\sqrt{x}$ as shown.

a Write down an expression for the area, A, of the swimming pool surface.

b The area, A, is to be $30\,\text{m}^2$. Write a quadratic equation that expresses this information.

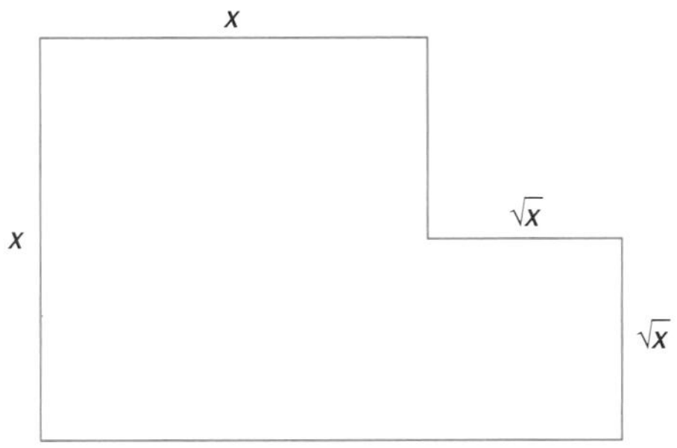

Not drawn to scale

c Find both the solutions of your equation in part **b**.

d Which of the solutions in part **c** is the correct value of x for the pool? State briefly why you made this choice. *M05q12*

6 a Given $x = 2.6 \times 10^4$ and $y = 5.0 \times 10^{-8}$, calculate the value of $w = x \times y$. Give your answer in the form $a \times 10^k$ where $1 \leq a < 10$ and $k \in \mathbb{Z}$.

b Which two of the following statements about the nature of x, y and w above are **incorrect**?

i $x \in \mathbb{N}$ ii $y \in \mathbb{Z}$ iii $y \in \mathbb{Q}$

iv $w < y$ v $x + y \in \mathbb{R}$ vi $\frac{1}{w} < x$ *M05q14*

7 In 2000 Herman joined a tennis club. The fees were £1200 a year. Each year the fees increase by 3%.

a Calculate, **to the nearest £1**, the fees in 2002.

b Calculate the **total** fees for Herman who joined the tennis club and remained a member for five years. *N04q7*

8 Anthony uses the formula

$$p = \frac{27q}{r + s}$$

to calculate the value of p when, correct to two decimal places, $q = 0.89$, $r = 1.87$ and $s = 7.22$.

a He estimates the value **without using a calculator**.

i Write down the numbers Anthony could use in the formula to estimate the value of p.

ii Work out the estimate for the value of p that your numbers would give.

b A calculator is to be used to work out the actual value of p. To what degree of accuracy would you give your calculator answer? Give a reason for your answer. *N00q14*

9 A woman deposits $100 into her son's savings account on his first birthday. She deposits $125 on his second birthday, $150 on his third birthday, and so on.

a How much money would she deposit into her son's account on his 17th birthday?

b How much in total would she have deposited after her son's 17th birthday? *M99q7*

Paper 2

1 i The nth term of an arithmetic sequence is given by $U_n = 63 - 4n$.

a Calculate the values of the first two terms of this sequence.

b Which term of the sequence is −13?

c Two consecutive terms of this sequence, u_k and u_{k+1}, have a sum of 34. Find k.

ii A basketball is dropped vertically. It reaches a height of 2 m on the first bounce. The height of each subsequent bounce is 90% of the previous bounce.

a What height does it reach on the eighth bounce?

b What is the total vertical distance travelled by the ball between the first and sixth time the ball hits the ground? *M04q2*

2 Ann and John go to a swimming pool.
They both swim the first length of the pool in 2 minutes.
The time John takes to swim a length is 6 seconds more than he took to swim the previous length.
The time Ann takes to swim a length is 1.05 times that she took to swim the previous length.

a i Find the time John takes to swim the third length.

ii Show that Ann takes 2.205 minutes to swim the third length.

b Find the time taken for Ann to swim a total of 10 lengths of the pool. *N03q2*

3 A National Lottery is offering prizes in a new competition.
The winner may choose one of the following.

Option one $1000 each week for 10 weeks.

Option two $250 in the first week, $450 in the second week, $650 in the third week, increasing by $200 each week for a total of 10 weeks.

Option three $10 in the first week, $20 in the second week, $40 in the third week continuing to double for a total of 10 weeks.

a Calculate the total amount you receive in the tenth week, if you select

i **option two** ii **option three.**

b What is the total amount you receive if you select option two?

c Which option has the greatest total value? Justify your answer by showing all appropriate calculations. *M02q2*

Theory of knowledge

As this is a rather long TOK entry, perhaps it can take place in the TOK lesson instead of the mathematical studies lesson.

Numbers and numerals (from ***Teacher support material–Theory of knowledge lessons from around the world*** © IBO, November 2000 Lesson 13)

Context

Most of us are so familiar with the system of numerals we use that it is hard for us to appreciate certain features of that very system. In order to highlight these features, this lesson asks students to design their own system of numerals. In this way, shortcomings and ambiguities in what they create can demonstrate more clearly the necessary characteristics of an effective system for representation and manipulation of numbers.

Aims

The aims of this lesson are:

- to distinguish between numbers and the symbols which represent them
- to make evident some of the assumptions embedded in our use of numerals
- to recreate and highlight some of the great leaps forward in number representation throughout the history of mathematics.

Group management

Divide participants into small groups.
Time must be allowed for sharing students' work across groups. Suggested time allocations are:

- 30 minutes for devising a system
- 30 minutes for presenting the systems to the other groups
- 20 minutes for summing up.

Focus activity

This is what you have to do.

1. Invent a series of symbols to represent numbers. These symbols should not be the same as any numeral system that you know. The number of different symbols and how they may be combined with one another (if at all) is entirely your choice.
2. Explain your numeral system to another group of students.
3. Show your audience how to represent:
   - three
   - forty-five
   - twenty
   - one hundred and seventeen.
4. Devise a series of problems for other students to solve, using your system of symbols.

Discussion questions

- What is the advantage of employing place value?
- Why does the number system which we generally use have base 10?
- What is the advantage of having a symbol for zero?
- Is there a difference between zero and nothing?

Links to other areas of TOK

- Can we think of mathematics as a language? Which features of language does it possess?
- Look closely at the following quotation. Do you agree with it?
  Numbers constitute the only universal language.
  Nathaniel West
- Many words refer to objects or classes of objects in the world (that is, they have a denotation). What do numerals denote or refer to?

Further Investigation

Research numeral systems from different parts of the world. Prepare responses to these questions.

- How many symbols are needed in each system?
- Does the system use a base? If so, what is it?
- Does it employ place value?
- Does it use a zero?
- Where exactly did each of these civilizations exist?
- What do the dates associated with the development of each number system suggest?

TOPIC 3

Sets, logic and probability

3.1 Basic concepts of set theory and set relations

3.1.1 Basic concepts

Mathematics deals with the classification of objects and describes connections between those objects. The objects might be everyday things like people, cars, foods etc, but might also be abstract things like numbers or algebraic expressions. One of the most fundamental ideas used for classification is that of a **set**. We have already met some sets of numbers in Topic 2. It is now time to say exactly what we mean by a set and to study methods of describing sets and relations between sets.

Definition of a set

In Topic 2 we saw that some special sets have their own names. The sets we met there included the natural numbers $\mathbb{N}$, the integers $\mathbb{Z}$, the rational numbers $\mathbb{Q}$ and the real numbers $\mathbb{R}$. We can attach a + or a – sign on the top right of these symbols to indicate only the positive or negative members (leaving out 0) of that set. Thus the negative integers would be $\mathbb{Z}^-$ and the positive reals would be $\mathbb{R}^+$.

Apart from such special names, the most common way of showing a set is to list or describe its members inside a pair of **set brackets**. For example, a set A of whole numbers from 1 to 7 inclusive would be written $A = \{1, 2, 3, 4, 5, 6, 7\}$. The objects in a set are called **elements**. If there are only a few elements, we can show them in a diagram. The elements can be almost anything, including people, objects, numbers, algebraic expressions, shapes or even other sets.

If an element belongs to a set, we only need to write it once. Also the order within the set does not matter. For example, $\{1, 1, 2, 2, 3, 3, 3, 4\} = \{1, 2, 3, 4\} = \{4, 2, 3, 1\}$ etc. The first of these is no different from the other two and it is a waste of time to write it in that way.

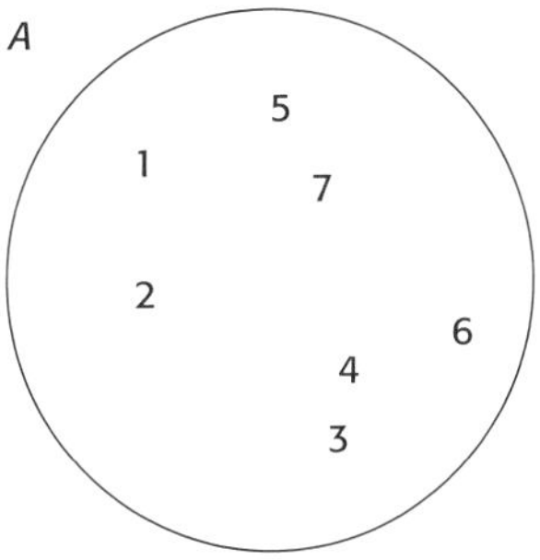

The set A of whole number from 1 to 7

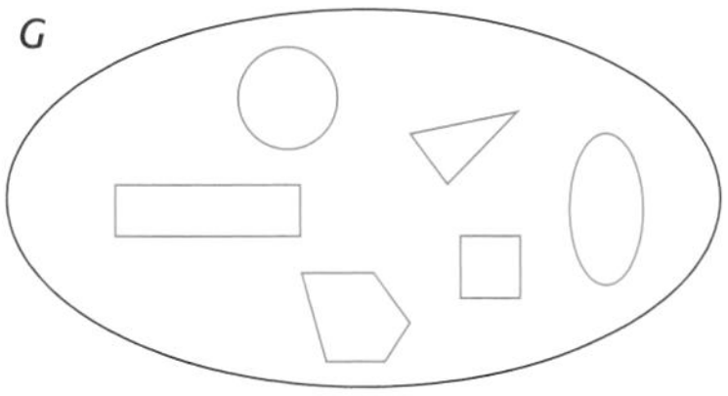

A set G of geometric shapes

Example 3.1.1a

State which of these sets are the same.

a $\{1, 3, 5, 7, 9\}$ b $\{1, 1, 5, 7, 7, 7, 9, 3\}$
c {odd numbers from 1 to 9 inclusive}
d $\{9, 7, 5, 3, 1, 1, 3, 5, 7, 9, 9\}$ e $\{7, 7, 5, 5, 3, 3, 1, 1\}$.

Solution
The first set contains all the odd numbers from 1 to 9. The order is unimportant and the number of appearances of each number is also unimportant. Hence all the sets are the same except for the fifth one, which has 9 missing.

Number of elements in a set

A set is **finite** if we can count the objects in it and the counting eventually comes to an end. The set A above, for example, contains seven elements. We could attempt to count the elements in $\mathbb{N}$, $\mathbb{Z}$ or even $\mathbb{Q}$, but the counting goes on for ever. We cannot even start to count the elements of $\mathbb{R}$ without immediately leaving some out. These number sets are said to be **infinite**.

If a set S is finite, we can write the number of elements it contains using the notation $n(S)$. For the sets A and G pictured, on page 115, $n(A) = 7$ and $n(G) = 6$.

Intervals

In Topic 2, we met the idea of a number line. Intervals of numbers on the real number line are often represented using curved brackets (parentheses), ( ) or square brackets, [ ]. Hence $[-2, 3]$ means *the set of all numbers from −2 to 3 inclusive*. Such a set is known as a **closed interval**. On the other hand $(-2, 3)$ *means the set of all the numbers from −2 to 3 but not including −2 and 3.* This is an **open interval**.

The two kinds of brackets can be mixed in one expression. For example $[1, 5)$ is *the set of all numbers from 1 to 5 including 1 but not 5.*

In these examples the notation means *all* the numbers in between, not just the whole numbers.

An **unbounded interval** is one that stretches to infinity in one or both directions. We can use the symbols $+\infty$ and $-\infty$ in the interval. The symbol '∞' must always appear next to an open interval bracket. This is because ∞ *is not a number* and should only be thought of as standing for a *tendency* to increase or decrease. So for example: $(-5, +\infty)$ is the set of all numbers starting at −5 and increasing indefinitely. The whole real number line can be written as $\mathbb{R} = (-\infty, +\infty)$.

The use of parentheses for an open interval should not be confused with their use to represent coordinates. Usually the meaning will be clear from the context.

Example 3.1.1b
Use interval notation to describe all the numbers between −360 and 360 inclusive.

Theory of knowledge

Is infinity a number? Is it ever correct to write $x = \infty$? Do some infinite sets have more elements than others? Give examples to back up your answer. If infinite sets are not the same size then what does "infinite" mean? Can you count the elements of some infinite sets? Explain your answer as clearly as you can. Can you think of an infinite set where it is not possible to count all the elements?

Solution
Both the end points are included so we use square brackets. The lesser value is entered first. Hence the answer is [−360, 360].

Example 3.1.1c
Use interval notation to describe all the numbers greater than an unknown value x and increasing indefinitely.

Solution
The task says 'greater than' but does not mention 'or equal to' so we use an open interval starting at x. 'Increasing indefinitely' means that we extend the interval to infinity and so that needs an open interval too. The answer is $(x, +\infty)$.

Containment and shared properties
The symbol $\in$ expresses containment of an element in a set. If we want to say an element is not in a set we use $\notin$.

Using our set $A = \{1, 2, 3, 4, 5, 6, 7\}$ above, we can say $1 \in \text{A}$ or $2, 3, 4 \in A$ but $19 \notin A$.

Similarly $-3 \notin \mathbb{Z}^+$ but $-3 \in \mathbb{Z}$. Also $\pi \in \mathbb{R}$ but $\pi \notin \mathbb{Q}$.

When elements of a set all share a common property, we express this fact using notation inside the set brackets. The most common symbols in use are |, : or /, all of which are read as *such that*. One example appeared in Topic 2: the set of rational numbers, $\mathbb{Q}$, is the set of numbers constructed from all possible ratios of integers, (avoiding 0 as a denominator).

$$\mathbb{Q} = \left\{ x = \frac{a}{b} : a, b \in \mathbb{Z}, b \neq 0 \right\}$$

Similarly, the set of even natural numbers can be extracted from the set of natural numbers as follows:

$\mathbb{N}_{even} = \left(n \in \mathbb{N} \mid \frac{n}{2} \in \mathbb{N} \right)$. (The set containing natural numbers with the property that when number is halved, the result is still a natural number.)

There can be more than one way to describe the same set. For example $\mathbb{N}_{even}$ above can also be described as $\mathbb{N}_{even} = \{2n \in \mathbb{N} \mid n \in \mathbb{N}\}$ (the set of numbers formed by doubling all the natural numbers).

Example 3.1.1d
Describe the interval $[a, b]$ using set brackets.

Solution
This is all real numbers between a and b inclusive. The answer is $\{x \in \mathbb{R} : a \leq x \leq b\}$.

Teachers and candidates should be aware that the bracket notations for an interval may vary in different parts of the world. For example, backward square brackets] [are sometimes used for an open interval instead of ( ). IB examiners are expected to be knowledgeable about valid local variations and will check them if uncertain.

Theory of knowledge
Do mathematical symbols such as $\in$, π, etc. have meaning in the same sense as words have meaning? Can mathematical symbols be ambiguous or vague like certain words, (for example: free, poor, good etc)?

Example 3.1.1e

Write as a set, all positive numbers which are exact multiples of 7.

Solution

We need to describe $7\times1, 7\times2$ etc. The answer is $\{7n \mid n \in Z^+\}$.

3.1.2 Subsets

Sometimes we need to refer to just some of the elements from within a set.

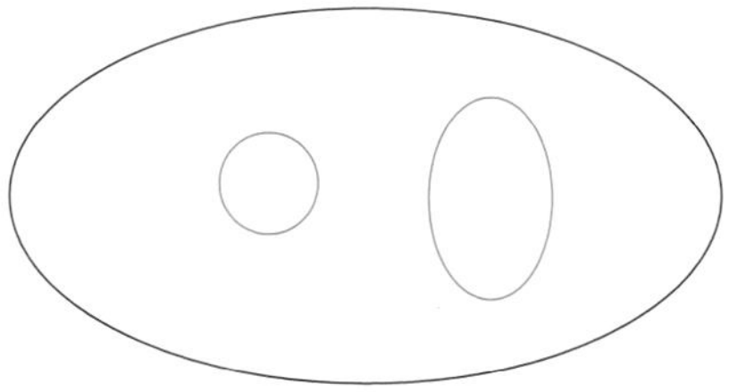

A subset of G with only curved shapes

A collection of elements all from within a particular set is called a **subset**.

A subset can also contain all the elements of the original set (or nothing at all) but it cannot contain new elements. To express a subset relation, we use the symbol "$\subset$" or, if we wish to allow possible equality, "$\subseteq$".

Suppose $B = \{2, 3\}$ and set A is the same as described on page 115. We can write $B \subset A$ (or $B \subseteq A$).

The set of integers is a subset of the real numbers so we can write $\mathbb{Z} \subset \mathbb{R}$. We can use the notation to show several subsets all nested within each other: $\mathbb{N} \subset \mathbb{Z} \subset \mathbb{Q} \subset \mathbb{R}$.

It is important to have a clear idea of the difference between containment, $\in$, of an element within a set and a subset relation, $\subset$, between sets.

In summary, a set S is a subset of a set T if every element of S also belongs to T. If $S \subset T$ and $T \subset S$ then the sets S and T are equal: $S = T$.

Example 3.1.2a

Use the symbols $\in$ and $\subset$ to write meaningful relations between the following:

$-4,\ 4,\ \{4\},\ \{-4\},\ \{-4,\ 4\},\ \{x \in \mathbb{Z} \mid x^2 = 16\}$

Solution

Lots of relations are possible, for example

$4 \in \{4\},\ 4 \in \{x \in \mathbb{Z}^+ \mid x^2 = 16\},\ -4 \in \{x \in \mathbb{Z} \mid x^2 = 16\},\ -4 \in \{-4,\ 4\},\ \{4\} \subset \{x \in \mathbb{Z}^+ \mid x^2 = 16\},$

$\{4\} \subset \{-4,\ 4\},\ \{-4,\ 4\} = \{x \in \mathbb{Z} \mid x^2 = 16\},\ \{-4,\ 4\} \subseteq \{x \in \mathbb{Z} \mid x^2 = 16\}$

Perhaps you can think of some more.

Example 3.1.2b

Decide which of the following statements about subsets are correct and which are incorrect.

a $[-3, 3] \subset \mathbb{R}$ b $[-3, 3] \subset \mathbb{Q}$ c $(-3, 3) \subset [-3, 3]$

d $\mathbb{N} \subset \mathbb{R}$ e $\{5, 6, 7\} \subset \mathbb{Q}$ f $(-\infty, +\infty) \subseteq \mathbb{R}$.

Solution

In each case look at the left-hand side and decide whether everything in that set also belongs to the set on the right-hand side.

a *Correct*. All numbers between -3 and 3 inclusive are certainly real.

b *Incorrect.* All numbers between −3 and 3 includes infinitely many irrational numbers. For example, one is $\sqrt{2}$. These are not members of $\mathbb{Q}$.

c *Correct.* The open interval on the left contains all numbers from −3 to 3, excluding −3 and 3 themselves. On the right, the closed interval contains all the same numbers as on the left but includes the end points as well.

d *Correct.* All natural numbers are also real.

e *Correct.* Each of the numbers in the set on the left can be written as a meaningful ratio, for example $5 = \frac{5}{1}$. Thus they are all rational.

f *Correct.* Here we have two different ways of showing the real numbers. The sets are equal and the symbol allows equality.

The empty set and the universal set

The only set that contains no elements at all is called the **empty set**.

- The empty set is written as $\varnothing$ or as { }.
- It is important to understand that $\varnothing$ is not the same as the number 0 or the set {0}. A number is not a set and the empty set contains *no elements*, not even the number 0.
- $\varnothing$ is a subset of all other sets.

Examples of $\varnothing$: The set of all pupils in your school who are 20 metres tall.
The set of all banks that offer 500% interest on savings accounts.
The set of all numbers which are both less than 2 and greater than 3.

When we are considering objects with a particular property, we often want to embed our sets of objects within a more general set for reference

That more general set is known as the **universal set**, *U*, for the problem we are considering.

The choice of U will depend on the context and should be made sensibly. For example, we might be considering the sets of people who like carrots or of people who like beans. In that case, a sensible choice for U might be the set of people who like vegetables, or at most, the set of all people. It would be ridiculous to let U be the set of all people and all computers because computers (so far) do not have any dietary preferences.

Example 3.1.2c

Decide what would be a sensible universal set to contain these sets.

a {−3, −2, −1, 0, 1, 2, 3}

b Solutions for y of the equation $ay - b = 0$, when a and b are positive whole numbers

c Three sets containing candidates in your school studying respectively IB Diploma Programme mathematical studies, French, and economics.

Solution

There can be several sensible choices for a universal set.

a The numbers considered are all integers. If we expect to be dealing only with integers throughout the problem, then the best universal set is probably $\mathbb{Z}$. Note that $\mathbb{Q}$ and $\mathbb{R}$ would be acceptable as well. $\mathbb{N}$ is definitely wrong, because it contains no negative numbers.

b The solution is $y = \frac{b}{a}$ and this is a ratio of whole numbers.

Hence choose $U = \mathbb{Q}$ or $\mathbb{Q}^+$.

c It depends on your school and on what you are investigating. Some possibilities are all IB Diploma Programme students in your school, all students at your school (if there are more than just IB students), all IB Diploma Programme students in your city, country, or the world etc.

> *Theory of knowledge*
>
> What is the difference between the empty set, $\varnothing$, the number zero and nothing at all? Can there be such a thing as a complete vacuum? Do you think we can consider the universe to be the ultimate universal set U containing everything?

3.1.3 Union and intersection

It is useful to look at how sets relate to each other. For example, what happens if we combine two sets? Can sets overlap?

The **union** of two or more sets is the set you get when you combine all the elements from the original sets into a single set. We use the symbol, $\cup$, for this operation.

Suppose $A = \{-7, 3, 32, \pi, 10^6, x, y\}$ and $B = \{-14, 8.3, 3, 32, \pi, 10^8\}$.

Then $A \cup B = \{-14, -7, 3, 8.3, 32, \pi, 10^6, 10^8, x, y$ is the union of A and B.

We can make a union with as many sets as we like and the order in which the sets are written is not important. For example $R \cup (S \cup T) = (R \cup S) \cup T = R \cup S \cup T$ and so on, for any sets R, S, T.

Another way to describe union is to say that if $x \in S \cup T$ then either $x \in S$ or $x \in T$ or possibly both (but x does not *have* to be in both).

If you are the sort of person who learns better from visual images, you might want to read this section in conjunction with Section 3.2 (page 126) on Venn diagrams. Look particularly at Examples 3.2.1a and b.

Examples 3.1.3a

Write down a single set that is equivalent to these unions.

a {All female humans} $\cup$ {All male humans}

b $\{1, 2, 3, 4\} \cup \{3, 4, 5, 6\}$

c $\mathbb{Q} \cup \{x \in \mathbb{R} \mid x \notin \mathbb{Q}\}$

d $[-8, 6) \cup [5, 8]$

e $[-8, 6) \cup [7, 8]$

Hint: part **e** is a trick question.

Solution

a This union includes everyone: {All humans}

b Just combine all the elements without repeating any: {1, 2, 3, 4, 5, 6}

c The union of all rational and all irrational numbers: $\mathbb{R}$.

d Two overlapping intervals. We do not care about the extent of the overlap. This is just all the numbers from −8 to 8 inclusive: [−8, 8]

e These intervals do not overlap. There is no very neat way of writing this as a single set. Either leave it as it is, or if you must write a single set, then it has to be $\{x \in \mathbb{R} : -8 \leq x < 6 \text{ or } 7 \leq x \leq 8\}$.

Example 3.1.3b

The set of Yuichi's friends is Y = {John, Li San, Pedro, Jacques, Maria, Annie}.

The set of Ingrid's friends is I = {Samantha, Ahmed, Phillipa, Yashar, Maria, Pedro}.

List the members of the sets $Y \cup I$ and explain in words what this set describes.

Solution

$Y \cup I$ = {John, Li San, Pedro, Jacques, Maria, Annie, Samantha, Ahmed, Phillipa, Yashar}; these are all the friends of *either* Yuichi *or* Ingrid (or both).

Intersection

The **intersection** of two sets is the set whose elements are *common to both* of them (the overlap). Intersection is denoted by the symbol $\cap$.

If the sets have no elements in common, then their intersection is the empty set $\varnothing$.

The idea of intersection can be extended to more than two sets by choosing the elements which are common to all the sets.

For the particular sets

$A = \{-7, 3, 32, \pi, 10^6, x, y\}$ and $B = \{-14, 8.3, 3, 32, \pi, 10^8\}$ that we used before, we have $A \cap B = \{3, 32, \pi\}$. We can take an intersection using as many sets as we like. The order does not matter. $(R \cap S) \cap T = R \cap (S \cap T) = R \cap S \cap T$ for any sets R, S, T.

If two sets do not overlap, we can still consider their intersection, but this will be the empty set as they have no element in common. In this case we write, for example, for sets S and T, $S \cap T = \varnothing$.

The empty set itself can be included in a union or intersection and satisfies the following rules:

- For any set S we have $S \cup \varnothing = S$ and $S \cap \varnothing = \varnothing$.
- Similarly if S is contained in some universal set U then $S \cup U = U$ and $S \cap U = S$.

Example 3.1.3.c

Referring to Yuichi and Ingrid's friends from the previous example, list the members of the sets $Y \cap I$ and explain in words what this set describes.

Solution

$Y \cap I =$ {Maria, Pedro}; these are the people who are friends with both Yuichi and Ingrid.

Example 3.1.3d

Wojtek's friends are in the set $W =$ {Maria, Yashar, John, Li San}. Y and I are the same sets from the previous examples.

a List the members of the set $W \cap Y$.

b Write down the set of people who are friends with all three of Yuichi, Ingrid and Wojtek and write this set in terms of Y, I and W.

Solution

a $W \cap Y =$ {John, Li San, Maria} b {Maria}, which is $W \cap Y \cap I$.

Examples 3.1.3.e

Simplify these intersections.

a $\{1, 2, 3\} \cap \{-1, 0, 1\}$ b $\{x, y, z\} \cap \{x, y, 1, 2\}$

c $\mathbb{N} \cap \mathbb{Q}$ d $\mathbb{Z} \cap \mathbb{Z}^+$ e $[2, 5] \cap [5, 9]$

Solutions

a The only common element is 1 so the answer is $\{1\}$.

b x and y appear in both sets. The answer is $\{x, y\}$.

c $\mathbb{Q}$ contains all of $\mathbb{N}$ so the answer is $\mathbb{N}$.

d $\mathbb{Z}^+$ contains all the positive integers but *not* 0 while $\mathbb{Z}$ is all the integers, positive, negative *and* 0, hence $\mathbb{Z}$ includes all members of $\mathbb{Z}^+$ and so the answer is $\mathbb{Z}^+$.

e The only number that the two intervals have in common is 5 and so the answer is $\{5\}$.

> *Theory of knowledge*
>
> Have you heard the old English proverb: *One swallow does not a summer make*? One example of something does not mean it is always true. If you think something is true in mathematics, it is necessary to prove it using general arguments and not just by finding an example where it happens to be true. However many examples you can find that support your claim, it only needs one counter-example that you were not aware of, for the claim to be proven false.

3.1.4 Complement of a set

The **complement** of a set S is the set containing elements which are *not* in S.

To make this idea meaningful we have to refer to a sensible universal set. For example, suppose we are studying sets containing members of different sports teams. Let the football team be the set F. Perhaps we want to list the people who are not in the football team. Clearly penguins and bottles of orange juice do not play sports, but it would be pointless to list them. Instead we should refer to a universal set containing all the sports players we are interested in. Then we can list those not in the football team, but still in the universal set. We use the notation F' or F^c for the complement.

Here's another example: If the set S consists of all people who like carrots and U is the set of all people, then S' is the set of all people who do not like carrots.

Example 3.1.4

Describe the irrational numbers using complement notation and state the universal set in this case.

Solution

$\mathbb{Q}$ is the set of rational numbers. Together with the irrational numbers, these make up the set $\mathbb{R}$. Thus we should choose $\mathbb{R}$ for our universal set and then the irrational numbers make up the set $\mathbb{Q}'$.

$\mathbb{Q} \cup \mathbb{Q}' = \mathbb{R}$ in this example, and, more generally, for any set A we have $A \cup A \rightarrow U$ and $A \cap A' = \varnothing$ always.

Relative complement

If we have more than one set inside our universal set, we might want to talk about the elements that are in one set but not another. Let us return to our sporting example. Suppose H is the set of hockey players and N the set of netball players. The players who play netball but not hockey are the players who are in N but not in H. This set is called the **relative complement** of H in N, often written as $N \cap H'$.

When we simply say 'complement' of A we mean the complement relative to the universal set.

There is some standard notation for relative complement.
$N \cap H'$ can be written as N/H or as $N - H$ (read as N minus H and described by some as 'set minus').
These two alternative notations for relative complement are not included in the official IB notation list for mathematical studies and will not appear in examinations, however, candidates may use them without penalty. Although in a sense $N \cap H'$ is a compound notation for a concept which is really basic and deserves its own notation, this form will be used in this book and candidates should get used to reading it as "N but not H" and thinking "relative complement".

Referring once again to the sets of numbers A and B used before:

$A = \{-7, 3, 32, \pi, 10^6, x, y\}$ and $B = \{-14, 8.3, 3, 32, \pi, 10^8\}$

we have $B \cap A' = \{-14, 8.3, 10^8\}$ while $A \cap B' = \{-7, 10^6, x, y\}$.

Compound relations

Here we consider more complicated structures using combinations of the basic relations. For example we might start to ask about sets such as $(S \cup T)'$ and $(S \cap T)'$ or we might introduce a third set V and ask about $S \cup (T \cap V)$, $(S \cap T) \cup V$ and so on. It is important to know which operation is being

Theory of knowledge

The use of the $B - A$ notation for relative complement is controversial. Some mathematicians (and teachers) feel uncomfortable with it. In fact there was rigorous debate whether to mention it at all in this book, not only because it is not on the IB notation list, but also because the use of 'minus' here is a bit dubious. The point is that 'minus' used in this way is actually rather different to the way it is used in normal arithmetic. For example, it is not an inverse operation for addition (which hasn't even been defined) or for union (which has). Despite that, it is a neat sort of notation and most people feel intuitively what is meant: 'In B but not in A'.
So the TOK question is: "Should we be allowed to 'abuse' notation in this way when everyone understands what is intended?"

Theory of knowledge

We must be very careful with brackets in mathematics. Are brackets in mathematics more important than brackets used in other areas of knowledge (for example, history, literature, science etc)?

applied first so careful use of brackets is essential. For example, the sets $(S \cup T)'$ and $(S \cap T)'$ are not usually the same. Similarly $S \cap (T \cup V)$ is not usually the same as $(S \cap T) \cup V$.

Expressions of the kind above can be rewritten using some rules called De Morgan's laws, but these laws are outside the scope of the mathematical studies syllabus. We will study such compound sets, using Venn diagrams, in the next section. Nevertheless, the interested reader is invited to explore the topic further by investigating De Morgan's laws.

Bertrand Russell (1872–1970)

No introduction to set theory is complete without a mention of Russell's antinomy (also known as Russell's paradox). It was mentioned early in this section that sets can have other sets as elements. The mathematician and philosopher, Bertrand Russell, posed a problem which occupied the mathematical community in the early days of set theory.

Russell's antinomy

Try to picture the set which contains all sets which do not contain themselves. Then ask the question: Does this set contain itself or not? You might see that there is a problem. Further investigation of this fascinating topic falls within the realm of theory of knowledge.

Theory of knowledge

What is an *antinomy*? (Don't confuse this word with the name of the metal antimony (Sb).)

Think of a barber who shaves men in his town. Suppose he shaves all the men in the town who do not shave themselves and shaves no other men. Does he shave himself or not? This analogy has been used as a rather quaint example of Russell's antinomy. Does the existence of this idea pose a problem for the very foundations of mathematics? Can we resolve the problem? Is it not really a problem at all? Should Russell just have grown a beard?

Exercise 3.1

1 **a** Which of these groups of items can be regarded as a set?

- **i** players for the Boston Red Sox
- **ii** members of the West Indies cricket team
- **iii** fast cars
- **iv** songs sung by Shakira
- **v** students in your school
- **vi** square roots of natural numbers
- **vii** members of a political party
- **viii** all sets whose elements are other sets
- **ix** all straight lines with negative gradient
- **x** laws passed by a parliament
- **xi** shades of the colour red
- **xii** species of fish in the river Nile
- **xiii** novels by the Egyptian writer Naguib Mahfouz.

b For each example in **a** which you consider to be a set, state whether the number of elements is finite or infinite.

c Use interval notation to describe all the numbers

- **i** from −80 to 150 inclusive
- **ii** less than 16.3
- **iii** greater than or equal to 16.3
- **iv** between 0 and 4π including 4π but not 0

d **i–iv** For each interval in part **c**, show the interval on a number line as described in Topic 2.

2 a Use set notation to describe these sets.

i the odd natural numbers

ii all natural numbers that have no remainder when divided by 5

iii all integer powers of 2 iv all possible real powers of 2

v all real numbers which have a well-defined square-root (There are several natural ways of writing this set.)

b Explain what is wrong with these statements.

i {potatoes, sweet potatoes, carrots, turnips} = { vegetables V | V grows under the ground}

ii $\{3\} = \{x : x^2 = 9\}$

3 a Write down two proper subsets of the interval [1, 9) which contain the number 7. (Note that *proper* means you must use $\subset$ rather than $\subseteq$.)

b State which of these subset relations are correct and which are incorrect.

i $(-1,1) \subset [-1,1]$ ii $[-1,1] \subset (-1,1)$ iii $(-1,1] \subset [-1,1]$

iv $0 \subset [-1,1]$ v $(-1,1] \subset [-1,1)$

c Write down a subset of the set {vegetables V | V grows under the ground}. Try to think of members of V that are grown in your country of origin. Repeat the question for a subset of V that grow in a country faraway instead.

4 A set of Jane's friends is J.

J = { Alison, Sarah, Paula, Samer, Malcolm, Vicki, Jill, Steven, Abebe}.

A set of Peter's friends is P.

P = { Sarah, Vicki, Jill, Abdullah, Mei-Li, Chuck, Abebe, Erasto}.

a Use set brackets to list all the people who are friends with both Jane and Peter. Write a name for this set in terms of J and P.

b Write out the names in the set $P \cap J'$ (friends of Peter but not Jane).

5 a Simplify these unions and intersections.

i $[0,1] \cup (1,3)$ ii $\mathbb{Z}^- \cup \{0\} \cup \mathbb{Z}^+$ iii $\mathbb{R} \cap [-6, 4]$

iv $\mathbb{N} \cap [6, 4]$ v $\{4.33, 3, 2, -1\} \cup (-1,5]$

vi $\mathbb{Q} \cap \{\pi, \sqrt{5}\}$ vii $\mathbb{N} \cup \mathbb{Z} \cup \mathbb{Q} \cup \mathbb{R}$ viii $\mathbb{N} \cap \mathbb{Z} \cap \mathbb{Q} \cap \mathbb{R}$

ix $\{2^n \mid n \in \mathbb{R} \text{ and } n \geq 0\} \cup \{2^n \mid n \in \mathbb{R} \text{ and } n < 0\}$

x $A \cup B \cup C \cup D$ where the sets contain IB students at your school studying
A: mathematical studies, B: mathematics SL,
C: mathematics HL and D: computer science.

b Let $S = [-1, 1]$ and $T = [0, 2]$. Simplify these expressions

i $S \cap T$ ii $S \cap T'$ iii $T \cap S'$ iv $(S \cap T') \cup (T \cap S') \cup (S \cap T)$

In general for any sets A, B, how can you simplify the expression $(A \cap B') \cup (B \cap A') \cup (A \cap B)$?

c For any set S what can you say about $S \cup S$ and $S \cap S$?

d (A bit harder) Show that for *any* sets S and T, if $S \subset T$ then

i $S \cup T = T$ ii $S \cap T = S$.

e Construct an example for each of the equalities in part **d** using sets A and B with your own choice of elements to *demonstrate* (note, not *prove*) these properties.

6 Write down $F \cap G'$ given sets F and G containing letters from various alphabets as follows:

F = {ä, é, ñ, å, Ç, Æ, µ, ß, Œ} and G = {Ç , s, ú, Æ, è}.

Hint for part **d**: part **i** goes as follows: Think about the left-hand side, (LHS). This is the set $S \cup T$. All the elements in this set consist of everything in S together with everything in T. This certainly means that we have everything in T, so we have everything in the set on the right-hand side, (RHS). Is there anything extra coming from S? No! S is a subset of T and so everything in S is also in T. There is nothing new. Hence the LHS and the RHS are the same and so $S \cup T = T$. You should be able to write something similar to prove the equality in part **ii**.

3.2 Venn diagrams and simple applications

John Venn (1834–1923)

So far we have only drawn pictures of individual sets, showing their contents. The English mathematician John Venn (1834–1923) was a prolific contributor to the developing areas of set theory and logic. One of his more famous achievements is the invention of the **Venn diagram**.

Venn was a man of many talents which included a knack for building machines. His machine for bowling cricket balls reportedly bowled out a top star in the Australian cricket team four times, when the team visited England in 1909!

The Venn diagram gives us a good method of picturing operations between sets such as the union and intersection as well as more complicated relations.

A Venn diagram consists of a box labelled U for the universal set. Subsets of U are drawn inside the box. Usually we use a circle for the subsets. The subsets should be labelled. If there is more than one subset, we use overlapping circles which should overlap in such a way as to show clearly, regions for all possible intersections. For example, if there are three circles, we must be able to see a region where all three overlap, as well as the three regions where only two of the circles overlap. If we know that an intersection is empty, we can show this by omitting the relevant overlap.

We can add useful information to the diagram in various ways.

- If the number of elements in the sets is small, then we can write them all in.
- We can write numbers on the diagram to show how many elements are in each region.
- We can use shading to show regions of union or intersection, complements etc.

In mathematical studies we restrict ourselves to a maximum of three subsets of the universal set.

Example 3.2.1a

Draw separate Venn diagrams containing a single set S and its complement S' inside a universal set U. Use shading to indicate S and S'.

Solution

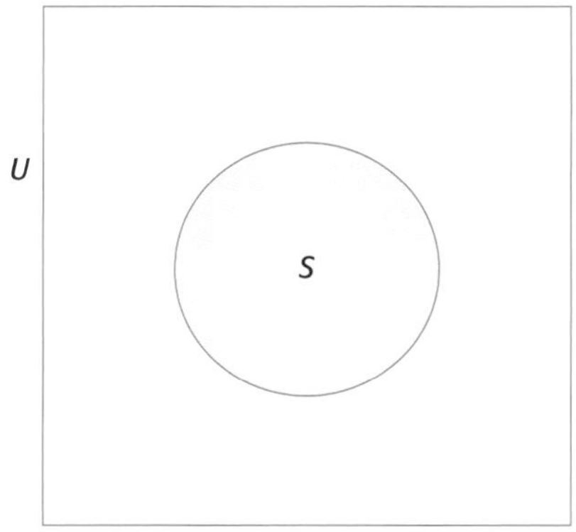

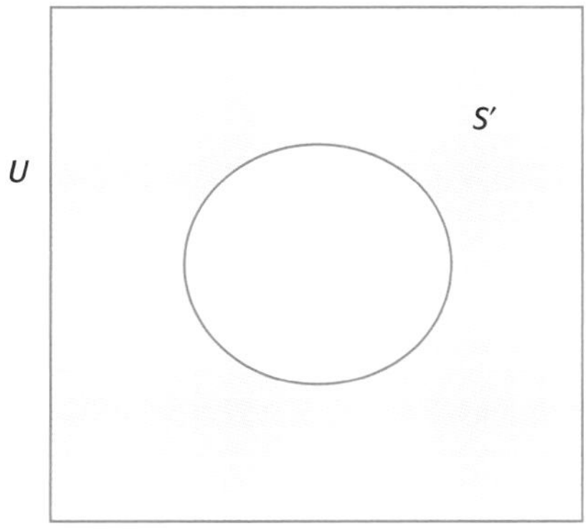

Example 3.2.1b

Draw separate Venn diagrams showing sets S and T, in which the union $S \cup T$ and the intersection $S \cap T$ are shaded.

Solution

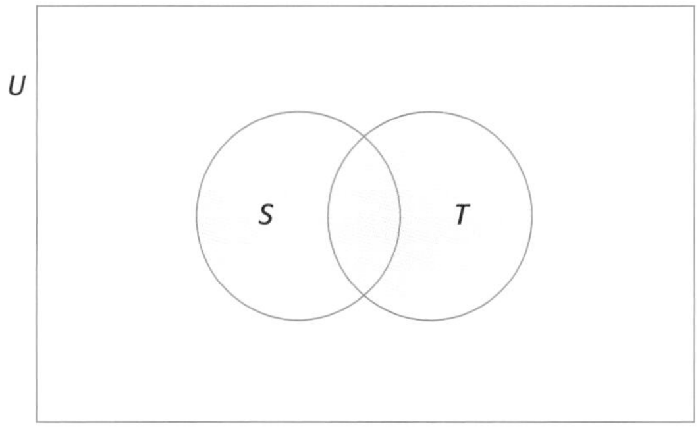

The union, $S \cup T$

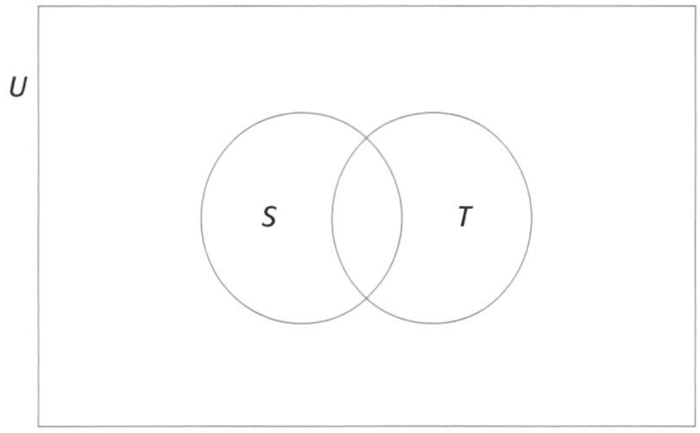

The intersection, $S \cap T$

The relative complement $(T \cap S')$ of S with respect to T contains all regions of T that are not shared with S. This region is shown in the following Venn diagram.

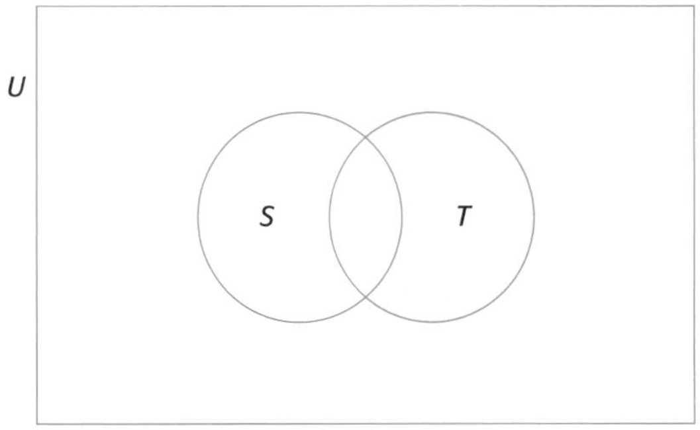

The relative complement of *S* in *T* : $(T \cap S')$

Example 3.2.1c

The Venn diagram shows three sets *S*, *T* and *V* containing symbols used in music and on playing cards. Write down the symbol(s) appearing in the set $S \cap (T \cap V')$. Copy the Venn diagram and shade this region.

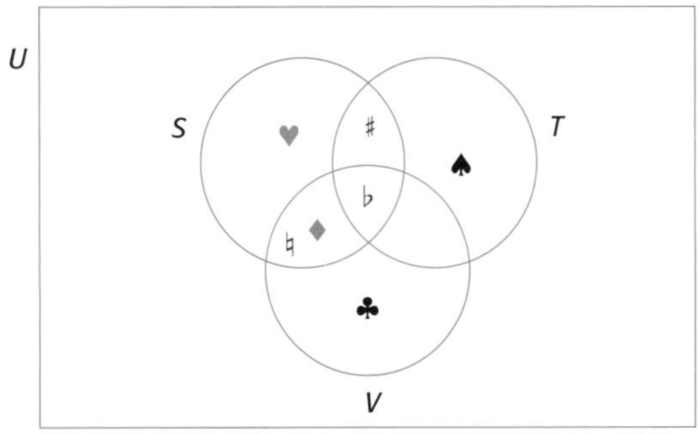

Solution

To solve this problem we break it into steps. First what are the members of $T \cap V'$? These are the elements that are in T but not V. Hence $T \cap V = \{♠, \#\}$. Now we decide if any of the elements of this set are also in S. The answer is just the symbol # and so this is the only element in $S \cap (T \cap V')$. The shading should be applied only to the region containing the # symbol.

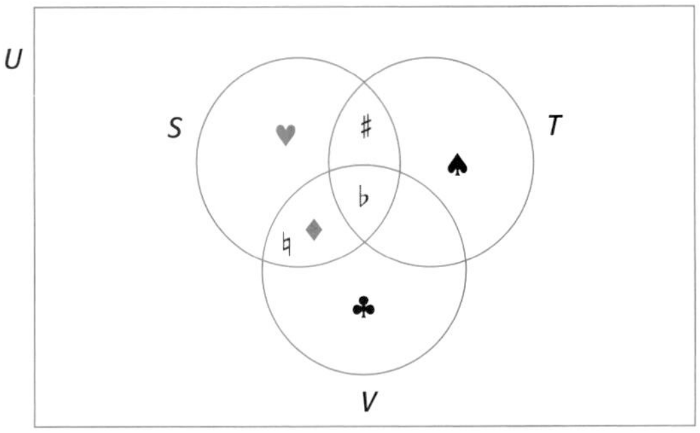

Example 3.2.1d

Use the symbols for union, intersection and so on to describe the shaded region in this diagram.

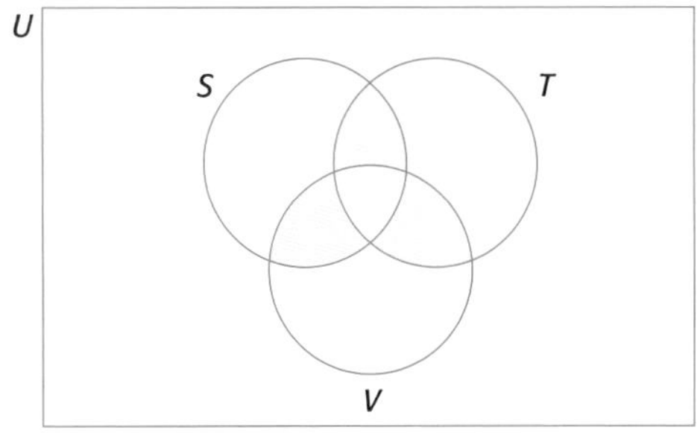

Solution
Consider just one of the shaded regions, for example the region common to S and V. Suppose the unshaded region in the middle was added to this region. That would give $S \cap V$. We could do this with all three pairs of regions and take the union to get $(S \cap V) \cup (S \cap T) \cup (T \cap V)$. This set contains all the shaded parts we want, but it still also contains the unwanted common intersection in the middle. This can be removed using the idea of complement so one answer is $((S \cap V) \cup (S \cap T) \cup (T \cap V)) \cap (S \cap T \cap V)'$.

As is sometimes the case, there is another way of writing this set. You should convince yourself that the expression $((S \cap V) \cap T') \cup ((S \cap T) \cap V')) \cup ((T \cap V) \cap S'))$ is just as good. Perhaps you can find even more ways.

Venn diagrams can be used in a way which we will find useful when we study probability problems.

Example 3.2.1e
This diagram shows the numbers of students taking the IB diploma in a school and also shows the numbers attending lessons in certain subjects.

The classes are represented by sets S: {Mathematical studies}, G: {German} and M: {Music}

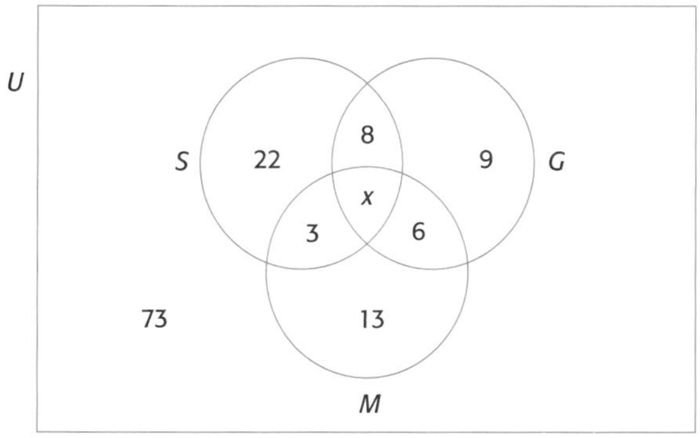

The number of IB students studying

a none of the three subjects is 73
b mathematical studies but neither German nor music is 22
c mathematical studies and music but not German is 3
d music and German but not mathematical studies is 6
e German is $6 + 8 + 9 + x = 23 + x$
f all three subjects is x
g music or German but not both is $8 + 9 + 3 + 13 = 33$
h mathematical studies or music but not German is $22 + 3 + 13 = 38$

If there are 136 students in this sample, find the value of x.

Solution
The total of all the values in the diagram is $134 + x$. Hence $134 + x = 136$ so $x = 2$.

Example 3.2.1f
Use the Venn diagrams in Example 3.2.1b to show that $n(S \cup T) = n(S) + n(T) - n(S \cap T)$.

Solution
$n(S \cup T)$ is the total of all elements in S and T together.
Imagine counting all the elements in S then all the elements in T then adding these answers.
This give $n(S) + n(T)$. But in doing this, we have counted the elements in the intersection $n(S \cap T)$ twice. These have to be removed once. Hence $n(S \cup T) = n(S) + n(T) - n(S \cap T)$.

Exercise 3.2

1 Draw correctly shaded Venn diagrams for these sets (if they are non-empty). Each diagram must have three intersecting sets S, T and V inside a universal set. If the set is empty, then just say so.

a $(S \cup T \cup V)'$ b $(S \cup T) \cap V'$ c $S \cap (T \cup V)'$
d $V' \cap (T \cap (S \cup V')')$ e $T' \cap V$ f $S \cap (T \cup V)'$
g $(S \cup V \cup T) \cap (S \cap T \cap V)$ h $T \cap \varnothing'$

2 Let $A = \{1, 2, 3, 4, 5, 8\}$, $B = \{1, 3, 5, 9\}$ and $C = \{2, 3, 5, 6, 7\}$. Take a universal set consisting of the whole numbers from 1 to 12 inclusive. Draw a Venn diagram showing these sets with their elements all positioned correctly.

3 Draw a Venn diagram showing the relationship between the sets $\mathbb{N}$, $\mathbb{Z}$, $\mathbb{Q}$, and $\mathbb{R}$. Place the following numbers in their correct positions in the diagram: $1, -1, \frac{4}{3}, \pi$.

4 Write down an expression to describe the shaded area in Diagram 1.

5 Write down an expression to describe the shaded area in Diagram 2.

6 This problem can be done after you have studied Topic 4.

Let U be the set of all quadratic expressions with variable x.

A is the set of quadratic expressions whose graphs have a minimum.

B is the set of quadratic expressions whose graphs have a maximum.

C is the set of all quadratic expressions which have at least one real zero.

Draw a Venn diagram showing these sets. If a region of intersection of any pair of sets is empty then do not draw those sets intersecting. In each of the remaining regions, write two quadratic expressions which properly belong in that region.

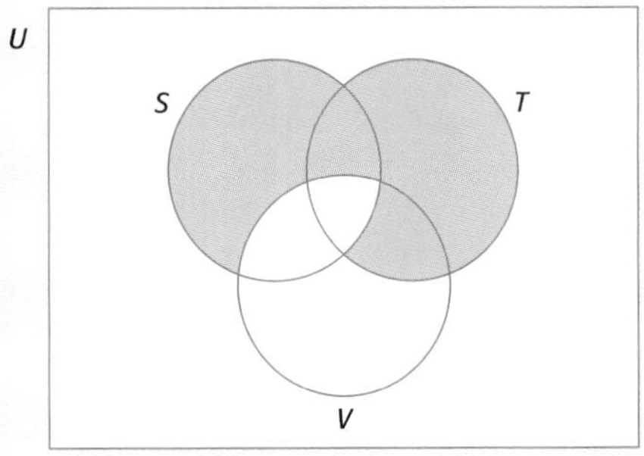

Diagram 1

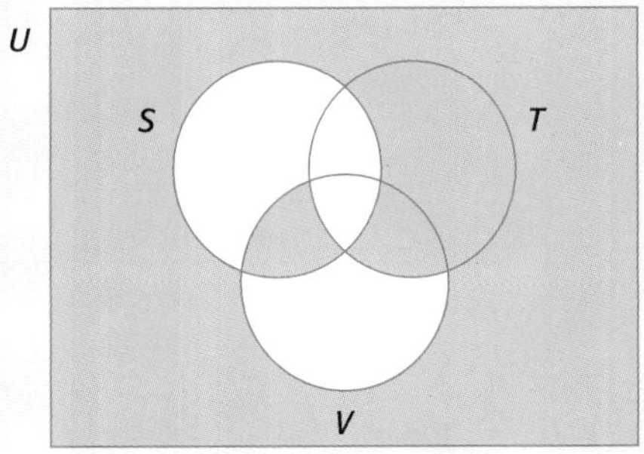

Diagram 2

3.3 Sample space for events

One important area of mathematics where set theory is used is in the study of probability.

A **sample space** is a set consisting of all possible results of a trial or experiment, or all events associated with a particular process. An individual outcome or result is also called an **event**.

Example 3.3.1a

A fair coin is tossed **a** once **b** twice **c** three times. What is the sample space?

Solution

a The coin can land with heads up or tails up. Label these outcomes h and t. The sample space is the set $T_1 = \{h, t\}$.

b For two tosses the first result can be a head or a tail and then the second result can also be a head or a tail. The possible outcomes are hh, ht, th and tt. The sample space is

$T_2 = \{hh, ht, th, tt\}$.

c For three tosses, $T_3 = \{hhh, hht, hth, thh, tht, tth, htt, ttt\}$.

Suppose you were to collect data (perhaps for an internal assessment project) relating gender (m or f) and height in metres for some people in your school. A single item from the collected data would consist of a pair that you could write, for example (f, 1.63). The sample space might look like this:

H = {(m, 1.73), (m, 1.98), (f, 1.63), (m, 1.80), (f, 1.45),(f, 1.69), (f, 1.72),(m, 1.50), (m, 1.86), (f, 1.61)}.

Investigation: Notice that $n(T_1) = 2^1$ while $n(T_2) = 4 = 2^2$ and $n(T_3) = 8 = 2^3$. What value do you think $n(T_6)$ will have? What about $n(T_i)$ for any i? Now suppose that we no longer care about the order in which the heads or tails land but only the number of each. How does this affect the sample space? Can you convince yourself that a new expression for $n(T_i)$ is $n(T_i) = i + 1$?

Example 3.3.1b

A six-sided die is thrown **a** once **b** twice. Write down the sample spaces. How many elements are in each of these sample spaces?

c How many elements do you think there will be if you throw three times?

Solution

a {1, 2, 3, 4, 5, 6}: 6 elements $(= 6^1)$

b {(1, 1), (1, 2), (1, 3), (1, 4), (1, 5), (1, 6), (2, 1), (2, 2), (2, 3), (2, 4), (2, 5), (2, 6) (3 ,1), (3, 2), (3, 3), (3, 4), (3, 5), (3, 6) (4, 1), (4, 2), (4, 3), (4, 4), (4, 5), (4, 6) (5, 1), (5, 2), (5, 3), (5, 4), (5, 5), (5, 6) (6, 1), (6, 2), (6, 3), (6, 4), (6, 5), (6, 6)}
36 elements $(= 6^2)$

c Each throw has 6 possible outcomes so there are $216 (= 6^3)$ elements.

When only two outcomes are possible and each outcome means the other cannot occur, the outcomes are called **mutually exclusive** and we refer to the events as **complementary**.

Example 3.3.1c

A couple has twin babies. Write down the sample space for the genders of the babies if they are **a** non-identical twins **b** identical twins.

Solution

Here we do not care about the order of the births. Let M be the event male and F female.

a Non-identical: both could be either gender so {MM, FM, FF}.

b Identical: they must be the same gender so {MM, FF}.

Example 3.3.1d

Write down some pairs of complementary events in the form of sample spaces.

Solution

Flipping a coin {heads, tails}, gender of a person {male, female}, state of well-being {alive, dead}, result of an examination {pass, fail} occupancy of a house {at home; away}, numerical comparison $(x \geq y,\ x < y)$ etc. Can you think of some more?

Exercise 3.3

1 What is the sample space for your score when throwing a single dart at a normal dart board? What is a suitable universal set for these scores? Which scores between 0 and 60 can you *not* achieve with one dart?

2 In a tennis game a player receives a score of 15 for the first point won, 30 for the second, 40 for the third point, then wins the game with the fourth point, unless both players have 40. In that case the next point won is called advantage (call this "a") then the player with the advantage either wins the game at the next point or the score falls back to 40 all if the point is lost. Write down a sample space S of pairs (for example, (15, 30)) for all the score combinations in a single game of tennis. What is $n(S)$?

3 A simple game involves throwing a six-sided die. A score of an odd number means you win, an even number means you lose.

- **a** Write down the sample space, G, for the game.
- **b** Write down a subset, W (winning numbers), of G.
- **c** Express the subset, L (losing numbers), in terms of W and list the members of L.
- **d** Show the situation in a Venn diagram.

4 **a** Write down the sample space, T, for the pairs of numbers that can be achieved by throwing two dice together (without worrying about order). What is the value of $n(T)$?

- **b** Suppose S is a subset of T consisting of pairs that add up to 8. Write down S and state the value of $n(S)$.

3.4 Basic concepts of symbolic logic, propositions

3.4.1 Basic concepts of symbolic logic

The word "logic" derives from the Greek *logos* meaning "word" For a mathematician, logic deals with the conversion of word statements to symbolic form and the use of that symbolic form to make deductions and create proofs. The rules of logic are useful in other areas of science such as electronics, as well as being of interest to philosophers.

Mathematical logic deals with basic statements called **propositions**.

A proposition can be **true**, T, or **false**, F, but might be **indeterminate**.

An indeterminate proposition is one which may be true or false depending on context, for example, the proposition "Your garden has no trees" can be true or false depending on whom you are speaking to. It is therefore indeterminate.

"Freddy is handsome" is an indeterminate proposition because it expresses only an opinion, which might be different for other speakers.

The items referred to in a proposition should be well-defined. For example "$x > 3$" by itself is not a proposition because it does not tell us what x is.

Questions, exclamations and orders are not propositions.

In the study of mathematical logic, we concentrate on propositions which have a well-defined *truth value*, that is, they are true or false.

There is a direct analogy between the two values, T or F, for a proposition and the nature of complementary events (see Section 3.3), for example, "girl" or "not girl" for a new baby. This analogy is exploited in the design of electronic circuits where a component of a circuit can have current or no current.

Example 3.4.1

Which of these are propositions? For each proposition identified, discuss whether it is true, false or indeterminate.

a All dogs have tails.

b My little sister is cute.

c Today it is snowing here.

d i $3^2 = 9$ ii $4^2 = 21$ iii π is a rational number

e How many people are there in the world?

f Quadratic equations contain at most a second power of the variable.

g $y = 6$

h For all $x \in \mathbb{R}$, $x^2 \geq 0$.

Solution

a This is a proposition. Some dog species do not have tails so it is false.

b This is a proposition but it depends on peoples' opinions so is indeterminate.

c This is a proposition but you will have to decide on its truth, or not, at the time you read this. It is indeterminate.

Theory of knowledge

Investigate *logical fallacies*. Find some examples of logical fallacies and discuss how we use them in everyday conversations.

d i Yes, a proposition which is true.
  ii A proposition which is false since $4^2 = 16$.
  iii Also a false proposition as π is irrational.

e A question is never a proposition.

f This is a proposition and it is true.

g This is not a proposition. We are not told what y is.

h Here we are told that x is a real number and this is a true proposition.

To avoid writing the words in a proposition all the time, we label propositions with letters. Usually we use p, q, r, s and so on, for example,

p: The wind is blowing.
q: I will lose my hat.

In this example, there might be a connection between the propositions. We now show how to write such connections.

In mathematical studies SL the maximum number of propositions used at once is three, often p, q and r.

Exercise 3.4

1 Which of the following are propositions? For each proposition identified, discuss whether it is true or false.

a The world is flat.
b Many cats like to eat fish.
c Why is the sky blue?
d i $x \in \mathbb{Z}$ ii $[-5, 5] \subset \mathbb{Z}$ iii π is a real number.
e Leave me alone!
f Logic is difficult.

3.5 Compound statements, use of Venn diagrams

3.5.1 Compound statements

All relations between two or more propositions are called **compound statements** or **propositions**.

Symbols used to connect propositions are called **connectives**.

Implication

For the propositions, p and q, on the previous page, you might want to say "*If* the wind is blowing *then* I will lose my hat". This is called **implication**. The symbol for this is an arrow *in the direction* of implication.

We write it as $p \Rightarrow q$ or sometimes as $q \Leftarrow p$.

Equivalence

If an implication works in both directions, it is called an **equivalence.**

The symbol for equivalence is $\Leftrightarrow$.
For example,

r: Fiona is a good footballer.
s: Fiona scores lots of goals.

We can write $r \Leftrightarrow s$. This amounts to saying that Fiona being a good footballer and Fiona scoring lots of goals are equivalent. It is also the same as $r \Rightarrow s$ together with $r \Leftarrow s$.

In Topic 2, Section 2.1.3, we introduced the mathematical relation **if and only if**. This relation is identical to the equivalence relation discussed here. $r \Leftrightarrow s$ means r holds *if and only* if s holds. For example "n is even if and only if n is divisible by 2" = "n is even $\Leftrightarrow n$ is divisible by 2".

The use of *therefore* when writing words for an implication is wrong! Consider the following: "If Antonia is wearing her hat, then the sun is shining." The meaning here is that if Antonia is wearing her hat then we can predict that the sun is shining, however, the sun is *not* shining *because* Antonia is wearing her hat. To say "therefore the sun is shining" is incorrect.

Always use "*if.... then*" when writing words for an implication.

Sometimes the meaning of the implication does allow an alternative form. We might write

"All cats like fish". This is short for "If you are a cat then you like fish."

George Boole (1815–1864) and Mary Boole (1832–1916)

George Boole (1815–1864) was a pioneer in the field of mathematical logic. One of his most important achievements was to write logic statements in mathematical form. George's wife, Mary (1832–1916) was 17 years his junior. She was also a talented mathematician considering herself to be a "mathematical psychologist". She made important contributions in the fields of mathematics and science education. Despite their age difference, the marriage was very successful and lasted until George's premature death.

Negation

If we deny a proposition we are saying it is *not* true. The negation of p is written $\neg p$ or $\sim p$ or p' or $\overline{p}$.

For the propositions used in the last example we have:

$\neg r$: Fiona is not a good footballer and
$\neg s$: Fiona does not score lots of goals

Example 3.5.1a

Using the propositions p and q

p: The wind is blowing
q: I will lose my hat

write in symbolic form the proposition "If the wind is not blowing, then I will not lose my hat."

Solution
Use the negated forms of p and q with the correct direction of implication:
$\neg p \Rightarrow \neg q$

Conjunction, disjunction and exclusive disjunction

These are fancy names for the ideas of "**and**", "**or**" and "**or but not both**".

Each has its own symbol.

- Conjunction (= **and**) with symbol $\wedge$
- Disjunction (= **or, possibly both**) with symbol $\vee$
- Exclusive disjunction (= **or but not both**) with symbol $\veebar$

The use is straightforward:

p: Mischa has a dog.
q: Yuri has a cat.

$p \wedge q$: Mischa has a dog and Yuri has a cat.

$p \vee q$: Mischa has a dog or Yuri has a cat (and both could be true).

$p \veebar q$: Either Mischa has a dog or Yuri has a cat, but **not** both are true.

Example 3.5.1b

Given w: I am wearing shorts
s: I am going to swim
r: I am going to run

Write in words these propositions.

a $w \wedge s$ **b** $w \Rightarrow r \vee s$ **c** $s \veebar r$ **d** $\neg s \Leftrightarrow r$ **e** $w \wedge \neg(s \vee r)$

Solution

a I am wearing shorts and I am going to swim.

b If I am wearing shorts then I am going to swim or to run.

c I am going to swim or to run, but not both.

d If I am not going to swim, then I am going to run and if I am going to run then I am not going to swim.

e I am wearing shorts and I am not going to run or to swim (*neither/nor* is acceptable).

Notice that some compound propositions have the same meaning (are equivalent).

For example $(p \Rightarrow q) \wedge (q \Rightarrow p)$ is the same as $p \Leftrightarrow q$,
$\neg(p \vee q)$ is the same as $\neg p \wedge \neg q$.
$\neg(p \wedge q)$ is the same as $\neg p \vee \neg q$.

To understand why these propositions are equivalent an example helps.

Take p: The sun is shining
and q: It is raining

Then $\neg p \wedge \neg q$ becomes "the sun is not shining **and** it is not raining"
while $p \vee q$ becomes "the sun is shining or it is raining (or both)"
and hence $\neg(p \vee q)$ is "**neither** is the sun shining **nor** is it raining"
which is clearly the same as "the sun is not shining **and** it is not raining."

Warning: The use of brackets is important when a compound proposition could be ambiguous without their use. For example, if we write $\neg p \wedge q$ this is different from $\neg(p \wedge q)$.
It's like the difference between $-x+y$ and $-(x+y)$ in algebra.
If you have even the smallest doubt, use brackets to ensure you have the right meaning.
If you misjudge this, extra, unneeded brackets might be redundant but they won't be wrong.

3.5.2 Analogy with set theory, Venn diagrams

There is a direct analogy between the operations of logic and those of set theory.

- The 'or' connective, $\vee$, corresponds to the union, $\cup$, of sets.
- The 'and' connective, $\wedge$, corresponds to the intersection, $\cap$, of sets.
- The 'negation' connective, $\neg$, corresponds to taking the complement, $'$, of a set.

We will be able to use these analogies to help us draw more difficult Venn diagrams after we have explored the idea of a truth table in the next section.

The **truth set** of a proposition p is the set of items satisfying the condition(s) described by p.

Example 3.5.2a

p: I have three dogs, called Fido, Rover and Shandy.
Represent proposition p in a truth set P on a Venn diagram.

Solution

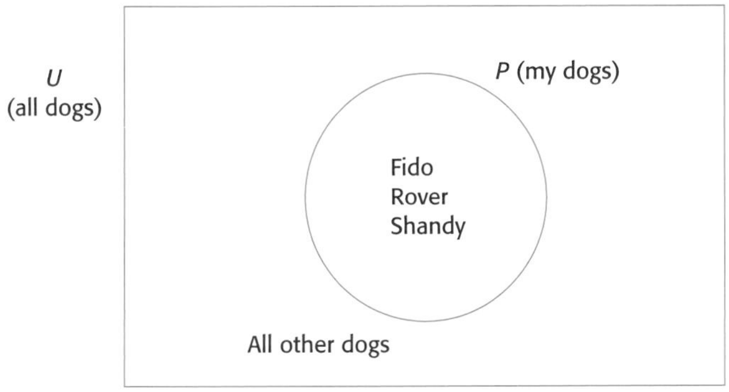

The truth set of proposition p

Example 3.5.2b

Draw a Venn diagram showing the truth sets for the propositions:

p: n is an even natural number less than 17
q: n is an odd natural number less than 17
r: n is a positive multiple of 3 which is less than 20.
Take U to be the set of integers from 1 to 20 inclusive.

Solution

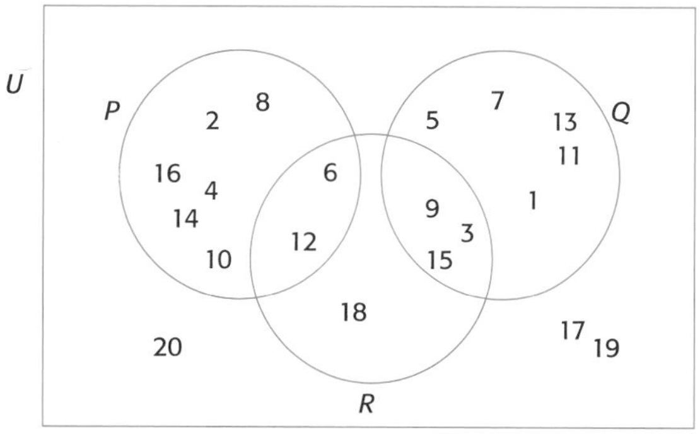

Example 3.5.2c

Using the truth sets in Example 3.5.2, part **b** above, list the truth sets for the propositions

a $p \wedge q$

b $p \wedge r$

c n is a positive multiple of 3 which is less than 20 but is not an even number less than 17

d $\neg(p \vee r)$

Write proposition **d** in words.

Solution

a We need the set $P \cap Q$ which is empty: $\varnothing$

b The set $P \cap R = \{6, 12\}$

c $\{3, 9, 15\}$

d $(P \cup Q)' = \{17, 18, 19, 20\}$

In words: n is an integer between 1 and 20 which is neither an even number less than 17 nor an odd number less than 17.

3.5.3 Knowledge and use of the "exclusive disjunction" and the difference between it and "disjunction"

It is important to be clear about the difference between union and disjunctive union. Some further examples will help.

We will use these propositions:

p: I will go swimming q: I will go shopping

Example 3.5.3a

Write in words the propositions **a** $p \vee q$ **b** $p \veebar q$
stating in each case whether you will do both the activities.

Solution

a $p \vee q$: I will go swimming or I will go shopping (I might do both).

b $p \veebar q$: I will go swimming or I will go shopping but I will not do both.

Example 3.5.3.b

a Explain why proposition **a** $(p \veebar q)\ (p \wedge q)$ must be represented by a column of F's in a truth table.

b What changes if we replace $(p \veebar q)$ with $p \vee q$?

Solution

a $p \veebar q$ stands for proposition p or q but not both while $p \wedge q$ stands for both p and q at once.

Clearly these cannot both occur at the same time so the proposition in **a** must be all false.

(It is called a **contradiction** – more on this later.)

b If we change the first bracket to $p \vee q$, then both p and q are now allowed at the same time. The conjunction must now just be the same as the second bracket: $(p \vee q) \wedge (p \wedge q) = p \wedge q$.

There is an operation in set theory known as **disjunctive union**, corresponding to the **disjunctive or** of logic, but this set operation is not included in the mathematical studies syllabus. You can speculate as to how it is defined as an exercise.

Exercise 3.5

1 Use the propositions: p: The mathematics teacher is very strict.
q: I work hard in the mathematics class.
r: I will do well in my mathematics test.

a Write in words the propositions
i $\neg p$ ii $p \wedge q \wedge r$ iii $\neg(p \vee q)$ iv $(p \Rightarrow q) \wedge (q \Rightarrow r)$

b Write in symbols the propositions
i If the mathematics teacher is very strict then I will not do well in my mathematics test.
ii Either the mathematics teacher is not very strict or I work hard in the mathematics class but not both.
iii If the mathematics teacher is very strict and I work hard in the mathematics class then I will do well in my mathematics test.

2 Consider the proposition
s: If the wind is strong then either I will lose my hat or I will leave my hat at home but not both of these.
Name and write down three simple propositions which make up s then express s in symbolic form in terms of your three basic propositions.

3 Draw a Venn diagram showing the truth sets P, S and T for the propositions
$p: x \geq 0, x \in \mathbb{Z}$

$s: x = 0$

$t: x \leq 0, x \in \mathbb{Z}$

Take U to be the set of all real numbers.

a Express S in terms of P and T.

b Express each of the sets

i $(P \cup T)'$

ii S'

in terms of one of the number sets $\mathbb{N}$, $\mathbb{Z}$, $\mathbb{Q}$, $\mathbb{R}$.

4 The Venn diagram shows the truth sets P and Q for some propositions p and q about integers.

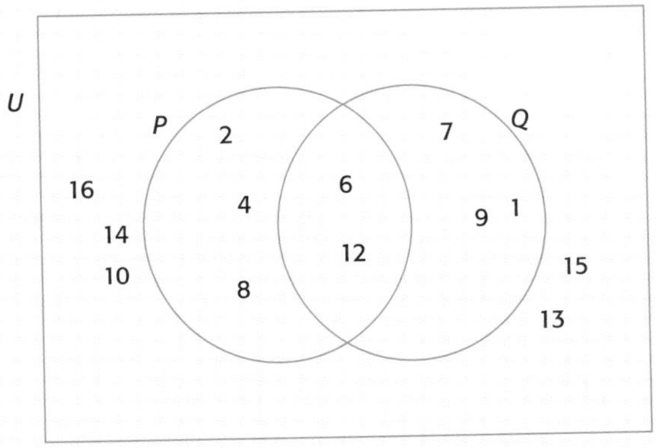

List the members of the truth sets for these compound propositions.

a $p \vee q$ b $p \veebar q$ c $p \wedge q$

d $(p \veebar q) \vee (p \wedge q)$ e $\neg p \veebar q$

f $\neg q \veebar p$ g $\neg(p \veebar q)$

Comment if any of these are the same.

3.6 Truth tables, contradiction and tautology

3.6.1 Truth tables

If a proposition is not indeterminate then it is either *true* (T), or *false* (F). True and false are complementary events. For two propositions, they are either both true or both false or one is true and one false: TT, FF, TF or FT. This reminds us of the situation of a coin flipped twice: {*hh*, *tt*, *th*, *ht*}. A truth table is a good way to show the possibilities

for two propositions p, q

p	*q*
T	T
T	F
F	T
F	F

or for three propositions p, q, r.

p	*q*	*r*
T	T	T
T	T	F
T	F	T
T	F	F
F	T	T
F	T	F
F	F	T
F	F	F

To find the truth table entries for complicated propositions, we build them up in simple stages filling in the columns for smaller parts of the proposition step by step.

Example 3.6.1a

Draw a truth table for propositions p and q adding extra columns for the compound statements $\neg p$, $p \wedge q$, $p \vee q$ and $p \veebar q$.

Solution

If p is true, then $\neg p$ is false. $p \wedge q$ can only be true if both p and q are true at the same time. Similar considerations also allow us to judge when $p \vee q$ and $p \veebar q$ are true.

p	q	$\neg p$	$p \wedge q$	$p \vee q$	$p \veebar q$
T	T	F	T	T	F
T	F	F	F	T	T
F	T	T	F	T	T
F	F	T	F	F	F

This table tells us a great deal of information. For example, it says that if p and q are both true, then

$p \wedge q$ is also true while $p \veebar q$ is false.

Example 3.6.1b
Draw a truth table involving three propositions p, q and r. Construct a column for the compound proposition $(p \vee q) \wedge (\neg p \vee r)$.

Solution
We need columns for p, q and r and a final column for the answer. In between, it is helpful to include columns for parts of the proposition, $\neg p$, $p \vee q$ and $\neg p \vee r$.

p	q	r	$\neg p$	$p \vee q$	$\neg p \vee r$	$(p \vee q) \wedge (\neg p \vee r)$
T	T	T	F	T	T	T
T	T	F	F	T	F	F
T	F	T	F	T	T	T
T	F	F	F	T	F	F
F	T	T	T	T	T	T
F	T	F	T	T	T	T
F	F	T	T	F	T	F
F	F	F	T	F	T	F

Notice how the first three columns were constructed using 4 T's and 4 F's, then two at a time, then one at a time. If there were four propositions we would start with 8 T's and so on.

Example 3.6.1c
Use a truth table to prove the statement in Section 3.5 that $\neg(p \vee q)$ is the same as $\neg p \wedge \neg q$.

Solution
The table will have to have columns for both of the compound propositions and these columns must turn out to be identical. It will help to include columns for $\neg p$, $\neg q$ and $p \vee q$ as well. You can see that the fifth and seventh columns are the same, demonstrating that the propositions in those columns are equivalent.

Truth values can be found using the GDC list facility. T is represented by '1' and F by '0'.

p	q	$\neg p$	$\neg q$	$\neg p \wedge \neg q$	$p \vee q$	$\neg(p \vee q)$
T	T	F	F	F	T	F
T	F	F	T	F	T	F
F	T	T	F	F	T	F
F	F	T	T	T	F	T

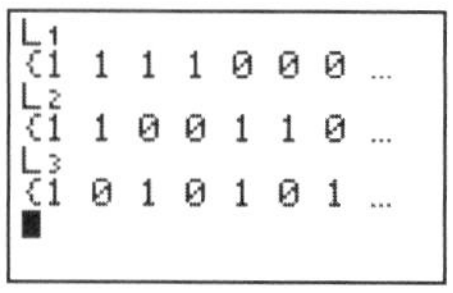

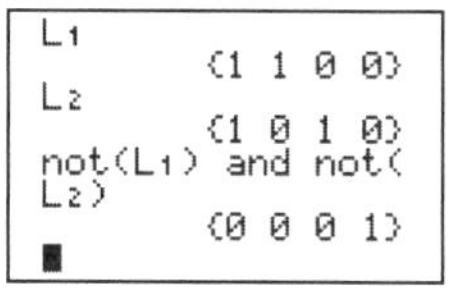

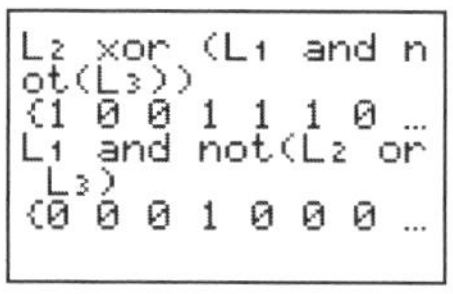

3.6.2 Logical contradiction and tautology

Contradiction

Suppose p is : All cows are yellow.
Consider the proposition $p \wedge \neg p$.

This is saying: All cows are yellow and not all cows are yellow.

Clearly p and $\neg p$ cannot both be true at the same time. This statement is an example of a **logical contradiction**. Here is the truth table.

p	$\neg p$	$\neg p \wedge \neg p$
T	F	F
F	T	F

The proposition, that all cows are yellow is nonsense but this has nothing to do with the fact that we found a contradiction above. The contradiction arises only because we tried to assert two mutually contradictory things at the same time.

A logical contradiction must necessarily be represented by an entire column of F's.

Similarly an entire column of F's always stands for a logical contradiction.

Example 3.6.2a

Show that $(\neg p \vee q) \wedge (\neg q \wedge p)$ is a logical contradiction.

Solution

Construct the truth table. To keep the table compact, take $s = (\neg p \vee q) \wedge (\neg q \wedge p)$.

p	q	$\neg p$	$\neg q$	$\neg p \vee q$	$\neg q \vee p$	s
T	T	F	F	T	F	F
T	F	F	T	F	T	F
F	T	T	F	T	F	F
F	F	T	T	T	F	F

Since we know that the fifth column is the same as $p \Rightarrow q$ while the sixth column is the proposition p and not q, both propositions cannot be true at the same time and so s is a contradiction.

Tautology

A proposition represented by an entire column of T values is called a **tautology**. Such a proposition is **always true**, no matter what. The simplest example of a tautology is the proposition $p \vee \neg p$. Using the previous example, it is always true that all cows are yellow or not all cows are yellow.

p	$\neg p$	$p \vee \neg p$
T	F	T
F	T	T

Example 3.6.2b

Show that $(p \wedge q) \vee (\neg p \vee \neg q)$ is a tautology.

Solution

p	q	$\neg p$	$\neg q$	$p \wedge q$	$\neg p \vee \neg q$	$(p \wedge q) \vee (\neg p \vee \neg q)$
T	T	F	F	T	F	T
T	F	F	T	F	T	T
F	T	T	F	F	T	T
F	F	T	T	F	T	T

Since we know that the seventh column contains entirely T values it represents a tautology.

More on the analogy with Venn diagrams

The analogy between logic connectives and set operations mentioned in Section 3.5 can be exploited to help us draw difficult Venn diagrams. Start with a simple case. Suppose you want to shade the set $P \cap Q'$. Represent P and Q by propositions p and q respectively. Fill in the truth table for $p \wedge \neg q$.

p	q	$\neg q$	$p \wedge \neg q$
T	T	F	F
T	F	T	T
F	T	F	F
F	F	T	F

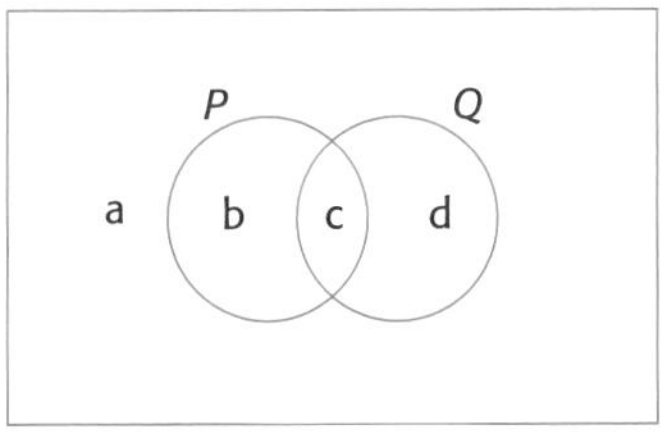

Now look at regions of the Venn diagram. To describe what to do, we have put labels **a–d** in the regions (but you do not have to do this).

Look at the region labelled **b**. It is a subset of P but not of Q. Hence assign the truth values T to p and F to q. Now look across the table to the last column. Here we see a T. This tells us we should shade region **b**.

What about the region labelled **a**? It is not a subset of P or of Q. Hence assign the truth values F to p and F to q. Now look across the table to the last column. Here we see an F. This tells us we should not shade region **a**. Just treat each region the same until all the shadings are decided.

Example 3.6.2c

Construct a truth table with three propositions p, q, r to help shade the region $P' \cap (Q \cap R)$ in a Venn diagram with sets P, Q, R.

Solution

p	q	r	$\neg p$	$q \wedge r$	$\neg p \wedge (q \wedge r)$
T	T	T	F	T	F
T	T	F	F	F	F
T	F	T	F	F	F
T	F	F	F	F	F
F	T	T	T	T	T
F	T	F	T	F	F
F	F	T	T	F	F
F	F	F	T	F	F

There is one T in the last column so only one region to shade, corresponding to P false but Q and R both true.

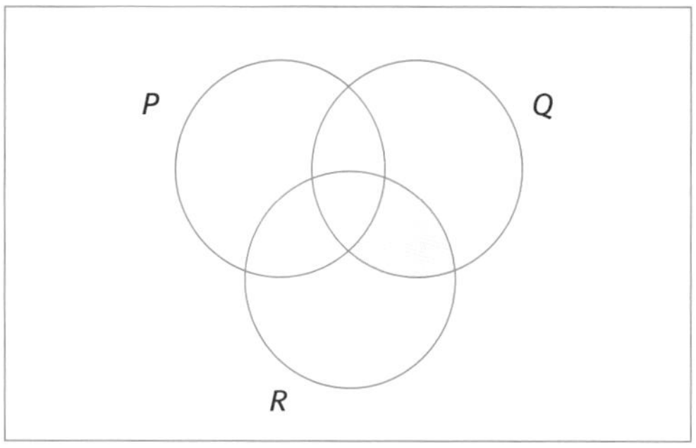

The Venn diagram associated with a tautology must represent T values in every region. The diagram will be completely shaded and this means that tautology is associated with the universal set U.

Similarly, a contradiction has a diagram with no shading and this represents the empty set, $\varnothing$.

Exercise 3.6

1 Show that the proposition $\neg p \vee [p \vee (q \wedge \neg q)]$ is a tautology.

2 Show that $\neg[p \wedge (\neg p \vee q)] \vee q$ is a tautology.

3 Show that $\neg q \wedge [p \wedge (q \vee \neg p)]$ is a contradiction.

3.7 Implication and related concepts

3.7.1 Definition of implication

This is the truth table for implication.

p	q	$p \Rightarrow q$
T	T	T
T	F	F
F	T	T
F	F	T

Part of this table is quite obvious.

If p is to imply q, then when p is true, so must q be. Similarly, if p implies q and q is false, then p cannot be true. Hence in the second line of the table, $p \Rightarrow q$ must be false. The problem appears when we ask what happens if p is false. From a false assertion, it is possible to conclude anything, in particular, that $p \Rightarrow q$ is true. Hence we assign the truth values T in these cases.

An example might help.

Suppose you have bought a CD using the internet.

The seller says "If you send me the money first" (that's statement p) "then I will send you the CD" (that's statement q).

There are four routes that the events might take.

i You send the money, the seller sends the CD. Then the seller's statement is true.

ii You send the money, the seller does not send the CD. The statement is false.

iii and **iv** You do not send any money. Now it is irrelevant whether the seller sends the CD or not.
We give the benefit of the doubt here and suppose the seller's statement is true and you would have received the CD if you had paid.

This might seem a bit contrived to the beginner, but it certainly should be clear, at least, that there is no contradiction in doing this.

Some mathematicians like to simply **define** implication by saying that p implies q is **the same as** the proposition $\neg p$ or q. That is: $p \Rightarrow q = \neg p \vee q$.

Example 3.7.1a

Show that the truth table values for the two propositions above are the same.

Solution

p	q	$\neg p$	$\neg p \vee q$	$p \Rightarrow q$
T	T	F	T	T
T	F	F	F	F
F	T	T	T	T
F	F	T	T	T

Example 3.7.1b

Show that the proposition $((p \veebar q) \vee (p \wedge q)) \Rightarrow (p \vee q)$ is a tautology.

Solution

Set $s = ((p \veebar q) \vee (p \wedge q)) \Rightarrow (p \vee q)$.

p	q	$p \veebar q$	$p \wedge q$	$(p \veebar q)$ $(p \wedge q)$	p q	s

p	q	$p \veebar q$	$p \wedge q$	$(p \veebar q)\ (p \wedge q)$	$p\ q$	s
T	T	F	T	T	T	T
T	F	T	F	T	T	T
F	T	T	F	T	T	T
F	F	F	F	F	F	T

In this example, the column for s is true in every case so s is a tautology. Notice also that the fifth and sixth columns are actually equal. We will discuss events of this kind further in the next section.

Example 3.7.1c

Show that $(p \vee (p \wedge q)) \Rightarrow p$ is a tautology while $(p \vee (p \wedge q)) \Rightarrow q$ is not.

Solution

p	q	$p \wedge q$	$p\ (p \wedge q)$	$p \vee (p \wedge q) \Rightarrow p$	$p\ (p \wedge q) \Rightarrow q$
T	T	T	T	T	T
T	F	F	T	T	F
F	T	F	F	T	T
F	F	F	F	T	T

The fifth column exhibits a tautology but the last column does not.

3.7.2 Converse, inverse and contrapositive

Starting with a simple implication, $p \Rightarrow q$, we can construct related propositions.

- The **converse** is obtained by swapping the propositions in the implication: $q \Rightarrow p$.
- The **inverse** is obtained by negating the propositions in the implication: $\neg p \Rightarrow \neg q$.
- The **contrapositive** is obtained by both negating and swapping the propositions: $\neg q \Rightarrow \neg p$.

Note that the contrapositive of the inverse is the converse.

Example 3.7.2a

Consider s: If my shoes are too small then my feet hurt.

Write in words, the converse, the inverse and the contrapositive of s.

Solution

The *converse*: If my feet hurt, then my shoes are too small.

The *inverse*: If my shoes are not too small, then my feet do not hurt.

The *contrapositive*: If my feet do not hurt, then my shoes are not too small.

Example 3.7.2b

Draw a truth table to compare the values of an original proposition with its converse, inverse and contrapositive. What do you notice?

Solution

p	q	$\neg p$	$\neg q$	implication $p \Rightarrow q$	converse $q \Rightarrow p$	inverse $\neg p \Rightarrow \neg q$	contrapositive $\neg q \Rightarrow \neg p$
T	T	F	F	T	T	T	T
T	F	F	T	F	T	T	F
F	T	T	F	T	F	F	T
F	F	T	T	T	T	T	T

Notice that some columns are the same. This is not the first time we have met identical columns so we now discuss this a little more.

3.7.3 Logical equivalence

Given the connectives introduced so far, there are infinitely many ways that we could form propositions from a pair p, q. On the other hand there are only $2^4 = 16$ different ways of entering T and F values in a table with four rows. This means that many different compound propositions have the same truth values. When this happens, we say the propositions are **equivalent**.

The symbol $\Leftrightarrow$ is used for equivalence.

$p \Leftrightarrow q$is a shorter way of writing $(p \Rightarrow q) \wedge (q \Rightarrow p)$.

Here is the truth table for equivalence.

p	q	$p \Rightarrow q$	$q \Rightarrow p$	$p \Rightarrow q$ $(=(p \Rightarrow q) \wedge (q \Rightarrow p))$
T	T	T	T	T
T	F	F	T	F
F	T	T	F	F
F	F	T	T	T

For a pair of propositions to be logically equivalent, they should both be true or both false at the same time. This is the same as saying that they have the same truth values (identical columns).

Example 3.7.3a

Consider the pairs of propositions

i p and $\neg(\neg p)$ ii $\neg p \vee q$ and $p \Rightarrow q$

a By comparing the truth values convince yourself that they are logically equivalent.

b In both cases construct the truth table for the equivalence to exhibit the appropriate last column.

For example, in **i** the last column should be $p \Leftrightarrow \neg(\neg p)$.

Solution

a i

p	p	$\neg(\neg p)$
T	F	T
F	T	F

ii

p	q	$\neg p$	$\neg p \vee q$	$p \Rightarrow q$
T	T	F	T	T
T	F	F	F	F
F	T	T	T	T
F	F	T	T	T

In both cases the columns for the propositions in question are the same so they are logically equivalent.

b i

p	p	$\neg(\neg p)$	$p \Rightarrow \neg(\neg p)$	$\neg(\neg p) \Rightarrow p$	$p \Rightarrow \neg(\neg p)$
T	F	T	T	T	T
F	T	F	T	T	T

ii The table here is shortened by using the last two columns from **a ii** as a starting point.

$\neg p \vee q$	$p \Rightarrow q$	$(\neg p \Rightarrow \vee q) \Leftrightarrow (p \Rightarrow q)$
T	T	T
F	F	T
T	T	T
T	T	T

Example 3.7.3b

Construct Venn diagrams corresponding to the truth tables for
i implication $p \Rightarrow q$ ii equivalence $p \Leftrightarrow q$.

Solution

i

p	q	$p \Rightarrow q$
T	T	T
T	F	F
F	T	T
F	F	T

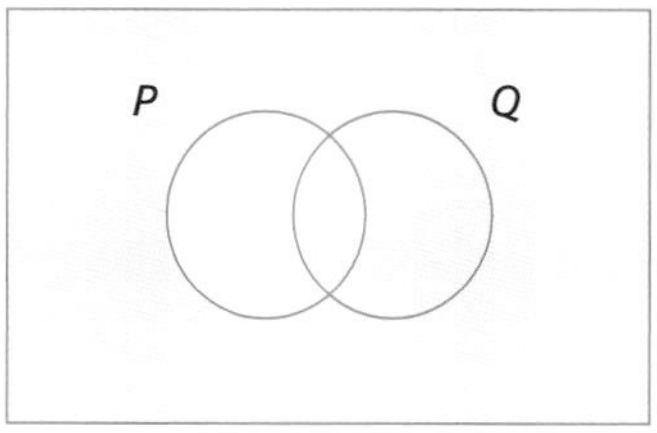

ii

p	q	$p \Rightarrow q$
T	T	T
T	F	F
F	T	F
F	F	T

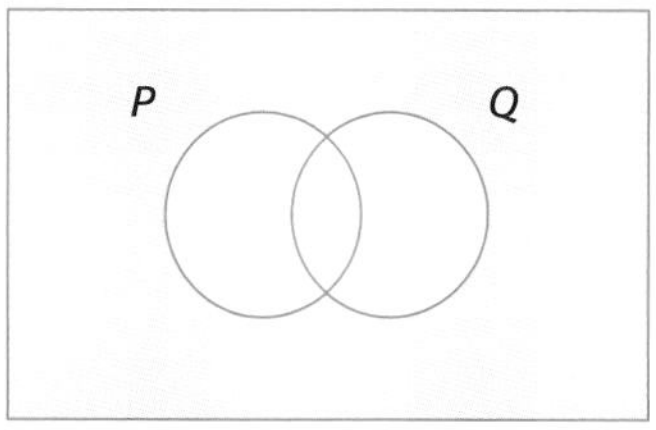

Exercise 3.7

1 Consider p: If I am spending ages on the phone then I am not studying enough.

Write in words, the converse, the inverse and the contrapositive of p.

2 Show that the proposition $((p \Rightarrow q) \wedge (q \Rightarrow r)) \Rightarrow (p \Rightarrow r)$ is a tautology.

Think of a set of three propositions to illustrate this result.

3 Draw a truth table and associated Venn diagram for the proposition $(p \Rightarrow q) \veebar (q \Rightarrow p)$.

Write down an expression involving set operations for the shaded area.

4 Construct the truth table for the proposition $(p \Rightarrow q) \Rightarrow \neg r$.

Draw a Venn diagram representing this proposition.

Use the fact that $p \Rightarrow q$ can also be written as $\neg p \vee q$ along with the result that $\neg(\neg p \vee q)$ is the same as $p \wedge \neg q$ to give an expression for the shaded area of the diagram in terms of set operations.

5 Translate the compound set operation $(S \cup T') \cap [(T \cap R') \cup (S \cap R)]$ to a compound logic proposition. Hence construct a truth table and use the table to help you draw a Venn diagram for this set.

By observing your diagram, simplify the set expression.

6 Decide using a truth table whether $p \Rightarrow p$ is a tautology and whether $p \Rightarrow \neg p$ is a contradiction.

Can you understand what is happening here?

7 Repeat question 6 using equivalence instead of implication.

3.8 Introduction to probability

3.8.1 Equally likely events

Cast your mind back to the discussion about flipping a coin. For a single flip the sample space was the set $\{h, t\}$. Assuming the coin is fair, the results head or tail are equally likely. This means that if we flip several times, about half the results will be heads and half tails. There is a randomness in the flipping and in practice the proportions might not be exactly half of each, but the more flips we make, the closer the ratio will get to a half. We say that the **probability** of the result "head" is $\frac{1}{2}$.

Similarly the probability of result "tail" is also $\frac{1}{2}$.

For a six-sided die, the sample space is $\{1, 2, 3, 4, 5, 6\}$. Again assuming a fair die, all the results are equally likely and there are six possibilities, so we say the probability of each single result is $\frac{1}{6}$.

In general, for n equally likely events, the probability of each individual event is $\frac{1}{n}$.

Did you know that...?

The word "probability" derives from the Latin *probare* (to prove, or to test). Informally, "probable" is one of several words applied to uncertain events or knowledge, being closely related in meaning to "likely, risky, hazardous", and "doubtful." "Chance, odds", and "bet" are other words expressing similar notions.

3.8.2 Probability of an event

Not all events are equally likely. Suppose you have a box of shoes of different colours. The colours are: 7 black, 4 white,

2 red, 6 blue, 5 brown. Put your hand in the box and take a shoe without looking. What is the probability of picking a red shoe?

There are 2 red and a total of 24 shoes in the box. The probability of picking red is $\frac{2}{24} = \frac{1}{12}$.

We turn this into a general formula as follows. The box is a universal set U. Then $n(U) = 24$. The red shoes form a subset $A \subset U$. The probability of picking a member of A is $P(A)$.

Then $P(A) = \frac{n(A)}{n(U)}$

Example 3.8.2

A drawer in my cupboard contains 53 socks. 13 are single socks and 40 are members of a pair. If I pick a sock at random, what is the probability of taking a single sock?

Solution

U is the drawer. $n(U) = 53$. Subset $S \subset U$ is the subset of single socks. $n(S) = 13$.

$$P(S) = \frac{n(S)}{n(U)} = \frac{13}{53}$$

3.8.3 Probability of a complementary event

Andrei Nikolaevich Kolmogorov (1903–1987)

Andrei Nikolaevich Kolmogorov (1903–1987) was a pioneer in giving probability a sound mathematical treatment. He was a professor at Moscow State University and received the honour of "Academician of the USSR Academy of Sciences" in 1939.

Theory of knowledge

"The theory of probability as a mathematical discipline can and should be developed from axioms in exactly the same way as geometry and algebra."

Andrei Nikolaevich Kolmogorov

In the last example, there are 40 socks that are members of a pair. The set of pairs of socks is the complement S' of S.

The probability of picking one of the socks from a pair is $P(S') = \frac{40}{53}$.

Notice that $P(S) + P(S') = \frac{13}{53} + \frac{40}{53} = \frac{13+40}{53} = \frac{53}{53} = 1$.

We often think of this relation re-ordered as

$P(S') = 1 - P(S)$

This is a general formula for all subsets.

Something else useful has emerged from this. Events in S or S' are complementary events. (Either you pick a single sock or you do not. There are no other possibilities.) This means that $S \cup S' = U$.

Since you *have* picked a sock the probability of picking a sock is

$$P(U) = \frac{n(U)}{n(U)} = 1.$$

This is the same 1 as appears on the right in the formula for P(S′) above.

Similarly, the set $S \cap S' = \varnothing$, the empty set.

(You cannot pick a sock that is both single and part of a pair.)

Hence $P(S \cap S') = P(\varnothing) = \frac{n(\varnothing)}{n(U)} = \frac{0}{1} = 0.$

- When an event is **certain** it has a probability of 1.
- When an event is **impossible** it has a probability of 0.

Example 3.8.3a

A normal six-sided die is thrown once. Calculate the probability that the top face is

a a six b not a six c a seven d no number

e a whole number from 1 to 6 inclusive.

Solution

The die is normal so it has faces marked 1 to 6. The sample space for the throw is $T = \{1, 2, 3, 4, 5, 6\}$ with $n(T) = 6$. Each face is equally likely. Hence the probability of each is $\frac{1}{6}$.

a $P(6) = \frac{1}{6}$ b $P(6') = 1 - \frac{1}{6} = \frac{5}{6}$ c 7 is impossible, $P(7) = 0$

d A number will show so P(none) = 0 e $P(1, 2, 3, 4, 5, 6) = P(T) = 1$

Example 3.8.3b

A six-sided die is weighted so that it is twice as likely to fall with 6 facing up than any other number. The die is thrown once. Calculate the probability that the top face is

a 6 b not 6 c 5 or 6.

Solution

Each of the numbers 1 to 5 has an equal probability of appearing. There are 5 of these numbers.

The 6 has twice the probability of the others so it is assigned weight 2 when adding up the probability values. 5 + 2 = 7 so we divide certainty (probability 1) into 7 equal parts and assign 1 part to each of faces 1 to 5 and 2 parts to 6. Hence

a $\frac{2}{7}$ b $\frac{5}{7}$ c $\frac{1+2}{7} = \frac{3}{7}$

Theory of knowledge

"God does not play dice with the universe" is a famous quote from the eminent scientist Albert Einstein. Einstein did not like the developments in atomic theory occurring in the early part of the 20th century, in which nature seemed to be inherently probabilistic (not just that we couldn't measure things exactly). This revolution in thought, known as *Quantum Mechanics* had an enormous impact in the early 20th century, not just on science but also on philosophy, art, poetry, architecture and music. See if you can find out more about this impact in your own area of interest. From the end of 2007, predicted new results from the CERN research centre in Geneva could have an even bigger impact on the way we think about the Universe. Exciting times are ahead. Try an Internet search for "Extra dimensions".

Exercise 3.8

A normal pack of 52 playing cards contains 4 suits, spades (♠), hearts (♥), diamonds (♦) and clubs (♣), consisting of values 2 to 10, jack, queen, king, ace. Each pack has one of each card value.

1 If a card is chosen at random from the pack find the probability that it is

a the queen of hearts b a king c not a diamond
d from a red suit (hearts or diamonds)
e a club but not an ace or a two.

2 Suppose two cards are chosen without replacement. What is the probability that

a the first is an ace and the second a king?
b the first is an ace and the second a king of the same suit?
c if five cards are drawn without replacement, what is the probability of getting a flush (all five in the same suit).

3 A die is made in the form of a tetrahedron, with faces showing a square, a circle, a parabola and a straight line. When thrown, each face is equally likely to fall face down. Calculate the probability of the downward face being

a a circle b not a curved shape
c not a square, a straight line or a circle.

4 A tray of area 100 cm^2 with raised edges is divided into three randomly shaped patches coloured red, blue and green . The areas of the patches are red 32 cm^2, blue 48 cm^2, green 20 cm^2. A very small ball is thrown into the tray and allowed to come to rest. Find the probability that the ball comes to rest

a in the red patch b not in the green patch
c in the red or the green patch.
(Assume that the ball never lies exactly across a boundary.)

Question 4 is not as silly as it seems. Irregular areas can actually be measured this way, usually using a computer. The unknown area is embedded in a known area and a random number generator generates a large number of coordinates of points inside the known area.

The ratio $\dfrac{\text{Number of points in unknown area}}{\text{Total number of points generated}} \times \text{Known area}$ is an estimate of the unknown area.

This is a simple application of a method called the *Monte-Carlo* method used throughout science and engineering.

3.9 Venn and tree diagrams in probability

3.9.1 Venn diagrams

When combinations of events become more complicated we use Venn diagrams to picture the structure and to help with finding probabilities. Consider the following problem.

Example 3.9.1a
Year 12 at a certain IB World School has 38 candidates.

One candidate anticipated mathematics and is not taking the subject in Year 12.

Of the other 37 candidates:

- 17 take mathematical studies (MS)
- 14 take mathematics SL
- 6 take mathematics HL.

A Year 12 candidate is picked at random. Find the probability that this candidate takes mathematical studies.

Solution
The problem is elementary and we could just write down the answer, but we use it to show how to use a Venn diagram. The diagram shown is labelled with the *numbers* of candidates in each set (rather than the actual candidates). There is no intersection of any of the sets.

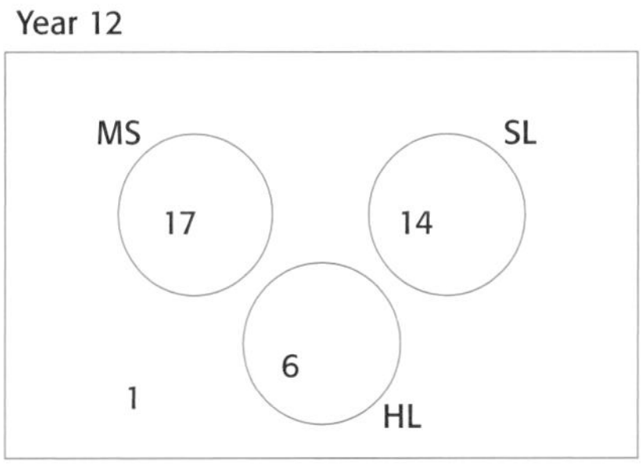

There are 17 members of set MS and 38 candidates in total so the probability is $\frac{17}{38}$.

We do not always have all the information. Sometimes we have to work out some of the numbers to put in the diagram.

Example 3.9.1b
In an IB World School the total number of candidates taking Diploma Programme

- mathematical studies is 22
- German is 16
- French is 21.

Of these

2 are taking all three subjects

9 are taking mathematical studies and French

7 are taking mathematical studies and German

6 are taking German only.

a Fill in this information on a Venn diagram.

b Find how many candidates are taking

  i German and French but not mathematical studies

  ii French only

  iii mathematical studies only.

c If there are 49 IB candidates in this school, calculate how many are doing none of the subjects.

Fill in the rest of the information on the Venn diagram.

d If a candidate is chosen at random from the IB candidates, calculate the probability that this candidate takes i just one of the subjects

ii German or mathematical studies.

e If we are told that this candidate *does* do German, what is the probability that s/he also takes mathematical studies?

Solution

a The 2 can be put in the central intersection and the 6 in the "German only" region immediately. There are 9 candidates taking mathematical studies and French but this includes the 2 already there.
We need to enter $9 - 2 = 7$ in the remaining mathematical and French region.
The 5 is obtained in a similar way.

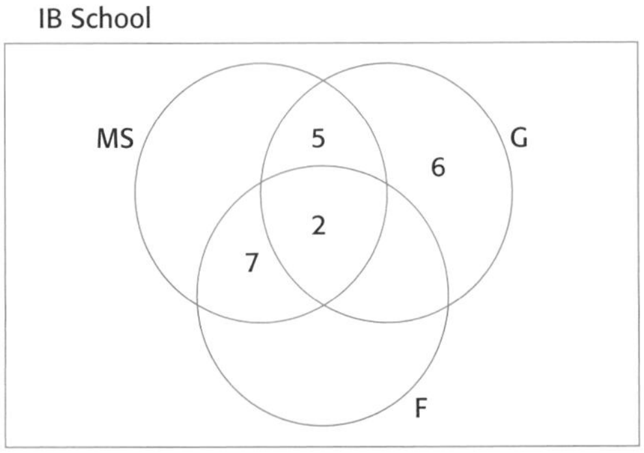

b i The total for German is 16. The missing entry for German and French but not mathematics must be $16 - (2 + 5 + 6) = 3$

ii Now we can also find the number for French only. It is $21 - (7 + 2 + 3) = 9$

iii For mathematics only there are $22 - (5 + 7 + 2) = 8$ candidates

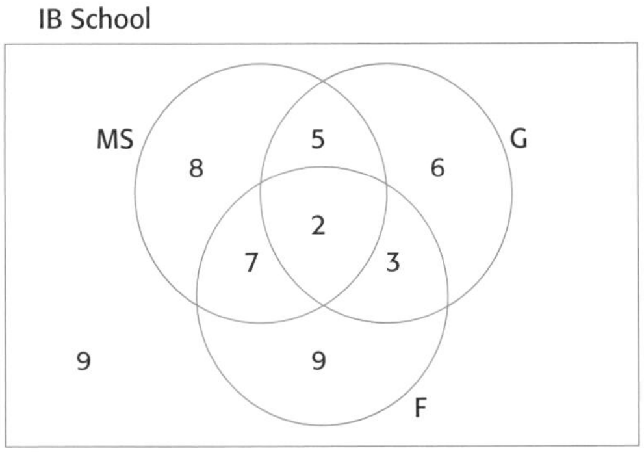

c We now have a total of $8 + 5 + 6 + 7 + 3 + 2 + 9 = 40$ candidates and there are 49 in the school.
The difference is 9, the number of candidates taking none of the three subjects.

d i The total number of candidates is 49. Count those taking only one subject. The answer is $6 + 8 + 9 = 23$. The probability must be $\frac{23}{49}$.

ii The number doing German or mathematical studies is $n(G \cup MS) = 8 + 7 + 2 + 5 + 3 + 6$ (or just $22 + 3 + 6$) $= 31$ so the probability is $\frac{31}{49}$.

e Now we know this candidate is one of the 16 doing German we can ignore other candidates.
Of these 16 there are $n(G \cap MS) = 7$ also doing mathematics. Hence the probability is $\frac{7}{16}$.
This is called a **conditional probability** and will be studied further in a later section.

3.9.2 Tree diagrams

We can represent **alternative** events along with their probabilities using a **tree diagram**.

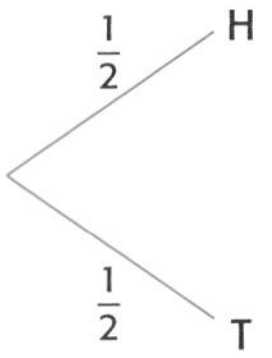

The simplest tree diagram has only two branches. For example the results heads (H) or tails (T) of flipping a fair coin once are shown in this diagram. The probability of each event is written next to the branch for that event.

If the coin is flipped twice, there are four possible outcomes. The tree is extended to a second level of branches, one pair attached to the branch for a head thrown first and one pair to that for a tail first.

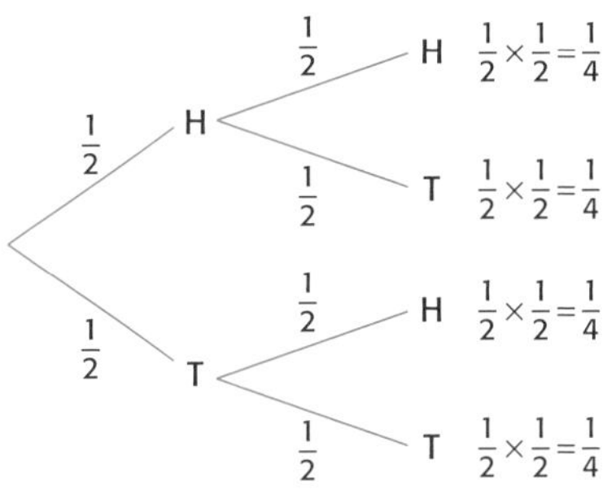

The probability of obtaining the sequence of events along a set of branches in the order represented by those branches is obtained by multiplying all the probabilities along the path taken. Hence the probability of obtaining a head followed by a tail comes from following the top branch first, then taking the lower branch.

The probability of getting this result is $\frac{1}{2} \times \frac{1}{2} = \frac{1}{4}$.

The combined probabilities have been added to the diagram.

Notice that these combined probabilities add up to 1.

$\frac{1}{4} + \frac{1}{4} + \frac{1}{4} + \frac{1}{4} = 1.$

(It is certain that if the coin is flipped twice, one of the possible final results will be what happens.)

Theory of knowledge

Why is order important in some areas of mathematics and not in others?

The probability of obtaining a head and a tail *regardless of order* can also be obtained from the diagram, by adding the two probabilities corresponding to those events: $\frac{1}{4} + \frac{1}{4} = \frac{1}{2}$.

Example 3.9.2

The probability that Jasmine wakes up by 7:30 am is 0.6.

If Jasmine wakes up by 7:30, then the probability that she will catch the school bus is 0.9.

If she sleeps later than 7:30, the probability that she will catch the bus is 0.4.

Draw a tree diagram representing this situation. Calculate the probability that Jasmine will catch the bus on any one day. If Jasmine goes to school on 280 days in the year, how many of these days will Jasmine catch the bus?

Solution

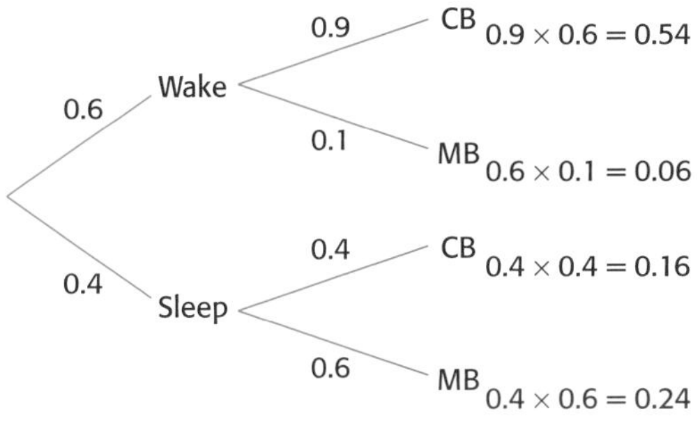

The first branch must represent waking by 7:30.

The probability of doing this is 0.6 so the probability of not doing so is $1 - 0.6 = 0.4$.

(They are complementary events.)

Similarly catching or not catching the bus are complementary. The second layer of branches will represent this event using the probabilities 0.9, 1 – 0.9 = 0.1, 0.4 and 1 – 0.4 = 0.6.

"Catch bus" and "miss bus" are abbreviated to CB and MB.

The probability that Jasmine catches the bus is given by 0.54 + 0.16 = 0.7.

To find the number of days out of a total of 280 we multiply 280 by the probability 0.7 to get $0.7 \times 280 = 196$ days.

Tree diagrams can get cluttered and hard to draw if the number of branches at each step is more than about 3 or if the number of steps is bigger than 3. (Try drawing one for three throws of a die for instance.) In Section 3.10 we look at formulae to help us calculate probabilities when the tree diagram is too cumbersome.

3.9.3 Solution of problems using "with replacement" and "without replacement"

Consider again the coloured shoes used in Section 3.8.2 on page 149. To make the tree diagrams easier we will put the 2 red, 6 blue and 4 white shoes in a separate box. We will pick a shoe at random, then *replace* it in the box and pick again.

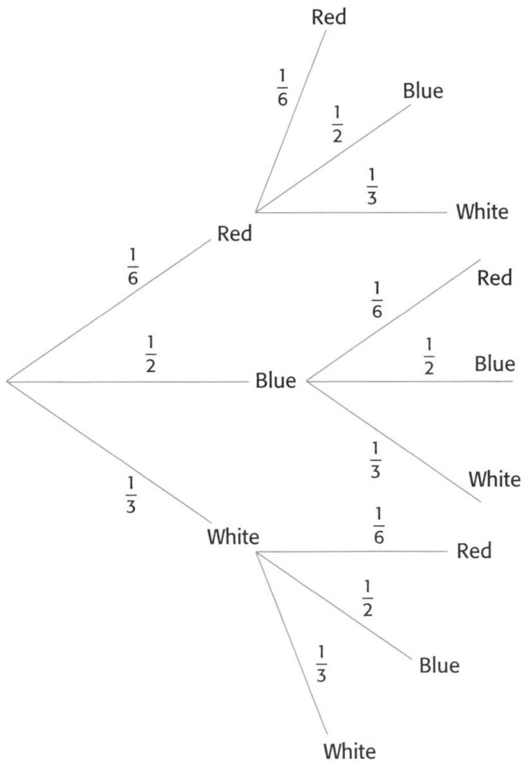

There are 12 shoes in the box for the first selection and because we replaced the first shoe, there are 12 for the second selection too. The probabilities at both stages of the process are

red $\frac{2}{12} = \frac{1}{6}$, blue $\frac{6}{12} = \frac{1}{2}$ and white $\frac{4}{12} = \frac{1}{3}$.

A tree diagram is feasible for this situation.

We can calculate many probabilities based on this diagram.

For example, the probability of getting 2 red shoes is $\frac{1}{6} \times \frac{1}{6} = \frac{1}{36}$. The probability of getting a blue first then a white is $\frac{1}{2} \times \frac{1}{3} = \frac{1}{6}$ while the probability of getting a blue and a white in either order is $\frac{1}{2} \times \frac{1}{3} + \frac{1}{3} \times \frac{1}{2} = \frac{1}{3}$.

Example 3.9.3

Suppose the shoe taken at the first selection is *not* replaced.

Draw a tree diagram for this situation. Calculate the probability of ending up with

a 2 red shoes b a pair of matching shoes

c shoes of different colours.

Solution

A similar diagram can be used even if the first shoe is *not* replaced.

Now at the second selection, there is one less shoe in the box. That makes 11.

We have to use 11 in the denominator for the right-hand branches of the tree.

Also, whichever colour is taken the first time is not replaced so there is one less of that colour to count in the numerator. Here is the tree.

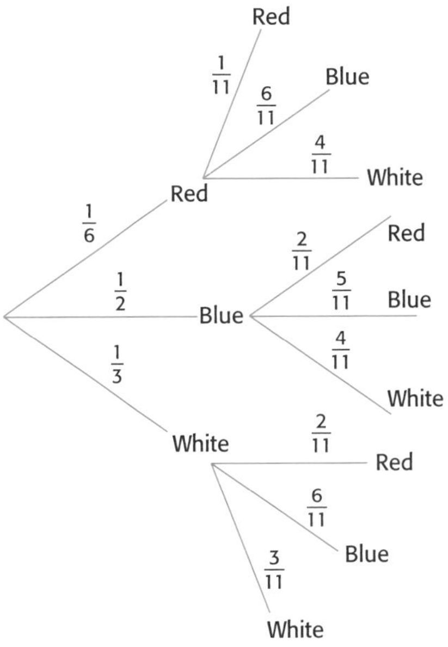

a the probability of getting 2 red shoes is

$$P(\text{two red}) = \frac{1}{6} \times \frac{1}{11} = \frac{1}{66}$$

b Here we need to add the contributions from following the tree for all possible matching pairs.

We get $P(\text{pair}) = \frac{1}{6}\times\frac{1}{11}+\frac{1}{2}\times\frac{5}{11}+\frac{1}{3}\times\frac{3}{11}=\frac{1+15+6}{66}=\frac{22}{66}=\frac{1}{3}$

c This can be done using the complement probability of **b**. "Non-matching shoes" is the complement event of getting a pair. Hence using the result in **b**

$$P(\text{non-matching}) = 1 - P(\text{pair}) = 1-\frac{1}{3}=\frac{2}{3}$$

Exercise 3.9

1 Sebastian will play either Hans or Markus in the tennis finals.

The probability that he will play Hans is 0.72.

If he plays Hans his past record indicates that his probability of winning is 0.35.

If he plays Markus, his probability of winning is 0.63.

a Find the probability that Sebastian will play Markus.
b Draw a tree diagram for the situation.
c Calculate the probability that Sebastian will win the final.

2 There are 78 students in a college music department.

2 students sing in a choir (C) and play in a jazz band (J) but are not in an orchestra (O).

27 students play in a jazz band and an orchestra

15 students do all three activities

48 sing in a choir but

13 sing in a choir *only*

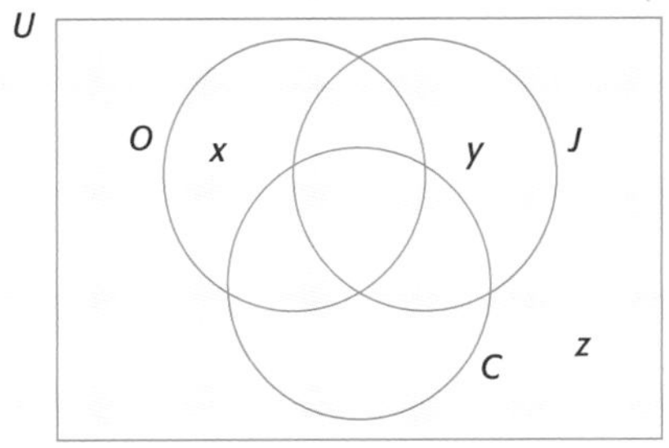

a Find how many students
i sing in a choir and play in an orchestra but not a jazz band
ii play in a jazz band and an orchestra but are not in a choir.

b Copy the Venn diagram and fill in all the information you have found so far.

c A total of 73 students do at least one of the activities and also there are 18 who do none or exactly one of the activities, but are not in the jazz band.
Write down three equations for x, y and z.

d Find x, y and z. How many students do exactly one of the activities?

e A student is chosen at random from the college.
Find the probability that this student
i does all three activities
ii sings in a choir or plays in both a jazz band and an orchestra.

f Given that a student belongs to $J \cup C$, find the probability that she also belongs to O.

3 Julie is deciding whether to visit Andrew, Martha or George.
The probability that she will visit Andrew is 0.7 and the probability that she will visit Martha is 0.2.
Julie will take her friend either to the cinema or to eat in a café. If Julie visits Andrew, the probability that they will go to the cinema is 0.6 but if she visits Martha it is 0.3. If she visits George it is equally likely that they will go to a cinema or a café.
a Draw a tree diagram for this situation and fill in all the probabilities.
b Find the probability that Julie will
i visit George and go to the cinema
ii eat in a café
iii eat in a café if she does not visit Andrew.

4 The tree diagram represents the probability that Eliza will work on her Spanish homework tonight, depending on whether it rains or not. The probabilities of events are shown on the diagram.

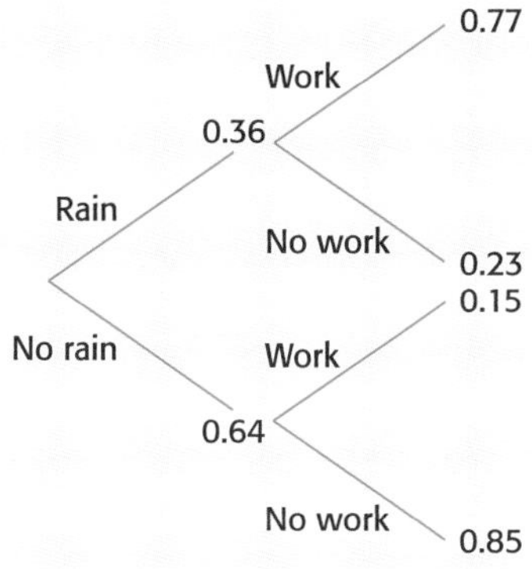

a Write down the probability that it will rain.
b Calculate the probability that Eliza will *not* do the homework.
c Find the probability that it will not rain and Eliza will do the work or that it will rain and she will not do the work.

3.10 Laws of probability, types of events

3.10.1 Combined events: $P(A \cup B) = P(A) + P(B) - P(A \cap B)$

In order to prove the above equation we will use two pieces of information that we have found earlier in this chapter. Firstly, when looking at Venn diagrams we saw that

$n(A \cup B) = n(A) + n(B) - n(A \cap B)$ (See Example 3.2.1f on page 130.)

then when looking at probabilities we found that $P(A) = \dfrac{n(A)}{n(U)}$.

Putting both of these together we get:

$$\frac{n(A \cup B)}{n(U)} = \frac{n(A)}{n(U)} + \frac{n(B)}{n(U)} - \frac{n(A \cap B)}{n(U)}$$

$$\Rightarrow P(A \cup B) = P(A) + P(B) - P(A \cap B)$$

Example 3.10.1a
In a class of 30 students, 15 own a bicycle, 10 own a car and 2 own both. Find the probability that a randomly chosen student owns either a bicycle or a car.

Solution

$$\text{P(bicycle)} = \text{P}(B) = \frac{15}{30} = \frac{1}{2}$$

$$\text{P(car)} = \text{P}(C) = \frac{10}{30} = \frac{1}{3}$$

$$\text{P}(B \cap C) = \frac{2}{30} = \frac{1}{15}$$

Therefore, $\text{P}(B \cup C) = \text{P}(B) + \text{P}(C) - \text{P}(B \cap C) = \frac{1}{2} + \frac{1}{3} - \frac{1}{15} = \frac{23}{30}$

Example 3.10.1b
In a group of 80 children the probability that a child has blonde hair is 0.23 and the probability that a child has blue eyes is 0.48. Given that the probability that a child has either blonde hair or blue eyes is 0.63, find the probability that a child has both blonde hair and blue eyes.

Solution
Let P(blonde hair) = $\text{P}(A) = 0.23$, P(blue eyes) = $\text{P}(B) = 0.48$ and $\text{P}(A \cup B) = 0.63$, then

$0.63 = 0.23 + 0.48 - \text{P}(A \cap B)$

So, $\text{P}(A \cap \text{B}) = 0.23 + 0.48 - 0.63 = 0.08$

Example 3.10.1c
In an Indian restaurant the probability that a customer orders nan is 0.68, the probability that he orders both nan and rice is 0.32 and the probability that he orders neither is 0.13. Find the probability that the customer orders only rice.

Solution
Let P(nan) = $\text{P}(A) = 0.68$ and P(rice) = $\text{P}(B)$.

P(nan and rice) = $\text{P}(A \cap B) = 0.32$

If the probability that a customer orders neither dish is 0.13, then the probability that a customer orders either one or the other is 0.87, that is, $\text{P}(A \cup B) = 0.87$

So, $0.87 = 0.68 + \text{P}(B) - 0.32$

$\Rightarrow \text{P}(B) = 0.87 - 0.68 + 0.32 = 0.51$

3.10.2 Mutually exclusive events $\text{P}(A \cup B) = \text{P}(A) + \text{P}(B)$

Mutually exclusive events are events which cannot happen together. For instance, when tossing a coin you cannot get a head *and* a tail or when choosing a card from a pack of playing cards you cannot choose a card that is red *and* black.

In mutually exclusive events $P(A \cup B) = P(A) + P(B)$

Example 3.10.2a

Furcan is an excellent marathon athlete. The probability that he will come in first place in the school marathon is 0.56. The probability that he comes in second place is 0.38. Calculate the probability that Furcan finishes the marathon in either first or second place.

Solution

Let $\text{P}(A)$ = probability of Furcan winning the marathon = 0.56

$\text{P}(B)$ = probability of second place = 0.38

Obviously, Furcan cannot be both first and second, therefore $\text{P}(A \cap B) = 0$ so, $\text{P}(A \cup B) = 0.56 + 0.38 = 0.94$

Example 3.10.2b

Albie has a pack of 52 playing cards. He chooses a card at random.

$\text{P}(A)$ = the probability that the card is a diamond.

$\text{P}(B)$ = the probability that the card is a black card.

$\text{P}(C)$ = the probability that the card is a court card (king, queen or jack)

a Find $\text{P}(A)$, $\text{P}(B)$ and $\text{P}(C)$.

b Are A and B mutually exclusive?

c Are A and C mutually exclusive?

Solution

a $\text{P}(A) = \frac{13}{52} = \frac{1}{4}$ $\quad \text{P}(B) = \frac{26}{52} = \frac{1}{2}$ $\quad \text{P}(C) = \frac{12}{52} = \frac{3}{13}$

b Since diamonds are red cards, $\text{P}(A \cap B) = 0$, therefore A and B are mutually exclusive.

c Since there are three court cards that are diamonds,

$\text{P}(A \cap C) = \frac{3}{52} \neq 0$, therefore A and C are not mutually exclusive.

Example 3.10.2c

You are given that $\text{P}(A) = 0.35$, $\text{P}(B) = 0.28$ and $\text{P}(A \cap B) = 0.62$. Are the events A and B mutually exclusive?

Solution

$\text{P}(A) + \text{P}(B) = 0.35 + 0.28 = 0.63$

$0.63 \neq 0.62$

Therefore $\text{P}(A \cap B) \neq 0$

So the events A and B are *not* mutually exclusive.

3.10.3 Independent events $P(A \cap B) = P(A) \times P(B)$

Independent events are events that are not dependent on each other but where there is a probability that they could still occur. For instance, the probability that a student's surname is Lee and the probability that a student has a brother named Bruce are two unconnected events and yet it is possible that a student could have the surname Lee and have a brother called Bruce. These are known as independent events.

For independent events $P(A \cap B) = P(A) \times P(B)$

Example 3.10.3a

The probability that I will be late for school on any one day is 0.05. The probability that I will eat rice for lunch is 0.23. Given that the two events are independent, find

a the probability that I am late for school and eat rice for lunch

b the probability that I am late for school or eat rice for lunch.

Solution

a Let $P(A)$ = probability of being late = 0.05 and

$P(B)$ = probability of eating rice = 0.23.

Since these two events are independent,

$P(A \cap B) = P(A) \times P(B) = 0.05 \times 0.23 = 0.0115$

b $P(A \cup B) = P(A) + P(B) - P(A) \times P(B)$

$= 0.05 + 0.23 - 0.0115$

$= 0.2685$

Example 3.10.3b

Let $U = \{1, 2, 3, 4, 5, 6, 7, 8, 9, 10, 11, 12, 13, 14, 15, 16, 17, 18, 19, 20\}$

A is the set of multiples of 3 contained in U and B is the set of multiples of 4 contained in U,

a Write down the elements of sets A and B.

b If a number is chosen at random, find the probability that

i it belongs to set A

ii it belongs to set B

iii it belongs to both A and B.

c Are sets A and B independent?

Solution

a $A = \{3, 6, 9, 12, 15, 18\}$, $B = \{4, 8, 12, 16, 20\}$

b i $P(A) = \frac{6}{20} = \frac{3}{10}$

ii $P(B) = \frac{5}{20} = \frac{1}{4}$

iii $P(A \cap B) = \frac{1}{20}$

c $\quad P(A) \times P(B) = \frac{3}{10} \times \frac{1}{4} = \frac{3}{40} \neq \frac{1}{20}$

So, A and B are *not* independent.

Example 3.10.3c

We have two sets A and B.
We are given that $P(A) = 0.44$, $P(B) = 0.25$ and $P(A \cup B) = 0.58$.

a Find $P(A \cap B)$.

b Are A and B independent? Give a reason for your answer.

c Are A and B mutually exclusive? Give a reason for your answer.

Solution

a Using the formula $P(A \cup B) = P(A) + P(B) - P(A \cap B)$, we get

$$0.58 = 0.44 + 0.25 - P(A \cap B)$$

$$P(A \cap B) = 0.44 + 0.25 - 0.58 = 0.11$$

b $P(A) \times P(B) = 0.44 \times 0.25 = 0.11 = P(A \cap B)$
So, A and B are independent since $P(A \cap B) = P(A) \times P(B)$.

c $P(A \cap B) \neq 0$, therefore A and B are not mutually exclusive.

3.10.4 Conditional probability $P(A|B) = \frac{P(A|B)}{P(B)}$

Conditional probability is when the probability of one event happening depends on another event previously having taken place.

$P(A|B)$ means "the probability of event A occurring, *knowing that* event B has occurred".

We use the formula $P(A|B) = \frac{P(A \cap B)}{P(B)}$

This can be demonstrated using a tree diagram where

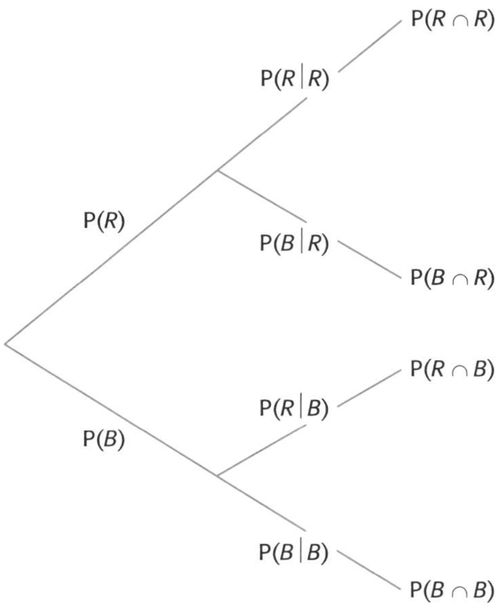

$P(R)$ is the probability of choosing a red item,
$P(B)$ is the probability of choosing a blue item,
$P(R | R)$ is the probability of choosing a red item, *knowing that* you have already chosen a red item, and $P(R \cap R)$ is the probability of choosing two red items.

From the tree diagram we see that:

$P(R \cap R) = P(R) \times P(R | R)$

So, $P(R | R) = \frac{P(R \cap R)}{P(R)}$

Also, $P(B \cap R) = P(R) \times P(B | R)$

So, $P(B | R) = \frac{P(B \cap R)}{P(R)}$

Similarly for the other two branches.

Example 3.10.4a

In a class of 30 students, 12 like to play golf, 17 like to play hockey and 3 play neither game.

Find **a** the probability that a student chosen at random likes to play golf

b the probability that a student chosen at random likes to play both golf and hockey

c the probability that a student chosen at random likes to play hockey knowing that the student likes to play golf.

Solution

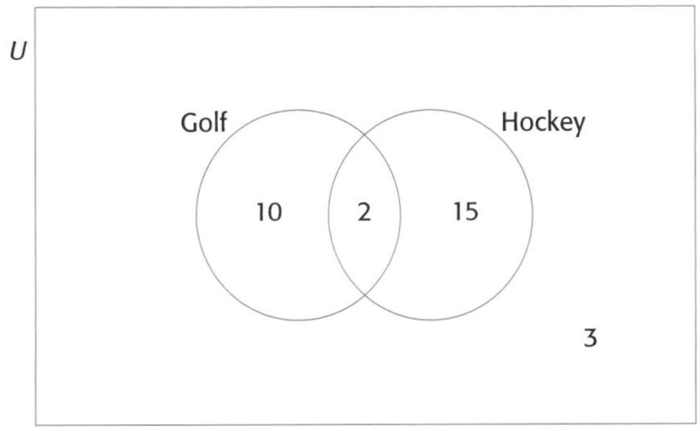

a $\text{P(golf)} = \frac{12}{30} = \frac{2}{5}$

b $\text{P(golf} \cap \text{hockey)} = \frac{2}{30} = \frac{1}{15}$

c $\text{P(hockey | golf)} = \frac{\text{P(hockey} \cap \text{golf)}}{\text{P(golf)}} = \frac{\frac{1}{15}}{\frac{2}{5}} = \frac{1}{6}$

Example 3.10.4b

The probability that Janie passes a mathematics test is 0.73. The probability that she passes a French test is 0.65. The probability that she passes neither is 0.08. She has a mathematics and French test on the same day. Her teacher told her that she has passed the French test. What is the probability that she has also passed the mathematics test?

Solution

Let $\text{P}(M) = 0.73$, $\text{P}(F) = 0.65$ and $\text{P}(M \cup F)' = 0.08$.

$\text{P}(M \cup F) = 1 - 0.08 = 0.92$

Using the formula $\text{P}(M \cup F) = \text{P}(M) + \text{P}(F) - \text{P}(M \cap F)$, we get

$0.92 = 0.73 + 0.65 - \text{P}(M \cap F)$

So, $\text{P}(M \cap F) = 0.73 + 0.65 - 0.92 = 0.46$

Therefore, the probability that Janie passes her mathematics test knowing that she has passed her French test is

$$\text{P}(M|F) = \frac{P(M \cap F)}{P(F)} = \frac{0.46}{0.65} = 0.708$$

Example 3.10.4c

The probability that Alison has a soft boiled egg for breakfast is 0.54. The probability that she travels to school by car is 0.86 if she has had a soft boiled egg for breakfast and 0.63 if she has not.

a Find the probability that Alison travels to school by car.

b If Alison travels to school by car, find the probability that she had a soft boiled egg that morning.

Solution

a $$\begin{aligned}\text{P(car)} &= \text{P(egg)} \times \text{P(car)} + \text{P(not egg)} \times \text{P(car)} \\ &= 0.54 \times 0.86 + 0.46 \times 0.63 \\ &= 0.7542\end{aligned}$$

b $$\begin{aligned}\text{P(egg} \mid \text{car)} &= \frac{P(\text{egg} \cap \text{car})}{P(\text{car})} \\ &= \frac{0.54 \times 0.86}{0.7542} \\ &= 0.616\end{aligned}$$

Exercise 3.10

1 The probability of having coffee, C, for breakfast is 0.89. The probability of eating toast, T, for breakfast is 0.69. The probability of having both toast and coffee is 0.73. Find the probability of having either coffee or toast or both for breakfast.

2 Given that $P(A) = 0.19$, $P(B) = 0.54$ and $P(A \cup B) = 0.41$, find $P(A \cap B)$.

3 The universal set in the adjacent diagram contains 60 items.

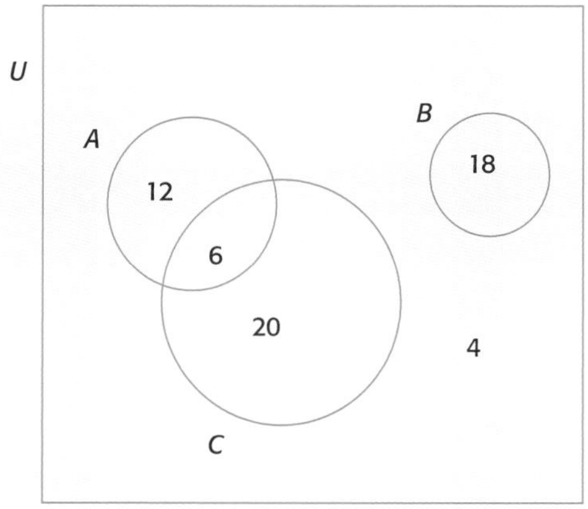

a Find $P(A)$, $P(B)$ and $P(C)$.

b Write down the sets that are mutually exclusive and give a reason for your answer.

4 The probability that the sun shines is 0.87. The probability that I am late for school is 0.06. If the two events are independent, find the probability that the sun shines and I am late for school.

5 A wheel has numbers 1–20. An arrow is placed in the centre of the wheel so that it can spin freely. The arrow is spun once. What is the probability that the arrow lands on

a an even number, E

b a multiple of 6, M

c a factor of 12, F

d a prime number, P?

Consider the events E and F.

e Are they independent, mutually exclusive or neither?

Consider the events M and P.

f Are they independent, mutually exclusive or neither?

6 A bag contains 6 green discs and 5 red discs. Two discs are chosen, at random and without replacement, from the bag.
Find the probability that

a the first disc is red and the second is green
b both discs are the same colour
c both discs are green, given that they are the same colour.

7 The probability that a newborn kitten is female is 0.56. If the kitten is female, the probability that it is tabby is 0.78. If the kitten is male, the probability that it is tabby is 0.39.
Find the probability that
a a kitten chosen at random is tabby
b a kitten is male, knowing that it is *not* a tabby.

8 In a group of 100 students, the probability of being smaller than average is 0.32. If the student is smaller than average then the probability of being overweight is 0.43 and if not, then the probability of being overweight is 0.38.
Find the probability that
a a student chosen at random is overweight
b a student chosen at random is not small, given that the student is overweight.

9 The table shows the results for 80 students in some school subjects.

	Mathematics	**English**	**History**	**Totals**
Grade 4–7	26	24	13	63
Grade 1–3	12	3	2	17
Totals	38	27	15	80

a Find the probability that a student chosen at random has a grade 4–7 in history.
b Find the probability that a student took the mathematics exam.
c Given that a student took the mathematics exam, find the probability that the student obtained a grade 1–3.

10 In a pack of 52 playing cards, let A be the event of choosing an ace, B the event of choosing a black card, C the event of choosing a club and D the event of choosing a diamond.

a Write down $P(A)$, $P(B)$, $P(C)$ and $P(D)$.
b Find $P(A|B)$
c Find $P(C|B)$
d Write down any events that are independent, giving a reason for your answer.
e Write down any events that are mutually exclusive, giving a reason for your answer.

Past examination questions for topic 3

Paper 1

1 The Venn diagram shows the universal set of real numbers $\mathbb{R}$ and some of its important subsets:

$\mathbb{Q}$: the rational numbers
$\mathbb{Z}$: the integers
$\mathbb{N}$: the natural numbers.

Write the following numbers in the correct position in the diagram.

$-1,\ 1,\ \pi,\ \frac{7}{16},\ 3.333\dot{3},\ \sqrt{3}$.

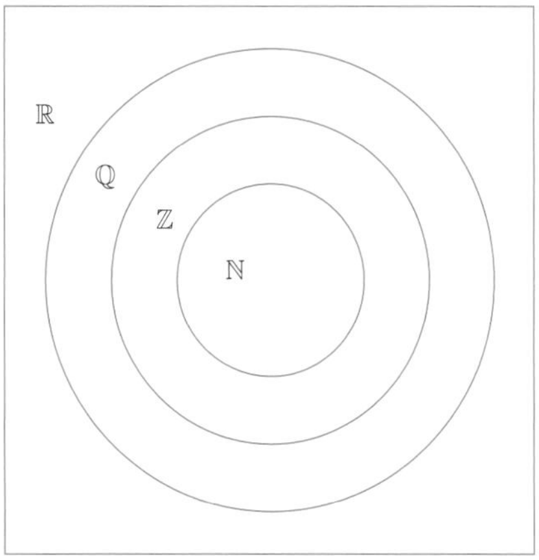

M06q1

2 Given the statements
p: The sun is shining
q: I am wearing my hat

a Write down, in words, the meaning of $q \Rightarrow \neg p$.

b Complete the truth table

p	q	$\neg q$	$q \Rightarrow \neg p$
T	T		
T	F		
F	T		
F	F		

c Write in symbols, the converse of $q \Rightarrow \neg p$.

M06q4

3 In a group of fifteen students, three names begin with the letter B and four begin with a G. The remaining eight names begin with A, C ,D, E, F ,H, I and J respectively.
The fifteen names are placed in a box. The box is shaken and two names are drawn out.

Find the probability that

a both names begin with any letter except G or B
b both names begin with the same letter
c both names begin with the letter H. *M06q8*

4 On a certain game show, contestants spin a wheel to win a prize, as shown in the diagram. The larger angles are 40° and the smaller angles are 20°.

Find the probability that a contestant
a will not win a prize
b will win a holiday in Greece (GH)
c will win a washer/dryer (WD), given that he knows that he has won a prize
d will win a holiday to Greece *or* a washer/dryer. *M00q10*

5 Events A and B have probabilities $P(A) = 0.4$, $P(B) = 0.65$ and $P(A \cup B) = 0.85$.

a Calculate $P(A \cap B)$.
b State with a reason whether events A and B are independent.
c State with a reason whether events A and B are mutually exclusive. *Specimen Paperq13*

6 Two propositions p and q are defined as follows:
p: the number ends in zero
q: the number is divisible by 5.

a Write in words
  i $p \Rightarrow q$
  ii the converse of $(p \Rightarrow q)$.
b Write in symbolic form
  i the inverse of $(p \Rightarrow q)$
  ii the contrapositive of $(p \Rightarrow q)$. *M01q12*

Paper 2

1 Let U be the set of all positive integers from 1 to 21 inclusive.

A, B and C are subsets of U such that:

A contains all the positive integers that are factors of 21
B is the set of multiples of 7 contained in U
C is the set of odd numbers contained in U.

a List all the members of set A.

b Write down all the members of

i $A \cup B$

ii $C' \cap B$.

Find the probability that a member chosen at random from A is also a member of $A \cap B \cap C$. *M06q2(i)*

2 A school jazz band contains three different musical instruments – saxophone (S), clarinet (C) and drums (D). Students in the band are able to play one, two or three different instruments. In a class of 40 IB students, 25 belong to the jazz band. Out of these 25

3 can play all three instruments
5 can play the saxophone and clarinet *only*
5 can play *at least* the clarinet and the drums
7 can play *at least* the saxophone and drums
16 can play the saxophone
12 can play the clarinet.

a Draw a Venn diagram and clearly indicate the number of students in each region.

b Show that the number of students who play drums *only* is 5.

c Find the probability that a student chosen at random from the IB class plays only the saxophone.

d Fine the probability that a student chosen at random from the IB class plays either the clarinet or drums or both.

e Given that a student plays the saxophone, find the probability that he also plays the clarinet. *N04q1*

3 On a particular day 100 children are asked to make a note of what they drank that day. They are given three choices, water (W), coffee (C) or fruit juice (F).

1 child drank only water.
6 children drank only coffee.
8 children drank only fruit juice.
5 children drank all three.
7 children drank water and coffee only.
53 children drank coffee and fruit juice only.
18 children drank water and fruit juice only.

a Represent this information on a Venn diagram.
b How many children drank none of the drinks listed?
c A child is chosen at random. Find the probability that the child drank
 i coffee
 ii water or fruit juice but not coffee
 iii no fruit juice, given that the child did drink water.
d Two children are chosen at random. Find the probability that both children drank all three drinks. *N03q2*

4 Let F be the set of all families that have exactly two children. Assuming P(boy) = P(girl) = 0.5, find the probability that a family chosen at random from F has exactly
 a two boys
 b two boys, if it is known that the first child is a boy
 c two boys, if it is known that there is a boy in the family. *N01q1(ii)*

Theory of knowledge

Applications of probability theory to everyday life

Two major applications of probability theory in everyday life are in risk assessment and in trade on commodity markets. Governments typically apply probability methods in environmental regulation where it is called "pathway analysis", and are often measuring well-being using methods that are stochastic in nature, and choosing projects to undertake based on statistical analyses of their probable effect on the population as a whole. It is not correct to say that statistics are involved in the modelling itself, as typically the assessments of risk are one-time and thus require more fundamental probability models, for example "the probability of another 9/11". A law of small numbers tends to apply to all such choices and perception of the effect of such choices, which makes probability measure a political matter.

A good example is the effect of the perceived probability of any widespread Middle East conflict on oil prices – which have ripple effects in the economy as a whole. An assessment by a commodity trade that a war is more likely versus less likely sends prices up or down, and signals other traders of that opinion. Accordingly, the probabilities are not assessed independently nor necessarily very rationally. The theory of behavioural finance emerged to describe the effect of such groupthink on pricing, on policy, and on peace and conflict.

Who decides when to "take a risk"?

How sure of the outcome do they have to be?

Who pays the price when it fails?

Do we really learn from our mistakes? Look at examples from economics and history.

TOPIC 4

Functions

4.1 Concept of a function as a mapping

4.1.1 Mappings

A **mapping** consists of two sets and a rule for assigning to each element in the first set one or more elements in the second set.
We say that A is **mapped** to B and write this as $m: A \to B$.

Consider these two sets. S is the set of students and G is the set of possible grades for a mathematics test. Each student is awarded a particular grade for the test.

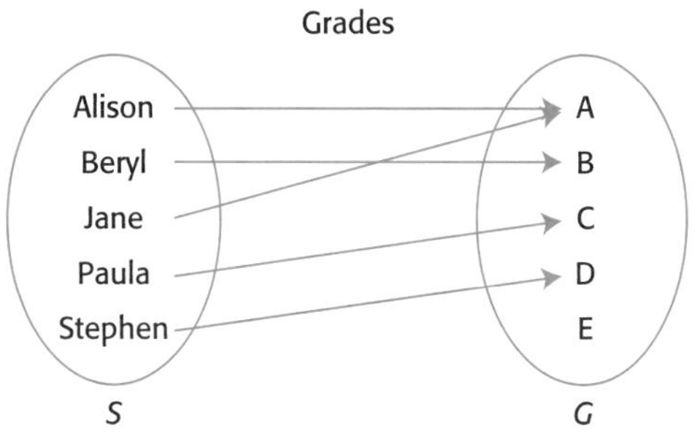

In this example, each element in set S is assigned to an element in set G. It is not possible for a student to be awarded more than one grade for the test but note that more than one student can be awarded the same grade.

4.1.2 Domain and range

A mapping consists of two sets. The first set is called the **domain** and the second set is called the **codomain**. An element in the domain is mapped to its **image** in the codomain.

If $a \in A$ and $b \in B$ and $m: A \to B$, we say that a is mapped onto b and b is the image of a. In the example, "Grades" above, the image of Paula is C.

Every element in the domain has an image but not every element in the codomain is an image of an element in the domain. The elements in the codomain that are images are a subset of the codomain. This subset is called the **image** set. The image set is often called the **range** of a mapping.

Example 4.1.2

Let $A = \{1, 2, 3\}$ and $B = \{2, 3, 4, 5, 6\}$, and let $m: A \to B$ be defined by the rule $a \to 2a$. Draw the mapping and identify the domain, the codomain, the image set and the range.

Solution

The domain is {1, 2, 3}.

The codomain is {2, 3, 4, 5, 6}.

The image set is {2, 4, 6}.

The range is {2, 4, 6}.

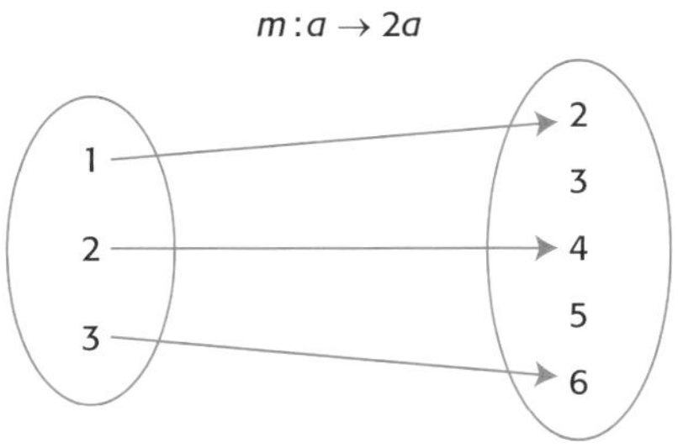

4.1.3 Functions

A **function** from set A to set B is a mapping where each element in the domain is assigned to one, and only one, element in the codomain.

The many-to-one and the one-to-one mappings are examples of functions.

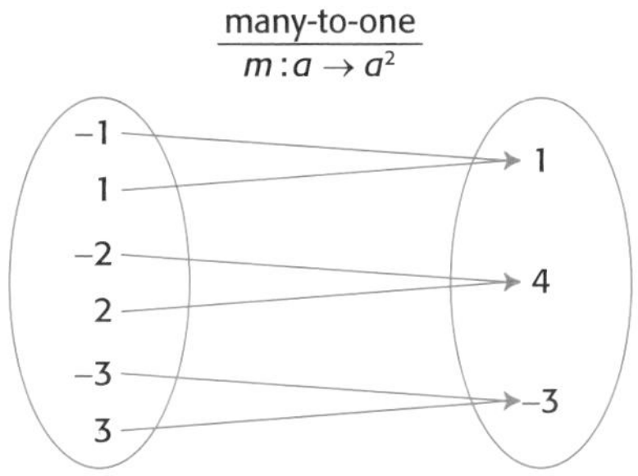

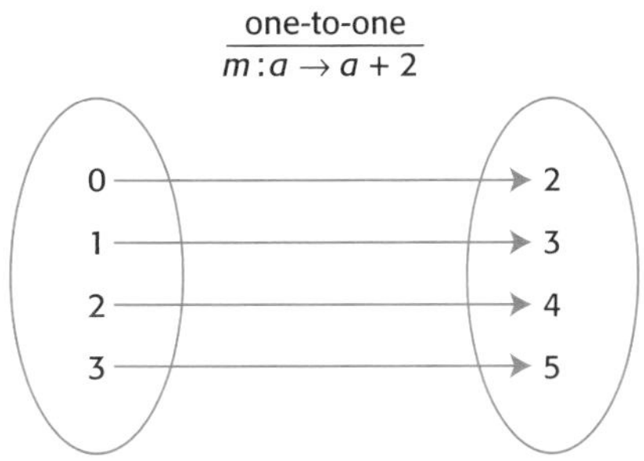

There are several ways to state that a function f maps an element, a, in the domain to an element, b, in the codomain:

- f maps a into b
- the value of f at a is b
- the image of a under f is b

Example 4.1.3a

Let the domain of a function f be $A = \{1, 2, 3\}$ and the codomain be $B = \{2, 3, 4, 5, 6, 7, 8\}$.

Let the mapping from A to B be defined by $f : x \to 3x - 1$. Show this mapping on a diagram and identify its type. State the domain, codomain and the range.

Solution

Type: the image of 1 is 2

the image of 2 is 5

the image of 3 is 8

∴ it is a one-to-one mapping

The domain is {1, 2, 3}.

The codomain is {2, 3, 4, 5, 6, 7, 8}.

The range is {2, 5, 8}.

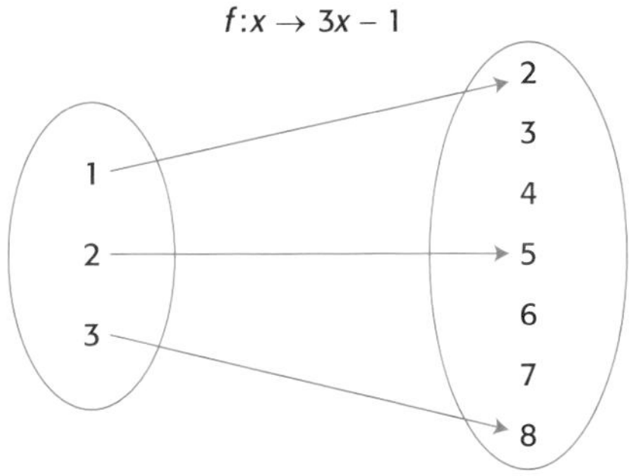

Notation

Functions involving real numbers use a particular notation. The following examples are in common use.

$f : x \rightarrow x^2$ ($x \in \mathbb{R}$) which means that f is a function that maps any real number x into x^2.

Or, alternatively, $f(x) = x^2 \quad (x \in \mathbb{R})$.

A function does not always need to be called f. We can use any letter.

Finding the values in the range

Example 4.1.3b

If $g(x) = x^2 + 2x - 3$, find $g(-1)$, $g(0)$ and $g(3)$.

Solution

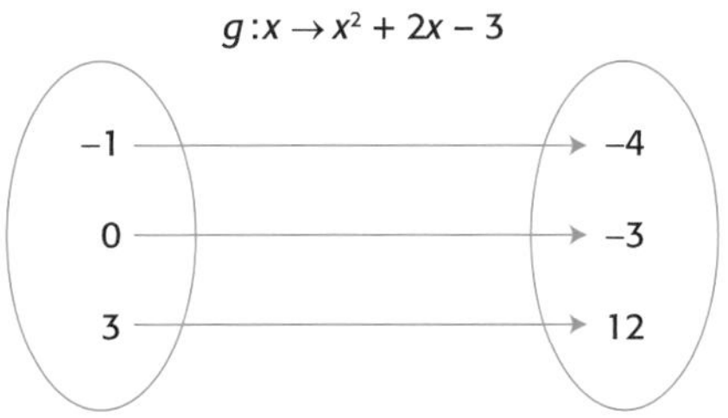

To find $g(-1)$, we substitute -1 for x.

Hence $g(-1) = (-1)^2 + 2(-1) - 3 = -4$

And similarly, $g(0) = (0)^2 + 2(0) - 3 = -3$ and $g(3) = (3)^2 + 2(3) - 3 = 12$.

So, $g(x) = x^2 + 2x - 3$ can also be written as $y = x^2 + 2x - 3$.

The letter y is often used to represent the elements in the range.

4.1.4 Graphs of functions

A pair (x, y), where the first element is in the domain and the second is in the range, is known as an **ordered pair**.

Consider the following:

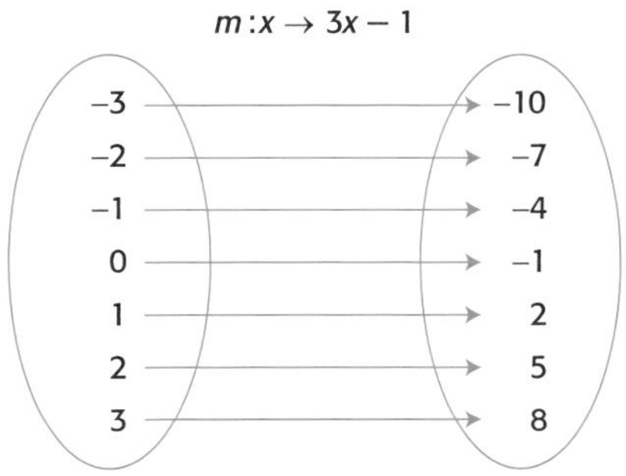

This is a one-to-one mapping, that is, it is a function. If the domain of the function involves all real numbers between -3 and $+3$ it can be written as

$f(x) = 3x - 1, \ -3 \leq x \leq 3, \ x \in \mathbb{R}$

The ordered pairs (x, y) can be written in the form of a table.

x	−3	−2	−1	0	1	2	3
$f(x)$	−10	−7	−4	−1	2	5	8

We can plot the ordered pairs on a set of axes and draw a line through the points. The line indicates that x can take all numeric values between −3 to +3 ($x \in \mathbb{R}$).

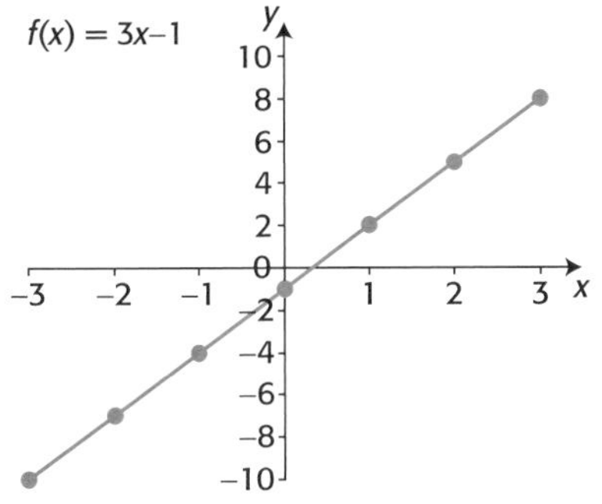

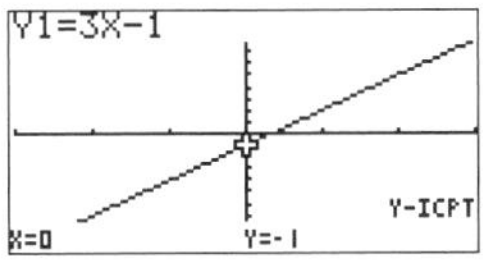

Unless stated otherwise, $x \in \mathbb{R}$.

If the domain of a function is stipulated, the line is often drawn as follows:

a For an inclusive domain, for example, $a \le x \le b$, the line is drawn within the given domain with closed circles at the end of the line.

b For an exclusive domain, for example, $a < x < b$, the line is drawn within the given domain with open circles at end of line.

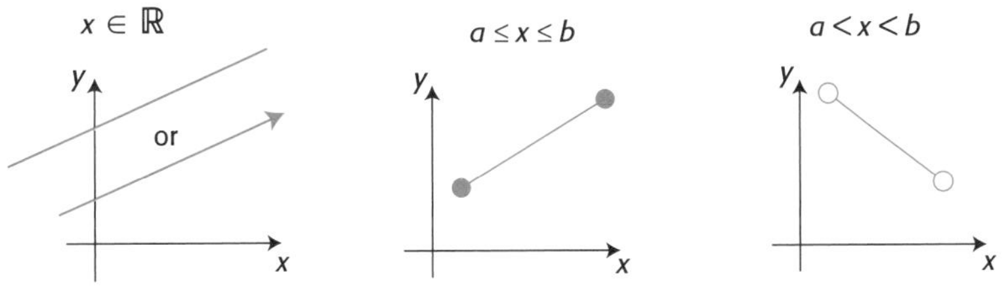

In practice, we often omit the closed circles for an inclusive domain. The line is simply drawn to the end points of the given domain.

The vertical line test: Consider the graphs above. If we draw a vertical line through any point on the x-axis, it will cross the graph exactly once. This means that for each x-value there is one, and only one, y-value. This corresponds to the definition of a function, that is, each element in the domain (x-value) maps onto only one element in the range (y-value). The vertical line test can be used to determine if a relation (mapping) is a function.

Functions are generally defined on the set of real numbers, $\mathbb{R}$, but they can also be defined on sets other than $\mathbb{R}$, in particular, the set of natural numbers, $\mathbb{N}$.

Example 4.1.4a

A sequence is defined by $t_n = 3 + 4(n - 1)$, $n \in \mathbb{N}$, $n > 0$. Plot the graph of t_n for $1 \leq n \leq 6$.

Solution

n	1	2	3	4	5	6
t_n	3	7	11	15	19	23

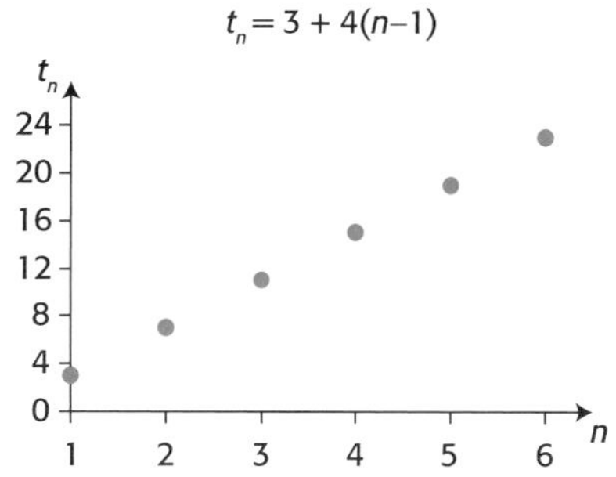

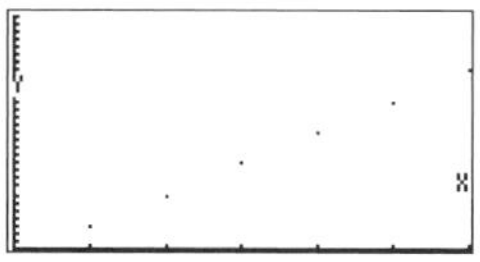

The vertical line test shows that $t_n = 3 + 4(n - 1)$ is a function. A line does not join the points because the function is defined for natural numbers only ($n \in \mathbb{N}$).

Graphs of functions may also be curved.

Example 4.1.4b

Draw the graph of $g(x) = x^2$, $-3 \leq x \leq 3$.

Solution

We first construct a table of values for $y = x^2$, $-3 \leq x \leq 3$.

x	−3	−2	−1	0	1	2	3
y	9	4	1	0	1	4	9

Then we plot these points and draw a smooth curve through them.

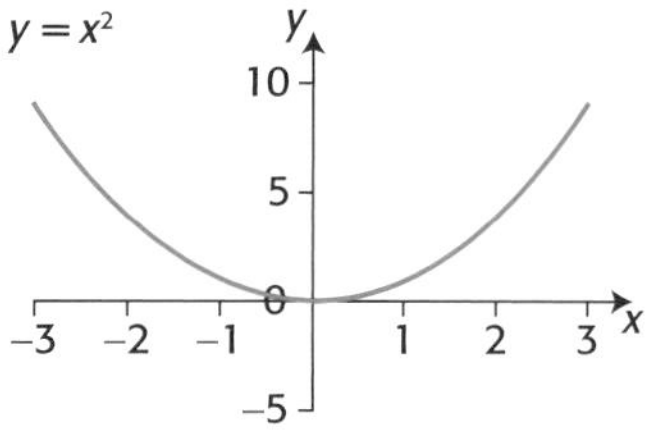

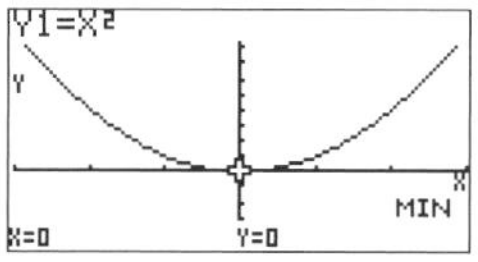

Theory of knowledge

What do you "see" or "think" when you look at a graph?
Do you see a useful drawing or a pretty picture?
Do you admire the shape and symmetry?
Do you question its accuracy?
Will you check that it is correct?
Do you wonder if it will be helpful?
Do you think about the person who drew it?
Was it a man or woman? Tall or small?
What is their nationality? What colour of eyes do they have?
If you daydream about the person rather than the accuracy and usefulness of the graph, does that make you any less mathematical?

Exercise 4.1

1 List the elements in the domain, codomain and the range for each of these mappings. State which mappings are functions.

a

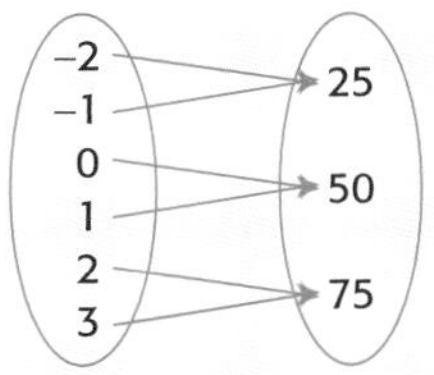

b

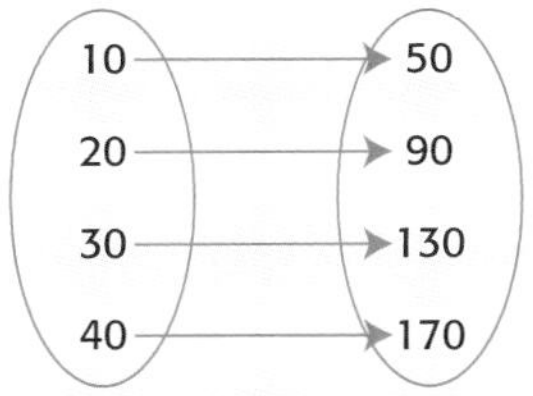

c

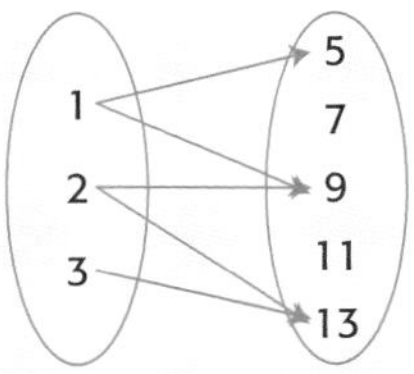

2 Let $A = \{-2, -1, 0, 1, 2\}$ and $B = \{1, 2, 3, 4, 5, 6, 7, 8, 9\}$. Draw the diagrams of these mappings for A into B.

a $m: a \to a^2$ b $m: a \to 3a - 1$ c $m: a \to 5 - a$

3 Find the images of −3, 0 and 3 under these mappings.

a $m: x \to 5x - 2$ b $f(x) = 3x^2 - 2$ c $g: x \to 3 - 2x$

4 Write each of these using function notation.

a f assigns to each real number two less than its double.
b w assigns to each real number 3 more than the square of the number.
c g assigns to each real number 3 times the cube of the number.

5 Find $f(-2)$, $f(0)$, $f(2)$ for each of these mappings.

a $f(x) = 7x - 5$ b $f(x) = -x^2 + 1$ c $f(x) = \sqrt{(x-1)^2}$
d $f(x) = \dfrac{1}{x}$ e $f(x) = x^2 + 3x - 10$ f $f(x) = 10 - 3x - 2x^2$

6 Write down the domain and range for each of these functions.

a

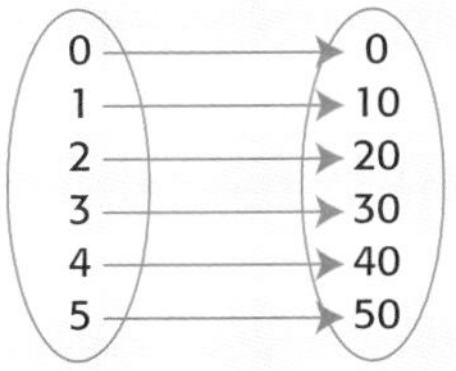

b

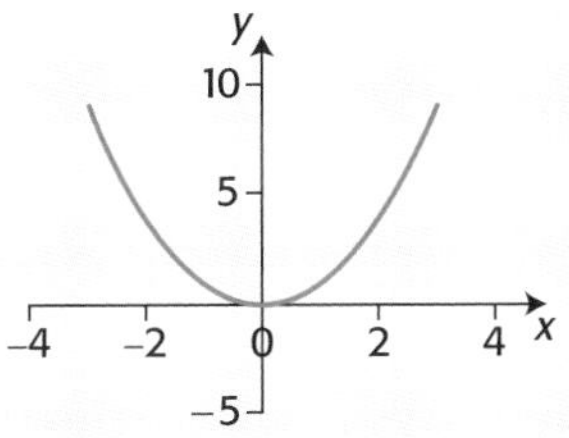

c

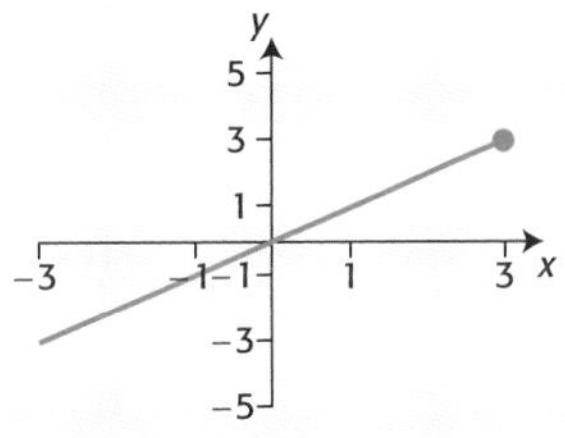

d $f(x) = -2x^2 - 1, -3 \le x \le 3$ e $f(x) = \sqrt{x}$

7 Construct a table of values for each of these functions, $-2 \le x \le 2$.

a $f(x) = 2x - 3$ b $g(x) = 2x^2 + 1$ c $S(x) = x(x - 1)$

8 Draw the graphs of these functions, $-3 \le x \le 3, x \in \mathbb{R}$.

a $m: x \to 3x + 1$ b $g(x) = 5 - x$ c $y = (1 - x)^2$

4.2 Linear functions and their graphs

Linear functions are described by $f: x \to mx + c$ or $f(x) = mx + c$.

They have an input variable (usually x) that is raised only to the first degree. First degree means that the highest exponent in the function is 1.

The graph of the function $f(x) = 3x - 1$, drawn in the previous section, shows that the ordered pairs (x, y) form a straight line. Graphs of functions of the first degree are always straight lines and hence the term *linear*.

4.2.1 Evaluating linear functions

To evaluate a function, we substitute the given value for the input variable.

Example 4.2.1a

Given $f(x) = 2x + 3$,
find $f(25)$

Solution

$f(25) = 2(25) + 3 = 53$

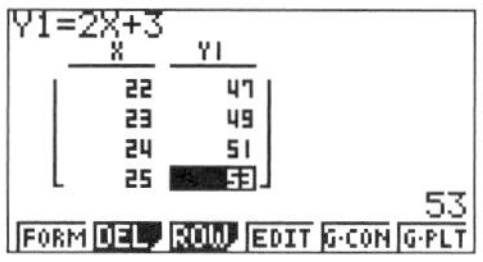

Example 4.2.1b

The total cost (\$) of producing x hundred pairs of jeans is given by $C(x) = 5000 + 22x$

Determine

a the cost of producing 2000 pairs of jeans

b the average cost per pair if 2000 pairs are produced.

Solution

a Total cost = 5000 + 22(2000) = \$49 000

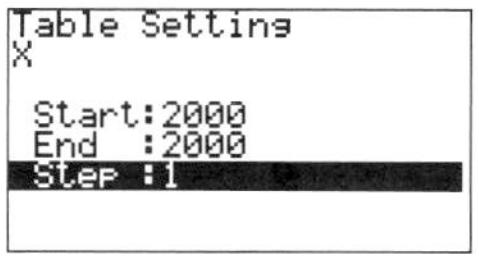

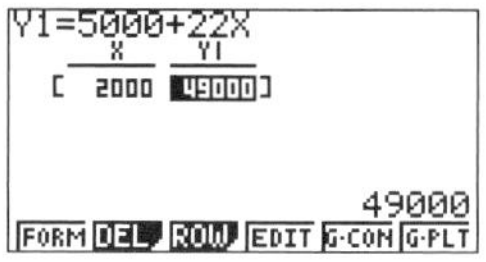

b Average cost per pair 49 000 ÷ 2000
= \$24.50

4.2.2 Finding the value of the input variable

To find the value of the input variable for a given value of the function, we equate the function to the given value.

Example 4.2.2a

Given $f : x \to 5x - 3$,
find x when $f(x) = 62$

Solution

$5x - 3 = 62$

$5x = 65$

$x = 13$

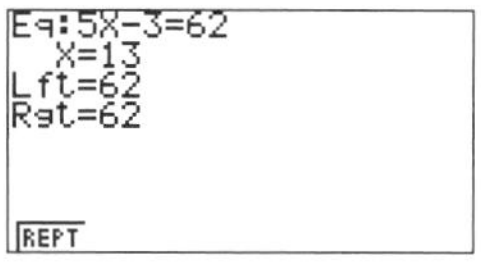

Example 4.2.2b

The normal air temperature T_H (°C), t thousand metres above sea level, is given by the function $T_H = T_{SL} - 2.9t$, where T_{SL} is the temperature at sea level.

If the temperature at the top of a mountain, 2600 m above sea level, is 14 °C, determine the temperature in a nearby coastal town.

Solution

$14 = T_{SL} - 2.9(2.6)$

$T_{SL} = 14 + 7.54 = 21.54$

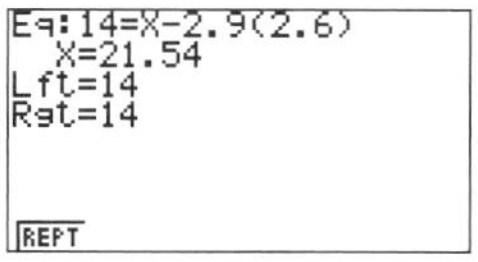

The temperature in the coastal town is 21.5 °C.

> *Theory of knowledge*
>
> *Consider the words "degree" and "root".*
> *Do they have only one meaning in mathematics?*
> *Do they have different meanings in other areas of knowledge?*
> *Can you think of other words that have a different meaning in other areas of knowledge (for example "normal")?*

4.2.3 Finding the function from a given table of values

To find the function, $f(x) = mx + c$, we first find the gradient m. Then we substitute an ordered pair $(x, f(x))$ to find the value of the constant c.

Example 4.2.3a

Find the function $f(x)$ described by this table of values.

x	−2	−1	0	1	2
f(x)	−7	−4	−1	2	5

→ → → →
+3 +3 +3 +3

Solution

There is a constant increase of 3 units in the value of the function for every one unit increase of the input variable, so the gradient is $+\frac{3}{1}$.

∴ The function has the form $f(x) = 3x + c$.

When $x = 0$, $3(0) + c = -1$.
Hence $c = -1$.
The function is $f(x) = 3x - 1$

Example 4.2.3b

These data represent the quantity of goods, $Q(p)$, purchased by consumers at certain prices, $\$p$. Find the demand function, $Q(p)$, for these goods.

$p	1.00	1.10	1.20	1.30	1.40
Q(p)	70	61	52	43	34

→ → → →
− 9 − 9 − 9 − 9

Solution

There is a constant decrease of 9 units purchased for every \$0.10 rise in the price. The gradient is

$$\frac{-9}{0.10} = \frac{-90}{1}$$

so the function has the form $Q(p) = -90p + c$.
When $x = 1$, $-90(1) + c = 70$
$\therefore\ c = 160$

Demand function is $Q(p) = -90p + 160$

Hint: the equation of a straight line can be found using linear regression. We place the x-values in one list and the corresponding y-values in a separate list.

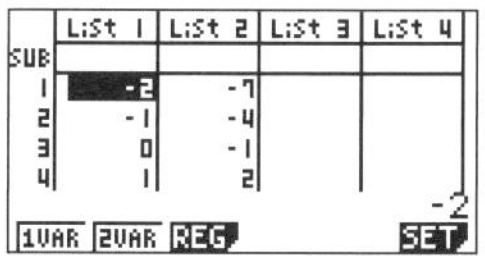

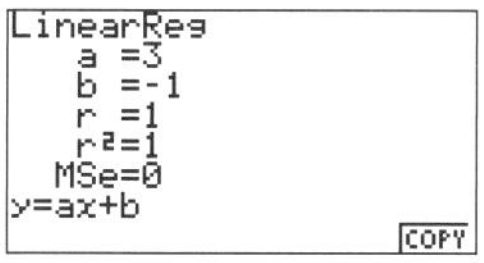

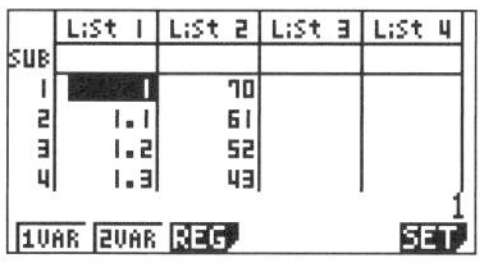

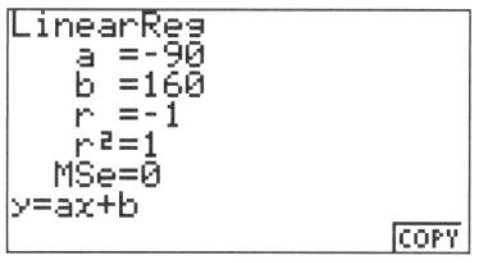

4.2.4 Common points

Linear functions which have different x-coefficients (that is different gradients) will have one point in common. To find this point, we equate the functions and solve for the input variable. To find the corresponding value of the function, we substitute the input variable into one of the functions.

(See also Section 2.7 – simultaneous equations)

Example 4.2.4a

Given $f(x) = 5x + 17$ and $g(x) = 3x - 5$, find the point common to both.

Example 4.2.4b

The demand function for a child's toy is given by $D(p) = 20\,000 - 150p$, where p is the price of the toy (in pesos). The supply function for the child's toy is $S(p) = 12\,000 + 100p$.

Solution

$f(x) = g(x) \Rightarrow 5x + 17 = 3x - 5$

$2x = -22$

$x = -11$

Substitute for x in $f(x)$

$f(-11) = 5(-11) + 17 = -38$

The common point is $(-11, -38)$

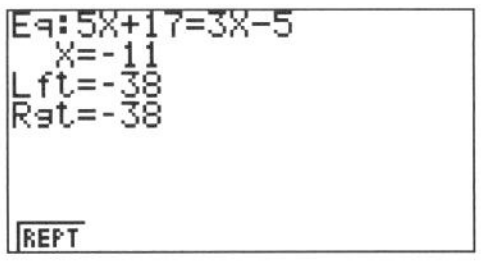

Determine the equilibrium price and quantity for the child's toy.

Solution

$D(p) = S(p) \Rightarrow 20000 - 150p = 12\,000 + 100p$

$8000 = 250p$

$p = 32$

Substitute for p in $D(p)$ $20\,000 - 150(32) = 15\,200$

The equilibrium quantity is 15 200 toys when the price is 32 pesos.

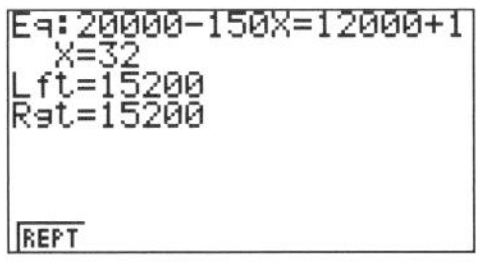

4.2.5 Graphs of linear functions

Linear functions can be represented graphically on the Cartesian plane by straight lines. The equation of a straight line is usually written in its gradient-intercept form, $y = mx + c$, where m is the gradient of the line and c is the y-intercept. Unless stated otherwise, $x \in \mathbb{R}$.

(See also Topic 5 – coordinate geometry)

Example 4.2.5a

Consider the function $f(x) = 3x - 1$. Using a table of values, draw the graph of $y = f(x)$ for $-2 \leq x \leq 2$.

Solution

x	−2	0	2
f(x)	−7	−1	5

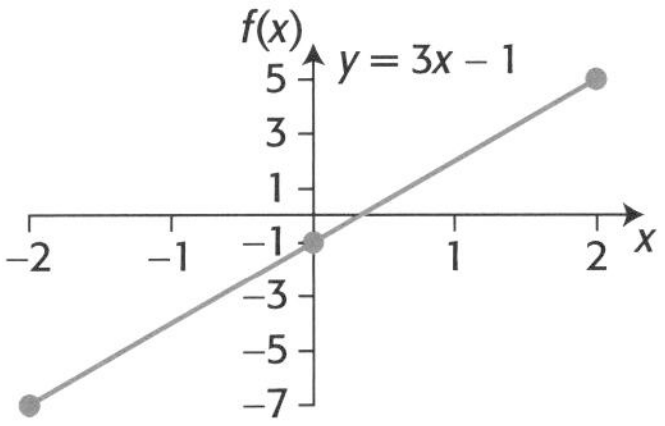

Example 4.2.5b

A local rock band charges a flat fee of €200 plus €75 per hour to play.

a Write a function in the form $C(t) = mt + c$ to express the cost of hiring the band to play for t hours.

b Using a table of values, draw the graph of $y = C(t)$, $0 \leq t \leq 8$.

c Use the graph to estimate the cost of hiring the band to play for $5\frac{3}{4}$ hours.

Solution

a $C(t) = 75t + 200$

t	0	4	8
C(t)	200	500	800

b

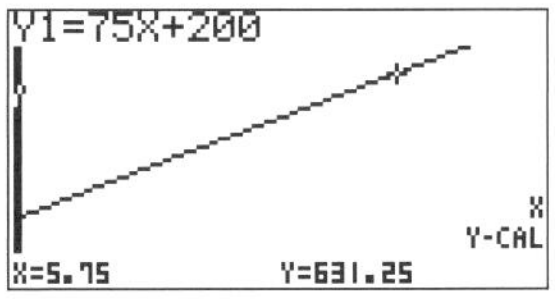

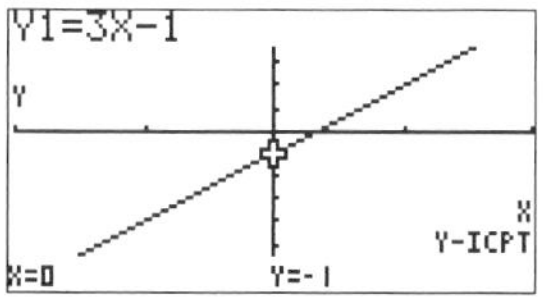

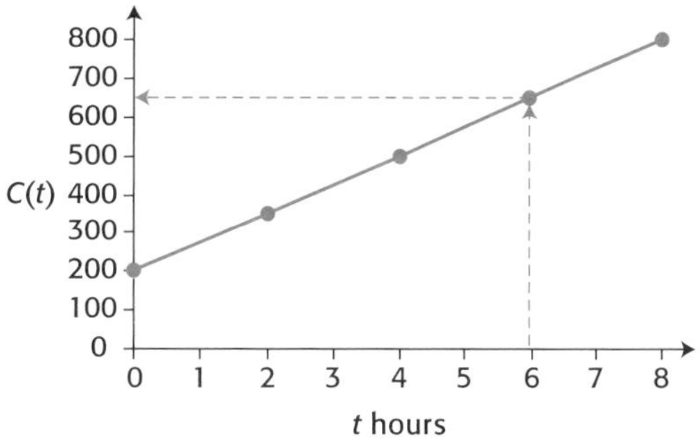

c Cost of hire is ≈ €650

4.2.6 Intersecting lines

Example 4.2.6

Where do the graphs of $f(x) = x + 4$ and $g(x) = 3x + 2$ meet?

(See also Section 2.7 and Topic 5, page 250)

Solution

Draw the graphs of $y = f(x)$ and $y = g(x)$ on the same axes. Draw (dotted) lines on the graph from the point of intersection to the axes.Write down the x- and y-coordinates of the point of intersection.

x	−2	0	2
f(x)	2	4	6

x	−2	0	2
g(x)	−4	2	8

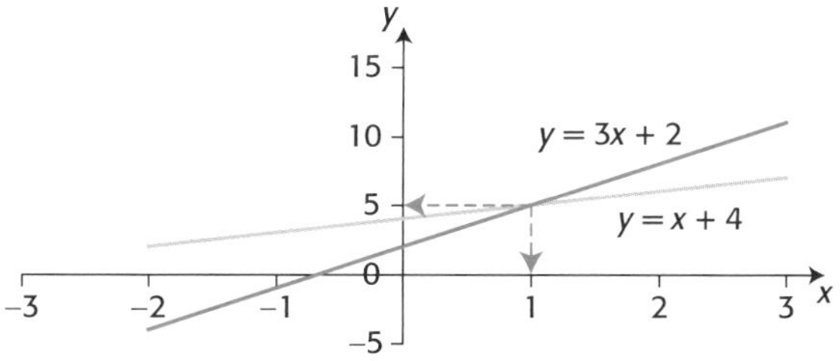

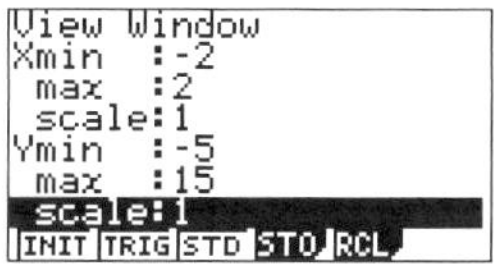

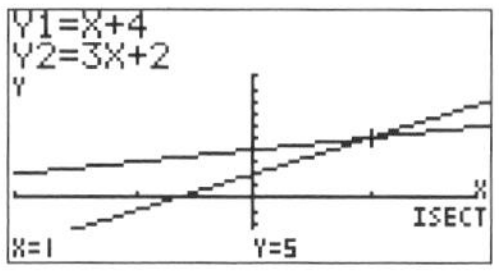

The coordinates of the point of intersection are $(1, 5)$.

The algebraic solution is: $f(x) = g(x) \Rightarrow 3x + 2 = x + 4$

$$2x = 2$$

$$x = 1$$

Substitute for $x = 1$ in $f(x)$, that is find $f(1)$

$$f(1) = 3(1) + 2$$

$$= 5$$

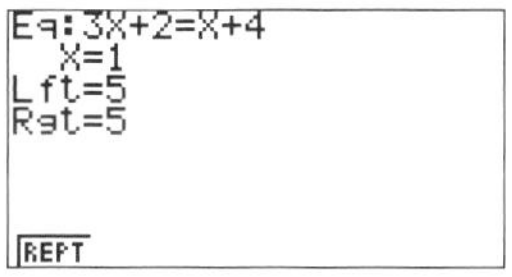

The point of intersection is $(1, 5)$.

Theory of knowledge

Linear functions with different gradients have a single point in common. What can be said about lines that either have no point in common or have an infinite number of common points?

4.2.7 Further applications of linear functions

Example 4.2.7a

Temperature conversion graphs

This formula converts the temperature from °C to °F.

$$F = \frac{9}{5}C + 32$$

a Draw the graph of F, $0 \leq C \leq 100$

b Use the graph to determine the temperature in °F when it is 30 °C.

c Use the graph to change 200 °F to °C.

Solution

a

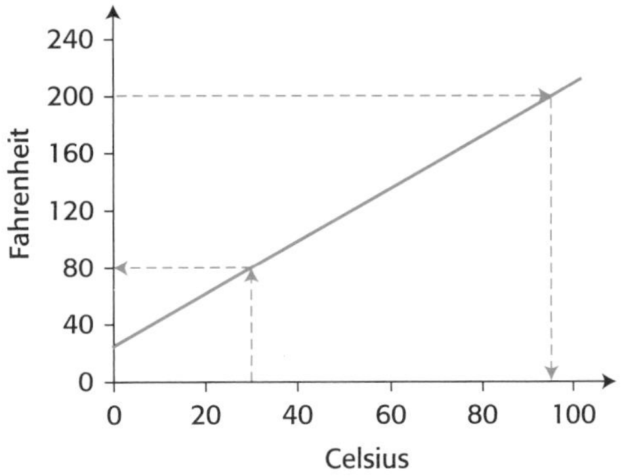

b 30 °C ~ 85 °F c 200 °F ~ 95 °C

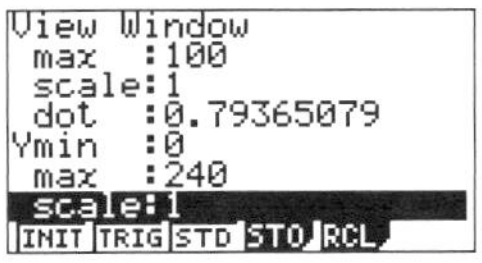

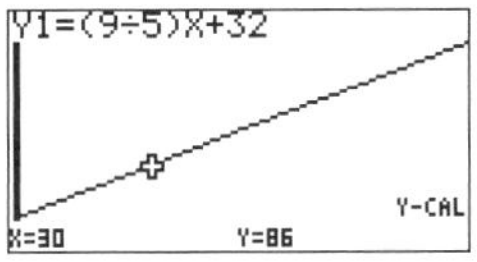

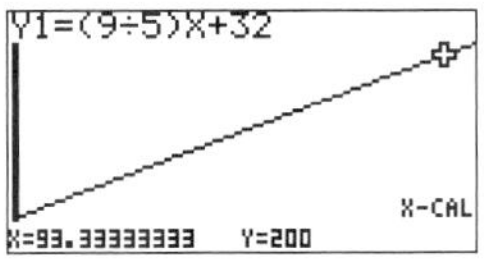

The algebraic solutions are:

b $F = \frac{9}{5}(30) + 32 = 86\,°\text{F}$

c $200 = \frac{9}{5}C + 32 \Rightarrow C = \frac{5}{9}(200 - 32)$

$= 93.3\,°\text{C}$

Example 4.2.7b

Demand/supply functions

The demand and supply functions for a particular item are given by

$D(p) = 14\,000 - 140p$ and $S(p) = 210p$, where p is the selling price ($) of the item.

Using both algebraic and graphical methods, find:

a the equilibrium price for this item

b the excess demand for this item when the selling price is $25

c the excess supply of this item when it sells for $45.

Solution

Algebraic: a $D(p) = S(p)$ when $14\,000 - 140p = 210p$

$$350p = 14\,000 \qquad p = \$40$$

b Excess demand is $D(p) - S(p)$

$$(14\,000 - 140 \times 25) - (210 \times 25) = 5250 \text{ units}$$

c Excess supply is $S(p) - D(p)$

$$(210 \times 45) - (14\,000 - 140 \times 45) = 1750 \text{ units}$$

Graphical:

\$p	0	20	50
D(p)	14000	11200	7000

\$p	0	20	50
S(p)	0	4200	10500

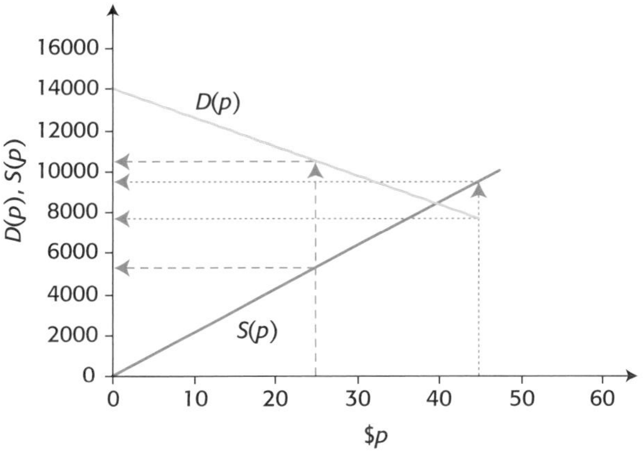

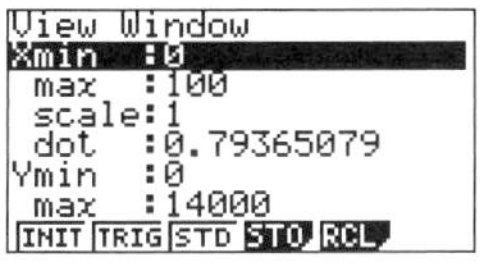

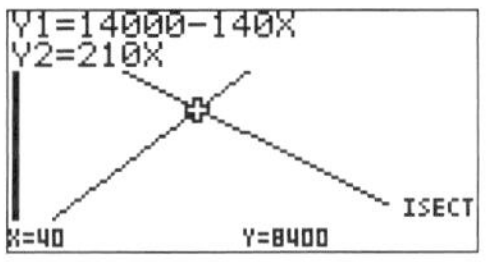

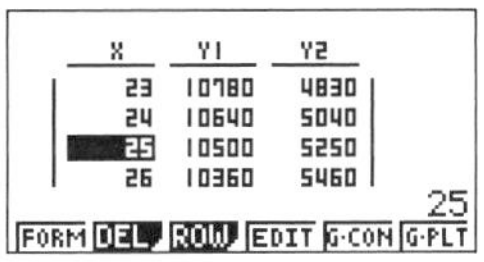

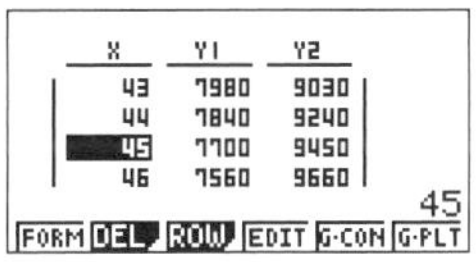

a Equilibrium price is \$40.

b Excess demand at \$25 is 10 500 – 5250 = 5250 units

c Excess supply at \$45 is 9450 – 7700 = 1750 units

Example 4.2.7c

Revenue/cost functions

A local firm produces and sells globbies. Each globbie sells for \$3. The cost of producing x globbies is given by the function $C(x) = 250 + 2x$.

a Determine the revenue function, $R(x)$.

b Draw the graphs of the cost and revenue functions for $0 \leq x \leq 500$.

c Use your graph to find

i the "break even point" (where revenue = cost) for the firm.

ii the profit made when 400 globbies are produced and sold.

Solution

a The revenue function is $R(x) = 3x$

b

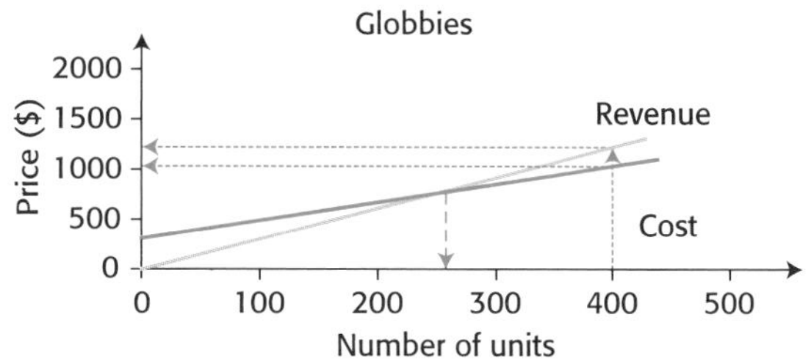

c i The firm breaks even when 250 globbies are produced and sold.

ii Profit = Revenue – Cost = 1200 – 1050 ≈ \$150

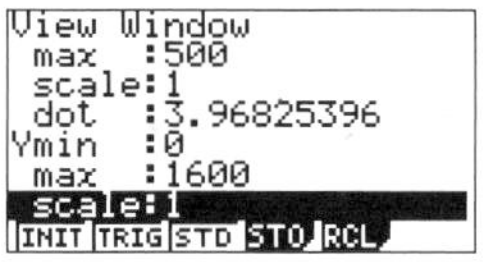

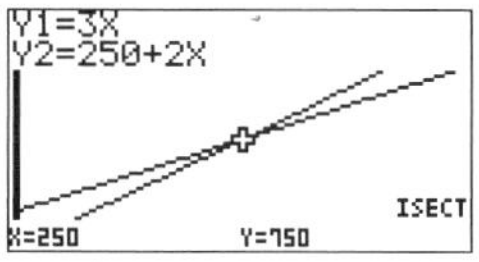

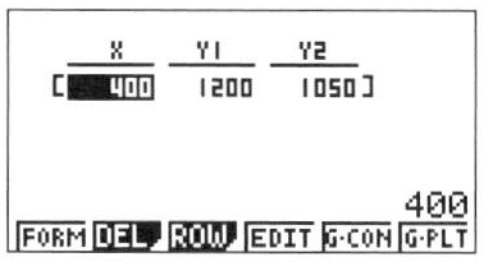

The algebraic solutions are:

c i Break even when $3x = 250 + 2x \Rightarrow x = 250$

The break even point is (250, 750).

ii Profit $= 3(400) - ((250 + 2(400))$

$= 1200 - 1050$

$= \$150$

Exercise 4.2

1 Write down the domain and range for these functions.

a

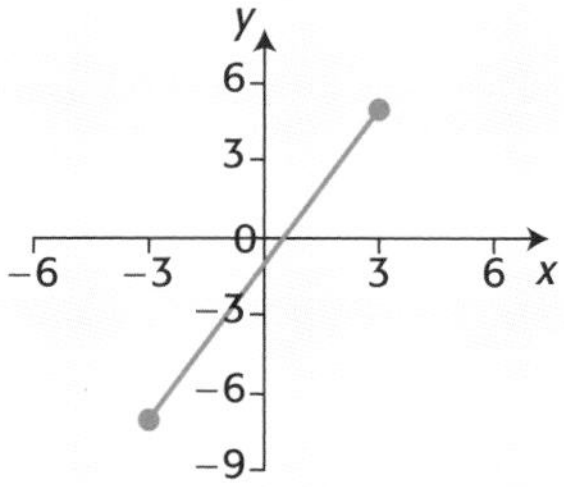

b

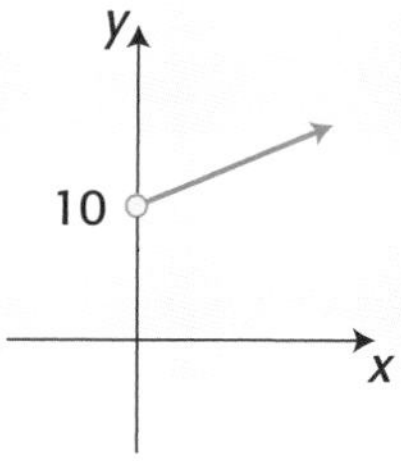

c

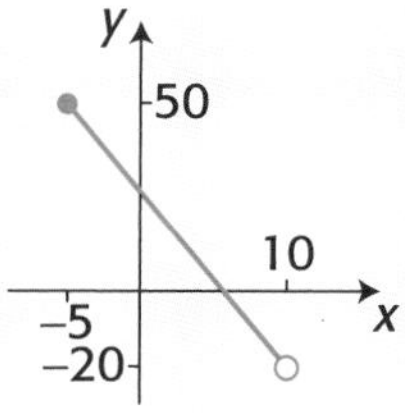

2 Construct a table of values and draw the graphs of these functions in the given domains.

a $f(x) = 2x + 1, -2 \le x \le 2$ b $g(x) = 20 - 3x, -10 \le x \le 10$

c $h(x) = \frac{1}{2}x + 2, -3 \le x \le 3$ d $y = \dfrac{80 + x}{5}, -100 \le x \le 100$

3 On the same axes, sketch the graphs of the lines which pass through these points. Clearly label the axes and intercepts.

a (0, 5) and (3, 0) b (0, –2) and (4, 0) c (0, 5) and (–3, 0)

4 Write down the gradients and y-intercepts for these functions.

a $y = 7x + 3$ b $3y = 2x - 9$ c $2y - 5x = 4$

d $1 - y = 4x$ e $2y = 6$ f $110x + 5y = 330$

5 Find the gradients and *y*-intercepts for each of these functions and hence write the equations of the straight lines in the form $Ax + By = C$.

a

x	−2	−1	0	1	2
f(x)	−9	−4	1	6	11

b

x	−2	−1	0	1	2
g(x)	7	5	3	1	−1

c

x	−2	−1	0	1	2
f(x)	−1	0.5	2	3.5	5

d

x	−2	−1	0	1	2
g(x)	15	12.5	10	7.5	5

6 Determine the coordinates of the point of intersection of each pair of functions.

a
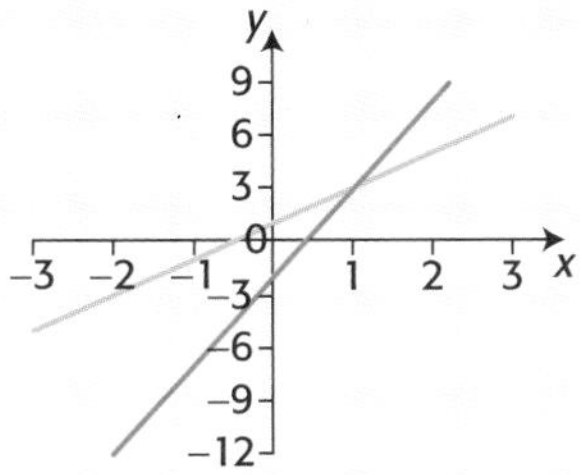

b
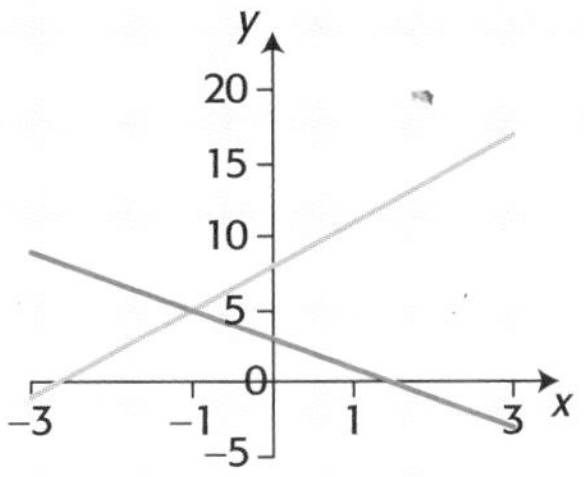

c
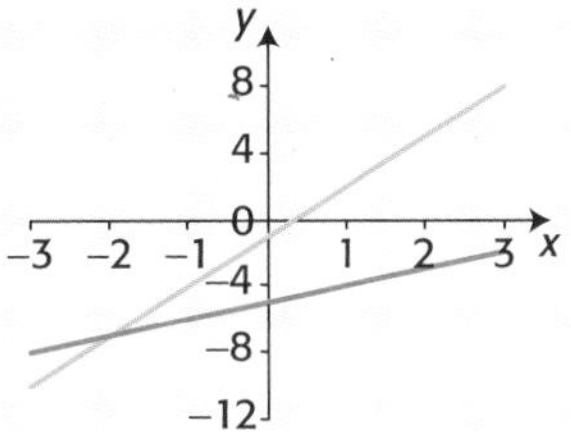

7 Draw the graphs of these pairs of lines and hence find the points where the graphs of the functions meet. Label the axes clearly.

a $y = 2x + 3$
$y = 5x - 6$

b $y = 9 - 2x$
$y = 2x - 1$

c $2y - x = -9$
$4y + 3x = 7$

d $0.5x + y = 2.5$
$0.3x + 0.4y = 1.0$

8 a A taxi charges a flat fee of 7500 rupiah (INR) plus 4400 rupiah per kilometre travelled.
- i Find the cost of a trip of 10 km and a trip of 60 km.
- ii Draw the graph of the cost of hiring a taxi.
- iii Use your graph to find the cost of a 50 km trip. Clearly show the method on the graph.

b A company's weekly profit, £*P*, can be found by $P = 25n - 60w - 2000$, where *n* represents the number of items sold and *w* represents the number of workers employed.
- i Find the profit when 10 workers make 120 items in a week.
- ii If 25 workers are employed, how many items must be made and sold for the company to break even?

c The formula to convert temperature from °F to °C is $C = \frac{5}{9}(F - 32)$.
- i Draw the graph of the temperature conversion function for the domain, $-100 < °F < 250$.
- ii Use your graph to find the temperature in °C when it is 100 °F, 0 °F and −32 °F.

d The cost ($*C*), of producing a book, is given by $C(x) = 46\,000 + 11.50x$, where *x* represents the number of books sold. Each book sells for $30.
- i Write down the revenue function $R(x)$.
- ii On the same axes, draw the graphs of the cost and revenue functions.

iii Determine the profit if
- 1000 books
- 5000 books are sold.

iv How many books must be sold to break even?

e The quantity demanded, $D(p)$, and quantity supplied, $S(p)$, of a particular item at a certain price (€p) are given by the functions, $D(p) = 5000 - 24p$ and $S(p) = 56p$.
Find
i the equilibrium price and quantity for this item
ii the excess quantity demanded when the price is €40
iii the excess quantity supplied when the price is €80.

f A water tank contains 12 500 litres. When the tap is fully open the water flows from the tank at the rate of 150 litres per minute.
i Write a linear function in the form $V(t) = mt + c$ to represent the volume of water remaining in the tank after t minutes if the tap is fully opened.
ii Determine the volume of water remaining in the tank after 15 minutes.
iii How long will it take for the volume of water remaining in the tank to reach 2000 litres?

4.3 Graphs of quadratic functions

A **quadratic function** is one which can be written in the form $f(x) = ax^2 + bx + c$ where a, b, and c are constants and $a \neq 0$. A quadratic function is a second-degree polynomial since the highest exponent in the function is 2.

4.3.1 The graph of the quadratic function

The graph of a quadratic function is a parabola.

If $a > 0$, the parabola opens upwards ∪. If $a < 0$, the parabola opens downwards ∩

The graph of $f(x) = x^2$

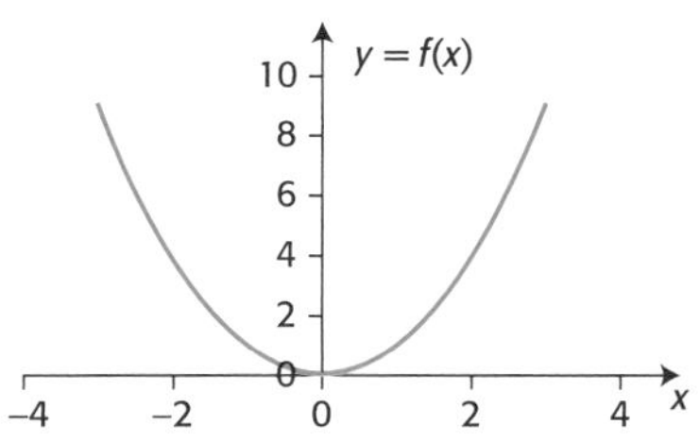

The graph of $f(x) = -x^2$

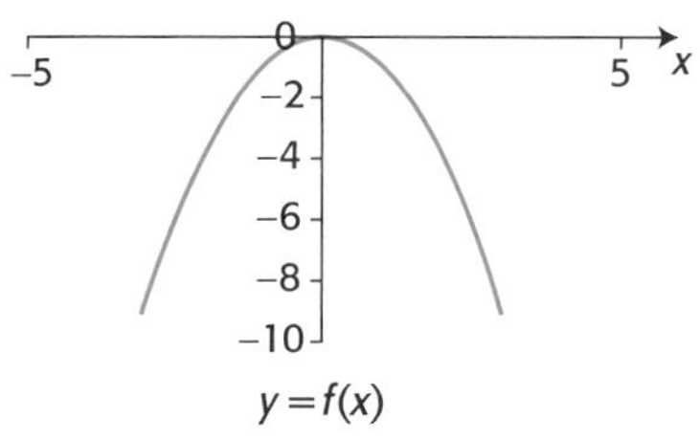

If we use x and y as variables, the equation of a quadratic function takes the form $y = ax^2 + bx + c$.

4.3.2 The axis of symmetry

Each parabola is symmetric about a vertical line called the axis of symmetry. In the example above, note that each parabola is symmetric about the y-axis. The y-axis is the axis of symmetry for $f(x) = x^2$ and $f(x) = -x^2$.

For the general quadratic, $y = ax^2 + bx + c$, the equation of the axis of symmetry is

$$x = \frac{-b}{2a}.$$

If there are two x-intercepts, the axis of symmetry passes through the midpoint of these intercepts. If there is only one x-intercept, the axis of symmetry passes through this point.

Example 4.3.2a

Find the equation of the axis of of symmetry of the function $f(x) = x^2 + 2x + 1$.

Show the axis clearly on a sketch of the graph of $y = f(x)$.

Solution

The axis of symmetry is $x = \frac{-b}{2a}$

$$= \frac{-2}{2 \times 1} = -1$$

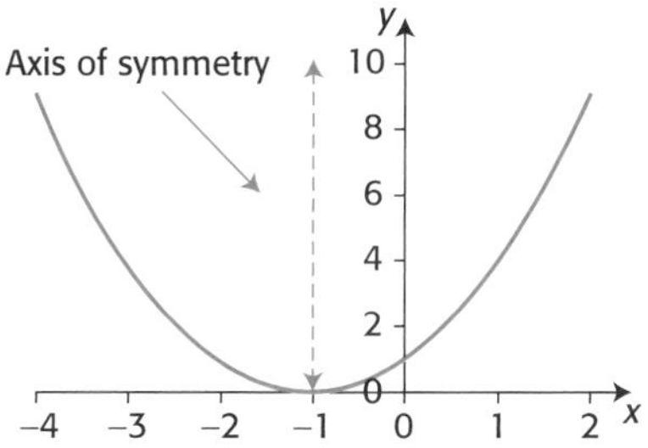

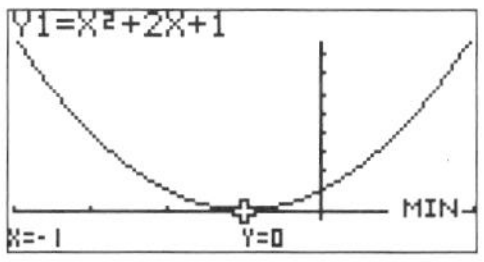

Example 4.3.2b

Find the equation of the axis of symmetry of the function $g(x) = -x^2 + 3x + 10$.

Show the axis clearly on a sketch of the graph of $y = g(x)$.

Solution

The axis of symmetry is $x = \frac{-b}{2a}$

$$= \frac{-3}{2 \times -1} = 1.5$$

or using the axis intercepts

$$x = \frac{-2+5}{2} = 1.5$$

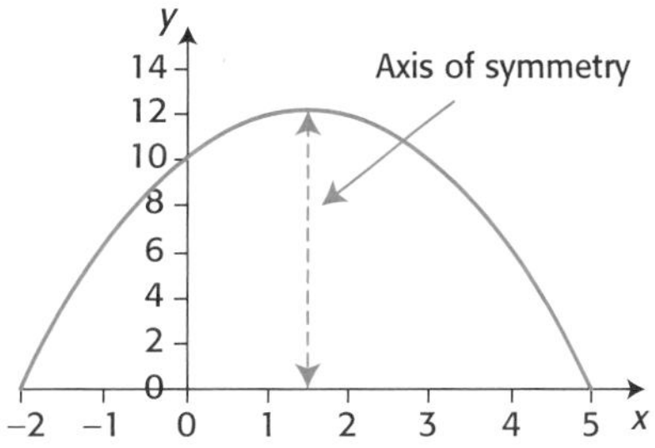

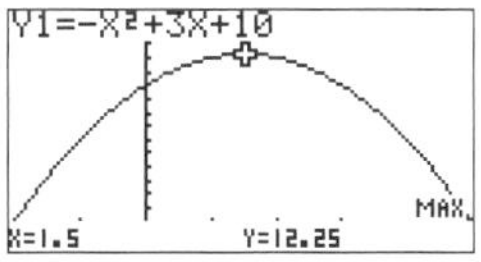

4.3.3 The vertex

The **vertex** is the lowest (if $a > 0$) or highest (if $a < 0$) point on the graph of a quadratic function.

The vertex lies on the **axis of symmetry**. To find the y-coordinate of the vertex, we substitute the value $x = \frac{-b}{2a}$ into the function.

So the vertex is at $\left(\frac{-b}{2a}, f\left(\frac{-b}{2a}\right)\right)$.

In Example 4.3.2a above, the vertex of the graph of $f(x) = x^2 + 2x + 1$ has coordinates $(-1, f(-1))$ and since $f(-1) = (-1)^2 + 2(-1) + 1 = 0$, the coordinates of the vertex are $(-1, 0)$.

Example 4.3.3a

Draw the graph of $f(x) = 2x^2 - 4x + 5$, $-2 \le x \le 2$. Determine the equation of the axis of symmetry and the coordinates of the vertex.

Solution

x	−2	−1	0	1	2
f(x)	21	11	5	3	5

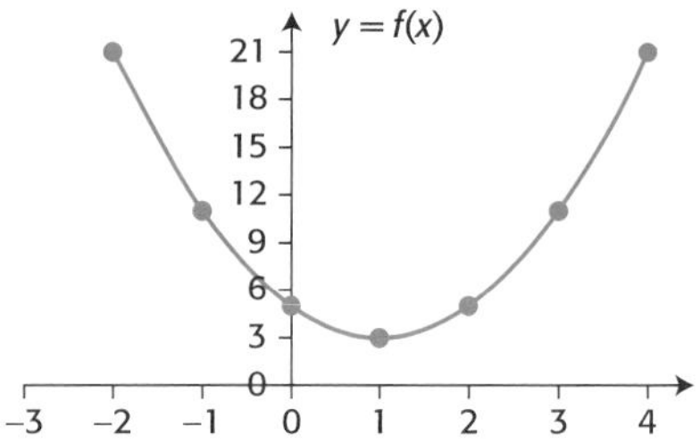

The equation of the axis of symmetry is $x = \dfrac{-b}{2a}$

$= \dfrac{+4}{2 \times 2} = +1, \quad x = 1$

$f(1) = 2(1)^2 - 4(1) + 5 = 3$

The coordinates of the vertex are $(1, 3)$.

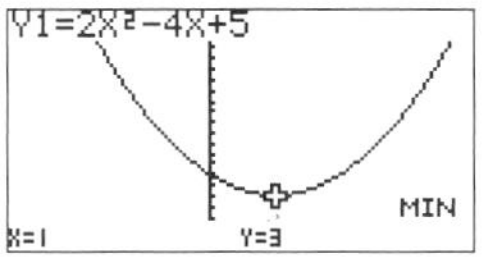

Example 4.3.3b

Draw the graph of $f(x) = -x^2 - 3x + 4$ for $-5 \le x \le 2$.

Determine the coordinates the vertex.

Solution

x	−5	−3	−1	0	1	2
f(x)	−6	4	6	4	0	−6

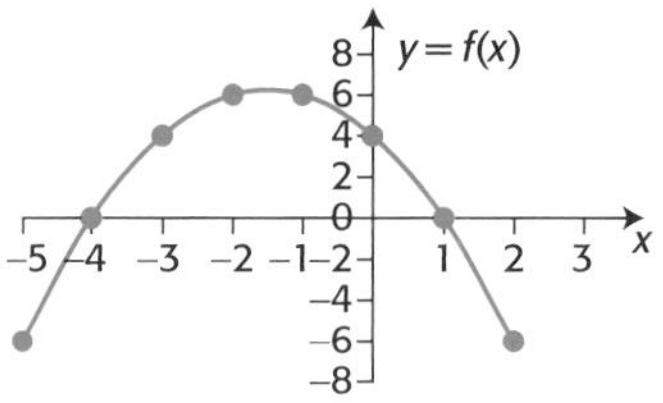

The equation of the axis of symmetry is $x = \dfrac{+3}{2 \times -1} = -1.5$

$f(-1.5) = -(-1.5)^2 - 3(-1.5) + 4$
$= 6.25$

The coordinates of the vertex are $(-1.5, 6.25)$.

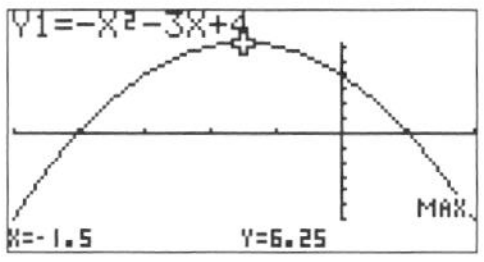

Maximum and minimum points

The vertex is either the highest or lowest point on the graph of a quadratic function.

If a, the coefficient of x^2, is **positive**, the vertex is the lowest point on the graph.

The y-coordinate of the vertex is the **minimum** value of the function.

If a is **negative**, the vertex is the highest point on the graph. The y-coordinate of the vertex is the **maximum** value of the function.

The absolute value of a influences the position of the vertex and the pitch (steepness) of the parabola.

Compare these three graphs. The higher the value of *a*, the steeper the pitch of the curve.

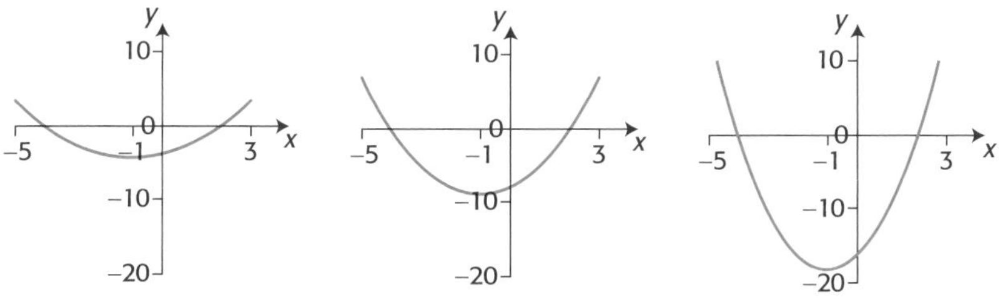

$a = 0.5$ $a = 1$ $a = 2$

4.3.4 The axis intercepts

(See also Section 2.7)

Notice that in each of the three graphs below, there is only one *y*-intercept because they are functions. In the first example there is only one *x*-intercept; in the second, there are no *x*-intercepts and in the third example there are two *x*-intercepts.

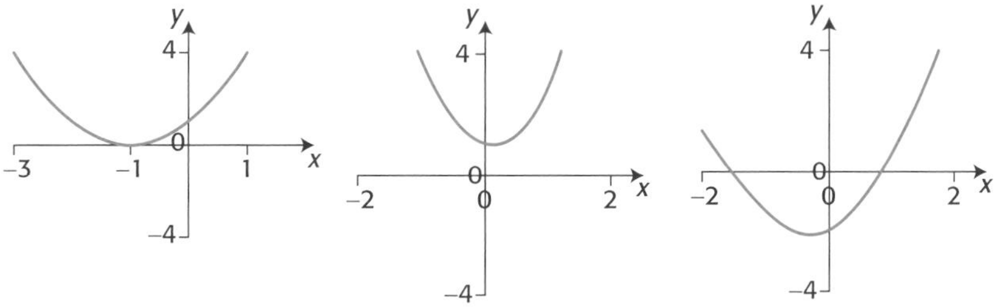

A quadratic function has one *y*-intercept and can have one, two or no *x*-intercepts.

The *y*-intercept

The point of intersection of a function graph with the *y*-axis occurs when $x = 0$.
The *y*-coordinate of this point is called the ***y*-intercept**.

It has the value of the constant, *c*, when the function is in the form $f(x) = ax^2 + bx + c$, that is,

$f(0) = a(0)^2 + b(0) + c = c$.

The *y*-intercept is at $(0, c)$.

The *x*-intercept

The point of intersection of a line with the *x*-axis occurs when $y = 0$.
The *x*-coordinate of this point is called the ***x*-intercept**.

To find the *x*-intercepts of a quadratic expression (in *x*), we set the expression to zero, and *solve the resulting quadratic equation for x*.

"Uh yeah, Homework Help Line? I need to have you explain the quadratic equation in roughly the amount of time it takes to get a cup of coffee."

Example 4.3.4a

Find the x-intercepts of $f(x) = x^2 + 2x + 2$.

Solution

Set the expression to zero

$x^2 + 2x + 2 = 0$.

Solve for x.

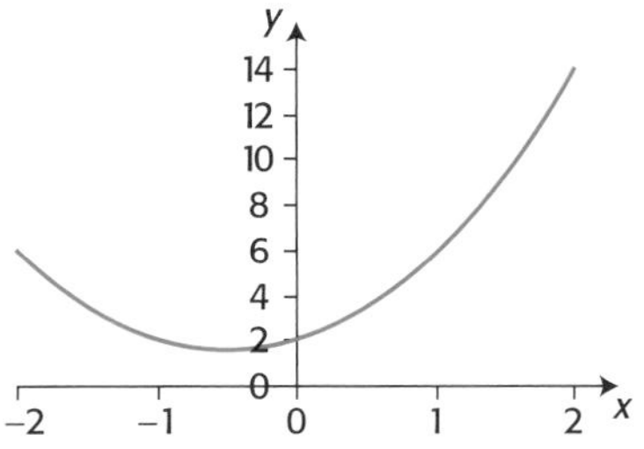

The equation cannot be solved (no real solutions). There are no x-intercepts. The graph of this function will not cut or touch the x-axis.

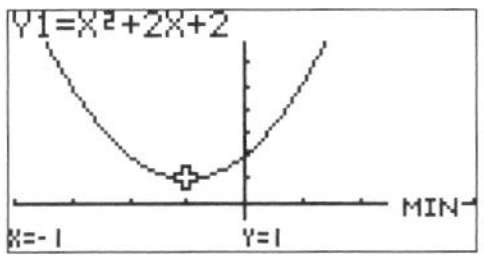

Example: 4.3.4b

Find the x-intercepts of $h(x) = x^2 + 2x + 1$.

Solution

Set the expression to zero

$x^2 + 2x + 1 = 0$.

Solve for x.

Factorize.

$(x + 1)(x + 1) = 0 \qquad x = -1$

or use the quadratic formula:

$$x = \frac{-2 \pm \sqrt{(-2)^2 - (4 \times 1 \times 1)}}{2 \times 1}$$

$$x = \frac{-2 \pm 0}{2} = -1$$

There is only one solution The graph of the function will touch the x-axis at $(-1, 0)$.

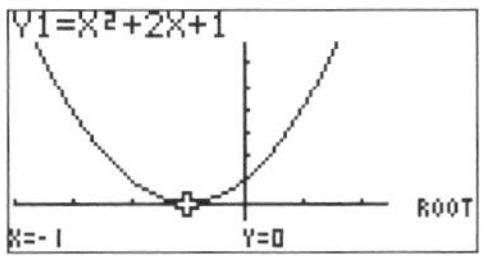

Example 4.3.4c

Find the x-intercepts of $f(x) = x^2 + 2x - 3$.

Solution

Set the expression to zero.

$x^2 + 2x - 3 = 0$.

Solve for x.

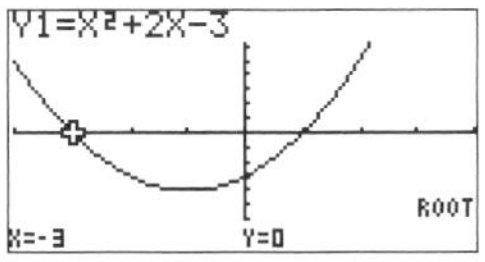

Factorize.

$(x + 3)(x - 1) = 0$

$x = 1, -3$

Or use the quadratic formula.

$$x = \frac{-2 \pm \sqrt{2^2 - (4 \times 1 \times -3)}}{2 \times 1}$$

$$x = \frac{-2 \pm \sqrt{16}}{2} = 1, -3$$

The graph of the function will cut the x-axis at $(1, 0)$ and $(-3, 0)$.
The x-intercepts are 1 and −3.

4.3.5 Graphing a quadratic function

The graph of a quadratic function is a **parabola**. The sign of a, the coefficient of x^2, determines whether the parabola curves up or down.

Example 4.3.5

Find the vertex and the axis intercepts for $f(x) = x^2 + 6x + 5$.
Graph the parabola by drawing a smooth curve through these points.

Solution

Note that the value of a is positive. The graph curves upwards.

The axis of symmetry is $x = \dfrac{-b}{2a} = \dfrac{-6}{2\times1} = -3$

$f(-3) = (-3)^2 + 6(-3) + 5 = -4$

The vertex is $(-3, -4)$.

The y-intercept occurs when $x = 0$. The y-intercept is 5.

The x-intercepts occur when $y = 0$.

$x^2 + 6x + 5 = 0$

$(x + 5)(x + 1) = 0, \quad x = -5, -1$ The x-intercepts are -5, and -1.

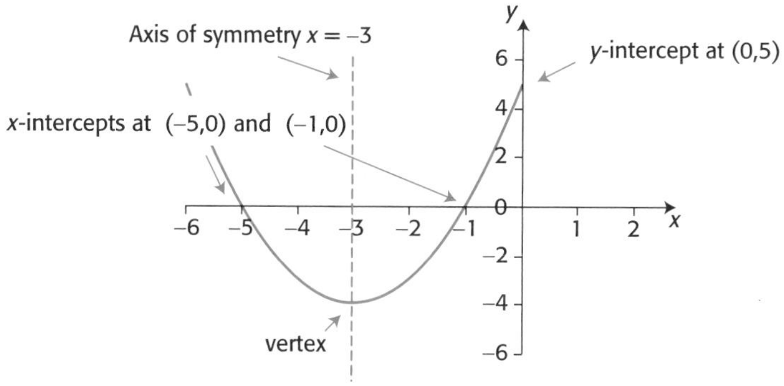

4.3.6 The factor form of a quadratic function: $y = a(x - h)(x - k)$

Quadratic functions whose graphs intersect the x-axis can be written in the form

$y = a(x - h)(x - k)$, where h and k are the x-intercepts of the function.

The constant, ahk, is the y-intercept.

The axis of symmetry is halfway between the intercepts, that is, at $x = \dfrac{h+k}{2}$.

If the curve touches the x-axis, then $k = h$ and the factor form becomes

$y = a(x - h)(x - h)$ or $y = a(x - h)^2$.

Example 4.3.6a

Find the intercepts and the vertex of $y = 2(x - 2)(x + 4)$.

Sketch the graph of the function.

Example 4.3.6b

Find the intercepts and vertex of $y = 3(x + 2)^2$.
Sketch the graph of the function.

Solution

The x-intercepts are 2 and –4.

The y-intercept is

$2 \times (0 - 2) \times (0 + 4) = -16$

The axis of symmetry is

$x = \dfrac{2 - 4}{2} = -1.$

At $x = -1$

$y = 2(-1 - 2)(-1 + 4) = -18$

The vertex is $(-1, -18)$.

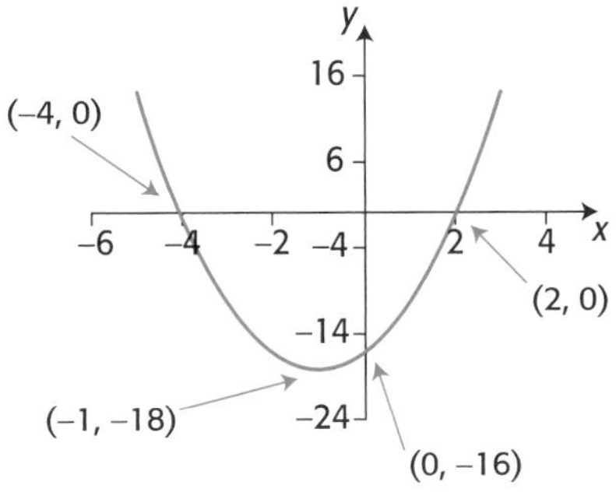

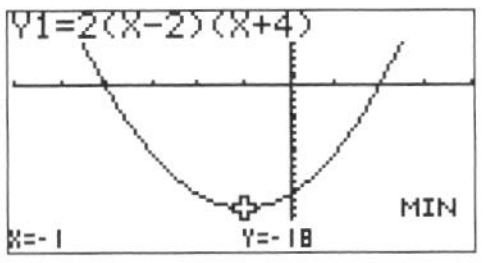

Solution

The graph touches the x-axis at $(-2, 0)$.

The y-intercept is

$3 \times (0 - 2)^2 = 12$

The axis of symmetry is

$x = -2.$

At $x = -2$

$y = 3(-2 + 2)^2 = 0$

The vertex is $(-2, 0)$.

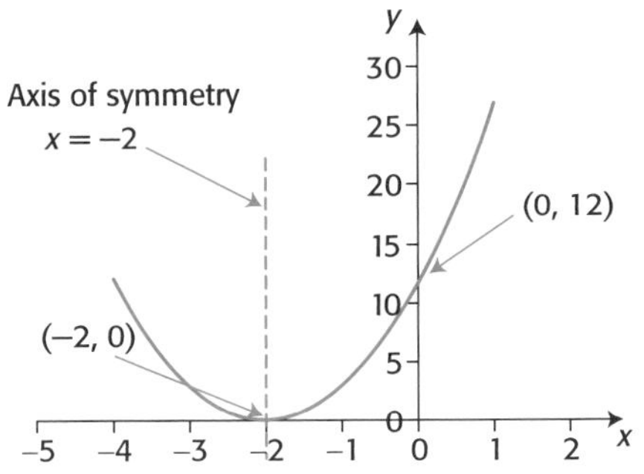

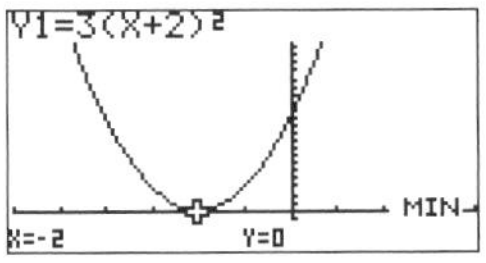

Finding the equation of a parabola from its graph

Example 4.3.6c ***Example 4.3.6d***

Determine the equations of these graphs of quadratic functions.

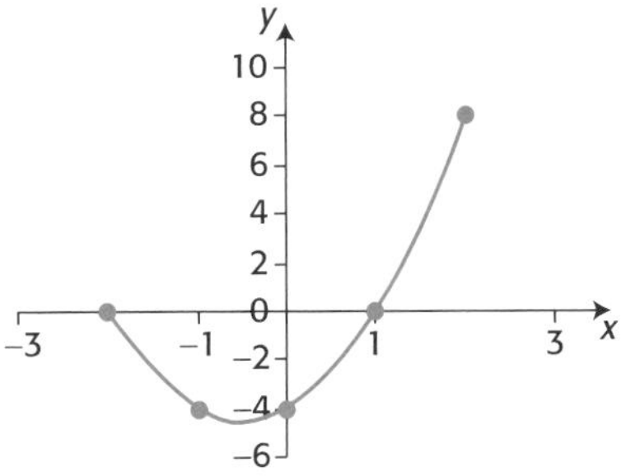

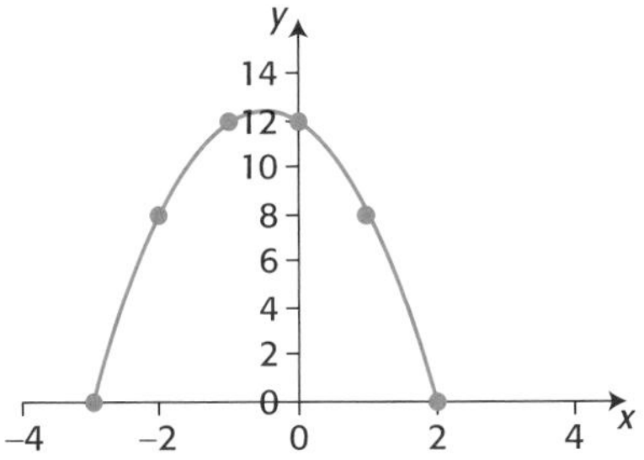

Solution

x-intercepts are –2 and 1.

Factor form: $y = a(x + 2)(x - 1)$

y-intercept is –4.

$a \times (0 + 2) \times (0 - 1) = -4$

so $a = 2$

Equation is $y = 2(x + 2)(x - 1)$

or $y = 2x^2 + 2x - 4$

Solution

x-intercepts are –3 and 2.

Factor form: $y = a(x + 3)(x - 2)$

y-intercept is 12

$a \times (0 + 3) \times (0 - 2) = 12$

so $a = -2$

Equation is $y = -2(x + 3)(x - 2)$

or $y = -2x^2 - 2x + 12$

Hint: If three points on a parabola are known, we can use a quadratic regression to find the equation of the quadratic function. In both of these examples the intercepts can be used.

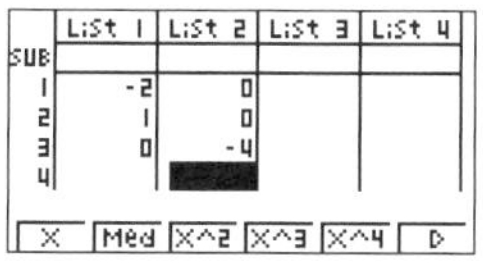

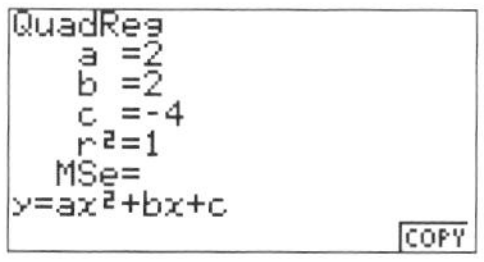

$y = 2x^2 + 2x - 4$

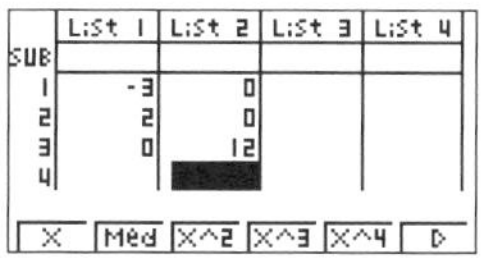

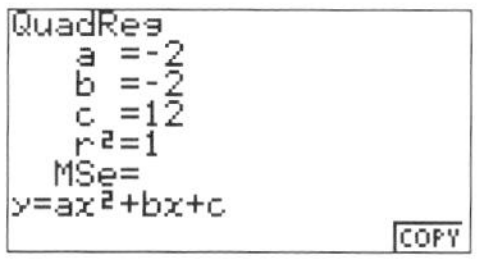

$y = -2x^2 - 2x + 12$

4.3.7 Applications of quadratic functions

Applications using quadratic functions generally involve establishing a quadratic equation first and then considering the intercepts or the vertex as possible solutions.

Example 4.3.7a

A rectangular fence has a perimeter of 40 m.

a Find the maximum area that could be enclosed by this fence.

b Write down the dimensions of the fence that maximize the area.

Solution

Let the width be w. The length is therefore $\frac{1}{2}(40 - 2w) = 20 - w$

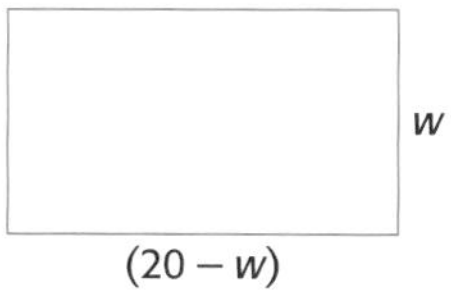

a Area = $w \times l$

$A = w(20 - w)$, a quadratic function.

The maximum is at vertex (10, 100).

Maximum area is 100 m^2.

b Maximum area occurs when $l = w = 10$ m.

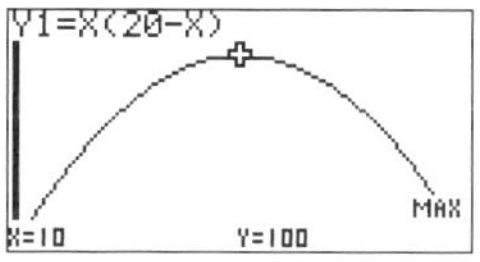

Show that: Sometimes, the function and the answer are known, and we are required to validate the answer. This type of question uses the words, "Show that …" It generally requires an algebraic solution.

Example 4.3.7b

A firm's output (P tonnes) is given by $P = -60 + 23w - w^2$, where w is the number of workers employed.

a Draw a sketch of the production function for $0 \le w \le 20$.

b *Show that* the maximum possible production is 72.25 tonnes.

Solution

a This is the graph of the production function.

b The maximum value is found at the vertex.

The axis of symmetry is $x = \frac{-b}{2a} = \frac{-23}{-2} = 11.5$

$P(11.5) = -60 + 23(11.5) - (11.5)^2 = 72.25$ tonnes

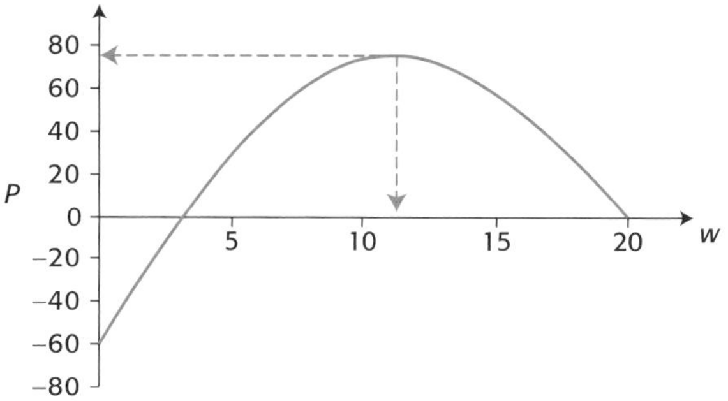

Example 4.3.7c

The profit function (in yen) for a particular business is given by $P(n) = -0.014n^2 + 67n - 50\ 000$, where n represents the number of units sold each week ($n < 3800$).

Find algebraically

a the weekly loss for the business if it remains closed

b the profit when 1200 units are sold.

Use your GDC to find

c the break-even point for the business

d the maximum weekly profit.

Solution

a $n = 0$ $\quad P(0) = -0.014(0)^2 + 67(0) - 50\ 000$

The business loses 50 000 yen each week.

b $P(1200) = -0.014(1200)^2 + 67(1200) - 50\ 000$

The profit when 1200 units sold is 18 040 yen.

(GDC)

c Break-even point is when $P(n) = 0$

$-0.014n^2 + 67n - 50\ 000 = 0 \qquad n = 926$

926 units must be sold for the business to break even.

d Maximum weekly profit is 30 161 yen when 2393 units are sold.

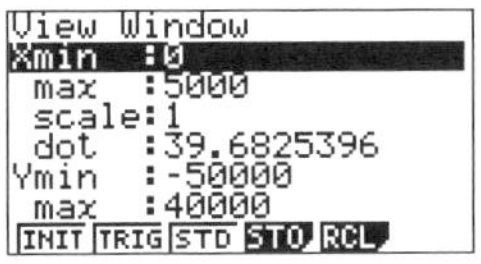

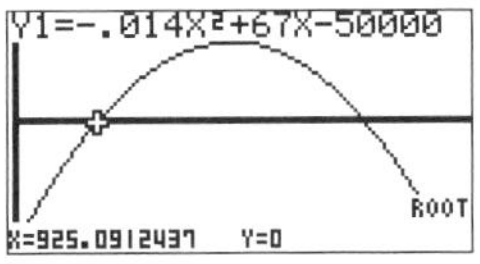

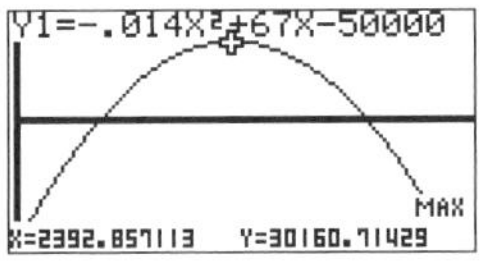

Exercise 4.3

1 State whether the graphs of these functions open upwards or downwards. Find the axis of symmetry for each function.

a $f(x) = x^2 + 2x - 8$ b $y = x^2 - 12x + 27$ c $S = 2t^2 - 8t + 28$

d $g(x) = 6 - 5x - x^2$ e $3x - x^2 = -10 + y$ f $C = 10n^2 - 5n - 30$

g $y = (x - 2)(x - 5)$ h $S = (t - 2)^2$ i $f(x) = (3 - x)^2$

2 Find the vertex of each of these quadratic functions. In each case, state whether the vertex is a maximum or minimum point.

a $y = x^2 + 2x - 8$ b $f(x) = x^2 - 4x + 2$ c $y = 2x^2 - 5x - 3$

d $f(x) = 1 + x - x^2$ e $S = 3t^2 - 2t - 4$ f $P(x) = (5 - x)^2$

g $A = n^2 - 9$ h $g(x) = (x + 5)^2 - 3$ i $y = -2x^2 + 20x - 50$

3 Find the axis intercepts and the vertex of each of these functions and hence sketch their graphs.

a $f(x) = x^2 - 7x + 6$ b $h(x) = x^2 + 3x - 10$ c $S = 4t^2 - 12t - 40$

d $y = -x^2 + x + 6$ e $y = -\dfrac{x^2}{2} + x + 4$ f $P(x) = -x^2 - 3x + 10$

g $y = (x - 3)^2$ h $g(x) = -(3 - 2x)^2$ i $y = 2x^2 - 7x - 15$

4 Write these expressions in factor form. Remove any common factors first.

a $x^2 + 5x + 6$ b $x^2 + 9x + 20$ c $2x^2 + 14x + 12$
d $6 + t - t^2$ e $2x^2 + 14x + 24$ f $x^2 - 3x - 40$
g $14 + 5x - x^2$ h $18 - 2x^2$

5 Find the equations of each of these quadratic functions. Write the equations in the form $y = ax^2 + bx + c$.

a

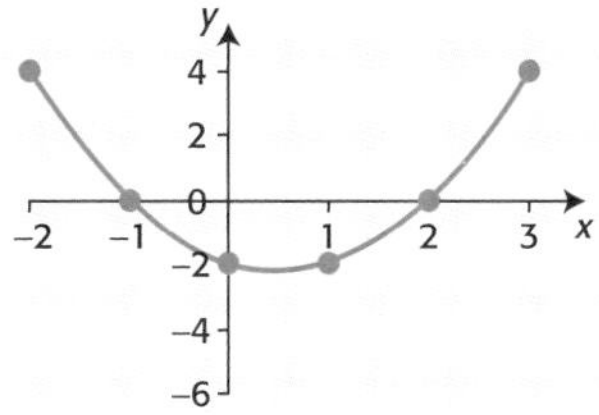

b

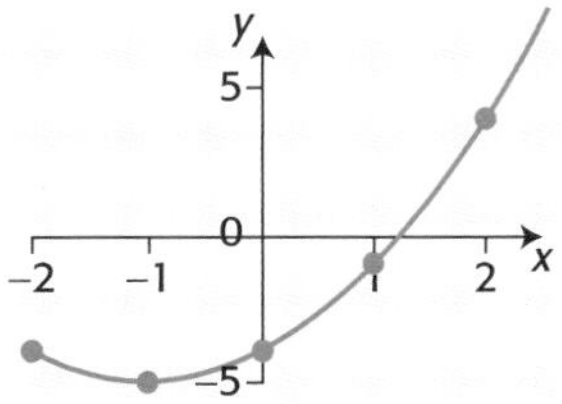

c

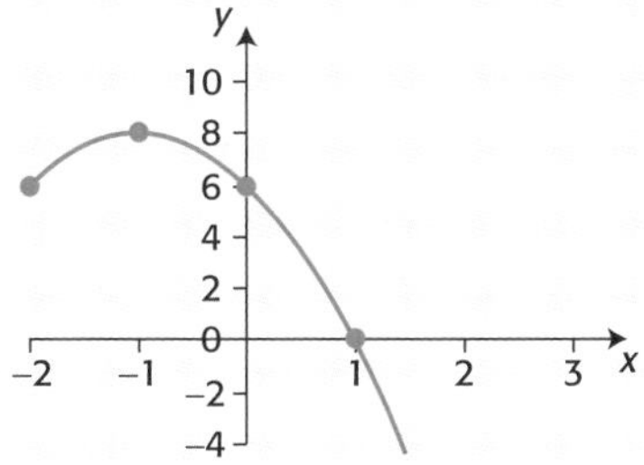

d

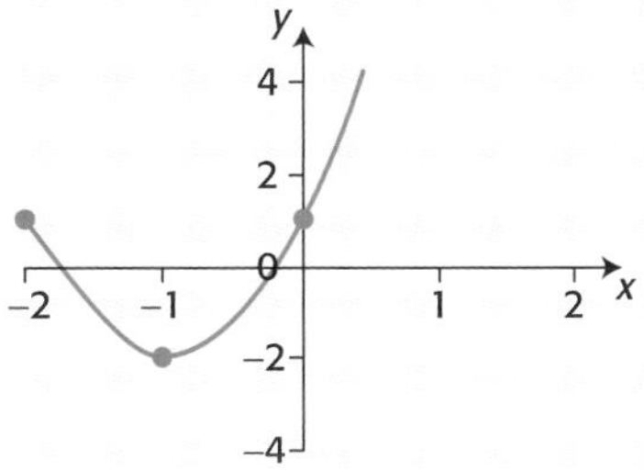

6 a The x-intercepts of a function are −1 and 5. The y-intercept is 10.
i Find the equation of the function in the form $y = ax^2 + bx + c$.
ii Find the maximum value of the function.

b The vertex of a function is (−2, 3) and the y-intercept is 15.
i Find the equation of the function in the form, $y = ax^2 + bx + c$.
ii Find the minimum value of the function.

7 a A firm's profit function, \$$P$, is $P = -x^2 + 20x - 60$, where x represents the quantity of goods sold per day.
i Draw the graph of the profit function for $0 \le x \le 25$, $0 \le y \le 50$.
ii Find the profit when 10 units and when 20 units are sold.
iii Determine the maximum profit for the firm and the quantity of goods sold to make that profit.

b A fence encloses a rectangular yard on three sides. The fence has a total length of 56 m.
i Write down expressions for the width and length of the yard, in terms of x.
ii Find the maximum possible area of the yard, and the corresponding dimensions.

c The path of an object launched into the air is parabolic. Its vertical height (in metres) above the ground after t seconds is given by $H(t) = 50t - t^2$.
i Find the height of the object after 10 seconds and after 40 seconds.

ii Draw a sketch of the function for $0 \leq t \leq 50$.
iii Find the maximum height reached by the object.
iv How long was the object in the air?

d The first term of an arithmetic sequence is 17 and the common difference is 13. The sum of the first n terms is 5390.
i Show that the solution for n can be found by the equation $13n^2 + 21n - 10\,780 = 0$.
ii Solve the equation and hence find the number of terms whose sum is 5390.

e The number of bacteria, N, in a culture t minutes after the start of an experiment is given by $N = 150 + 69t + 3t^2$.
i Determine the number of bacteria present at the start of the experiment.
ii Find the number of bacteria present after 5 minutes.
iii Find the time required for the number of bacteria to reach 2000.
iv Sketch the graph of N against t for $0 \leq t \leq 25$.

f A rectangle has a perimeter of 36 cm. If we let x represent the width of one side of the rectangle
i find the possible values which x can take
ii find the length of the rectangle in terms of x
iii show that the area A of the rectangle is given by $A(x) = x(18 - x)$
iv sketch the graph of A against x
v find the dimensions of the rectangle which has the largest possible area for the given perimeter.

4.4 Graphs and properties of exponential functions

4.4.1 Evaluating exponential expressions

An **exponential expression** is comprised of a base number a, which is a positive constant, together with a variable exponent b.

Finding individual values of an exponential expression is not difficult when both a and b are small integers.

Example 4.4.1a
Evaluate a^b for $a = 2$ and $b = 3$

Solution
$2^3 = 2 \times 2 \times 2 = 8$

Example 4.4.1b
Evaluate a^{b-3} for $a = 5$ and $b = 5$

Solution
$5^{5-3} = 5^2 = 5 \times 5 = 25$

However, the computation of a^b becomes difficult when a is not an integer, even for small integer values of b. The computation

becomes extremely difficult when b is not an integer.
For example, evaluate

$2.36^{1.5}$

Understanding the nature of exponential expressions is best done by careful observation of the graphs of exponential functions.

4.4.2 The graph of $f(x) = a^x$

The **exponential function** f with a base a is defined by $f(x) = a^x$, $a > 0, a \neq 1, x \in \mathbb{R}$.

Example 4.4.2a

$a > 1$

Consider the function $f(x) = 2^x$.

Example 4.4.2b

$0 < a < 1$

Consider the function

$f(x) = \left(\frac{1}{2}\right)^x$.

a Construct tables of values for $-3 \leq x \leq 3$.

b Draw the graphs of $y = f(x)$, $-3 \leq x \leq 3$, $0 < y < 8$.

c Observe the nature and properties of the graphs.

Solutions

a $f(x) = 2^x$

x	−3	−2	−1	0	1	2	3
y	0.125	0.25	0.5	1	2	4	8

$f(x) = \left(\frac{1}{2}\right)^x$

x	−3	−2	−1	0	1	2	3
y	8	4	2	1	0.5	0.25	0.125

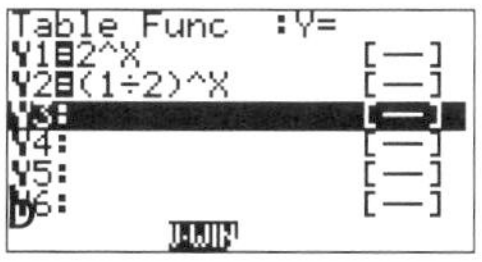

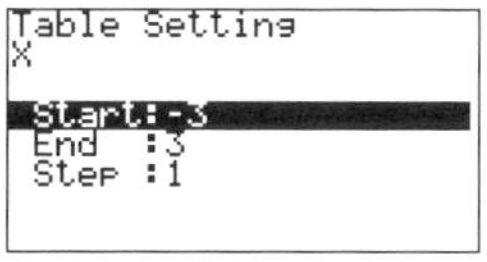

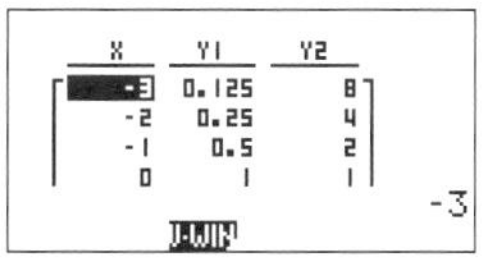

b

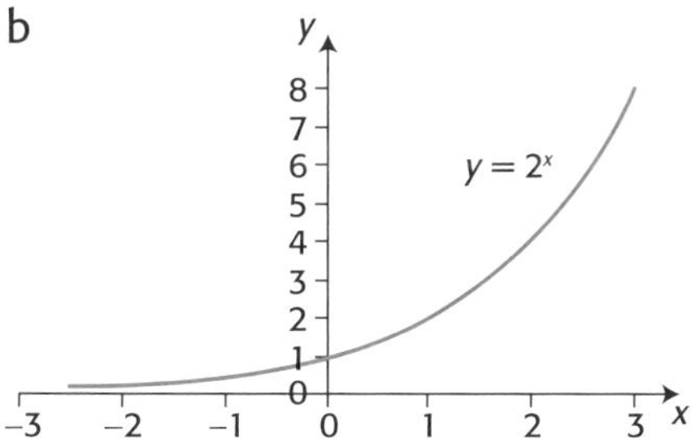

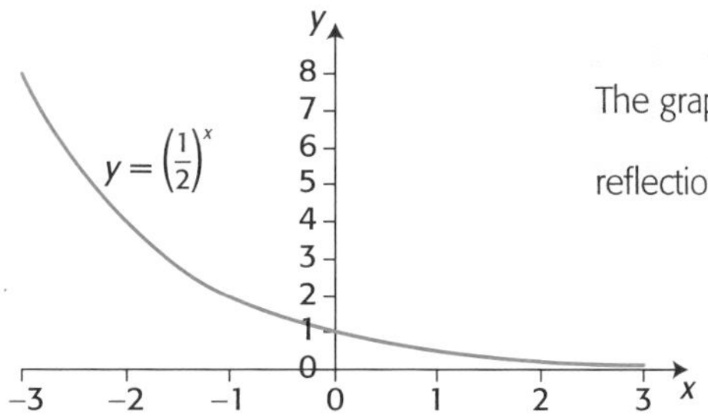

The graph of $y = \left(\frac{1}{2}\right)^x$ is the reflection in the y-axis of $y = 2^x$.

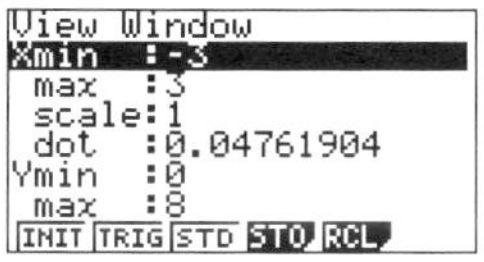

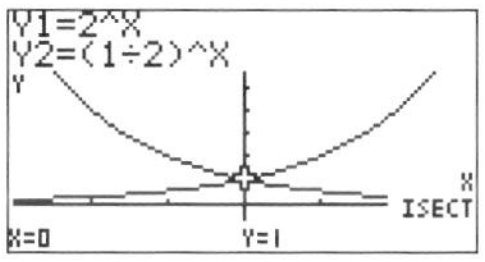

c

$y = 2^x$

$y = \left(\frac{1}{2}\right)^x$

The domain is real numbers between −3 and +3 and the range is $y > 0$.

The graph crosses the y-axis at (0, 1).

As the value of x increases, the value of y increases at an increasing rate.

As the value of x decreases, y approaches zero **asymptotically.** That is, as x gets smaller and smaller, the graph gets closer and closer to the x-axis ($y = 0$) without ever crossing it.

The domain is real numbers between −3 and +3 and the range is $y > 0$.

The graph crosses the y-axis at (0, 1).

As the value of x increases, y approaches zero (asymptotically).

As the value of x decreases, the value of y increases, at an increasing rate.

The graphs have the same shape but the first graph, ($a > 1$), is increasing as the value of x increases. This is an example of exponential growth.

The second graph, ($0 < a < 1$), is decreasing as the value of x increases. This is an example of exponential decay.

Theory of knowledge

Why do both graphs contain the point (0, 1)?
Why do both graphs have the x-axis as an asymptote?

Increasing values of a

All exponential functions have the same characteristics. The only thing that varies is the steepness of their graphs.

Example 4.4.2c

Consider the functions $y = 2^x$ and $y = 5^x$.

a On the same set of axes, draw the graphs for $-2 \leq x \leq 2$, $0 \leq y \leq 25$.

b Observe the nature and properties of the graph.

Solution

a

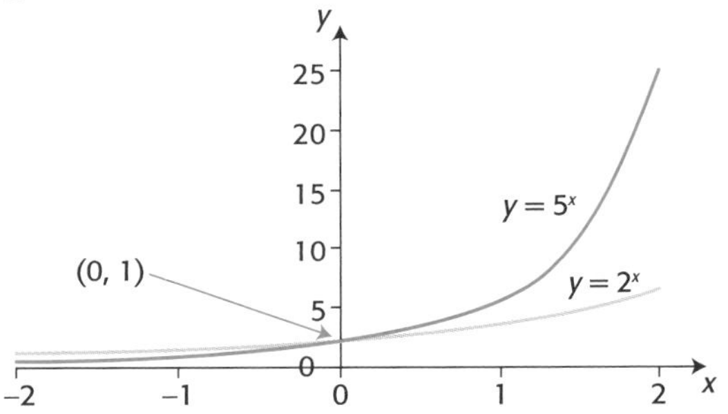

b

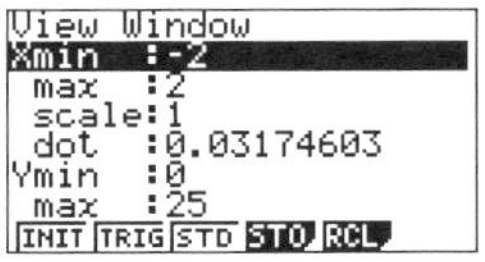

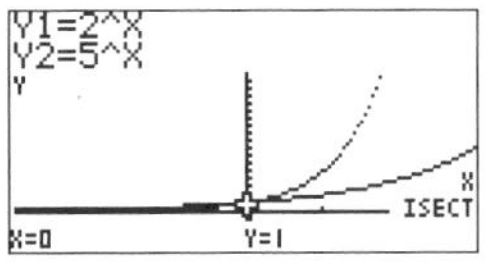

The two graphs share the characteristics already described. The only difference is that the graph of $y = 5^x$ rises more steeply than the graph of $y = 2^x$.

- As the value of a increases, the graph of $y = a^x$ becomes steeper.

4.4.3 The graph of $f(x) = a^{\lambda x}$

The steepness of the graph of an exponential function is also determined by the value of λ (the Greek letter, lambda).

Example 4.4.3a

$\lambda > 1$

Consider the functions $y = 2^x$

and $y = 2^{2x}$

Example 4.4.3b

$0 < \lambda < 1$

Consider the functions $y = 2^x$

and $y = 2^{0.5x}$

a On the same set of axes draw the graphs for $-2 \leq x \leq 2$.

b Observe the nature and properties of the graphs.

Solutions

a

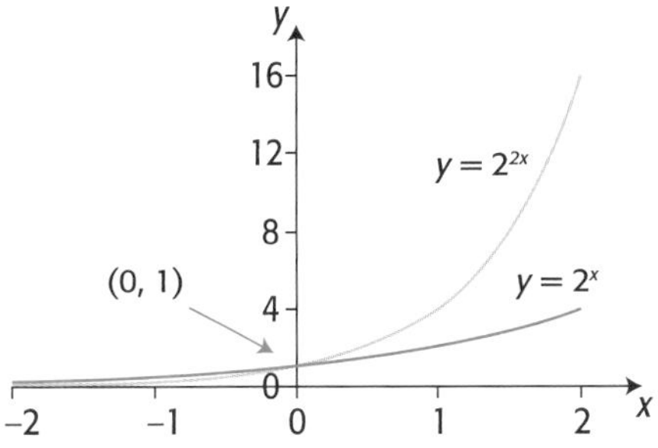

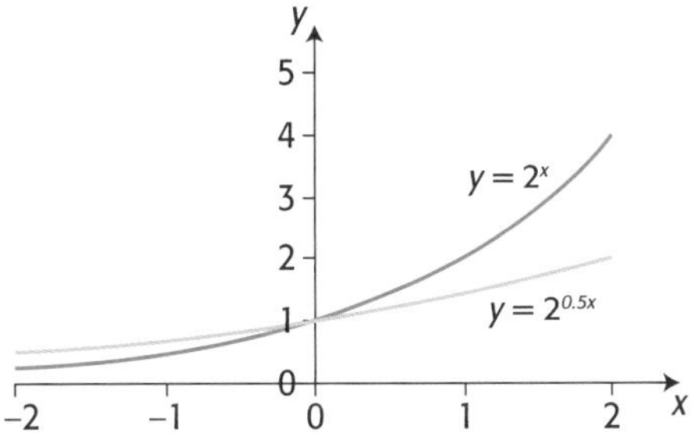

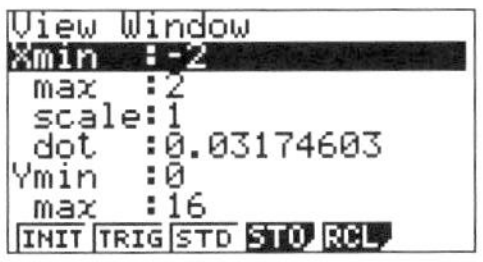

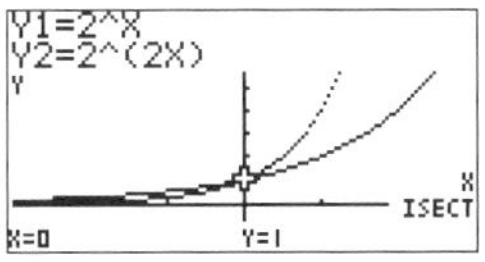

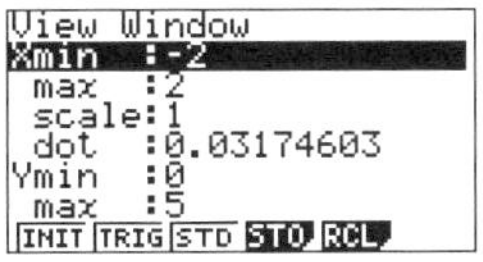

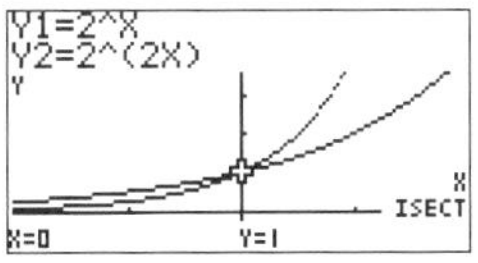

b In each case, both graphs cut the y-axis at (0, 1), are asymptotic to the x-axis and have the same shape.

The graph of $g(x) = 2^{2x}$ rises more steeply than the graph of $f(x) = 2^x$.

The graph of $g(x) = 2^x$ rises more steeply than the graph of $f(x) = 2^{0.5x}$.

- The graph of $f(x) = a^{\lambda x}$ becomes steeper as the value of λ increases.

Example 4.4.3c

$\lambda < 0$

Consider the functions $y = 2^{-x}$ and $y = 2^x$.

a On the same set of axes, draw the graphs for $-2 \leq x \leq 2,\ 0 \leq y \leq 5$.

b Observe the nature and properties of the graphs.

Solution

a

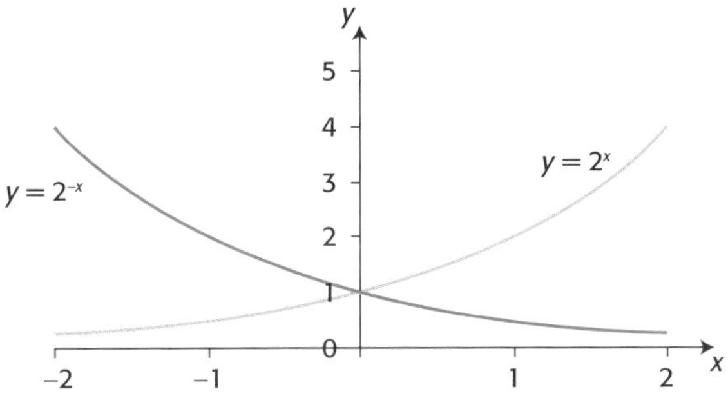

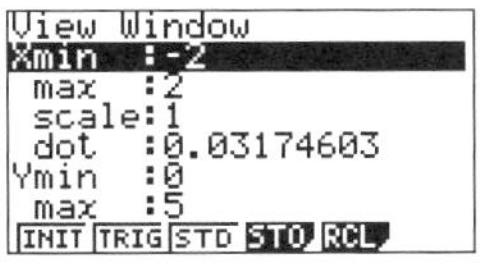

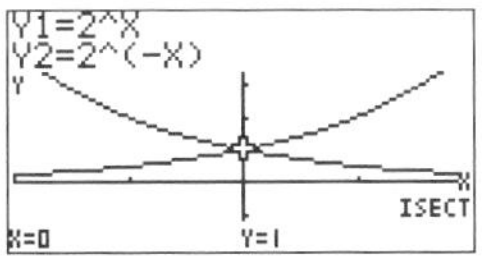

Theory of knowledge

Note that the graph of $y = 2^{-x}$ is the same as the graph of $y = \left(\frac{1}{2}\right)^x$. Why? Is this true for all values of a?

b Both graphs cut the y-axis at $(0, 1)$, are asymptotic to the x-axis, and have the same shape. The graph of $y = 2^{-x}$ is the reflection of the graph of $y = 2^x$ in the y-axis.

4.4.4 The graph of $f(x) = ka^{\lambda x}$

Example 4.4.4a

$k > 1$

Consider the functions $y = 5 \times 2^x$ and $y = 2^x$

Example 4.4.4b

$0 < k < 1$

Consider the functions $y = \frac{1}{5} \times 2^x$ and $y = 2^x$

a On the same set of axes draw the graphs for $-2 \leq x \leq 2$.

b Observe the nature and properties of the graphs.

Solutions

a

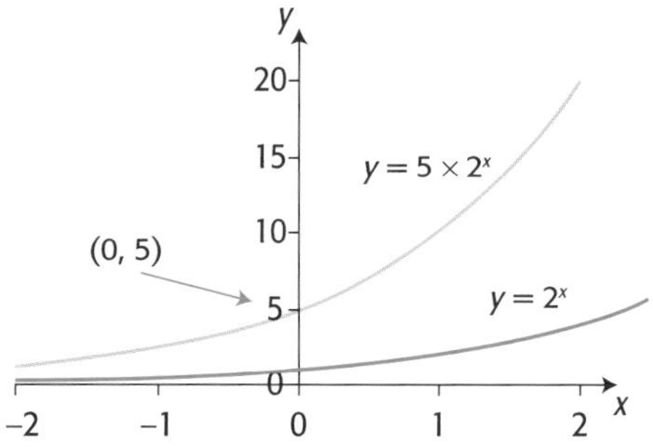

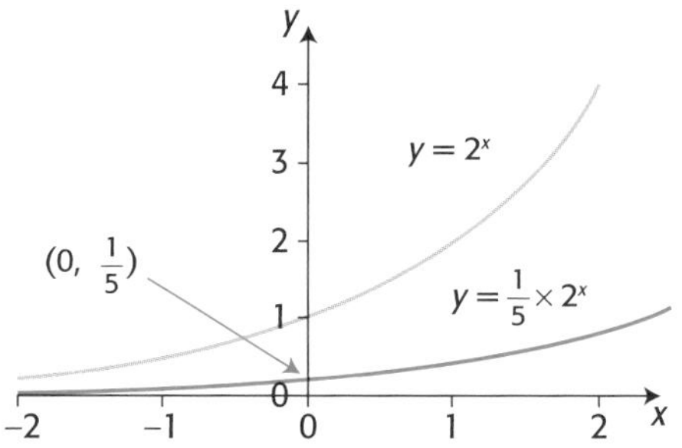

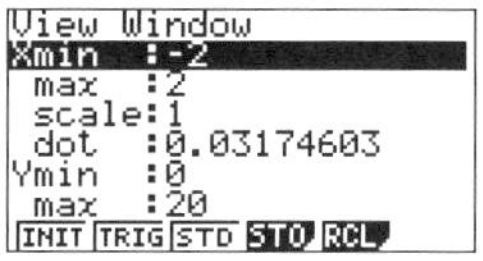

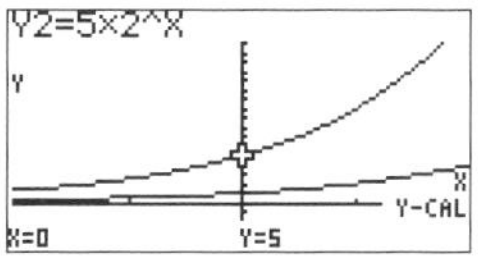

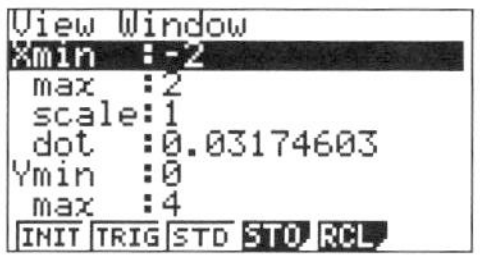

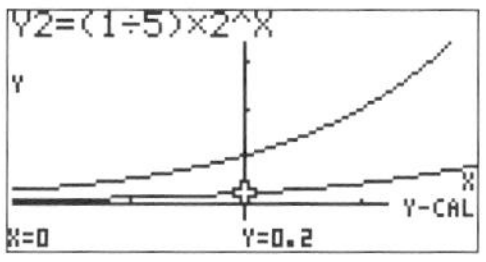

b The graphs of $y = 5 \times 2^x$ and $y = \frac{1}{5} \times 2^x$ have the same shape as all exponential graphs. They are asymptotic to the x-axis.

The graph of $y = 5 \times 2^x$ has its y-intercept at $(0, 5)$.

For every value of x, the value of y for $y = 5 \times 2^x$ is five times the corresponding y-value for $y = 2^x$.

The graph of $y = \frac{1}{5} \times 2^x$ has its y-intercept at $(0, \frac{1}{5})$.

For every value of x, the value of y for $y = \frac{1}{5} \times 2^x$ is one fifth the corresponding y-value for $y = 2^x$.

Example 4.4.4c

$k < 0$

Consider the functions $y = -2 \times 2^x$ and $y = 2 \times 2^x$.

a On the same set of axes, draw the graphs for $-2 \leq x \leq 2$, $-8 \leq y \leq 8$.

b Observe the nature and properties of the graphs.

Solution

a

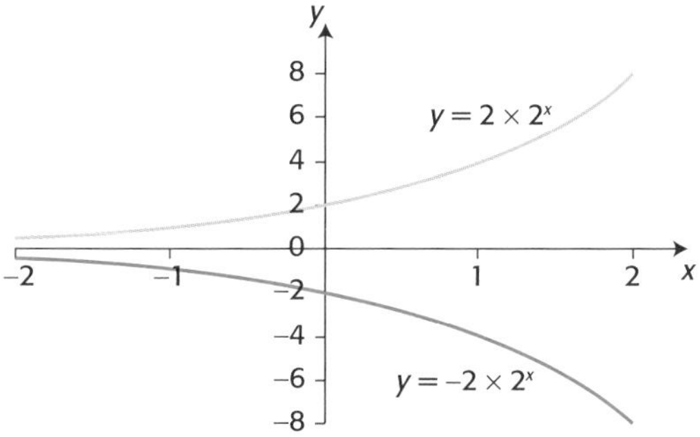

b The graph of $y = -2 \times 2^x$ is a reflection of $y = 2 \times 2^x$ in the x-axis. It does not pass through $(0, 2)$ in this case, its y-intercept is $(0, -2)$.

For every value of x, the value of y for $y = -2 \times 2^x$ is the negative of the corresponding y-value for $y = 2 \times 2^x$.

4.4.5 The graph of $f(x) = ka^{\lambda x} + c$

Example 4.4.5a

$c > 0$

Consider the functions $y = 2^x$ and $y = 2^x + 1$.

Example 4.4.5b

$c < 0$

Consider the functions $y = 2^x$ and $y = 2^x - 1$.

a On the same set of axes draw the graphs for $-2 \le x \le 2$.

b Observe the nature and properties of the graphs.

Solution

a

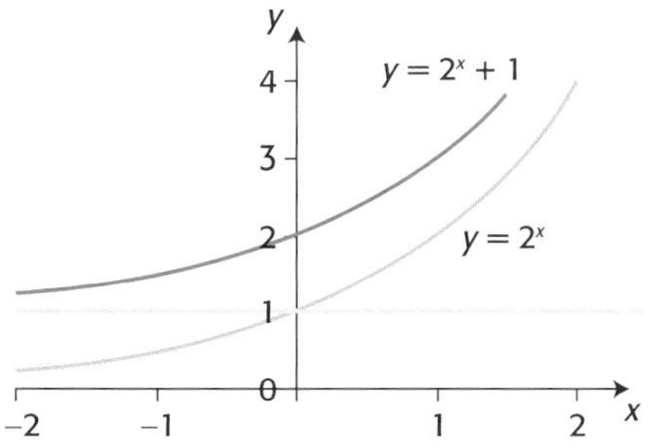

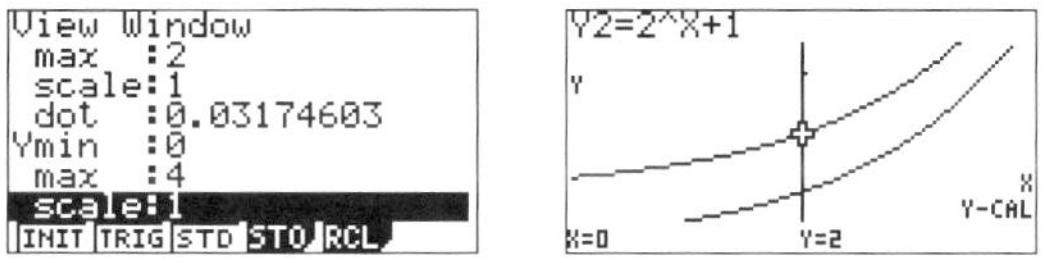

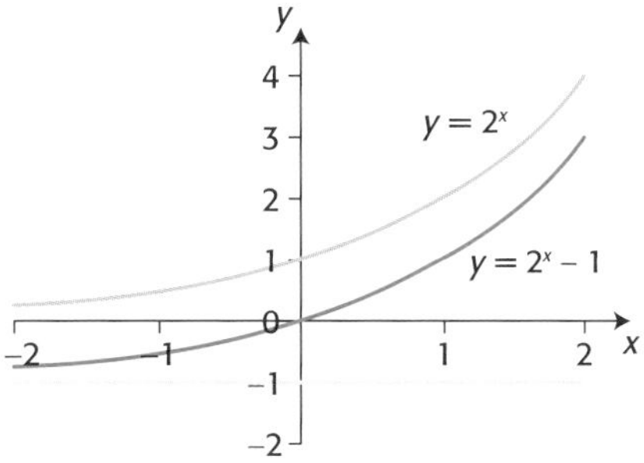

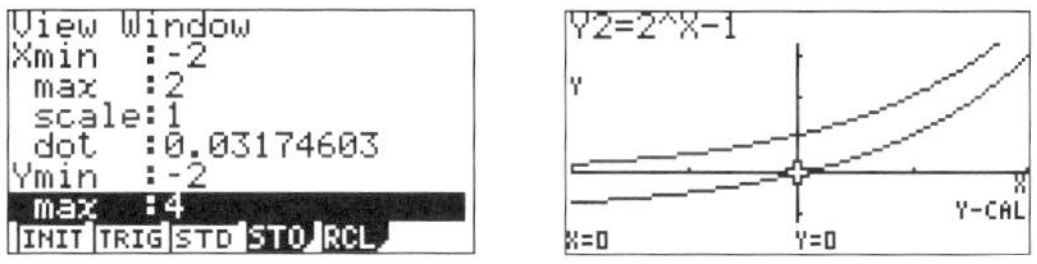

b The graph of $y = 2^x + 1$ passes through (0, 2) whereas the original graph of $y = 2^x$ passes through (0, 1).

The asymptote of the graph of $y = 2^x + 1$ is now at $y = 1$.

The graph of $y = 2^x - 1$ passes through (0, 0) whereas the original graph of $y = 2^x$ passes through (0, 1).

The asymptote of the graph of $y = 2^x - 1$ is now at $y = -1$.

For every value of x, the value of y for $y = 2^x + 1$ is one unit greater than the corresponding y-value for $y = 2^x$.

For every value of x, the value of y for $y = 2^x - 1$ is one unit less than the corresponding y-value for $y = 2^x$.

4.4.6 Evaluating exponential functions

To evaluate any function, we substitute the given value for x.

Example 4.4.6a

For $f(x) = 3^x$, find **a** $f(1)$

b $f(-5)$

Solution

a $f(1) = 3^1 = 3$

b $f(-5) = 3^{-5} = 0.004\,12$

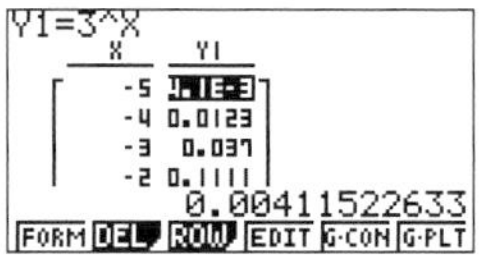

Example 4.4.6b

Find $P(1)$ given $P(x) = 20 \times 5^{2x+1}$

Solution

$P(1) = 20 \times 5^{2\times1+1} = 20 \times 5^3$

$= 2500$

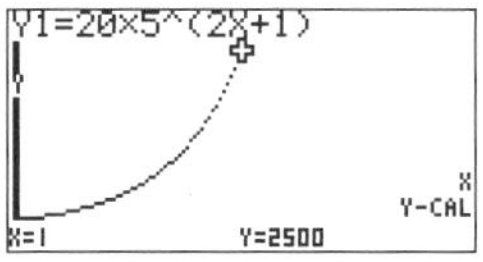

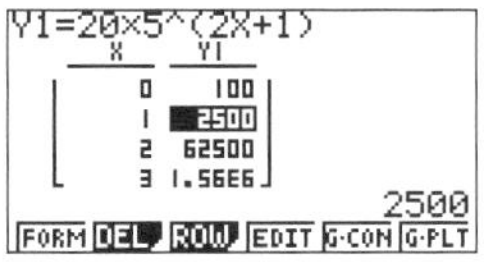

4.4.7 Finding exponential functions from their graphs

Example 4.4.7a

The function whose graph is drawn in fig. 1 is of the form $f(x) = a^x$. Find the value of a and hence write down the equation of $f(x)$.

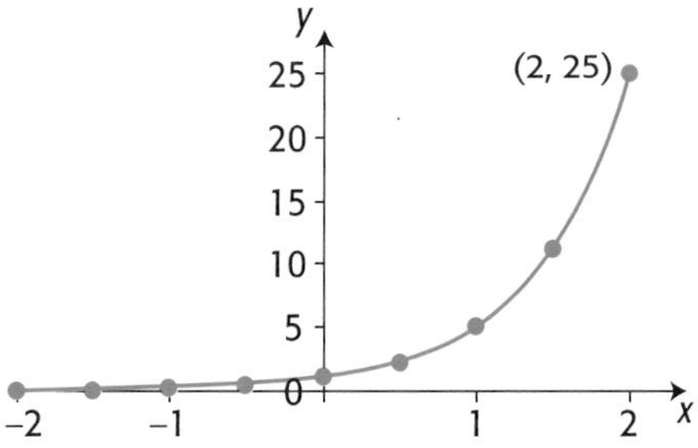

Figure 1

Solution

Substitute (2, 25) into $f(x)$.

$25 = a^2 \therefore a^2 = 25 \ a = 5 \ (a > 0)$

The function is $f(x) = 5^x$.

Example 4.4.7b

The function whose graph is drawn in fig. 2 is of the form $g(x) = ka^x$. Write two equations in k and a and solve them to find the values of k and a and hence find the equation of $g(x)$.

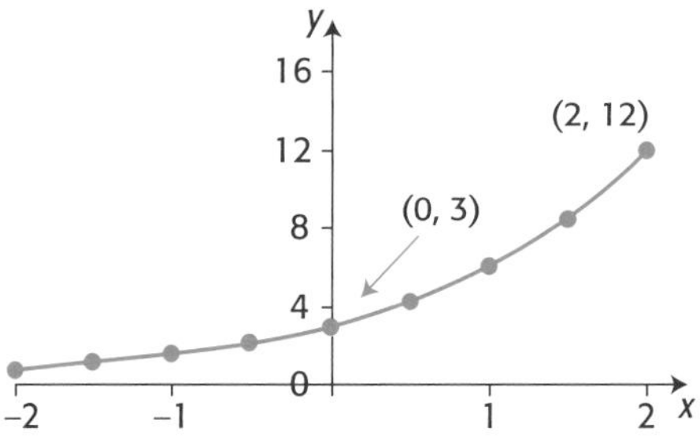

Figure 2

Solution

Substitute (0, 3) into $g(x)$.

$3 = ka^0 \quad \therefore k = 3$

Now substitute (2, 12) into $y = ka^x$.

$12 = 3 \times a^2$

$a^2 = 4 \qquad \therefore a = 2 \ (a > 0)$

The function is $g(x) = 3 \times 2^x$.

4.4.8 Applications of exponential functions

Compound interest

$$Fv = C\left(1+\frac{r}{100\,k}\right)^{kn}$$

Fv = final value
C = initial investment
r = percentage rate
k = number of compound periods per year
n = number of years

Example 4.4.8a

A sum of €2000 is invested at 6% per annum nominal interest, compounding monthly.

Find the value of the investment after

a 6 months b 2 years.

Solution

a $Fv = 2000\left(1+\frac{6}{1200}\right)^{6}$

$= €2060.76$

b $Fv = 2000\left(1+\frac{6}{1200}\right)^{24}$

$= €2254.32$

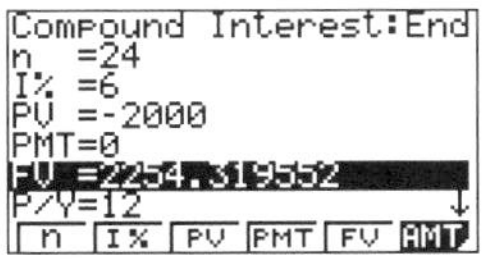

Exponential growth

Example 4.4.8b

The population increase on a Pacific island since 1970 can be modelled by the function

$P = 2500 \times 1.2^{0.3t}$, where t is the number of years after 1970.

a Draw the population graph for $0 \leq t \leq 50$ and $0 \leq P \leq 40\,000$.

b Use the graph to estimate

i the population of the island in 1970

ii the population of the island in 2000

iii the first year that the population of the island will be greater than 34 000.

Solution

a

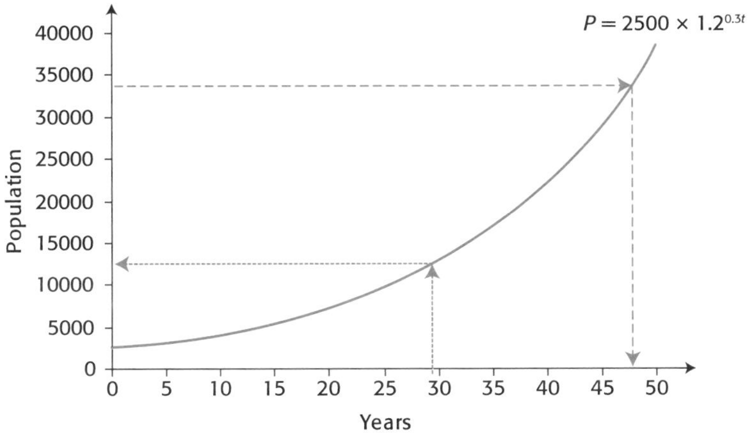

b i 2500 ii 13 000 iii $t = 48$, that is in the year 2018

Example 4.4.8c

For the population function in Example 4.4.7b above, find exact answers to parts **b i, ii** and **iii**.

Solution

i $P = 2500 \times 1.2^{0.3 \times 0} = 2500$

ii $P = 2500 \times 1.2^{0.3 \times 30} = 2500 \times 1.2^{9} = 12\ 900$

iii $2500 \times 1.2^{0.3t} = 34\ 000 \quad t = 47.7$, that is in the year 2018

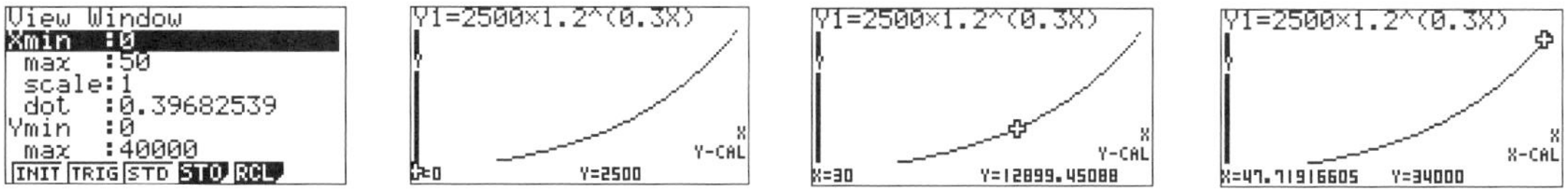

Exponential decline (decay)

Example 4.4.8d

The temperature (°C) of a pot of water, removed from the stove after boiling, is given by $T = 18 + 82(0.9)^{t}$, where t is the number of minutes after the pot is removed from the stove.

a Draw the graph of temperature against time for $0 \le t \le 50$.

b Find the temperature of the water at the moment it is removed from the stove.

c Calculate the temperature of the water 10 minutes after being removed from the stove.

d How many minutes will have elapsed before the temperature of the water is at 40 °C?

e Find the minimum possible temperature of the water.

Solution

a

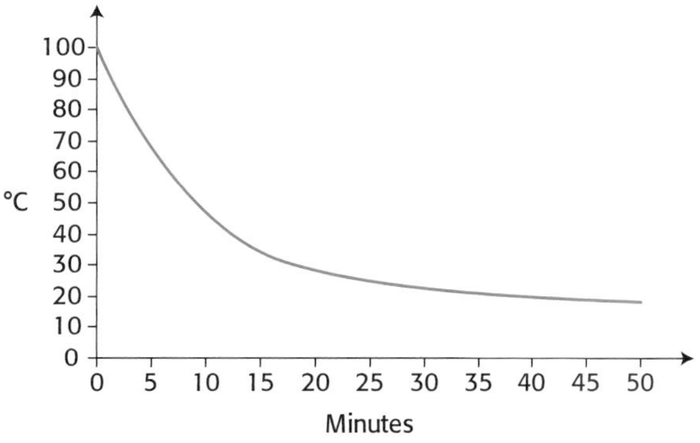

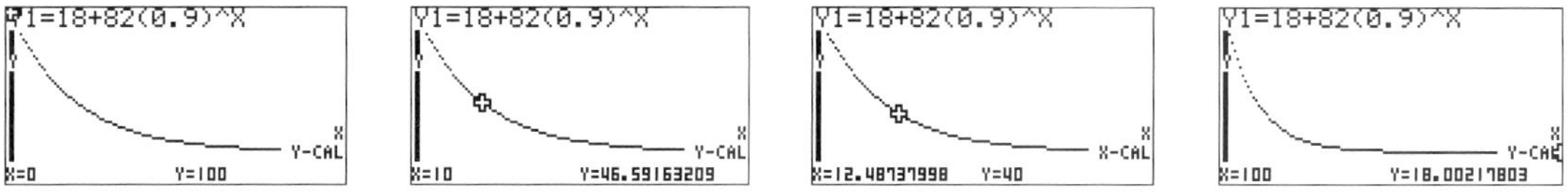

b $T = 18 + 82(0.9)^{0} = 100\ °C$

c $T = 18 + 82(0.9)^{10} = 46.6\ °C$

d $18 + 82(0.9)^{t} = 40$, $t = 12.5$ minutes

e Temperature approaches 18 °C.

Exercise 4.4

1 Evaluate **a** 2^4 **b** 5^0 **c** 3^{-1} **d** $125^{-0.25}$ **e** $64^{0.5}$
f 6×10^{-2}

2 Evaluate these exponential expressions for
i $x = -3$ **ii** $x = 0$ **iii** $x = 3$

a 5^x **b** -5^x **c** 5^{-x} **d** $5^{0.2x}$ **e** $5^{-0.2x}$ **f** -5^{-x}

3 Evaluate these exponential functions for
i $x = -5$ **ii** $x = 0$ **iii** $x = 5$

a $f(x) = \left(\frac{1}{2}\right)^x$ **b** $g(x) = 2\left(\frac{2}{3}\right)^{x+1}$ **c** $h(x) = \left(\frac{2}{5}\right)^{2x}$

d $y = 8^{2x-3}$

4 **a** Given $y = k \times 2^x$ and $y = 48$ when $x = 2$, find the value of k.

b Given $f(x) = 5a^{2x}$ and $f(3) = 1220.7$, find the value of a.

5 Construct a table of values for $-2 \le x \le 2$ and hence draw the graphs of

a $y = 2^{x+1} - 1$ **b** $f(x) = \left(\frac{1}{2}\right)^{2x-1}$

6 On the same set of axes, draw the graphs of $y = 3 = \left(\frac{1}{2}\right)^{0.5x+1}$ and

$y = x + 3$ for a domain of $-3 \le x \le 3$ and range of $0 \le y \le 10$.
Find the point of intersection of the two graphs.

7 The diameter of a tree (in cm) is given by $d = 3 \times 2.5^{0.12t}$ where t represents the number of years after planting.

Calculate
a the number of years taken for the diameter of the tree to double
b the diameter of the tree after 15 years.

8 **a** The growth of the population of a colony of bacteria is given by $P(t) = 200(40)^{\frac{t}{8}}$, where t represents hours.

i Determine the initial population of the colony.
ii Find the number of hours it will take for the population to double.
iii Find the number of hours it will take for the population to reach 1 000 000.
iv Calculate the number of bacteria present after 24 hours.

b The area of land covered by rainforest in a country is decreasing by 3.5% per annum. There are 1.5 million hectares of rainforest at the moment.

i Determine the area of land that will be covered by rainforest in ten years' time.
ii Write an exponential function which can be used to determine the area of land, L, that will be covered by rainforest after t years.
iii Draw the graph of the function for the first ten years.
iv Calculate the number of years it will take for the rainforest to cover less than 0.5 million hectares.

c $1000 is invested at 8% nominal interest, with quarterly compounds. Use the compound interest formula to find
   i the value of the investment after one year and after five years.
   ii the number of years it will take for this investment to double.
   iii the annual rate of interest required for the doubling period to be one year less than your answer to part **ii**.

d The value of a car has depreciated 15% per annum. After four years, the car was worth £7600.
   i Find the original value of the car.
   ii Find the number of years it will take for the value of the car to fall below £4000.
   iii Calculate the value of the car after ten years. If the owner sells the car after ten years for £3500, determine the profit/loss made against the book value of the car.

Leonhard Euler (1707–1783)

Leonhard Euler (1707 –1783) was a Swiss mathematician and physicist. He is considered to be the pre-eminent mathematician of the 18th century and one of the greatest of all time; he is certainly one of the most prolific, with collected works filling over 70 volumes. Euler developed many important concepts and proved numerous lasting theorems in diverse areas of mathematics, from calculus to number theory to topology. In the course of this work, he introduced much of modern mathematical terminology, defining the concept of a **function**, and its notation, such as **sin**, **cos**, and **tan** for the trigonometric functions.

4.5 Graphs and properties of the sine and cosine functions

We can use **sine** and **cosine** curves to model real life situations which demonstrate periodic or 'wave' behaviour such as tides, sound and hours of daylight.

Students should be familiar with the nature of the graphs of functions, for $0 \leq x \leq 360°$ in particular, and be aware of the effects of changes to the values of a, b, and c in the functions $f(x) = a\sin bx + c$ and $f(x) = a\cos bx + c$; $a, b, c, \in \mathbb{Q}$.

It is useful to remember these values of **sin** and **cos** that are equal to 0 and 1.

$\sin 0° = 0$	$\cos 0° = 1$
$\sin 90° = 1$	$\cos 90° = 0$
$\sin 180° = 0$	$\cos 270° = 0$
$\sin 360° = 0$	$\cos 360° = 1$

Note that calculators should always be set to "degrees".

4.5.1 The basic sine and cosine curves, $y = f(x)$, $0° \le x \le 360°$

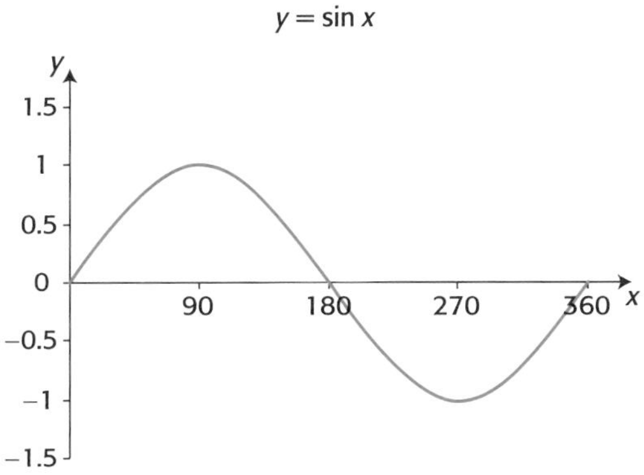

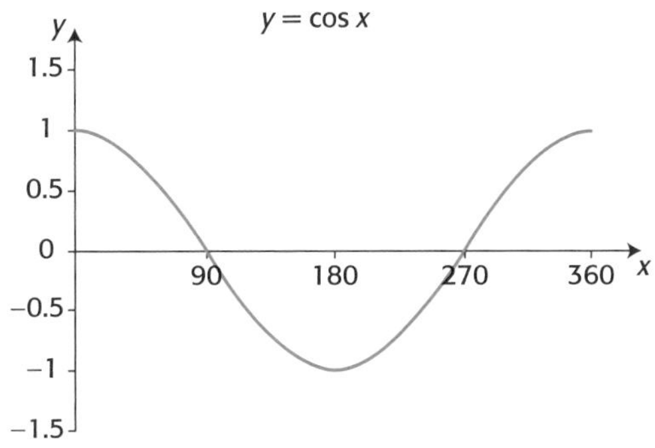

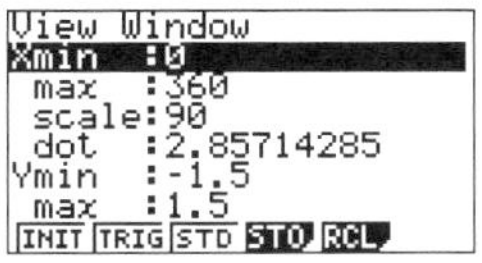

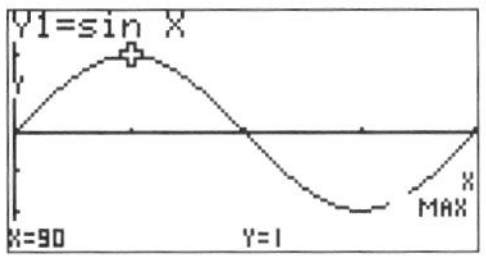

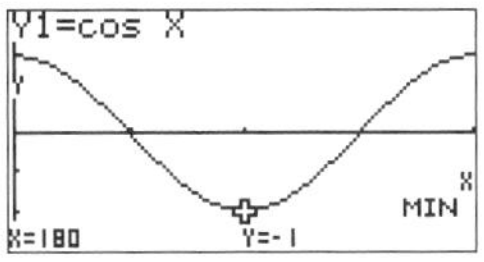

The values of sin x and cos x lie between –1 and +1.
Each curve returns to its original position at 360°.

The sine curve begins at the midpoint of its range, (0, 0).
The cosine curve begins at its maximum, (0, 1).

The amplitude and period

The **amplitude** of a periodic function is half the distance between the maximum and minimum values of the function.
The **period** of a function is the distance along the horizontal axis of one full cycle, that is the number of degrees required for the function to repeat.

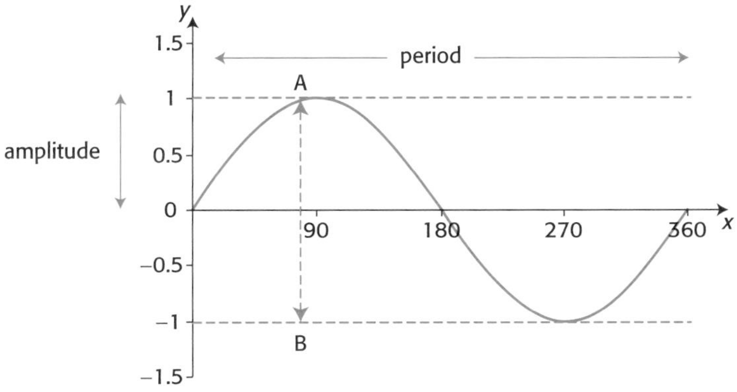

The amplitude is half the distance from A to B = 1. The period is 360°.

4.5.2 The effect of a in $y = a\sin x$ and $y = a\cos x$

In the following examples, the dotted lines are the graphs of sin x and cos x.

Example 4.5.2a

$a > 1$

Draw the graphs of $y = 2\sin x$ and $y = 2\cos x$ for $0° \le x \le 360°$.
Determine the period and amplitude for each function.

Solution

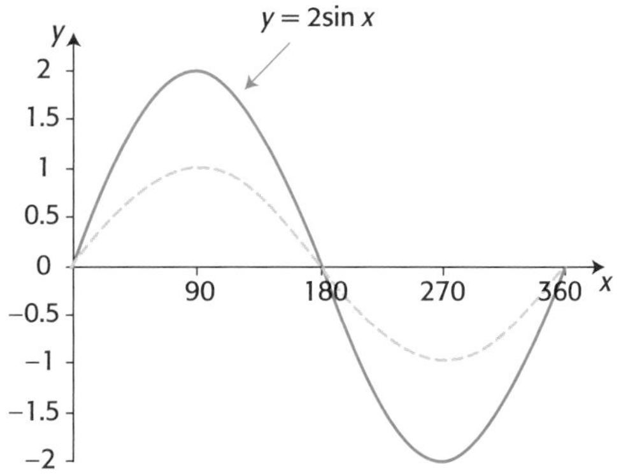

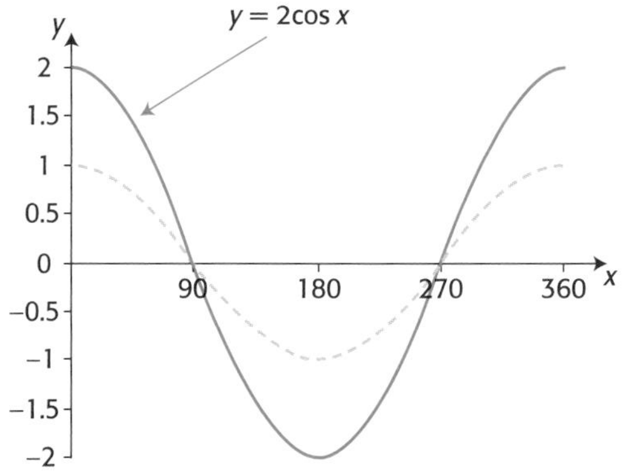

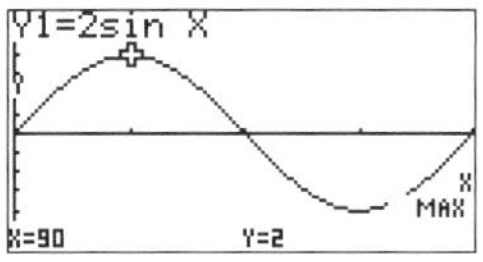

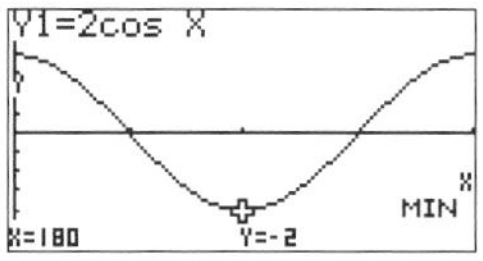

The sine curve begins at mid-range (0, 0), and has a maximum at (90°, 2). The cosine curve begins at its maximum (0, 2).

The period is 360°. The amplitude of both $y = 2\sin x$ and $y = 2\cos x$ is 2.

Example 4.5.2b

$0 < a < 1$

Draw the graphs of $y = \frac{1}{2}\sin x$ and $y = \frac{1}{2}\cos x$ for $0° \leq x \leq 360°$.

Determine the period and amplitude for each function.

Comment on the effect of the value of a in the functions $y = a\sin x$ and $y = a\cos x$.

Solution

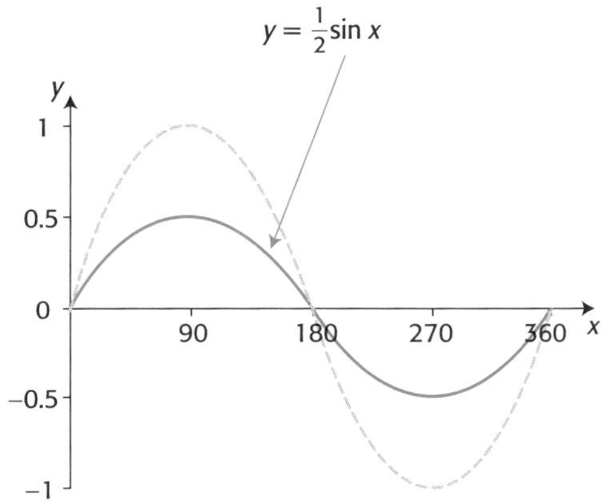

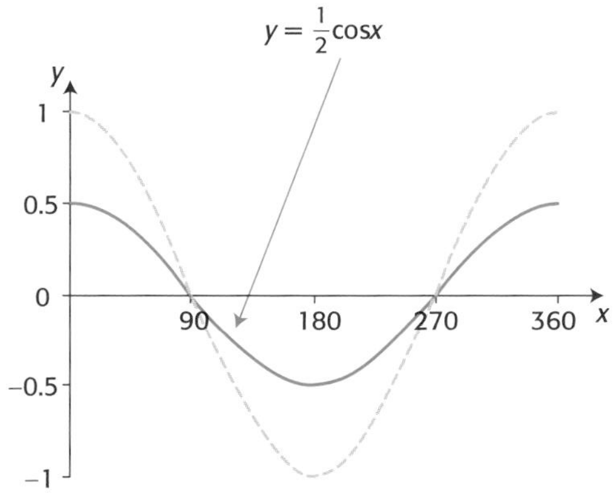

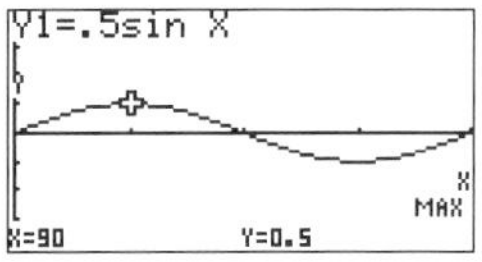

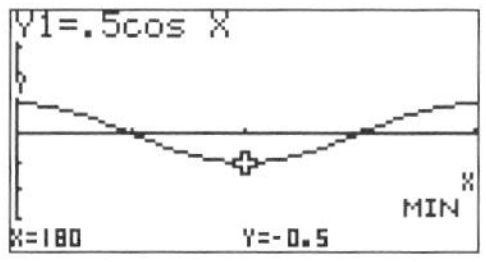

The sine curve begins at (0, 0) and has a maximum at (90, 0.5).

The cosine curve begins at its maximum (0, 0.5).

The period is 360°. The amplitude of both $y = 0.5\sin x$ and $y = 0.5\cos x$ is 0.5.

- The amplitude of the functions $y = a\sin x$ and $y = a\cos x$ is the value of a.

Example 4.5.2c

$a < 0$

Draw the graphs of $y = -\sin x$ and $y = -\cos x$ for $0° \le x \le 360°$. Determine the value of the period and amplitude for each function. Note the effect of $-a$ in the functions $y = a\sin x$ and $y = a\cos x$.

Solution

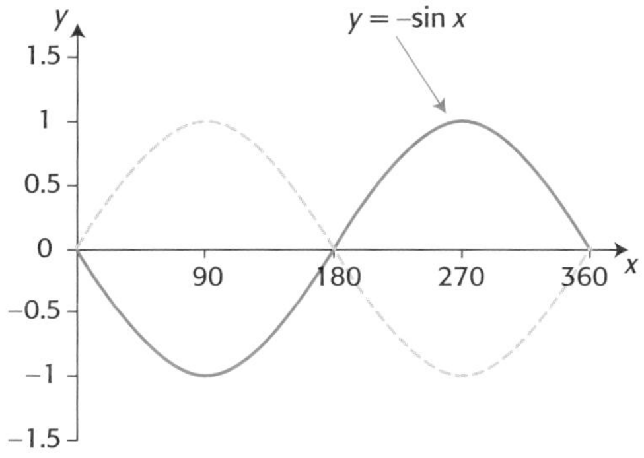

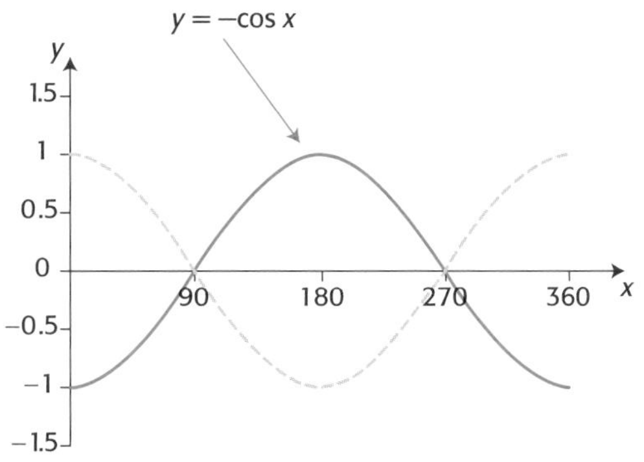

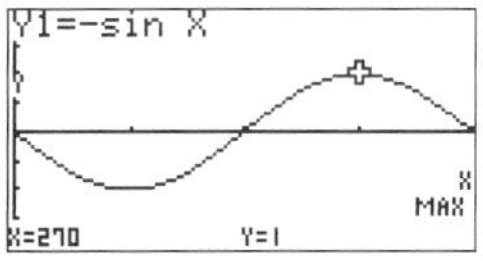

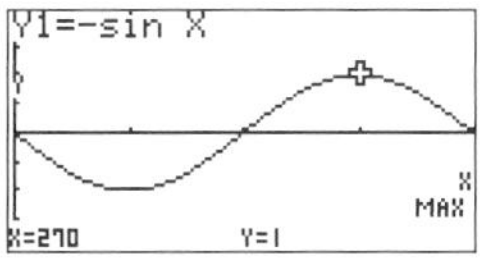

The amplitude is 1. The period is 360°.

When the value of a is negative, the graphs of $y = \sin x$ and $y = \cos x$ are reflected in the x-axis.

- The amplitude and period of the functions $y = a\sin x$ and $y = a\cos x$ are not affected by a negative value of a.

4.5.3 The effect of b in $y = \sin bx$ and $y = \cos bx$

Example 4.5.3a

Draw the graphs of $y = \sin 2x$ and $y = \cos 2x$ for $0° \le x \le 360°$. Determine the period and amplitude for these curves.

Solution

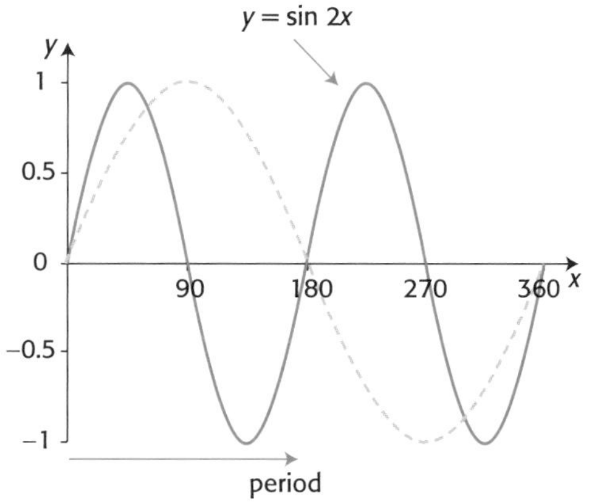

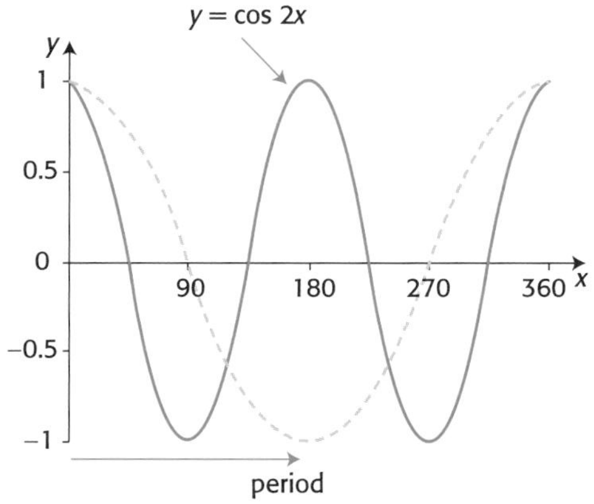

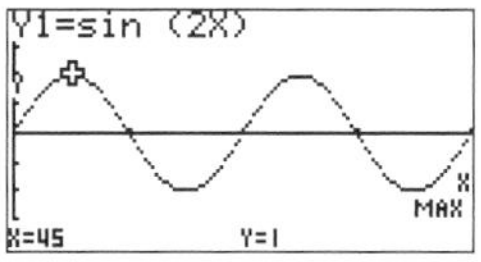

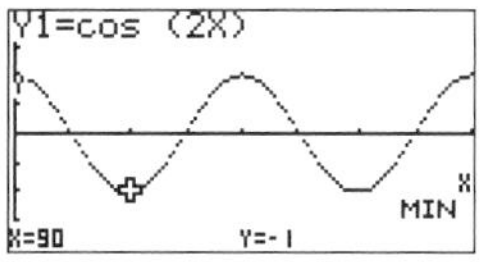

The amplitude is 1 (the value of *a* is 1). The period is 180°.

The period for $y = \sin 2x$ and $y = \cos 2x$ is half the period for $y = \sin x$ and $y = \cos x$.

When $b = 2$, the period is $360 \div 2 = 180°$.

Example 4.5.3b

Draw the graphs of $y = \sin 0.5x$ and $y = \cos 0.5x$ for $0° \leq x \leq 720°$. Determine the period and amplitude for these curves.

Solution

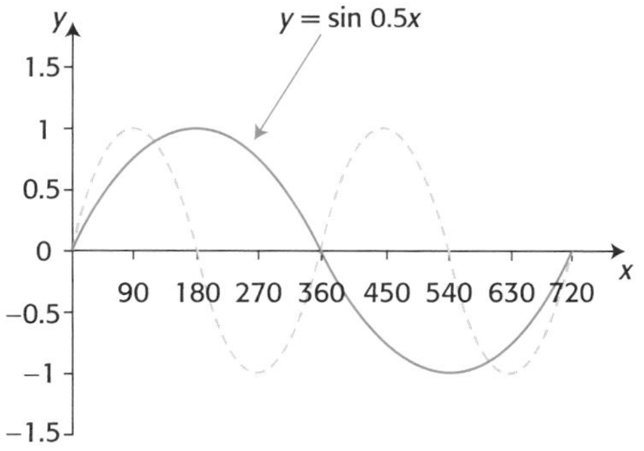

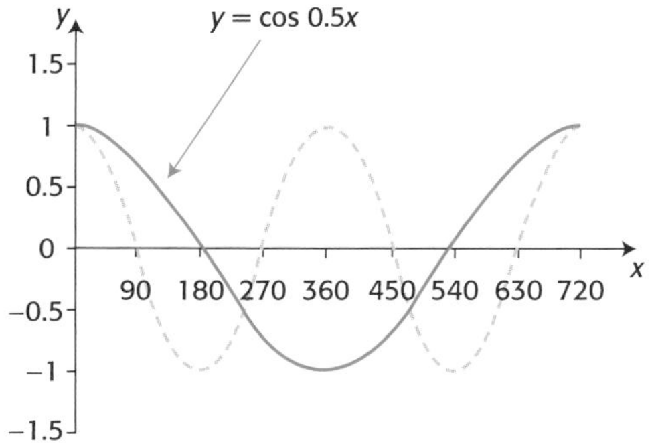

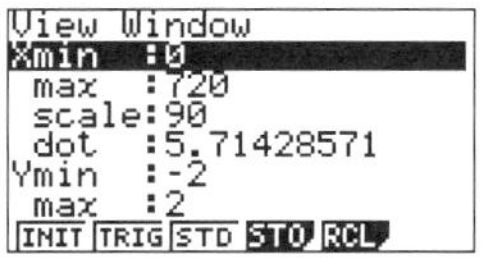

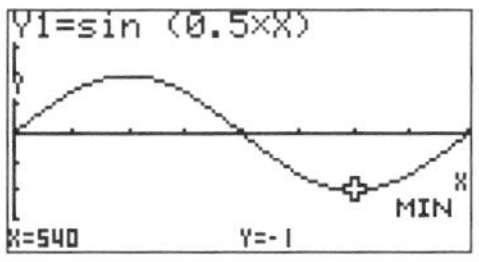

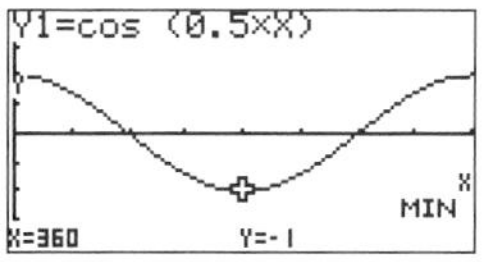

The amplitude is 1 (the value of *a* is 1). The period is 720°.

The period for $y = \sin 0.5x$ and $y = \cos 0.5x$ is double the period for $y = \sin x$ and $y = \cos x$.

When $b = 0.5$, the period is $360 \div 0.5 = 720°$.

The **period** of the functions $y = \sin bx$ and $y = \cos bx$ is $\frac{360}{b}$. The value of b does not affect on the amplitude of the functions.

The period is always expressed as a positive value.

4.5.4 The effect of c in $y = \sin x + c$ and $y = \cos x + c$

Example 4.5.4a

Draw the graphs of $y = \sin x + 1$ and $y = \cos x + 1$ for $0° \leq x \leq 360°$. Observe the effect that the value of c has on the curves.

Solution

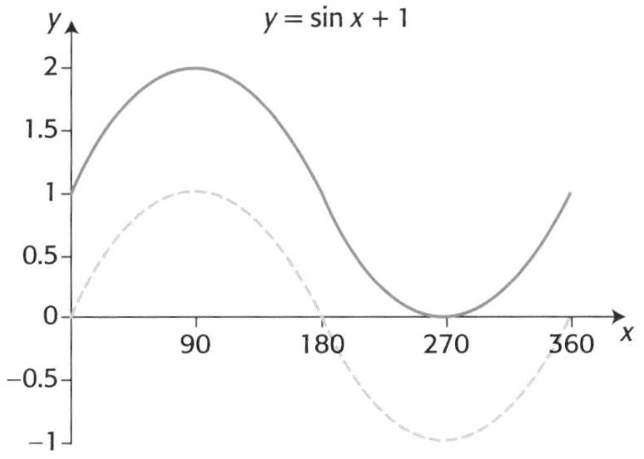

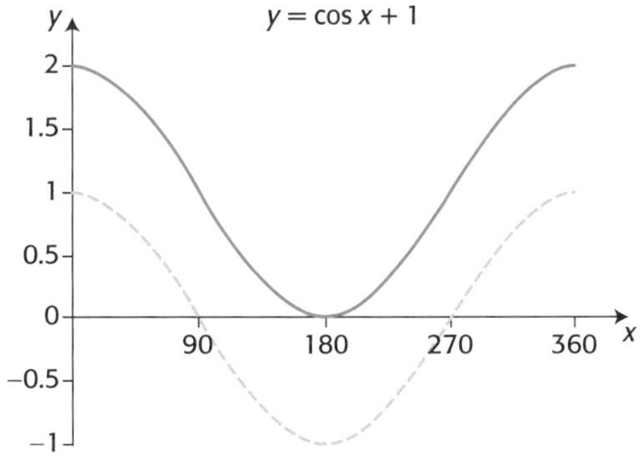

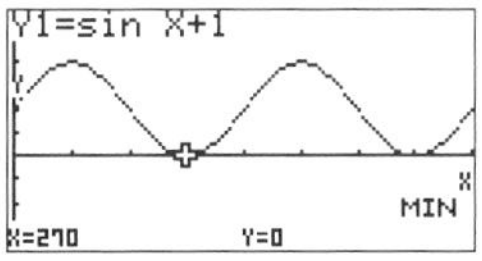

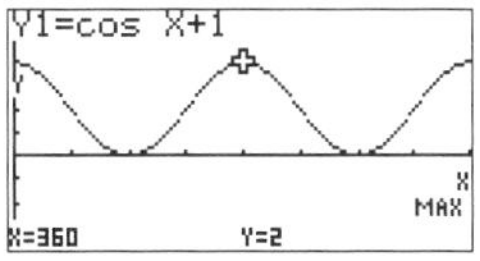

The curves have been shifted vertically by $c = 1$ unit. The period and amplitude of the functions are not affected.

Example 4.5.4b

Draw the graphs of $y = \sin x - 2$ and $y = \cos x - 2$ for $0° \leq x \leq 360°$. Observe the effect that the value of c has on the curves.

Solution

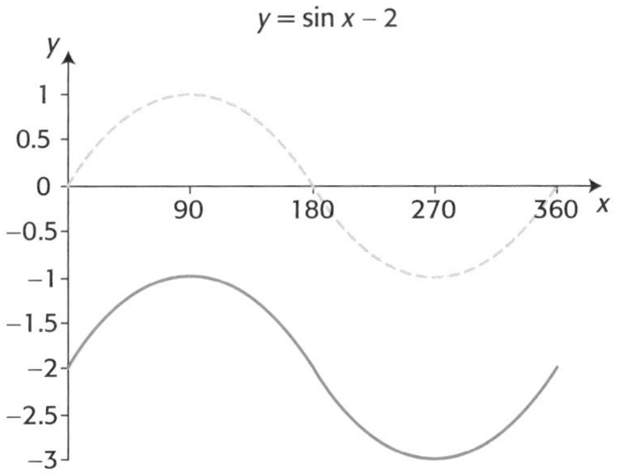

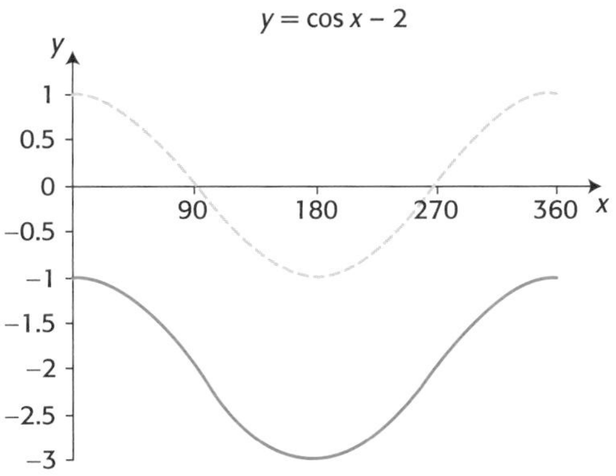

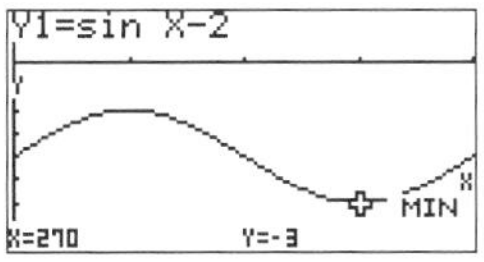

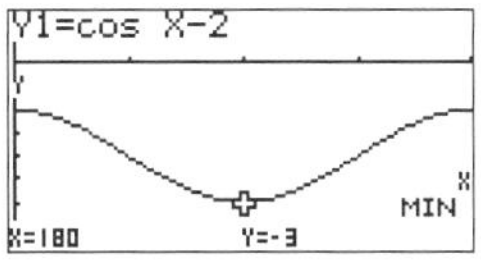

The curves have been shifted vertically by $c = -2$ units.
The period and amplitude of the functions are not affected.

When a number, c, is added to or subtracted from the functions $y = \sin x$ and $y = \cos x$ the curves are shifted (translated) vertically by the value of c.

4.5.5 The graphs of $y = a\sin bx + c$ and $y = a\cos bx + c$

Example 4.5.5a

Determine the amplitude, period and the vertical shift for the function $f(x) = 2\sin 2x - 1$, $0° \le x \le 360°$.
Sketch the graph of $y = f(x)$.

Solution

$a = 2$ – amplitude is 2
$b = 2$ – period is $360 \div 2 = 180°$.
$c = -1$ – shift down by 1 unit.
Sine curve starts at $(0,-1)$

$f(x) = 2\sin 2x - 1$

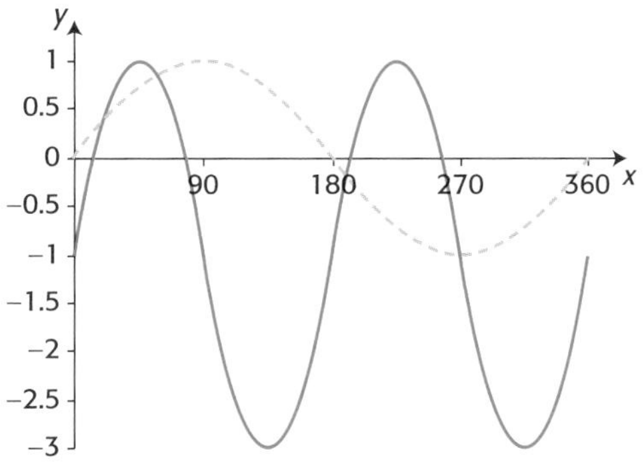

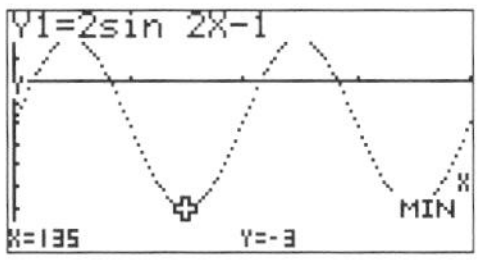

Example 4.5.5b

Determine the amplitude, period and the vertical shift for the function

$g(x) = 0.5\cos 0.5x + 2$, $0° \le x \le 720°$.

Sketch the graph of $y = g(x)$.

Solution

$a = 0.5$ – amplitude is 0.5
$b = 0.5$ – period is $360 \div 0.5 = 720°$.
$c = +2$ – shift up by 2 units.
Cosine curve starts at $(0, 2.5)$

$g(x) = 0.5 \cos 0.5x + 2$

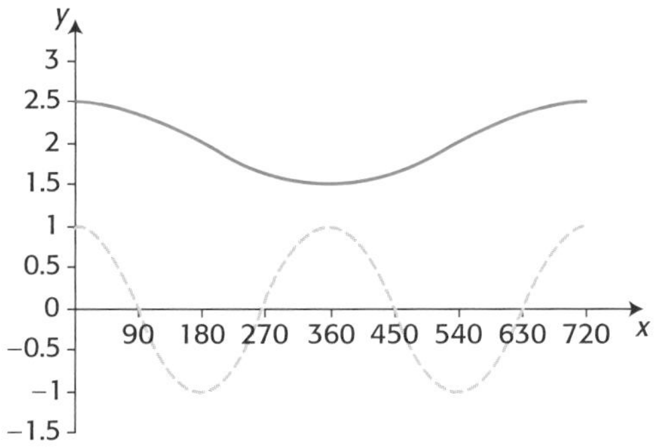

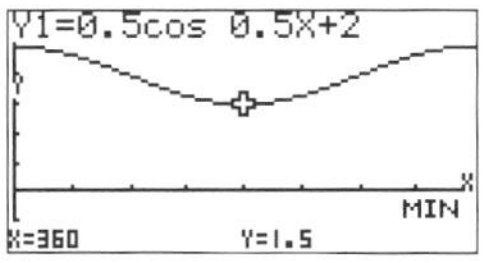

Example 4.5.5c

Determine the amplitude, period and the vertical shift for the function $f(x) = 0.2\sin 3x + 1$, $0° \le x \le 360°$.
Sketch the graph of $y = f(x)$.

Solution

$a = 0.2$ – amplitude is 0.2
$b = 3$ – period is $360 \div 3 = 120°$.
$c = +1$ – shift up by 1 unit.
Sine curve starts at $(0, 1)$

Example 4.5.5d

Determine the amplitude, period and the vertical shift for the function $g(x) = -3\cos x - 2$, $0° \le x \le 360°$. Sketch the graph of $y = g(x)$.

Solution

$a = -3$ – amplitude is 3
$b = 1$ – period is $360 \div 1 = 360°$.
$c = -2$ – shift down by 2 units.
Cosine curve starts at $(0, -5)$ Negative sign means it starts at its minimum.

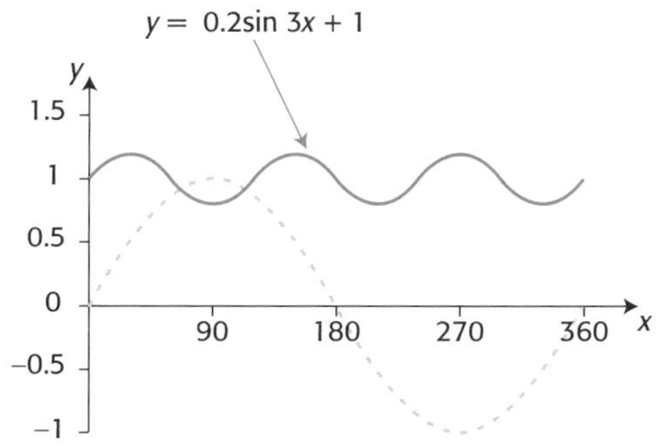

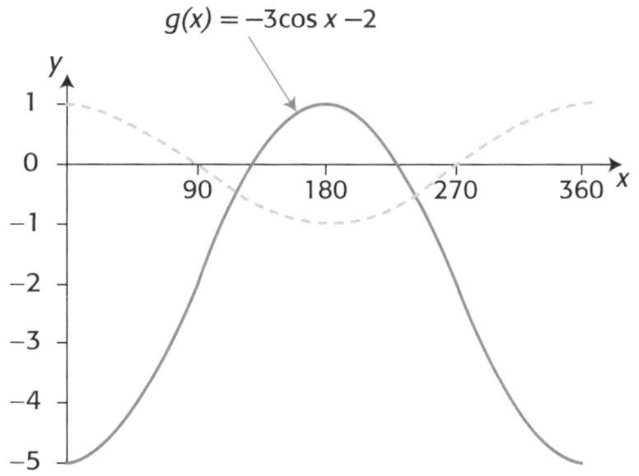

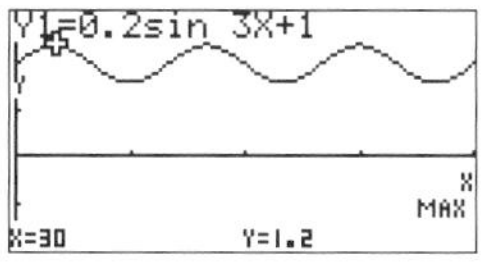

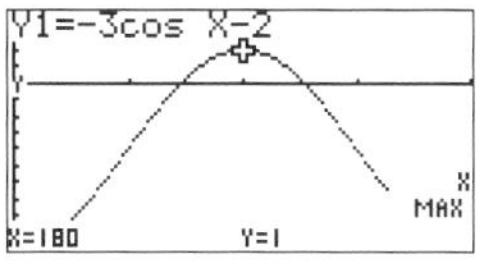

4.5.6 Finding the function from its graph

Determine the amplitude, period and vertical shift for the following functions and hence write the equation of the curves in the form $y = a\sin bx + c$ or $y = a\cos bx + c$.

Example 4.5.6a

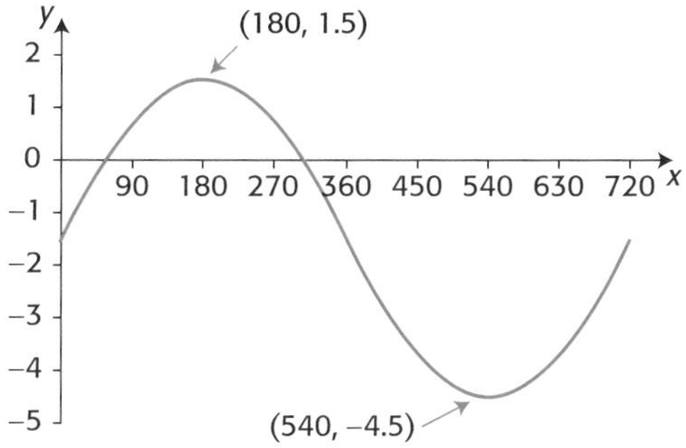

Solution

This is a sine curve: starts at mid-value.

amplitude: $a = \dfrac{1.5-(-4.5)}{2} = 3$

period is 720°, $\dfrac{360}{b} = 720$, $b = 0.5$

vertical shift is −1.5, $c = -1.5$

The function is

$y = 3\sin 0.5x - 1.5$

Example 4.5.6b

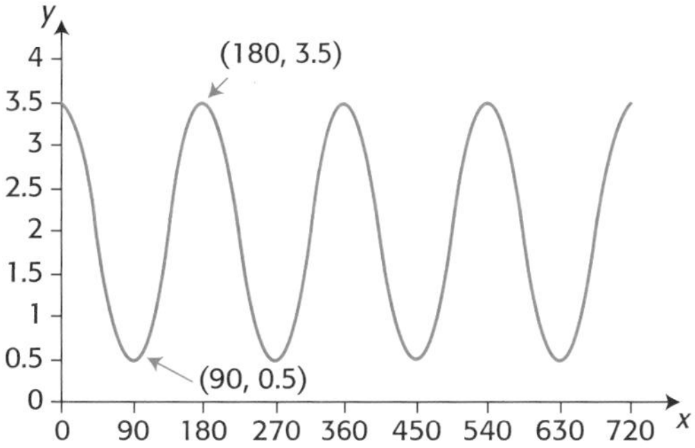

Solution

This is a cosine curve: starts at maximum.

amplitude: $a = \dfrac{3.5 - 0.5}{2} = 1.5$

period is 180°, $b = 2$

vertical shift is +2, $c = 2$

The function is

$y = 1.5\cos 2x + 2$

4.5.7 Evaluating sine and cosine functions

There will be only one value of $f(x)$ for each value of x.

Example 4.5.7a

Find $f(30)$ for $f(x) = 2\sin x - 0.5$

Solution

$f(30) = 2\sin 30° - 0.5$

$(GDC) = 0.5$

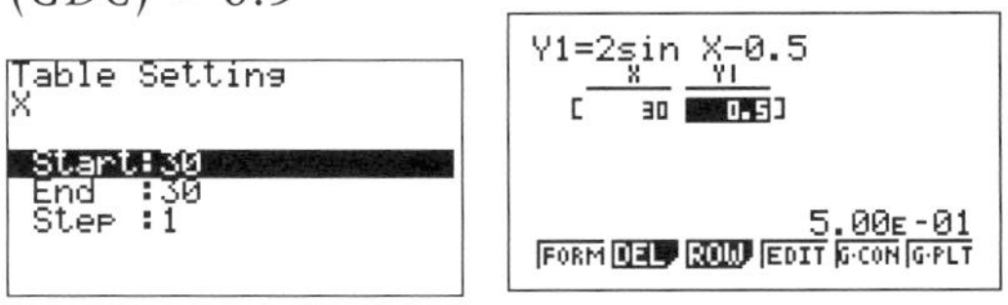

Example 4.5.7b

Find $g(120)$ for $g(x) = 0.5\cos 2x + 5$

Solution

$f(120) = 0.5\cos 240° + 5$

$(GDC) = 4.75$

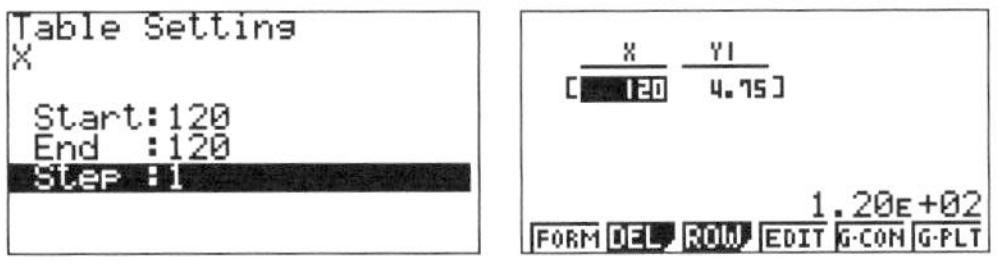

There may be more than one value of x for each value of $f(x)$ for any given domain.

Example 4.5.7c

Solve for x given

$2\sin x - 0.5 = 1,$
$0° \le x \le 450°$

Solution Graphical

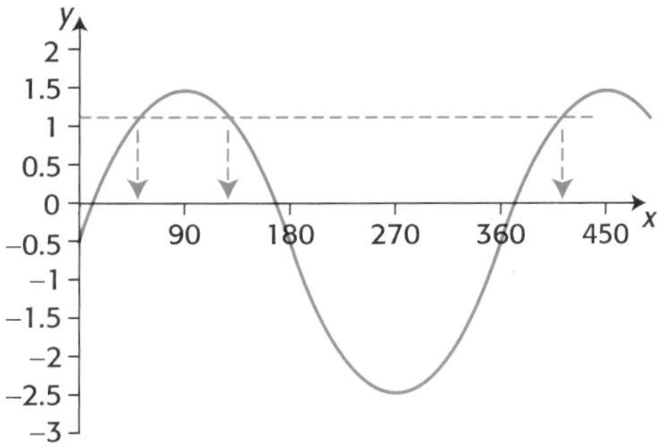

$x \approx 50°$ or $130°$ or $410°$

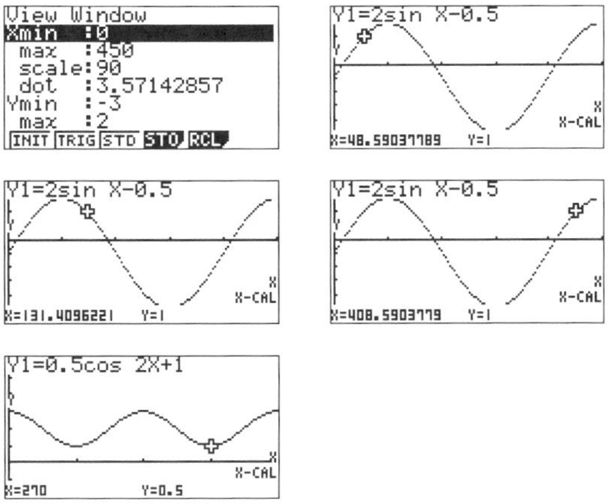

Example 4.5.7d

Solve for x given

$0.5\cos 2x + 1 = 0.5,$
$0° \le x \le 360°$

Solution Graphical

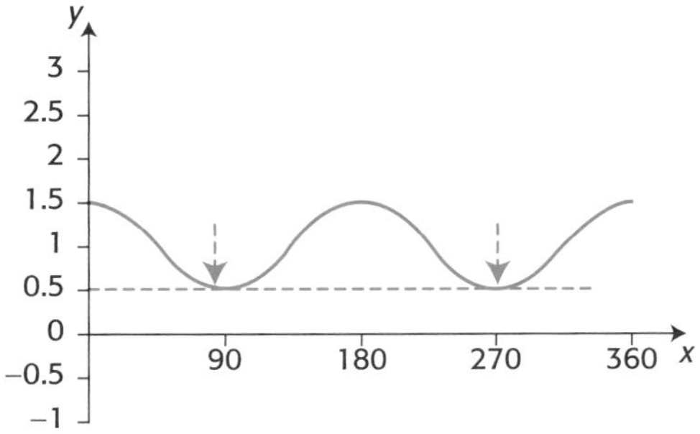

$x \approx 90°$ or $270°$

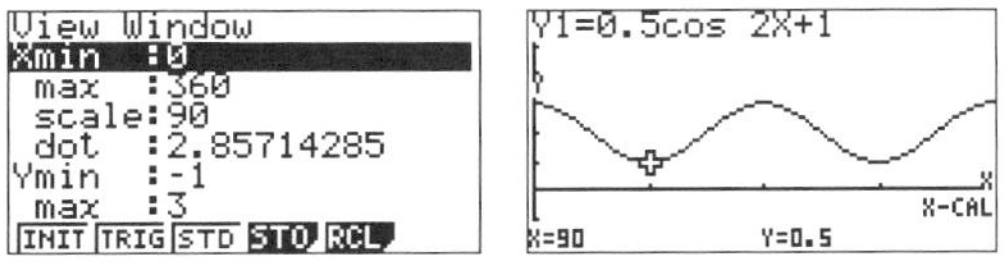

4.5.8 Negative domain

Example 4.5.8a

Draw the graphs of $y = \sin x$ and $y = \cos x$ for $-360° \le x \le 0°$. Determine the amplitude and period of each function.

Solution

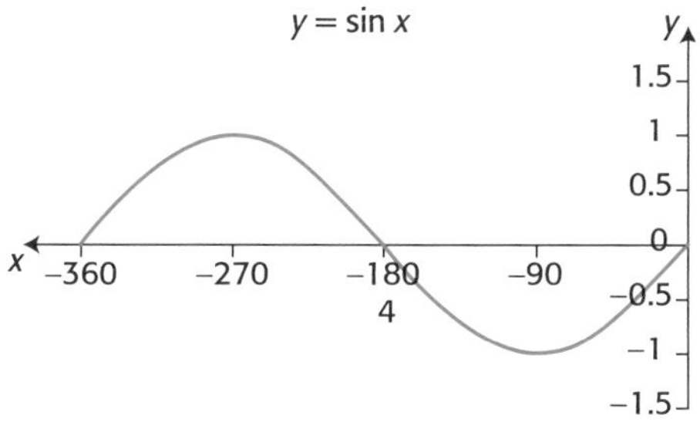

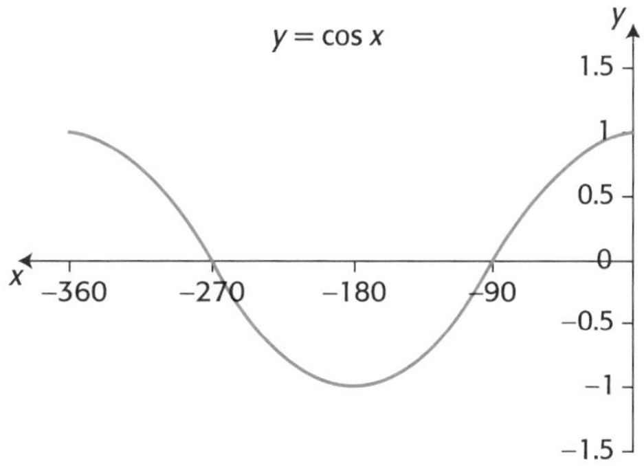

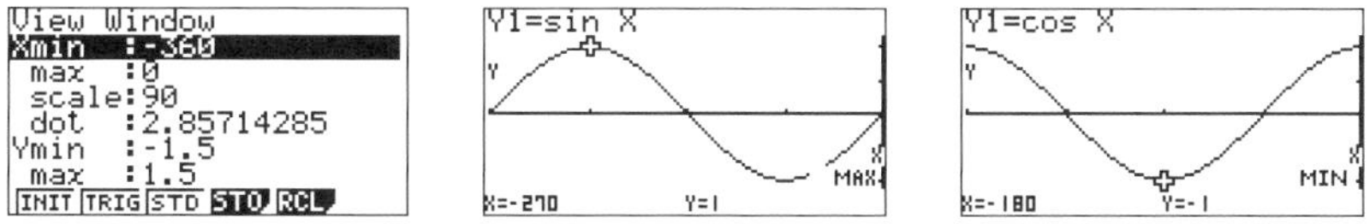

The amplitude is 1. The period is 360°. The negative domain has no effect.

Example 4.5.8b

Solve for x given

$0.4\sin 0.5x + 1 = 0.75$,

$-720° \le x \le 0°$

Solution

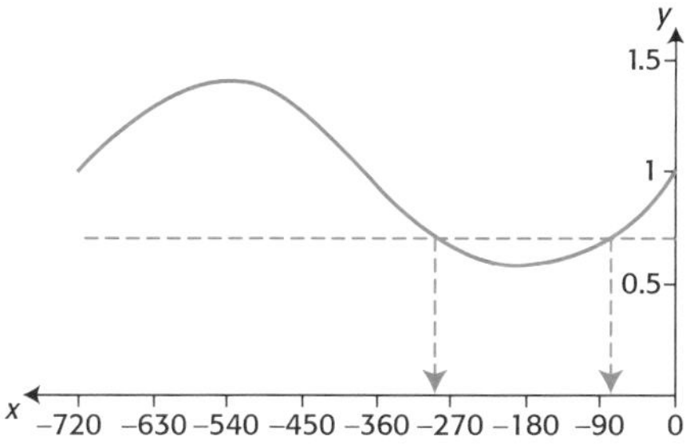

$x \approx -75°$ or $-280°$

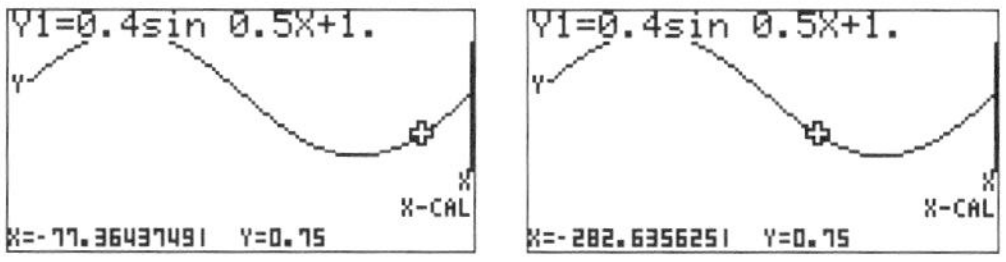

Example 4.5.8c

Solve for x given

$2\cos x - 1 = -2$,

$-360° \le x \le 0°$

Solution

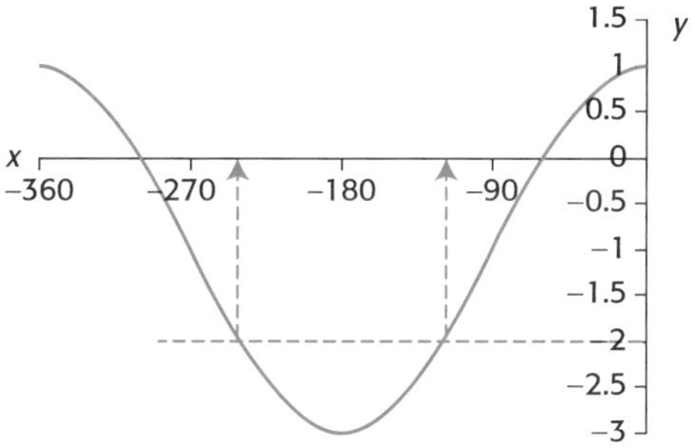

$x \approx -120°$ or $-240°$

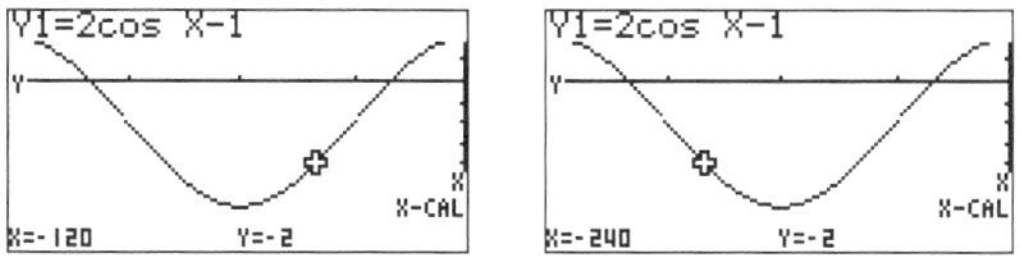

4.5.9 Applications for sine and cosine functions

Example 4.5.9a

The height of the tide in a local harbour was measured. It was found that the water rose 3 m above the mean sea level and fell 3 m below the mean sea level in one complete cycle over a 12-hour period.

a Plot the graph to represent the height of water with respect to the mean sea level over an 18-hour period.

b Assuming that the height of water with respect to mean sea level can be modelled by a sine function, determine the amplitude, period and the amount of vertical shift for this function.

c Determine the equation of the sine function in the form $H(t) = a\sin bt + c$, with bt measured in degrees and where t is in hours. $H(t)$ is the height with respect to mean sea level.

d If the mean sea level occurred at midnight, use the equation of the sine function to determine the height of the tide at 8 am.

Solution

a We construct a table of values and plot the graph.

Time, *t* hours	0	3	6	9	12	15	18
H(*t*), metres	0	3	0	−3	0	3	0

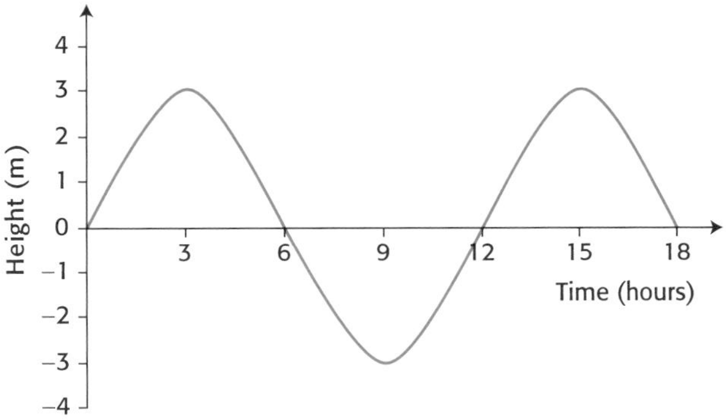

b Amplitude is

$$\frac{\max - \min}{2} = \frac{3 - -3}{2} = 3,\ b = \frac{360}{\text{period}} = \frac{360}{12} = 30°$$

Vertical shift is 0.

c The equation of the sine function is $H(t) = 3\sin 30t$

d 8 am is 8 hours after midnight.... $H(8) = 3\sin(30 \times 8) = -2.60$ m, that is 2.60 m below mean sea level.

Check answers using GDC.

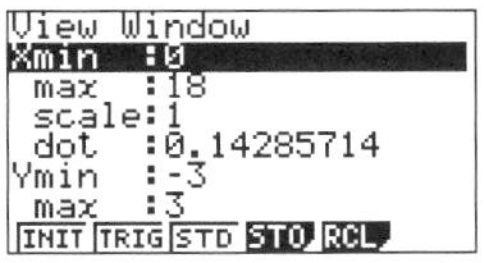

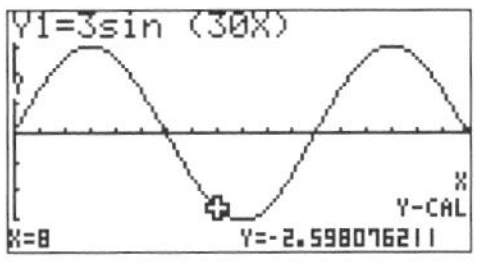

Example 4.5.9b

The air temperature (°C) for a 24-hour period at a particular location can be modelled by the function $T = 25 - 5\cos 10t$, where t is the number of hours after midnight and $10t$ is measured in degrees.

a Complete the table of values for the temperature function.

Time, *t* hours	0	4	8	12	16	20	24
Temp, °C	20			27.5			27.5

b Draw the graph of T for $0° \le t \le 24°$.

c Use the graph to estimate
  i the temperature at 10 am
  ii the time interval when the temperature was greater than 28 °C
  iii the increase in temperature between 10 am and midday
  iv the time of day when the temperature was at its maximum.

Solution

a

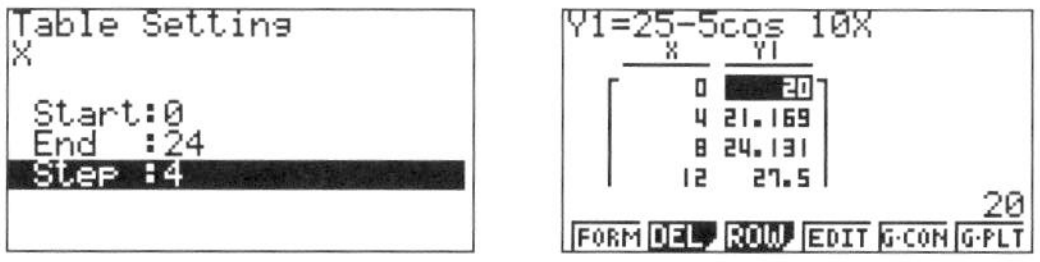

Time, *t* hours	0	4	8	12	16	20	24
Temp, °C	20	21.2	24.1	27.5	29.7	29.7	27.5

b

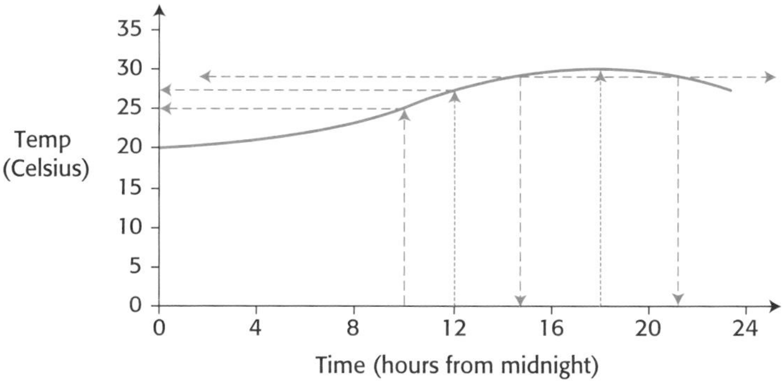

c i Temperature at 10 am ≈ 25 °C
  ii Temperature > 28 °C between 2:30 pm and 9:30 pm.
  iii Increase in temperature of 27.5 – 25 = 2.5 °C
  iv Maximum temperature at 6 pm.

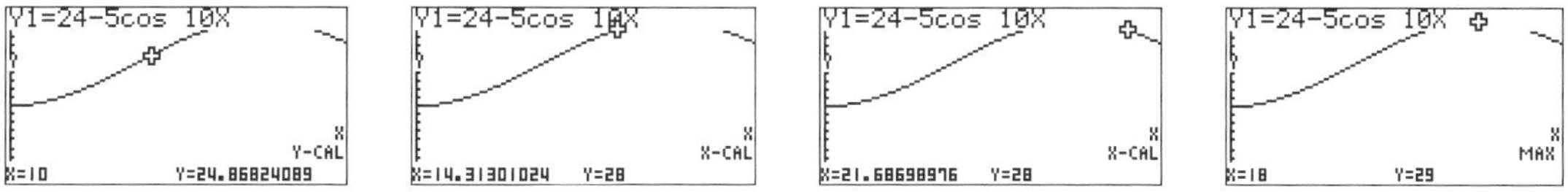

Exercise 4.5

1 State the amplitude and period for each of these functions.

a $f(x) = \sin x + 3$ b $y = 2\sin x$ c $g(x) = 2\cos x - 1$

d $f(x) = 0.5\sin x$ e $0.2\cos x - 3 = y$ f $g(x) = \sin 3x$

g $y = -2\cos 0.5x + 1$ h $f(x) = -\sin 2x$ i $g(x) = -0.5\cos 2x$

2 Sketch the graphs of these functions for $0° \le x \le 360°$.

a $y = \sin 3x$ b $y = 2\sin x - 1$ c $y = -2\sin x + 1$

d $y = \cos 3x$ e $y = 2\cos x - 1$ f $y = -2\cos x + 1$

3 Sketch the graphs of these functions for $-360° \le x \le 0°$.

a $y = \sin 3x$ b $y = 2\sin x - 1$ c $y = -2\sin x + 1$

d $y = \cos 3x$ e $y = 2\cos x - 1$ f $y = -2\cos x + 1$

4 Determine the equations of the sine or cosine functions shown in these diagrams.

a

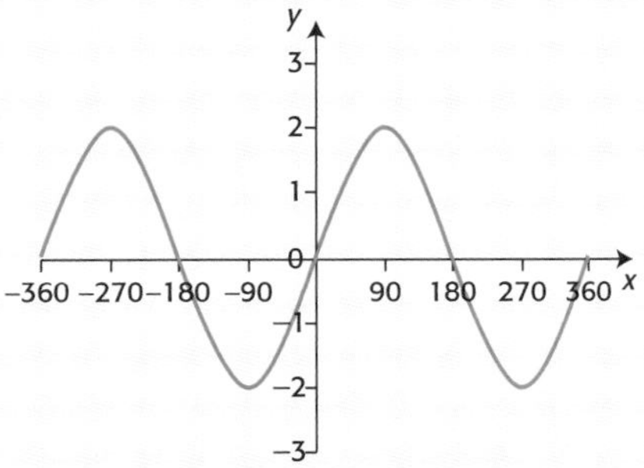

b

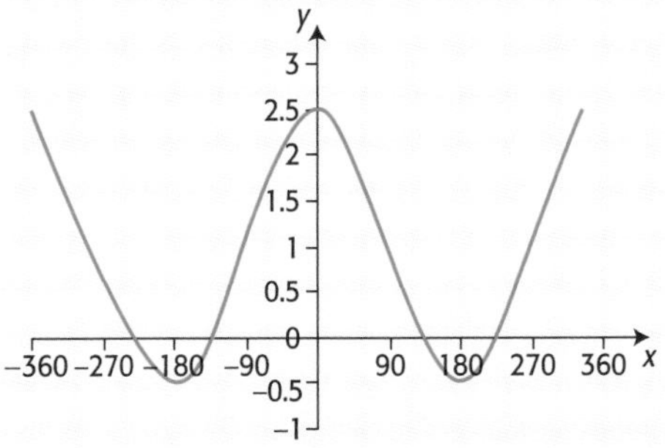

c

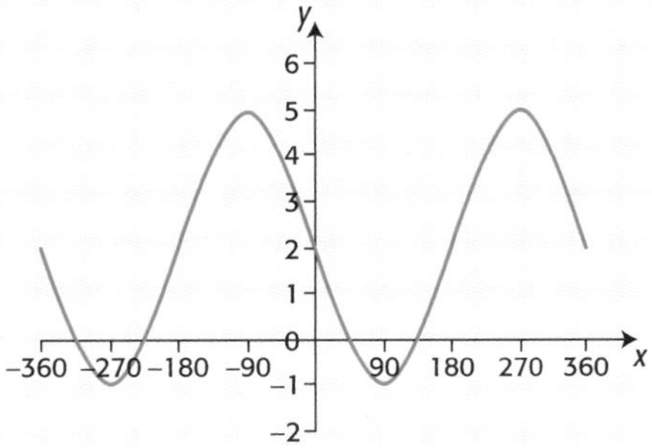

d

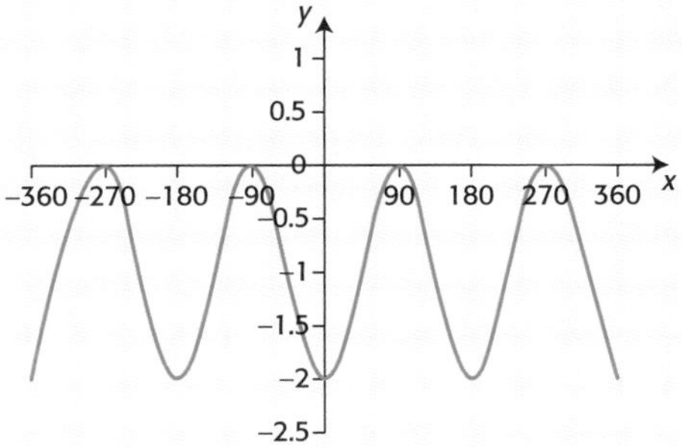

5 State the range for each of the graphs in question 4.

6 Use a graphic display calculator to find all solutions in the given domain for each of these functions.

a $\sin 0.5x + 1 = 0.75, \ -90° \le x \le 90°$

b $3\cos x - 2 = -1, \ 0° \le x \le 540°$

c $-1.5\sin x + 2 = 0.75, \ -360° \le x \le 0°$

d $-2\cos 2x = -1, \ -180° \le x \le 180°$

e $\sin 2x + 1 = 0.2, \ -180° \le x \le 180°$

f $1.5\cos 3x = 0, \ -180° \le x \le 0°$

7 a As a wave passes a jetty pylon, the depth of water, D metres, at time t seconds is given by the function, $D = 0.2\sin 50t + 5, t > 0$.

i Find the maximum and minimum depth of water. When do these occur?

ii Find the first time the depth of water reaches a mark on the pylon which is exactly 5 metres above the seabed.

b For the function $f(x) = -3\cos 2x + 1.2$, write down

i the amplitude

ii the period

iii the minimum value.

iv Find the values of x for which $f(x) = 0, \ 180° \le x \le 360°$.

c The position of a particle at time t seconds is given by the function $y = 3 - \sin 1.5t, t > 0$.

i Find the maximum and minimum values of y and the times when they first occur.

ii Determine the amplitude of the particle's movement.

iii Sketch the graph of y. State the period for one cycle of the particle's movement.

d The number of caribou in a national park can be modelled by the function, P, which has the equation, $P = 1000 - 650\sin 15t$, where t is the number of weeks since the discovery of a virus, $0 < t \leq 52$.

- **i** What was the original population of caribou in the park?
- **ii** Determine the population after 10 weeks and after 50 weeks.
- **iii** Determine the minimum number of caribou present in the first six months after the discovery of the virus and the time when that occurs.
- **iv** When does the population first return to its original size?
- **v** For how many weeks is the population less than 500?

e The temperature, T (°C), of the sea water at a certain location is given by the function $T = 14 - 3\cos 15t$, $0° < t \leq 24°$.
The initial temperature is measured at 6 am.

- **i** Find the initial temperature of the water.
- **ii** Determine the temperature of the water at midday.
- **iii** Find the maximum temperature of the water and the time of day when it occurs.
- **iv** Find the interval of time for which the temperature of the water is below 12 °C.

4.6 Accurate graph drawing

Important points to remember:

- Graphs should always be drawn in pencil.
- A ruler should be used for drawing the axes.
- Axes must be labelled.

  For graphs of $y = f(x)$, the horizontal axis is labelled x and the vertical axis is labelled y.

  For functions involving time, the horizontal axis is labelled t. The vertical axis is labelled as the function of time, for example, $H(t)$ or $P(t)$.
- Scales are usually given.

  For example: Use 1 cm to represent 10 units on the horizontal axis and 1cm to represent 20 units on the vertical axis.

  Mark and label the scale accurately on the axes.
- Coordinates of points are usually determined using a table. These points should be plotted carefully and exactly.
- For linear functions, use a ruler to draw the straight line. All other functions are curved. Draw them neatly through all the plotted points. If an error is made, erase the mistake and redraw.
- Domains are often stated. Make sure the drawn line or curve does not exceed the given domain.

4.7 Use of a GDC to sketch and analyse some simple, unfamiliar functions

Some important features of a graph include its general shape, maximum and minimum points, x- and y-intercepts and asymptotes. For the following functions, all of the above features are expected to be found/observed using a graphic display calculator. Algebraic explanations will not be assessed.

4.7.1 Polynomials

Cubic polynomials: $f(x) = ax^3 + bx^2 + cx + d,\ a \neq 0.$

Example 4.7.1a

Sketch the graph of
$y = x^3$ for $-2 \leq x \leq 2$.

Find the x- and y-intercepts.

Solution

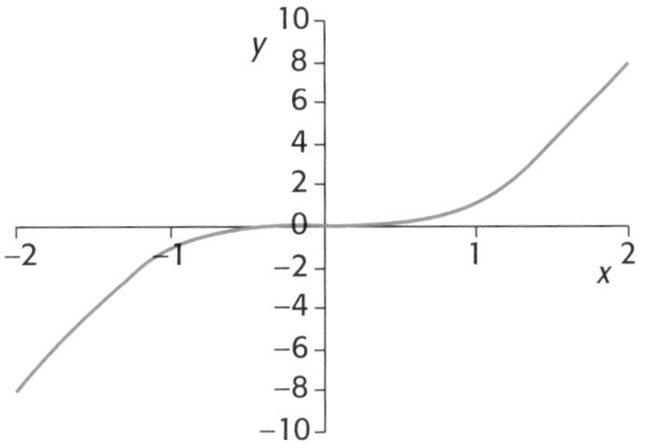

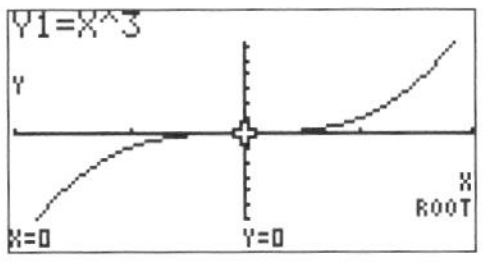

The x-intercept is 0.
The y-intercept is 0.

Example 4.7.1b

Sketch the graph of
$y = (x - 2)(x - 1)(x + 3)$ for $-3 \leq x \leq 2$.

Find

a the x- and y-intercepts

b the maximum and minimum points.

Solution

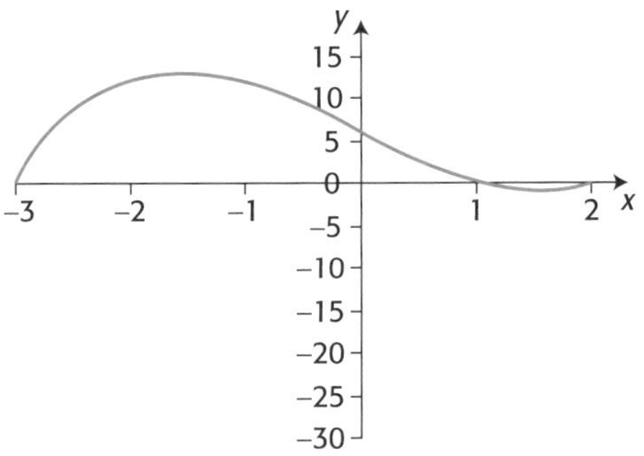

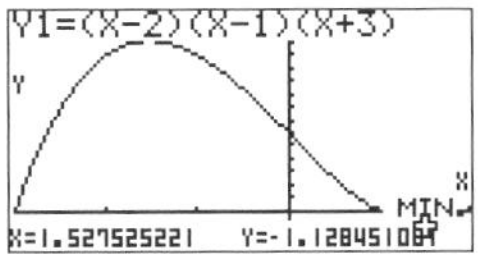

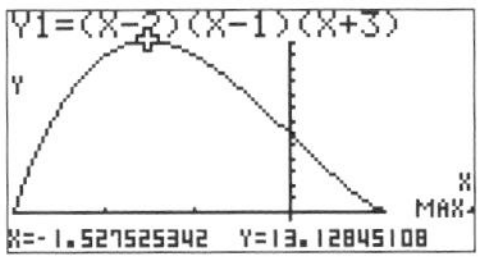

The x-intercepts are -3, 1, 2.
The y-intercept is 6.
Maximum at $(-1.53, 13.1)$
Minimum at $(1.53, -1.13)$

Quartic polynomials: $f(x) = ax^4 + bx^3 + cx^2 + dx + e,\ a \neq 0.$

Example 4.7.1c

Sketch the graph of
$y = x^4$ for $-2 \leq x \leq 2$.
Find the x- and y-intercepts.

Example 4.7.1d

Sketch the graph of
$y = (x - 1)^2(x + 1)(x + 2)$
for $-2.5 \leq x \leq 2$.
Find **a** the x- and y-intercepts
b the minimum point.

Solution

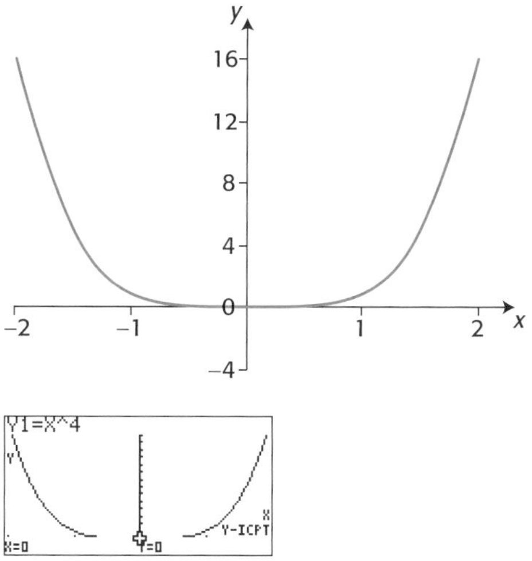

The x-intercept is 0.
The y-intercept is 0.

Solution

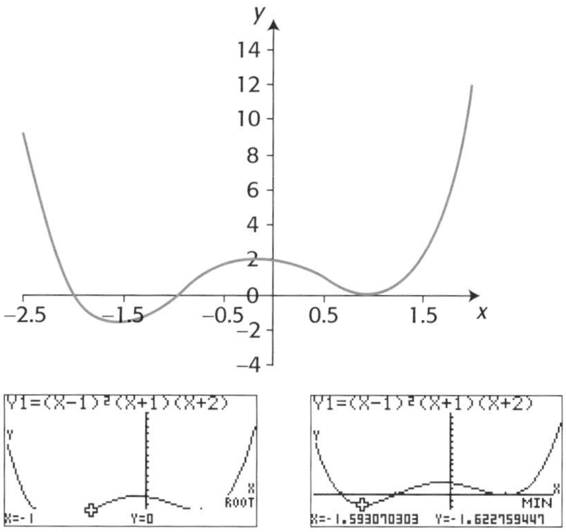

The x-intercepts are −2, −1, 1.
The y-intercept is 2.
Minimum at (−1.59, −1.62)

4.7.2 Maximum/minimum points in a given domain

The maximum or minimum value in a given domain can occur at a turning point or at an end point of the stated domain. The value of the function at the end points of a given domain needs to be checked.

Consider the function $y = (x - 2)(x - 1)(x + 3)$ in Example 4.7.1b above. For the given domain of $-3 \leq x \leq 2$, the minimum value of the function is −1.13 when $x = 1.53$. Suppose the domain for the function had been given as $-4 \leq x \leq 3$. Would this affect the maximum or minimum value of the function?

Example 4.7.2

Sketch the graph $y = (x - 2)(x - 1)(x + 3)$ for $-4 \leq x \leq 3$. Find the maximum and minimum values of the function in the given domain.

Solution

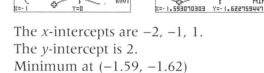

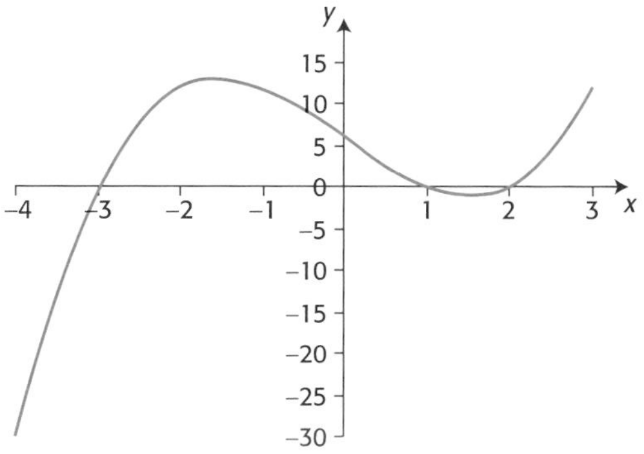

Check the end points of the domain.

When $x = -4$, $y = -30$. The minimum value of the function (−30) occurs at the end point of the domain.

When $x = 3$, $y = 12$. The maximum value of the function still occurs at the turning point $(-1.53, 13.1)$.

4.7.3 Rational functions $y = \dfrac{1}{ax + c}$

Rational functions have **asymptotes**, that is values which the function approaches but never equals. The asymptotic behaviour of rational functions can be summarized by considering the nature of the graph of the function (y) as the value of the input variable (x) gets very large or very small (tends to infinity) and also as the value of x approaches the vertical asymptote(s).

Example 4.7.3a

Consider the function $y = \dfrac{1}{x}$.

Using your GDC

a Write down the equations of any asymptotes.

b Discuss the behaviour of the graph as it approaches its asymptotes.

Solution

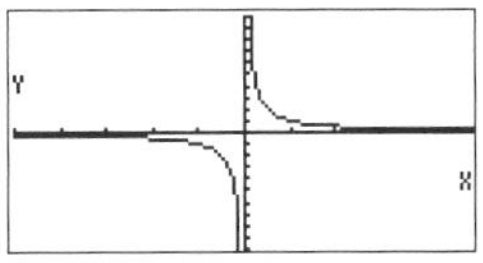

a Horizontal asymptote when y = 0.
Vertical asymptote when $x = 0$

b Discussion
as $x \to +\infty$, $y \to 0$ from above
as $x \to -\infty$, $y \to 0$ from below
as $x \to 0$ from the right, $y \to +\infty$
as $x \to 0$ from the left, $y \to -\infty$

Example 4.7.3b

Consider the function $y = \dfrac{1}{x - 2}$.

Using your GDC

a Write down the equations of any asymptotes.

b Discuss the behaviour of the graph as it approaches its asymptotes.

Solution

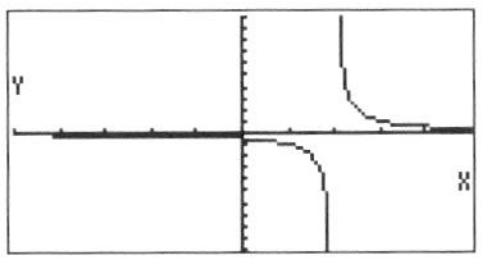

We need to be careful when using the graphic display calculator to study asymptotic behaviour. In normal view windows, the asymptotes may not be obvious.

a Horizontal asymptote when $y = 0$.
Vertical asymptote when $x = 2$

b Discussion
as $x \to +\infty$, $y \to 0$ from above
as $x \to -\infty$, $y \to 0$ from below
as $x \to 2$ from the right, $y \to +\infty$
as $x \to 2$ from the left, $y \to -\infty$

Use of the "zoom" facility is often useful. In many cases, the view window will need to be adjusted to clearly see the asymptotes.

Consider the following screens.

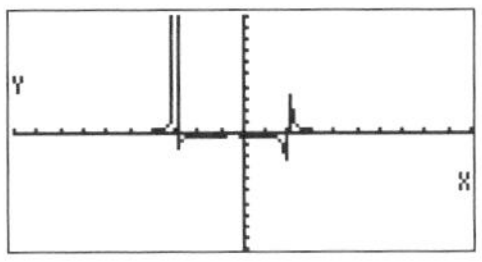

standard view screen asymptotes not clear

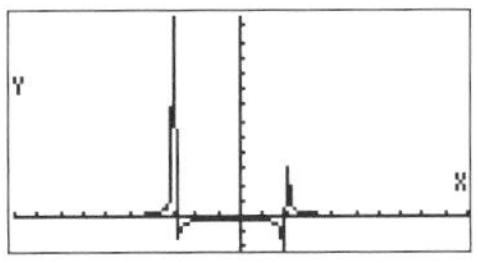

"zoom auto"

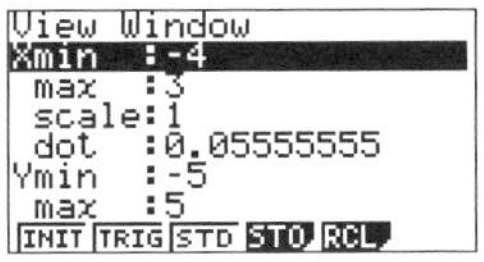

set view window

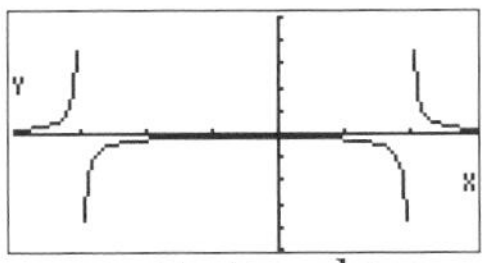

asymptotes clear

4.7.4 Logarithmic functions

Example 4.7.4a

Consider the function
$y = \log x$.

a Draw a sketch of $y = \log x$ for $0 < x \le 5$.

b Find the x-intercept.

c Write down the equation of the vertical asymptote.

Solution

a

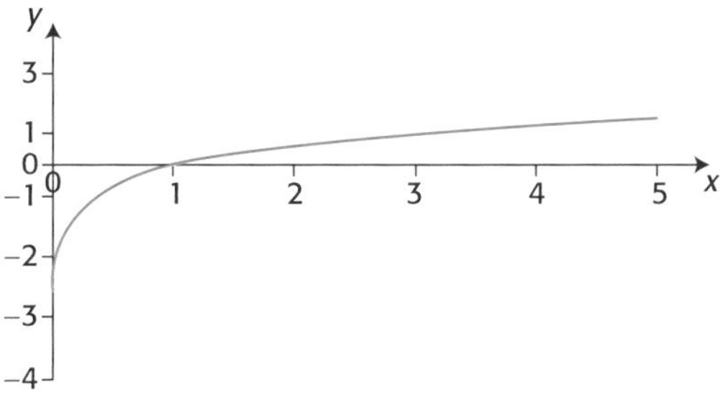

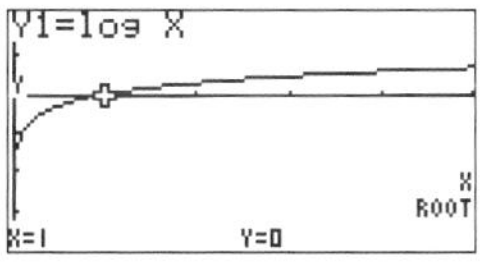

b The x-intercept is 1.

c The vertical asymptote is $x = 0$.

Example 4.7.4b

Consider the function
$y = 3\log(2x)$.

a Draw a sketch of $y = 3\log(2x)$ for $0 < x \le 5$.

b Find the x-intercept.

c Write down the equation of the vertical asymptote.

Solution

a

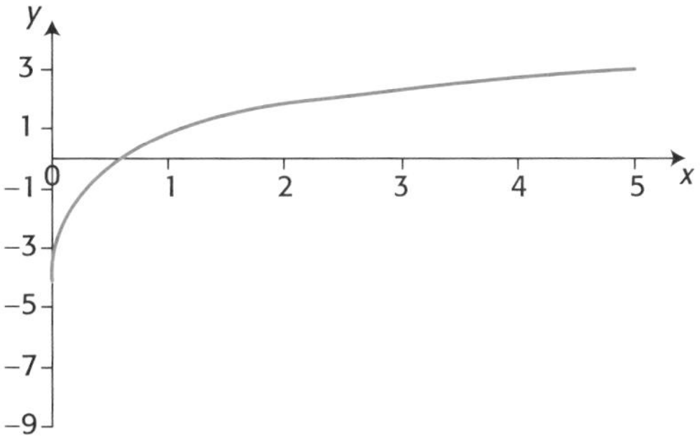

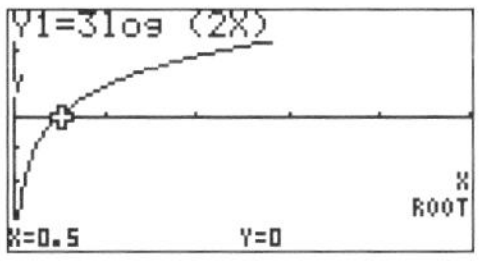

b The x-intercept is 0.5.

c The vertical asymptote is $x = 0$.

4.7.5 Some unfamiliar functions

Some examples of more complicated functions are given below. Students should attempt to analyse their own versions of more complex functions to gain practice in determining the intercepts, maximum and minimum values and asymptotes. In each case, the graphic display calculator is expected to be used in the analysis of the graphs of these functions.

Example 4.7.5a

$y = x\log x$

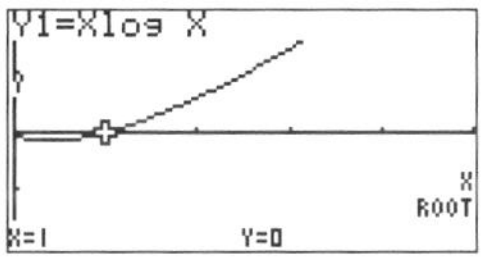

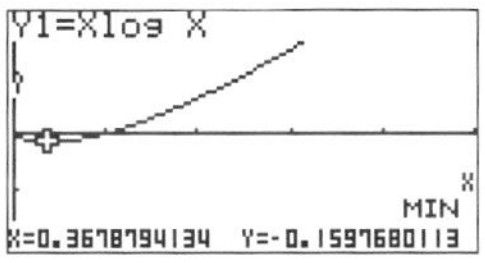

x-intercept is 0.

minimum at (0.368, −0.160)

Example 4.7.5b

$$y = \frac{1}{\sqrt{x}}$$

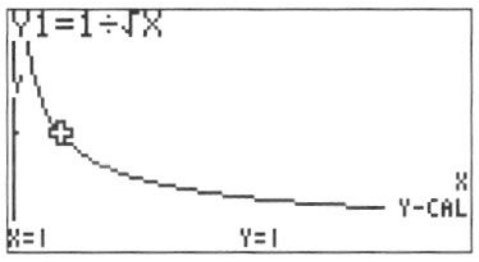

Vertical asymptote, $x = 0$

Horizontal asymptote, $y = 0$ Discussion:
as $x \to +\infty$, $y \to 0$ from above, as $x \to 0$, $y \to +\infty$

Example 4.7.5c

$$y = \frac{x+1}{x^2 - 2x - 8}$$

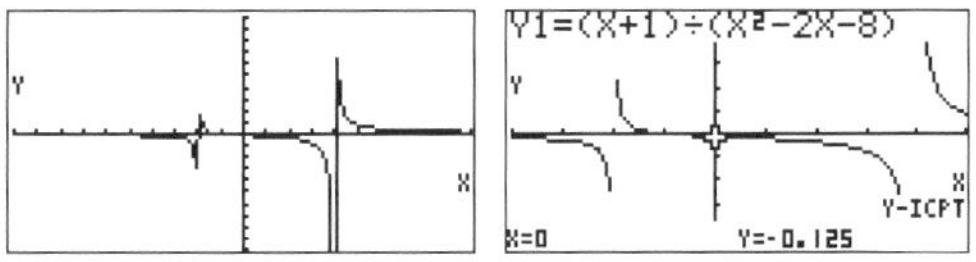

The first screen is a standard view-screen. The asymptotic behaviour is not clear. The second screen clearly identifies the behaviour of the graph as it approaches the asymptotes.

x-intercept at $(-1, 0)$
y-intercept at $(0, -0.125)$

Vertical asymptotes are $x = -2$ and $x = 4$.
Horizontal asymptote is $y = 0$.

Discussion

as $x \to +\infty$, $y \to 0$ from above
as $x \to -\infty$, $y \to 0$ from below
as $x \to -2$ from the right, $y \to +\infty$
as $x \to -2$ from the left, $y \to -\infty$
as $x \to 4$ from the right, $y \to +\infty$
as $x \to 4$ from the left, $y \to -\infty$

Example 4.7.5d

$$y = (x - 3) \times 2^x$$

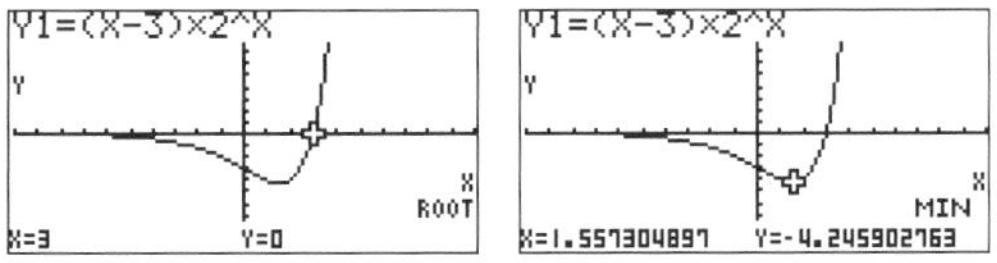

x-intercept is 3
y-intercept is -3
Minimum at $(1.56, -4.25)$

Horizontal asymptote is $y = 0$.

Discussion

as $x \to +\infty$, $y \to +\infty$
as $x \to -\infty$, $y \to 0$ from below

Exercise 4.7

Using the graphic display calculator, find the main features of the graphs of these functions. Discuss asymptotic behaviour where appropriate.

1 $y = x^3 - 4.5x^2 + 2$

2 $y = x^3 - 0.5x^2 - 4x + 2$

3 $y = x^4 - 6x^3 + 13x^2 - 12x + 4$

4 $y = \dfrac{1}{3x - 1}$

5 $y = \dfrac{1}{x^2 - 3x + 2}$

6 $y = \dfrac{2^x}{5x}$

7 $y = \sqrt{x}(x - 2)$

8 $y = \dfrac{x}{\log x}$

9 $y = (\sin x)^2$, $-180^\circ \le x \le 180^\circ$.

4.8 Using a GDC to solve equations involving unfamiliar functions

It is recommended that solutions are found by finding the point(s) of intersection of the graphs of functions. In many cases, using "Solver" would be appropriate but, given that more than one solution may exist, this might lead to some solutions not being found.

Example 4.8

Solve for x in each of these functions.

a $x - 2 = \dfrac{1}{x}$

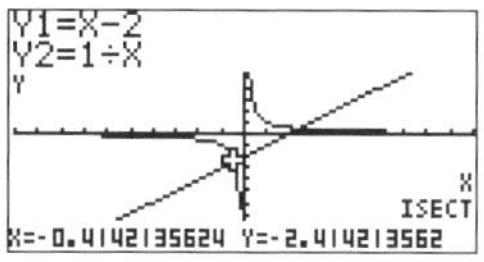

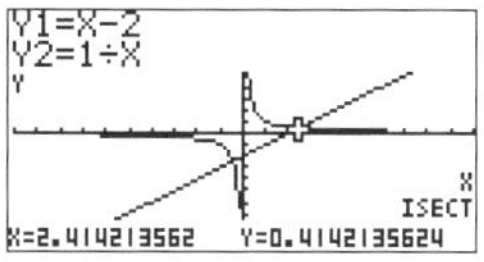

$x = -0.414$ or 2.41

b $5x = 3^x$

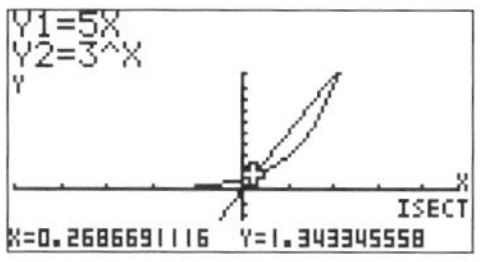

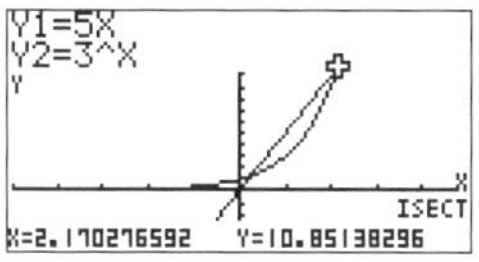

$x = 0.269$ or 2.17

c $\log x = x - 5$

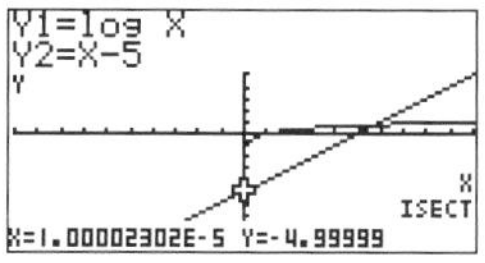

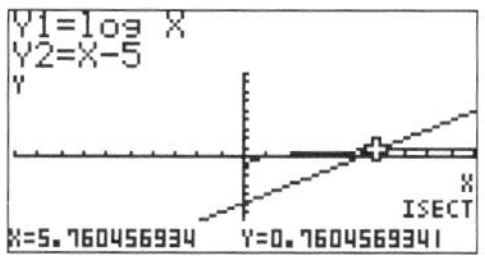

$x = 1.00 \times 10^{-5}$ or 5.76

d $x^3 - 2x^2 + 5x - 6 = 7$

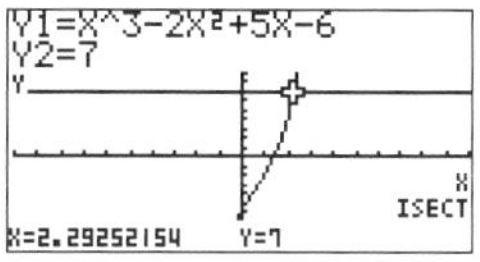

$x = 2.29$

e $\sin x = \dfrac{1}{x}$, $-180° \leq x \leq 180°$.

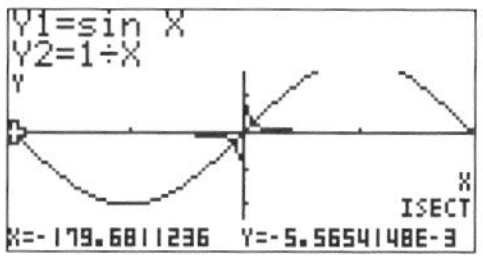

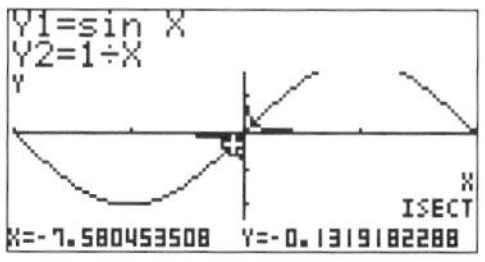

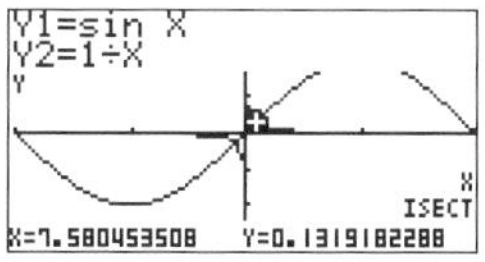

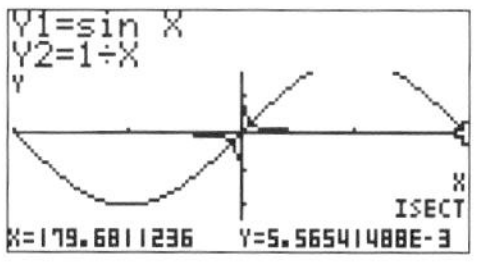

$x = \pm 179.68°, \pm 7.58°$

Exercise 4.8

Solve for x in each of these functions.

a $x^3 - 4.5x^2 + 2 = 1.2$

b $x\log x = 3$

c $\dfrac{1}{3x-1} = x$

d $\dfrac{1}{x^2 - 3x + 2} = 2$

e $3x + 5 = 5^x$

f $3\log 2x = x - 3$

g $\dfrac{x-2}{x-5} = 2 - 3x$

h $2\sqrt{x} - 3 = \dfrac{1}{2x}$

i $(\sin x)^3 = \dfrac{1}{x}$, $-90° \leq x \leq 90°$

Past examination questions for topic 4

Paper 1

1 The function $Q(t) = 0.003t^2 - 0.625t + 25$ represents the amount of energy in a battery after t minutes of use.

a State the amount of energy held by the battery immediately before it was used.

b Calculate the amount of energy available after 20 minutes.

c Given that $Q(10) = 19.05$, find the average amount of energy produced per minute for the interval $10 \le t \le 20$.

d Calculate the number of minutes it takes for the energy to reach zero. *M06q7*

2 Two functions $f(x)$ and $g(x)$ are given by

$$f(x) = \frac{1}{x^2 + 1}, \qquad g(x) = \sqrt{x}, \quad x \ge 0.$$

a Sketch the graphs of $f(x)$ and $g(x)$ together on the same diagram using values of x between −3 and 3, and values of y between 0 and 2. You must label each curve.

b State how many solutions exist for the equation

$$\frac{1}{x^2 + 1} - \sqrt{x} = 0.$$

c Find a solution of the equation given in part **b**. *M06q15*

3 The graph represents the function $y = 4\sin 3x$.

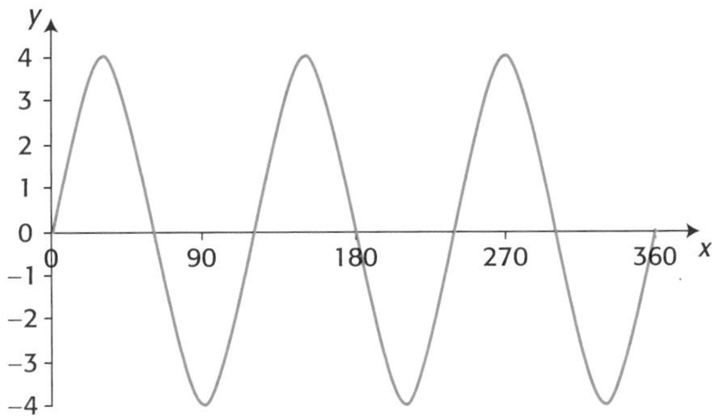

a i Write down the period of the function.

ii Write down the amplitude of the function.

b Draw the line $y = 2$ on the diagram.

c Using the graph, or otherwise, solve the equation $4\sin 3x = 2$ for $0° \le x \le 90°$. *N05q6*

4 The equation $M = 90\ 2^{-\frac{t}{20}}$ gives the amount, in grams, of radioactive material held in a laboratory over t years.

a What was the original mass of the radioactive material?

The table lists some values for M.

t	60	80	100
M	11.25	v	2.8125

b Find the value of v.

c Calculate the number of years it would take for the radioactive material to have a mass of 45 grams. *M05q5*

5 The diagram shows the graph of $y = c + kx - x^2$, where k and c are constants.

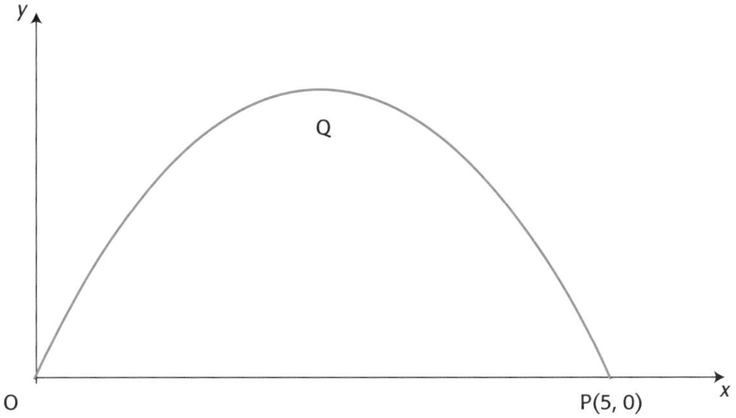

a Find the values of k and c.

b Find the coordinates of Q, the highest point on the graph. *M05q6*

6 These are the graphs of three trigonometric functions. The x-variable is measured in degrees, with $0° \le x \le 360°$. The amplitude a is a positive constant with $0 < a \le 1$.

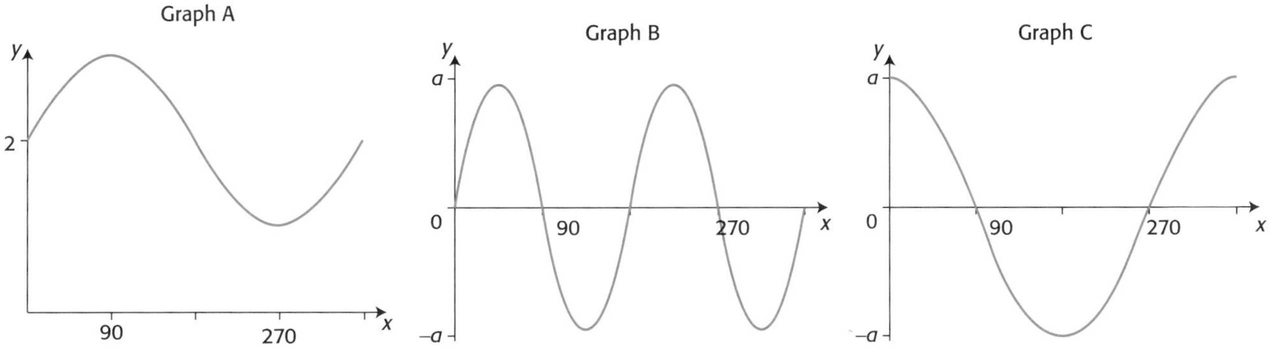

a Copy the table and write the letter of the graph next to the function representing that graph.

Function	**Graph label**
$y = a\cos x$	
$y = a\sin 2x$	
$y = 2 + a\sin x$	

b State the period of the function shown in graph B.

c State the range of the function $2 + a\sin x$ in terms of the constant a. *M05q8*

7 The graph of the function $f : x \rightarrow 30x - 5x^2$ is given in the diagram.

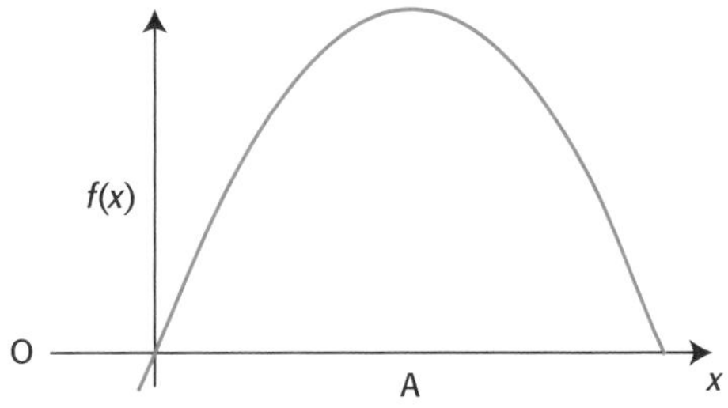

a Factorize fully $30x - 5x^2$.
b Find the coordinates of the point A.
c Write down the equation of the axis of symmetry. *N03q4*

8 Consider the function $f(x) = 2\sin x - 1$ where $0° \leq x \leq 720°$.

a Write down the period of the function.
b Find the minimum value of the function.
c Solve $f(x) = 1$. *N03q10*

9 The conversion formula for temperature from the Fahrenheit (F) to the Celsius (C) scale is given by $C = \frac{5(F-32)}{9}$

a What is the temperature in degrees Celsius when it is 50° Fahrenheit?

There is another temperature scale called the Kelvin(K) scale.

The temperature in degrees is given by $K = C + 273$.

b What is the temperature in **Fahrenheit** when it is zero degrees on the Kelvin scale? *M03q2*

10 The diagram shows a chain hanging between two hooks A and B.

The points A and B are at equal heights above the ground. P is the lowest point on the chain. The ground is represented by the x-axis. The x-coordinate of A is –2 and the x-coordinate of B is 2. Point P is on the y-axis.

The shape of the chain is given by $y = 2^x + 2^{-x}$ where $-2 \leq x \leq 2$.

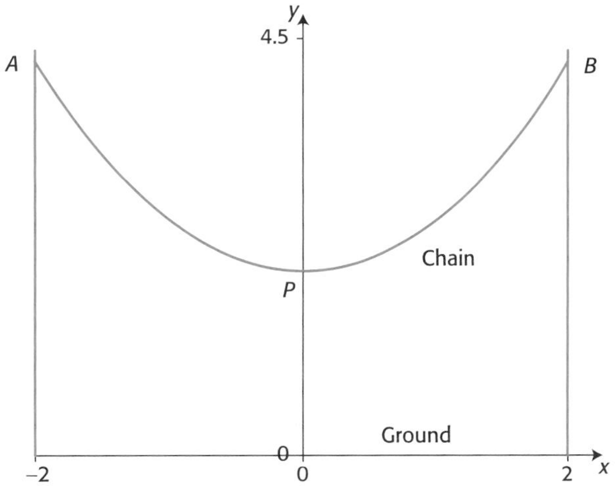

a Calculate the height of the point P.
b Find the range of y. Write your answer as an interval or using inequality symbols. *M03q12*

Paper 2

1

a A function $f(x)$ is defined by
$f(x) = 2x^2 - 10x + 60, -5 \le x \le 8$.

x	−5	0	2	5	8
f(x)	160	*a*	*b*	60	108

i Write down the values of a and b.
ii Using the values in the table, draw the graph of $f(x)$ on a set of coordinate axes. Use a scale of 1 cm to represent 1 unit on the horizontal axis and 1 cm to represent 20 units on the vertical axis.
iii Show that the coordinates of the vertex of the graph are (2.5, 47.5).
iv State the values of x for which the function is increasing.

b A second function $h(x)$ is defined by
$h(x) = 80, \quad 0 \le x \le 8$.

i On the same axes used for part **a**, draw the graph of $h(x)$.
ii Find the coordinates of the points at which $f(x) = h(x)$.
iii Find the vertical distance from the vertex of the graph of $f(x)$ to the line $h(x)$. *M06q1*

2 A function is represented by the equation $f(x) = 3(2)^x + 1$.
The table of values of $f(x)$ for $-3 \le x \le 2$ is given below.

x	−3	−2	−1	0	1	2
f(x)	1.375	1.75	*a*	4	7	*b*

a Calculate the values of a and b.
b On graph paper, draw the graph of $f(x)$, for $-3 \le x \le 2$, taking 1cm to represent 1 unit on both axes.
The domain of the function $f(x)$ is the real numbers, $\mathbb{R}$.
c Write down the range of $f(x)$.
d Using your graph, or otherwise, find the approximate value of x when $f(x) = 10$. *N05q1*

3 A small manufacturing company makes and sells x machines each month. The monthly cost, C, in dollars, of making x machines is given by

$$C(x) = 2600 + 0.4x^2.$$

The monthly income, I, in dollars, obtained by selling x machines is given by

$$I(x) = 150x - 0.6x^2.$$

a Show that the company's monthly profit can be calculated using the quadratic function

$$P(x) = -x^2 + 150x - 2600.$$

b The maximum profit occurs at the vertex of the function $P(x)$. How many machines should be made and sold each month for a maximum profit?

c If the company does maximize profit, what is the selling price of each machine?

d Given that $P(x) = (x - 20)(130 - x)$, find the smallest number of machines the company must make and sell each month in order to make a positive profit. *M05q1*

4 The graph shows the tide heights, h metres, at time t hours after midnight, for Tahini Island.

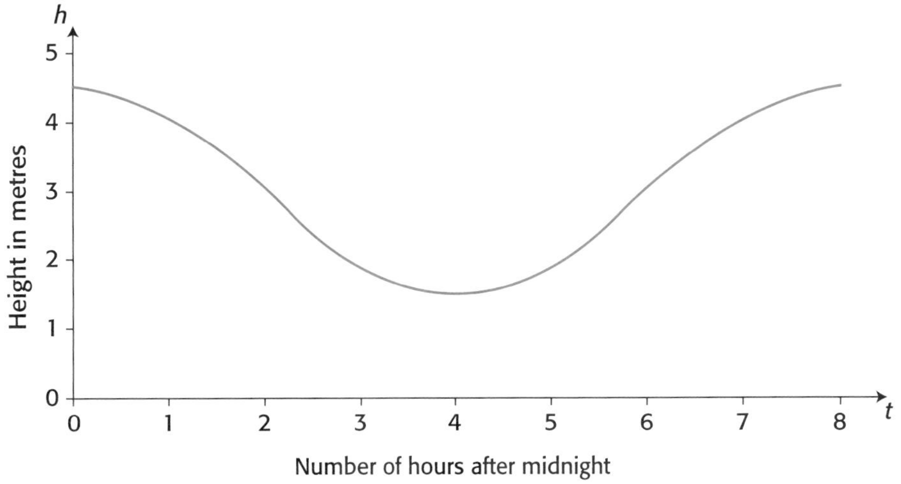

a Use the graph to find
  i the height of the tide at 03:15
  ii the times when the height of the tide is 3.5 metres.

b The best time to catch fish is when the tide is **below** 3 metres. Find this best time, giving your answer as an inequality in t.

Due to the location of Tahini Island, there is little variation in the pattern of tidal heights. The maximum tide height is 4.5 metres and the minimum tide height is 1.5 metres. The height h can be modelled by the function

$$h(t) = a \cos bt^0 + 3$$

c Use the graph to find the values of the variables a and b.

d Hence **calculate** the height of the tide at 13:00.

e At what time would the tide be at its lowest point in the second 8-hour period? *M02q5*

Theory of knowledge

Is there a non-mathematical mind?

"Right and Wrongheadedness" Overcoming Math Anxiety by S Tobias, Haughton Mifflin Co., 1978

Some people epitomized by the character in the following passage by Philip Roth are so distracted by what Henry James called the "felt life" that they cannot concentrate on the essential mathematical information as it is presented. They are so fascinated with detail with the "people" part of the issues, that abstracting the numbers and the ratios from a problem wrenches them from their real interests. Roth tells of Nathan, a sickly and feverish young boy, whose father tried to sharpen his mind by giving him arithmetic problems to solve. As Roth tells it, the father would announce a problem like this:

"Marking down" he would say, not unlike a recitation student announcing the title of a poem. "A clothing dealer, trying to dispose of an overcoat cut in last year's style, marked it down from the original price of \$30 to \$24. Failing to make a sale, he reduced the price to \$19.20. Again he found no takers, so he tried another price reduction and this time sold it.... All right, Nathan, what was the selling price if the last markdown was consistent with the others?" Or, "Making a chain". "A lumberjack has six sections of chain, each consisting of four links. If the cost of cutting open a link......" and so on.

The next day, while my mother whistled Gershwin and laundered my father's shirts, I would daydream in my bed about the clothing dealer and the lumberjack. To whom had the haberdasher finally sold the overcoat? Did the man who bought it realize that it was cut in last year's style? If he wore it to a restaurant, would people laugh? And what did "last year's style" look like anyway? "Again he found no takers", I would say aloud, finding so much to feel melancholy about in that idea. I still remember how charged for me was that word "takers". Could it have been the lumberjack with his six sections of chain who, in his rustic innocence, had bought the overcoat cut in last year's style? And why suddenly did he need an overcoat? Invited to a fancy ball? By whom?......

My father was disheartened to find me intrigued by fantasies and irrelevant details of geography and personality and intention, instead of the simple beauty of the arithmetic solution. He did not think that was intelligent of me and he was right. The young Roth (this tale must be autobiographical) was fascinated by the possibilities of personality. To call his mind "non-mathematical" is to miss what he represents: a beacon in a continuum of human curiosity in search of meaning.

Questions

1 If you are intrigued by the "people" part of a problem does that necessarily make you less mathematical?
2 Is mathematics gender related?
3 Is it the case that there are different types of people with different mindsets in mathematics? And is this also the case in the other areas of knowledge?

TOPIC 5

Geometry and trigonometry

5.1 Coordinates, points and lines in two dimensions

5.1.1 The Cartesian plane

René Descartes (1596–1650)

Cartesian means relating to the French mathematician and philosopher René Descartes, who, among other things, worked to merge algebra and Euclidean geometry. This work was influential in the development of analytic geometry, calculus, and cartography. The idea of this system was developed in 1637 in two writings by Descartes. In part two of his *Discourse on Method* he introduces the new idea of specifying the position of a point or an object on a surface, using two intersecting axes as measuring guides. In *La Géométrie*, he further explores these concepts.

- In order to plot points in the plane we need to draw a pair of perpendicular axes, one horizontal, the ***x*-axis**, and the other vertical, the ***y*-axis**.
- The two axes divide the plane into four **quadrants** that are numbered in an anticlockwise direction.
- The point where both axes meet is called the **origin** and is labelled O.
- Once we have drawn both axes, we set a **scale** for each one.The scales do not have to be the same on both axes. The numbers on the scales increase in a positive direction.
- The position of any point in the Cartesian plane can be indicated in terms of its **coordinates**. Each point has two coordinates: the ***x*-coordinate**, the **horizontal** step from O, and the ***y*-coordinate**, the vertical step from O.

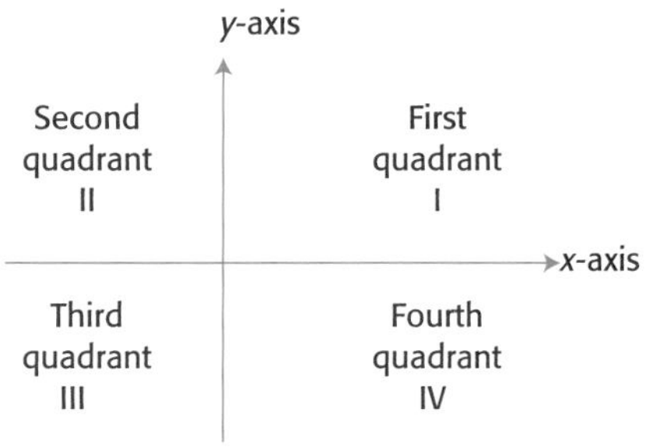

Notation

$P(a, b)$ indicates that

- the x-coordinate or **first** coordinate of P is a
- the y-coordinate or **second** coordinate of P is b.

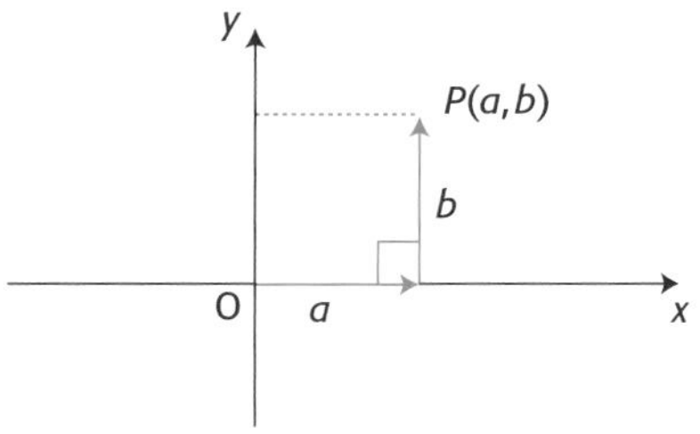

Each time you draw a pair of axes remember to label them.

Example 5.1.1a

Write down the coordinates of the points A, B, C, D, E and F.

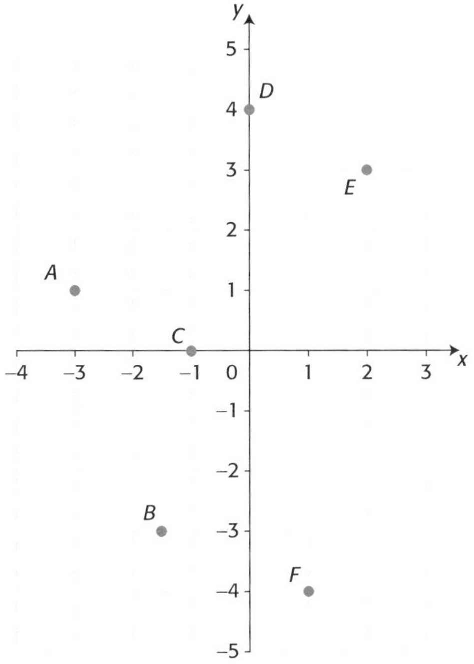

Solution

$A(-3, 1)$, $B(-1.5, -3)$, $C(-1, 0)$, $D(0, 4)$, $E(2, 3)$, $F(1, -4)$

Example 5.1.1b

Plot the points $A(4, 1)$ and $B(1, 4)$.

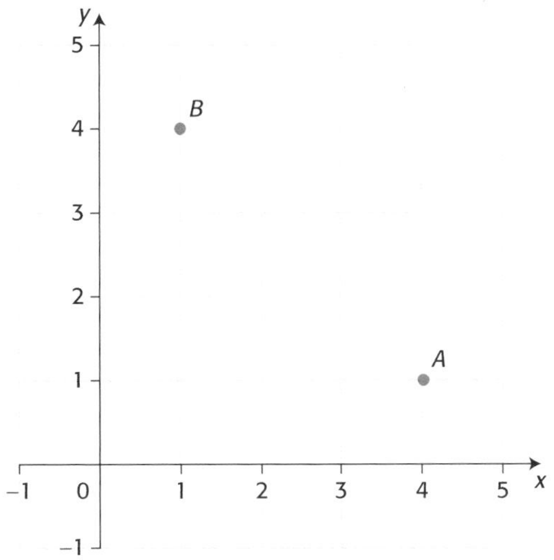

Solution

As we can see in this example points $A(4, 1)$ and $B(1, 4)$ do not have the same position in the Cartesian plane. We conclude that the *order* in which we write the coordinates of a point matters.

Example 5.1.1c

a Draw a set of axes for $-3 \le x \le 4$ and $-3 \le y \le 4$.

b Draw the line passing through the points $A(-2, -2)$ and $B(3, 3)$. Label it L.

c Write down the coordinates of two other points lying on L.

d Decide whether the point $C(1, 2)$ lies on L.

e The point $D(-1, b)$ lies on L. Write down the value of b.

Solution

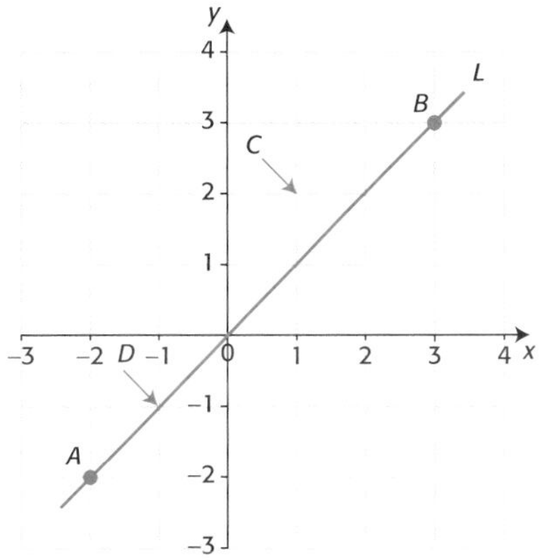

a and b on graph

c For example $(0, 0)$ and $(2, 2)$

d C does not lie on L.

e $b = -1$

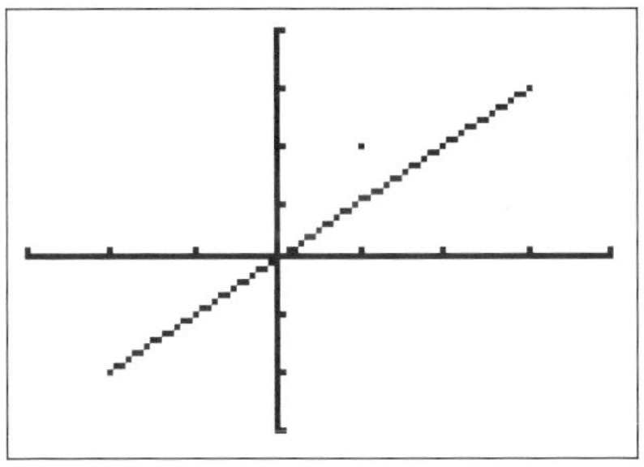

5.1.2 Midpoints

The **midpoint**, M, of a line segment AB is the point which is halfway between the points A and B.

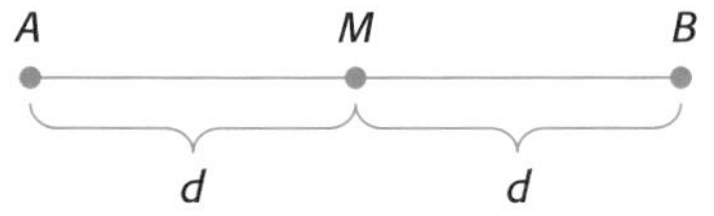

Example 5.1.2a

Consider the points $A(1,8)$ and $B(3,14)$.

a Calculate M, the coordinates of the midpoint of AB.

b Represent A, B and M on a pair of axes.

Solution

a Let $M\ (a, b)$ then

$\rightarrow a$ is halfway between the x-coordinates of the points A and B, so $a = \dfrac{1+3}{2} = 2$

$\rightarrow b$ is halfway between the y-coordinates of the points A and B, so $b = \dfrac{8+14}{2} = 11$

$\therefore M$ is $(2, 11)$.

b

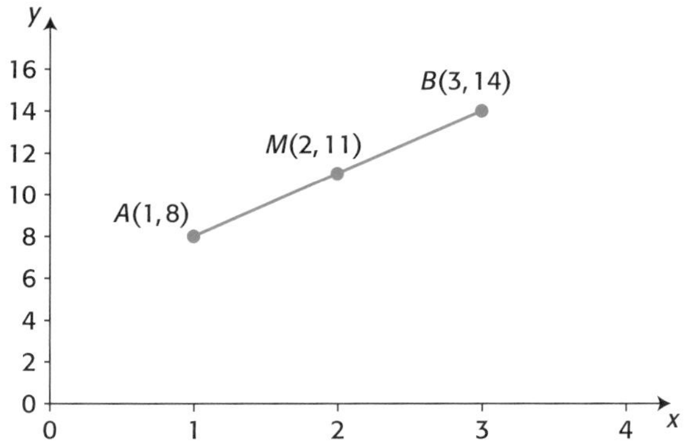

Example 5.1.2b

Points $A(-2,-1)$, $B(3,-1)$ and $C(3,4)$ are three of the vertices of square $ABCD$.

a Plot A, B and C on a pair of coordinate axes.

b i Write down the coordinates of point D.

ii Plot D on the same pair of axes. Hence draw $ABCD$.

c Find the coordinates of the point M at which the diagonals of the square $ABCD$ meet.

As A and B are different points we need to differentiate between their coordinates when noting them. To do this we use different **suffixes**: x_1 and x_2, y_1 and y_2.

If $A(x_1, y_1)$ and $B(x_2, y_2)$ then the midpoint of the segment AB has coordinates $\left(\dfrac{x_1+x_2}{2}, \dfrac{y_1+y_2}{2}\right)$.

Solution

a On graph

b i $D(-2,4)$ ii On graph

c The points where the diagonals meet is halfway between two opposite vertices of the square. A and C are opposite therefore

x-coordinate of $M = \dfrac{-2+3}{2} = 0.5$

y-coordinate of $M = \dfrac{-1+4}{2} = 1.5$

$\therefore M$ is $(0.5, 1.5)$.

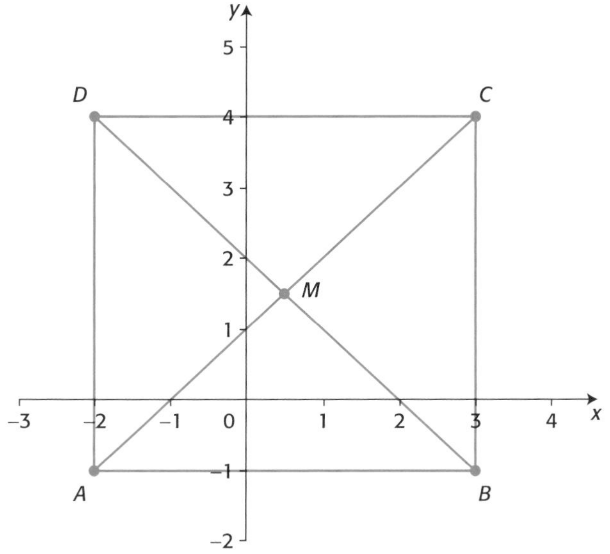

5.1.3 Distances between points

The distance between two points is the length of the line segment joining them.

Example 5.1.3a

Find the distance between the points

a $A(-2, 3)$ and $B(4, 3)$

b $C(2, 2)$ and $D(4, 8)$.

Solution

a

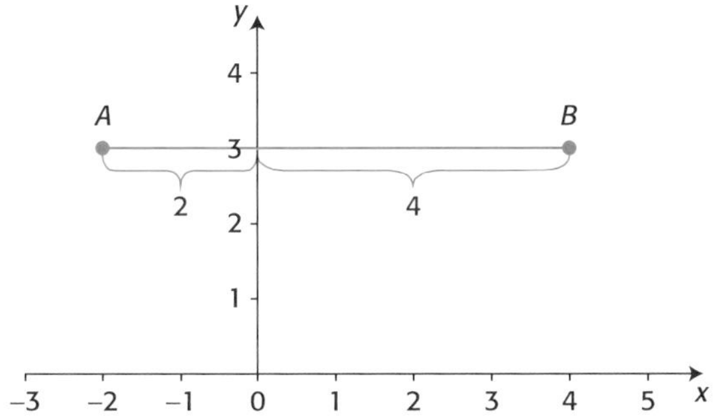

As AB is horizontal then the distance from A to B is $2 + 4 = 6$.

b

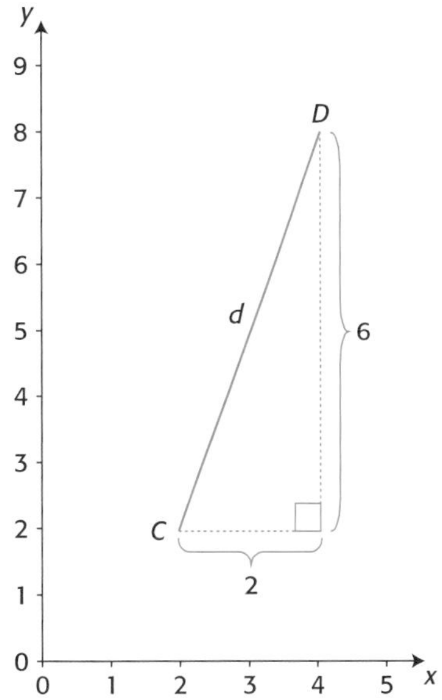

The distance from C to D is the length d of the hypotenuse of the right-angled triangle with sides parallel to the axes.

$d = \sqrt{2^2 + 6^2}$

$d = \sqrt{40}$

Each time we need to find the distance d between two points we can either draw a diagram as we have done above or use a formula.

If A is (x_1, y_1) and B is (x_2, y_2), this diagram will help us to find this formula.

- d represents the length of the hypotenuse of the right-angled triangle, ABC, where AC and BC are parallel to the axes.
- $(x_2 - x_1)$ is the *x-step*
- $(y_2 - y_1)$ is the *y-step*

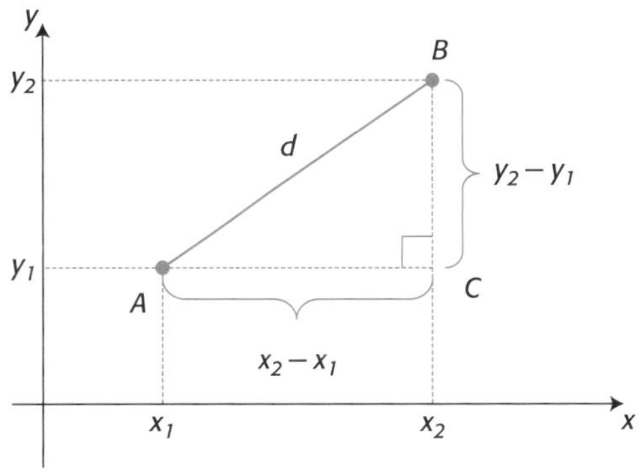

In general, if A is (x_1, y_1) and B is (x_2, y_2) then the distance from A to B is $d = \sqrt{(x_2 - x_1)^2 + (y_2 - y_1)^2}$

This formula is useful when
- the segment is neither horizontal nor vertical
- we are given the distance and have to find the coordinates of one of the points.

Example 5.1.3b

a Find the distance between $D(-1, 4)$ and $E(2, -3)$.

b Find the length of the line FG where $F(3, 1)$ and $G(3, -4)$.

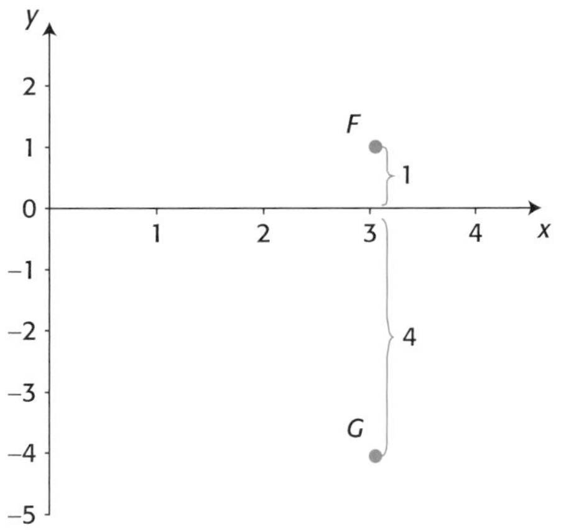

Solution

a $\left.\begin{array}{l} x_1 = -1; x_2 = 2 \\ y_1 = 4; y_2 = -3 \end{array}\right\} \Rightarrow$

$DE = \sqrt{(2 - -1)^2 + (-3 - 4)^2} = \sqrt{58} = 7.62 \text{ (3 s.f.)}$

b The segment FG is vertical (parallel to the y-axis) so $FG = 1 + 4 = 5$ (see figure alongside)

Example 5.1.3c

a Plot the points $A(-1, -2)$, $B(4, -2)$, $C(4, 2)$ and $D(-1, 2)$.

b What type of quadrilateral is $ABCD$?

c Find the area of $ABCD$.

a

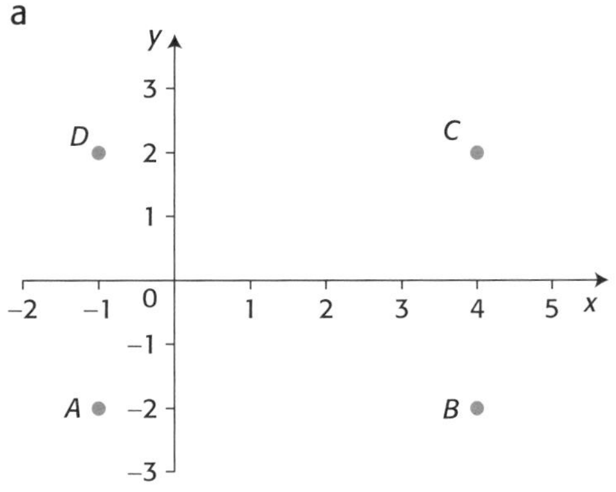

Solution

b $ABCD$ is a rectangle.

c Base = 5 (horizontal line) Height = 4 (vertical line)

Area = $b \times h = 5 \times 4 = 20$

Example 5.1.3d

The distance between $A(1, 2)$ and $B(a, -1)$ is 5. Find all the possible values of a.

Solution

Substituting into the distance formula

$$5 = \sqrt{(a - 1)^2 + (-1 - 2)^2}$$

$$25 = (a - 1)^2 + 9$$

$$16 = (a - 1)^2 \therefore a - 1 = 4 \text{ or } a - 1 = -4$$

$$\therefore a = 5 \text{ or } a = -3$$

In this type of question it is useful to draw a sketch to better understand what you are asked to find.

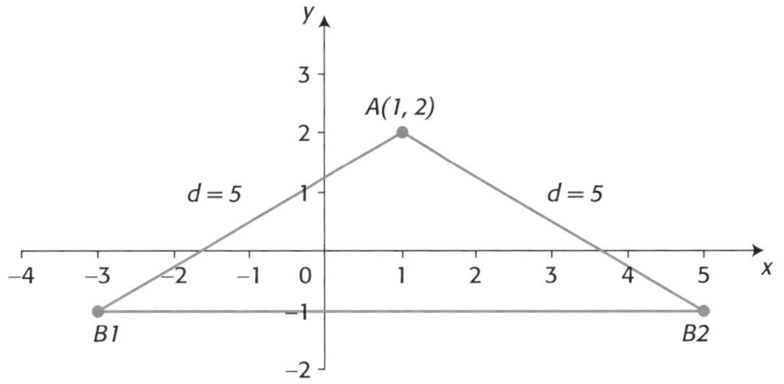

Exercise 5.1

1 Write down the coordinates of the points A, B, C, D and E plotted on the pair of axes in the diagram.

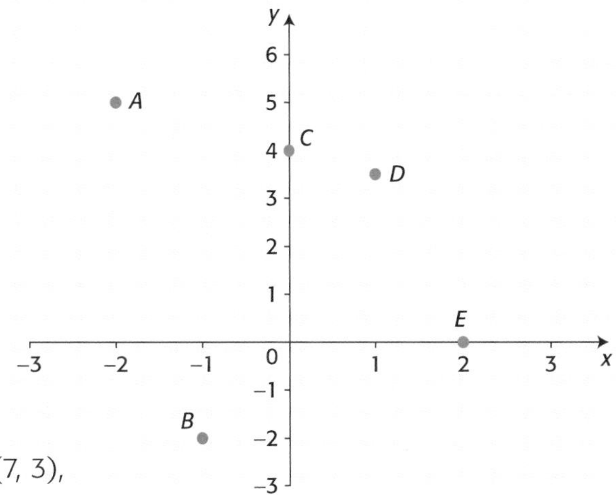

2 a Plot the points $A(1, -1)$, $B(1, 1)$, $C(-1, 1)$ and $D(-1, -1)$.
b What kind of quadrilateral is $ABCD$?
c Find
i the area of $ABCD$
ii the perimeter of $ABCD$.

3 Calculate the distance between

a $A(1, 4)$ and $B(1, -3)$
b $A(1, 4)$ and $C(-5, 4)$
c $A(1, 4)$ and $D(-2, 6)$.

4 a Plot and label the points $A(-1, 1)$, $B(-2, 5)$ and $C(7, 3)$, using the same scale for both axes. Join the points.
b Calculate the length of the lines AB and AC.
The length of CB is $\sqrt{85}$.
c Show that ABC is a right-angled triangle.

5 Consider the parallelogram $ABCD$ where A is $(-3, -1)$, B is $(-3, 4)$ and C is $(5, 6)$.

a Calculate the coordinates of the midpoint, M, of the line AC.
b Find the coordinates of the fourth vertex, D, of the parallelogram.

6 a The distance between the points $A(4, 5)$ and $B(4, y)$ is 8. Find all the possible values of y.
b The distance between the points $A(4, 5)$ and $B(x, -3)$ is 10. Find all the possible values of x.

7 a Draw the line passing through the points $A(-1, 6)$ and $B(0, 4)$. Label it L.
b $P(1, y)$ lies on L. Write down the value of y.
c Calculate the distance from A to P. Give your answer correct to 4 significant figures.

5.2 Lines, their properties and relationships

5.2.1 Gradient of a line

Example 5.2.1a

Two buildings, A and B, have the same type of water tank with a capacity of 540 litres.

a When the consumption is constant in building A this volume of water lasts 6 hours. The graph shows the water consumption for building A.

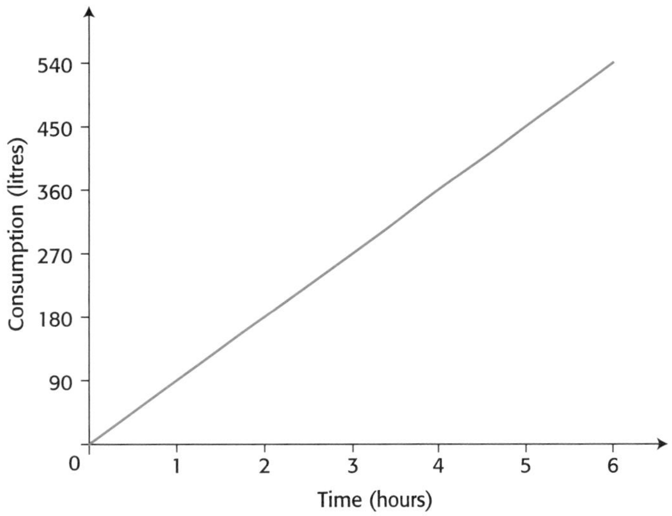

We see from this graph that in building *A* the volume of water that has been consumed increases by 90 litres per hour.

b When the consumption is constant in building B this volume of water lasts 9 hours. This graph shows the water consumption for building B.

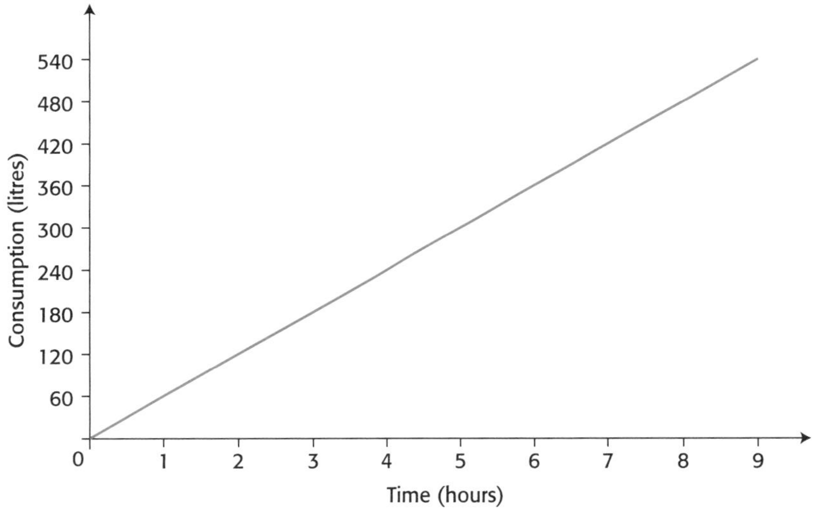

We see from this graph that in building *B* the volume of water that has been consumed increases by 60 litres per hour.

This graph shows the consumption of water for both buildings.

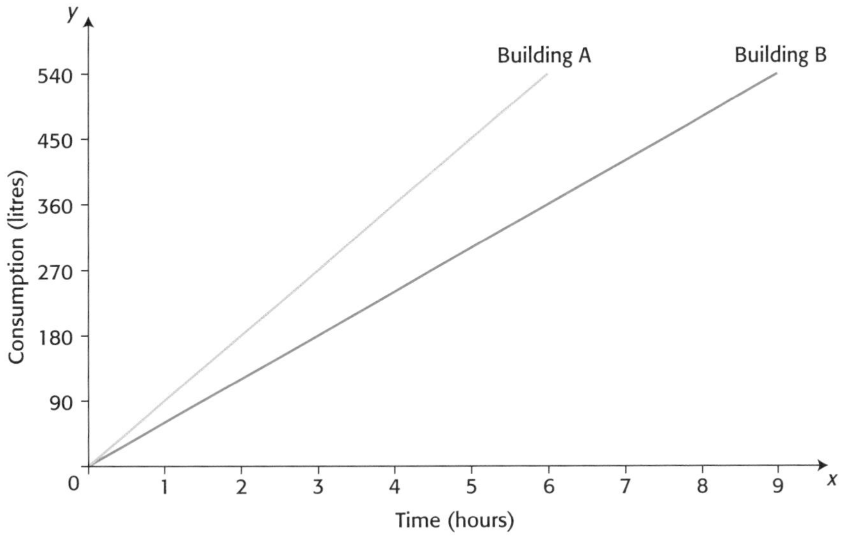

The line for building A has a steeper **slope** than the line for building B. The **gradient** of each line tells us how steep it is.

c This graph shows the amount of water remaining in building A when the consumption is constant.

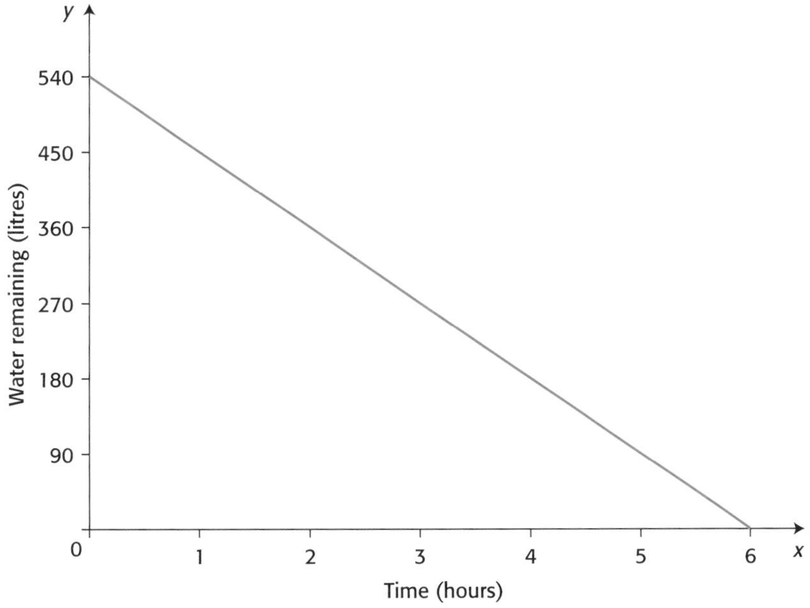

In this example the amount of water left decreases by 90 litres per hour.

Using this definition we can say that the gradient of the line in

The **gradient** of a line is the number that tells us by how much y increases or decreases each time that x increases by 1 unit.

a is 90 $\left(90 = \frac{90}{1} = \frac{180}{2} = \ldots = \frac{540}{6} = \frac{\text{vertical step}}{\text{horizontal step}}\right)$,

b is 60 $\left(60 = \frac{60}{1} = \frac{120}{2} = \ldots = \frac{540}{9} = \frac{\text{vertical step}}{\text{horizontal step}}\right)$,

c is –90 $\left(-90 = \frac{-90}{1} = \frac{-180}{2} = \ldots = -\frac{540}{6} = \frac{\text{vertical step}}{\text{horizontal step}}\right)$.

In the next diagram the points $A(x_1, y_1)$ and $B(x_2, y_2)$ have been plotted. From the diagram it can be seen that the vertical step is $y_2 - y_1$ and the horizontal step is $x_2 - x_1$.

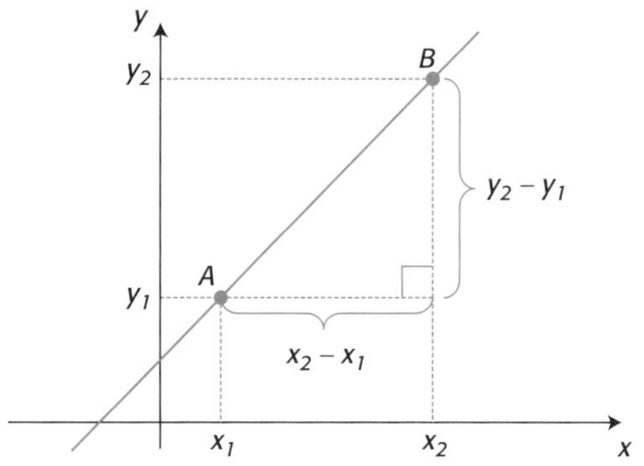

The **gradient**, m, of the line passing through $A(x_1, y_1)$ and $B(x_2, y_2)$ can be determined by the quotient $m = \dfrac{y_2 - y_1}{x_2 - x_1}$.

Note that the same suffix order is used in both the top and the bottom of the gradient formula.

Example 5.2.1b

For each of these pairs of points

a $A(1, 3)$ and $B(3, 7)$ b $A(-2, 4)$ and $B(2, 2)$

c $A(-4, 5)$ and $B(2, 5)$ d $A(-1, 2)$ and $B(-1, 4)$

i sketch the line joining them and label the points A and B

ii find the gradient of the line.

Solution

a i

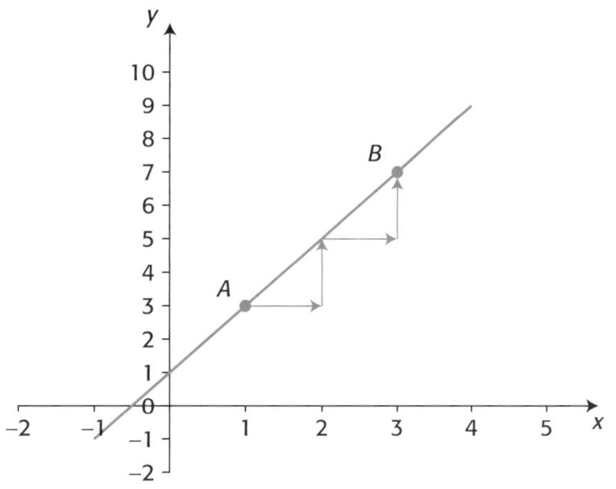

ii We take $x_1 = 1$, $x_2 = 3$, $y_1 = 3$, $y_2 = 7$ and substitute into the gradient formula

$$m = \frac{y_2 - y_1}{x_2 - x_1} = \frac{7 - 3}{3 - 1} = \frac{4}{2} = 2$$

The gradient is *positive* and each time that *x increases* 1 unit, *y increases* 2 units.

b i

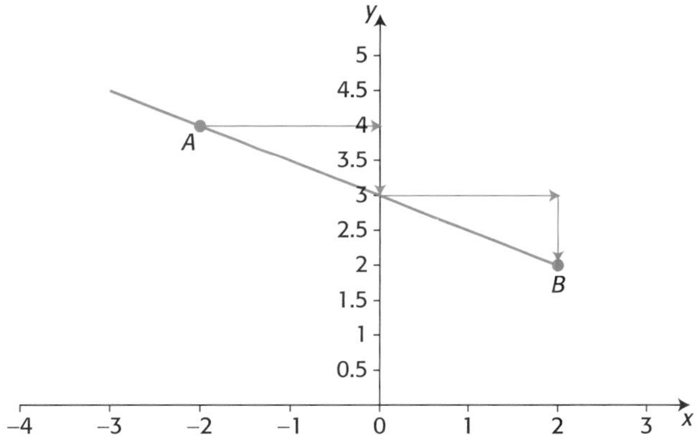

ii $x_1 = -2$, $x_2 = 2$, $y_1 = 4$, $y_2 = 2$

$$m = \frac{y_2 - y_1}{x_2 - x_1} = \frac{2 - 4}{2 - -2} = \frac{-2}{4} = -\frac{1}{2}$$

The gradient is *negative*. Each time that *x increases* 2 units, *y decreases* 1 unit or, similarly, each time that *x increases* 1 unit, *y decreases* $\frac{1}{2}$ unit.

c i

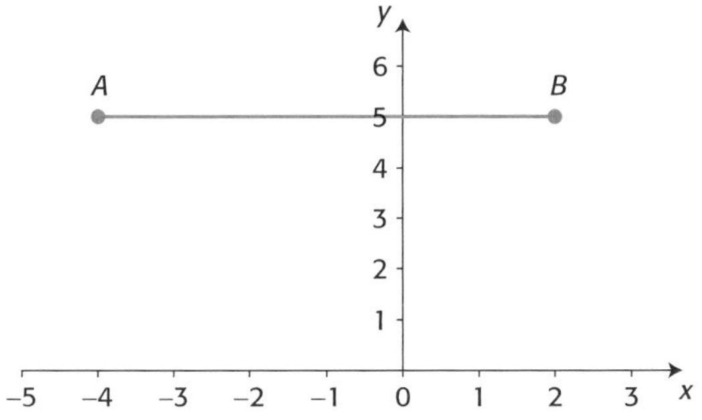

ii $x_1 = -4$, $x_2 = 2$, $y_1 = 5$, $y_2 = 5$

$$m = \frac{y_2 - y_1}{x_2 - x_1} = \frac{5-5}{2--4} = \frac{0}{6} = 0$$

This is a *horizontal* line. As x increases, y remains *constant*. The line is parallel to the x-axis. Each time x increases by 1 unit, y neither increases nor decreases.

d i

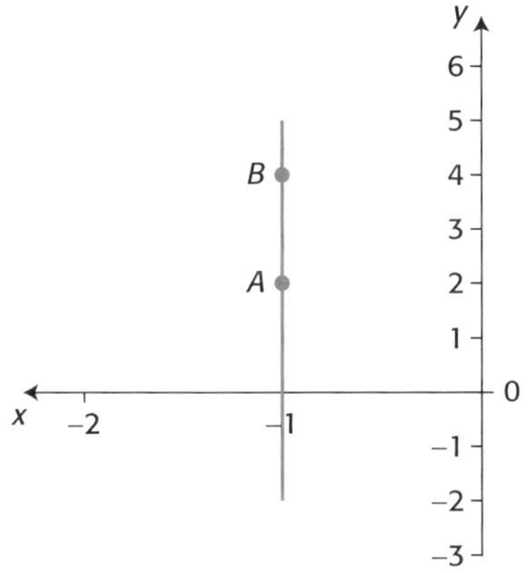

ii $x_1 = -1$, $x_2 = -1$, $y_1 = 2$, $y_2 = 4$

$$m = \frac{y_2 - y_1}{x_2 - x_1} = \frac{4-2}{-1--1} = \frac{2}{0}$$

This operation is not meaningful as it is not possible to divide by zero. We say the gradient is **not defined** or **not well-defined**.

Therefore the gradient of this line *is not defined.*

This a vertical line. It is parallel to the y-axis. (Loosely speaking as a line gets steeper and becomes vertical, its gradient increases without limit and becomes infinite.)

We conclude from the previous examples that

- **vertical** lines have no gradient
- for a non-vertical line the gradient can be positive, zero or negative.

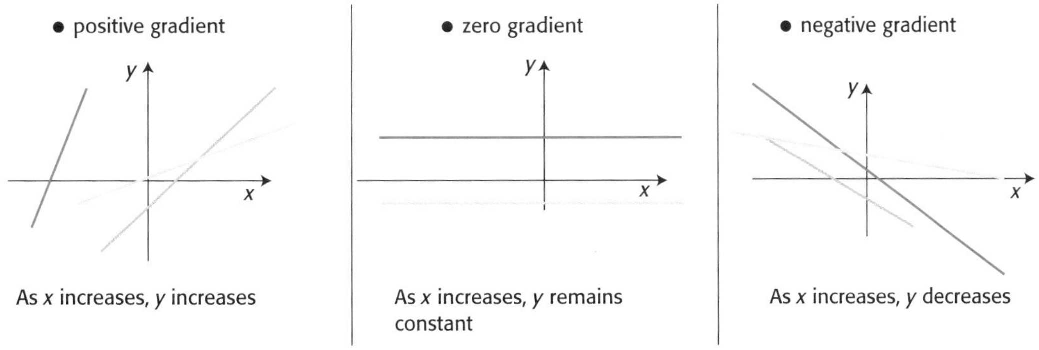

Example 5.2.1c

a Draw a line with gradient $-\frac{3}{5}$ and through $A(4, 2)$.

b Draw a line with gradient 2 through $B(-3, 0)$.

Solution

a We first plot the point $A(4, 2)$.

Gradient $= -\frac{3}{5} = \frac{y\text{-step}}{x\text{-step}}$ so "each time x increases 5 units, y decreases 3 units".

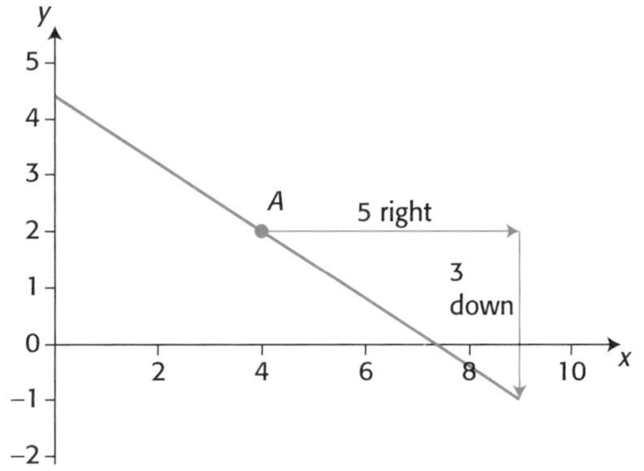

b We first plot the point $B(-3, 0)$.

Gradient $= 2 = \frac{2}{1} = \frac{y\text{-step}}{x\text{-step}}$ so "each time x increases 1 unit, y increases 2 units".

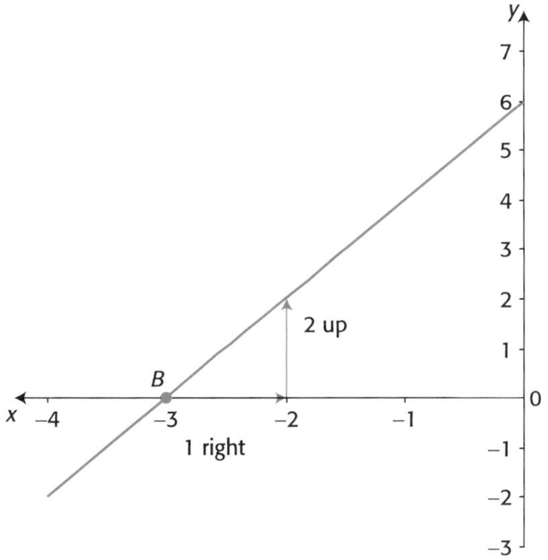

Example 5.2.1d

Find the gradient of each of these lines by using their graphs.
Explain the meaning of the gradient for each line.

a

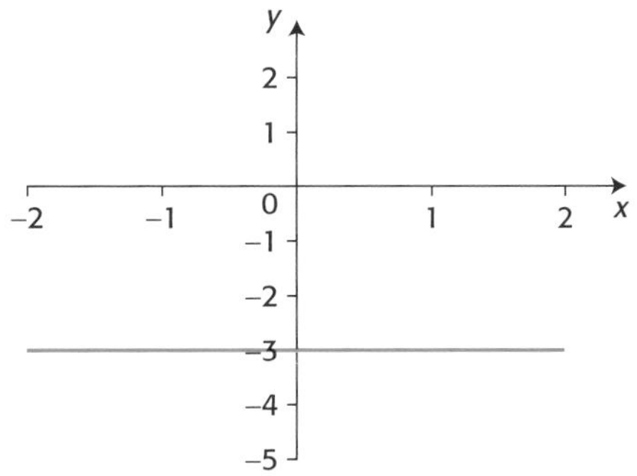

b

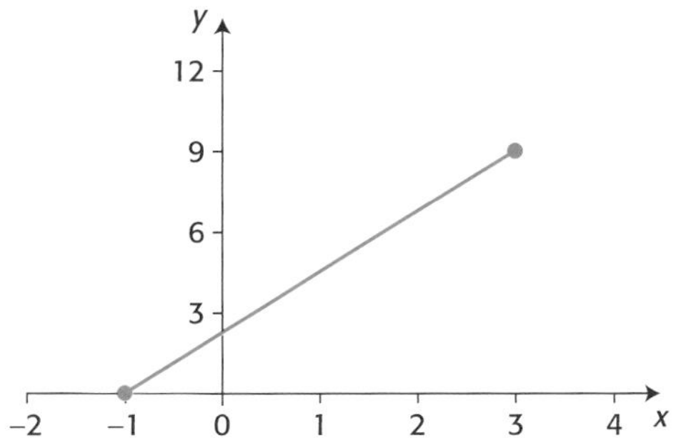

c

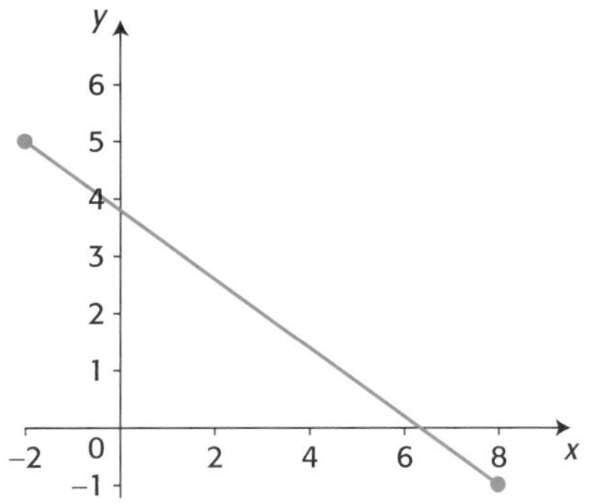

d

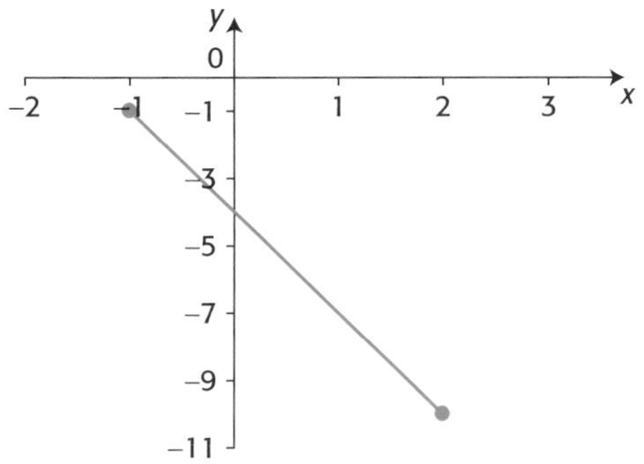

Solution

a As this is a horizontal line, its gradient is zero. As x increases, y remains constant.

b This line passes through the points $(-1, 0)$ and $(3, 9)$.

$$\text{Gradient} = \frac{9 - 0}{3 - -1} = \frac{9}{4}$$

The gradient is $\frac{9}{4}$. This means that each time that x increases 4 units, y increases 9 units.

c This line passes through the points $(-2, 5)$ and $(8, -1)$.

$$\text{Gradient} = \frac{-1 - 5}{8 - -2} = -\frac{6}{10} = -\frac{3}{5}.$$

The gradient is $-\frac{3}{5}$. This means that each time that x increases 3 units, y decreases 5 units.

d This line passes through the points $(-1, -1)$ and $(2, -10)$.

$$\text{Gradient} = \frac{-10 - -1}{2 - -1} = -\frac{9}{3} = -3.$$

The gradient is -3. This means that each time that x increases 1 unit, y decreases 3 units.

5.2.2 Parallel lines

Two lines are **parallel** if, and only if, they have the same gradient.

This means that

- If two lines are parallel then they have the same gradient.
- If two lines have the same gradient then they are parallel.

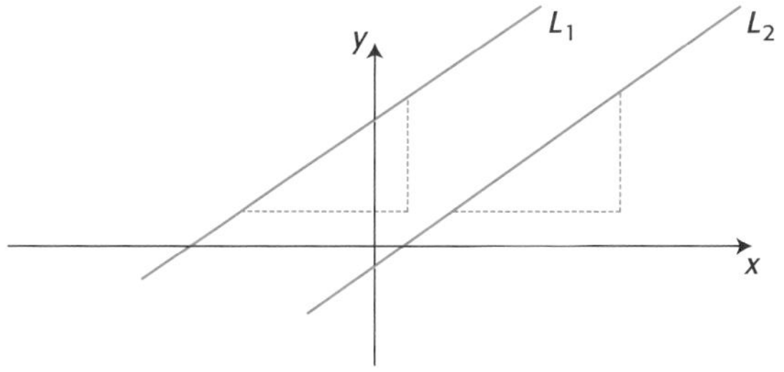

In symbols $L_1 // L_2$ reads "L_1 is parallel to L_2".

Example 5.2.2

Line L_1 passes through the points $A(-1, 3)$ and $B(2, 6)$.

a Draw L_1 and label it.

b Write down the gradient of L_1.

c Draw and label a line L_2 passing through the origin and parallel to L_1.

Point D lies on L_2 and its x-coordinate is 2.5.

d Write down the y-coordinate of D.

Solution

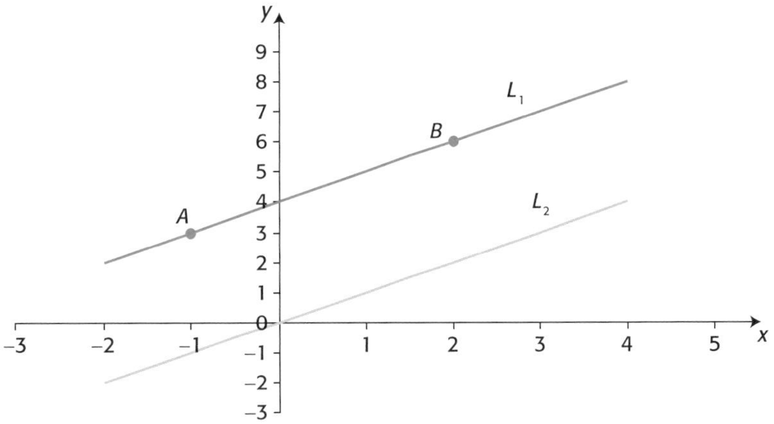

a On graph.

b gradient of $L_1 = m = \dfrac{6-3}{2--1} = 1$

c On graph.

d As $L_1 \parallel L_2$, the gradient of $L_2 = 1$ and L_2 passes through the points $(0, 0)$ and $(2.5, a)$

then $1 = \dfrac{a-0}{2.5-0} \Rightarrow a = 2.5$

5.2.3 Perpendicular lines

We have already said that the x-axis and the y-axis are perpendicular. So now we are interested in finding existing relationships between perpendicular lines when there is the same scale on both axes.

Let's draw line L_1 through O and $B(3, 2)$ and the line L_2 through O and $C(-2, 3)$. We see that these two lines are perpendicular.

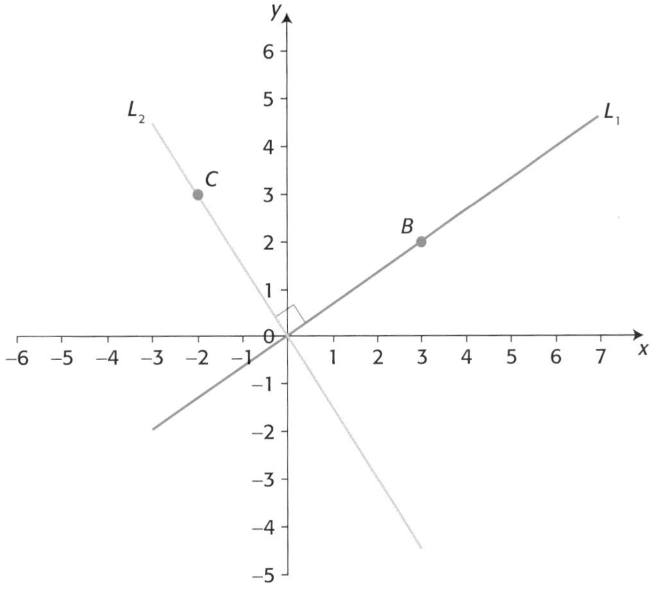

Notation $L_1 \perp L_2$ reads "L_1 is perpendicular to L_2".

Gradient of L_1 is $m_1 = \dfrac{2}{3}$ and the gradient of L_2 is $m_2 = -\dfrac{3}{2}$.

Each gradient is the **negative reciprocal** of the other.

This means that their product is $-1\left(\dfrac{2}{3} \times -\dfrac{3}{2} = -1\right)$.

Two lines are **perpendicular** if, and only if, their gradients are **negative reciprocals** of each other.

Example 5.2.3a

Write down the gradient of all lines perpendicular to the line with gradient

a 2 b $-\frac{3}{4}$

c m with $m \neq 0$ (that is, the line is not horizontal)

Solution

a Since the gradients of two perpendicular lines are negative reciprocals, the gradient of all lines perpendicular to the one given is $-\frac{1}{2}$. Check: $-\frac{1}{2} \times 2 = -1$

b The gradient of all lines perpendicular to the one given is $\frac{4}{3}$.
Check: $-\frac{3}{4} \times \frac{4}{3} = -1$

c The gradient of all lines perpendicular to the one given is $-\frac{1}{m}$.
Check: $m \times -\frac{1}{m} = -1$

Example 5.2.3b

The points $A(3, c)$ and $B(-1, 0)$ lie on the line L which is perpendicular to all lines with gradient –0.5. Find the value of c.

Solution

Let m be the gradient of L then

$$m = \frac{0 - c}{-1 - 3} = \frac{-c}{-4} = \frac{c}{4}$$

Now $\frac{c}{4} \times -0.5 = -1$ (product of gradients of two perpendicular lines is -1) $\therefore c = 8$

5.2.4 Equations of lines

(See also Topic 4.2 on page 175)

Collinear points are points which lie on the same straight line. Any two points are collinear because there is always a straight line that passes through them.

In this diagram A, B and C are collinear as they all lie on the line L. However D, E and F are not collinear.

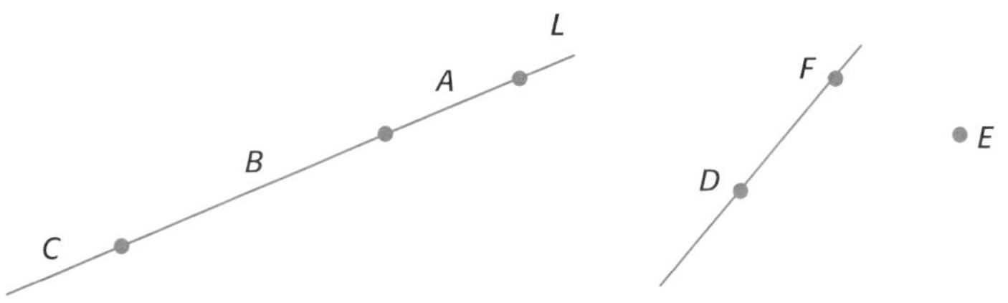

Clearly, as A, B and C lie on the same line then gradient of AB = gradient of CB = gradient of AC.

Example 5.2.4a

$A(1, 1)$, $B(2, -3)$ and $C(4, t)$ are collinear. Find the value of t.

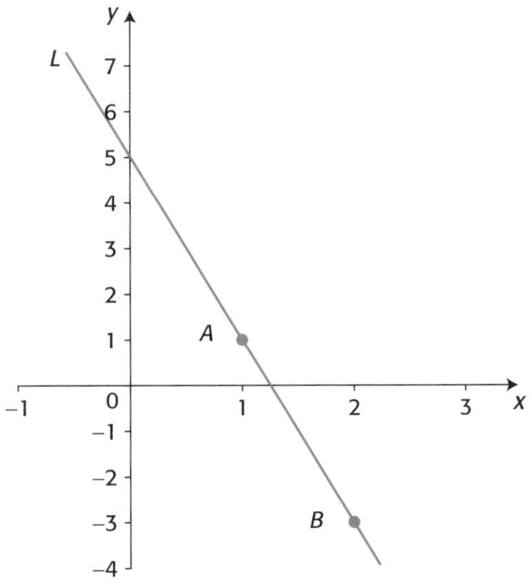

Solution

If A, B and C are collinear then if m = gradient of AB, the gradient of $BC = m$.

$$m = \frac{-3 - 1}{2 - 1} = -4 \text{ and gradient of } BC = \frac{t - -3}{4 - 2}$$

then
$$-4 = \frac{t+3}{2}$$
$$-8 = t + 3$$
$$-11 = t$$

Example 5.2.4b

The line L joins the points $A(1, 1)$ and $B(2, -3)$ as in the diagram above.

Let $P(x, y)$ be any point on L.

a Find an equation linking x and y.

b Hence write down the coordinates of two points lying on L different from A and B.

c Decide whether the point $T(0.5, 3.5)$ lies on L. Justify your answer.

Solution

a Let m be the gradient of L then $m = -4$ (same as Example 5.2.4*a*). P lies on L therefore gradient of BP is -4.

Gradient of $BP = \dfrac{y - -3}{x - 2}$ then

$$\frac{y + 3}{x - 2} = -4 \text{ and}$$

$$y + 3 = -4(x - 2)$$

The equation $y+3=-4(x-2)$ is the equation of the line L. This means that the coordinates of *any* point on L satisfy this equation.

b Let $x = 3$, then substituting this value in the equation of L

$$y+3=-4(3-2)$$

$$\therefore y=-4-3=-7$$

therefore the point $(3,-7)$ lies on L.

Let $x = -1$, then substituting into the equation of L

$$y+3=-4(-1-2)$$

$$\therefore y=12-3=9$$

therefore another point lying on L is $(-1,9)$.

To find points lying on a line we just give x any real value and substitute for x in the equation of the line to find y.

c If $T(0.5, 3.5)$ lies on L, then its coordinates will satisfy the equation of L. Therefore if we substitute 0.5 for x in the equation, the value of y should be 3.5.

$$y + 3 = -4(0.5 - 2)$$

$$\therefore y=6-3=3 \neq 3.5$$

therefore T does not lie on L.

Remember: a point lies on a line *if and only if* the coordinates of the point satisfy the equation of this line.

The equation of a line L is an equation that links the coordinates of *any* point on L.

Let $A(s, r)$ be a point on the line L and m be the gradient of L. If $P(x, y)$ is any point on L then the equation of L is $\dfrac{y-r}{x-s} = m$ and this can be rewritten as $y - r = m(x - s)$ and then written in the form $y = mx + c$ or $ax + by + d = 0$.

Example 5.2.4 c

Write down the equation of the line passing through $A(1, 1)$ and $B(2, -3)$ in the form

a $y=mx+c$

b $ax+by+d=0$

Solution

We have already found the equation of this line in the previous example. The equation is $y+3=-4(x-2)$. Now we have to write this equation in the required form.

a $y+3=-4(x-2)$

$\therefore y=-4x+8-3$

$\therefore y=-4x+5$

Here $m = -4$ and $c = 5$.

b $4x+y-5=0$

Here $a = 4$, $b = 1$ and $d = -5$

Any multiple of this equation is also correct.

Examples

$-4x-y+5=0$

$8x+2y-10=0$

Example 5.2.4d

a Find the equation of the line which has gradient 3 and passes through the point $A(1,4)$. Write the equation in the form $y=mx+c$.

b Find the equation of a line that passes through the points $C(-2,3)$ and $D(1,-1)$. Write down the equation in the form $ax+by+d=0$ where $a\in\mathbb{Z}$, $b\in\mathbb{Z}$, $d\in\mathbb{Z}$.

Solution

a Let $P(x, y)$ be any point on L, then the equation of L is

$y=3x+c$ and $(1,4)$ satisfies this equation so $4=3\times1+c \quad \therefore c=1$.

The equation is $y=3x+1$.

b $m=\dfrac{-1-3}{1--2}=-\dfrac{4}{3}$, so if $P(x, y)$ is any point on the line then

$$\frac{y\;-\;-1}{x\;-\;1} = -\frac{4}{3}$$

$$3(y+1) = -4(x-1)$$

$$3y+3 = -4x+4$$

so the required equation is $4x+3y-1=0$.

Equations of horizontal and vertical lines

Example 5.2.4e

The line passing through the points $(2,-1)$ and $(2,3)$ is vertical. Every point lying on this line has an x-coordinate of 2.

The equation of the line is $x = 2$.

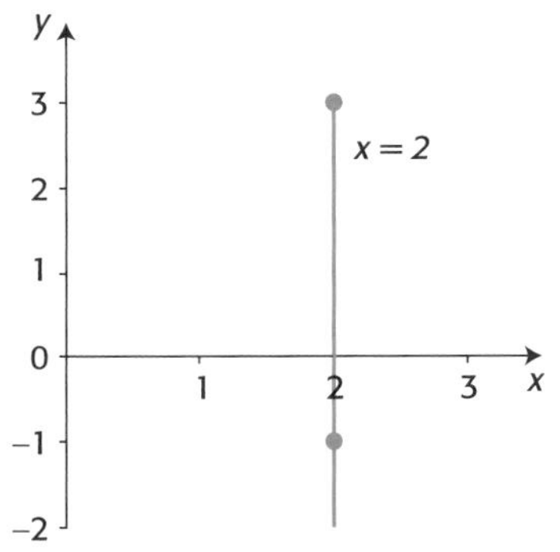

Example 5.2.4f

The line passing through the points $(-2,-3)$ and $(2,-3)$ is horizontal. Every point lying on this line has a y-coordinate of -3.

The equation of the line is $y = -3$.

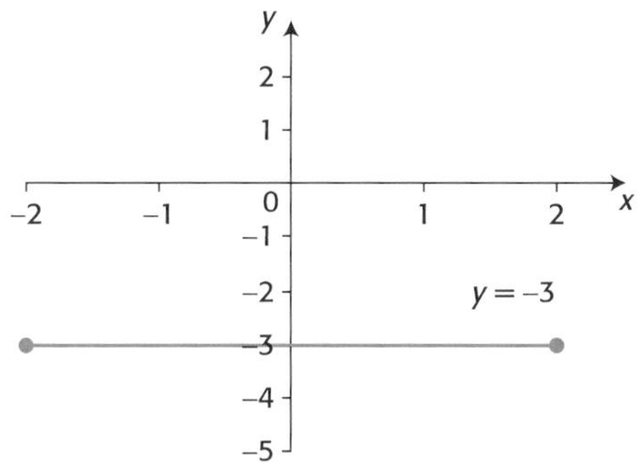

In general: The equation of any vertical line is of the form $x = k$ where k is a constant. The equation of any horizontal line is of the form $y = c$ where c is a constant.

5.2.5 Axis intercepts

- The point of intersection of a line with the y-axis occurs when $x = 0$. The y-coordinate of this point is called the **y-intercept**.
- The point of intersection of a line with the x-axis occurs when $y = 0$. The x-coordinate of this point is called the **x-intercept**.

If the equation is given in the form $y = mx + c$ then $(0, c)$ is the point of intersection with the y-axis and the y-intercept is c.

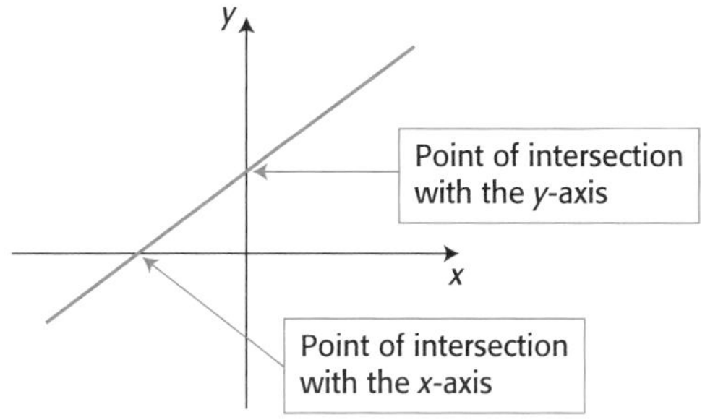

Example 5.2.5a

The equation of a line L is $3x + 2y + 6 = 0$.

a Write down the y-intercept of L.
b Write down the x-intercept of L.
c Hence draw L.

Solution

a Let $x = 0$, $3 \times 0 + 2 \times y + 6 = 0 \therefore y = -3$
b Let $y = 0$, $3x + 2 \times 0 + 6 = 0 \therefore x = -2$
c

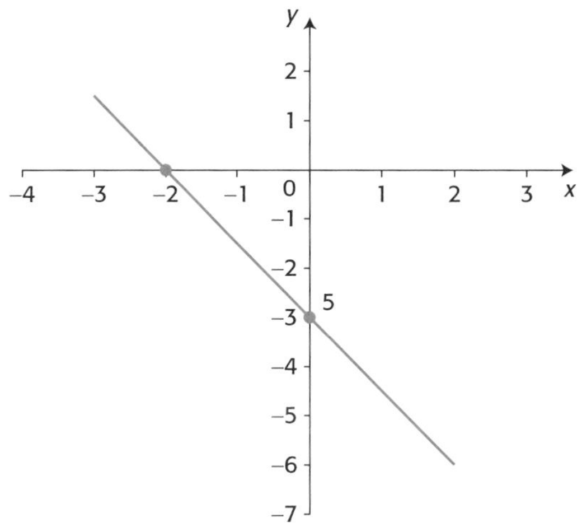

Example 5.2.5.b

The equation of a line L is $-3x + y = 0$.

a Write down the x- and y-intercepts.
b Write down the gradient.
c Hence draw L.

Solution

a L intercepts both axes at the point $(0, 0)$, so both intercepts are 0.
b The equation can be written in the form $y = 3x$. So the gradient is 3.
c

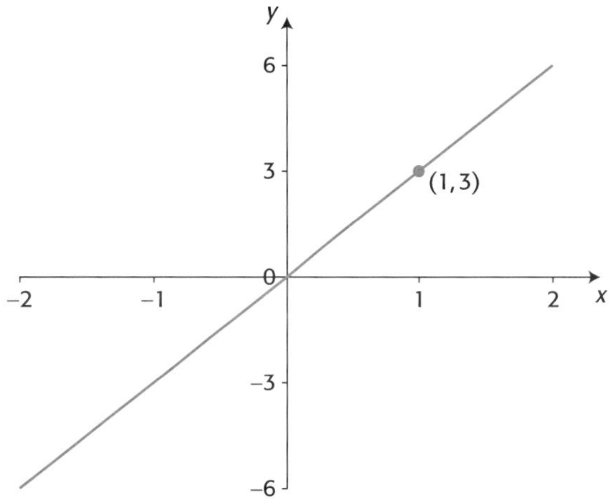

Example 5.2.5c

Find the equations of the lines shown in these diagrams. Write them in the form $y = mx + c$.

a

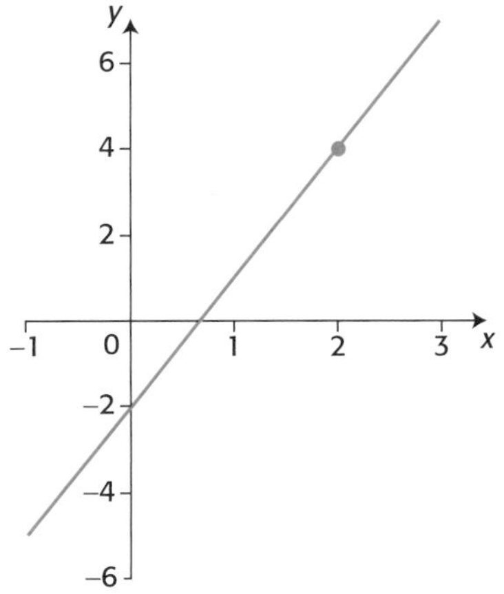

b

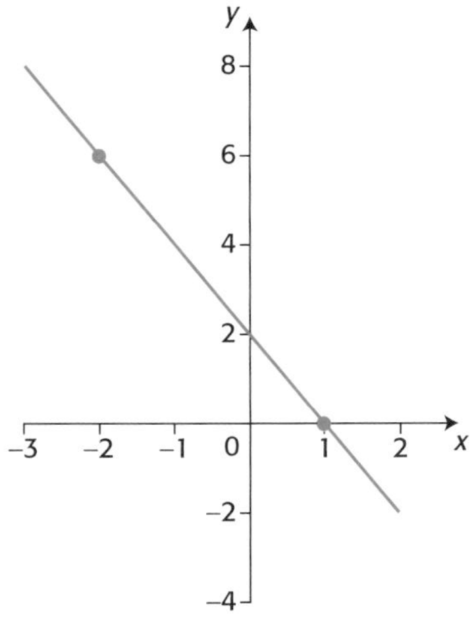

a The y-intercept is $c = -2$. The equation so far is $y = mx - 2$. We still need to find the gradient of the line. The point $(2, 4)$ lies on the line so

$4 = m \times 2 - 2$

$m = 3$

and the equation of the line is $y = 3x - 2$.

b The line passes through the points $(-2, 6)$ and $(1, 0)$. The gradient of the line is $m = \dfrac{0-6}{1--2} = -2$

so $y = -2x + c$.

Substitute $(-2, 6)$ into the equation.

$6 = -2(-2) + c$

$2 = c$

and the equation is $y = -2x + 2$.

5.2.6 Point of intersection of lines in the plane

If two lines L_1 and L_2 are parallel then they have the same gradient and they may be

- different lines

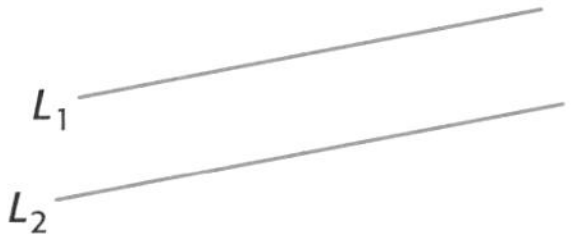

There is no point of intersection. Different y-intercepts.

or

- coincident lines

There is an infinite number of points of intersection.

If two lines L_1 and L_2 are not parallel then they have different gradients and they meet at just one point.

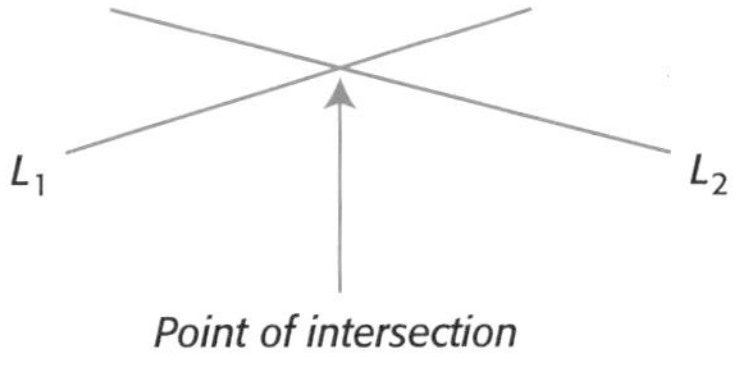

Example 5.2.6a

Show that the lines L_1: $x + 2y = 6$ and L_2: $3x + 6y = -1$ never meet.

Solution

If they never meet they are parallel but not coincident.

Gradient of L_1	Gradient of L_2
$2y = 6 - x$	$6y = -1 - 3x$
$y = 3 - \frac{1}{2}x$	$y = -\frac{1}{6} - \frac{1}{2}x$
$m_1 = -\frac{1}{2}; c_1 = 3$	$m_2 = -\frac{1}{2}; c_2 = -\frac{1}{6}$

These lines never meet as they have the same gradient but different y-intercepts.

Example 5.2.6b

a i Draw the lines $L_1 : y - x = 1$ and $L_2 : y + 1.5x - 6 = 0$ on a pair of Cartesian axes.

ii Mark the point of intersection. Label it P.

b Find the coordinates of P by algebraic means.

Solution

a i Axis intercepts of L_1 are at $(0, 1)$ and $(-1, 0)$ and axis intercepts of L_2 are at $(0, 6)$ and $(4, 0)$.

ii On graph.

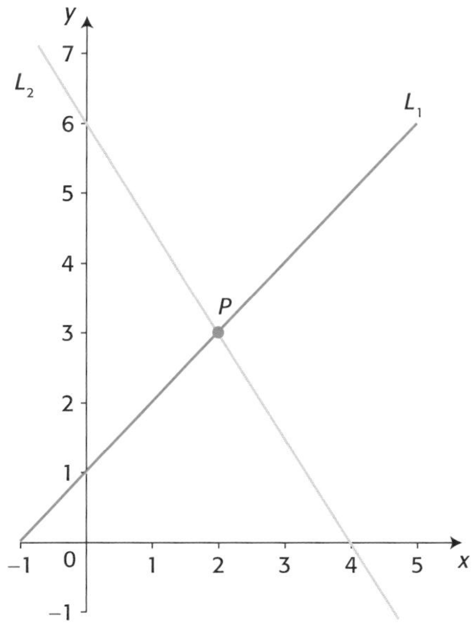

b Solve the pair of simultaneous equations

$$\begin{cases} y - x = 1 \\ y + 1.5x - 6 = 0 \end{cases}$$

$y = 1 + x$ and $y = 6 - 1.5x$

$1 + x = 6 - 1.5x$

$2.5x = 5$

$x = 2$ and $y = 3$

Therefore P is the point $(2, 3)$.

Example 5.2.6c

Find, using a GDC, the point of intersection of the lines
$y = -2x - 1$ and $6x - 2y + 8 = 0$.

Solution

Method 1

Rearrange both equations into the form $y = mx + c$ and proceed by considering the lines as the graphs of the linear functions $f(x)$ and $g(x)$. Draw a graph of these functions and find the intersection point as explained in Topic 4, Section 4.2.6, page 179.

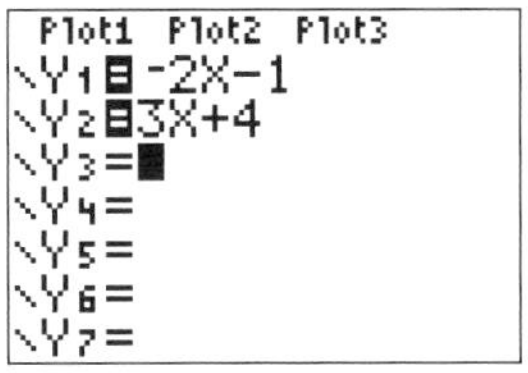

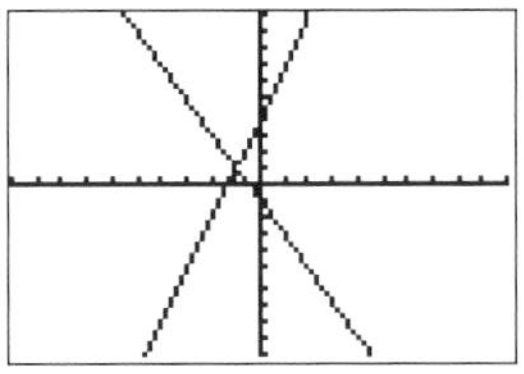

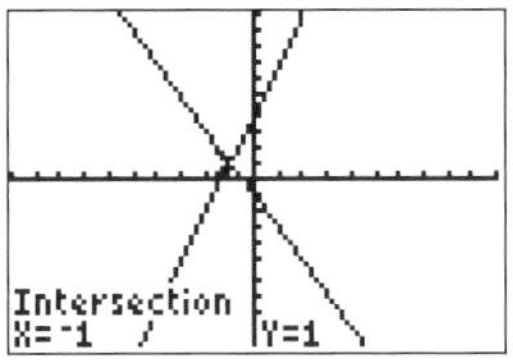

The lines intersect at the point $(-1, 1)$.

Method 2

Solve the pair of simultaneous equations

$$\begin{cases} 2x + y = -1 \\ 6x - 2y = -8 \end{cases}$$ as explained in Topic 2, Section 2.7, page 104.

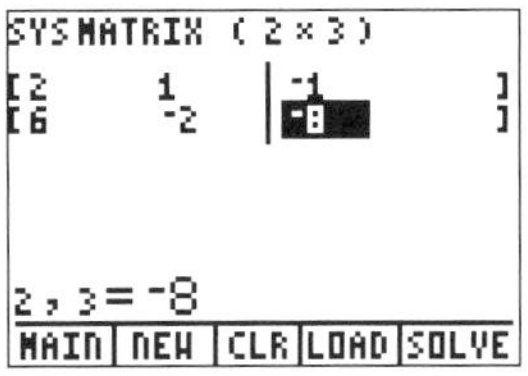

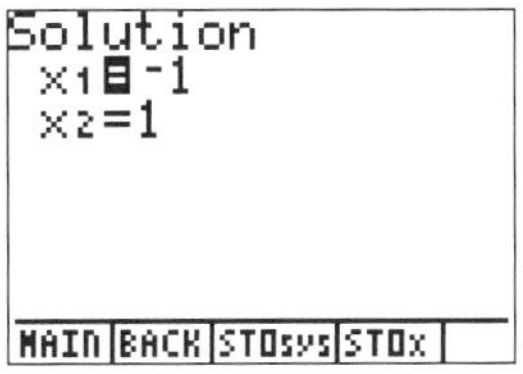

The lines intersect at $x = -1$, $y = 1$.

Exercise 5.2

1 Find the gradient and the y-intercept of these lines.

a $2x + 3y - 8 = 0$ b $3(x - 2y) = 6$ c $2y = 5$ d $\frac{2x - 6y}{3} = 4 + y$

2 Draw a line

a passing through $(1, 2)$ and with gradient -3

b passing through $(-1, 3)$ and gradient $\frac{1}{2}$

c that cuts the x-axis at $(2, 0)$ and with y-intercept -2.

3 Write down the equation of the lines from question 2.

4 Consider the line L: $y = -\frac{1}{3}x + 5$.

a The points $A(5, r)$ and $B(s, 0)$ lie on L. Write down the values of r and s.

b Write down the gradient of all lines perpendicular to the line L.

c Write down the equation of L in the form $ax + by + d = 0$ where $a \in \mathbb{Z}$, $b \in \mathbb{Z}$, $d \in \mathbb{Z}$.

5 Decide whether the points $A(0, -5)$, $B(3, 1)$ and $C(-1, -6)$ are collinear. Justify your answer.

6 Write down the equations of these lines.

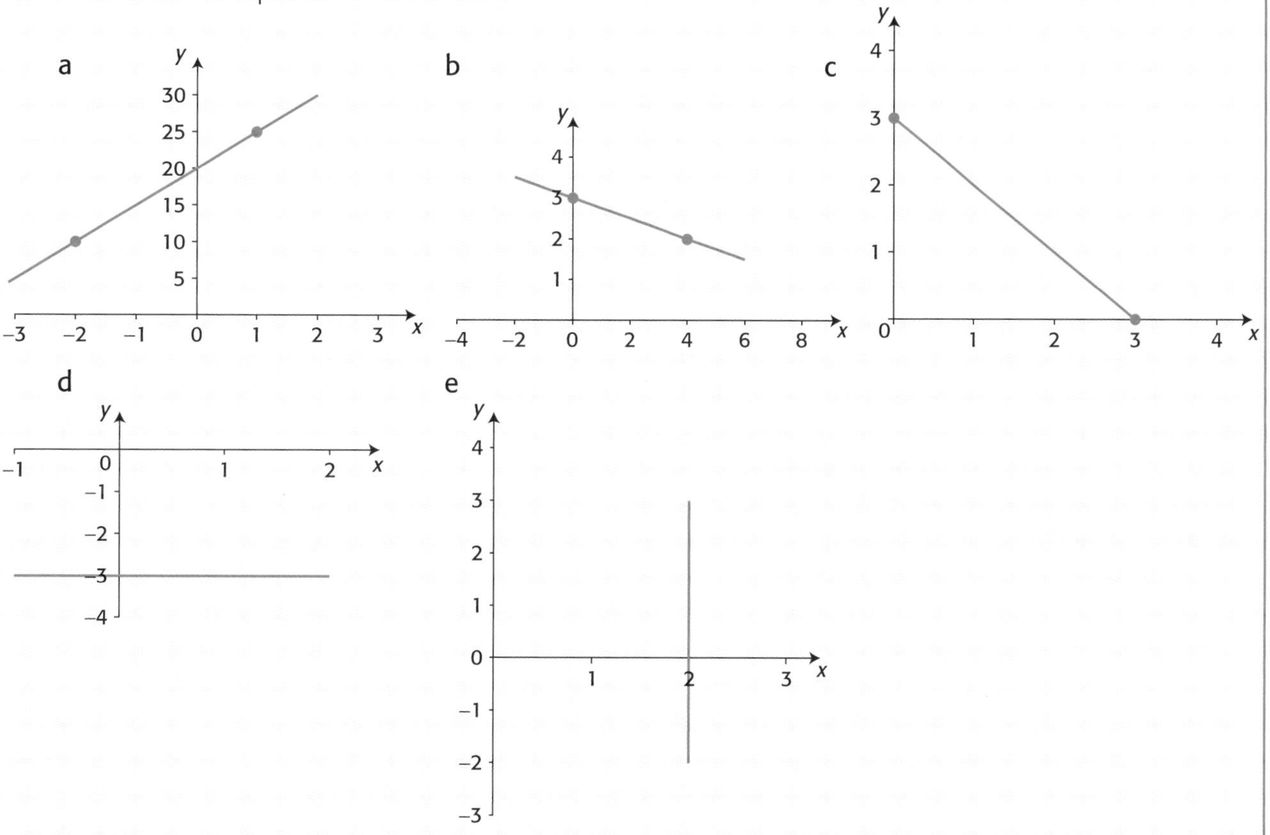

7 Draw each line on a different pair of Cartesian axes.

a $y = \frac{1}{4}x - 3$ b $3x - 5y + 30 = 0$ c $3y = 6$ d $x = -1$

8 Given the points $A(2, 5)$ and $C(-2, 3)$

a find the coordinates of the midpoint, M, of the line AC

b write down the equation of the line L that passes through the point M and is perpendicular to the line AC. (L is said to be the perpendicular bisector of AC.)

Point $B(-1, 6)$ lies on L. $ABCD$ is a rhombus.

c Find the coordinates of the point D.

9 Use your GDC to find the point of intersection of the lines $y = 4x + 9$ and $7x + 2y + 12 = 0$.

10 Which one of these four pairs of lines meet at

a just one point
b no point
c an infinite number of points?

i $y - 2x + 1$ and $3y + 6x - 8$ ii $\dfrac{y-3}{x-1} = 6$ and $y = 6x - 3$

iii $3y + 5 - 0$ and $x - 2$ iv $x - 2y + 8 = 0$ and $\dfrac{1}{2}x - y = 0$

11 This is the plan of a village. Lines L_1: $y = -x + 3$ and L_2: $y = x - 5$ represent the only two roads in the area.

On both axes 1 unit represents 1 km.

Family Alvarez lives at A.
There is a mall at M.

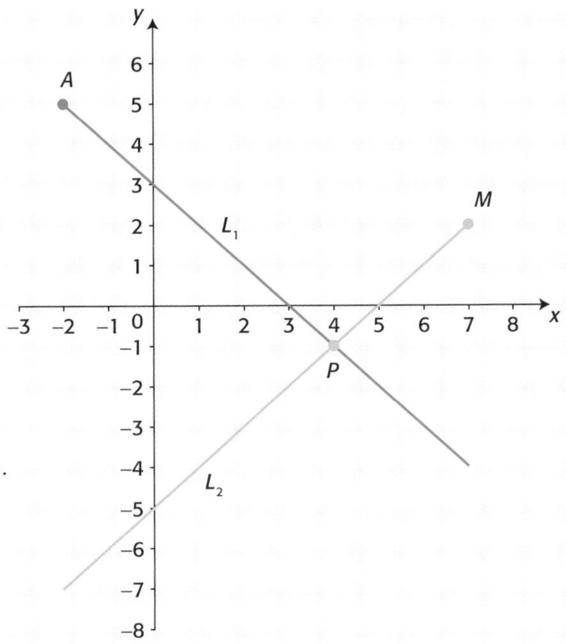

a Find the coordinates of the point of intersection of the roads, P.
b Calculate the distance that family Alvarez have to **drive** each time they go to the mall.
c Show that L_1 and L_2 are perpendicular lines.
d A school, S, is located on the road L_2. P is exactly halfway between M and S. Find the coordinates of S.
e Calculate the area enclosed by the points M, P and A.

12 The equation of the straight line, L_1, is $3x - 5y + 4 = 0$.

a Find the equation of the line L_2, parallel to L_1 that passes through the origin, O.

L_1 intersects the y-axis at A.

b Write down the coordinates of A.

L_3 passes through A and is parallel to the horizontal axis.

c Write down the equation of L_3.

5.3 Right-angled trigonometry, sine, cosine and tangent

5.3.1 The sides of a right-angled triangle

In any right-angled triangle, the **hypotenuse** is the side which is opposite the right angle. The way we refer to each of the other two sides depends on the angle of the triangle that is being considered.

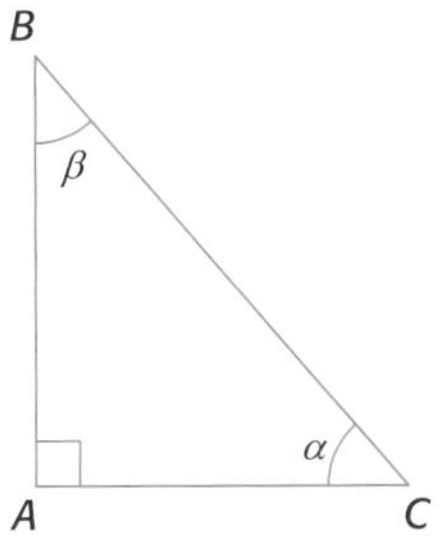

If we consider the triangle ABC in the diagram.

a For angle α

- side AB is **opposite** angle α
- side AC is **adjacent** to angle α

Did you know that...?

"Trigonometry has an enormous variety of applications. The ones mentioned explicitly in textbooks and courses on trigonometry are its uses in practical endeavours such as navigation, land surveying, building, and the like. It is also used extensively in a number of academic fields, primarily mathematics, science and engineering."
www.answers.com

b For angle β

- side AB is adjacent to angle β
- side AC is opposite angle β

Example 5.3.1

For each of these triangles write down using letters

a the hypotenuse (Hyp)

b the side opposite angle θ (Opp)

c the side adjacent to angle θ (Adj).

The **hypotenuse** is the **longest side** of a right-angled triangle.
Greek: teinein – to stretch or span
hypoteinein – to subtend.

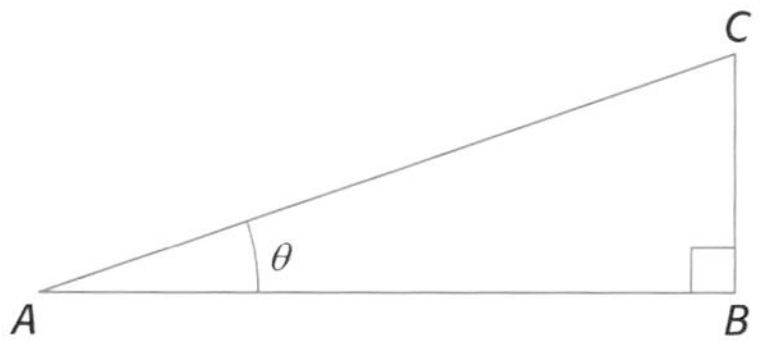

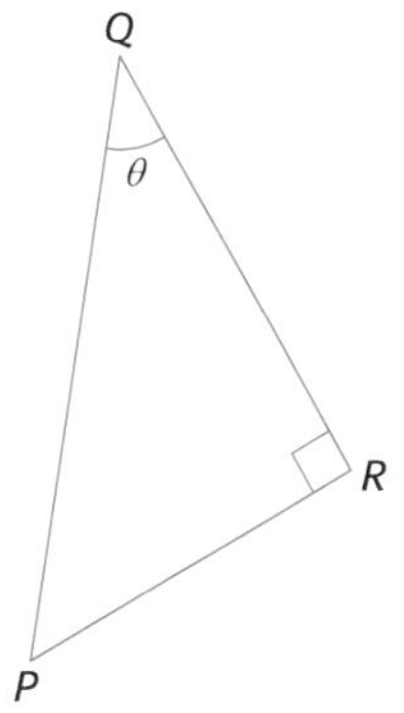

Solution

a Hyp = AC

b Opp $\theta = BC$

c Adj $\theta = AB$

a Hyp = QP

b Opp $\theta = PR$

c Adj $\theta = RQ$

5.3.2 The trigonometric ratios

Consider the triangles ABC and ADE in the diagram. They have the same angles and corresponding sides are proportional (triangles with this property are called similar triangles).

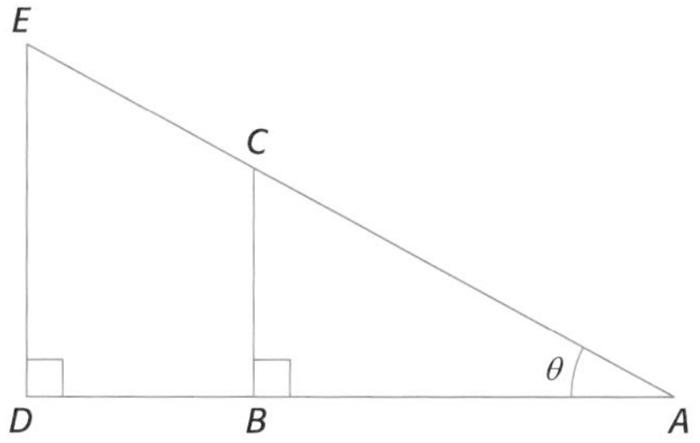

We have then $\frac{\text{ED}}{\text{DA}} = \frac{\text{CB}}{\text{BA}}$; $\frac{\text{DA}}{\text{AE}} = \frac{\text{BA}}{\text{AC}}$; $\frac{\text{ED}}{\text{AE}} = \frac{\text{CB}}{\text{AC}}$

The three ratios written above can also be written as

$\frac{\text{Opp}}{\text{Adj}}$, $\frac{\text{Adj}}{\text{Hyp}}$ and $\frac{\text{Opp}}{\text{Hyp}}$ respectively.

- ED and CB are opposite angle θ
- DA and BA are adjacent to angle θ
- AE and AC are the hypotenuses of the triangles.

In any triangle that is **similar** to triangle *ABC* these **ratios** will remain constant.

In any right-angled triangle we define the three **trigonometric ratios** as

$$\sin\theta = \frac{\text{Opp}}{\text{Hyp}},\ \cos\theta = \frac{\text{Adj}}{\text{Hyp}},\ \tan\theta = \frac{\text{Opp}}{\text{Adj}}$$

"$\sin\theta$" is read "sine of θ"

"$\cos\theta$" is read "cosine of θ"

"$\tan\theta$" is read "tangent of θ"

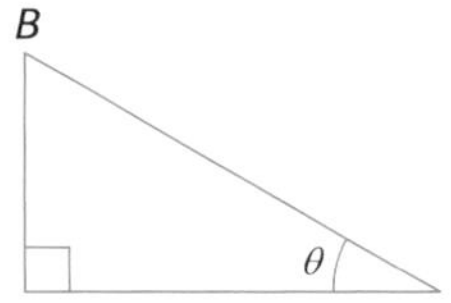

The trigonometric ratios are used to find angles and sides of triangles.

Example 5.3.2a

Write down the three trigonometric ratios for the angle marked β in terms of the sides of the triangle.

a

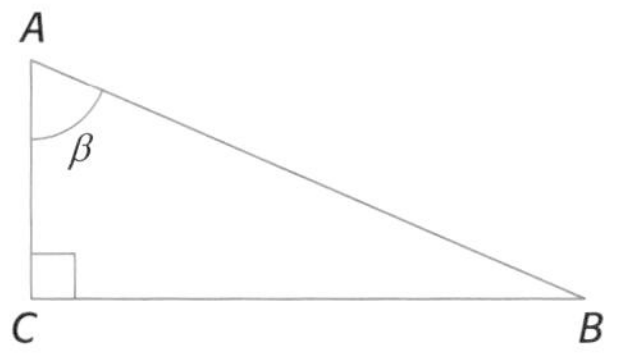

b

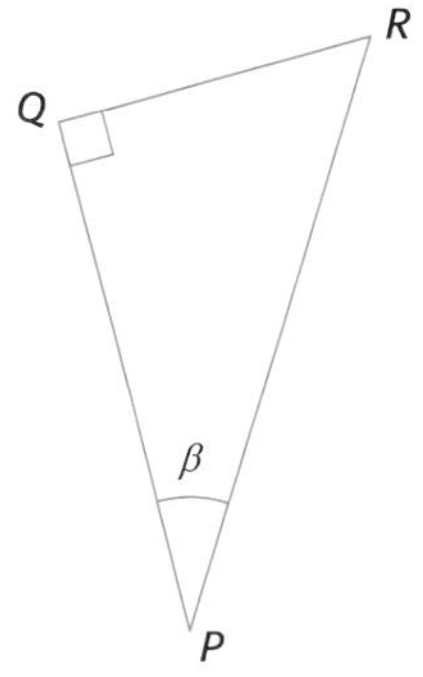

Solution

a $\sin\beta = \frac{\text{CB}}{\text{AB}};\ \cos\beta = \frac{\text{AC}}{\text{AB}};\ \tan\beta = \frac{\text{CB}}{\text{AC}}$

b $\sin\beta = \frac{\text{QR}}{\text{RP}};\ \cos\beta = \frac{\text{QP}}{\text{RP}};\ \tan\beta = \frac{\text{QR}}{\text{QP}}$

Example 5.3.2b

Write down the value of

i $\sin\alpha$ ii $\cos\alpha$ iii $\tan\alpha$

for each of these triangles.

a

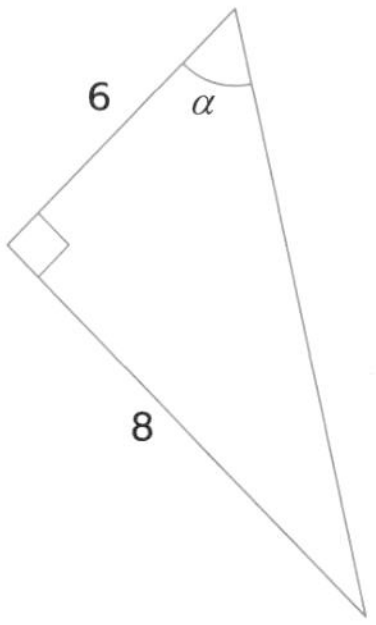

b

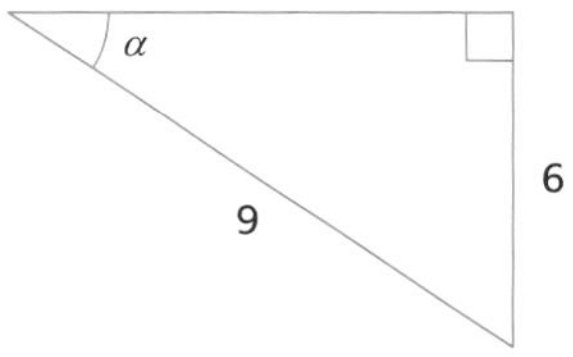

Solution

a To find $\sin\alpha$ and $\cos\alpha$ we need to find the length of the hypotenuse. Let the hypotenuse be x then using Pythagoras' theorem

$x^2 = 6^2 + 8^2 \quad \therefore x^2 = 100 \quad \therefore x = 10$
and

i $\sin\alpha = \dfrac{8}{10}$ or $\sin\alpha = 0.8$

ii $\cos\alpha = \dfrac{6}{10}$ or $\cos\alpha = 0.6$

iii $\tan\alpha = \dfrac{8}{6}$ or $\tan\alpha = 1.33(3\text{ s.f.})$

b To find $\sin\alpha$ and $\cos\alpha$ we need to find the length of the side adjacent to α. Let the side adjacent to α be x then using Pythagoras' theorem

$x^2 + 6^2 = 9^2 \quad \therefore x^2 = 81 - 36 \quad \therefore x = \sqrt{45}$
and

i $\sin\alpha = \dfrac{6}{9}$ or $\sin\alpha = 0.667(3\text{ s.f.})$

ii $\cos\alpha = \dfrac{\sqrt{45}}{9}$ or $\cos\alpha = 0.745(3\text{ s.f.})$

iii $\tan\alpha = \dfrac{6}{\sqrt{45}}$ or $\tan\alpha = 0.894(3\text{ s.f.})$

Example 5.3.2c

a Draw a right-angled triangle ABC whose right angle is at A and in which the tangent of $\hat{B}$ is 3. Label all the angles.

b Draw a right-angled triangle PQR whose right angle is at P and in which the sine of $\hat{Q}$ is 0.8. Label all the angles.

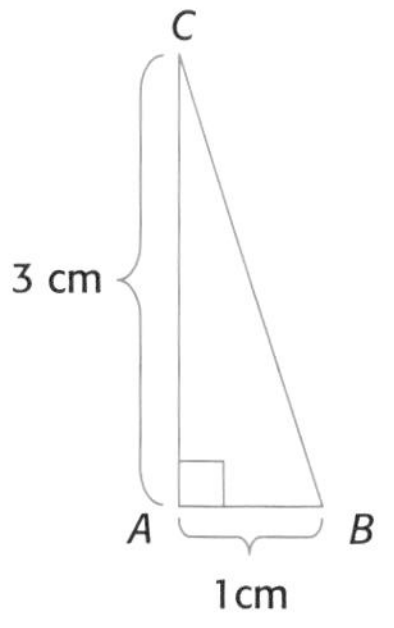

Solution

a $\tan\hat{B} = \dfrac{\text{Opp}}{\text{Adj}}$ and $\tan\hat{B} = 3 \;\therefore \dfrac{\text{Opp}}{\text{Adj}} = 3.$

Therefore we can take Opp = 3 and Adj = 1 or Opp = 6 and Adj = 2 and so on.

b $\sin\hat{Q} = \dfrac{\text{Opp}}{\text{Hyp}}$ and $\sin\hat{Q} = 0.8$ then $\dfrac{\text{Opp}}{\text{Hyp}} = 0.8 = \dfrac{8}{10}$

Therefore we can take Opp = 8 and Hyp = 10 or Opp = 4 and Hyp = 5 and so on.

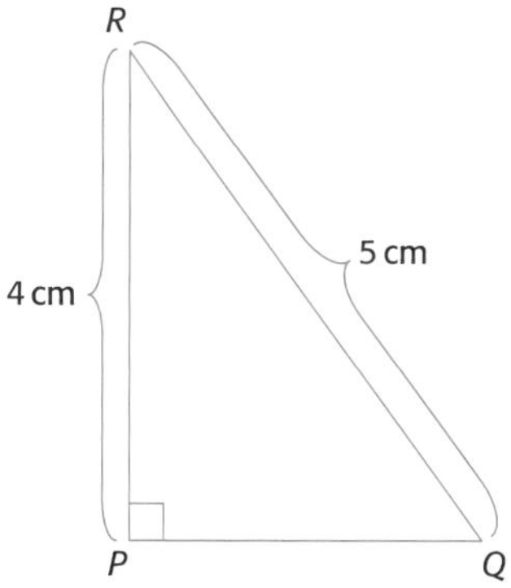

Example 5.3.2d

Write down a trigonometric equation to link angle α and side x.

a

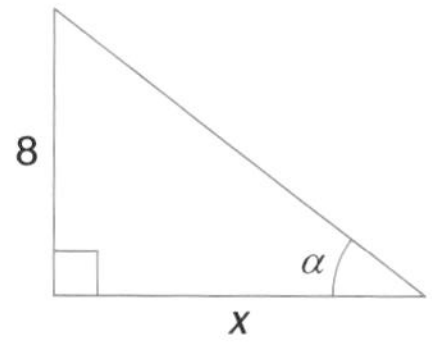

b

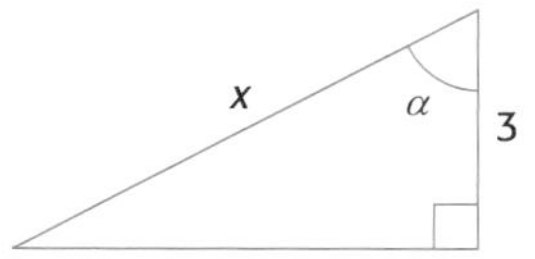

c

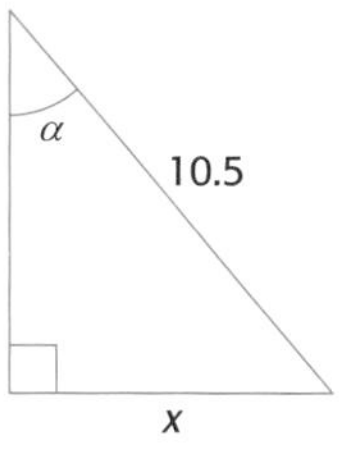

Solution

a $\tan\alpha = \dfrac{8}{x}$ b $\cos\alpha = \dfrac{3}{x}$ c $\sin\alpha = \dfrac{x}{10.5}$

5.3.3 Finding the sides of a right-angled triangle

If we are given an *acute angle* and a *side* of a right-angled triangle we can find the lengths of the other two sides using the appropriate trigonometric ratios.

Example 5.3.3

Find the lengths of the unknown sides in these triangles.

a

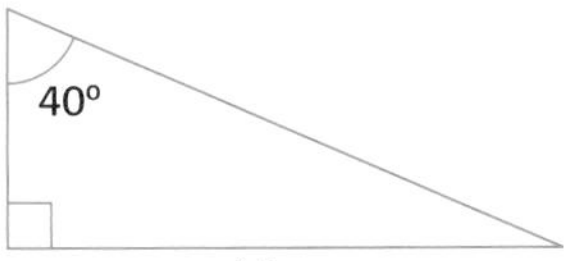

b

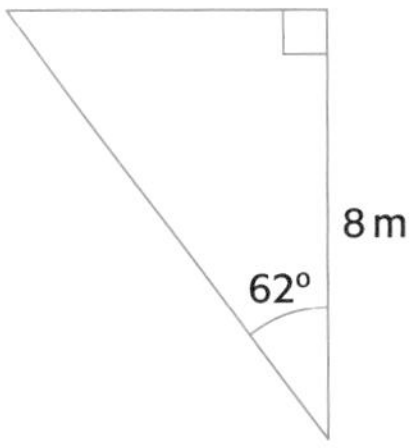

c

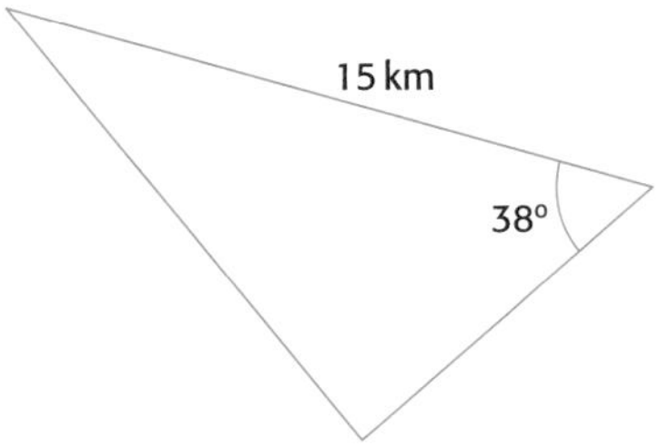

Note that the triangles are not labelled. It is useful to label the vertices *A, B, C.*

Remember to set up your GDC in degrees.

Solution

a

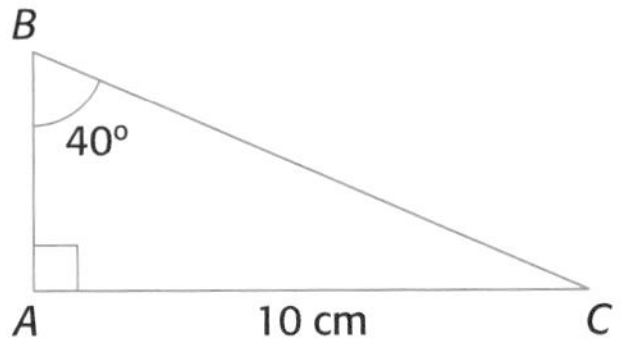

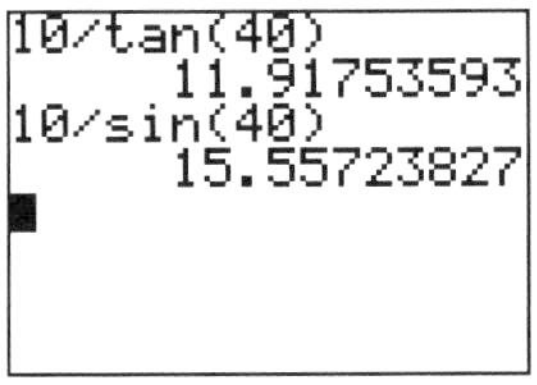

$$\tan 40° = \frac{10}{AB} \quad \therefore AB = 11.9 \text{ cm (3 s.f.)}$$

$$\sin 40° = \frac{10}{BC} \quad \therefore BC = 15.6 \text{ cm (3 s.f.)}$$

Once we had calculated AB we could have found BC by using Pythagoras.

$$BC^2 = AB^2 + AC^2$$

$$BC^2 = \left(\frac{10}{\tan 40°}\right)^2 + 10^2$$

$$BC = 15.6 \text{ cm (3s.f.)}$$

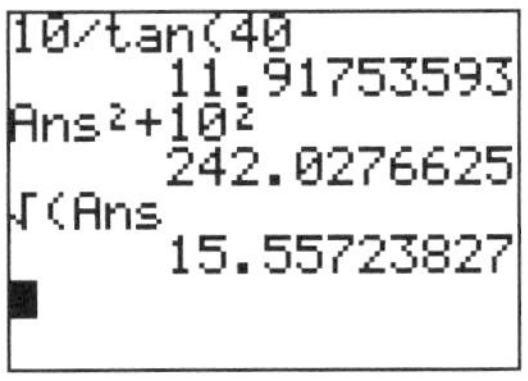

b

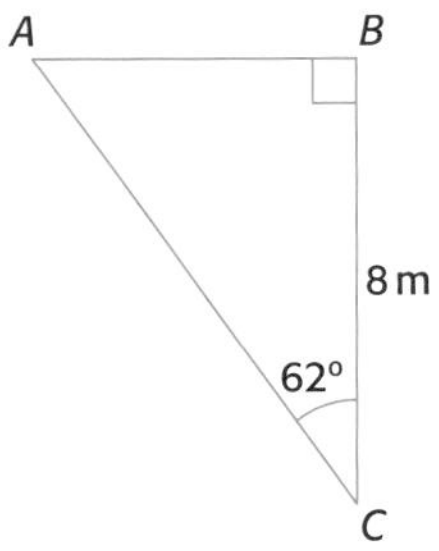

$$\cos 62° = \frac{8}{AC} \quad \therefore AC = 17.0 \text{ m (3s.f.)}$$

$$\tan 62° = \frac{AB}{8} \quad \therefore AB = 15.0 \text{ m (3s.f.)}$$

c

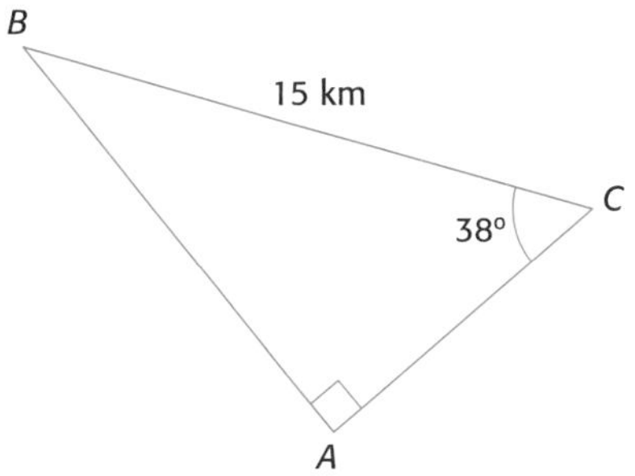

$$\cos 38° = \frac{AC}{15} \quad \therefore AC = 11.8 \text{ km (3s.f.)}$$

$$\sin 38° = \frac{AB}{15} \quad \therefore AB = 9.23 \text{ km (3s.f.)}$$

5.3.4 Finding the angles in a right-angled triangle

If we are given two sides of a right-angled triangle we can find the sizes of the angles of the triangle by using appropriate trigonometric ratios.

Example 5.3.4a

Find the sizes of the acute angles in this triangle.

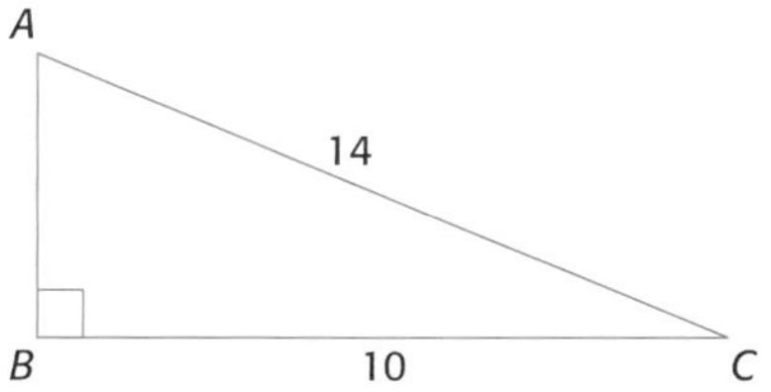

Solution

- Find $\hat{C}$.

  $\cos \hat{C} = \frac{10}{14}$ so we need to find the acute angle $\hat{C}$ that satisfies this equation. We write $\hat{C} = \cos^{-1}\left(\frac{10}{14}\right)$

  To find $\hat{C}$ we use the calculator.

  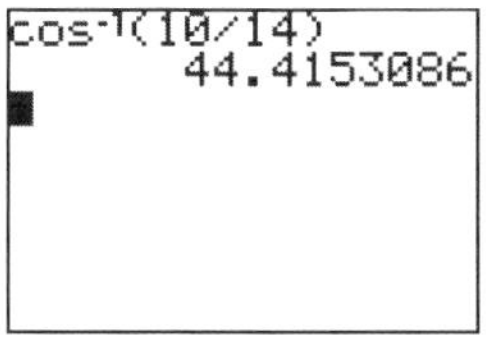

  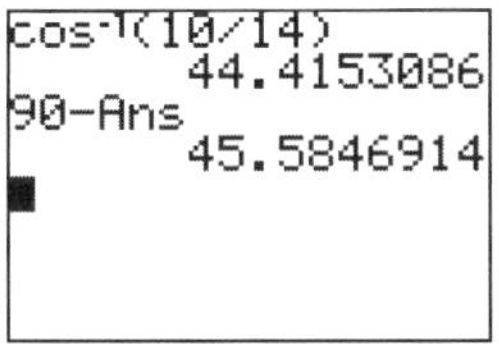

  So $\hat{C}$ = 44.4° (3 s.f.)

- Find $\hat{A}$.

  $\hat{A} + \hat{B} + \hat{C} = 180°$ and $B = 90° \quad \therefore A + C = 90° \quad \therefore \hat{A} = 90 - \hat{C}$

  So $\hat{A}$ = 45.6° (3 s.f.)

$\cos^{-1}\frac{10}{\sqrt{4}}$ means "$\hat{C}$ is the angle with a cosine of $\frac{10}{14}$".

Example 5.3.4b

Find the angles marked ϕ. Give your answers to the nearest degree.

a

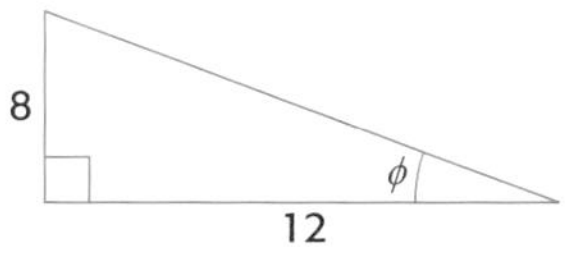

b

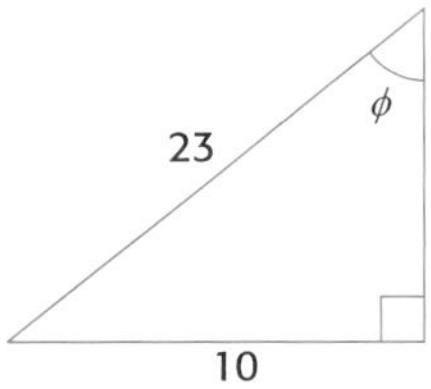

Solution

$\tan \phi = \dfrac{8}{12}$ so we need to find the acute angle ϕ that satisfies this equation.

We write $\phi = \tan^{-1}\left(\dfrac{8}{12}\right)$ and use the calculator to find ϕ.

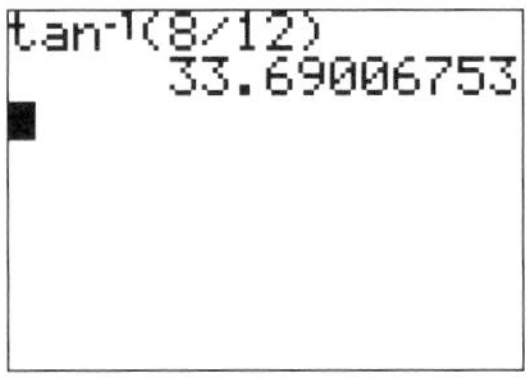

so ϕ = 34° to the nearest degree.

Solution

$\sin \phi = \dfrac{10}{23}$ so we need to find the acute angle ϕ that satisfies this equation.

We write $\phi = \sin^{-1}\left(\dfrac{10}{23}\right)$ and use the calculator to find ϕ.

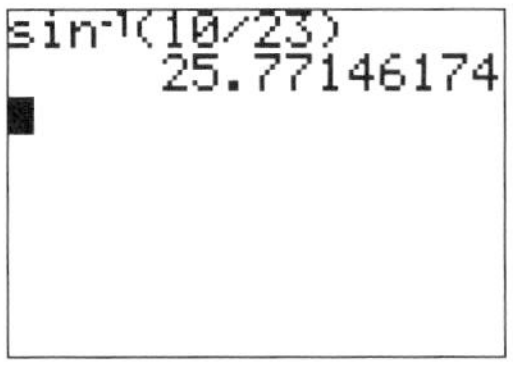

so ϕ = 26° to the nearest degree.

5.3.5 Right-angled triangles hidden in other shapes

Right-angled triangles can be used to find sides and angles in other geometric figures. Many figures, for example, triangles that are not right-angled, rectangles, squares, circles, rhombuses and so on, have right-angled triangles hidden inside them or forming parts of them. We just need to work out where they are and use right-angled trigonometry to find the unknown measurements.

Some examples are shown here.

Isosceles triangle

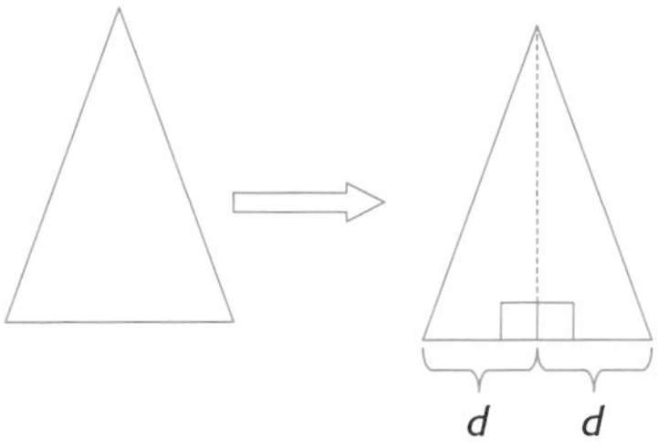

"Any" triangle

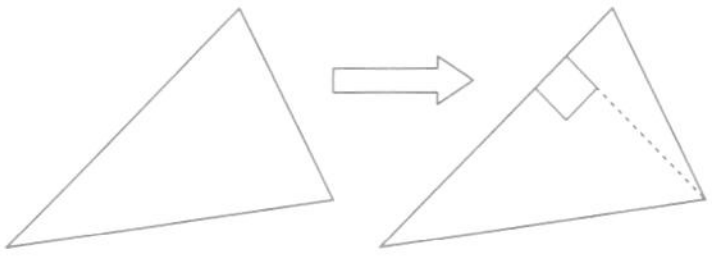

Rectangle

Rhombus

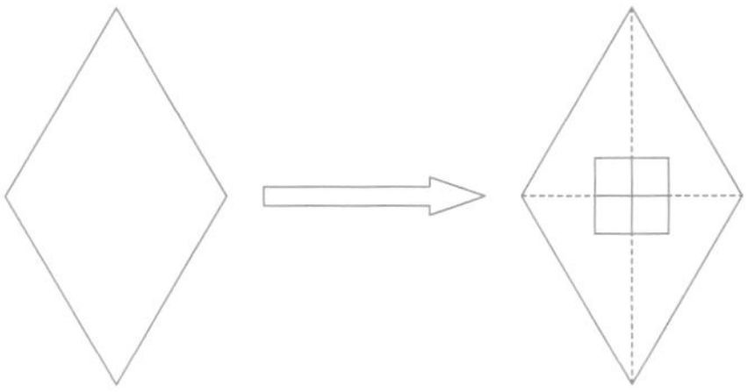

Trapezium

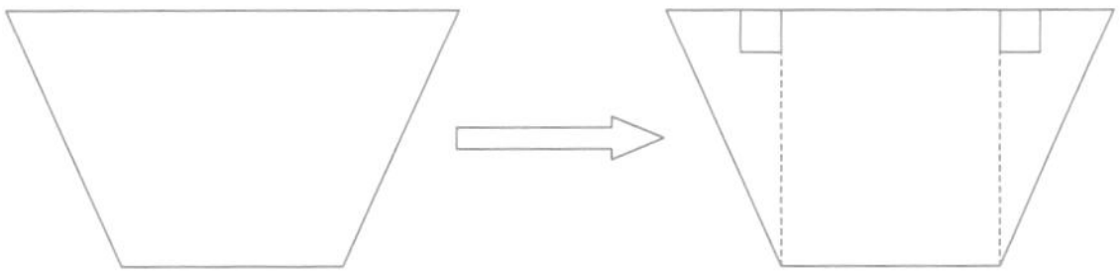

Parallelogram

Circle with centre O and radius r

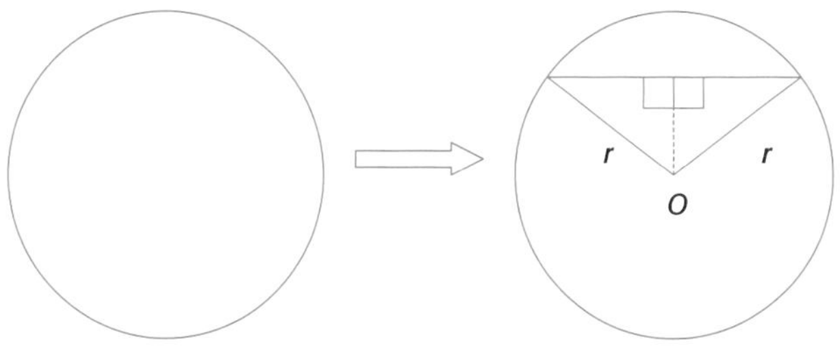

Example 5.3.5a

The length of the shorter side of a rectangular park is 30 m. The park has a path joining two opposite corners. The path is 50 m long.

a Make a diagram to represent the information given.

b Calculate the size of the angle β that the path makes with the longer side of the park.

Solution

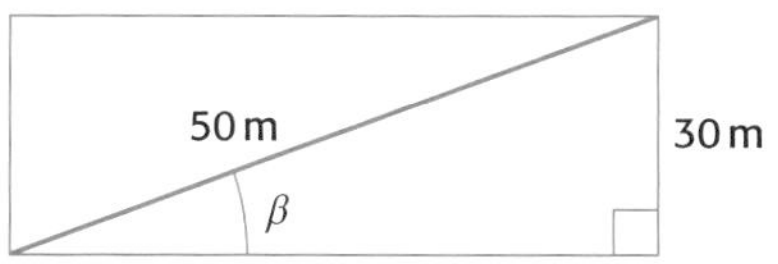

$$\sin\beta = \frac{30}{50}$$

$$\beta = \sin^{-1}\left(\frac{30}{50}\right)$$

$$\beta = 36.9°\,(3\text{s.f.})$$

Example 5.3.5b

The size of the larger angle of a rhombus is 140° and the length of its shorter diagonal is 12 cm.

a Make a diagram to show all the information given.

b Calculate the length of the longer diagonal.

Solution

a

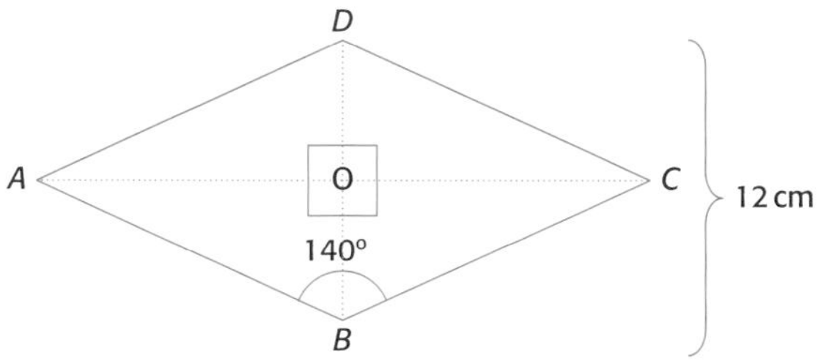

b The longer diagonal is AC.

The diagonals of the rhombus bisect each other at right angles and bisect the angles at each vertex, so $OB = 6$ cm and $O\hat{B}A = 70°$.

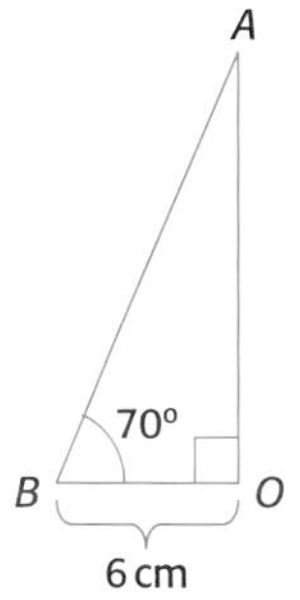

$$\tan 70° = \frac{OA}{OB}$$

$$\therefore \tan 70° = \frac{OA}{6}$$

$$\therefore OA = 6 \times \tan 70°$$

and $AC = 2 \times OA = 33.0\text{ cm}$ (3 s.f.)

Example 5.3.5c

In this diagram A and C represent two villages connected by the road AC. At B there is a hospital that is 33 km away from A. It has been decided to construct a new road from B to AC.

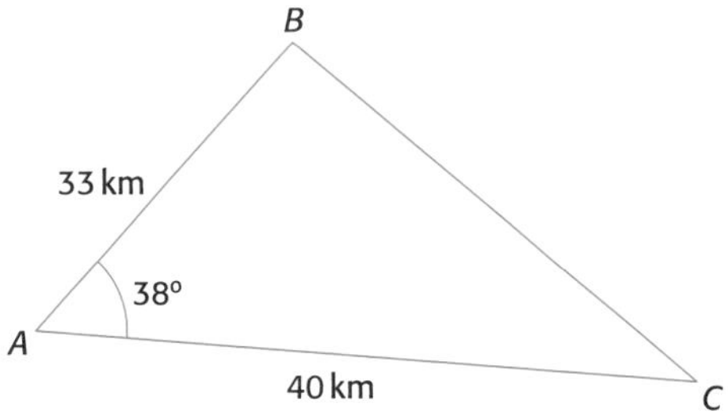

a Calculate the length of the new road if it is to be **as short as possible**. Give your answer to the nearest km.

b Which one of the two villages will be closer to the new road? Justify your answer.

Solution

a The shortest distance between B and the line AC is the **perpendicular distance**. Consider the line perpendicular to AC through B then let O be the point of intersection of these two lines. We are looking for OB.

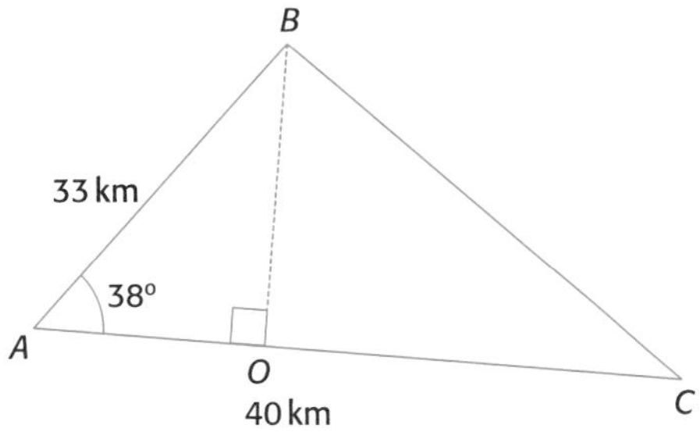

The shortest distance between a point and a line is the perpendicular distance.

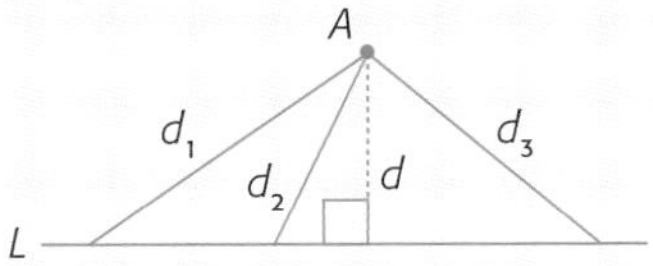

The perpendicular distance between A and the line L is d.

$$\sin 38° = \frac{OB}{33}$$

$\because OB = 33 \times \sin 38°$

$\therefore OB = 20$ km to the nearest km.

b Find AO and OC and compare their lengths.

$\cos 38° = \dfrac{AO}{33}$ $\quad \therefore AO = 26.0$ km (3s.f.)

and $OC = 40 - AO$ $\quad \therefore OC = 14.0$ km (3s.f.) so C will be closer to the new road as $OC < AO$.

Exercise 5.3

1 Find the angle marked ϕ in these right-angled triangles.

a

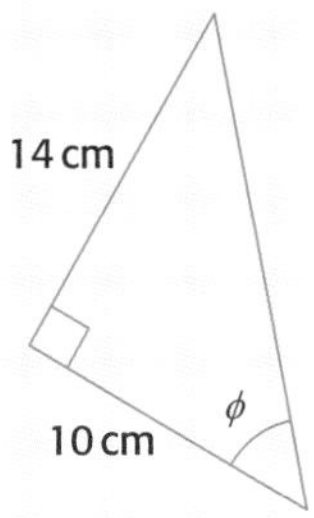

b

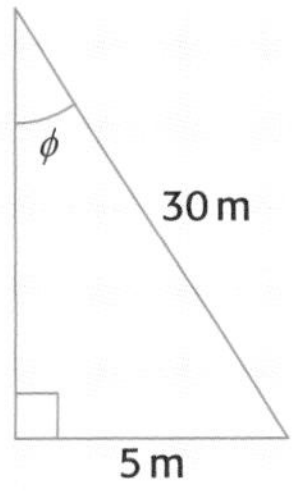

c

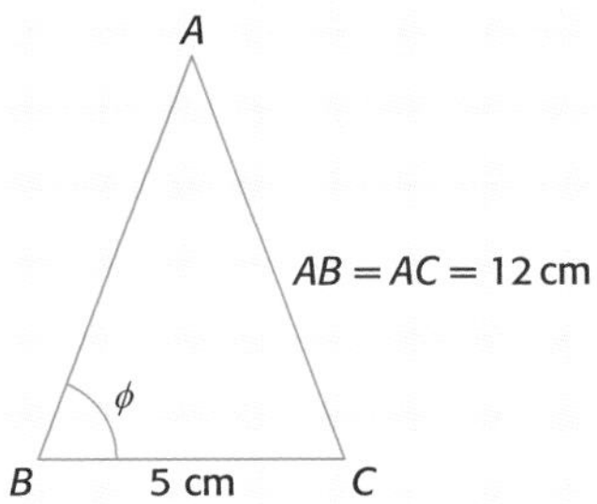

2 Find all the unknown sides and angles in these triangles.
Give your answers correct to 2 decimal places.

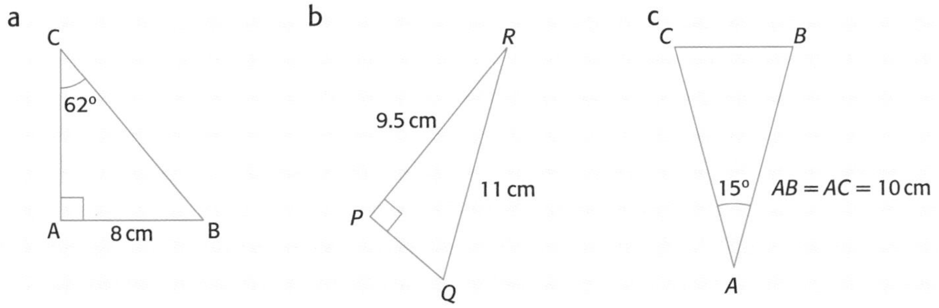

3 Draw a right-angled triangle ABC whose right angle is at A and in which the cosine of $\hat{B}$ is 0.6. Label all the vertices.

4 A ship starts its journey at A and sails 30 km due east until it reaches B, and then 50 km due north until it reaches C.

- a Represent this information in a clear and labelled diagram.
- b Calculate the distance from A to C.
- c Calculate the size of the angle CAB.

5 A ladder rests against a wall. The angle that the ladder makes with the wall is 35°. The ladder is 10 m long. Calculate how far the top of the ladder is from the ground. Give your answer to the nearest metre.

6
- a Plot the points $A(0, 3)$ and $B(-6, 0)$ on a pair of Cartesian axes. Use the same scale on both axes.
- b Join the points A, B and O, where O is the origin.
- c Calculate the tangent of the angle ABO.
- d Calculate the gradient of line AB.
- e Comment on the results obtained in parts **c** and **d**.

7
- a The shorter diagonal of a rhombus is 5 cm and the longer is 12 cm. Calculate the size of the larger angle of the rhombus.
- b The two diagonals of a rectangle make an angle of 140° and the width of the rectangle is 6 cm. Calculate the length of the rectangle.

8 Find x. Give your answer to the nearest degree.

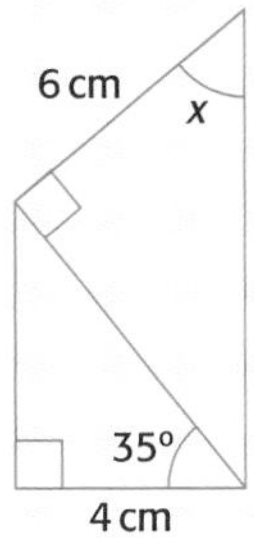

5.4 The sine and cosine rules, area of a triangle

We have seen in the previous section that if we are given any right-angled triangle and two of its sides or one of its acute angles and one side, we can find all the other measurements of the triangle.

Now we are going to study, in particular, those triangles that are *not* right-angled.

To make the notation and formulae simpler we label the triangles in the following way.

We label the side opposite

- angle A as "*a*"
- angle *B* as "*b*"
- angle *C* as "*c*".
- We say that "angle *A* is **included** between sides *c* and *b*" or "angle C is included between sides *a* and *b*" and so on.

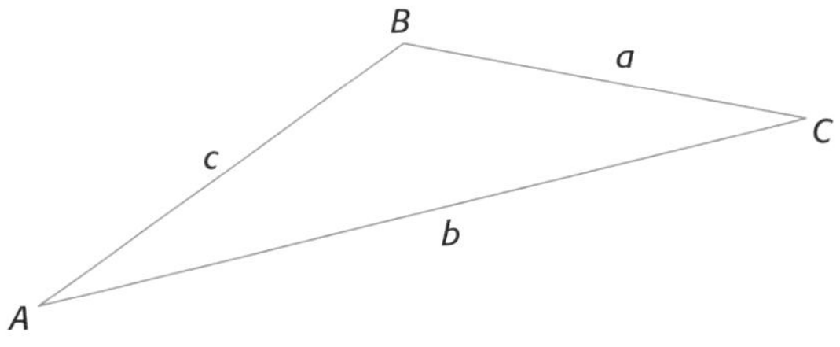

5.4.1 The sine rule

If we are given

- two sides of a triangle and a non-included angle
- two angles and one side

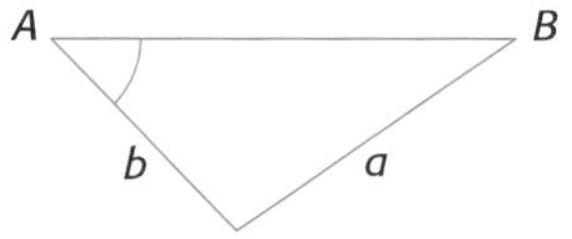

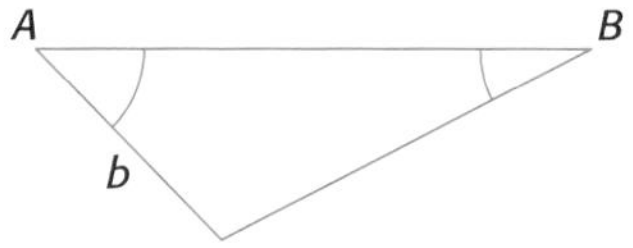

then we can find the other measurements of the triangle.

> In any triangle *ABC* with angles *A*, *B* and *C* and opposite sides *a*, *b* and *c* respectively, the **sine rule** applies.
> $$\frac{a}{\sin A} = \frac{b}{\sin B} = \frac{c}{\sin C}$$

Example 5.4.1a

Find the length of *CB*. Give your answer correct to the nearest millimetre.

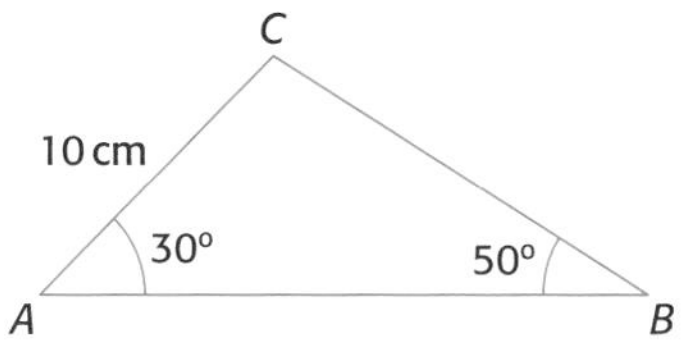

Solution

$$\frac{10}{\sin 50^\circ} = \frac{CB}{\sin 30^\circ} \quad \therefore CB = \frac{10 \times \sin 30^\circ}{\sin 50^\circ}$$

$\therefore CB = 6.5$ cm or 65 mm to the nearest millimetre.

Example 5.4.1b

Find the size of the angle marked x in this triangle.

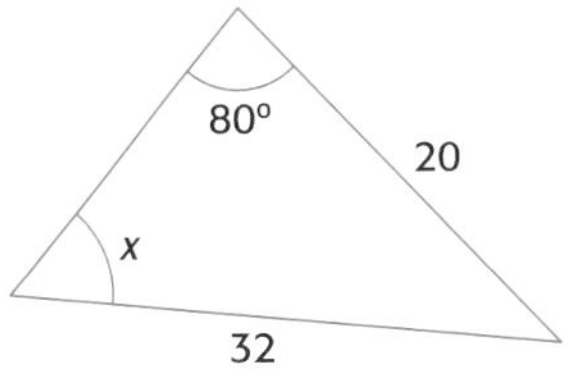

Solution

$$\frac{32}{\sin 80–} = \frac{20}{\sin x} + \sin x = \frac{20 \times \sin 80–}{32}$$

$+ \sin x = 0.6155\ldots$

but this equation has two solutions that lie between 0 and 180°, $x = 38.0°$ and $x = 142°$, both correct to 3 s.f. However, the only possible value for x is 38.0° as $142° + 80° > 180°$ which is impossible.

The two solutions of the equation $\sin x = 0.6155$ can be seen clearly by finding the points of intersection of the functions

$f(x) = \sin x$ and $g(x) = 0.6155$

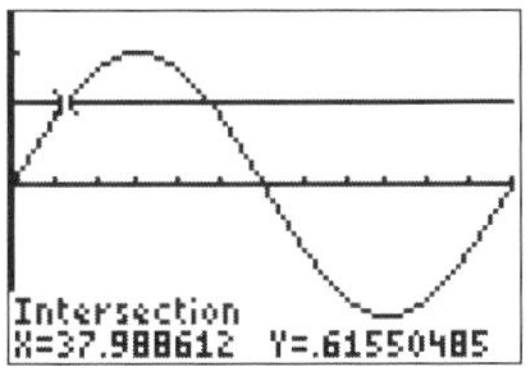

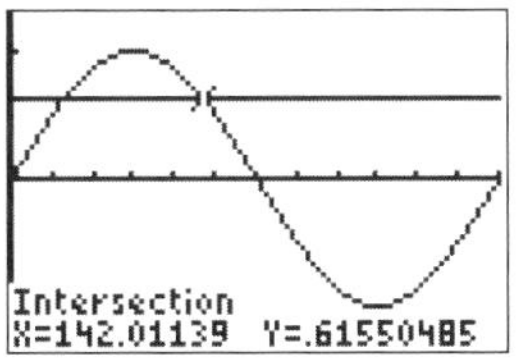

5.4.2 The cosine rule

If we are given

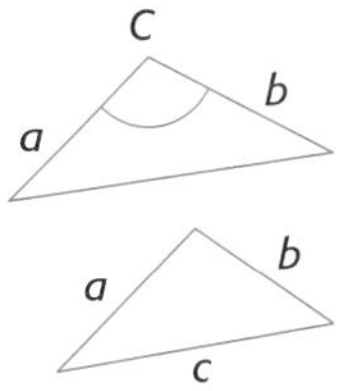

- two sides and the included angle or
- three sides of a triangle then we can find all the measurements of the triangle.

In any triangle ABC with angles A, B and C and opposite sides a, b and c respectively, the **cosine rule** applies.

$a^2 = b^2 + c^2 - 2bc\cos A$

This formula can be rearranged to

$$\cos A = \frac{b^2 + c^2 - a^2}{2bc}.$$

The first version of the cosine rule is useful when we need to find a side and the second version when we need to find an angle.

Example 5.4.2a

Find the length of AC.

Give your answer correct to 2 decimal places.

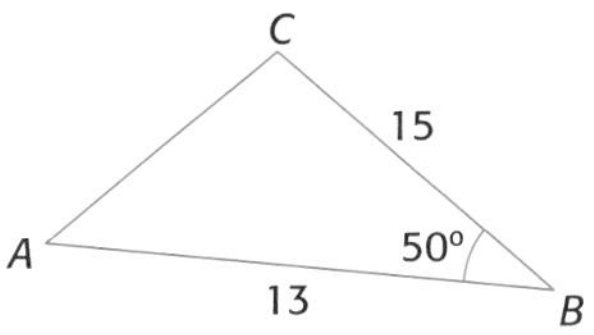

Solution

$AC^2 = 13^2 + 15^2 - 2 \times 13 \times 15 \times \cos 50°$

$AC = \sqrt{13^2 + 15^2 - 2 \times 13 \times 15 \times \cos 50°}$

$AC = 11.97$ correct to 2 d.p.

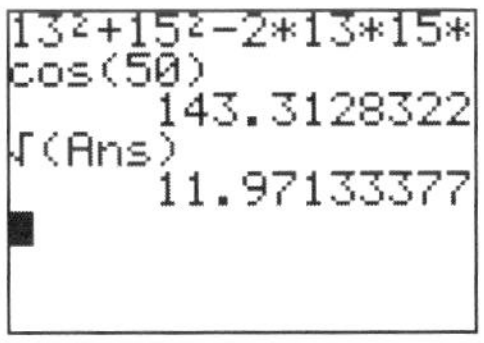

Example 5.4.2b

Find the size of $A\hat{C}B$. Give your answer to the nearest degree.

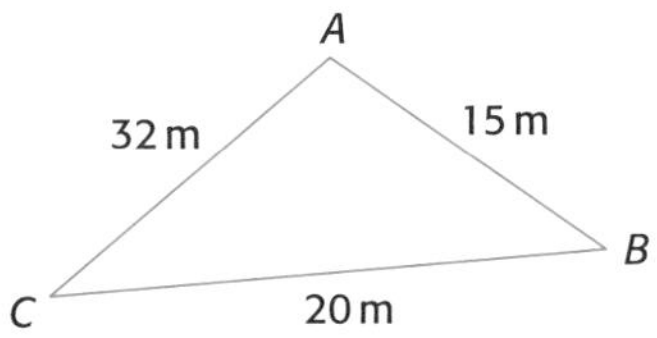

Solution

$$\cos C = \frac{32^2 + 20^2 - 15^2}{2 \times 32 \times 20}$$

$$C = \cos^{-1}\left(\frac{32^2 + 20^2 - 15^2}{2 \times 32 \times 20}\right)$$

$C = 20°$ correct to the nearest degree.

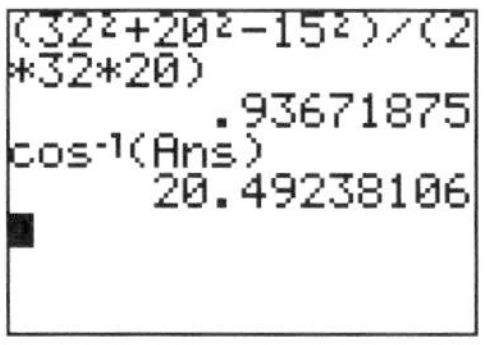

Example 5.4.2c

The positions of the towns P, Q and R are is such that P is 10 km due south from Q. R is somewhere to the west of the line PQ, 9 km from R and 13 km from P.

a Draw a clear, labelled diagram to represent this information.

b Calculate the size of angle $P\hat{Q}R$.

Solution

a

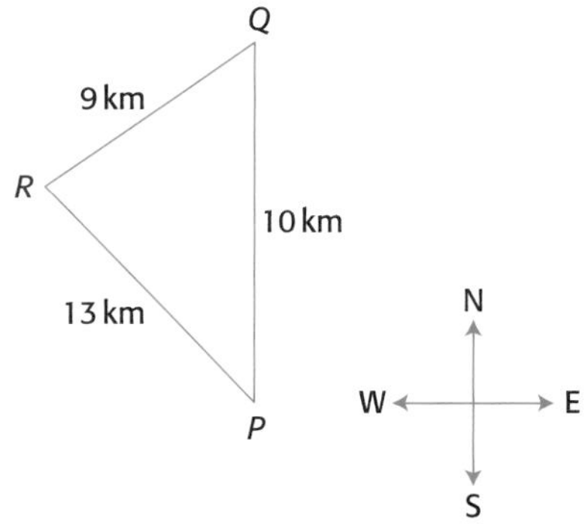

b $$\cos P\hat{Q}R = \frac{9^2 + 10^2 - 13^2}{2 \times 9 \times 10}$$

$$P\hat{Q}R = \cos^{-1}\left(\frac{9^2 + 10^2 - 13^2}{2 \times 9 \times 10}\right)$$

$$P\hat{Q}R = 86.2° \text{ (3 s.f.)}$$

5.4.3 Area of a triangle

We already know the formula

$$\text{Area of a triangle} = \frac{1}{2}(b \times h)$$

where b is the length of one side of the triangle and h is the length of the corresponding height.

This formula can only be used when we are given the length of one side and its corresponding height. However, there may be situations where we are not given this information but we still can calculate the area of the triangle.

Remember that a triangle has three heights, one height per side. Some examples are shown here.

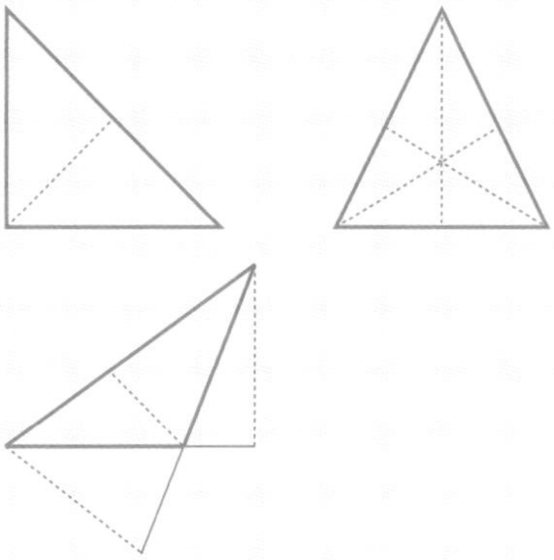

Example 5.4.3a

Calculate the area of the triangle *ABC*.

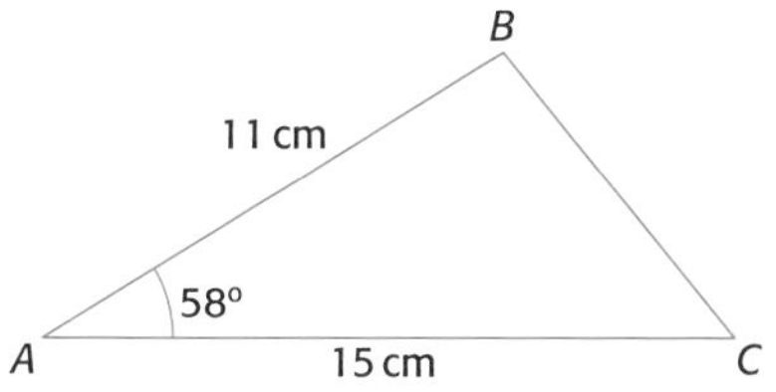

Solution

We need to find the height corresponding to side *AC* and then substitute into the formula

Area of a triangle $= \frac{1}{2}(b \times h)$

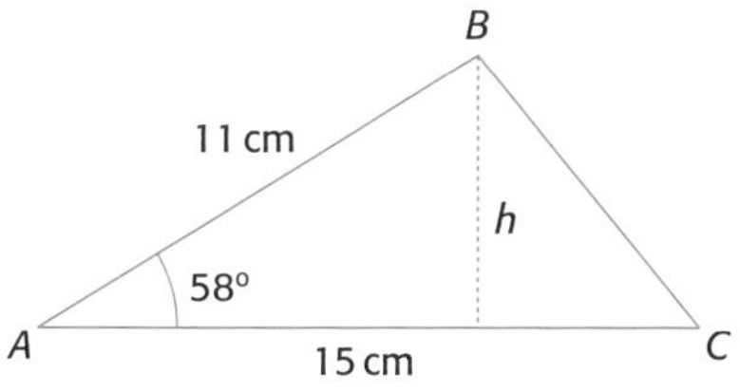

$\sin 58== \frac{h}{11} + h = 11 \times \sin 58=.$

Substituting this value into the formula of the area of the triangle we obtain

Area of a triangle $= \frac{1}{2}(b \times h) = \frac{1}{2}(15 \times 11 \times \sin 58^\circ)$

Area of a triangle $= 70.0\ \text{cm}^2$ (3 s.f.)

In general

In any triangle *ABC* with angles *A*, *B* and *C* and opposite sides *a*, *b* and *c* respectively, this rule applies.

Area of triangle $= \frac{1}{2}bc \sin A$

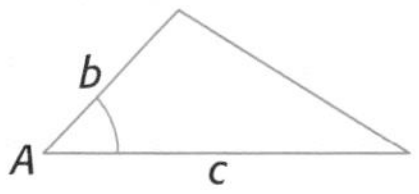

Example 5.4.3b

Calculate the area of the triangle *ABC* given in Example 5.4.2b.

Solution

We already know that $\hat{C} = 20.49\ldots$. Substituting into the new formula

Area $= \frac{1}{2} \times 32 \times 20 \times \sin 20.49^\circ$

Area $= 112\ \text{m}^2$ (3 s.f.)

Example 5.4.3c

Louis and Sahmia are in the park next to the fountain, at *F*. Louis starts to walk due west straight to the lake, at *L*, 120 m away. Sahmia turns through an angle of 100° and walks towards the statue, at *S*, 90 m away.

a Make a clear, fully labelled diagram to represent this information.

b Calculate how far apart Louis and Sahmia are when they are at the fountain and at the statue respectively.

c Calculate the area of the park enclosed by the three points *F*, *L* and *S*. Give your answer to the nearest square metre.

Solution

a

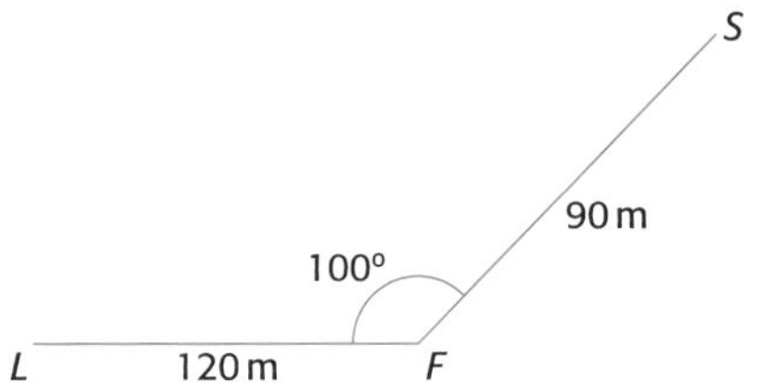

b $LS^2 = 120^2 + 90^2 - 2 \times 120 \times 90 \times \cos 100°$

$LS = \sqrt{120^2 + 90^2 - 2 \times 120 \times 90 \times \cos 100°}$

$LS = 162\text{ m}\ (3\text{s.f.})$

c $\text{Area} = \frac{1}{2} \times 120 \times 90 \times \sin 100°$

Area = 5318 m^2 correct to the nearest square metre

Exercise 5.4

1 Find the sides marked with letters.

a

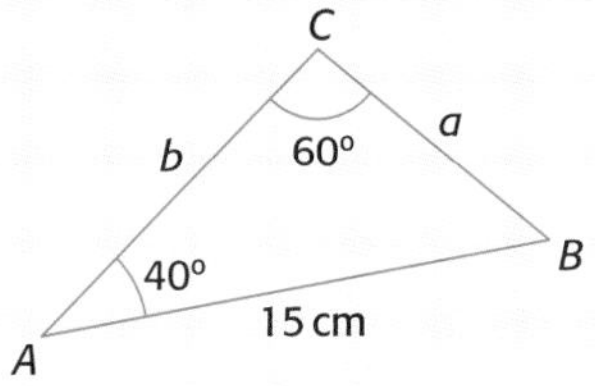

b

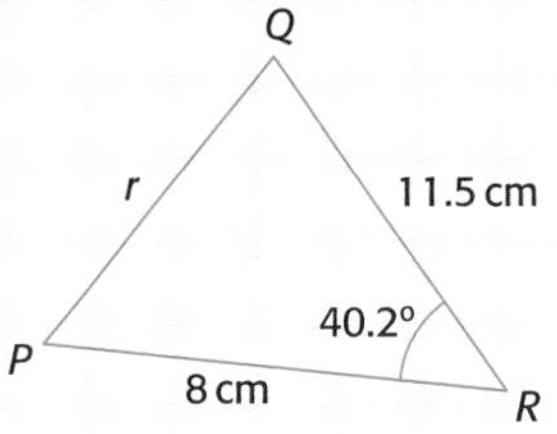

2 i Find the angle marked *β*.

ii Hence find the area, *A*, of each triangle.

a

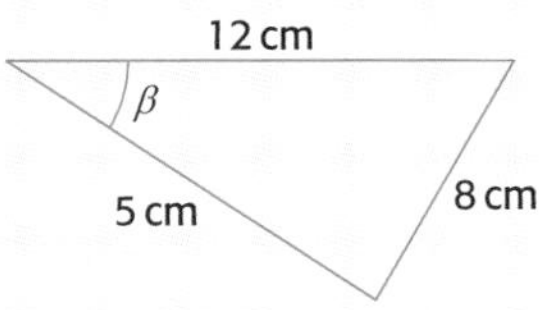

b

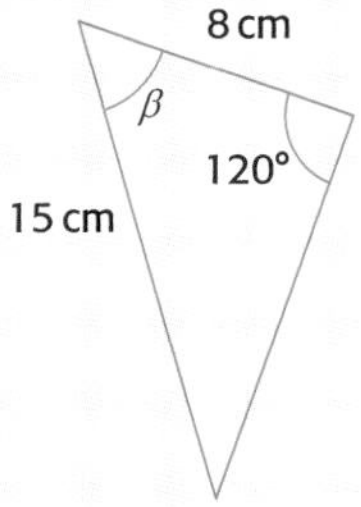

3 The area of the triangle *ABC* is 30 cm^2, *AB* = 12 cm and *AC* = 10 cm. Find $\hat{A}$ if it is known that $\hat{A}$ is an acute angle.

4 In this diagram, *O* is the centre of the circle with diameter 12 cm. *AB* is a chord (this means that *AB* is a line with *A* and *B* lying on the circumference) of length 6.5 cm.

a Find the size of the angle *OAB*.

b Find the area of the triangle *OAB*.

c Find the shaded area.

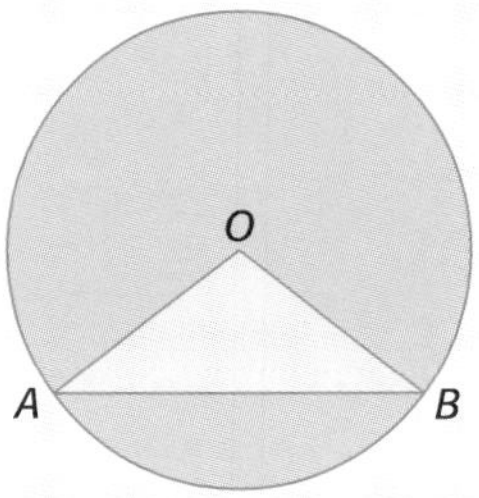

5 *AC* and *BD* bisect each other at *O*, making an angle of 130°. *AC* = 8 cm and *BD* = 5 cm.

a Make a clear, labelled diagram to represent this information.

b Write down what type of quadrilateral *ABCD* is.

c Calculate the area of *ABCD*.

6 There are three points $A(1, 1)$, $B(3, 7)$ and $C(5, 3)$.

a Plot and label A, B and C on a pair of Cartesian axes. Use the same scale on both axes.
b Draw the lines AB and AC.
c Calculate the lengths of AB, AC and CB.
d Calculate the angle between the lines AB and AC.

5.5 Geometry of solids

5.5.1 Solids with all the faces plane

Prisms

A **prism** is a solid figure whose end faces have the same *size* and *shape* and are parallel to one another, and in which each of the other sides is a parallelogram.

A **solid** is a **three-dimensional** shape.

Here are some examples of prisms.

Cube   Cuboid   Triangular prism   Hexagonal prism

Architecture and prisms.

Pyramid

A pyramid is a solid figure with a polygonal base and triangular faces that meet at a common point. This point is called the "apex".

Here are some examples of pyramids.

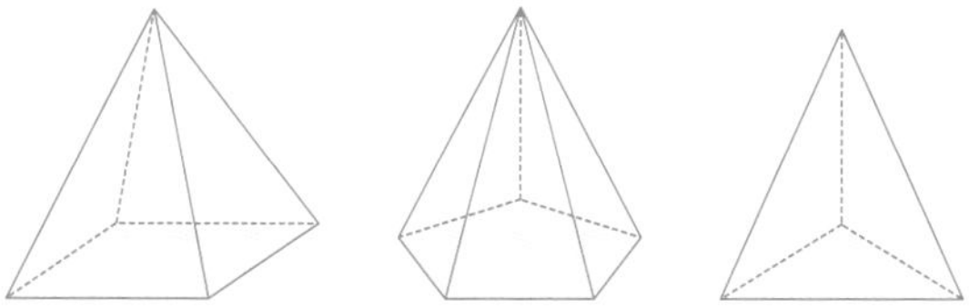

The base is a square.   The base is a pentagon.   The base is triangle.

The pyramids of Cheops and Chephren, are the largest in Egypt.

5.5.2 Solids with at least one curved face

The cone, the cylinder and the sphere are solids with at least one curved face.

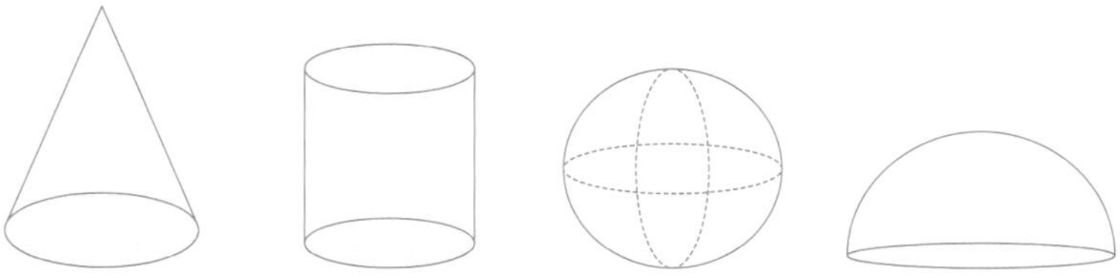

Cone: one plane face and one curved face   Cylinder: two plane faces and one curved face   Sphere: one curved face   Hemisphere: one curved face and one plane face

5.5.3 Surface area of solids with all faces plane

The surface area of a solid is the sum of the areas of all its faces.

Example 5.5.3

Calculate the surface areas of these solids.

a

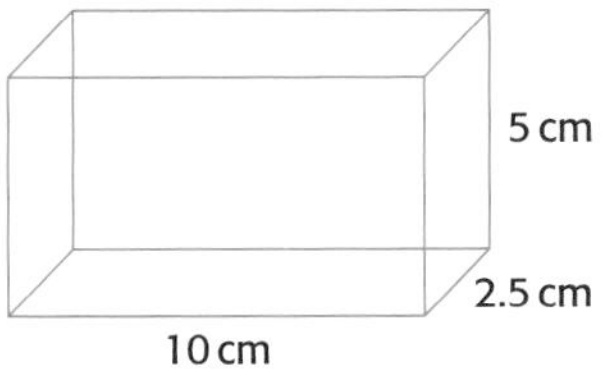

b

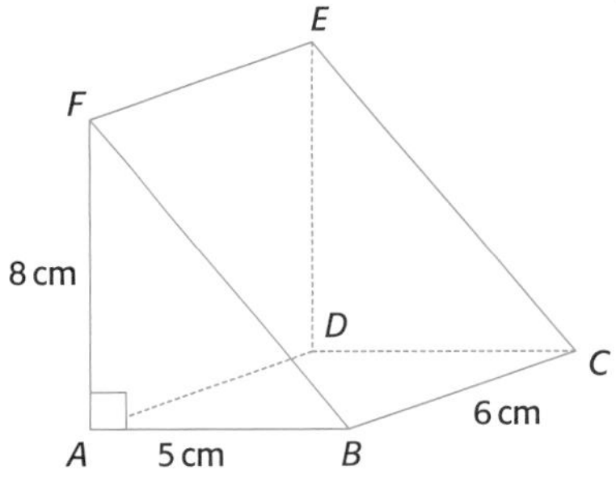

c

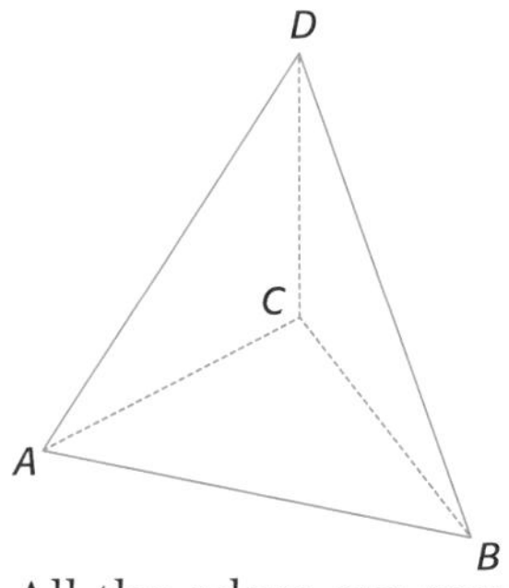

All the edges are equal and $AB = 6\,\text{cm}$.

Solution

a This cuboid has 6 rectangular faces.

$$\left.\begin{array}{l}2 \text{ faces of } 2.5\text{ cm} \times 5\text{ cm}\\ 2 \text{ faces of } 2.5\text{ cm} \times 10\text{ cm}\\ 2 \text{ faces of } 10\text{ cm} \times 5\text{ cm}\end{array}\right\} \Rightarrow \begin{aligned}\text{Surface area} &= 2\times 2.5\times 5 + 2\times 2.5\times 10 + 2\times 10\times 5\\ &= 175\text{ cm}^2\end{aligned}$$

b This triangular prism has 5 faces: 2 triangular faces and 3 different rectangular faces.

A_1 = Area of triangular face $= \frac{1}{2} \times 5\text{ cm} \times 8\text{ cm} = 20\text{ cm}^2$

A_2 = Area of rectangular face ADEF $= 8\text{ cm} \times 6\text{ cm} = 48\text{ cm}^2$

A_3 = Area of rectangular face ABCD $= 5\text{ cm} \times 6\text{ cm} = 30\text{ cm}^2$

A_4 = Area of rectangular face BCEF $= 6\text{ cm} \times \text{BF}$ and

$\text{BF}^2 = 5^2 + 8^2 \therefore \text{BF} = \sqrt{89}$

$A_4 = 6\text{ cm} \times \sqrt{89}\text{ cm}$

Area total $= 2A_1 + A_2 + A_3 + A_4 = 175\text{ cm}^2$ (3 s.f.)

c This tetrahedron has 4 equal faces. Each face is an equilateral triangle.

The size of each angle is 60° so

Area $\triangle$ ABC $= \frac{1}{2} \times 6 \times 6 \times \sin 60°$

Total surface area $= 4 \times$ Area $\triangle ABC = 62.4\,\text{cm}^2$ (3 s.f.)

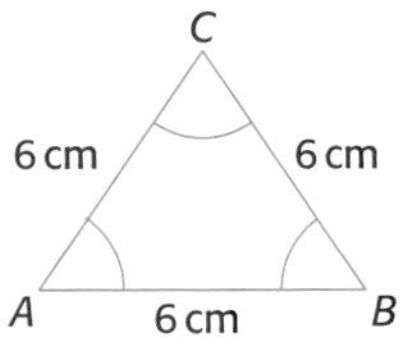

5.5.4 Surface area of solids with at least one curved face

For these solids there are formulae for calculating the area of the curved face.

Solid	Area of curved face
Cone 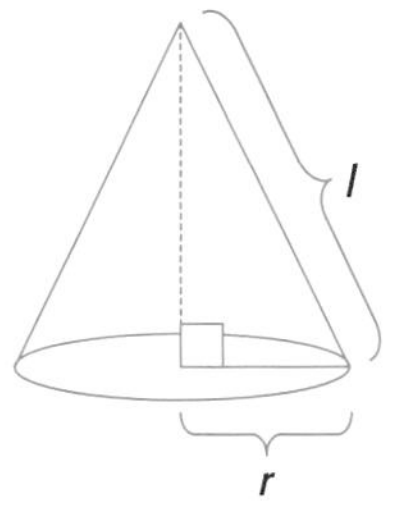	$A = \pi r l$ where r is the radius, l is the slant height.
Sphere 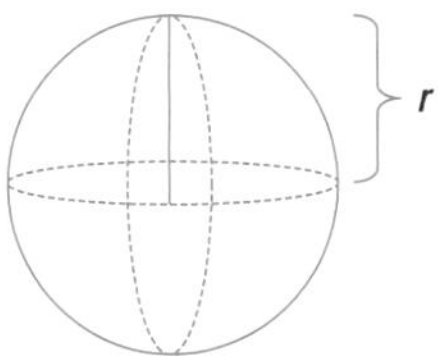	$A = 4\pi r^2$ where r is the radius
Cylinder 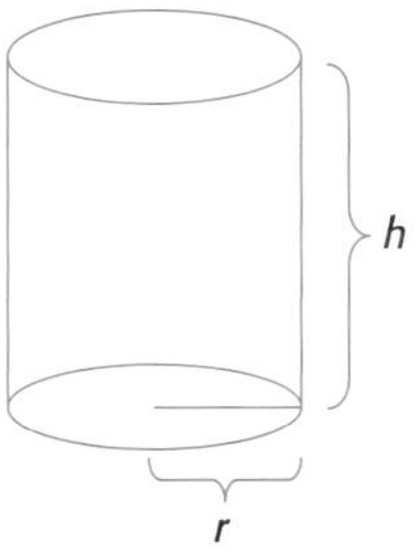	$A = 2\pi r h$ where r is the radius, h is the height

Example 5.5.4

Calculate the total surface area of each solid.

a

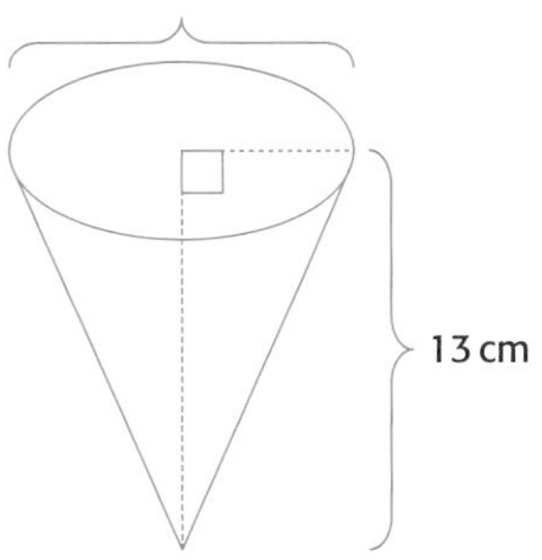

b

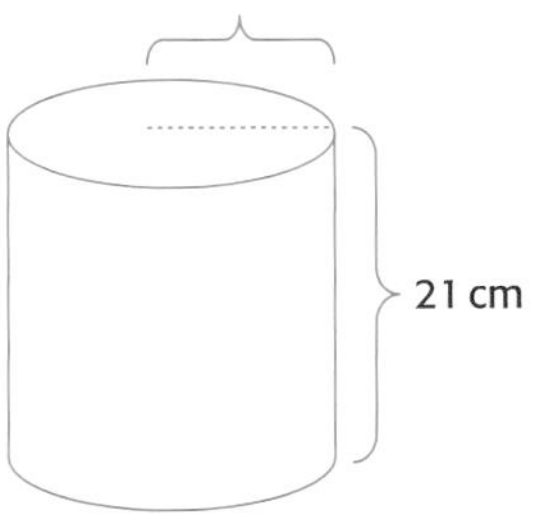

c

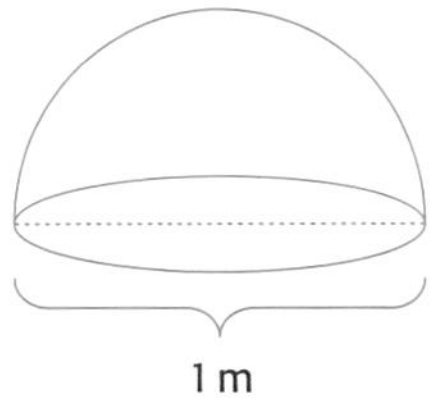

Solution

a This cone has two faces, one plane and one curved.

Let A_1 be the area of the plane face (a circle) and A_2 be the area of the curved face.

$A_1 = \pi \times 4.5^2$
$A_2 = \pi \times 4.5 \times l$

We need to find the **slant height**, l.

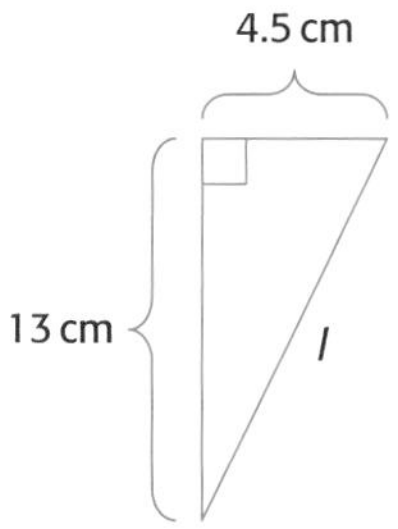

$$l^2 = 4.5^2 + 13^2$$
$$l = \sqrt{189.25}$$
$$\text{so total area} = A_1 + A_2 = 258\,\text{cm}^2 (3\text{ s.f.})$$

b This cylinder has three faces, two plane and one curved.
Total area $= 2 \times (\pi \times 13.6^2) + 2\pi \times 13.6 \times 21 = 2960\,\text{cm}^2$ (3 s.f.)

c This hemisphere has two faces, one plane and the other curved.
Let A_1 be the area of the plane face and A_2 be the area of the curved face.
$A_1 = \pi \times 0.5^2$
$A_2 = \frac{1}{2}(4\pi \times 0.5^2)$
Total area $= A_1 + A_2 = 2.36\,\text{m}^2$ (3 s.f.)

5.5.5 Volume

The **volume** of a solid is the amount of space it occupies. Volume is measured in cubic units: cubic centimetres, cubic metres, and so on.

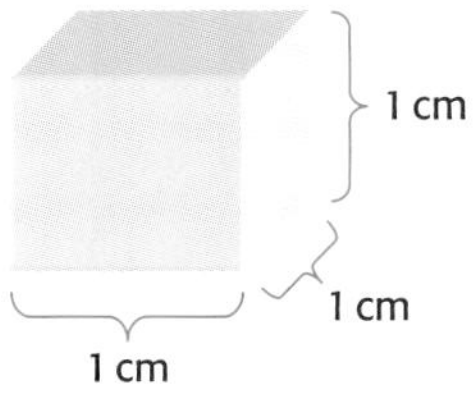

For example, **one cubic centimetre** is the space occupied by a cube with an edge of 1 cm.

Similarly **one cubic metre** is the space occupied by a cube with and edge of 1 m.

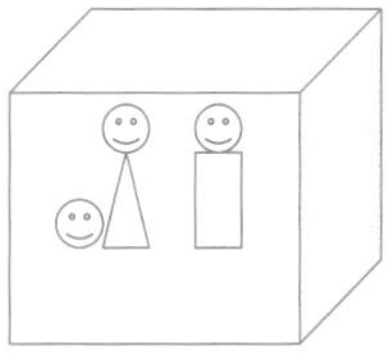

Find out how many students in your mathematical studies class can get into a 1 m^3 box. Use this information to estimate the volume of a person.

Prisms

To calculate the volume of a prism we need first to identify its **base**. We have said in the definition that prisms have two **ends** that are equal in shape and size and are parallel. Either of these ends can be taken as the base of the prism. In the prisms in these diagram both possible bases are shaded.

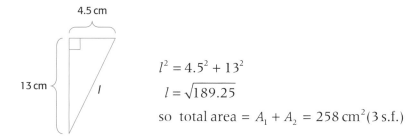

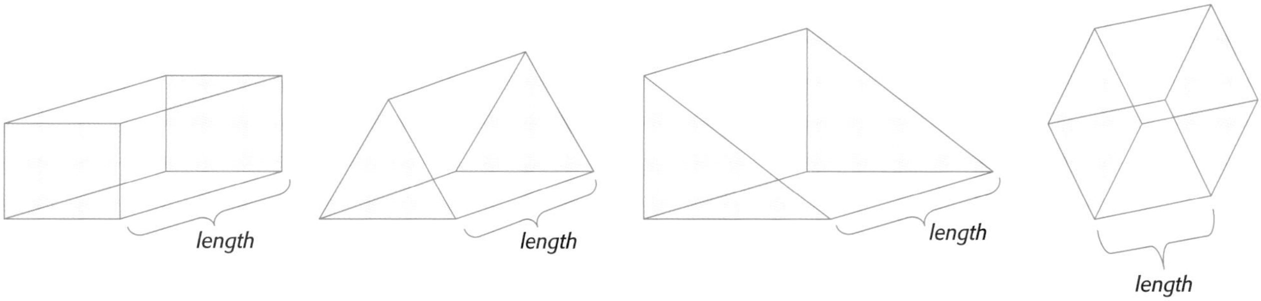

Volume of prism – Area of base = length
The cuboid is a special case of this formula.

Volume of a cuboid = $l \times w \times h$, where l is the length, w is the width, h is the height.

Example 5.5.5a

Calculate the volumes of these solids.

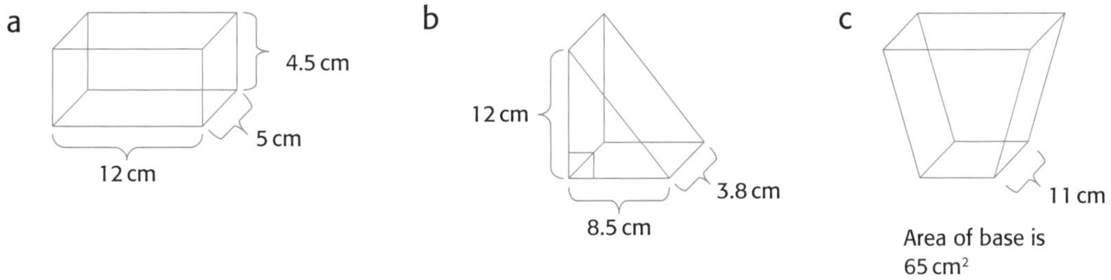

Solution

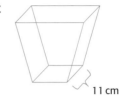

a This is a cuboid with l = 12 cm, w = 5 cm, h = 4.5 cm so

Volume of a cuboid = 12 cm × 5 cm × 4.5 cm = 270 cm^3

b This is a triangular prism.

Volume of prism = Area of base × length

$$= \frac{1}{2}(8.5\text{ cm} \times 12\text{ cm}) \times 3.8\text{ cm}$$

$$= 193.8\text{ cm}^3$$

c The base is a trapezium.

Volume of prism = Area of base × length = 65 cm^2 × 11 cm = 715 cm^3

Cylinder

The cylinder has two parallel "ends" that are equal in shape and size. To calculate the volume of a cylinder we use the same formula as for the volume of a prism.

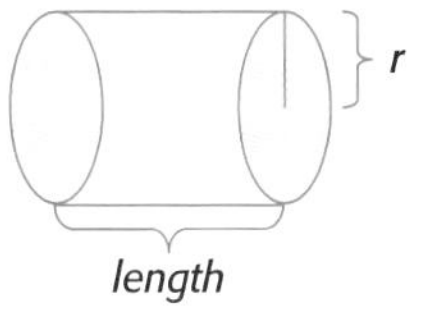

However, the *length* of the cylinder can be also taken as the *height*, so we can write the formula for the volume of the cylinder like this.

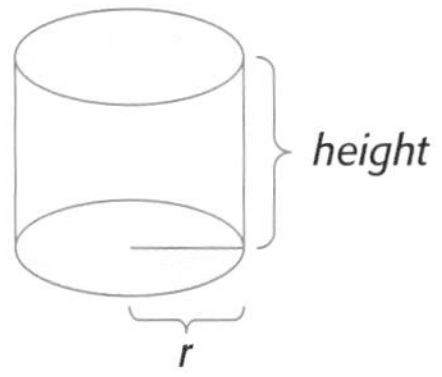

Volume of a cylinder = Area of base × length or
Volume of a cylinder = $\pi r^2 h$, where r is the radius, h is the height.

Example 5.5.5b

The volume of a cylinder is 250π cm^2 and its height is 10 cm. Calculate the diameter of its base.

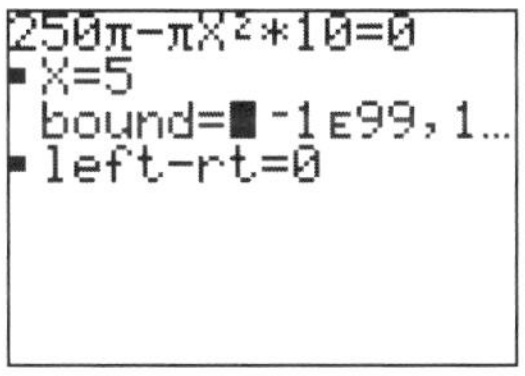

Solution

$250\pi = \pi r^2 \times 10 \therefore r^2 = \dfrac{250\pi}{10\pi} \therefore r^2 = 25 \therefore r = 5\text{ cm}$ or using the GDC

Let d be the diameter of the base then $d = 2 \times r = 2 \times 5 = 10$ cm

Example 5.5.5c

Find a method to calculate the volume of the upper part of the building in the photo.

Oscar Niemeyer museum at Curitiva, Brazil

Solution

Observe that the upper part of the building looks like an eye. It has two parallel plane faces (front and back) and two curved faces (top and bottom). Assume that the area of each plane face is C, and the length is L then

Volume of the "eye" = $C \times L$

Pyramids and cones

The volumes of these solids can be calculated using the following formula.

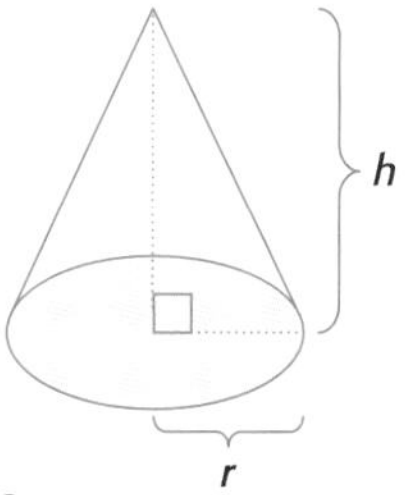

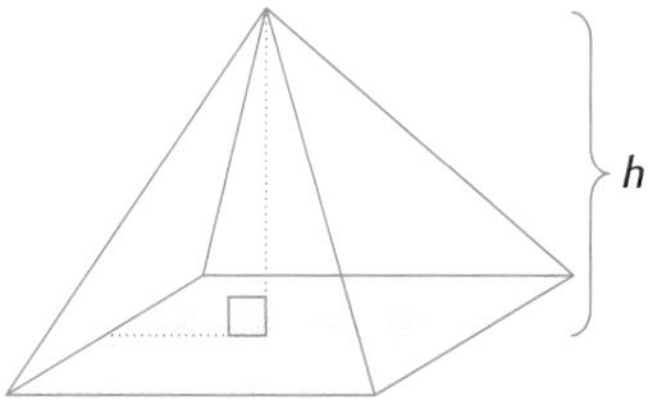

$$\text{Volume} = \frac{1}{3}(\text{Area of base} \times \text{height})$$

and in the case of the cone this can be written as

Volume of the cone $= \frac{1}{3}(\pi r^2 h)$, where r is the radius and h is the height.

- the volume of a cone is *one third of the volume of a cylinder* with the same base and height as the cone.
- the volume of a pyramid is *one third of the volume of a cuboid* with the same base and height as the pyramid.

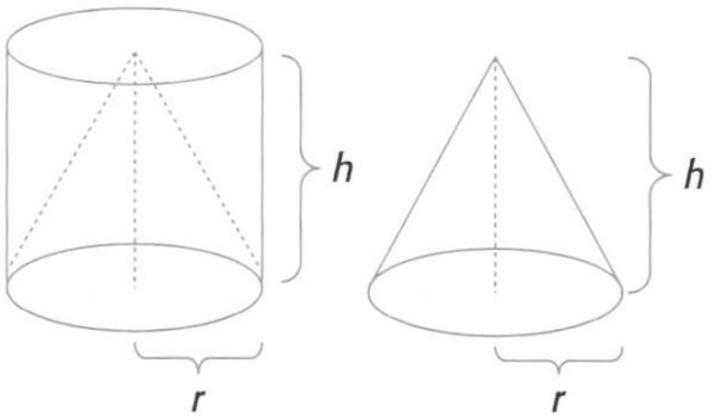

Volume of cone $= \frac{1}{3} \times$ Volume of cylinder

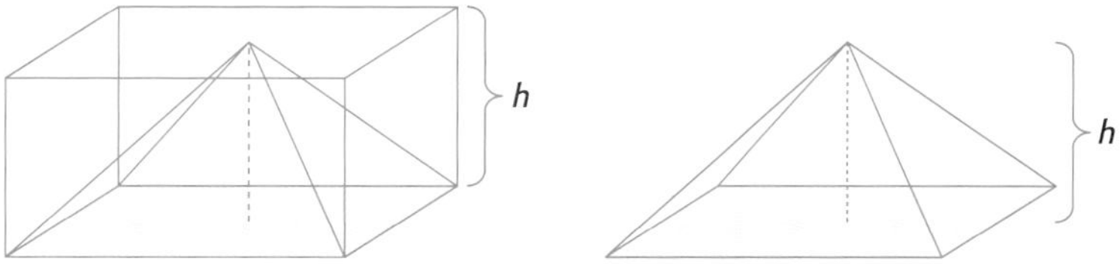

Volume of pyramid $= \frac{1}{3} \times$ Volume of cuboid

Example 5.5.5 d

Write down a formula to calculate the volume, V, of each solid.

a

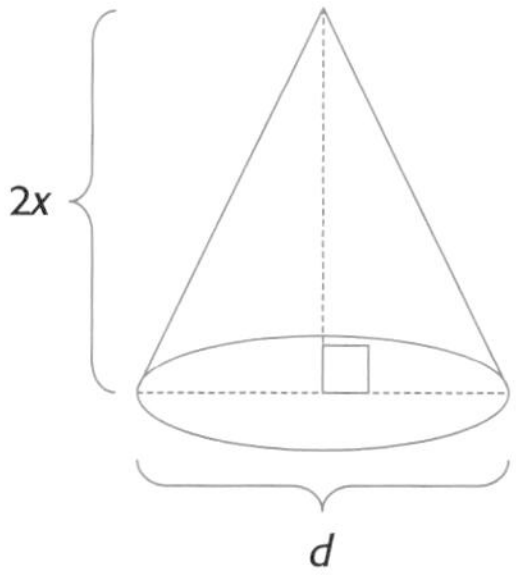

b

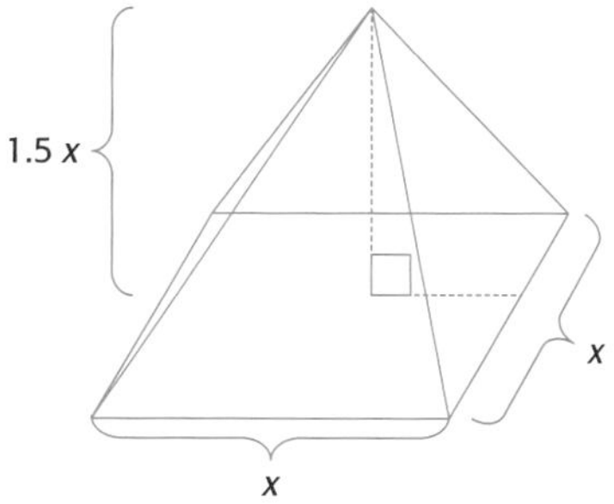

Solution

a $V = \frac{1}{3}\left(\pi\left(\frac{d}{2}\right)^2 2x\right) \therefore V = \frac{1}{6}\pi d^2 x$

b $V = \frac{1}{3}(x^2 \times 1.5x) \therefore V = 0.5x^3$

Sphere

Volume of a sphere $= \frac{4}{3}\pi r^3$, where r is the radius

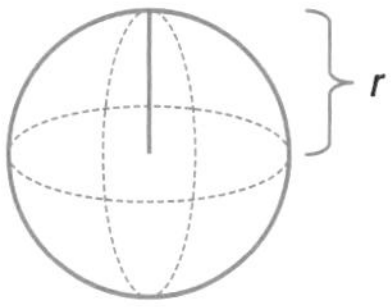

Example 5.5.5e

a Find the radius of a spherical ball if its volume is 4189 cm³. Give your answer correct to the nearest cm.

b Use your answer to part **a** to calculate the cost of covering the ball with leather if 1 square metre of leather costs \$100. Give your answer correct to 2 d.p.

Solution

a $4189 = \frac{4}{3}\pi r^3 \therefore r = \sqrt[3]{\frac{4189 \times 3}{4\pi}} \therefore r = 10 \text{ cm}$

or using the GDC

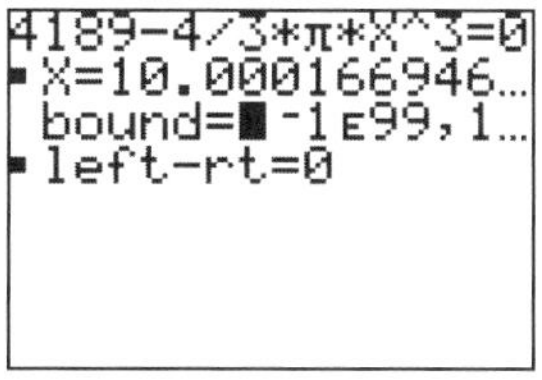

b We need to find the surface area of the ball.

$A = 4\pi r^2 = 4\pi \times 10^2 = 400\pi \text{ cm}^2$

$1 \text{ m}^2 = 10000 \text{ cm}^2$ and

$\frac{10000 \text{ cm}^2}{\$100} = \frac{400\pi \text{ cm}^2}{\$x} \therefore x = \frac{400\pi \times 100}{10000} \therefore x = \$12.57$

How would you estimate the volume of this log?
Can you think of different ways of doing it?

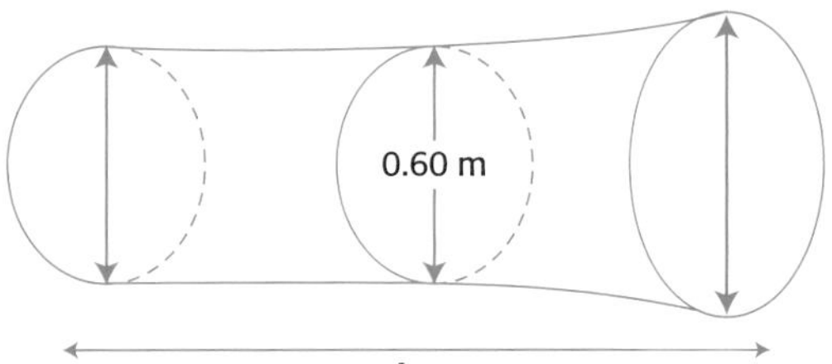

> **Theory of knowledge**
>
> Often there are several different ways to arrive at an answer. For each method there will be advantages and disadvantages. Sometimes one method may be more accurate than another. How do you decide which method is best? Do you think that everyone would agree with you? Are there different ways to arrive at an answer in other areas of knowledge such as the sciences, history, aesthetics?

5.5.6 Lengths of lines joining vertices with vertices, vertices with midpoints and midpoints with midpoints

Look at the cuboid in the diagram. We may need to calculate some particular lengths, for example, AC, AE, and so on. To do so we need to find the appropriate right-angled triangles containing the line segments, AC, AE, and so on.

Example 5.5.6

In the cuboid $ABCDEFGH$, $AB = 12$ cm, $BE = 10$ cm and $BC = 4.5$ cm.

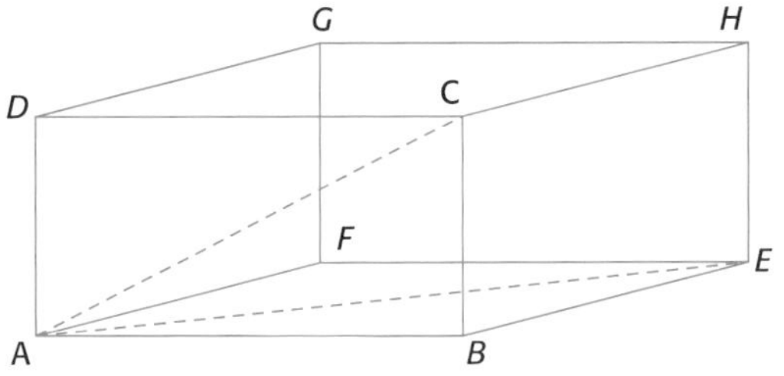

a Calculate the length of

i AE ii AC iii BG.

Let M be the midpoint of CH and R be the midpoint of GH.

b Calculate the distance from

i M to R ii M to A.

Solution

a i To calculate AE we can work with the right-angled triangle ABE.

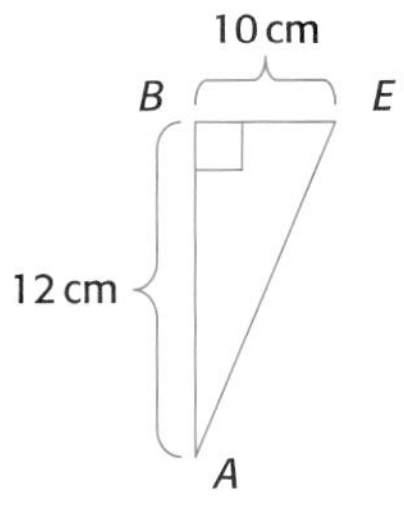

$$AE^2 = AB^2 + BE^2$$
$$AE^2 = 12^2 + 10^2$$
$$AE = \sqrt{244}\text{ cm} = 15.6\text{ cm (3 s.f.)}$$

ii To calculate AC we can work with the right-angled triangle $\triangle ABC$.

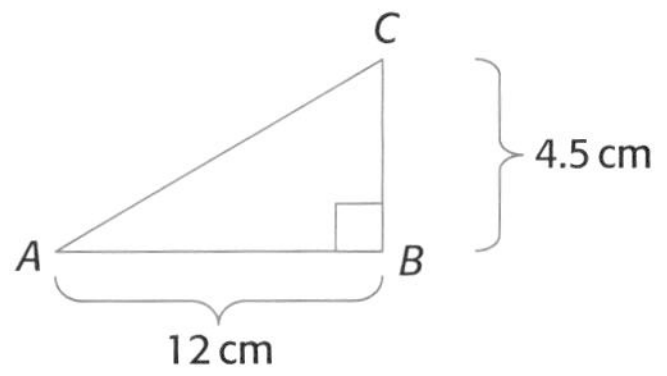

$$AC^2 = AB^2 + BC^2$$
$$AC^2 = 12^2 + 4.5^2$$
$$AC = \sqrt{164.25}\text{ cm} = 12.8\text{ cm (3 s.f.)}$$

iii To calculate BG we can work with the right-angled triangle $\triangle BFG$.

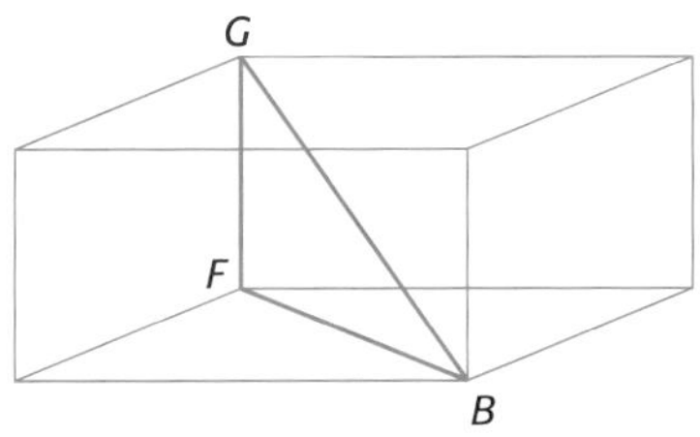

$FB = AE = \sqrt{244}$ cm as they are both the diagonals of the rectangle ABEF and $FG = BC = 4.5$ cm

$$GB^2 = GF^2 + FB^2$$
$$GB^2 = 4.5^2 + (\sqrt{244})^2$$
$$GB = 16.3\text{ cm}$$

b i $\triangle RHM$ is clearly right-angled as $\hat{H} = 90°$, $MH = 5\,\text{cm}$, $HR = 6\,\text{cm}$ so

$RM^2 = MH^2 + HR^2 \therefore RM^2 = 5^2 + 6^2$

$\therefore RM = 7.81\,\text{cm}$

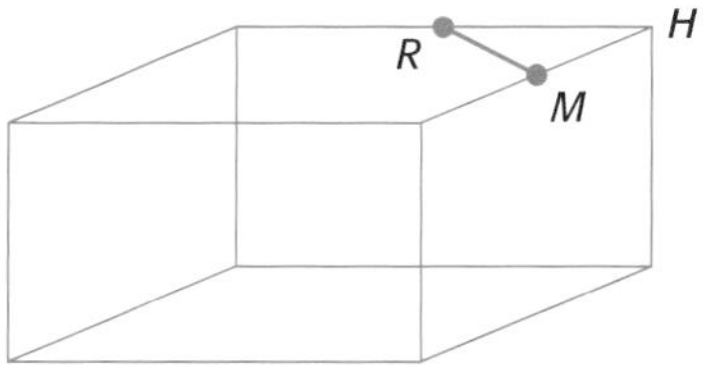

ii In this case we can work with the right-angled triangle APM, where PM is the perpendicular line through M to the face $ABEF$ of the cuboid.

$PM = 4.5$ cm and AP can be found as it is the hypotenuse of $\triangle ABP$.

$AP^2 = AB^2 + BP^2$

$AP^2 = 12^2 + 5^2$ so $AP = 13\,\text{cm}$ and

$AM^2 = AP^2 + PM^2$

$AM^2 = 13^2 + 4.5^2$

$AM = 13.8\,\text{cm}$ (3 s.f.)

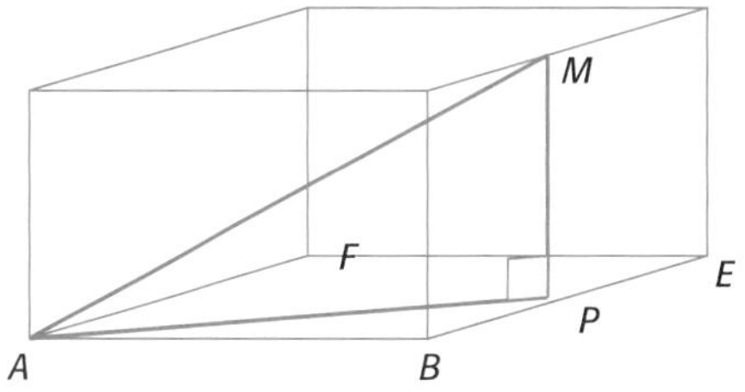

5.5.7 Sizes of angles between two lines and between lines and planes

Consider the line AB which intersects with the plane q at the point B. To calculate the size of the angle α that this line makes with q, drop a perpendicular line from A. This line meets q at the point P.

- The angle between the line AB and the plane q is the angle between the lines AB and BP.

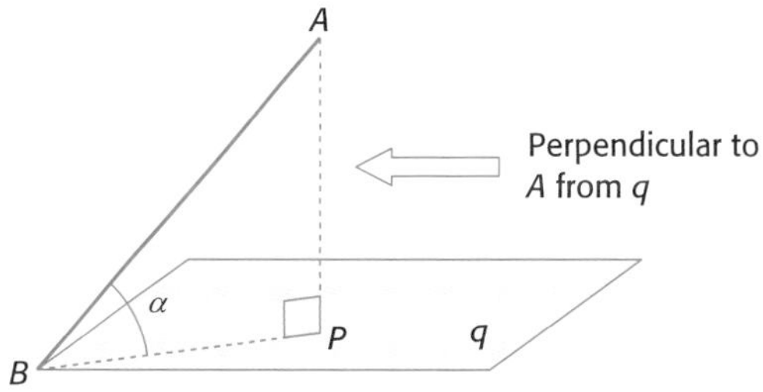

Example 5.5.7a

Copy the cuboid shown in the diagram and mark the angle that

a the plane $ABEF$ makes with the line AH. Label this angle α.

b the plane $ADGF$ makes with the line GB. Label this angle β.

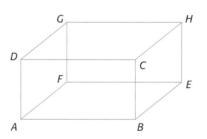

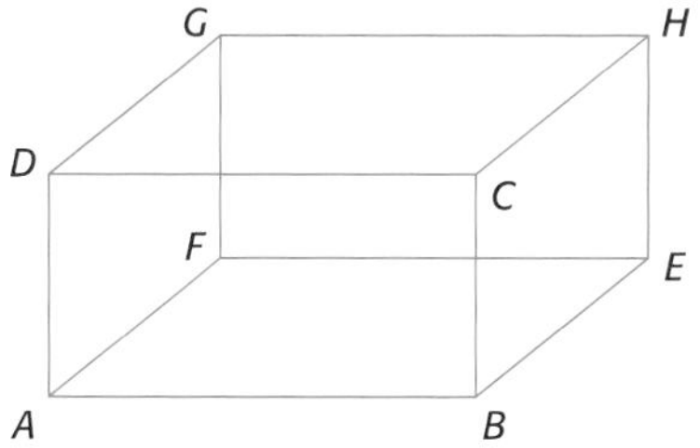

Solution

a

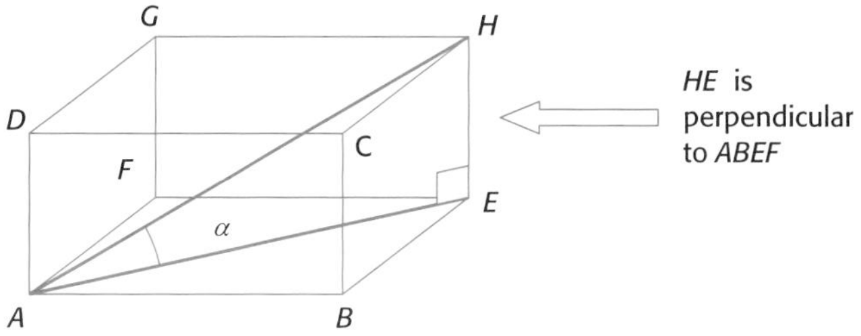

b

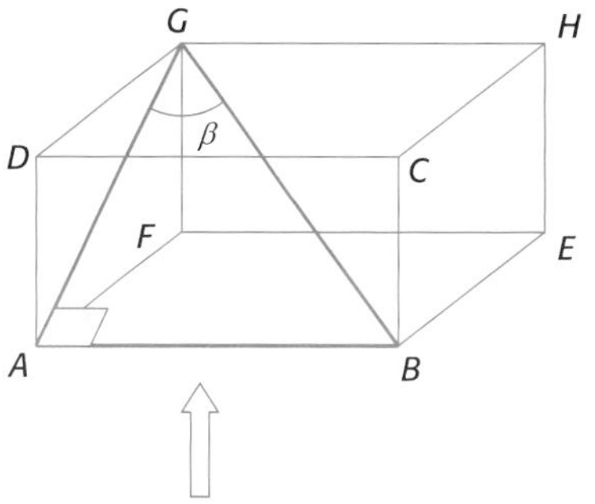

AB is perpendicular to *ADGF*.

Example 5.5.7b

Solution

This cuboid is the one used in Example 5.5.6.

Find

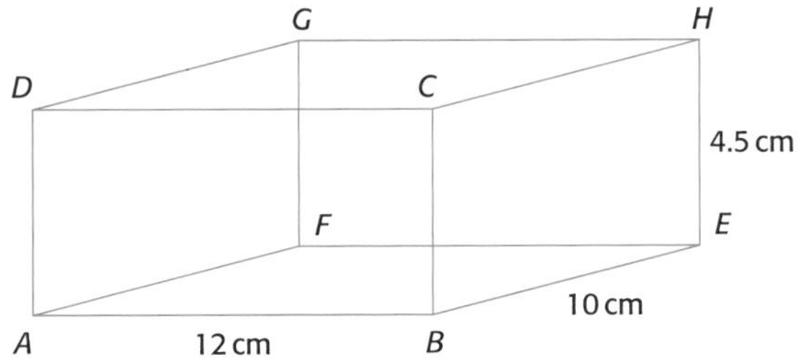

a $A\hat{E}B$

b the angle made between the plane $ABEF$ and the line AH

c the angle made between the lines AC and CE.

Solution

a Consider $\triangle ABE$ which is right-angled at B. The required angle is $A\hat{E}B$. $\therefore$ $\tan A\hat{E}B = \dfrac{12}{10}$

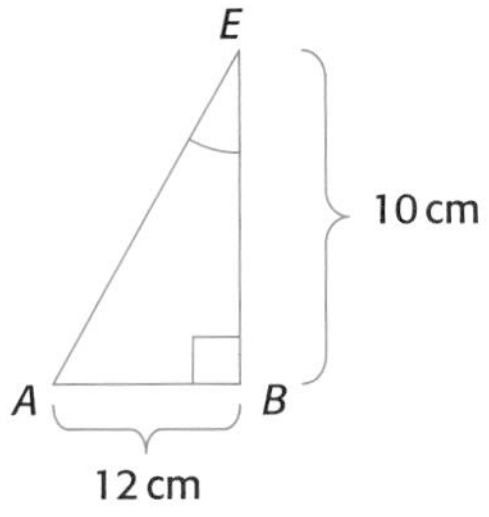

$$A\hat{E}B = \tan^{-1}\left(\frac{12}{10}\right)$$

$$A\hat{E}B = 50.2°$$

b The angle we are looking for is $H\hat{A}E$ so we work with the right-angled triangle HAE where $HE = 4.5$ cm. We have found that $AE = \sqrt{244}$ cm, so

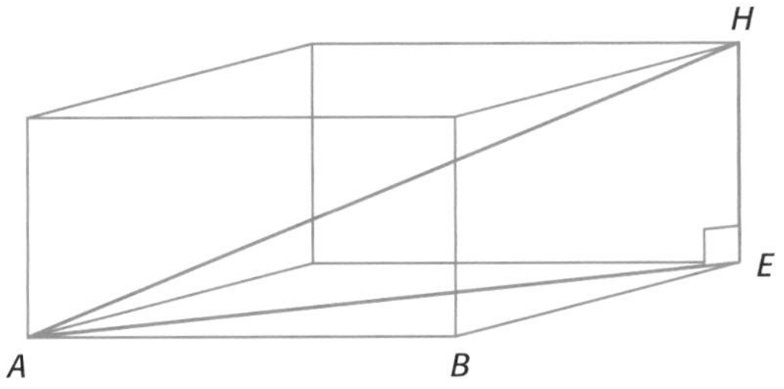

$$\tan H\hat{A}E = \frac{4.5}{\sqrt{244}}$$

$$H\hat{A}E = \tan^{-1}\left(\frac{4.5}{\sqrt{244}}\right)$$

$$H\hat{A}E = 16.1°$$

c The angle between the lines AC and CE is $A\hat{C}E$ so we work with the triangle ACE where $AC = \sqrt{164.25}$ cm and $AE = \sqrt{244}$ cm. We find CE by using Pythagoras in the right-angled triangle CBE.

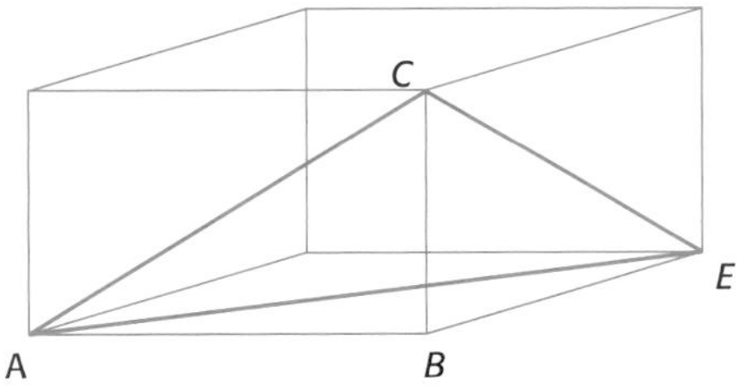

$\therefore CE^2 = 4.5^2 + 10^2 \therefore CE = \sqrt{120.25}$ cm

Now we need to find $A\hat{C}E$.
In ΔACE and using the cosine rule we have

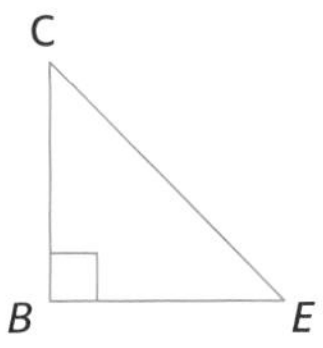

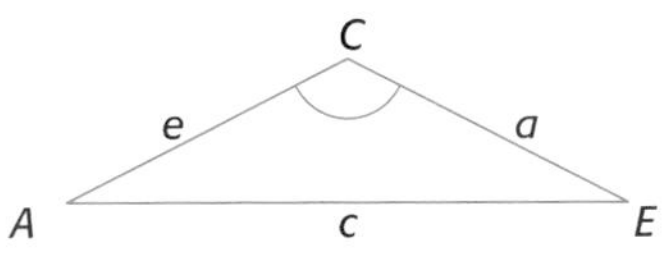

$$\cos A\hat{C}E = \frac{a^2 + e^2 - c^2}{2ae}$$

$$\cos A\hat{C}E = \frac{\left(\sqrt{120.25}\right)^2 + \left(\sqrt{164.25}\right)^2 - \left(\sqrt{244}\right)^2}{2 \times \sqrt{120.25} \times \sqrt{164.25}}$$

$$A\hat{C}E = \cos^{-1}\left(\frac{\left(\sqrt{120.25}\right)^2 + \left(\sqrt{164.25}\right)^2 - \left(\sqrt{244}\right)^2}{2 \times \sqrt{120.25} \times \sqrt{164.25}}\right)$$

$$A\hat{C}E = 81.7°$$

Example 5.5.7c

In the square-based pyramid in the diagram all the edges are 10 cm long and M is the midpoint of BC.

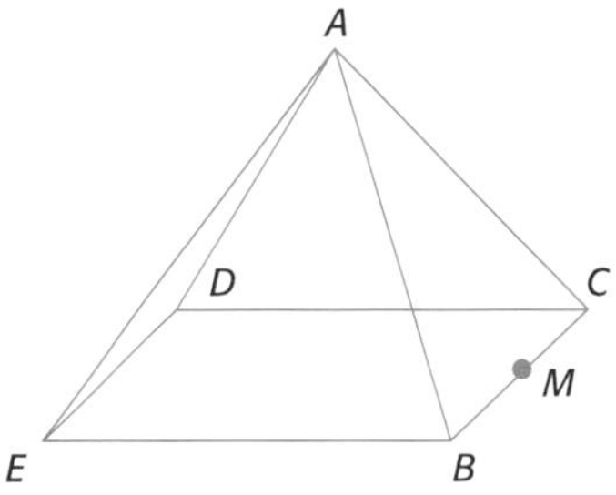

a Find the length of AM.

b Find the angle that AM makes with the base of the pyramid.

Solution

a AM is the height of the triangle ABC and $BM = 5$ cm as M is the midpoint of BC.

In the right-angled triangle AMC,

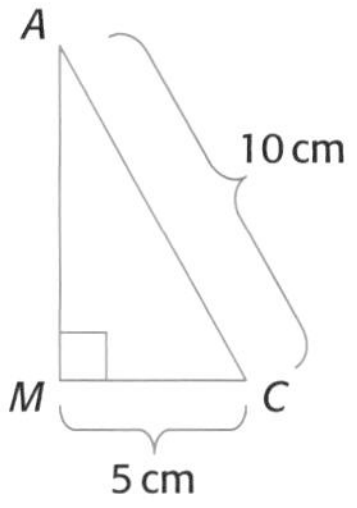

$AM^2 + 5^2 = 10^2 \therefore AM = \sqrt{75} = 8.66$ cm (3 s.f.)

b We drop the perpendicular from A to the base and label point O where this line meets the base. We are looking for $O\hat{M}A$.

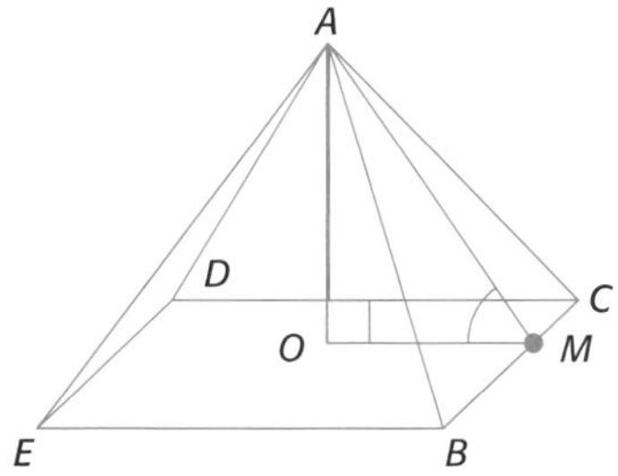

In $\Delta OMA, OM = 5$ cm and $AM = \sqrt{75}$ cm

$\therefore \cos O\hat{M}A = \dfrac{5}{\sqrt{75}} \quad \therefore O\hat{M}A = \cos^{-1}\left(\dfrac{5}{\sqrt{75}}\right)$

$\therefore O\hat{M}A = 54.7°$

Exercise 5.5

1 Copy and complete this table. Tick (✓) the last column if the solid is a prism.

Solid	Number of faces	Number of plane faces	Number of curved faces	Prism
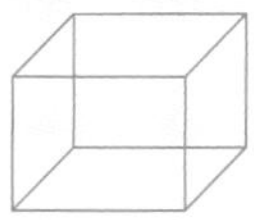				

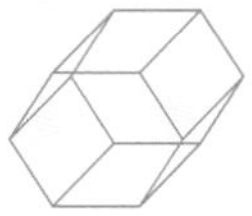				
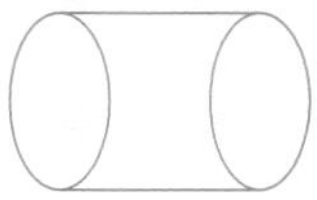				
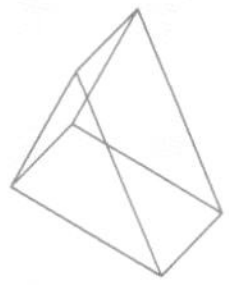				
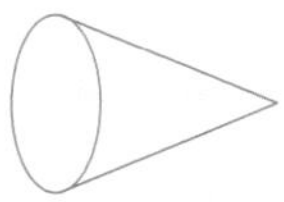				

2 Calculate the surface areas and the volumes of these solids.

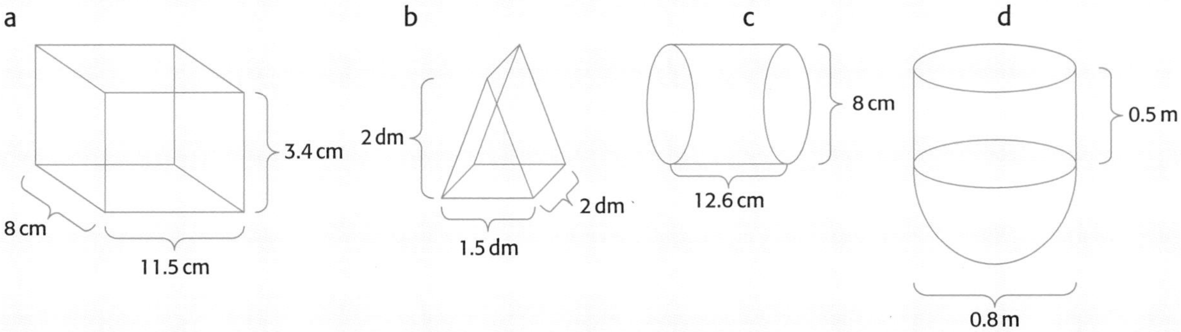

3 The volume of a sphere is $1000\,\text{cm}^3$. Calculate the length of its radius. Give your answer correct to the nearest cm.

4 A rectangular box is to be made from cardboard. The length of the cardboard is 35 cm and the width is 28 cm. Four squares with side of x cm are to be cut from the corners as shown in the diagram.

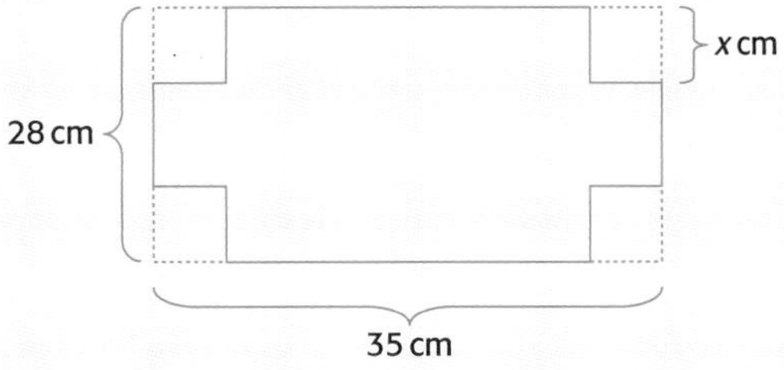

a Write down an expression in terms of x for the volume of the box.

b The volume of the box is $1914\,\text{cm}^3$ and it does not have a top.

i Use your GDC to find the value of x.

ii Use your answer to part **i** to find the surface area of the box.

5 Consider the cuboid shown in the diagram. Draw different sketches to mark the angles made between

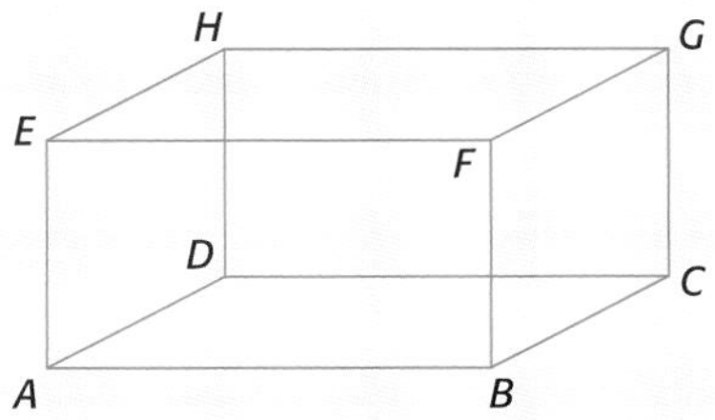

a the face *AEHD* and the line *AG*

b the plane *ABFE* and the line *CF*

c the lines *HB* and *BG*

d the planes *BDHF* and *DHEA*

e the plane *CDHG* and the line *HM* where *M* is the midpoint of *FG*.

6 The height and the diameter of the base of a cylindrical glass are 15 cm and 6 cm respectively.

a Calculate the volume of the glass. Give your answer to the nearest cm^3.

b Calculate the surface area of the glass.

7 The slant height of a cone is 10 cm and the angle that the slant height makes with the base is α. The height of the cone is 8 cm.

a Make a clear and labelled sketch to represent the given information.

b Find the radius of the base.

c Calculate the size of α. Give your answer to the nearest degree.

8 *OABCD* is a squared-based pyramid. *O* is vertically above the centre of the base, *X*. $AB = 1$ m and $OX = 1$ m.

a Calculate the height of one of its triangular faces.

b Hence calculate the surface area of the pyramid.

c Calculate the angle that any of the triangular faces makes with the base of the pyramid.

d Calculate the angle that *OB* makes with the plane *ABCD*.

e Find the volume of the pyramid.

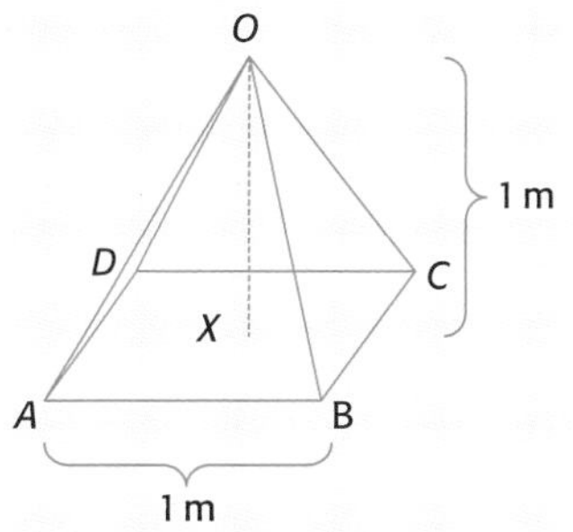

Past examination questions for topic 5

Paper 1

1 The diagram shows a circle of radius R and centre O. A triangle AOB is drawn inside the circle. The vertices of the triangle are at the centre, O, and at two points, A and B, on the circumference. Angle AOB is $110°$.

a Given that the area of the circle is $36\pi\,\text{cm}^2$, calculate the length of the radius R.

b Calculate the length of AB.

c Write down the side length, L, of a square which has the same area as the given circle.

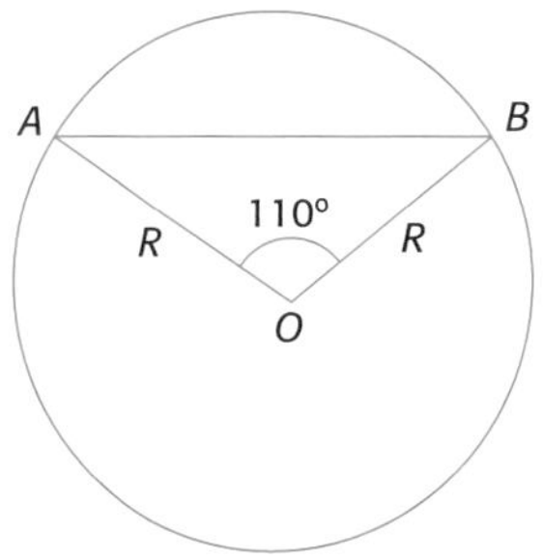

M06q9

2 $OABCD$ is a square-based pyramid of side 4 cm as shown in the diagram. The vertex D is 3 cm directly above X, the centre of square $OABC$. M is the midpoint of AB.

a Find the length of XM.

b Calculate the length of DM.

c Calculate the angle between the face ABD and the base $OABC$.

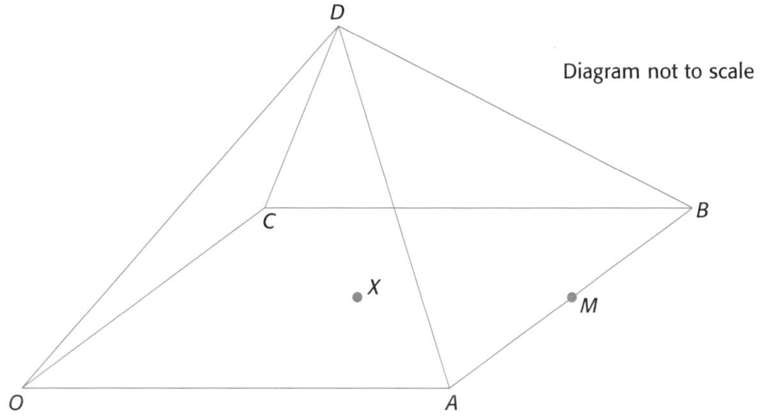

N04q9

3 The diagram shows the side view of a tent. The side of the tent, AC, is 6 m high. The ground AB slopes upwards from the bottom of the tent at point A, at an angle of 5° from the horizontal. The tent is attached to the ground by a rope at point B, a distance of 8 m from its base.

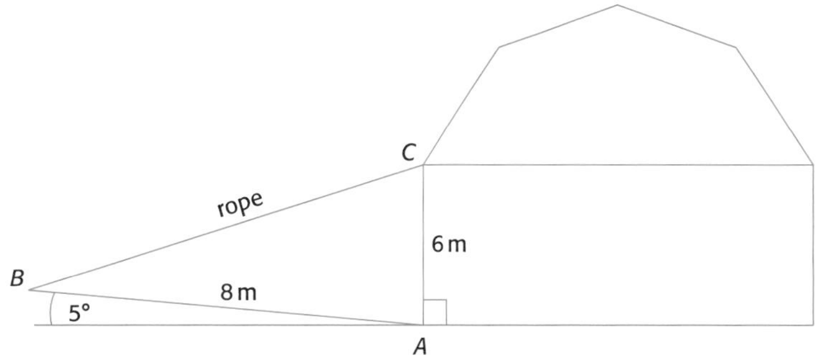

a Calculate the angle BAC.
b Calculate the length of the rope, BC.
c Calculate the angle, CBA, that the rope makes with the sloping ground. *M04q14*

4 The figure shows two adjacent triangular fields, ABC and ACD, where $AD = 30$ m, $CD = 80$ m, $BC = 50$ m, angle $ADC = 60°$ and angle $BAC = 30°$.

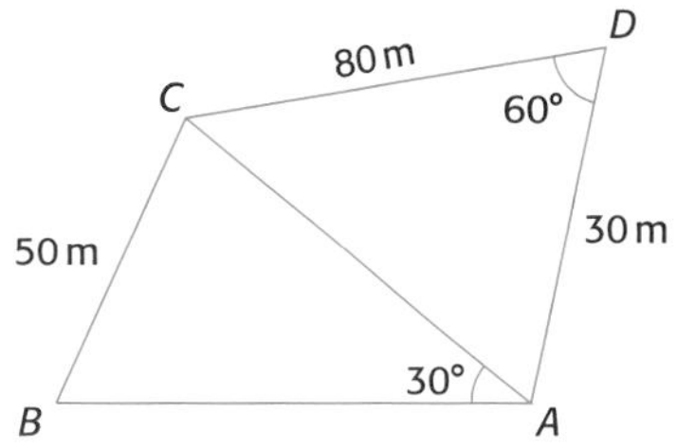

Diagram not to scale

a Using triangle ACD calculate the length AC.
b Calculate the size of angle ABC. *M03q8*

5 A is the point $(2, 3)$ and B is the point $(4, 9)$.
a Find the gradient of the line segment AB.
b Find the gradient of a line perpendicular to the line segment AB.
c The line $2x + by - 12 = 0$ is perpendicular to the line segment AB. What is the value of b? *M01q11*

6 The diagram shows the points P, Q and M. M is the midpoint of PQ.
a Write down the equation of the line PQ.
b Write down the equation of the line through M which is perpendicular to the line PQ.

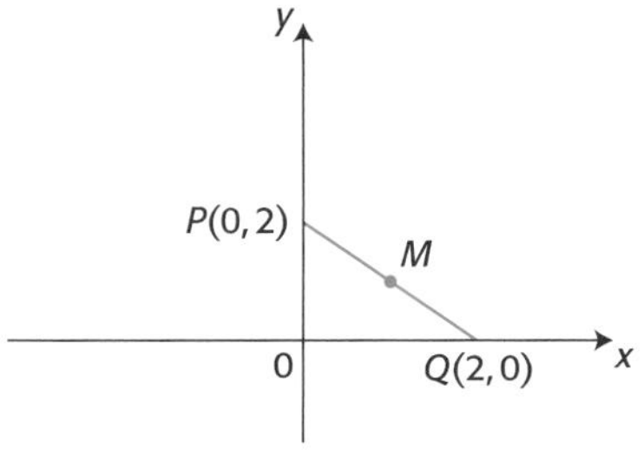

M00q5

Paper 2

1 The figure below shows a rectangular prism with some side lengths and diagonal lengths marked.
$AC = 10$ cm, $CH = 10$ cm, $EH = 8$ cm, $AE = 8$ cm.

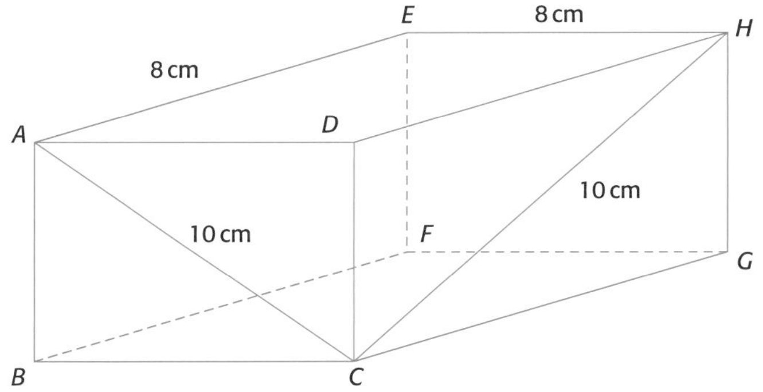

a Calculate the length of AH.
b Find the size of angle ACH.
c Show that the total surface area of the rectangular prism is 320 cm^2.
d A triangular prism is enclosed within the planes $ABCD$, $CGHD$ and $ABGH$. Calculate the volume of this prism. *M06q2(ii)*

2 The vertices of quadrilateral $ABCD$ as shown in the diagram are $A(-8, 8)$, $B(8, 3)$, $C(7, -1)$ and $D(-4, 1)$.

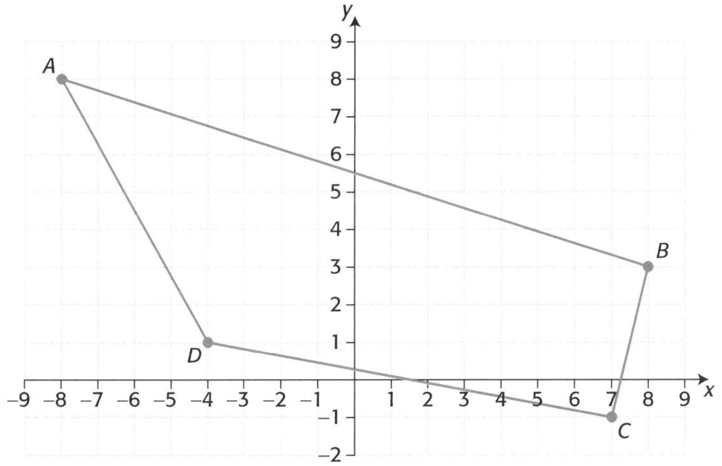

The gradient of line AB is $-\frac{5}{16}$.

a Calculate the gradient of line DC.
b State whether or not DC is parallel to AB and give a reason for your answer.

The equation of the line through A and C is $3x + 5y = 16$.

c Find the equation of the line through B and D expressing your answer in the form $ax + by = c$, where a, b and $c \in \mathbb{Z}$.

The lines AC and BD intersect at point T.

d Calculate the coordinates of T. *N05q3*

3 i a A farmer wants to construct a new fence across a field. The plan is shown in the diagram. The new fence is indicated by a dotted line. Calculate the length of the fence.

b The fence creates two sections of land. Find the area of the smaller section of land, *ABC*, given the additional information shown below.

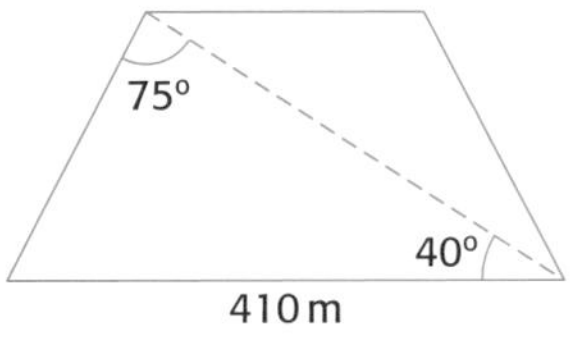

Diagram not to scale

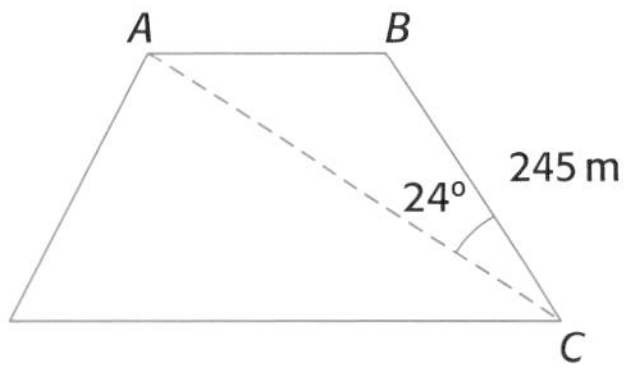

Diagram not to scale

ii Find the volume of this prism. *M05q2*

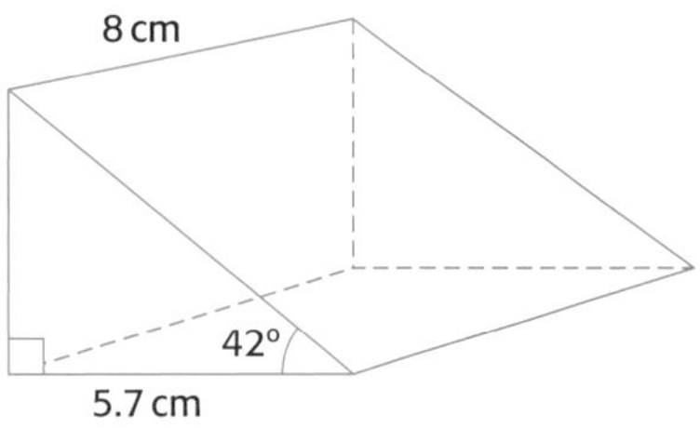

Diagram not to scale

4 The line L_1 shown on the set of axes below has equation $3x+4y=24$. L_1 cuts the x-axis at A and the y-axis at B.

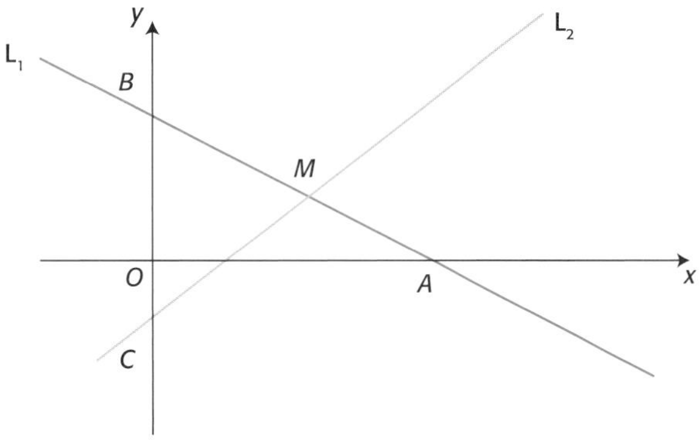

Diagram not to scale

a Write down the coordinates of A and B.

M is the midpoint of the line segment AB.

b Write down the coordinates of M.

The line L_2 passes through the point M and the point $C(0,-2)$.

c Write down the equation of L_2.

d Find the length of

i MC

ii AC.

e The length of AM is 5. Find

i the size of angle CMA

ii the area of the triangle with vertices C, M and A. *M00q4*

Theory of knowledge

How much mathematics can there be? (from *The Mathematical Experience* – Davis & Hersh)

With billions of bits of information being processed every second by machine, and with 200 000 mathematical theorems of the traditional, hand-crafted variety produced annually, it is clear that the world is in a Golden Age of mathematical production. Whether it is also a golden age for new mathematical ideas is another question altogether.

It would appear from the record that mankind can go on and on generating mathematics. But this may be a naïve assessment based on linear (or exponential) extrapolation, an assessment that fails to take into account diminution due to irrelevance or obsolescence. Nor does it take into account the possibility of internal saturation. And it certainly postulates continuing support from the community at large.

The possibility of internal saturation is intriguing. The argument is that within a fairly limited mode of expression or operation there are only a very limited number of recognizably different forms, and while it would be possible to proliferate these forms indefinitely, a few prototypes adequately express the character of the mode. Thus, although it is said that no two snowflakes are identical, it is generally acknowledged that from the point of view of visual enjoyment, when you have seen a few, you have seen them all.

In mathematics, many areas show signs of internal exhaustion – for example, the elementary geometry of the circle and the triangle, or the classic theory of functions of a complex variable. While one can call on the former to provide five-finger exercises for beginners and the latter for application to other areas, it seems unlikely that either will ever again produce anything that is both new and startling within its bounded confines.

It seems certain that there must be a limit to the amount of living mathematics that humanity can sustain at any time. As new mathematical specialties arise, old ones will have to be neglected.

All experience so far seems to show that there are two inexhaustible sources of new mathematical questions. One source is the development of science and technology, which makes ever new demands on mathematics for assistance. The other source is mathematics itself. As it becomes more elaborate and complex, each new, completed result becomes the potential starting point for several new investigations. Each pair of seemingly unrelated mathematical specialties poses an implicit challenge – to find a fruitful connection between them.

Although each special field in mathematics can be expected to become exhausted, and although the exponential growth in mathematical production is bound to level off sooner or later, it is hard to foresee an end to all mathematical production, except as part of an end to mankind's general striving for more knowledge and more power. Such an end to striving may indeed come about some day. Whether this end would be a triumph or a tragedy, it is far beyond any horizon now visible.

1 Do you agree that the growth in mathematical production will level off? What about in other areas of knowledge?
2 Do you think that mankind will ever stop striving for more knowledge and more power?

TOPIC 6

Statistics

Here are just a few examples of the type of statistics that we see every day in newspapers, magazines, on the television, etc.

- The amount of sunlight reaching the planet could be down 10%.
  www.omidyar.net/group/users-env
- In Bangladesh infant mortality is about 60 per 1000 live births.
  www.indexmundi.com/bangladesh
- Experts estimate that orangutans could be extinct in the world in as few as ten years.
  www.orangutan.com

In many of our school subjects too, statistical tests are used to make sense of data collected.

It may be in biology, in geography, in economics or in any other subject.

Of course, researchers have to make sure that the data they collect are random and free from bias in any way if they want to use them to draw meaningful conclusions.

6.1 Discrete and continuous data

6.1.1 Discrete data

Discrete data can be *counted*, for example the number of heads we get when a coin is tossed, the number of red cars in a parking lot, the number of visitors to the Blue Mosque in one hour, etc.

6.1.2 Continuous data

Continuous data can be *measured*, for example weight, height, time etc. We can express continuous data to a required number of significant figures.

Theory of knowledge

It is important that timing is as accurate as possible in the Olympic Games. However, the same degree of accuracy is not necessary when timing the cooking of a pot of rice. So, how important is it to have "exact" values?
What do you understand by an "exact value"?
Can a measurement ever be exact?
Is it more important to be "exact" in mathematics and science than it is in other areas of knowledge such as science, history, aesthetics?

Time is an example of continuous data.

Did you know that...?

OMEGA was the first company to be entrusted with the official timekeeping of all disciplines at the Los Angeles Games in 1932, using chronographs and stop-watches developed by its subsidiary Lémania. OMEGA's association with the Olympic Games led to well over half a century of pioneering developments in the field of sports timekeeping.

One of the first innovations by the brand was the world's first independent, portable and water-resistant photoelectric cell (1945). This was later followed by the world's first photofinish camera, the Racend OMEGA Timer (1949), which was a major innovation that solved the problem of grouped arrivals in track events. At the Helsinki 1952 Olympic Games, OMEGA became the first company ever to use electronic timing in sport, with the OMEGA Time Recorder (OTR), which was homologated by the International Amateur Athletics Federation on the basis of a rating certificate from the Observatory of Neuchâtel that proved it was accurate to within 0.05 seconds in 24 hours.

At the same Olympic Games in 1952, OMEGA was also awarded the Olympic Cross of Merit for "exceptional services to the world of sport". In 1961, OMEGA invented the Omegascope, which allowed the time of each competitor followed by the TV camera to be superimposed on the TV screen. The 1966 European Athletics Championships in Budapest marked a turning point in sports timekeeping, since they were the first European Championships at which the electronically recorded times were taken as the official times. It was the OTR and the Omegascope that recorded this unique moment. One year later, OMEGA introduced its "contact pads" for swimming competitions.

This simple new technology reacted only to the touch of the swimmers and was not affected by water splashes. Such contact pads have been used ever since at all the world's major swimming events. In 1990, the brand opened up sports timekeeping to the mass market with the launch of the Scan'O'Vision – a low-cost and popular version of the photofinish camera. OMEGA's most recent development in timing technology was to bring sports timekeeping into the internet age with live timing of swimming events, which allows anyone with internet access to view swimming and diving competition results in real time on the OMEGA Timing Internet site, www.omegatime.com

Exercise 6.1

1 State whether these data are discrete or continuous.

a the heights of bonsai trees
b the shoe sizes of the All Blacks rugby team
c the numbers of differently coloured hats on a market stall
d the weights of baby chimpanzees
e the times taken for students to write their extended essays
f the marks obtained by grade 9 in a French test
g the weights of ten pieces of sushi
h the hours of sunlight each day during the month of July
i the scores obtained when a dart is thrown 50 times
j the times taken for students in grade 11 to walk 500 m
k the number of visitors to Niagara Falls each day
l the exact weights of 100 one-kg bags of rice.

6.2 Frequency tables and polygons

6.2.1 Frequency tables

When we have a large amount of data it is easier to interpret if the data are set up in a table or a graph.

Example 6.2.1

60 students from grade 9 are asked to pick a number between 0 and 4. The result is shown here. Represent the information in a frequency table.

1 4 3 3 4 0 2 2 1 3 3 4 0 4 3 4 3 2 0 1

2 3 4 3 3 0 3 4 2 0 1 2 3 3 4 1 0 3 4 2

0 3 4 4 2 3 3 1 0 4 2 3 2 4 1 0 3 4 3 2

Solution

We can make a **tally chart** and represent the information in a **frequency table.**

Number chosen	Tally	Frequency
0	𝍸 IIII	9
1	𝍸 II	7
2	𝍸 𝍸 I	11
3	𝍸 𝍸 𝍸 IIII	19
4	𝍸 𝍸 IIII	14
	Total	60

We can also represent these data on a graph.

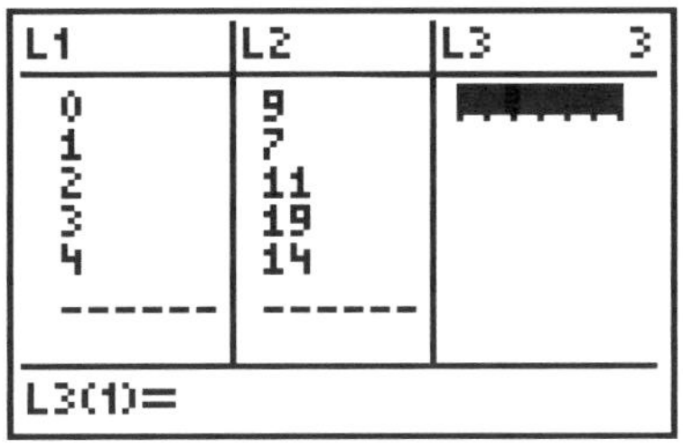

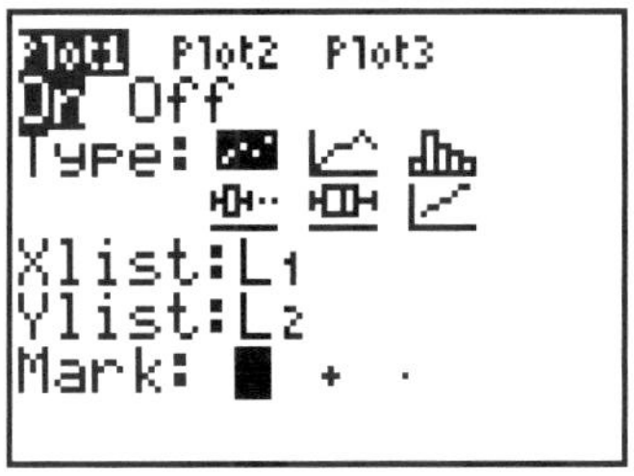

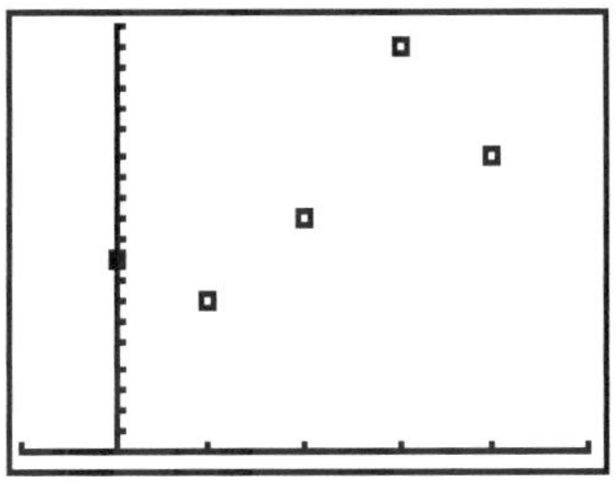

(window –1–5 on *x*-axis, 0–20 on *y*-axis)

Remember that the data are **discrete** so do *not* connect the points on the graph.
(This mistake is seen frequently in internal assessment projects.)

6.2.2 Frequency polygons

When we have **continuous** data, we can represent them on a **frequency polygon.**

Example 6.2.2

The table shows the time, in minutes to the nearest three minutes, of 100 telephone calls made from the same cell phone in one month.

Time (*t* minutes)	Frequency
3	42
6	23
9	8
12	5
15	6
18	4
21	4
24	2
27	6

Draw a frequency polygon to represent this information.

Solution

With the x-axis representing the time and the y-axis the frequency, we can plot the points in the table and join them up with a smooth curve. We can also use a GDC and enter the time in List 1 and the frequency in List 2 and use the Stat Plot function as shown here.

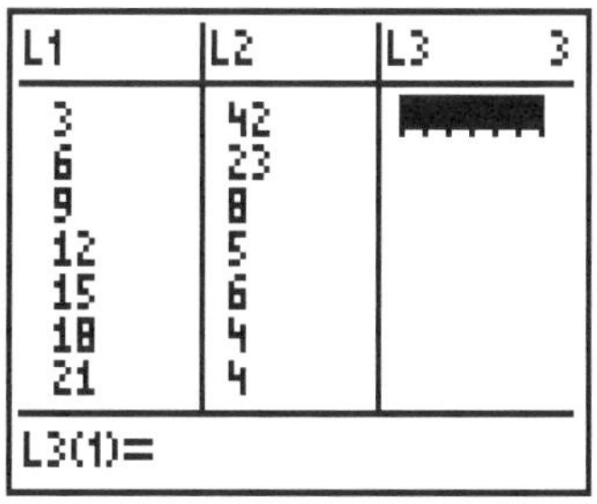

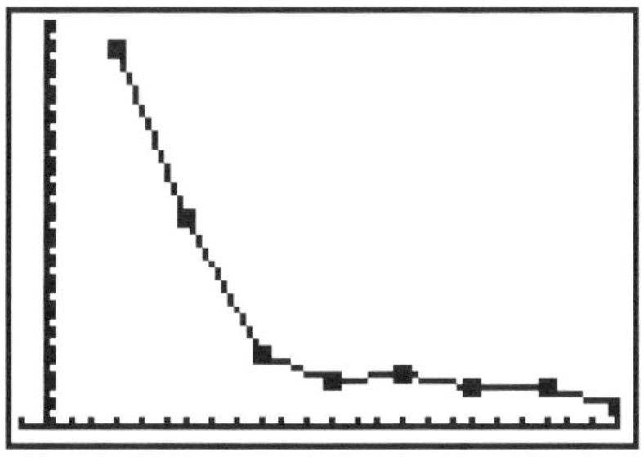

(window –1–30 on x-axis and 0–45 on y-axis)

6.3 Grouped and continuous data, histograms and stem plots

6.3.1 Grouped discrete or continuous data: frequency tables

When there are a lot of data spread over a wide range, then it is useful to **group** the data. There should be between 5 and 15 equal groups. We can group both discrete and continuous data.

Example 6.3.1

A survey was carried out in a shopping mall to find out how old people were when they passed their driving test. 150 people between the ages of 15 and 92 were interviewed and the results are shown in the table.

Age (x)	**Frequency**
$15 \leq x < 25$	85
$25 \leq x < 35$	33
$35 \leq x < 45$	14
$45 \leq x < 55$	8
$55 \leq x < 65$	3
$65 \leq x < 75$	4
$75 \leq x < 85$	2
$85 \leq x < 95$	1

Remember that ages are continuous data and we have to be careful to include all possible ages.

Theory of knowledge

When conducting a survey how can you ever be sure that the replies given are true?

In example 6.3.1 the people questioned could have lied about their real age and/or the age when they passed their driving test.

What type of questionnaire was used and to whom was it distributed?

What difference could this make to the results?

6.3.2 Mid-interval values

To find the mid-interval value we halve the sum of the end values. In the example above, the midpoint of the first class interval is $(15 + 25) \div 2 = 20$.

6.3.3 Upper and lower boundaries

For discrete data, the lower value in the class interval is the lower boundary and the upper value is the upper boundary.

For continuous data, the boundaries are calculated by averaging the upper value of one group with the lower value of the next group.

Example 6.3.3

Consider this table of data.

Age	Frequency
0–9	8
10–19	12
20–29	15
30–39	6

The upper boundary of the first interval is $(9 + 10) \div 2 = 9.5$. This is also the lower boundary of the second interval. So, the boundaries would now be $0 \leq \text{age} < 9.5$ and $9.5 \leq \text{age} < 19.5$, and so on.

6.3.4 Frequency histograms

A frequency histogram is a useful way to represent data in a visual manner.

A **frequency histogram** has *equal* class-intervals. There should be no spaces between the bars.

The ages in Example 6.3.1 can be represented in a frequency histogram as shown here.

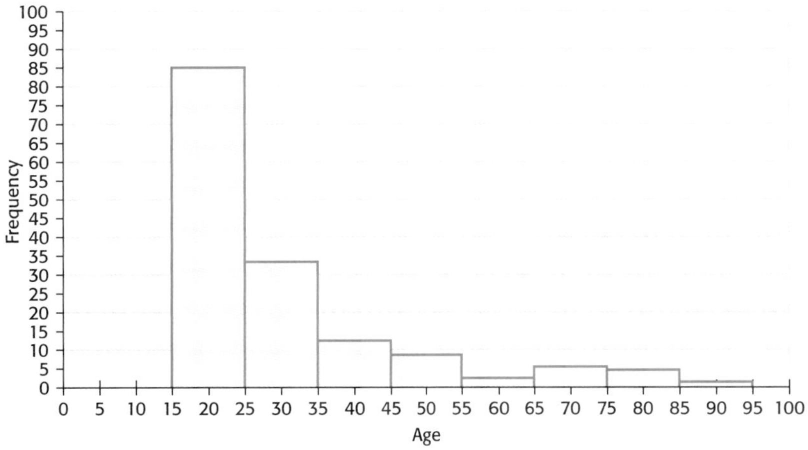

6.3.5 Stem and leaf diagrams (stem plots)

We can also represent data in a **stem and leaf diagram**.

Example 6.3.5

These are the times taken (in minutes) for 50 batteries to run out.

450	451	453	453	455	462	465	466	469	470	470	477	478	483	484	485	488	489
489	491	495	495	498	500	501	502	504	504	505	506	508	511	511	513	514	517
519	521	522	522	536	537	539	541	544	544	546	547	548	548				

These data can be represented in a stem and leaf diagram.

Stem	Leaf
45	0 1 3 3 5
46	2 5 6 9
47	0 0 7 8
48	3 4 5 8 9 9
49	1 5 5 8
50	0 1 2 4 4 5 6 8
51	1 1 3 4 7 9
52	1 2 2
53	6 7 9
54	1 4 4 6 7 8 8

Key 45 | 0 = 450 minutes

This is an example of an **ordered** stem and leaf diagram because the values are in ascending order.

It can also be useful to compare two sets of data using a "back-to-back" stem and leaf diagram.

This diagram below shows the grades obtained in an English exam and a mathematics exam for the 30 students in an IB1 class.

English		Mathematics
	1	9
	2	7 9
8 6 6	3	3 8 9
9 7 4	4	0 5 7 9
8 6 6 4 3 0	5	1 3 5 5 8
9 7 7 4 1	6	1 2 4 5 7 9
9 8 8 5 3	7	4 6 8
9 8 5 5 3 3	8	1 3 7 9
1 0	9	3 8

Key 6|3|8 = 36 for English and 38 for mathematics

Exercise 6.3

1 The numbers of goals scored by Sparta during their last 30 games were

0 1 3 0 0 2 1 1 0 2 3 0 1 2 2 5 0 2 1 1 4 3 2 1 0 1 2 3 6 0

Represent this information in a frequency table and draw a graph.

2 The numbers of sixes obtained when twelve dice were rolled 50 times were

3 4 4 6 7 2 8 4 6 5 7 6 5 11 5 6 0 7 6 5 1 5 6 10 7 5 6 7 4 9
5 6 7 2 9 10 3 8 5 7 9 1 6 5 7 0 12 5 8 6

Represent this information in a frequency table and draw a graph.

3 The heights (to the nearest cm) of 85 bean plants after five weeks are given in the table.

Height	2	3	4	5	6	7	8	9
Frequency	8	21	24	13	8	6	3	2

Draw a frequency polygon to represent this information.

4 Sixty people were asked how often they travelled by metro (underground railway) each month. The results are shown below.

6	26	14	30	4	42	56	20	22	48	20	8	34	10	4	68	10	28	46	30
10	24	8	22	20	80	20	38	48	6	8	38	42	68	22	76	16	62	6	12
60	16	74	36	8	32	44	14	12	46	76	50	16	44	16	40	38	10	32	64

a Represent this information in a grouped frequency table.
b Draw a histogram to represent the information graphically.

5 The weekly wages for 100 farm workers in India are shown in the table.

Weekly wage (*w*)	**Frequency**
$10 \le w < 15$	3
$15 \le w < 20$	12
$20 \le w < 25$	14
$25 \le w < 30$	35
$30 \le w < 35$	21
$35 \le w < 40$	5
$40 \le w < 45$	8
$45 \le w < 50$	2

Draw a histogram to display this information.

6 The number of hours of CAS completed in the first term for the 80 students at Idowell School is recorded below.

16	32	8	41	27	103	19	31	42	6	21	51	13	38	72	80	20	61	82	73
10	36	25	39	65	101	10	49	43	78	33	25	27	26	83	94	14	53	51	85
45	41	92	18	29	63	81	102	32	29	9	18	15	32	41	52	33	100	28	48
19	98	39	19	40	73	61	97	19	73	63	64	41	27	70	28	55	88	46	95

Represent this information in an ordered stem and leaf diagram.

7 This stem and leaf diagram shows the number of *SMS* messages that eighty IB1 students sent in one week.

Stem	**Leaf**
0	4 5 8 8 8 9 9
1	2 2 4 5 5 7 8 9 9
2	0 0 0 1 1 2 3 4 5 7 7 8 8 9
3	0 1 1 1 2 2 3 4 5 5 6 9
4	1 1 2 2 2 3 4 4 4 6 6 7 8 9 9
5	0 1 2 5 5 6 7 9
6	2 2 3 4 4 5 6 8
7	1 4 8 9
8	0 5 9

Key 0 | 4 = 4 messages

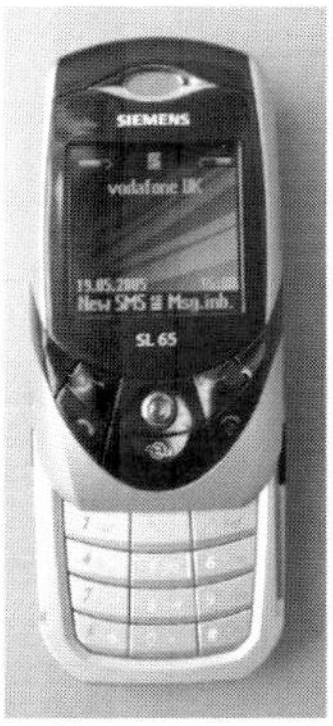

a Represent the information in a grouped frequency table.
b Draw a histogram to represent the information.

8 This stem and leaf diagram shows the percentage scores for an English and an economics test for 40 students in grade 8.

English test (leaf)	Stem	Economics test (leaf)
	2	5 9
9 9	3	6 8 9
8 3 2 0	4	5 6 7
9 9 6 4 3	5	1 2 3 3 3 5 8 9 9
8 7 7 3 1	6	0 0 1 2 2 4 5 5 6 8
9 6 5 4 3 1 0	7	3 4 5 6
8 5 3 2 2 1 1 0	8	0 1 1 1 2 8
9 8 5 4 4 3 2 2 1	9	3 5 6

Key 9|3|6 = 39% English and 36% economics

a Represent the information in separate grouped frequency tables.
b Draw histograms to represent the information.
c Comment on the shape of the histograms.

6.4 Cumulative frequency, box plots, percentiles and quartiles

6.4.1 Cumulative frequency tables

To construct a cumulative frequency table we write down the upper boundary of each class interval in one column and the corresponding cumulative frequency in another.

The **cumulative frequency** is the total frequency of all values less than or equal to a given value of the variable.

Example 6.4.1

Jun's parents decided to monitor the length of time that he spoke on the telephone. The table shows the time, to the nearest minute, that Jun spoke during his last 80 calls. Represent this information in a cumulative frequency table.

Number of minutes	Frequency
0–2	8
3–5	12
6–8	28
9–11	20
12–14	8
15–17	4

Solution

Number of minutes	Upper boundary	Frequency	Cumulative frequency
0–2	2.5	8	8
3–5	5.5	12	20
6–8	8.5	28	48
9–11	11.5	20	68
12–14	14.5	8	76
15–17	17.5	4	80

6.4.2 Cumulative frequency curves (ogives)

To draw the graph of the cumulative frequency (often called an ogive) we plot the value of the **upper boundary** with the **cumulative frequency** value.

The cumulative frequency curve for the data in Example 6.4.1*a* is shown in the diagram.

> **Did you know that...?**
>
> An ogive is a curved shape, figure, or feature.

The cumulative frequency is always shown on the *vertical* axis.

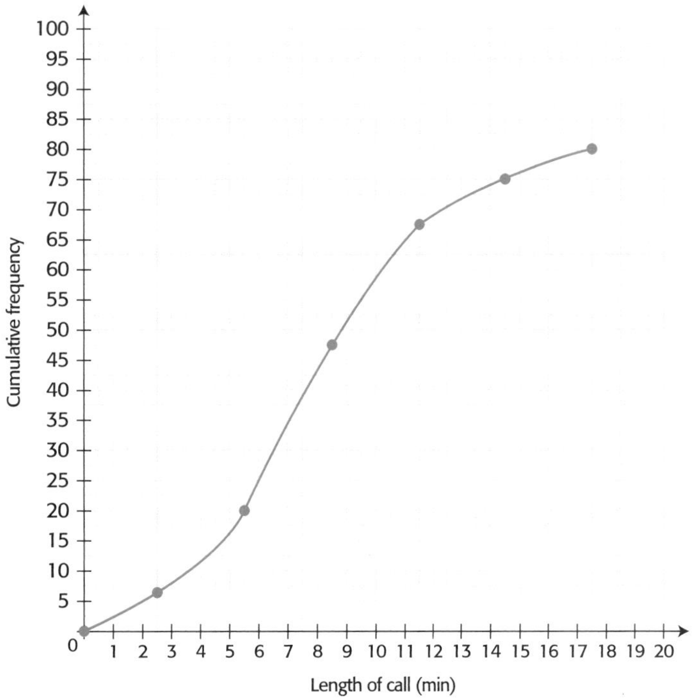

6.4.3 Percentiles and quartiles

We can use the cumulative frequency curve to find estimates of the percentiles and quartiles.

The **lower quartile,** Q_1, is found by reading the value on the curve corresponding to $n \div 4$ on the cumulative frequency axis, where n is the total frequency.

The **middle quartile** or **median** is found by reading the value on the curve corresponding to $n \div 2$ on the cumulative frequency axis.

The **upper quartile,** Q_3, is found by reading the value on the curve corresponding to $3n \div 4$ on the cumulative frequency axis.

The **percentiles,** say ***p*%**, are found by reading the value on the curve corresponding to $pn \div 100$ on the cumulative frequency axis.

The **interquartile range** is found by subtracting the lower quartile from the upper quartile.

For example, in the cumulative frequency diagram above

Lower quartile $\cong 5.5$ This is the value corresponding to $80 \div 4 = 20$

Median ≅ 7.8	This is the value corresponding to 80 ÷ 2 = 40
Upper quartile ≅ 10.2	This is the value corresponding to 3(80 ÷ 4) = 60
40th percentile ≅ 7.2	This is the value corresponding to 40% of 80 = 32

The inter-quartile range = 10.2 – 5.5 = 4.7

Example 6.4.3a

100 students attempt to complete a jigsaw puzzle. The time (in minutes) that it takes each one is shown in the table.

a Construct a cumulative frequency table.

b Draw a cumulative frequency graph.

c Use your graph to estimate

i the lower quartile

ii the median

iii the upper quartile

iv the interquartile range

v the 30th percentile.

Time in minutes, t	**Frequency**
$0 \le t < 1$	3
$1 \le t < 2$	5
$2 \le t < 3$	7
$3 \le t < 4$	9
$4 \le t < 5$	22
$5 \le t < 6$	28
$6 \le t < 7$	8
$7 \le t < 8$	5
$8 \le t < 9$	6
$9 \le t < 10$	4
$10 \le t < 11$	2
$11 \le t < 12$	1

Solution

a

Time less than	**Cumulative frequency**
$t < 1$	3
$t < 2$	8
$t < 3$	15
$t < 4$	24
$t < 5$	46
$t < 6$	74
$t < 7$	82
$t < 8$	87
$t < 9$	93
$t < 10$	97
$t < 11$	99
$t < 12$	100

b

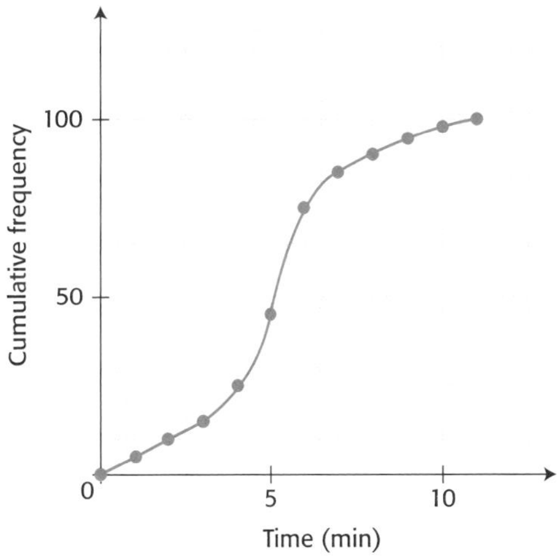

c i The lower quartile is approximately 4 (read the value on the horizontal axis corresponding to 25 on the vertical axis).

ii The median is approximately 5.2 (read the value on the horizontal axis corresponding to 50 on the vertical axis).

iii The upper quartile is approximately 6 (read the value on the horizontal axis corresponding to 75 on the vertical axis).

iv The interquartile range is 6 – 4 = 2

v The 30th percentile is approximately 4.4 (read the value on the horizontal axis corresponding to 30 on the vertical axis).

Example 6.4.3b

From this cumulative frequency graph find

a the median

b the interquartile range

c the 70th percentile.

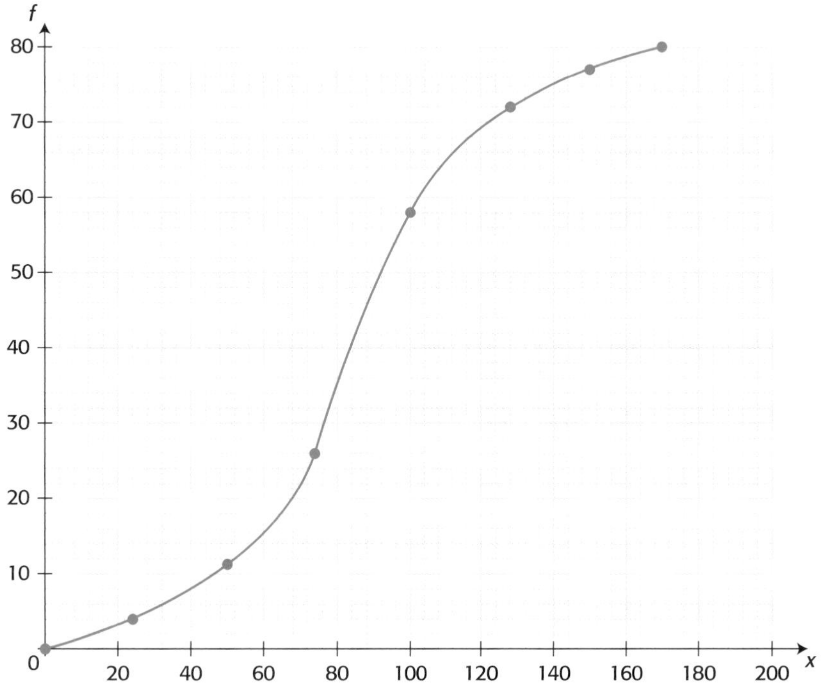

Solution

a The median is approximately 86 (the value corresponding to 40 on the vertical axis).

b The interquartile range is approximately 102 – 67 = 35 (the value corresponding to 60 on the vertical axis minus the value corresponding to 20 on the vertical axis).

c The 70th percentile is approximately 98 (the value corresponding to 70% of 80 = 56 on the vertical axis).

6.4.4 Box and whisker plots (box plots)

Another useful way to represent data is in a **box and whisker plot.**

A box and whisker plot looks like this.

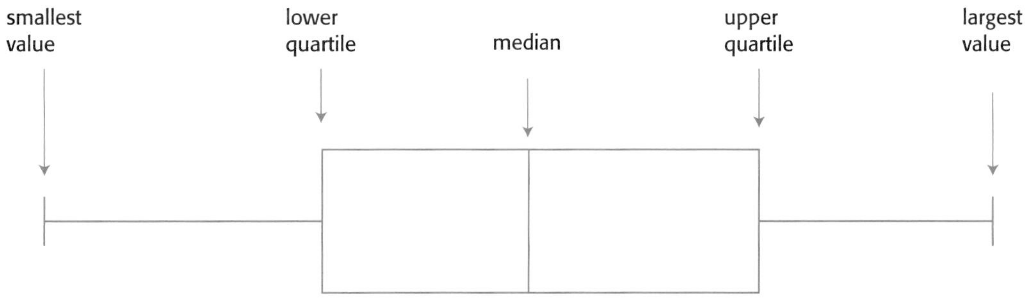

Referring to Example 6.4.1 on page 291, if we knew that Jun's shortest call was 1 minute long and his longest was 17 minutes , then the box and whisker plot would look like this.

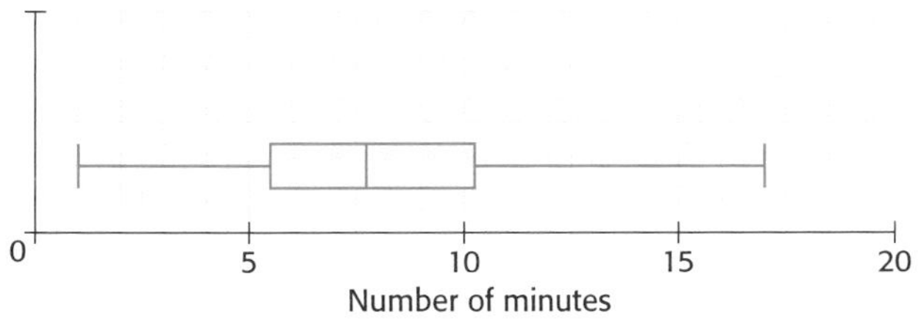

Outliers

An **outlier** is a value that is much smaller or much larger than the other values.
Normally we consider an outlier to be:

smaller than the lower quartile – 1.5 × the interquartile range

or larger than the upper quartile + 1.5 × the interquartile range.

Example 6.4.4a

1 This stem and leaf diagram shows the results, as percentages, for French and Spanish tests taken by 23 grade 7 students.

a Represent the information on box plots.

b Are there any outliers?

French		Spanish
	Stem	
8	2	
7 5 2	3	8 9
3 2 2 1	4	0 5
9 6 5 1 0	5	1 2 3 3 8 9
7 4 3 2	6	3 4 4 5
9 5 2 1	7	0 6 7 8
5 3	8	1 2 2
	9	4 8

Key 3 | 8 =38%

Solution

a **French:** lowest mark = 28, lower quartile is 6th entry = 42, median is 12th entry = 56, upper quartile is 18th entry = 71, largest mark = 85.

Spanish: lowest mark = 38, lower quartile is 6th entry = 52, median is 12th entry = 64, upper quartile is 18th entry = 78, largest mark = 98.

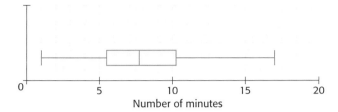

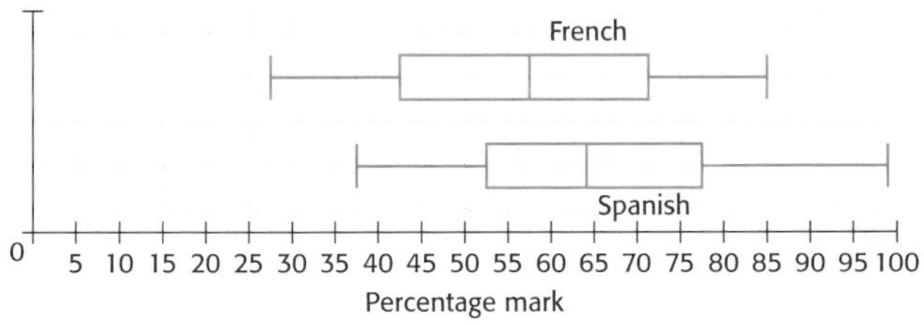

b Inter-quartile range for French = 71 – 42 = 29
$42 - 1.5 \times 29 = -1.5$ and
$71 + 1.5 \times 29 = 114.5$, so there are no outliers for French.

Interquartile range for Spanish = 78 – 52 = 26
$52 - 1.5 \times 26 = 13$ and
$78 + 1.5 \times 26 = 117$, so there are no outliers for Spanish.

Example 6.4.4b

The temperature each day, in degrees Celsius, at 12 noon in Tokyo in the month of July was

31° 28° 30° 27° 46° 32° 31° 28° 30° 27° 30° 31° 30° 30° 28° 29° 32°
27° 29° 30° 31° 30° 32° 27° 29° 30° 31° 30° 28° 31° 31°

a Represent this information in a box and whisker plot.

b Are there any outliers?

Solution

a First arrange the data in ascending order:

27° 27° 27° 27° 28° 28° 28° 28° 29° 29° 29° 30° 30° 30° 30° 30° 30° 30° 30°
30° 31° 31° 31° 31° 31° 31° 31° 32° 32° 32° 46°

Lowest entry = 27°, lower quartile is 8th entry = 28°,
median is 16th entry = 30°,
upper quartile is 24th entry = 31° and largest entry = 46°

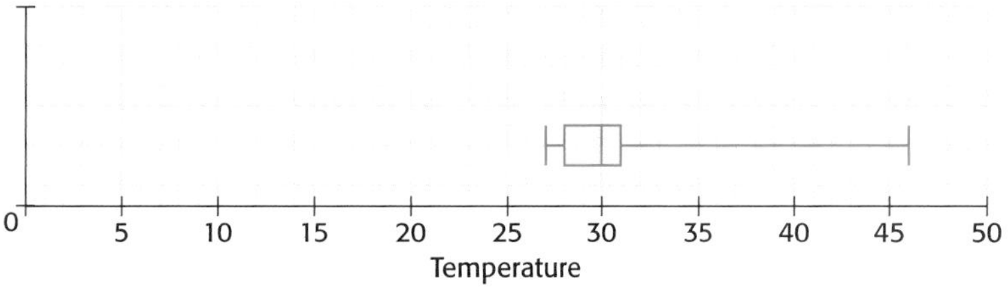

b Interquartile range = 31 – 28 = 3
$28 - 1.5 \times 3 = 23.5$
$31 + 1.5 \times 3 = 35.5$
Therefore 46° is an outlier as it is larger than 35.5.

"I'm a consultant they brought in to create some new buzzwords."

Theory of knowledge

"Stem and leaf diagram", "box and whisker plot".

Does the name given to a mathematical process make a difference to the result?

How do titles influence us in other areas of knowledge?

Exercise 6.4

1 This cumulative frequency graph shows the time taken for 200 students to complete a mathematical task.

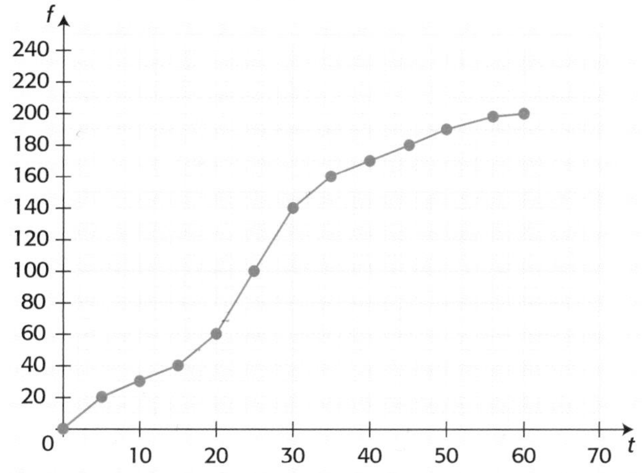

From the graph find **a** the median time

b the interquartile range

c the percentage of students who completed the task in under 22 minutes.

2 This cumulative frequency graph shows the heights, to the nearest cm, of 150 twelve-year-old children in Shanghai.

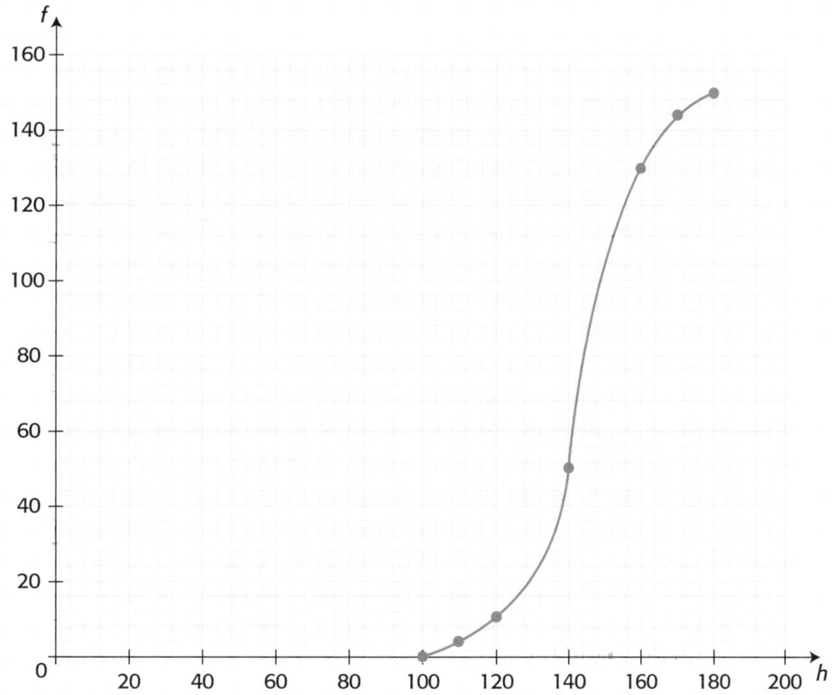

From the graph find **a** the median height

b the interquartile range

c the tenth percentile

d the percentage of children taller than 165 cm.

3 The box and whisker plot represents the number of attempts that it took a group of IB students to hit a bullseye in archery.

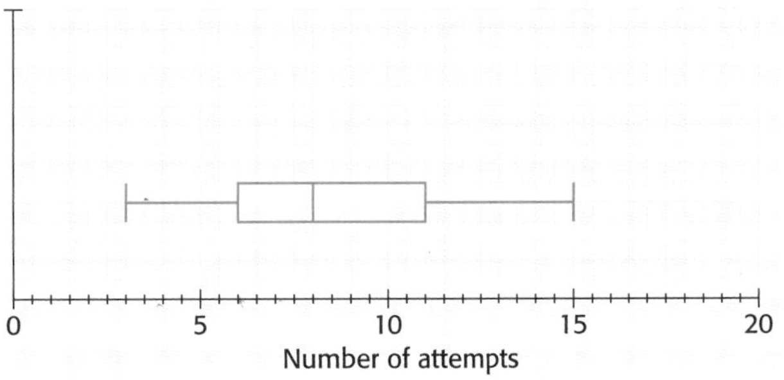

Write down
- a the lowest number of attempts
- b the median
- c the interquartile range
- d the largest number of attempts.

4 The table shows the number of piglets in a litter.

Number of piglets	Frequency
6	8
7	15
8	16
9	20
10	22
11	7
12	4
13	2
14	1

- a Find the lower quartile, the median and the upper quartile.
- b Draw a box and whisker plot to represent the information.
- c Are there any outliers?

5 A national park is open to visitors every day of the year. The number of visitors each day for a whole year was recorded and is shown in the table.

Number of visitors (n)	Frequency
$0 < n \leq 100$	55
$100 < n \leq 200$	36
$200 < n \leq 300$	58
$300 < n \leq 400$	98
$400 < n \leq 500$	52
$500 < n \leq 600$	30
$600 < n \leq 700$	21
$700 < n \leq 800$	15

- a Draw a cumulative frequency graph to represent this information.
- b Find the median and the interquartile range.
- c Find the percentage of days when there were more than 550 people visiting the park.

6 Hans studied an article in Die Welt. He recorded the number of words in each sentence as shown in the frequency table.

a Draw a cumulative frequency graph to represent this information.
b Write down the lower quartile, the median and the upper quartile.
c Given that the smallest sentence had one word and the largest had 43 words, draw a box and whisker plot.

Number of words	Frequency
1–5	6
6–10	28
11–15	36
16–20	14
21–25	10
26–30	3
31–35	1
36–40	0
41–45	2

7 The weights, in kilograms, of 200 students are given in the table.

Weight (w kg)	Frequency
$40 < w \leq 50$	38
$50 < w \leq 60$	63
$60 < w \leq 70$	41
$70 < w \leq 80$	20
$80 < w \leq 90$	12
$90 < w \leq 100$	13
$100 < w \leq 110$	8
$110 < w \leq 120$	5

a Draw a cumulative frequency diagram to represent this information.
b Write down
  i the median
  ii the interquartile range.
c Find the percentage of students who weigh less than 75 kg.
d Find the least weight of the heaviest 10% of the students.
e If the lightest student weighs 43 kg and the heaviest weighs 117 kg, then represent this information on a box and whisker plot.

8 The percentage scores for a biology test for 40 boys and 40 girls are given in the table.

Percentage score (s)	Boys (frequency)	Girls (frequency)
$10 \leq s < 20$	2	0
$20 \leq s < 30$	3	1
$30 \leq s < 40$	4	5
$40 \leq s < 50$	6	8
$50 \leq s < 60$	10	12
$60 \leq s < 70$	12	6
$70 \leq s < 80$	2	4
$80 \leq s < 90$	1	2
$90 \leq s < 100$	0	2

a Draw separate cumulative frequency diagrams for the boys and the girls.

b For each group write down

i the median ii the interquartile range.

c Given that the lowest grade for the boys was 18% and for the girls was 27% and that the highest grade for the boys was 84% and for the girls 96%, represent the information on box and whisker plots and comment on your diagrams.

6.5 Measures of central tendency for discrete and continuous data

6.5.1 Simple discrete data

Three measures of central tendency are the mode, the median and the mean.

Mode This is the value that is the *most frequent.*

Median This is the value that is in the *middle*.

Mean This is the *sum* of all the values divided by the *number* of values.

Remember: to find the median, write the values in ascending or descending order. (If you have n values, then the middle one is the ½ (n + 1)th entry.)

Example 6.5.1

Consider the numbers: 6 8 3 2 6 5 1 6 8.

Find the mode, median and mean.

Solution

Mode = 6, as this number occurs three times.

Put the numbers in order 1 2 3 5 6 6 6 8 8

There are nine numbers, therefore the $\frac{1}{2}(9 + 1)$th, that is the fifth number is the median.

Median = 6 as this is the fifth entry.

Mean = $(1 + 2 + 3 + 5 + 6 + 6 + 6 + 8 + 8) \div 9 = 45 \div 9 = 5$

Or we can use a GDC and the results are shown below.

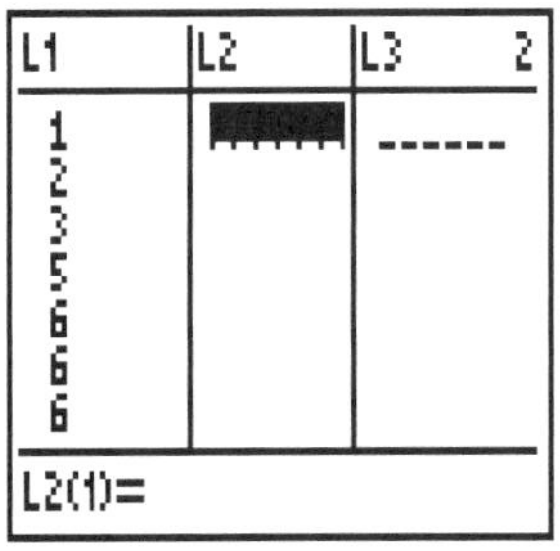

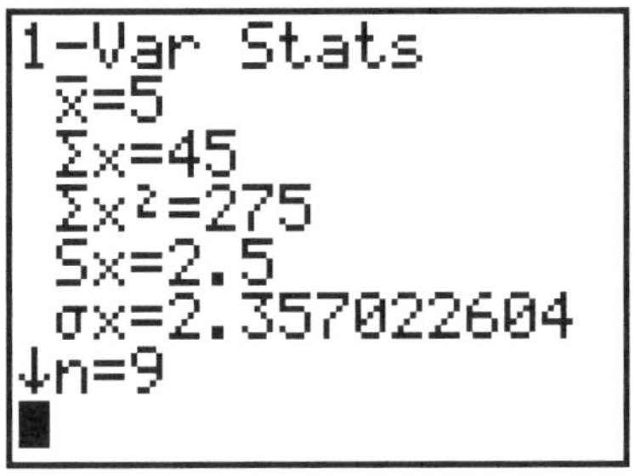

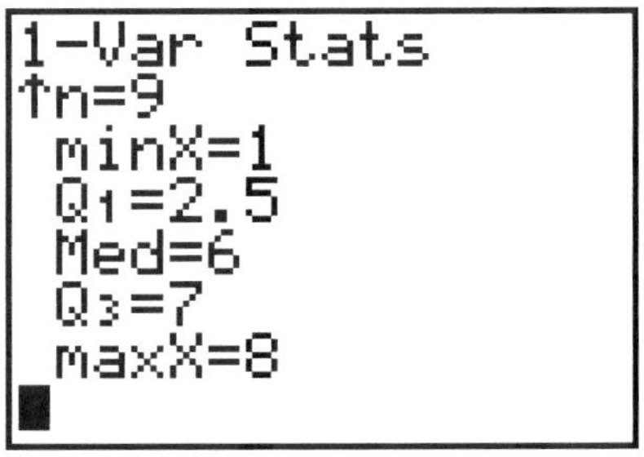

6.5.2 Sigma notation, Σ

Σ means "the sum of ".

For example, $\sum_{r=1}^{n} r$ means "the sum of all values of r from $r = 1$ up to and including $r = n$".

$$\sum_{r=1}^{n} r = 1 + 2 + 3 + \ldots\ldots + n$$

The values for r are *integer* values.

Example 6.5.2

Calculate $\sum_{r=1}^{4} r^2$

Solution

$$\sum_{r=1}^{4} r^2 = 1^2 + 2^2 + 3^2 + 4^2 = 1 + 4 + 9 + 16 = 30$$

6.5.3 Discrete data in a frequency table

To find the measures of central tendency for discrete data in a frequency table use these formulae.

Mode This is the value that has the *highest* frequency.

Median This is the *middle* value.

Mean This is found by first multiplying each value (x_i) by its corresponding frequency (f_i), then adding these products ($\sum(f_i x_i)$) and finally by dividing by the sum of the frequencies ($\sum(f_i)$).

Example 6.5.3

Calculate the mode, median and mean for these data.

x_i	f_i
1	4
2	9
3	23
4	25
5	16
6	3
Total	80

Solution

Mode = 4 because this value has the highest frequency.

Median = 4 It is the $\frac{1}{2}(80 + 1)$th = 40.5th value (between the 40th and 41st values). Both the 40th and 41st values are 4, therefore the median = 4.

The mean is found as follows:

x_i	f_i	x_if_i
1	4	4
2	9	18
3	23	69
4	25	100
5	16	80
6	3	18
Total	80	289

Mean = 289 ÷ 80 = 3.61

Or using the GDC

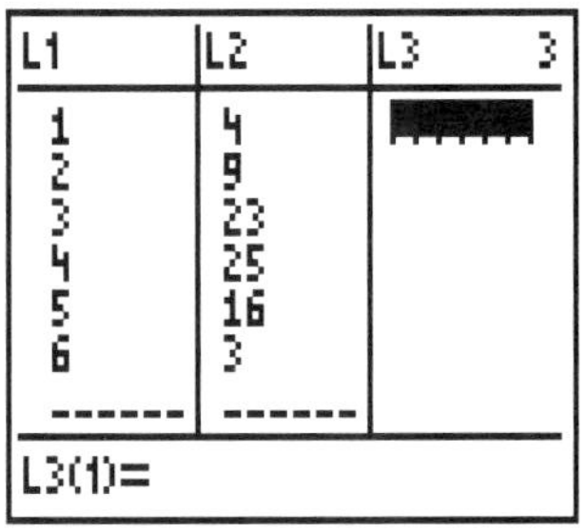

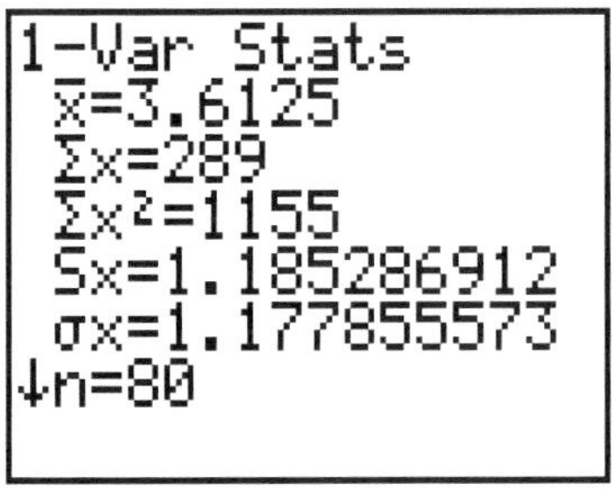

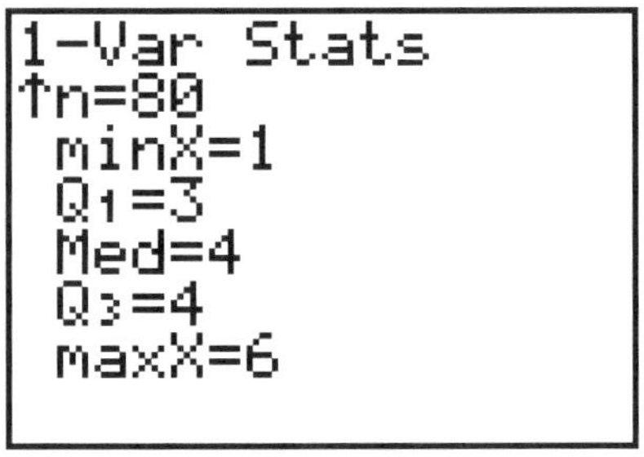

6.5.4 Grouped discrete or continuous data

When the data are grouped we can find the **modal group** (or modal class) and *estimated* values for the median and mean.

We cannot find *exact* values for the median and mean as we do not know the individual values of the data.

Modal group	This is the group or class interval that has the largest frequency.
Median	This is the *estimated* middle value.

The median can be found either from the cumulative frequency curve or from the table of values.

To find the median from a grouped frequency table follow these steps.

1. Identify the group the median is in.
2. Decide which entry in that group is the median value, a.
3. Find the total frequency of that group, b.
4. Find the width of that group, c.
5. Then use the formula:

   Median = (lower limit of group) + $(a \div b) \times c$

(This method is used in the next example.)

To find an *estimate* of the mean follow these steps.

1. Find the midpoint of each group, x_i.
2. Multiply each midpoint by its corresponding frequency, f_i.
3. Find the sum, $\Sigma x_i f_i$.
4. Divide this sum by the sum of the frequencies Σf_i.

   Mean = $\Sigma x_i f_i \div \Sigma f_i$

Example 6.5.4

The time, in minutes, taken to complete a roller blade marathon is given in the table.

Time (t minutes)	Frequency
$0 \le t < 50$	20
$50 \le t < 60$	61
$60 \le t < 70$	83
$70 \le t < 80$	90
$80 \le t < 90$	106
$90 \le t < 100$	62
$100 \le t < 110$	49
$110 \le t < 120$	29
Total	500

Calculate

a the modal group **b** an estimate of the median

c an estimate of the mean.

Solution

a The modal group = $80 \le t < 90$ as this group has the highest frequency.

b The median will be the time for the 250th roller blader.

Cumulative frequency of first three groups = 164 and 250 − 164 = 86

The median is in the group $70 \le t < 80$.

It is the 86th time in that group. So $a = 86$

The frequency of the group is 90, so $b = 90$

The width of the group is 10, so $c = 10$

So, median = 70 + 86 ÷ 90 × 10 = 79.6

c The mean is found as follows:

Midpoint, x_i	Frequency, f_i	$x_i f_i$
45	20	900
55	61	3355
65	83	5395
75	90	6750
85	106	9010
95	62	5890
105	49	5145
115	29	3335
Total	$\sum f_i = 500$	$\sum x_i f_i = 39\,780$

So, the mean = 39 780 ÷ 500

= 79.56

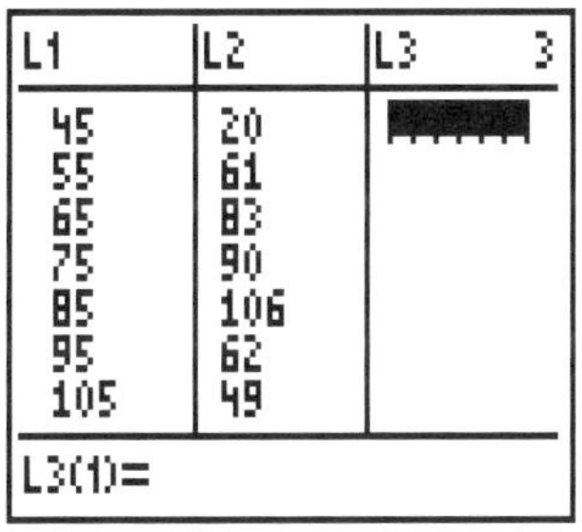

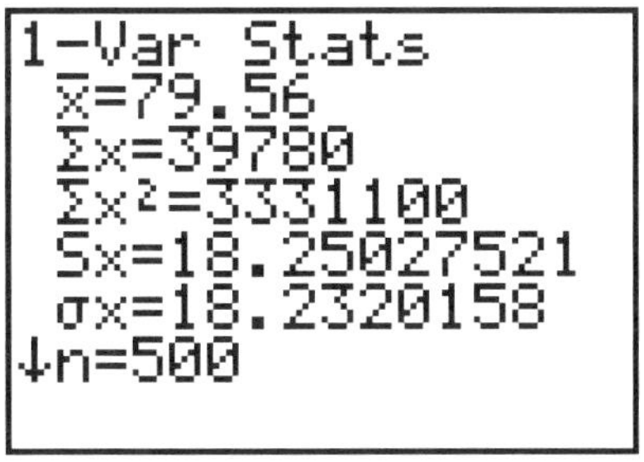

The GDC does *not* give us the correct value of the median for a grouped frequency table because we have entered the mid-values of each group into our list to find the mean. For a more accurate estimate of the median we would need to enter the end values of each group along with the corresponding frequencies.

Exercise 6.5

1 Calculate the mode, median and mean values of these data.

a 4 2 3 8 1 6 4 9 4 3 2

b

Number of piglets in litter	Frequency
8	8
9	16
10	21
11	33
12	40
13	19
14	7
15	6

2 The times taken for 30 camels to cover a distance of 10 km are given in the table.

Time taken (t min)	Frequency
$50 \le t < 55$	2
$55 \le t < 60$	8
$60 \le t < 65$	10
$65 \le t < 70$	3
$70 \le t < 75$	4
$75 \le t < 80$	2
$80 \le t < 85$	1

a Find the range of times for the modal group.

b Calculate an approximate value for the median and the mean times .

3 The mode of these data is 5, the median is 6 and the mean is 6.5.
Given that $s < t$, find the values of s and t.

1 1 2 3 s 5 5 7 8 9 10 t 12 12

4 The reaction times of 50 children for catching a ruler are shown in the table.

Time (s)	Frequency
$0 \le s < 1$	3
$1 \le s < 2$	10
$2 \le s < 3$	14
$3 \le s < 4$	15
$4 \le s < 5$	4
$5 \le s < 6$	2
$6 \le s < 7$	1
$7 \le s < 8$	1

a Write down the modal class.

b Find an approximation for the median and mean times.

5 The mean weight of the 21 students in grade 11 is 61 kg. When Martin joins the class the mean weight becomes 61.2 kg.
Calculate Martin's weight.

6 The mean weight of ten pineapples is 0.526 kg. Two more pineapples weighing 0.638 kg and 0.589 kg respectively are added to the other ten.
Calculate the mean weight of all twelve pineapples.

7 If 13 cats weigh an average of 5.6 kg and 12 dogs weigh an average of 12.2 kg, find the average weight of all 25 animals.

6.6 Measures of dispersion

6.6.1 Range

The **range** is found by subtracting the smallest value from the largest value.

Example 6.6.1

The numbers of eggs laid by ten chickens in one month are shown below. Find the range.

20 22 22 23 25 26 29 30 30 31

Solution

Range = 31 – 20 = 11

6.6.2 Interquartile range *(IQR)*

The **interquartile range** is found by subtracting the lower quartile, Q_1, from the upper quartile, Q_3.

$$IQR = Q_3 - Q_1$$

This value can be found from a cumulative frequency graph or algebraically.

Example 6.6.2a

Find the interquartile range of these numbers.

3 5 5 7 8 8 9 10 12 15 16

Solution

There are 11 numbers. Q_1 will be the $(11 + 1) \div 4$ = third number = 5

Q_3 will be the $3(11 + 1) \div 4$ = ninth number = 12

So the $IQR = 12 - 5 = 7$

Example 6.6.2b

Tom takes the lift up to the 12th floor every morning. He makes a note of the number of seconds that he waits for the lift each day for 100 days. The times are shown in the table. Draw a cumulative frequency curve and find the interquartile range.

Number of seconds, s	Frequency	Time less than	Cumulative frequency
$0 \le s < 15$	10	$s < 15$	10
$15 \le s < 30$	12	$s < 30$	22
$30 \le s < 45$	28	$s < 45$	50
$45 \le s < 60$	17	$s < 60$	67
$60 \le s < 75$	14	$s < 75$	81
$75 \le s < 90$	11	$s < 90$	92
$90 \le s < 105$	5	$s < 105$	97
$105 \le s < 120$	3	$s < 120$	100

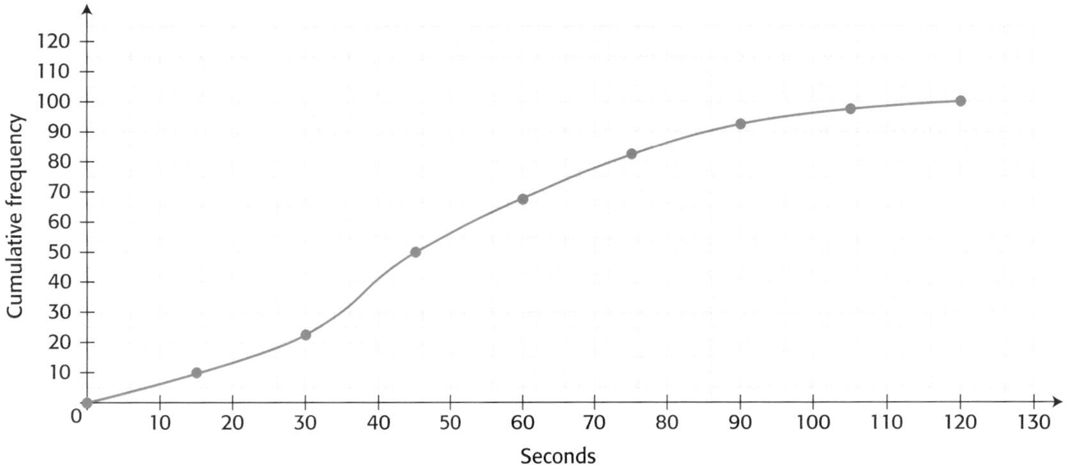

Solution

$Q_1 \approx 32 \qquad Q_3 \approx 69$

Therefore the $IQR = 69 - 32 = 37$

(Algebraically: $Q_1 \approx 30 + \frac{3}{28} \times 15 = 31.6,$

$Q_3 \approx 60 + \frac{8}{14} \times 15 = 68.6$

Therefore $IQR = 68.6 - 31.6 = 37$)

For grouped frequency tables, the GDC does *not* give us accurate values for Q_1 and Q_3 because we have entered the midpoint of each group in order to find the mean value.

6.6.3 Standard deviation

The standard deviation is a measure of dispersion that gives us an idea of how the values are related to the mean value. A small standard deviation implies that most values are close to the mean whereas a large standard deviation implies that the values have a large spread.

The standard deviation is often referred to as the "root-mean-squared-deviation" because we find the deviation of each entry from the mean, then we square these values, next we find the mean of the squared values and finally we take the square root of this answer.

"It's a new keyboard for the statistics lab. Once you learn how to use it, it will make the computation of the standard deviation easier."

Example 6.6.3a

Find the mean and standard deviation of these numbers.

3 4 4 7 10 11 12 12 13 15

Solution

Mean = (3 + 4 + 4 + 7 + 10 + 11 + 12 + 12 + 13 + 15) ÷ 10 = 91 ÷ 10 = 9.1

x_i	x_i – mean	$(x_i - \text{mean})^2$
3	−6.1	37.21
4	−5.1	26.01
4	−5.1	26.01
7	−2.1	4.41
10	0.9	0.81
11	1.9	3.61
12	2.9	8.41
12	2.9	8.41
13	3.9	15.21
15	5.9	34.81
Total		164.90

Standard deviation = $\sqrt{(164.9 \div 10)}$ = 4.06 (to 3 significant figures)

Or, using a GDC we get:

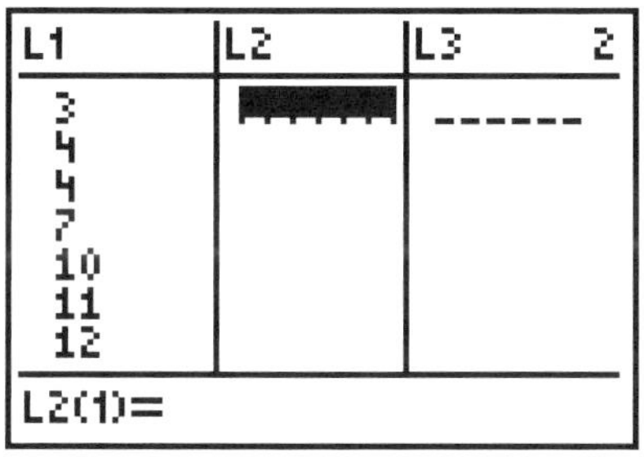

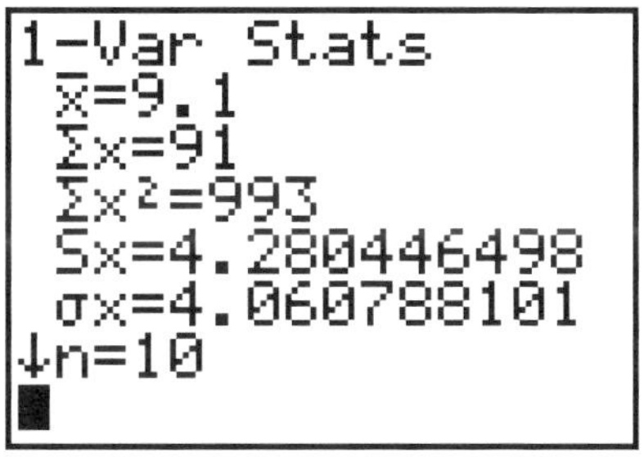

Example 6.6.3b
50 students (25 girls and 25 boys) were asked how many pairs of shoes they owned. The results are shown in the table.

Use your GDC to calculate the mean and standard deviation for the boys and girls separately and comment on your answer.

Boys	Number of pairs of shoes	Girls
0	4	2
0	5	4
6	6	3
13	7	6
5	8	3
1	9	2
0	10	2
0	11	2
0	12	1

Solution
Entering the data into a GDC we get these results.

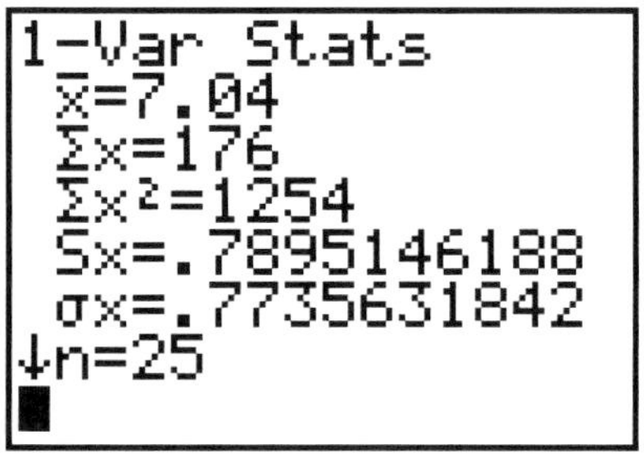

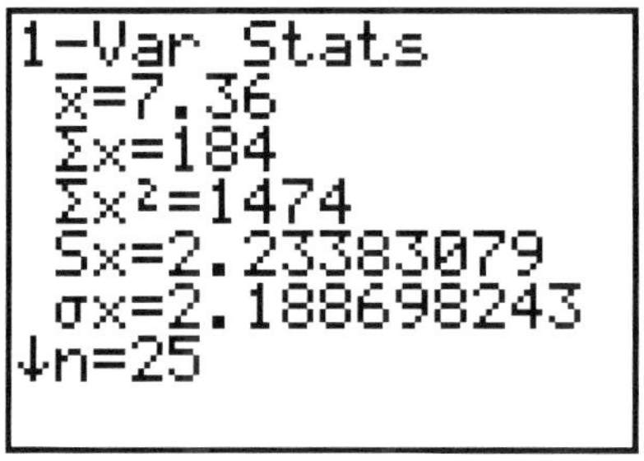

So, from the GDC: Boys' mean = 7.04 standard deviation = 0.774

Girls' mean = 7.36 standard deviation = 2.19

Both the boys and the girls have, on average, approximately 7 pairs of shoes. The standard deviation for the boys is small which implies that most boys have close to 7 pairs of shoes. However, the standard deviation for the girls is much larger which implies that some girls will have many fewer than 7 pairs of shoes and some will have many more.

It is often impossible to find the mean and standard deviation for a whole population. This could be due to time restrictions, financial or other reasons.

If we have, say, a random sample of 12 babies' heights from the UK, then the standard deviation of those 12 babies heights is σ_x on the TI GDC and s_n on the Casio. This is the one we use all the time in mathematical studies.

If we wanted to estimate the standard deviation of *all* the babies' heights in the UK, based on our random sample, then we would use s_x on the TI and s_{n-1} on the Casio.

Exercise 6.6

1 For these sets of data calculate
i the range
ii the interquartile range
iii the standard deviation.

a 4 2 3 8 1 6 4 9 4 3 2

b

Number of piglets in litter	Frequency
8	8
9	16
10	21
11	33
12	40
13	19
14	7
15	6

2 The times taken for 30 camels to cover a distance of 10 km are given in the table.

Time taken (t min)	Frequency
$50 \le t < 55$	2
$55 \le t < 60$	8
$60 \le t < 65$	10
$65 \le t < 70$	3
$70 \le t < 75$	4
$75 \le t < 80$	2
$80 \le t < 85$	1

a Draw a cumulative frequency graph to find the interquartile range of the times taken.

b Calculate an approximate value for the standard deviation of the times taken.

3 The reaction times of 50 children for catching a ruler are shown in the table.

Time (t seconds)	Frequency
$0 \le t < 1$	3
$1 \le t < 2$	10
$2 \le t < 3$	14
$3 \le t < 4$	15
$4 \le t < 5$	4
$5 \le t < 6$	2
$6 \le t < 7$	1
$7 \le t < 8$	1

a Draw a cumulative frequency graph to find the interquartile range.

b Find an approximation for the standard deviation of the time taken.

4 Calculate the mean and standard deviation for these data.

6 3 8 5 2 9 11 21 15 8

5 The number of telephone calls to a doctor's surgery was monitored every hour for a month. The data collected are shown in the table.

Number of calls per hour	**Frequency**
6	10
8	42
9	31
12	63
15	29
16	32
18	10
19	3
20	14
25	6

Calculate **a** the mean number of calls per hour

b the standard deviation.

6 The mean of these numbers is 33.

26 31 14 x 72 28 15

a Find the value of x.

b Calculate the standard deviation.

7 100 seedlings were measured and their heights are recorded in the table.

Height (cm)	**Frequency**
1–2	8
3–4	31
5–6	42
7–8	16
9–10	3

Calculate an approximation of the mean and the standard deviation of the heights.

8 The times taken for 30 students to complete this Sudoku puzzle are shown in the table.

1	6	4						2
2			4		3	9	1	
		5		8		4		7
	9				6	5		
5			1		2			8
		8	9				3	
8		9		4		2		
	7	3	5		9			1
4						6	7	9

Time (m minutes)	**Frequency**
$5 \le m < 10$	5
$10 \le m < 15$	3
$15 \le m < 20$	8
$20 \le m < 25$	7
$25 \le m < 30$	4
$30 \le m < 35$	1
$35 \le m < 40$	2

Find approximate values for the mean and standard deviation.

9 The percentage marks obtained in a mathematics test by the 30 boys and 30 girls at Fun Academy are shown in the table.

Girls' frequency	Percentage mark	Boys' frequency
0	0–10	3
0	11–20	1
0	21–30	2
4	31–40	1
5	41–50	5
8	51–60	10
10	61–70	2
3	71–80	1
0	81–90	3
0	91–100	2

a Calculate an estimated value for the mean and standard deviation for the girls and the boys separately.

b Comment on your answers.

6.7 Scatter diagrams and linear correlation

6.7.1 Scatter diagrams

Whenever we have two sets of data that we suspect may be related (that is, one set of data is dependent on the other), then there are various methods that we can use to check whether or not there is any correlation between the two sets. One of these is a scatter graph. We can plot the data on a scatter diagram with the independent variable on the horizontal axis and the dependent variable on the vertical axis. The pattern of dots will give us a visual picture of how closely, if at all, the variables are related. Some examples of variables that may be related are

- the number of road accidents and age of driver
- the time taken to complete a task and the training received
- the total score for the IB diploma and the number of hours of study
- the diameter of the trunk of a tree and the age of the tree
- the weight of a polar bear and its height.

Theory of knowledge

How do we know when two things are related? Can two sets of data have a very strong correlation and yet not be related? Try to think of your own examples. Can you find any examples in advertising?

Types of correlation

We can have a **positive correlation**. This implies that the dependent variable increases as the independent variable increases.

We can have a **negative correlation**. This implies that the dependent variable decreases as the independent variable increases.

We can have **no** correlation.

We can also have a *strong, moderate* or *weak* correlation.

Examples:

Positive correlation

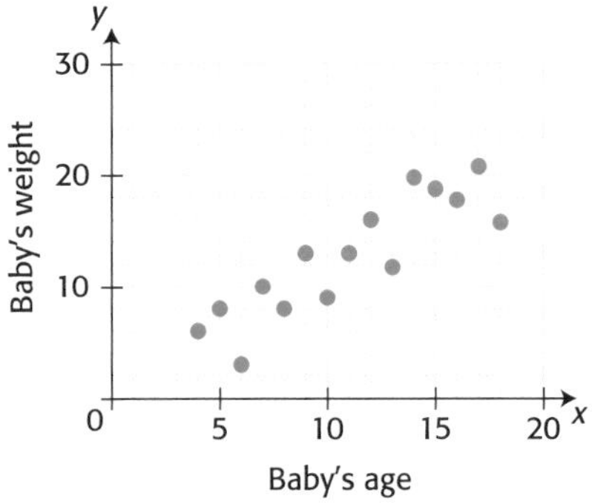

Negative correlation

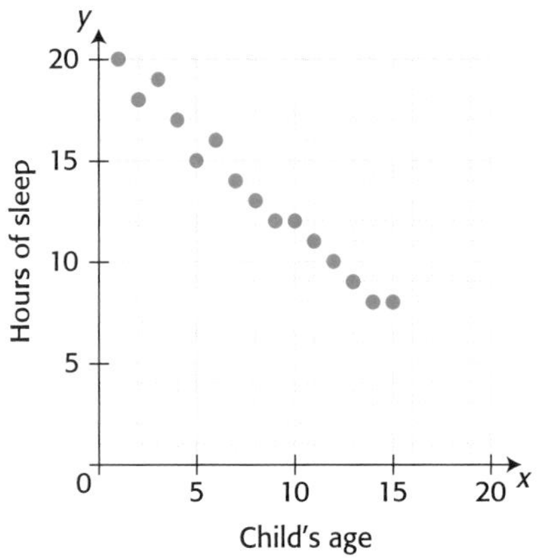

No correlation

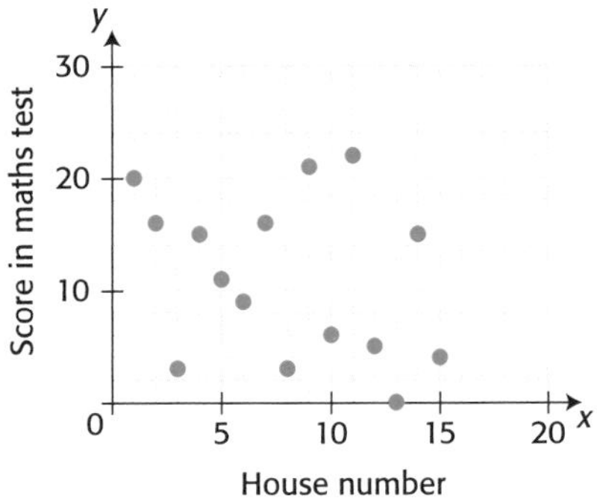

This diagram shows a *very strong positive* correlation.

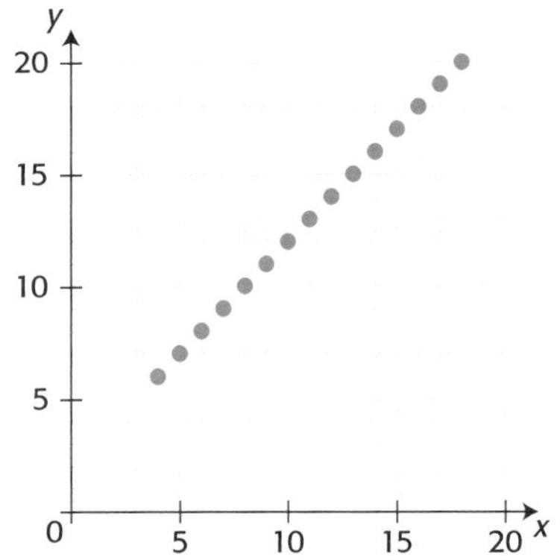

This diagram shows a *moderate positive* correlation.

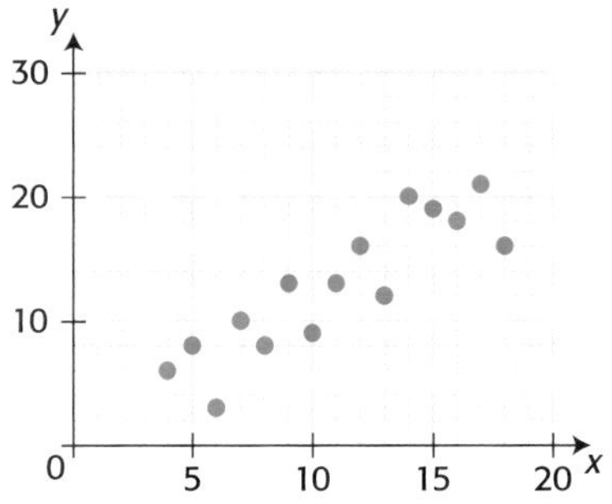

This diagram shows a *strong negative* correlation.

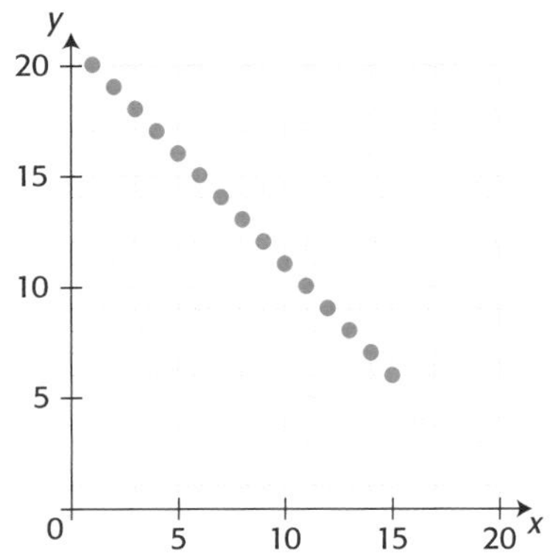

This diagram shows a *weak negative* correlation.

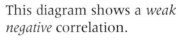

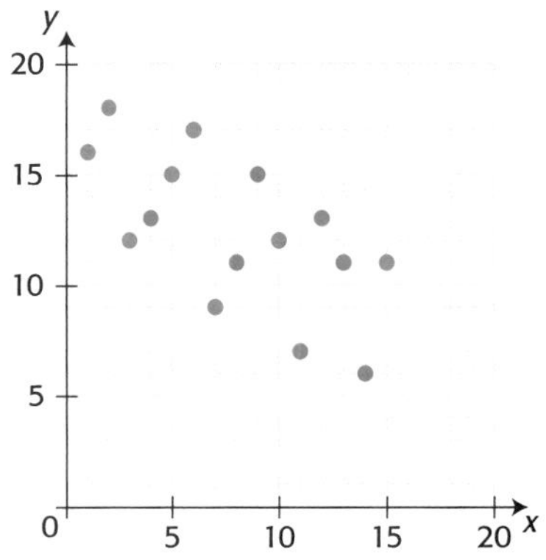

We can also have *linear* or *non-linear* correlations.

For example

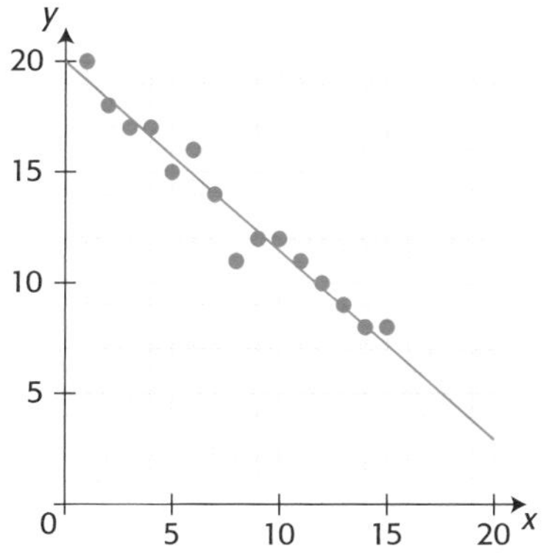

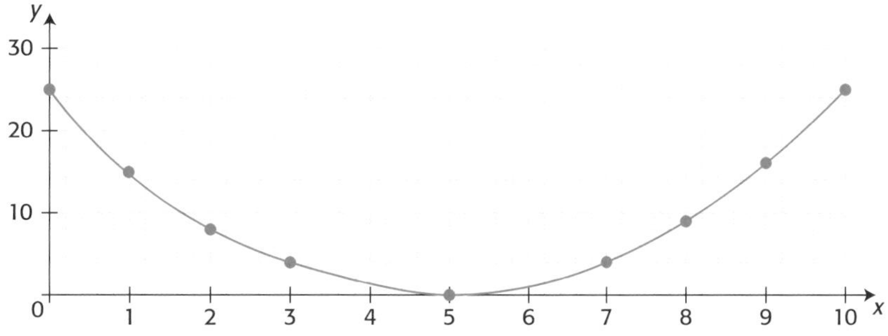

In the mathematical studies course, you will only need to learn about *linear* correlations, but you might want to use other types of correlations in your project.

Example 6.7.1a

It is thought that the number of observed specimens of jellyfish in an atoll depend on the temperature. A record of the temperature and the number of jellyfish was kept over a two-week period.

Plot these points on a scatter diagram and comment on the type of correlation.

Temperature (°C)	**Number of jellyfish**
20°	135
21°	138
22°	150
19°	135
24°	162
26°	201
31°	263
27°	221
24°	168
21°	155
21°	149
20°	152
21°	148
19°	124

Solution

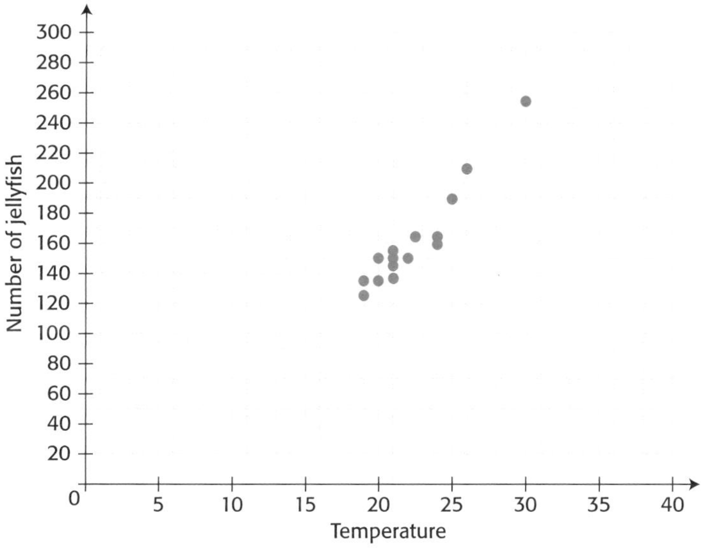

As we can see from the points plotted, there is a moderately strong linear correlation between the temperature and the number of jellyfish.

Example 6.7.1b

A mathematics teacher wanted to check if there was a correlation between her predicted grades and the actual grades of her mathematical studies class.

Draw a scatter diagram to represent the data and comment on the correlation.

Predicted grade	5	6	4	5	4	3	7	7	6	4	4	2	5	7
Obtained grade	4	7	2	3	5	5	7	6	7	5	3	4	3	5

Solution

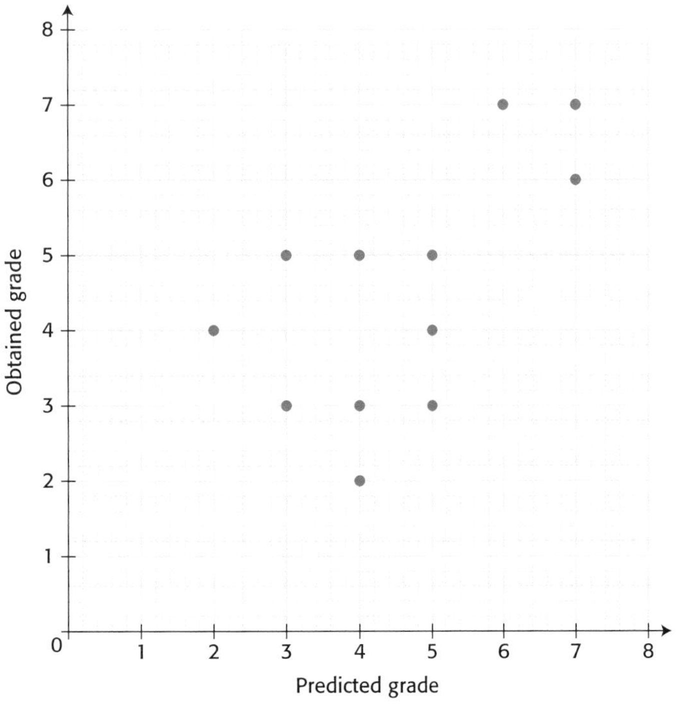

Here there is only a weak positive correlation.

6.7.2 Line of best fit, by eye, passing through the mean point

First of all we must find the *mean* of both sets of data and plot this point on the scatter diagram. Then we draw a line that passes through this plotted point and is close to all the other points with about an equal number of points above and below the line.

The line of best fit does not need to go through the origin and in most cases it will not.

Example 6.7.2

Consider the previous two examples. We will draw in the line of best fit by eye.

a The mean temperature is 22.6°, the mean number of jellyfish is 164. Plot this point on the scatter diagram and draw a best fit line through it.

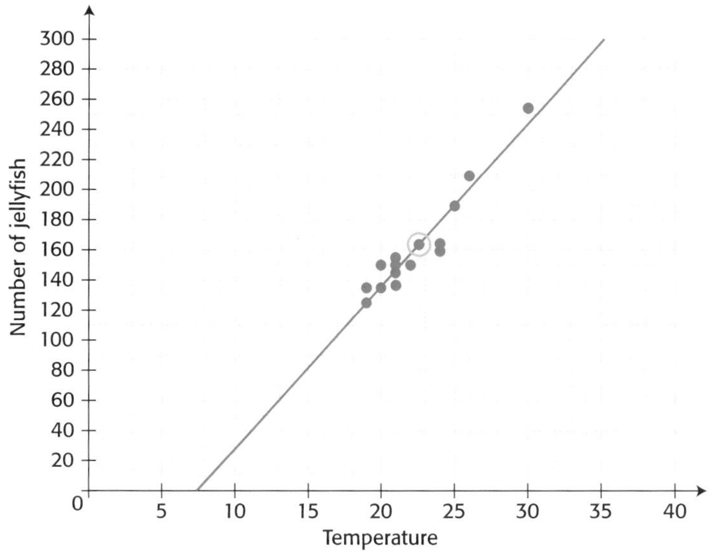

b The mean predicted grade is 4.93, the mean obtained grade is 4.71. Plot this point on the scatter diagram and draw a straight line through it.

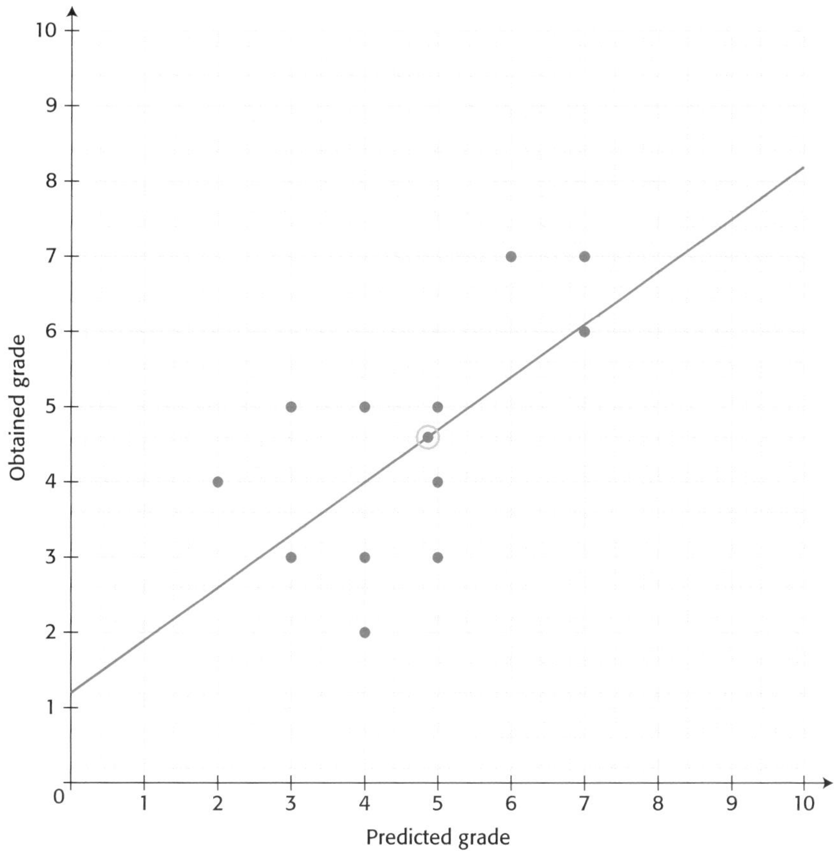

Exercise 6.7a

1 Consider these sets of data.

For each set **i** plot the points on a scatter diagram and discuss the type of correlation
ii find the mean of x and y
iii plot this point on your diagram and draw the line of best fit by eye.

a

x	18	20	23	25	29	33	35	38	39	42	45	50
y	48	46	43	40	35	34	32	28	26	23	21	15

b

x	2	8	8	12	15	16	16	17	19	20
y	10	1	3	15	16	8	1	19	7	3

c

x	2	4	6	8	10	12	14	16	18	20	22	24
y	16	17	20	23	24	25	25	29	33	32	35	39

d

x	15	16	17	18	19	20	21	22	23	24
y	1	5	2	8	12	12	11	14	20	21

e

x	1	2	3	4	5	6	7	8	9	10	11	12
y	22	20	18	16	17	15	13	13	11	9	10	8

2 This table gives the heights and weights of 16 randomly chosen students.

Height, x cm	148	151	152	158	159	160	160	163	168	172	173	175	181	183	185	185
Weight, y kg	42	48	53	54	62	61	84	63	60	76	92	70	73	85	79	91

a Plot the points on a scatter diagram and discuss the correlation.
b Find the mean of the heights and weights.
c Plot this point on your diagram and draw the line of best fit by eye.

3 Twenty couples were interviewed at random and asked their ages.
The table shows the results.

Age (male)	19	25	26	31	31	33	34	37	39	39	43	45	48	52	58	59	62	65	70	78
Age (female)	18	24	27	30	31	32	35	36	39	36	41	43	50	51	55	60	59	63	67	75

a Plot the points on a scatter diagram and discuss the correlation.
b Find the mean of the ages.
c Plot this point on your diagram and draw the line of best fit by eye.

4 The temperature of a cup of green tea was taken every minute and the results are shown in the table.

Time	0	1	2	3	4	5	6	7	8	9	10
Temperature (°C)	100	90	82	75	68	60	55	49	44	39	35

a Plot the points on a scatter diagram and discuss the correlation.
b Find the mean of the time and temperature.
c Plot this point on your diagram and draw the line of best fit by eye.

5 The table shows the weights of babies and the weights of their fathers.

Weight of father, kg	58	65	69	72	74	76	82	85	85	92
Weight of baby, kg	2.2	3.1	3.7	3.6	3.8	3.9	3.1	4.0	4.1	5.2

a Plot the points on a scatter diagram and discuss the correlation.
b Find the mean weights of the fathers and the babies.
c Plot this point on your diagram and draw the line of best fit by eye,

6.7.3 Bivariate data: the concept of correlation

Bivariate data is a set of data where there are two variables for each observation, for example, height and weight, distance and time.

6.7.4 Pearson's product-moment correlation coefficient; use of the formula $r = \dfrac{s_{xy}}{s_x s_y}$

Karl Pearson **(1857–1936) Biometrician, statistician and applied mathematician**. Karl Pearson read mathematics at Cambridge but made his career at University College, London. Pearson was an established applied mathematician when he joined the zoologist W. F. R. Weldon and launched what became known as biometry; this found institutional expression in 1901 with the journal *Biometrika*. Weldon had come to the view that "the problem of animal evolution is essentially a statistical problem" and was applying Galton's statistical methods. Pearson's contribution consisted of new techniques and eventually a new theory of statistics based on the Pearson curves, correlation, the method of moments and the chi-square test. Pearson was eager that his statistical approach be adopted in other fields and amongst his followers was the medical statistician Major Greenwood. Pearson created a very powerful school and for decades his department was considered the only place to learn statistics. Yule, Wishart and F. N. David were among the distinguished statisticians who started their careers working for Pearson. Amongst those who attended his lectures were the biologist Raymond Pearl, the economist H. L. Moore, the medical statistician, Austin Bradford Hill, and Jerzy Neyman; in the 1930s Wilks was a visitor to the department. Pearson's influence extended to Russia where Slutsky (see minimum chi-squared method) and Chuprov were interested in his work. Pearson had a great influence on the language and notation of statistics and his name often appears on the *Words* pages and *Symbols* pages—see, for example, population, histogram and standard deviation.

It is useful to know the strength of the relationship between any two sets of data that are thought to be related.

Pearson's product-moment correlation coefficient, r, is one way to find a numerical value that can be used to determine the strength of a linear correlation between the two sets of data.

The formula for Pearson's product-moment correlation for two sets of data x and y is

$r = \frac{s_{xy}}{s_x s_y}$ where s_{xy} is the **covariance** and s_x and s_y are the standard deviations of x and y respectively.

$$s_{xy} = \sum \frac{(x-\bar{x})(y-\bar{y})}{n} \quad \text{or} \quad \frac{\sum xy}{n} - \frac{\sum x}{n}\frac{\sum y}{n}$$

$$s_x = \sqrt{\frac{\sum (x-\bar{x})^2}{n}} \quad \text{or} \quad \sqrt{\left(\frac{\sum x^2}{n} - \bar{x}^2\right)}$$

$$s_y = \sqrt{\frac{\sum (y-\bar{y})^2}{n}} \quad \text{or} \quad \sqrt{\left(\frac{\sum y^2}{n} - \bar{y}^2\right)}$$

The covariance, s_{xy}, will always be given in exam questions where students are expected to use the formula to calculate the value of r. In other cases the students will be expected to use a GDC to find the value for r.

6.7.5 Interpretation of positive, zero and negative correlations

r can take all values between −1 and +1 inclusive.
$r = -1$ indicates a *perfect negative linear* correlation
$r = 0$ indicates *no linear* correlation and
$r = +1$ indicates a *perfect positive linear* correlation.
Interpretations of r for other values are roughly as follows:

Strong negative	Moderate negative	Weak negative	Very weak negative	Very weak positive	Weak positive	Moderate positive	Strong positive

−1   −0.75   −0.5   −0.25   0   0.25   0.5   0.75   1

The degree of correlation will depend on the number of pairs in the sample. The more pairs there are the lower the correlation coefficient has to be for a good fit. Another method to use would be the p-value on the GDC. A very small p-value (see page 330) means a strong correlation.

On your GDC you will also be given r^2, the coefficient of determination. This is an indication of how the variation in one set of data, y, can be explained by the variation in the other set of data, x. For example, if $r^2 = 0.866$, this means that 86.6% of the variation in set y is caused by the variation in set x. For this value of r^2, $r = 0.930$ which indicates a strong positive linear relationship.

Example 6.7.5a

These data are for the Dutch football first division and they show the position of the team and the number of goals scored. Given that $s_{xy} = -74.4$, calculate the correlation coefficient, r, and comment on this value.

Position	1	2	3	4	5	6	7	8	9	10	11	12	13	14	15	16	17	18
Goals	69	62	58	49	44	43	43	41	39	38	35	33	28	27	27	24	20	8

Solution

The covariance has been given so you can use either the formula for r or your GDC.

$s_{xy} = -74.4$

$s_x = 5.19$

$s_y = 14.8$

$r = s_{xy} \div (s_x s_y) = -0.969$

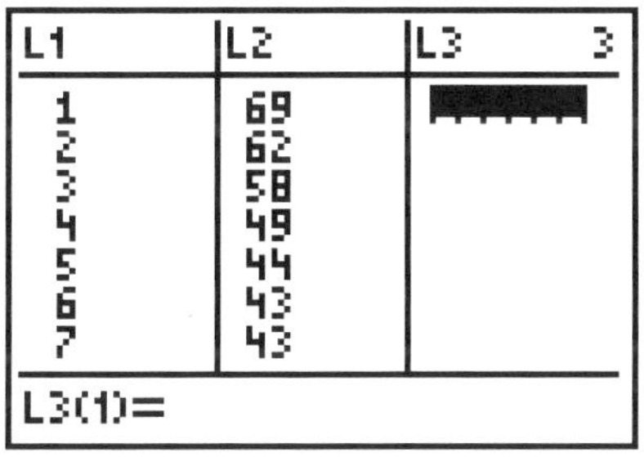

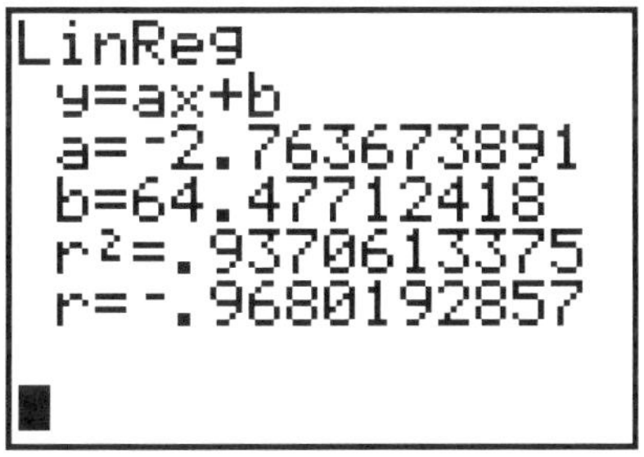

GDC value for $r = -0.968$

Comment: There is a very strong, negative linear correlation.

Example 6.7.5b

For the data in the table calculate the correlation coefficient, r, and comment on your result.

Height, *x* cm	148	151	152	158	159	160	160	163	168	172	175	183
Weight, *y* kg	42	48	53	54	62	61	63	65	59	70	73	75

Solution

Since the covariance has not been given, we must use a GDC to find the value of r.

From the GDC we get:

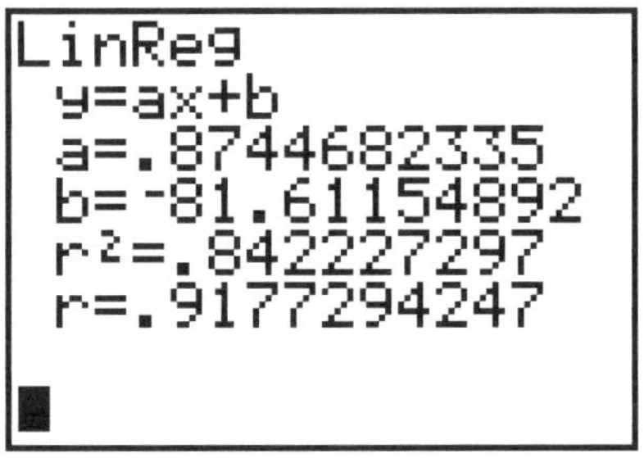

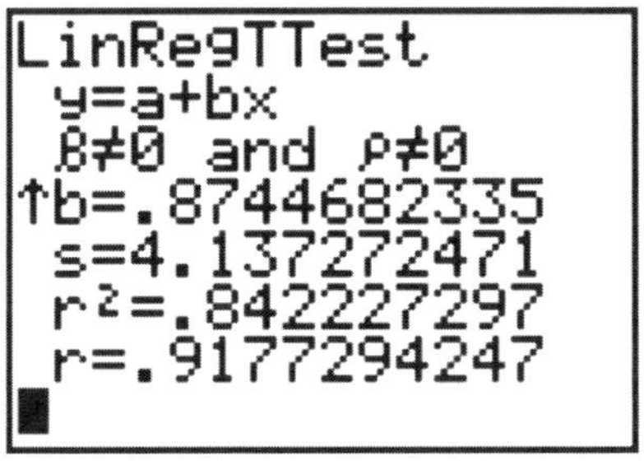

$r = 0.918$

This means that there is a very strong, positive linear relation between the two sets of data.

Or we could use the linear regression t-test on the GDC with the following result:

A t-test is another way to find a simple linear regression.

The value for r is obviously the same as before, but here we are also given the p-value. This is 0.000 025 8 which also indicates a very strong and positive correlation.

Example 6.7.5c

Ivan thought that there would be a strong correlation between the distance his classmates lived from the school and the time it took them to travel to school. After he gathered the information, Ivan calculated that the covariance was 20.3, the standard deviation of the time was 8.12 minutes and the standard deviation of the distance was 6.83 km.

Find the correlation coefficient, r, and comment on the result.

Solution

$$r = \frac{s_{xy}}{s_x s_y} = 20.3 \div (8.12 \times 6.83) = 0.366$$

There is only a weak, positive linear relationship between the distance travelled and the time taken.

Exercise 6.7b

1 State the type of relationship (positive / negative / linear / non-linear) and the strength of the correlation (perfect / strong / moderate / weak / none) shown in these diagrams.

a

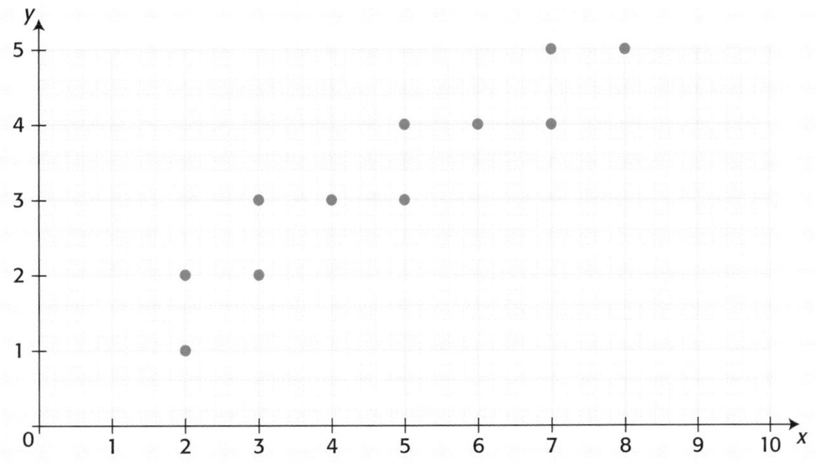

b

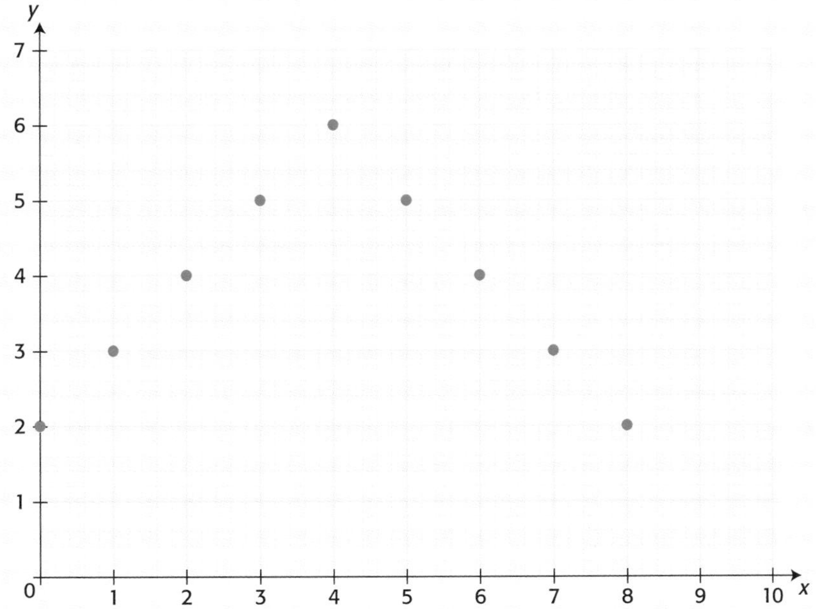

c

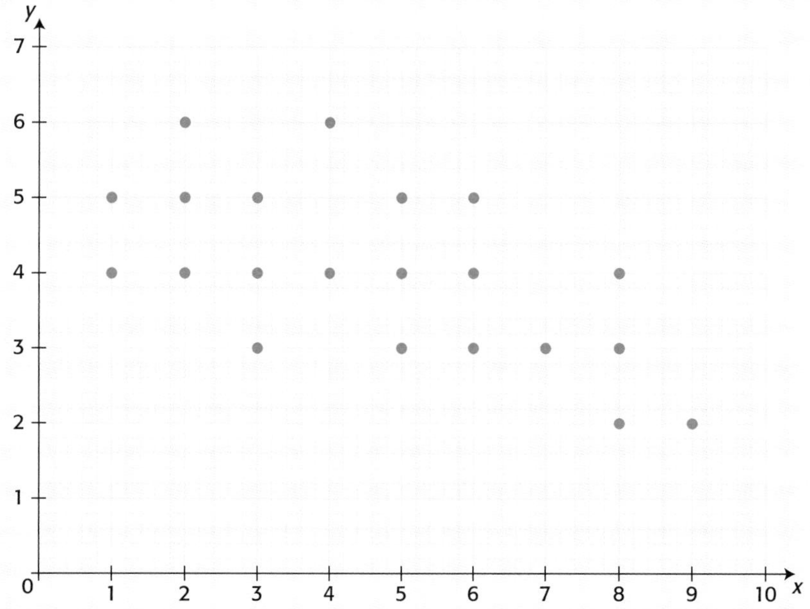

d

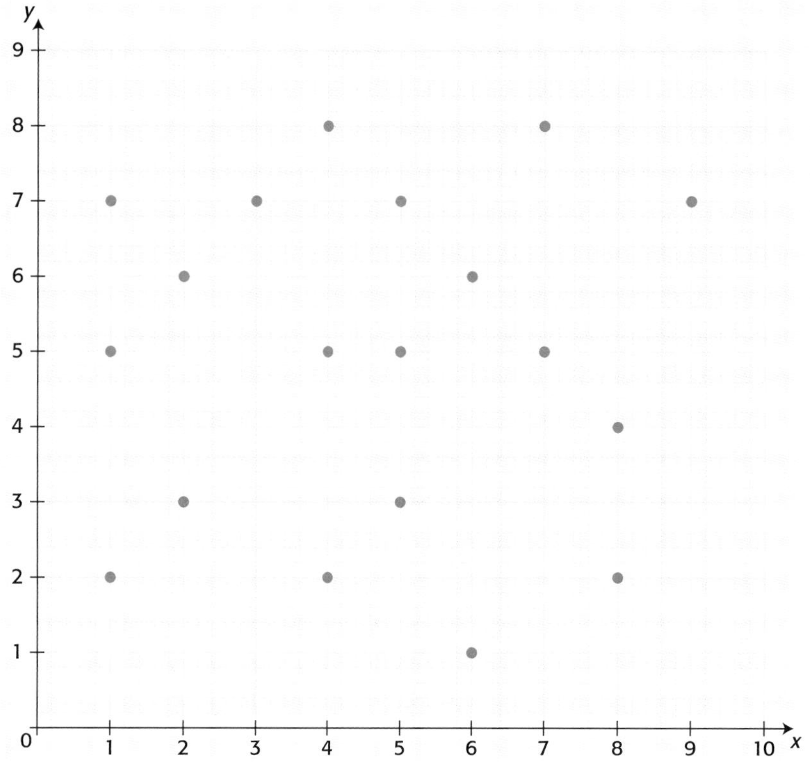

e

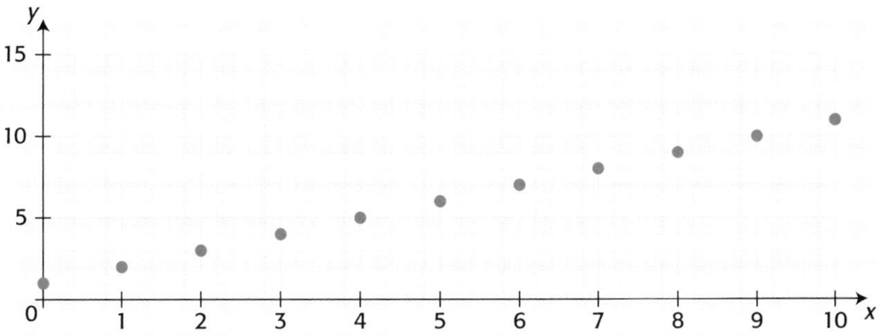
y
15
10
5
0
1
2
3
4
5
6
7
8
9
10
x

f

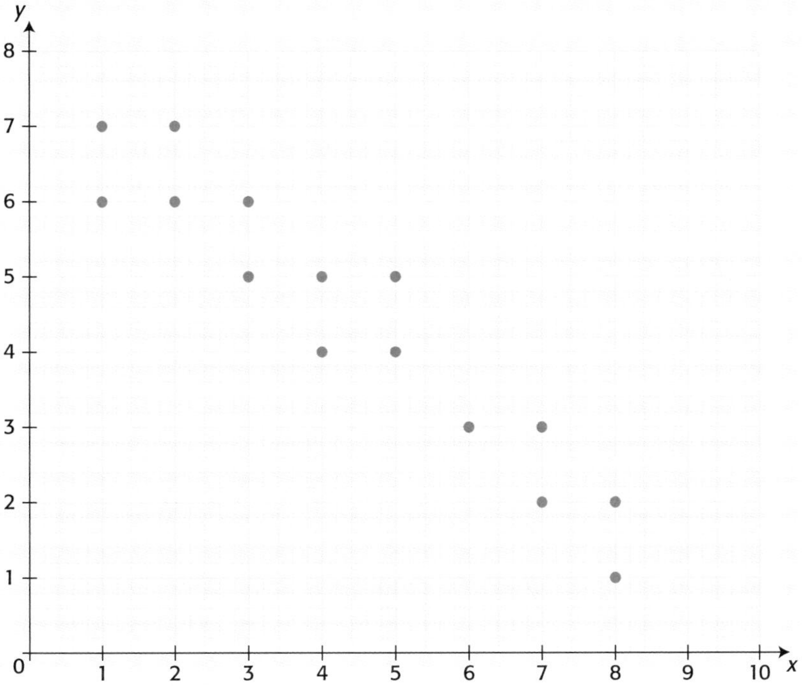
y
8
7
6
5
4
3
2
1
0
1
2
3
4
5
6
7
8
9
10
x

g

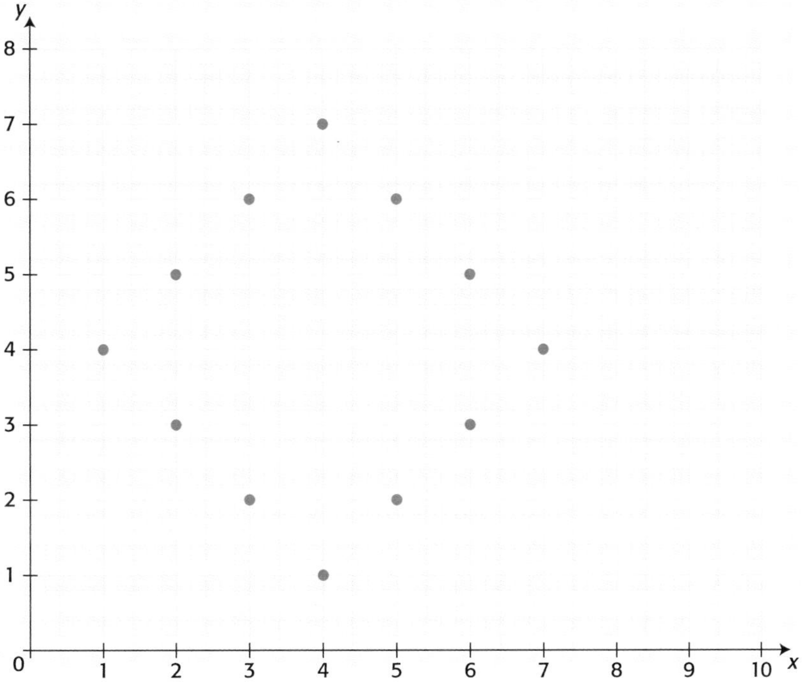
y
8
7
6
5
4
3
2
1
0
1
2
3
4
5
6
7
8
9
10
x

2 The table shows the temperature taken at various heights up a mountain. Calculate the correlation coefficient, *r*, and comment on your result.

Height, *h* metres	5000	10000	15000	20000	25000	30000	35000
Temperature, °C	5	−35	−48	−52	−55	−53	−60

3 The table shows the number of coins tossed and the number of heads obtained. Calculate the correlation coefficient, *r*, and comment on the result.

Number of coins	10	15	20	25	30	35	40	45	50	55	60	65	70	75
Number of heads	6	8	9	11	18	19	20	20	23	28	34	33	34	42

4 The GNP (gross national product) and life expectancy for several countries is shown in the table.

GNP $ per capita	41210	39640	24990	18700	11940	4820	1510	800	350	240
Life expectancy	77	80	78	77	72	71	68	60	52	41

Calculate the correlation coefficient, *r*, and comment on the result.

5 The percentage scores for a physics and mathematics test for each of the 16 students in grade 11 at Fun Academy are shown below.

Physics	23	34	39	43	47	52	58	67	73	75	79	82	86	88	91	97
Maths	31	33	42	55	48	51	60	71	72	69	83	84	89	96	92	89

Calculate the correlation coefficient, *r*, and comment on your result.

6 The average daily temperature in Malaga (Spain) in July was 35° with a standard deviation of 3°. An ice cream salesman calculated that, on average, he sold 800 ice creams a day with a standard deviation of 35. If the covariance for the temperature and the number of ice creams sold is 96, calculate the correlation coefficient, *r*, and comment on the result.

7 The distance travelled by bus between various places in Australia and the cost of the journey is given in the table.

Distance, km	10	20	25	40	50	70	75	80	95	120
Cost, AU$	5.00	7.80	8.00	9.20	11.00	13.00	13.50	13.80	15.00	16.80

Calculate the correlation coefficient, *r*, and comment on your result.

8 Twelve students were given 20 words in a different language to learn. They were allowed different lengths of time to study the words and the number of errors that they made is shown in the table.

Study time, min	1	2	3	4	5	6	7	8	9	10	11	12
Errors made	19	17	17	15	12	13	11	9	6	5	5	3

Calculate the correlation coefficient and comment on your result.

9 The mean height of the members of a team of basketball players is 205 cm with a standard deviation of 8.23 cm. The mean length of their hands is 20.8 cm with a standard deviation of 1.63 cm. Given that the covariance of their heights and hand lengths is 3.58, calculate the correlation coefficient, *r*, and comment on your result.

10 The ages of 15 people and the times it took them to complete a training circuit are shown in the table.

Age, years	15	16	18	19	20	23	25	28	32	35	40	47	51	55	62
Time, min	25	20	18	15	14	15	14	14	13	16	21	29	31	33	35

Given that the covariance is 85.0, calculate the standard deviation of the ages and the time taken and use the formula to find the correlation coefficient, *r*, and comment on your result.
Check your answer using your GDC.

6.8 The linear regression formula

The regression line for y on x, where y is the dependent variable, is also known as the least squares regression line. It is the line drawn through a set of points such that the sum of the squares of the distance of each point from the line is a minimum. The regression line is more accurate than the line of best fit by eye.

If there is a reasonably strong correlation, you can use the regression line of y on x to predict values of y for various values of x.

Do not use the regression line for prediction purposes for values beyond the region of the given data.

The formula for the regression line of y on x is

$$(y - \bar{y}) = \frac{s_{xy}}{(s_x)^2}(x - \bar{x})$$

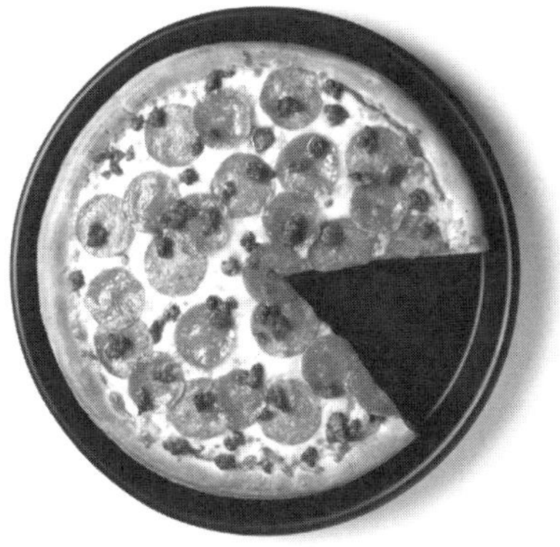

where $\bar{x}$ and $\bar{y}$ are the means of x and y, s_x is the standard deviation of x and s_{xy} is the covariance.

In exam questions you will always be given the value of s_{xy}.

Example 6.8a

Mr Lori was trying out a new recipe for pizza. The data in the table shows how popular the new pizza was.

Day number (x)	1	3	5	7	9	11	13	15	17	19
Number of pizzas sold (y)	10	15	19	24	30	37	43	45	53	60

a Write down
  i the correlation coefficient, r
  ii the regression line of y on x.

b Use your line to predict the number of pizzas sold on day 8 and on day 14.

c Can you predict how many pizzas will be sold on day 30?

Solution

Entering the data into List 1 and List 2 we find the following from the GDC.

```
LinReg
 y=ax+b
 a=2.757575758
 b=6.024242424
 r²=.9940555932
 r=.9970233664
```

a i From the GDC, $r = 0.997$

ii From the GDC, the equation of the regression line is $y = 2.76x + 6.02$

b On day 8, $y = 2.76(8) + 6.02 = 28.1$. Therefore approximately 28 pizzas are sold.

On day 14, $y = 2.76(14) + 6.02 = 44.7$. Therefore approximately 45 pizzas are sold.

c No, you cannot because 30 lies outside the range of data that you used to calculate the regression line.

Example 6.8b

Dave thought that there was a strong correlation between a child's IQ score and SAT score. The IQ scores of some children were

78 89 92 96 99 100 102 111 113 128 135

The SAT is a standardized test for US college entrance.

a Calculate

i the mean IQ

ii the standard deviation of the IQ.

b Given that their mean SAT score was 896 with standard deviation 174 and that $s_{xy} = 2740$, find the regression line of y on x.

c Use your regression line to estimate the SAT score of a child with an IQ score of

i 84 ii 123

Solution

Entering the data into List 1 and List 2 we find the following from the GDC.

```
1-Var Stats
 x̄=103.9090909
 Σx=1143
 Σx²=121589
 Sx=16.79556218
 σx=16.01394021
↓n=11
```

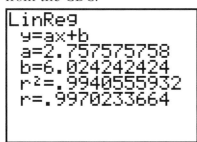

a i From the GDC, the mean IQ = 104

ii From the GDC, the standard deviation = 16.0

b $(y - 896) = 2740/16^2\ (x - 104)$

$y = 10.7x - 217$

c i $y = 10.7(84) - 217 = 681$

ii $y = 10.7(123) - 217 = 1099$

Exercise 6.8

1 Barrels of liquid detergent are kept in storage drums for some time before being bottled for sale. During this time there is some evaporation of the water content. An examination of a drum provided these results.

Storage time (weeks)	3	5	6	9	11	13
Evaporation loss (ml)	41	57	61	73	80	91

a Using your GDC find the correlation coefficient, r.

b Using your GDC find the equation of the regression line of y on x.

c Use your line to predict the evaporation loss after 10 weeks.

d Can you predict the evaporation loss after 20 weeks?

2 Each day, a factory records the number of items it produces and the total production cost in euros.
The results over a two-week period are shown in the table.

Items	25	32	35	45	21	39	56	43	51	48	29	33	41	47
Cost (€)	350	423	439	536	302	461	602	522	570	555	387	412	502	546

a Find the correlation coefficient, r.

b Find the equation of the regression line.

c Estimate the cost of producing 40 items.

3 Ten members of a sports club take part in a 100 metre hurdle race. The table shows the average number of hours per week that each member trains and the time taken to complete the race.

Training (hours)	12	6	9	3	21	8	10	5	6	25
Time (seconds)	15.3	16.2	15.9	18.4	14.2	14.8	14.1	16.1	16	13.8

a Find the correlation coefficient, r .

b Find the equation of the regression line.

c Estimate the time taken for a member who trains 18 hours a week.

4 The average number of hours of study per week by some students and their total score in the IB diploma are shown in the table.

Hours study	14	18	19	21	24	8	6	15	28	20	25	10
Total score	29	35	36	40	43	22	19	32	45	38	44	24

a Find the correlation coefficient, r.

b Find the equation of the regression line.

c Estimate the score for a student who studies 22 hours a week.

5 The weights and heights of 15 polar bears are given in the table.

Weight (kg)	330	346	362	384	391	432	457	466	482	543	569	587	653	689	745
Height (m)	1.54	1.58	1.63	1.78	1.98	2.11	2.17	2.24	2.31	2.39	2.43	2.48	2.52	2.57	2.74

a Find the correlation coefficient, r.
b Find the equation of the regression line.
c Estimate the height of a polar bear who weighs 500 kg.

6 The distance travelled by bus between various places in Australia and the cost of the journey are given in the table.

Distance, km	10	20	25	40	50	70	75	80	95	120
Cost, AU$	5.00	7.80	8.00	9.20	11.00	13.00	13.50	13.80	15.00	16.80

a Find the equation of the regression line.
b Estimate the cost of a 100 km journey.
c Can you use your line to estimate the cost for a journey of 200 km?

7 The GNP and life expectancy for several countries are shown in the table.

GNP $ per capita	41 210	39 640	24 990	18 700	11 940	4820	1510	800	350	240
Life expectancy	77	80	78	77	72	71	68	60	52	41

a Find the equation of the regression line.
b Estimate the life expectancy for a country whose GNP is $20 000.

8 The percentage scores for a physics and mathematics test for each of the 16 students in grade 11 at Fun Academy are shown in the table.

Physics	23	34	39	43	47	52	58	67	73	75	79	82	86	88	91	97
Maths	31	33	42	55	48	51	60	71	72	69	83	84	89	96	92	89

a Find the equation of the regression line.
b Estimate the mathematics score for a student who scores 60 on the physics test.

9 The table shows the number of coins tossed and the number of heads obtained.

Number of coins	10	15	20	25	30	35	40	45	50	55	60	65	70	75
Number of heads	6	8	9	11	18	19	20	20	23	28	34	33	34	42

a Find the equation of the regression line.
b Estimate the number of heads when 52 coins are tossed.
c Can you use your regression line to estimate the number of heads obtained when 120 coins are tossed?

10 The table shows the weights of babies and the weights of their fathers.

Weight of father, kg	58	65	69	72	74	76	82	85	85	92
Weight of baby, kg	2.2	3.1	3.7	3.6	3.8	3.9	3.1	4.0	4.1	5.2

a Find the equation of the regression line.
b Estimate the weight of a baby whose father weighs 80 kg.

6.9 The χ^2 test for independence, *p*-values

The χ^2 test for independence

We are often interested in finding out whether or not certain sets of data are related. Suppose we collect data on the favourite colour of car for men and women. We may want to find out whether favourite colour of car and gender are independent or related.

One way to do this is to perform a χ^2 **(chi-squared)** test for independence.

To set up the test

- We first set up a **null hypothesis**, H_0, and an *alternative hypothesis, H_1*.
  H_0 states that the data sets are independent.
  H_1 states that the data sets are not independent.
- Then we decide on the level of significance to use.
  The most common levels are 1%, 5% and 10%.
- We can put our data in tables.
  The elements in the table are our **observed** data and the table is known as a **contingency table**.
  The elements in the contingency table should be frequencies.

For example, the contingency table for the favourite colour of car could be:

	Black	White	Red	Blue	Total
Male	51	22	33	24	130
Female	45	36	22	27	130
Total	96	58	55	51	260

From the observed data we can calculate the **expected** frequencies.

Because our values are independent, we can use probabilities to calculate the expected values.

So, the expected number of men who like black cars is

$$\frac{130}{260} \times \frac{96}{260} \times 260 = 48$$

and the expected number of men who like white cars is

$$\frac{130}{260} \times \frac{58}{260} \times 260 = 29$$

Expected frequencies for the other colours are calculated in a similar way.

The expected table of frequencies looks like this.

	Black	White	Red	Blue	Total
Male	48	29	27.5	25.5	130
Female	48	29	27.5	25.5	130
Total	96	58	55	51	260

The GDC calculates the expected frequencies for you but you must know how to find them by hand in case you are asked to show one or two calculations in an exam question.

- The expected frequencies can *never* be less than 1.
- At most 20% of the expected frequencies can be between 1 and 5.
- At least 80% of the expected frequencies must be 5 or higher.
- If there are too many frequencies between 1 and 5 then it is possible to combine rows or columns.

Now we are ready to calculate the χ^2 value using the formula

$$\chi^2_{calc} = \sum \frac{(f_o - f_e)^2}{f_e}$$

f_o is the observed value
f_e is the expected value

$$\chi^2_{calc} =$$
$$\frac{(51-48)^2}{48} + \frac{(22-29)^2}{29} + \frac{(33-27.5)^2}{27.5} + \frac{(24-25.5)^2}{25.5} + \frac{(45-48)^2}{48} + \frac{(26-29)^2}{29}$$
$$+ \frac{(22-27.5)^2}{27.5} + \frac{(27-25.5)^2}{25.5} = 6.13$$

This value can also be found using the GDC as follows:

```
X²-Test
 Observed:■A]
 Expected:[B]
 Calculate Draw
```

```
X²-Test
 X²=6.130780933
 P=.1054179418
 df=3
```

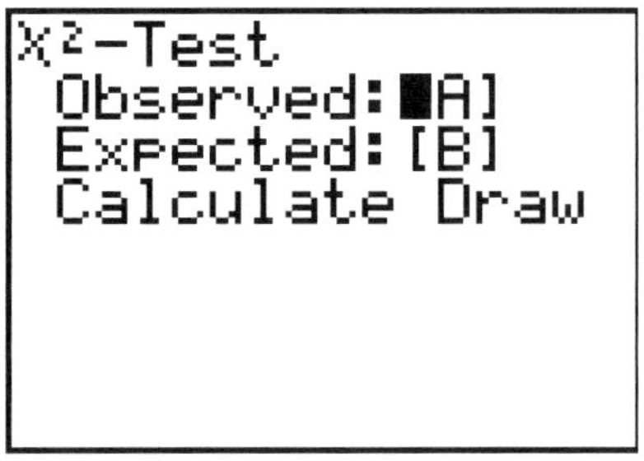

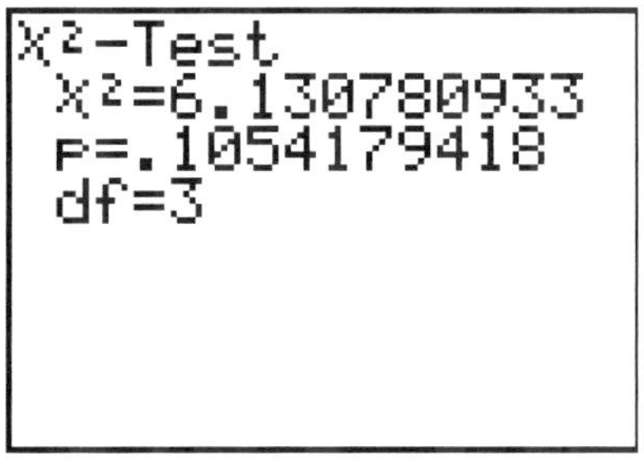

Now, the number of **degrees of freedom** must be calculated.

To find this value for a χ^2 test for independence, we multiply (number of rows − 1) by (number of columns − 1)

Degrees of freedom =
(rows − 1)(columns − 1)

So, in the example above,
number of degrees of freedom = (2 − 1) × (4 − 1) = 3

We use this value to look up the critical value in our information booklet, (see Topic 10, Section 10.5.1) or in any standard tables of chi-squared values that should be available in a library.

If we accept the null hypothesis at the 1% level we are saying that we are confident that 99% of the time the null hypothesis is correct.

At the 1% level, the critical value = 11.345
At the 5% level, the critical value = 7.815
At the 10% level, the critical value = 6.251

If the χ^2_{calc} value is **less than** the critical value then we **accept** the null hypothesis.
If the χ^2_{calc} value is **more than** the critical value then we **do not accept** the null hypothesis.

We can also use the *p*-value on the GDC to decide whether to accept or reject the null hypothesis.

If the p-value is *less* than the significance level then we ***do not accept*** the null hypothesis.
If the p-value is *more* than the significance level then we ***do accept*** the null hypothesis.

In the example above, at the 5% level, $6.13 < 7.815$, therefore we accept the null hypothesis that favourite colour of car is independent of gender. (Or from the GDC the p-value = 0.105 which is bigger than 0.05 therefore we accept the null hypothesis.)

Example 6.9a

A survey was conducted to find out which type of flower males and females preferred.

Eighty people were interviewed outside a florist shop and the results are shown in the table.

	Rose	**Carnation**	**Lily**	**Freesia**	**Total**
Male	16	10	5	8	39
Female	19	6	4	12	41
Total	35	16	9	20	80

Using the χ^2 test, at the 5% significance level, determine whether the favourite flower is independent of gender.

a State the null hypothesis and the alternative hypothesis.
b Show that the expected frequency for female and roses is approximately 17.9.
c Write down the number of degrees of freedom.
d Write down the χ^2_{calc} value at for this data.
e Comment on your result.

Solution

a H_0: Favourite flower type is independent of gender.

H_1: Favourite flower type is not independent of gender.

b $\frac{41}{80} \times \frac{35}{80} \times 80 = 17.9375$

So, expected frequency for roses and females is approximately 17.9.

c Degrees of freedom = $(2 - 1)(4 - 1) = 3$

d From the GDC, the $\chi^2_{calc} = 2.12$

The p-value = 0.548

e From the information booklet that you will use in the exam, the critical value is 7.815.

$2.12 < 7.815$ therefore we can accept the null hypothesis and conclude that favourite flower is independent of gender.

Or, using the p-value, $0.548 > 0.05$, therefore we can accept the null hypothesis.

Example 6.9b

300 people of different ages were interviewed and asked which genre of books they mostly read (fiction/non-fiction/science fiction). The results are shown in this table of observed frequencies.

Age/book type	Fiction	Non-fiction	Science fiction	Total
0–25 years	23	16	41	80
26–50 years	54	38	38	130
51+ years	29	43	18	90
Total	106	97	97	300

Using the χ^2 test, at the 5% significance level, determine whether the type of book is independent of age.

a State the null hypothesis and the alternative hypothesis.
b Show that the expected frequency for science fiction between 26–50 years is 42.
c Write down the number of degrees of freedom.
d Write down the χ^2_{calc} for this data.
e Comment on your result.

Solution

a H_0: The genre of book is independent of age.
H_1: The genre of book is not independent of age.

b $\frac{130}{300} \times \frac{97}{300} \times 300 = 42.0$

c $(3 - 1)(3 - 1) = 4$

d Using the GDC, the $\chi^2_{calc} = 26.9$

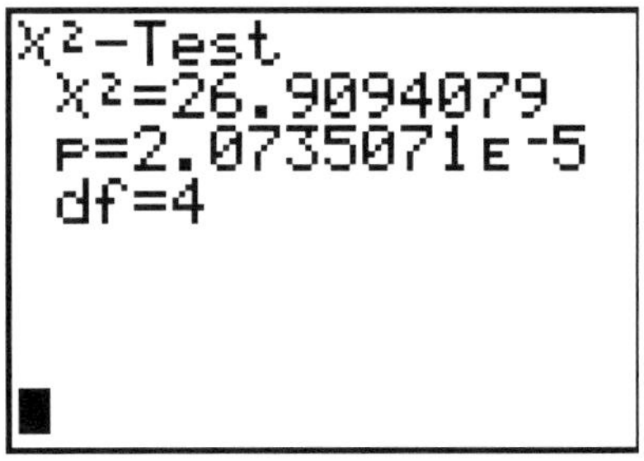

The critical value from the information booklet = 9.488
26.9 > 9.488, therefore we do *not* accept the null hypothesis.
So, the genre of book is *not* independent of age.

We could also use the p-value.
Since 0.000 020 7 < 0.05, we must reject the null hypothesis.
So, the genre of book is dependent on age.

Example 6.9c

Three different flavours of cat food were tested on different breeds of cats to see if there was any connection between favourite flavour and breed.

The results are shown in the table.

	Beef	**Chicken**	**Fish**	**Total**
Persian	4	16	8	28
Siamese	5	11	18	34
Manx	3	9	4	16
Burmese	6	3	13	22
Total	18	39	43	100

A χ^2 test at the 5% significance level is to be set up.

a State the null hypothesis and the alternative hypothesis.
b Write down the table of expected frequencies.
c Show that the number of degrees of freedom is 6.
d Write down the χ^2_{calc} for this data.
e Comment on your result for a 5% significance level.

Solution

a H_0: Preferred flavour of food is independent of breed.
H_1: Preferred flavour of food is not independent of breed.

b From the GDC the table of expected frequencies is:

```
[B]
[[5.0 10.9 12.0…
 [6.1 13.3 14.6…
 [2.9 6.2  6.9 …
 [4.0 8.6  9.5 …
```

So, the table of expected values is:

	Beef	**Chicken**	**Fish**	**Total**
Persian	5	11	12	28
Siamese	6	13	15	34
Manx	3	6	7	16
Burmese	4	9	9	22
Total	18	39	43	100

c The number of degrees of freedom = $(4 - 1)(3 - 1) = 6$

d From the GDC the $\chi^2_{calc} = 13.7$

```
χ²-Test
 χ²=13.74092401
 p=.0326679433
 df=6
```

e The p-value $= 0.0327 < 0.05$ therefore we do not accept the null hypothesis and we conclude that favourite flavour of cat food is dependent on breed.
Or, from the information booklet, the critical value = 12.592
$13.7 > 12.592$, therefore we reject the null hypothesis.

Example 6.9d

Sixty 18-year-old youths were asked what their favourite type of film was (adventure, crime, romantic, comedy, science fiction). The results are shown in the table.

	Adventure	Crime	Romantic	Comedy	Sci-fi	Total
Male	10	7	1	6	8	32
Female	2	8	14	3	1	28
Total	12	15	15	9	9	60

A χ^2 test at the 5% significance level is set up.

a State the null hypothesis and the alternative hypothesis.
b Write down the table of expected frequencies.
c Write down the number of degrees of freedom.
d Write down the χ^2_{calc} for this data.
e Comment on your result.

Solution

a H_0: The favourite type of film is independent of gender.
H_1: The favourite type of film is not independent of gender.
b From the GDC we get this table.

	Adventure	Crime	Romantic	Comedy	Sci-fi	Total
Male	6.4	8	8	4.8	4.8	32
Female	5.6	7	7	4.2	4.2	28
Total	12	15	15	9	9	60

Since 4 out of the 10 values are less than 5 we will combine "comedy" and "science fiction" to get this table.

	Adventure	Crime	Romantic	Comedy + Sci-fi	Total
Male	6.4	8	8	9.6	32
Female	5.6	7	7	8.4	28
Total	12	15	15	18	60

c Degrees of freedom = $(2 - 1)(4 - 1) = 3$
d From the GDC the $\chi^2_{calc} = 22.1$

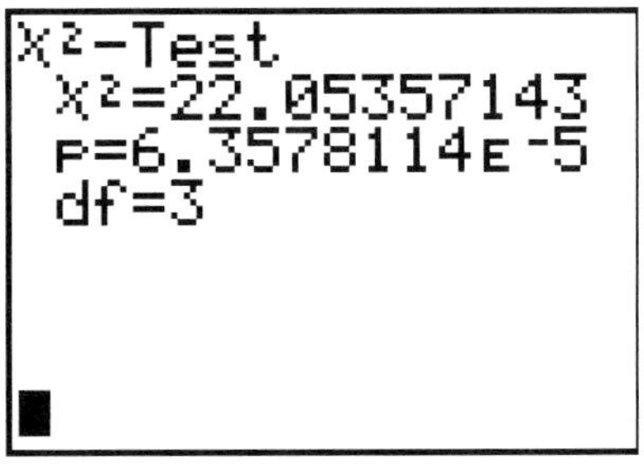

e The p-value = 0.000 063 6 which is less than 0.05, therefore we reject the null hypothesis and deduce that film preference is dependent on gender.

Or, from the information booklet, the critical value = 7.815. Since $22.1 > 7.815$, we do not accept the null hypothesis.

Exercise 6.9

1 Conrad had friends from many different countries. He was interested to find out if natural hair colour was related to nationality. Since Conrad attended a large international school which had students from over 45 countries, it was very easy for him to collect a lot of data. His observed data are shown in the table.

	Black	**Brown**	**Blonde**	**Ginger**	**Total**
American	12	24	19	3	58
Asian	54	16	1	2	73
European	21	36	45	17	119
Total	87	76	65	22	250

Test, at the 5% significance level, if there is a connection between hair colour and nationality.

a State the null hypothesis and the alternative hypothesis.
b Write down the table of expected frequencies.
c Write down the number of degrees of freedom.
d Write down the χ^2_{calc} for these data.
e Comment on your result.

2 Pedro was interested in finding out whether or not the number of hours per week spent watching the television had an influence on grades (GPA). He collected this information.

	Low GPA	**Average GPA**	**High GPA**	**Total**
0–9 hours	5	23	57	85
10–19 hours	13	55	32	100
> 20 hours	23	43	19	85
Total	51	121	108	270

Test, at the 5% significance level, if there is a connection between GPA and number of hours spent watching television.

a State the null hypothesis and the alternative hypothesis.
b Write down the table of expected frequencies and show that the expected frequency for 0–9 hours and a high GPA is 34.
c Show that the number of degrees of freedom is 4.
d Write down the χ^2_{calc} for these data.
e Comment on your result.

3 Wan-Gyu loves to fish. He wanted to find out if the number of fish he caught on average was related to the time of day that he went fishing. He kept a good record over 90 days and the results are shown in the table.

	Morning	**Afternoon**	**Evening**	**Total**
0–3 fish	8	5	12	25
4–6 fish	13	8	11	32
> 7 fish	9	17	7	33
Total	30	30	30	90

Test, at the 5% significance level, if there is a connection between time of day and number of fish caught.

a State the null hypothesis and the alternative hypothesis.
b Write down the table of expected frequencies.
c Write down the number of degrees of freedom.
d Write down the χ^2_{calc} for these data.
e Comment on your result.

4 Tim had a part-time job working at Luigi's ice cream shop. He decided to see if there was a relation between the temperature and the number of ice creams sold. His observations are shown in the table.

	< 21°	**21°–30°**	**> 30°**	**Total**
< 500 ice creams	14	23	12	49
500–750	10	31	17	58
> 750	8	26	9	43
Total	32	80	38	150

Test, at the 5% significance level, if there is a connection between temperature and number of ice creams sold.

a State the null hypothesis and the alternative hypothesis.
b Write down the table of expected frequencies.
c Write down the number of degrees of freedom.
d Write down the χ^2_{calc} for these data.
e Comment on your result.

5 Haruna wanted to find out the connection between the cost of a battery and the number of hours that it lasted. She tested several different types of batteries and her observed results are shown in the table.

Cost	**< 50 hours**	**50–75 hours**	**75–100 hours**	**> 100 hours**	**Total**
< \$2	12	18	21	15	66
\$2–\$4	10	15	23	20	68
> \$4	8	12	22	24	66
Total	30	45	66	59	200

Test, at the 5% significance level, if there is a connection between cost and hours.

a State the null hypothesis and the alternative hypothesis.
b Write down the table of expected frequencies and show that the expected frequency for a battery that cost more than \$4 and lasted less than 50 hours is 9.9.
c Write down the number of degrees of freedom.
d Write down the χ^2_{calc} for these data.
e Comment on your result.

6 Sonia wanted to find out if there was a connection between the type of art that people liked and their ages. She interviewed many visitors to an art gallery and her observed results are shown in the table.

Age	Modern	Landscape	Portrait	Still life	Total
< 20 years	15	8	4	6	33
20–40	12	15	7	9	43
> 40	8	18	12	6	44
Total	35	41	23	21	120

Test, at the 5% significance level, if there is a connection between age and preference for art type.

a State the null hypothesis and the alternative hypothesis.
b Write down the table of expected frequencies.
c Write down the number of degrees of freedom.
d Write down the χ^2_{calc} for these data.
e Comment on your result.

7 Naa Sakle decided to compare the average heights of different nationalities at her school. The data she collected are shown in the table.

Height(m)	European	African	Asian	American	Total
< 1.70	5	8	24	13	50
1.70–1.80	33	14	9	21	77
>1.80	12	8	2	11	33
Total	50	30	35	45	160

Test, at the 5% significance level, if there is a connection between nationality and height.

a State the null hypothesis and the alternative hypothesis.
b Write down the table of expected frequencies.
c Write down the number of degrees of freedom.
d Write down the χ^2_{calc} for these data.
e Comment on your result.

8 To find out if there was a relationship between blood pressure and smoking a survey was carried out and the observed data are shown in the table.

	High blood pressure	Not high blood pressure	Total
Smoker	79	37	116
Non-smoker	35	49	84
Total	114	86	200

The idea is to test the following hypotheses:
H_0: Blood pressure is independent of smoking.
H_1: Blood pressure is not independent of smoking.
This table shows the expected frequencies.

	High blood pressure	Not high blood pressure	Total
Smoker	*a*	*b*	116
Non-smoker	*d*	*c*	84
Total	114	86	200

a Show that the value for *a* is 66.1 and write down the values for *b*, *c* and *d*.
b Write down the number of degrees of freedom.
c Write down the χ^2_{calc} for these data .
d Comment on your result at the 5% significance level.

9 Sarah wanted to find out if the age that a baby first walked was related to gender. Her findings are set out in the table.

	< 10 months	10–14 months	> 14 months	Total
Male	15	43	12	70
Female	10	44	16	70
Total	25	87	28	140

Test, at the 5% significance level, if there is a connection between walking age and gender.

a State the null hypothesis and the alternative hypothesis.
b Show that the expected frequency of a male walking after 14 months is 14.
c Write down the number of degrees of freedom.
d Write down the χ^2_{calc} for these data .
e Comment on your result.

10 Charles decided to look into the question of fox hunting. 200 people were asked for their opinion and the results are shown in the table.

	Ban fox hunting	Allow fox hunting	Total
Rural	31	82	113
Urban	54	33	87
Total	85	115	200

Charles wanted to test the following hypotheses:

H_0: Fox hunting ban is independent of where a person lives.
H_1: Fox hunting ban is not independent of where a person lives.

a The table below sets out the elements required to calculate the χ^2 value for these data.

	f_o	f_e	$f_o - f_e$	$(f_o - f_e)^2$	$(f_o - f_e)^2 \div f_e$
Rural/ban	31	48	−17	289	6.02
Rural/allow	82	65	17	289	4.45
Urban/ban	54	37	17	289	7.81
Urban/allow	33	*a*	*b*	*c*	*d*

a Write down the values of *a*, *b*, *c* and *d*.
b Write down the χ^2_{calc} for these data.
c Write down the number of degrees of freedom.
d Using either the critical value of χ^2 for the 5% significance level or the *p*-value decide whether or not the null hypothesis should be accepted.

Theory of knowledge

Lying with statistics
(taken from *How to Lie with Statistics* by Darrell Huff)

Consider these test results.
Raw marks

Name	A	B	C	D	E	F	G	H	Total
Alan	100	30	47	72	40	75	30	47	441
Bart	90	38	43	60	20	65	48	70	434
Charlie	61	36	40	45	41	55	62	80	420
Derek	63	32	51	90	30	70	47	35	418
Edgar	56	55	41	82	45	40	49	41	409
Frank	80	45	49	64	65	45	38	20	406
George	23	47	45	55	60	80	32	60	402
Harry	40	35	52	70	56	20	60	65	398
Ian	85	40	60	40	28	51	55	30	389
John	72	54	50	10	25	35	66	75	387
Kyle	48	57	55	34	70	60	36	10	370
Lars	10	60	59	20	35	30	70	58	342

Scaled scores: high = 100, low = 0 and all others are scaled accordingly.

Name	A	B	C	D	E	F	G	H	Total
Lars	0	100	95	12	30	17	100	69	423
Kyle	42	90	75	30	100	67	15	0	419
John	69	80	50	0	10	25	90	93	417
Ian	83	34	100	38	6	52	62	29	414
Harry	33	17	60	75	72	0	75	79	411
George	14	57	25	56	80	100	5	71	408
Frank	78	50	45	68	90	42	20	14	407
Edgar	51	83	5	90	50	33	48	44	404
Derek	59	7	55	100	20	83	43	36	403
Charlie	57	20	0	44	42	58	80	100	401
Bart	89	27	15	63	0	75	45	86	400
Alan	100	0	35	77	40	92	0	53	397

Position rank scores: high = 1 and low = 12 etc.

Name	A	B	C	D	E	F	G	H	Total
Frank	4	6	7	5	2	8	9	11	52
Kyle	9	2	3	10	1	5	10	12	52
Charlie	7	9	12	8	6	6	3	1	52
Harry	10	10	4	4	4	12	4	4	52
Ian	3	7	1	9	10	7	5	10	52
Edgar	8	3	11	2	5	9	6	8	52
Derek	6	11	5	1	9	3	8	9	52
George	11	5	9	7	3	1	11	5	52
Bart	2	8	10	6	12	4	7	3	52
Alan	1	12	8	3	7	2	12	7	52
John	5	4	6	12	11	10	2	2	52
Lars	12	1	2	11	8	11	1	6	52

Eight test scores are shown for the twelve students.
The first set are the raw marks – *unscaled* scores.

Here the highest mark represents the best student in a subject. But notice that some subjects have higher marks than others.

The second set have been adjusted – *scaled* scores. Here the highest mark gets 100 and the lowest 0 – a line is drawn and the other marks are read from the line. This is quite an acceptable technique. Now we notice that the total scores are the reverse of the raw scores.

The third set are based on *position* – with the highest mark getting 1 and the lowest 12. Now all the students have a score of 52!

"Lies, damned lies, and statistics"
Benjamin Disraeli

There are many other ways to "lie with statistics" such as *false percentages, misleading graphs, comparing unlike data, etc.*

So, we must be careful when presented with statistical data.

Some questions we can ask are:

- Who is presenting the data? Are they knowledgeable in the area? Are they reliable and unbiased?
- How did they get their data? Who filled in the questionnaire or survey? Was it a random sample? Were the questions written in such a way that the answers were likely to be true?
- Is any relevant information missing? Are we only seeing what the person wants us to see? Is there another side to the story?
- Is the finding consistent with previous reports? Can we believe it?

1 What can you do to prove that a certain statistical result is trustworthy?

2 Are statistical results more likely to be true in certain areas of knowledge than in others?

Past examination questions for topic 6

Paper 1

1 This table shows the age distribution of the teachers who smoke at Laughlin High School.

Ages	Number of smokers
$20 \le x < 30$	5
$30 \le x < 40$	4
$40 \le x < 50$	3
$50 \le x < 60$	2
$60 \le x < 70$	3

a Calculate an estimate of the mean smoking age.

b Draw a histogram to represent the data. *M00q9*

2 This stem and leaf diagram gives the heights in cm of 39 school children.

Stem	Leaf
13	2, 3, 3, 5, 8
14	1, 1, 1, 4, 5, 5, 9
15	3, 4, 4, 6, 6, 7, 7, 7, 8, 9, 9
16	1, 2, 2, 5, 6, 6, 7, 8, 8
17	4, 4, 4, 5, 6, 6
18	0

Key 13 | 2 represents 132 cm

a i State the lower quartile height.
ii State the median height.
iii State the upper quartile height.
b Draw a box and whisker plot of the data. *Spec05q7*

3 These are the heights of some sunflowers in cm.

180 184 195 177 175 173 169 167 197 173 166
183 161 195 177 192 161 165

Represent the data by a stem and leaf diagram. *Spec05q29*

4 The table shows the number of children in 50 families.

Number of children	Frequency	Cumulative frequency
1	3	3
2	m	22
3	12	34
4	p	q
5	5	48
6	2	50
	Total, T	

a Write down the value of T.
b Find the values of m, p and q. *N99q2*

5 The mean of the ten numbers listed here is 5.5.
4, 3, a, 8, 7, 3, 9, 5, 8, 3
a Find the value of a.
b Find the median of these numbers. *M99q3*

6 Peter marked 80 exam scripts. He calculated the mean mark for the scripts to be 62.1.
Maria marked 60 scripts with a mean of 56.8.
Peter discovers an error in his marking. He gives two extra marks each to eleven of the scripts.

a Calculate the new value of the mean for Peter's scripts.

After the corrections have been made Peter and Maria put all their scripts together.

b Calculate the value of the mean for all the scripts. *M05q13*

7 The graph shows the cumulative frequency for the yearly incomes of 200 people.

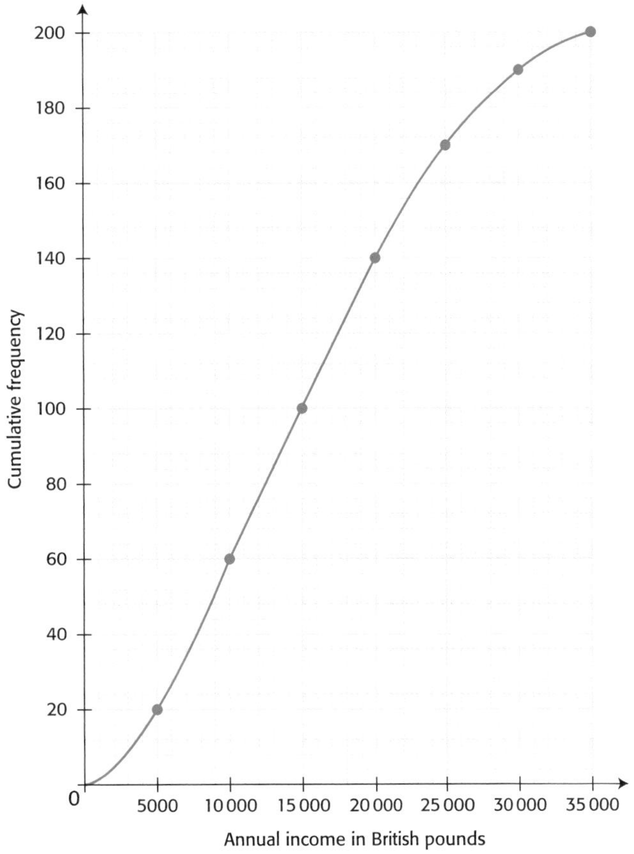

Use the graph to estimate

a the number of people who earn less than 5000 British pounds per year.

b The median salary of the group of 200 people.

c The lowest income of the richest 20% of this group. *M01q8*

8 The lengths and widths of ten leaves are shown on this scatter diagram.

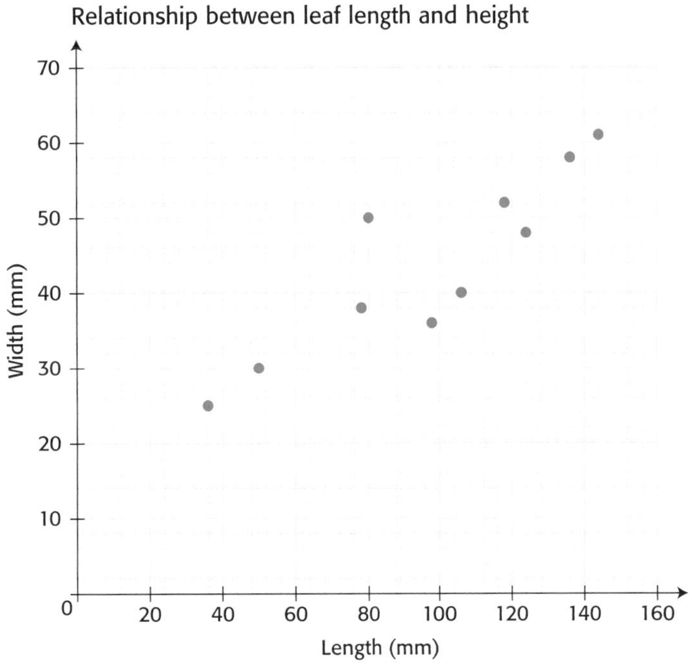

a Plot the point $M(97, 43)$ which represents the mean length and mean width.
b Draw a suitable line of best fit.
c Write a sentence describing the relationship between leaf length and leaf width for this sample. *M09q13*

9 Tom performs a chi-squared test to see if there is any association between the time to prepare for a penalty kick (short time, medium time, long time) and the outcome (scores a goal, does not score a goal). Tom performs this test at the 10% level.

a Write down the null hypothesis.
b Write down the number of degrees of freedom.
c The p-value for this test is 0.073. What conclusion can Tom make? Justify your answer. *Spec05q30*

Paper 2

1 The table shows the times, to the nearest minute, taken by 100 students to complete a mathematics task.

Time (t) minutes	11–15	16–20	21–25	26–30	31–35	36–40
Number of students	7	13	25	28	20	7

a Construct a cumulative frequency table with upper class boundaries 15.5, 20.5 etc.
b On graph paper, draw a cumulative frequency graph, using a scale of 2 cm to represent 5 minutes on the horizontal axis and 1 cm to represent 10 students on the vertical axis.
c Use your graph to estimate
 i the number of students that completed the task in less than 17.5 minutes
 ii the time it will take for $\frac{3}{4}$ of the students to complete the task. *N00q3*

2 The heights of 200 students are recorded in this table.

Height (h) in cm	**Frequency**
$140 \le h < 150$	2
$150 \le h < 160$	28
$160 \le h < 170$	63
$170 \le h < 180$	74
$180 \le h < 190$	20
$190 \le h < 200$	11
$200 \le h < 210$	2
Total	200

a Write down the modal group.
b Calculate an estimate of the mean and standard deviation of the heights.

The cumulative frequency curve for these data is drawn below.

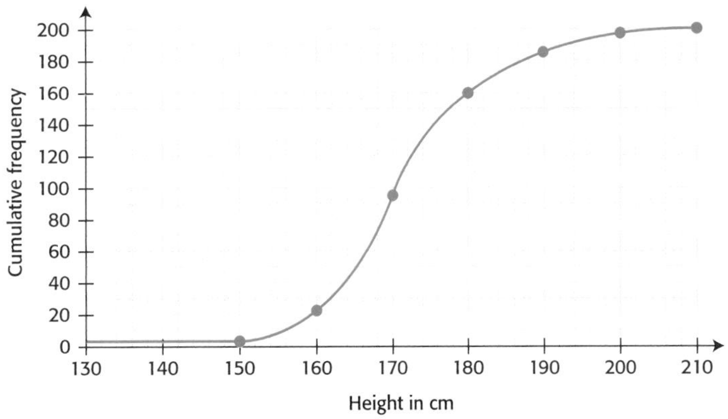

c Write down the median height.

d The upper quartile is 177.3 cm. Calculate the interquartile range.

e Find the percentage of students with heights below 165 cm. *N03q1*

3 A group of 25 females was asked how many children they each had. The results are shown in the histogram.

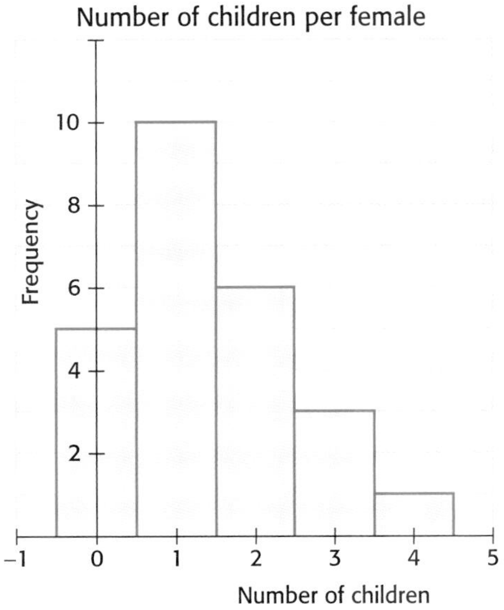

a Show that the mean number of children per female is approximately 1.4.

b Find the approximate standard deviation for these data.

c Another group of 25 females was surveyed and it was found that the mean number of children per female was 2.4 and the standard deviation was 2. Use the results from parts **a** and **b** to describe the differences between the number of children the two groups of females have.

d A female is selected at random from the first group. What is the probability that she has more than two children?

e Two females are selected at random from the first group. What is the probability that
  i both have more than two children?
  ii only one of the females has more than two children?
  iii the second female selected has two children given that the first female selected had no children? *M00q2*

4 The heights and weights of ten students selected at random are shown in the table.

Student	1	2	3	4	5	6	7	8	9	10
Height, *x* cm	155	161	173	150	182	165	170	185	175	145
Weight, *y* cm	50	75	80	46	81	79	64	92	74	108

a Plot this information on a scatter graph. Use a scale of 1 cm to represent 20 cm on the x-axis and 1 cm to represent 10 kg on the y-axis.
b Calculate the mean height.
c Calculate the mean weight.
d It is given that $s_{xy} = 44.31$.
  i By first finding the standard deviations of the heights, correct to two decimal places, show that the gradient of the regression line of y on x is 0.276.
  ii Find the equation of the regression line.
  iii Draw this line on your graph.
e Use your line to estimate
  i the weight of a student of height 190 cm.
  ii the height of a student of weight 72 kg.
f It is decided to remove the data for student number 10 from all calculations.
  Explain, *briefly*, what effect this will have on the line of best fit. *N01q7*

5 The sketches below represent scatter diagrams for the way in which variables x, y and z change over time t in a given experiment. They are labelled 1, 2 and 3.

1

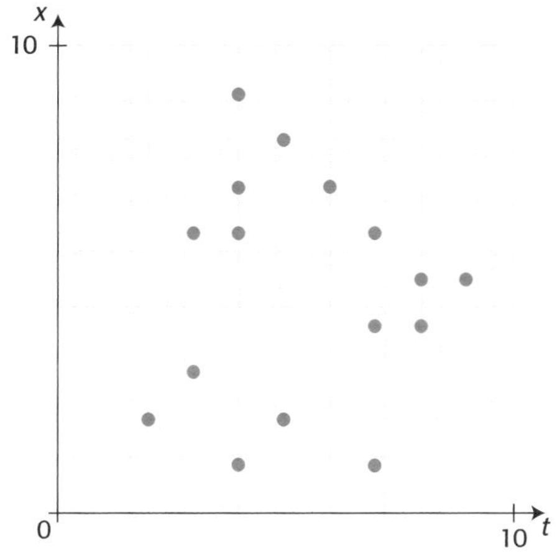

2

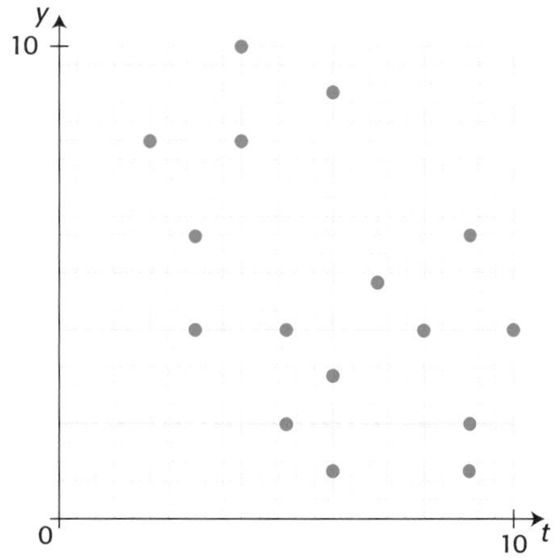

3

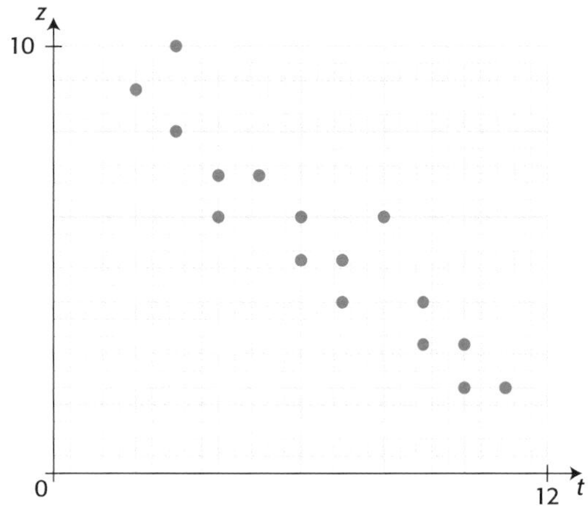

a State which of the diagrams indicate that the pair of variables
  i is not related
  ii shows strong linear correlation.

b A student is given a piece of paper with five numbers written on it. She is told that three of these numbers are the product moment correlation coefficients for the three pairs of variables shown in sketches 1, 2 and 3 above. The five numbers are

  0.9, −0.85, −0.20, 0.04, 1.60

  i For each sketch above, state which of the five numbers is the most appropriate value for the correlation coefficient.
  ii For the two remaining numbers, state why you reject them for this experiment. *N04q7*

6 It is decided to take a random sample of 10 students to see if there is a linear relationship between height and shoe size. The results are given in the table.

Height (x cm)	175	160	180	155	178	159	166	185	189	173
Shoe size (y)	8	9	8	7	10	8	9	11	10	9

a Write down the equation of the regression line of shoe size, y, on height, x, giving your answer in the form $y = mx + c$.

b Use your equation to predict the shoe size of a student who is 162 cm in height.

c Write down the correlation coefficient.

d Describe the correlation between height and show size. *Spec05q3*

7 For his mathematical studies project a student gave his classmates a questionnaire to complete. The results for the question on the gender of a student and the specific subjects taken by the student are given in the table.

	History	**Biology**	**French**	
Female	22	20	18	(60)
Male	20	11	9	(40)
	(42)	(31)	(27)	

This table gives the expected values.

	History	**Biology**	**French**
Female	p	18.6	16.2
Male	q	r	10.8

a Calculate the values of p, q and r.
The chi-squared test is used to determine if the choice of subject is independent of gender, at the 5% level of significance.

b i State a suitable null hypothesis.

ii Show that the number of degrees of freedom is two.

iii Write down the critical value of chi-squared at the 5% level of significance.

c The calculated value of chi-squared is 1.78. Do you accept the null hypothesis? Explain your answer. *N01q7*

8 In the small town of Joinville, population 1000, an election was held. The results are shown in the table.

	Urban voters	**Rural voters**
Candidate A	295	226
Candidate B	313	166

In parts **a** and **c** below we will use a chi-squared test to decide whether the choice of candidate depends on where the voter lives.

H_o: the choice of candidate is independent of where the voter lives.

a i Write down the alternative hypothesis.

ii Use the information above to fill in a and b in the table below.

Cell	f_o	f_e	$f_o - f_e$	$(f_o - f_e)^2$
1	295	317	−22	484
2	226	204	22	484
3	313	291	22	484
4	166	a	b	484

b i Calculate the chi-squared value.

ii Write down the number of degrees of freedom.

iii State the critical value at the 5% level of independence.

c i State your conclusion.

ii Why did you arrive at this conclusion? *M03q7*

TOPIC 7

Introductory differential calculus

In this topic we will discover why and how differential calculus was discovered, and what it means.

Newton (1687)

Approximately 325 years ago, Isaac Newton was investigating how to find the velocity (speed) of a moving body, which is the name for the rate of change of distance with respect to time. He was interested in finding this as an application of physics, and plotted the graphs of distance against time. The problem was how to find the velocity when the graph was a curve. Newton did not publish his findings until 1687, although he had been working on this topic for a number of years before then. He called it the *Method of Fluxions*.

Leibniz (1684)

And here comes the coincidence. Gottfried Wilhelm Leibniz was also working on the same idea at the same time but not from a practical point of view. Although Newton and Leibniz were in communication with each other, neither realized initially that they were both exploring the same topic, but approaching it from different directions. Leibniz published his account of differential calculus in 1684. He called it *Differential Calculus*.

The big question became "Who discovered differential calculus first?" Many people thought that Leibniz had plagiarized Newton's ideas and made them his own, just changing the notation. As you see from the paragraph above, Newton used a different name and he also used a different notation. In fact, Leibniz's name and notation are what we use today, as they are easier to use than Newton's. In 1715 the Royal Society credited the discovery of differential calculus to Isaac Newton. One year later, in 1716, Leibniz died. He was not to know that later on, the accusations made against him of plagiarism were dropped, and both men were then given credit.

An interesting side-effect of this controversy is that in England Leibniz's notation was not adopted for over a century, since it was regarded as a matter of national pride to use Newton's notation.

Uses of calculus

Calculus is all about how one quantity changes with another. So why should we be concerned about how one quantity changes with another? Do you travel by car? The speedometer tells you how your distance from a point is changing with respect to time at any instant. Here the y-axis is distance, the x-axis is time. When you were pushed back in your seat, you were accelerating, which is what happens when speed changes with time. In this case the y-axis is speed, the x-axis is time. Calculus will give the value of your acceleration at any instant.

Suppose a swimming pool was the perfect bottom half of a sphere. Now fill it up at a constant rate from a hosepipe. What is the rate at which the height of the water is changing? ("rate" means changing with respect to time). Does the height start increasing quickly and then slow down, or the opposite? Calculus can tell you the answer.

In China there is considerable concern about population control. There is a one child policy, that each woman can have only one child (this is a simplified example). To make such an important and courageous policy as this, someone must have been able to work out the difference that a policy of one child versus two children would have on the future population. Calculus can tell you the answer.

How quickly will bird flu spread through the world?

What speed must a rocket reach to go into orbit?

How quickly does water pressure increase as a scuba diver descends in the water?

Calculus can tell you the answers to all these problems. It was invented to solve problems, not to make life difficult for students!

We are now going to use Leibniz's notation $\left(\frac{dy}{dx}\right)$ to find the rate of change of a function $y = f(x)$ with respect to x. But first we need to recapitulate how to find the gradient of a straight line joining two points.

7.1 Gradient of a line through two points on a curve, tangent to a curve

7.1.1 The gradient of a line through two points (see also Topic 5)

Consider two points, $P(x_1, y_1)$ and $Q(x_2, y_2)$, lying on a graph.

The gradient of the line joining them is defined as $\dfrac{y_2 - y_1}{x_2 - x_1}$ or $\dfrac{rise}{run}$.

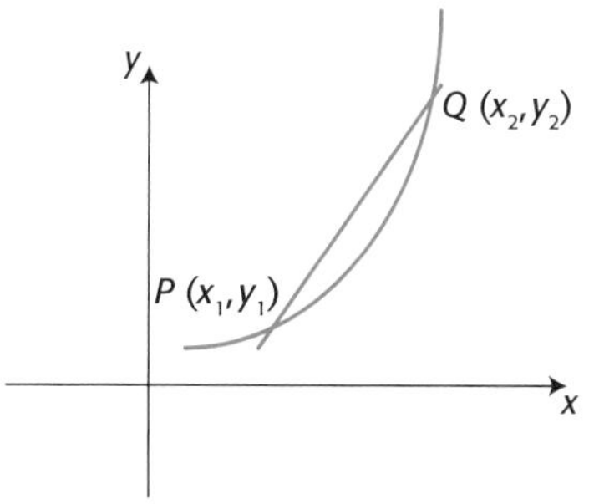

The gradient of line PQ is positive if, reading from left to right, the line goes up, and the gradient is negative if, reading from left to right, the line goes down

Example 7.1.1

Find the gradient of the line joining $A\ (-1, 3)$ to $B\ (4, 2)$.

Solution

First we draw a diagram.

In this diagram the line AB goes down and so the gradient of line AB is negative.

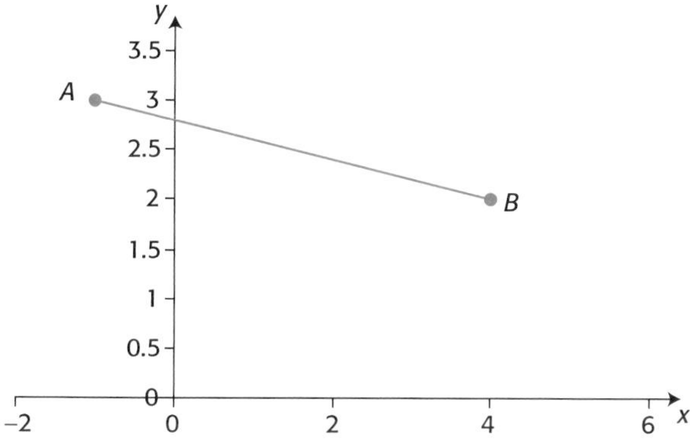

Here $A\ (-1, 3)$ is (x_1, y_1) so that $x_1 = -1$ and $y_1 = 3$

$B\ (4, 2)$ is (x_2, y_2) so that $x_2 = 4$ and $y_2 = 2$

The gradient is $\dfrac{y_2 - y_1}{x_2 - x_1} = \dfrac{2-3}{4--1} = -\dfrac{1}{5} = -0.2$

7.1.2 The gradient of a curve

This is essentially the problem Sir Isaac Newton was trying to solve when he wanted to find the *instantaneous* velocity of a moving object whose velocity was continually changing. Nowadays it could be equated to finding the instantaneous velocity of a car, say, which is driving in town and is continually speeding up and slowing down. He knew that on a graph of distance against time, distance ÷ time = velocity (speed) and also distance ÷ time = gradient so he set about trying to find the gradient when the graph was a curve.

To find the gradient we need a straight line. So what do we do when we have a curve, and wish to find the gradient at a point

P on the curve? Consider lines PQ_1, PQ_2, PQ_3, ... on the curve, where P is fixed and Q_1, Q_2, Q_3, ... are on the curve and moving closer and closer to P. The closer the Q's get to P, the closer the chords resemble a small piece of the curve. Finally the Q's will merge with P and there will only be one point of contact with the curve.

Such a line is called the **tangent** to the curve at P.

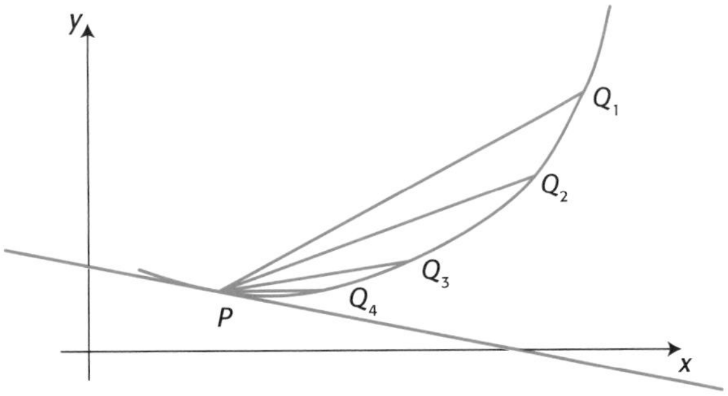

The gradients of the lines PQ_1, PQ_2, PQ_3, PQ_4, ... become closer and closer to the gradient of the tangent to the curve at P. This gradient of the tangent at P is called the gradient of the curve at P.

The gradient of the curve at P is the gradient of the tangent to the curve at P.

Question

Which of the following graphs have a line which is a tangent to the curve?

a

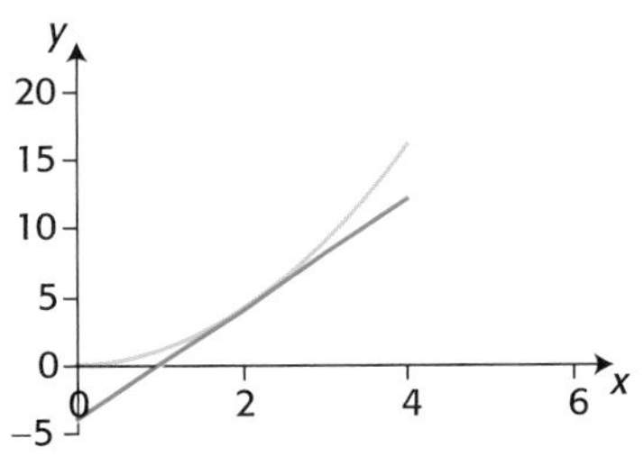

b

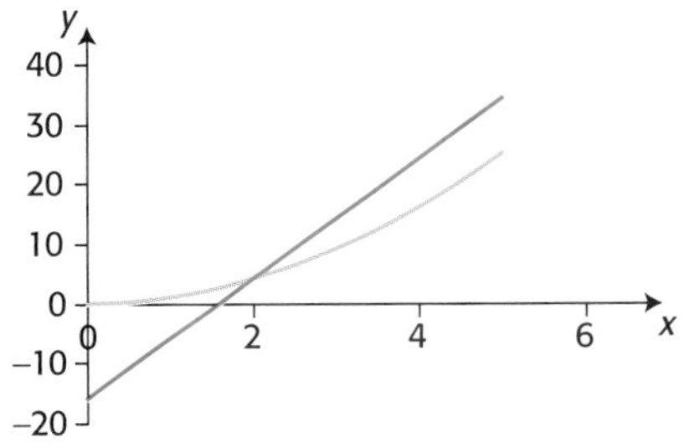

Solution

In graph **a** the line is a tangent; it touches the curve at one point.

Example 7.1.2

Consider the point $P(2, 5)$ on the curve whose equation is $y = x^2 + 1$. Let Q_1 be the point on the curve with x-coordinate 3. Let Q_2 be the point on the curve with x-coordinate 2.5. Let Q_3 be the point on the curve with x-coordinate 2.1. Let Q_4 be the point on the curve with x-coordinate 2.01. Consider the lines PQ_1, PQ_2, PQ_3 and so on. Find the gradient of each of these lines and estimate the gradient of the tangent at P.

Solution

A good idea is to sketch the curve on your graphics calculator. Then use it to find the y-coordinates of the Q's.

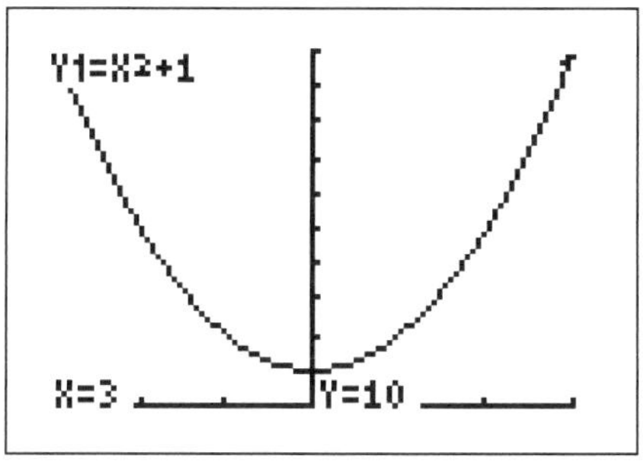

Copy the curve, mark in the points P and Q_1 and join them up with a straight line.

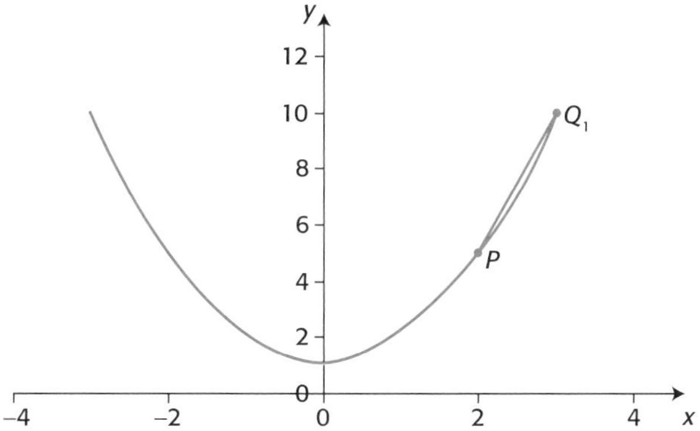

The curve is a parabola.

We find Q_1 is the point $(3, 10)$ and so the gradient of PQ_1 is $\dfrac{10-5}{3-2}=5$

Similarly we find Q_2 is the point $(2.5, 7.25)$ and so the gradient of PQ_2 is $\dfrac{7.25-5}{2.5-2}=4.5$

We find Q_3 is the point $(2.1, 5.41)$ and so the gradient of PQ_2 is $\dfrac{5.41-5}{2.1-2}=4.1$

We find Q_4 is the point $(2.01, 5.0401)$ and so the gradient of PQ_2 is $\dfrac{5.0401-5}{2.01-2}=4.01$

- It looks as though the gradient of the tangent to the curve at $P(2, 5)$ is 4. This is the same as saying that the gradient of the curve at P is 4.

We cannot calculate the gradient when Q is actually at P because then we would have gradient $= \dfrac{rise}{run} = \dfrac{0}{0}$ which cannot be evaluated. That is why we have to make Q get as close as we can to P, and see what happens.

We have established that finding the gradient of the curve at P is the same as finding the gradient of the tangent to the curve at P. The gradient is changing all the time so working through, as in the previous example, is clearly going to be time consuming if we

> *Theory of knowledge*
>
> If mathematics is about following rules and we know that $\dfrac{10}{10}=1$ and $\dfrac{4}{4}=1$ then why does $\dfrac{0}{0}\neq 1$?

have to find several different gradients. We need to see if there is a pattern.

- Consider the curve whose equation is $y = x^2 + 1$ again.

For the curve whose equation is $y = x^2 + 1$ we have found that when $x = 2$ the gradient of the curve is 4.

Let us see what the gradient is when $x = 3$. Take values of x close to 3, say 3.1, 3.01, 3.001, follow the steps through as in the previous example and see what happens.

When Q has x-coordinate 3.1 the gradient is $\dfrac{10.61-10}{3.1-3} = 6.1$.

When Q has x-coordinate 3.01 the gradient is $\dfrac{10.0601-10}{3.01-3} = 6.01$.

When Q has x-coordinate 3.001 the gradient is $\dfrac{10.006001-10}{3.001-3} = 6.001$.

- When $x = 3$ the gradient of the curve is 6.

Can you make a guess what the gradient of the curve might be when $x = 5$?

When $x = 5$ the gradient of the curve is 10. For this particular curve the gradient at any point is always twice the x-value.

- For the curve $y = x^2 + 1$ the gradient of the curve is $2x$.

Look back at the sketch of the curve.
We can see that it gets steeper as x increases, so a gradient of $2x$ is reasonable. Also when x is negative, the gradient is negative.

7.1.3 Finding the gradient of a curve at a general point P

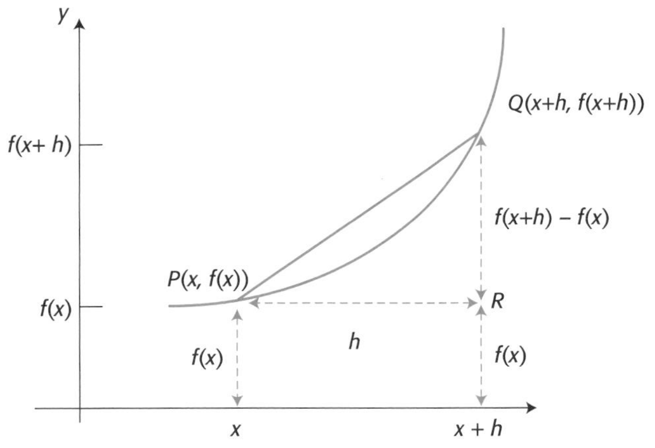

The gradient of PQ can be written as $\dfrac{QR}{PR} = \dfrac{f(x+h)-f(x)}{h}$

If we make Q move closer and closer to P then h will become closer and closer to zero.

We say that the gradient of the tangent at P is the limit of the above as $h \to 0$.

Gradient of the tangent at P is $\lim_{h\to 0}\left(\frac{f(x+h)-f(x)}{h}\right)$

We call the gradient of the curve or tangent at P the gradient function and write it as $f'(x)$ (pronounced f-dashed x).

Alternatively, we can say that the gradient of the curve whose equation is $y = f(x)$ is $\frac{\mathrm{d}y}{\mathrm{d}x}$ (pronounced dee-y by dee-x).

Combining these results we have the gradient

$$f'(x) = \lim_{h\to 0}\left(\frac{f(x+h)-f(x)}{h}\right) = \frac{\mathrm{d}y}{\mathrm{d}x}$$

Example 7.1.3

Estimate the gradient of the curve whose equation is $y = 2x^3 - 4x + 5$ at the point where $x = 1$, by taking h to be successively 0.1, 0.01, 0.001 in the formula $f'(x) = \lim_{h\to 0}\left(\frac{f(x+h)-f(x)}{h}\right)$ and filling in the table below.

Solution

Using the GDC we can type in the equation, use **CALCULATE VALUE** or **TABLE** and write down the y values for each x-value.

We must first find the y-value when $x = 1$.

***x*-value**	***y*-value**	**Gradient**
1	3	....
$h = 0.1$ so $x = 1.1$	3.262	**2.62** (3.262 − 3)/ 0.1
$h = 0.01$ so $x = 1.01$	3.020602	**2.0602** (3.020602 − 3)/ 0.01
$h = 0.001$ so $x = 1.001$	3.002006	**2.006** (3.002006 − 3)/ 0.001

The gradient of the curve at the point where $x = 1$ is the limit as $h \to 0$ and is 2.

Exercise 7.1

You will be expected to use your GDC for as many of these questions as are appropriate. A sketch diagram often helps, especially in deciding whether the gradient is positive or negative.

1 Find the gradient of the line joining the points $A(-1,2)$ and $B\ (2,7)$.

2 Find the gradient of the line joining the points P and Q on the curve $y = x^3$ where P has x-coordinate 0.5 and Q has x-coordinate 2.4.

3 Find the gradient of the line joining the points $X\ (2,6)$ and $Y\ (4.5,1)$.

4 Estimate the gradient of the curve whose equation is $y = 2x^3 + x - \frac{3}{x}$ at the point where $x = 2$, by taking h to be successively 0.1, 0.01, 0.001 in the formula $f'(x) = \lim_{h \to 0}\left(\frac{f(x+h) - f(x)}{h}\right)$ and filling in the table below.

Write your y-values correct to 4 decimal places and your gradient correct to 4 significant figures

***x*-value**	***y*-value**	**Gradient**
2		
$h = 0.1$ so $x = \ldots\ldots$		
$h = 0.01$ so $x = \ldots.$		
$h = 0.001$ so $x = \ldots.$		

5 The gradient of the line joining $P(-2, -4)$ and $Q(x, 5)$ is 3. Find x.

6 A is the point $(3, y)$. The gradient of the line AB is 4. Calculate y if B is the point $(5, 10)$.

7 Estimate the gradient of the curve given by $f(x) = x^2 - 2x + 5$ at the point where $x = 3$, by taking h to be successively 0.1, 0.01, 0.001 in the formula $f'(x) = \lim_{h \to 0}\left(\frac{f(x+h) - f(x)}{h}\right)$ and filling in the table below.

Write the y-values to as many significant figures as you consider necessary.

***x*-value**	***y*-value**	**Gradient = *f (x)***
3		
$h = 0.1$ so $x = \ldots\ldots$		
$h = 0.01$ so $x = \ldots.$		
$h = 0.001$ so $x = \ldots.$		

8 You are given that $f(x) = 4x^2 - 5x + 1$. Calculate the value of $\left(\frac{f(x+h) - f(x)}{h}\right)$ for $x = 2$ and $h = 0.005$.

9 You are given that $f(x) = x + \frac{1}{x}$. Calculate the value of $\left(\frac{f(x+h) - f(x)}{h}\right)$ for $x = 4$ and $h = 0.003$.

10 $f(x) = 2x^3 - 4x + 5$. Calculate the value of $\left(\frac{f(x+h) - f(x)}{h}\right)$ when $x = 1$ and $h = 0.001$ and then $h = 0.0001$ giving your answers correct to as many decimal places as the calculator gives. Estimate the value of the gradient at $x = 1$.

7.2 Derivatives of polynomials and negative powers of the variable

7.2.1 Formula for $\frac{dy}{dx}$ when $y = ax^n$

So far we have found in a previous example that when $y = x^2 + 1$ the gradient of the curve seems to be given by $\frac{dy}{dx} = 2x$.

You can also use your GDC to find the numerical value of the gradient at any particular point, see Topic 1.7, Example 1.7.1 on page 48.

Make sure you know how to do this for your particular GDC.

But what happens if we want to find a general formula for the gradient of any curve?

What would happen with $y = x^3$ for instance?

We could complete a table of values as we did in the previous section, and try and find a formula that works for each curve, but that is very time consuming. The general formulae are given below. The proofs are beyond the scope of this book.

- When $y = x^n$ where n is a real number

$$\frac{dy}{dx} = nx^{n-1}$$

- When $y = ax^n$ where a is a constant

$$\frac{dy}{dx} = nax^{n-1}$$

- When $y = u(x) \pm v(x)$ where u and v are functions of x

$$\frac{dy}{dx} + \frac{du}{dx} = \frac{dv}{dx}$$

Although these results are true for all real n, we shall only be using integer values of n, both positive and negative, and the value $n = 0$.

Other letters

So far we have only used the axes $y \uparrow$ and $x \longrightarrow$ to find out how y changes as x increases.

If we wanted to find the speed of a car say, as it accelerated away from traffic lights,

we would plot a graph with axes "distance s" $\uparrow$ (we tend to use s not d for distance) against "time t" $\longrightarrow$.

In this case we would be finding the velocity or speed at any point as the rate of change of distance with respect to time at that point, $\frac{ds}{dt}$ (pronounced dee-*s* by dee-*t*).

We could also find the rate of change of air pressure in a hot air balloon with respect to height above ground, with pressure on the *y*-axis and height on the *x*-axis and call it $\frac{dp}{dh}$.

Example 7.2.1a

Find **a** $\frac{ds}{dt}$ **b** $\frac{dy}{dx}$ for the graphs of these functions.

a $s = 6t^3 - 4t$ **b** $y = x^5 + 3x$

Solution

a First we must write all the *t* values as powers of *t*.

$s = 6t^3 - 4t^1$

We use the above results:

$$\frac{ds}{dt} = 6(3t^{3-1}) - 4\,(1t^{1-1}) = 18t^2 - 4$$

b $$\frac{dy}{dx} = 5(x^{5-1}) + 3(1x^{1-1}) = 5x^4 + 3$$

Differentiating a constant

This means finding the gradient of the line $y = k$ where k is a constant.

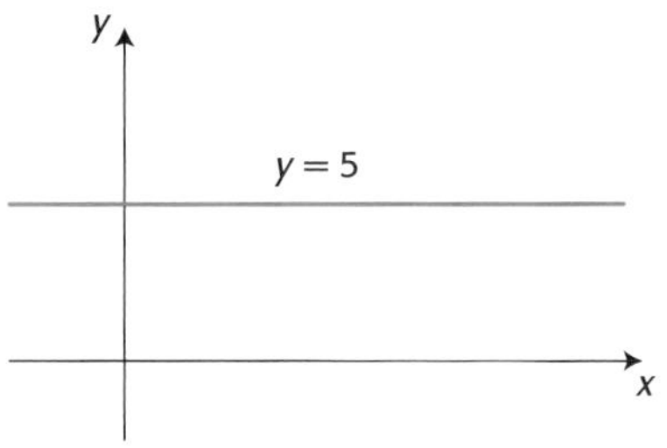

The line $y = 5$ (say) is a horizontal straight line parallel to the *x*-axis, and so its gradient or rate of change, is zero.

Now check to see if the formula agrees with this intuitive value.

We use the fact that

$$x^0 = 1$$

The equation $y = 5$ can be written as $y = 5 \times 1 = 5 \times x^0$ and using the results, we see that $\frac{dy}{dx} = 5(0x^{0-1}) = 0$ (because anything multiplied by $0 = 0$).

In fact, the graph of $f(x) = k$ where k is any constant always has a gradient of zero. This can be seen clearly on the diagram above. The gradient of the line is zero.

$$y = k, \ \frac{dy}{dx} = 0$$

Example 7.2.1b

Find $f'(x)$ for the curve given by $f(x) = 3x^4 - x^2 + 4$.

Solution

$f'(x) = 3(4x^{4-1}) - 2x^{2-1} + 0 = 12x^3 - 2x$

Example 7.2.1c

If $s = 5t^2 + 3t - 7$, find $\frac{ds}{dt}$.

Solution

We use the same rules for differentiating even though we have different letters.

$\frac{ds}{dt} = 10t + 3$

7.2.2 Negative powers of x (or t)

You will have met standard form, $a \times 10^k$ in Topic 2.
For example, a number like 0.0325 can be expressed as 3.25×10^{-2}.
Multiplying by 10^{-2} is the same as *dividing* by 10^2.

Conversely, dividing by 10^2 is the same as multiplying by 10^{-2}, dividing by 3^2 is the same as multiplying by 3^{-2} and more generally we have

$$\frac{1}{x^n} = x^{-n}, \quad \frac{2}{x^n} = 2x^{-n}, \quad \frac{1}{2x^n} = \frac{1}{2}x^{-n} \text{ or } \frac{x^{-n}}{2}$$

Also we have $\frac{1}{x^{-n}} = x^n$

It is important to note that there are many ways of expressing the same function. For instance, $\frac{3x^4}{7} = \frac{3}{7}x^4$ and $\frac{3}{7x^4} = \frac{3x^{-4}}{7} = \frac{3}{7}x^{-4}$.

Notice also that the 3 is always in the numerator and the 7 is always in the denominator.

The 3 and the 7 are *constants multiplied by a power of x* and stay exactly where they are, so that if the constant is on the top (numerator) it stays there and if it is on the bottom (denominator) it stays on the bottom.

Example 7.2.2a

Write these expressions in the form ax^n where a is a constant and n is an integer.

a $\frac{2}{x}$ b $\frac{4x}{3}$ c $\frac{2}{x^{-3}}$ d $\frac{4}{5x^2}$ e $\frac{2}{x^4}$ f $\frac{2}{5x^{-2}}$

Solution

a $2x^{-1}$ b $\frac{4x^1}{3}$ or $\frac{4}{3}x$ c $2x^3$ d $\frac{4x^{-2}}{5}$ or $\frac{4}{5}x^{-2}$

e $2x^{-4}$ f $\frac{2}{5}x^2$ or $\frac{2x^2}{5}$

Example 7.2.2b

Find the value of $\frac{ds}{dt}$ for the curve whose equation is $s = \frac{1}{t^2}$.

Solution
Firstly we need to express the equation of the curve as a power of t.

$s = \frac{1}{t^2}$ can be written as $s = t^{-2}$. Differentiating,

$$\frac{ds}{dt} = -2t^{-2-1} = -2t^{-3} \text{ or } \frac{-2}{t^3}.$$

Example 7.2.2c
For the curve whose equation is $f(x) = \frac{2}{3x} - 5x^3$, find the gradient function $f'(x)$.

Solution
Express the equation of the curve in powers of x. The 2 and the 3 are *constants multiplied by a power of x* and stay exactly where they are; the 2 in the numerator and the 3 in the denominator.

$$f(x) = \frac{2}{3}x^{-1} - 5x^3$$

$$f'(x) = \frac{2}{3}(-1x^{-1-1}) - 5(3x^{3-1}) = -\frac{2}{3}x^{-2} - 15x^2 \quad \text{or} \quad -\frac{2}{3x^2} - 15x^2$$

Example 7.2.2d
Find $\frac{dy}{dx}$ for the curve whose equation is $y = \frac{1}{3x} + 2x + 7$.

Solution
Express the equation in powers of x. Remember that *constants multiplied by a power of x* stay exactly where they are.

$$y = \frac{1}{3}x^{-1} + 2x^1 + 7$$

$$\frac{dy}{dx} = \frac{1}{3}(-1x^{-2}) + 2(1x^0) + 0 = -\frac{1}{3x^2} + 2 \quad (\text{since } x^0 = 1)$$

The 7 is a constant added (not multiplied) and its gradient is 0.

Example 7.2.2e
Find $\frac{dP}{dt}$ for the curve with equation $P = (t + 3)(2t - 1)$

Solution
Multiply out the brackets first.

$$P = 2t^2 - t + 6t - 3 = 2t^2 + 5t - 3$$

$$\frac{dP}{dt} = 4t^1 + 5 = 4t + 5$$

7.2.3 The second derivative

We may need to differentiate a function twice. We just perform the same operation twice over. The function is written $f''(x)$

(pronounced *f* double dashed *x*) or $\frac{d^2y}{dx^2}$ (pronounced dee-two-*y* by dee-*x*-squared). It has many uses but they are beyond the scope of this book.

The distance *s* travelled by a ball thrown upwards is a function of time *t*. The velocity of the ball at any time is given by $\frac{ds}{dt}$ and its acceleration is given by the *second derivative*, $\frac{d^2s}{dt^2}$.

Example 7.2.3

Find $\frac{d^2y}{dx^2}$ for the function $y = 3x^2 + 7x - 3$.

Here, $\frac{dy}{dx} = 6x^1 + 7x^0 = 6x + 7$ (since the constant –3 is not *multiplied by a power of x,* but subtracted, and so its derivative is zero).

Differentiate again and we have $\frac{d^2y}{dx^2} = 6$.

Exercise 7.2

You will need plenty of practice to become familiar with differentiating different powers of *x*, and switching to negative powers. In each case give your answer in the same format as the question, either as a constant divided by a positive power of *x*, or a constant multiplied by a negative power of *x*.

1 Write these functions in the form ax^n where *a* is a constant.

a $y = \frac{2}{5x}$ b $y = \frac{1}{2x^3}$ c $y = \frac{3}{4x^7}$ d $y = \frac{1}{3x}$ e $y = \frac{3}{5x^2}$

2 Express as a positive power of *x*:

a $y = 3x^{-2}$ b $y = \frac{2}{3}x^{-1}$ c $y = \frac{1}{x^{-2}}$ d $y = \frac{1}{4}x^{-2}$ e $y = \frac{4}{3x^{-1}}$

3 Find $\frac{dy}{dx}$ for each of these functions of *x*.

a $y = 2x^{-1}$ b $y = 9 + x^4$ c $y = 9x^3$ d $y = \frac{9}{x^3}$

4 Find the derived function $f'(x)$ for each of these functions.

a $f(x) = x^5 - 3x^3$ b $f(x) = 2x^2 + 3x$ c $f(x) = \frac{4}{x^2}$ d $f(x) = x^6 + 5x^4 + 1$

5 Differentiate these functions with respect to time. That means, find $\frac{dy}{dt}$.

a $y = \frac{4}{t^2}$ b $y = 3t^4 - 5t^2 + 3$ c $y = t^2 - 3t$ d $y = 1$

6 Find $\frac{ds}{dt}$ for each of these functions.

a $s = 4t^{-3} + 2t - 5$ b $s = \frac{4}{t^3} - 4t^2 - 2t$ c $s = \frac{7}{t} + 2t$ d $s = t + 3.6$

7 Find the derived function $\frac{dy}{dx}$ for these functions.

a $y = 2x^{-1} + 4x^{-2}$ b $y = \frac{2}{5}x$ c $y = \frac{3x^2}{4}$ d $y = 3x^{-2} + 6$

8 Find $\frac{d^2y}{dx^2}$ when $y = x^3 = 2x^2 + 5x = 4$.

9 Differentiate this function with respect to t.

$$f(t) = 4t^3 + 2t - 6 + \frac{3}{t}$$

10 Find the second derivative of the function $y = 3x^2 = 2x + 5$ with respect to x.

11 Find $\frac{ds}{dt}$ for the function $s = \frac{3}{7t} + 2t^2$.

12 The distance (s cm) of a radio controlled car from its start position after t seconds is given by the formula $s = 12t - t^2$.

a Find the distance after 5 seconds.

b Find how far it travels in the third second.

c Find its speed $\frac{ds}{dt}$ after 5 seconds.

7.3 Gradient of a curve and equation of a tangent to a curve

7.3.1 Gradients of curves for given values of x

Now we have a formula for finding the gradient of any curve which is an integral power of x, we can quickly find the gradient of a curve at any point on the curve that we choose.

- First we differentiate to find the gradient
- Then we substitute the particular value of x to find the gradient of the curve at that point

Remember, the gradient of the curve is also the gradient of the tangent to the curve at that point.

Example 7.3.1a

Find the gradient of the tangent to the curve whose equation is $y = \frac{4}{x} + 2$ at the point where $x = 2$.

Solution

We recognize that this is the same as finding $\frac{dy}{dx}$ and substituting $x = 2$.

We cannot differentiate until we have written y as a power of x.

$$y = 4x^{-1} + 2$$

Differentiate with respect to x.

$$\frac{dy}{dx} = (-1)4x^{-2} + 0 = -4x^{-2} = \frac{-4}{x^2}$$

Substitute $x = 2$.

$$\frac{dy}{dx} = \frac{-4}{2^2} = -1$$

Example 7.3.1b

A power boat moves in a straight line such that at time t seconds its distance s from a fixed point O on that line is given by $s = 2t^2 - 3t + 1$. Find the speed after 3 seconds.

Solution

We know speed is rate of change of distance with respect to time, which is $\frac{ds}{dt}$.

Differentiate with respect to t.

$\frac{ds}{dt} = 4t - 3$

Substitute $t = 3$.

Speed is $4(3) - 3 = 9\ ms^{-1}$ or 9 m/s

Example 7.3.1c

Find the value of the gradient of the curve whose equation is $y = (x - 3)(x + 2)$ at the point where it crosses the positive x-axis.

Solution

A quick sketch on your GDC will help.

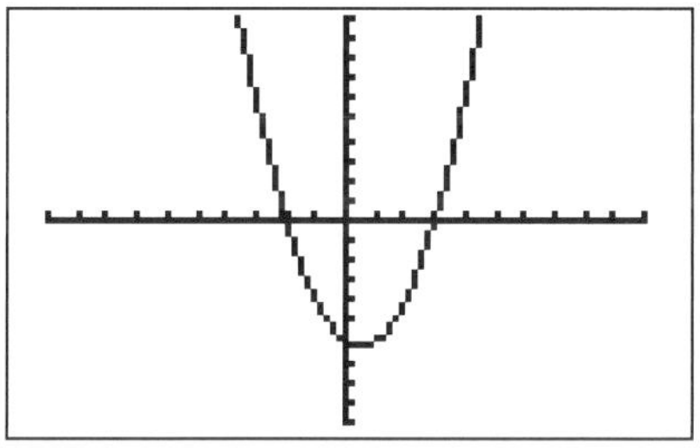

From your GDC you can see the curve crosses the positive x-axis at the point where $x = 3$.

Alternatively you could put y = 0 in the equation and obtain $x = 3$ or you could use your GDC and choose the positive value of x.

Multiply out the brackets.

$y = x^2 - x - 6$

Differentiate with respect to x.

$\frac{dy}{dx} = 2x - 1$

Substitute $x = 3$ to get $\frac{dy}{dx} = 2(3) - 1 = 5$

Gradient where the curve crosses the positive x-axis = 5

The next example shows you how to use your GDC to find the gradient of a curve without doing any calculation. If no method is specified, then this method is acceptable and indeed, expected. Sometimes you may be told to express the equation of the curve in powers of x and hence find the gradient, so it is important that you know how to find the gradient by either method.

Example 7.3.1d

The spread of untreated HIV in a certain village can be modelled by the equation $N = 4 \times 1.6^t$ where N is the number of people infected and t is the time in days after the first infection. Use your GDC to write down the rate of change after seven days.

Solution

The rate of change means the gradient, $\frac{\mathrm{d}N}{\mathrm{d}t}$. For the given N, calculating this is beyond the scope of this book but we can use the GDC to find it.

A quick sketch from $x = 0$ to $x = 10$ gives the picture, and the value of $\frac{\mathrm{d}N}{\mathrm{d}t}$ can be found as instructed in Topic 1.7. (We use $\frac{\mathrm{d}y}{\mathrm{d}x}$ on our GDC.)

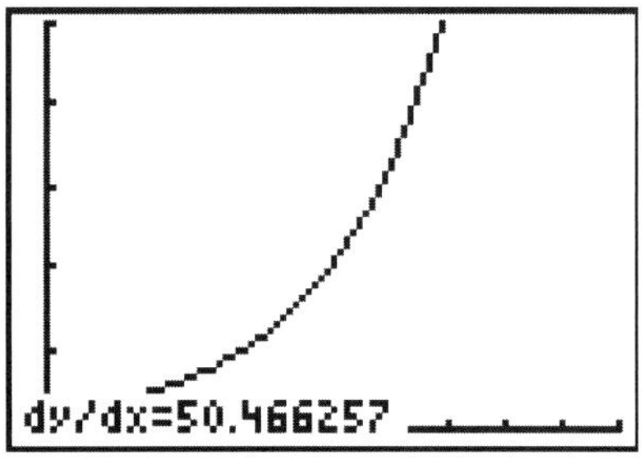

The rate of change after seven days is an increase of 50.47 people per day, say 50 people per day.

Example 7.3.1e

Find the gradient function $f'(x)$ of the curve $f(x) = 2x^3 - 7x^2 + 5x$.

Find also the gradient of the curve at the point where $x = 1$.

Solution

The question says "find", which means that relevant working must be shown. A quick sketch will help and you can use this to check your answer.

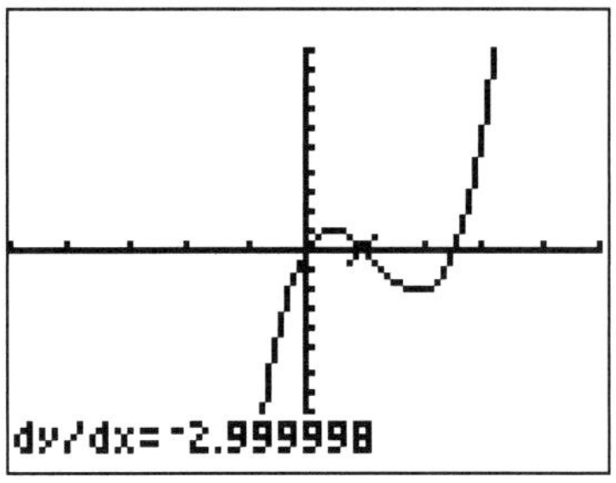

The equation is already written in powers of x.

Differentiate with respect to x.

$$\frac{\mathrm{d}y}{\mathrm{d}x} = 6x^2 - 14x + 5$$

Substitute $x = 1$.

$$\frac{dy}{dx} = 6 - 14 + 5 = -3$$

(you can see from the sketch that the gradient at the point where $x = 1$ is negative. This should warn you that something is wrong if you get a positive answer)

7.3.2 Values of x where the gradient is given

If the value of the gradient is given at a certain point, and also the equation of the curve is known, then we can differentiate the equation of the curve to get the gradient function and put this equal to the given value of the gradient. Then we solve for x.

Example 7.3.2a

The equation of a curve is given by $y = 5x^3 - 2x^2 - 4$.
Find the x-coordinates of the points where the gradient is 5.

Solution

"Gradient is 5" means $\frac{dy}{dx} = 5$.

Differentiating we get $\frac{dy}{dx} = 15x^2 - 4x$

So $15x^2 - 4x = 5$ which gives $15x^2 - 4x - 5 = 0$.

There are many different ways to solve this quadratic equation. One such way is to draw it and find the intersection with $y = 0$ (see Topic 1, page 28).

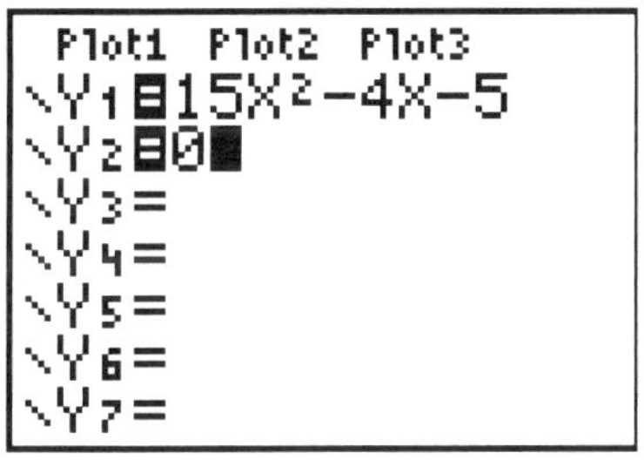

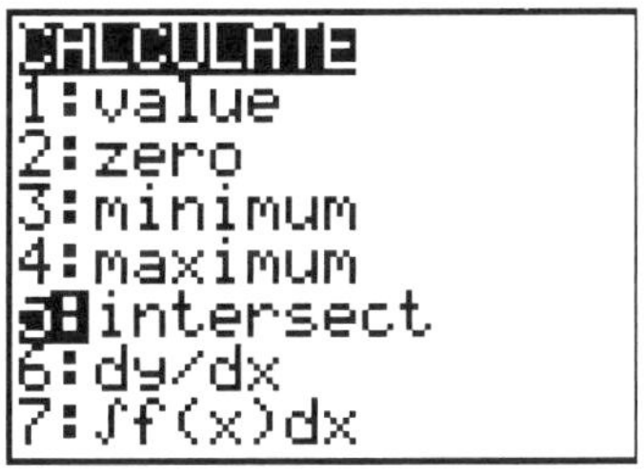

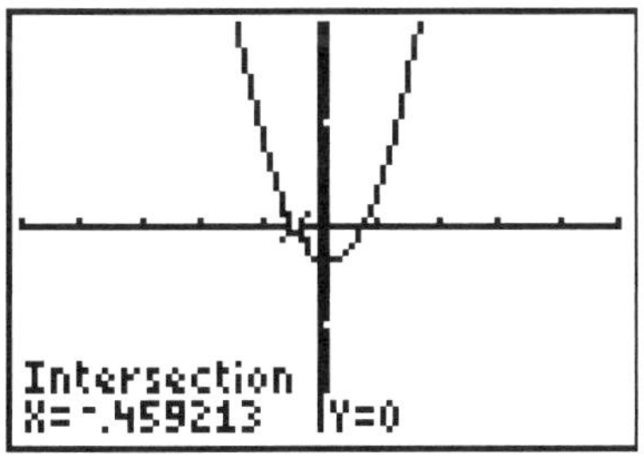

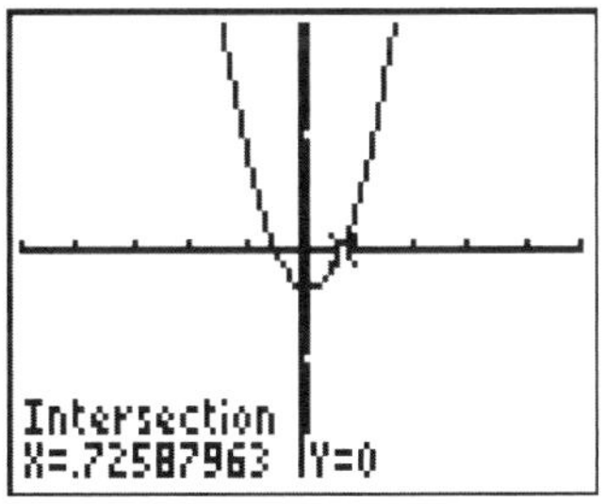

We obtain the x-coordinates, $x = -0.459$ and $x = 0.726$.

Example 7.3.2b

Raoul blows up a balloon. The volume, V cm³, after he has blown for t seconds is given by the formula

$$V = 100t^2 + 80t.$$

a Find how many seconds Raoul has been blowing when the rate of increase of the volume is 480 cm^3 per sec.

b The balloon bursts after Raoul has been blowing for 6 seconds. Find the greatest volume of the balloon, just before it bursts.

Solution

a Rate of increase of volume means $\frac{dV}{dt}$.

Differentiate with respect to t.

$$\frac{dV}{dt} = 200t + 80$$

Put this equal to 480.

$$200t + 80 = 480$$

$$t = 2$$

Raoul has been blowing for 2 seconds.

b Volume of balloon when $t = 6$

$$V = 100(6)^2 + 80(6) = 4080 \text{ cm}^3$$

7.3.3 Equation of the tangent at a given point

Now we know how to find the gradient of a curve at any point on the curve (it is equal to the gradient of the tangent at that point) we can find the equation of the tangent at any point by using $y = mx + c$ where m is the gradient of the tangent line.

Example 7.3.3

Find the equation of the tangent to the curve whose equation is $y = 3x^2 + \frac{2}{x} - 5$ at the point P where $x = 2$.

Solution

There are four things we need to do here. You will do these every time.

1 Find the gradient at the point P and call it m.

2 Find the y-coordinate of P as well as the x-coordinate. (Sometimes it will be given.)

3 Use $y = mx + c$ (the general equation of a straight line) and substitute for m, x and y for the point P which the line goes through, to find c.

4 Write the equation with your values of m and c.

So, here we go! Rewrite the equation in powers of x.

$$y = 3x^2 + 2x^{-1} - 5$$

Differentiate with respect to x.

$$\frac{dy}{dx} = 6x - 2x^{-1-1} = 6x - 2x^{-2} = 6x - \frac{2}{x^2}$$

1 To find the gradient at P, substitute $x = 2$.

$$\frac{dy}{dx} = 12 - \frac{2}{4} = 11\frac{1}{2}, \text{ so } m = 11\frac{1}{2}$$

2 To find the y-coordinate of P substitute $x = 2$ into the equation of the curve.

$$y = 3(2)^2 + \frac{2}{2} - 5 = 12 + 1 - 5 = 8$$

So P is the point $(2, 8)$.

3 Substitute into $y = mx + c$.

$$8 = 11\frac{1}{2} \times 2 + c \qquad \Rightarrow c = -15$$

4 Equation of required tangent is $y = 11\frac{1}{2}x - 15$.

You can check this using your GDC.

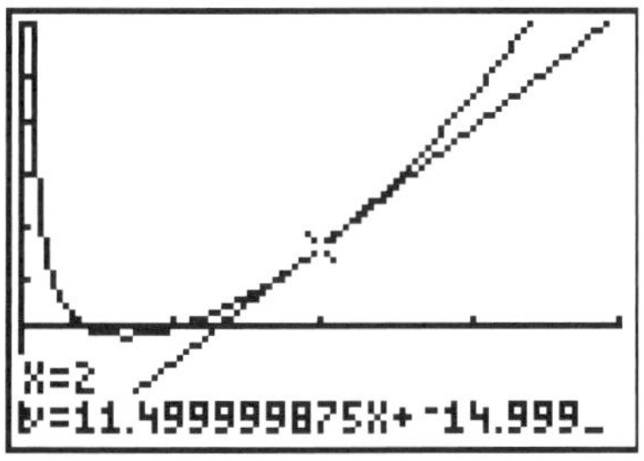

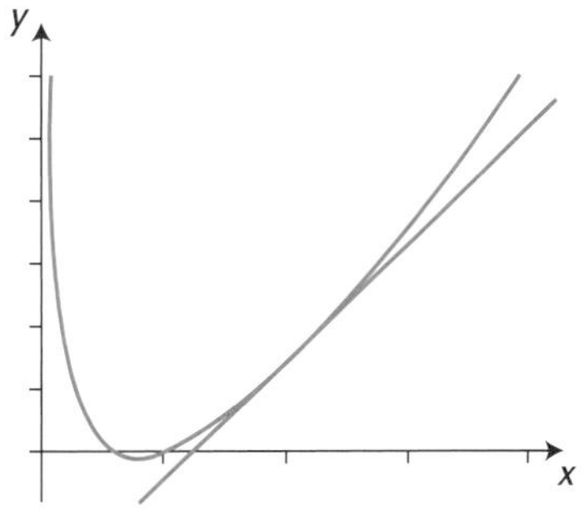

Exercise 7.3

You are expected to use your GDC in almost every question, and so should not be put off by unusual looking equations.

1 Find the equation of the tangent to the curve whose equation is $y = 5x^2 - 2x^4 + 3$ at the point where $x = 1$. Show all working and check your answer by GDC.

2 Find the equation of the tangent to the curve whose equation is $y = 2x^3 - 4x$ at the point where $x = 2$.

3 Find the value of x where the tangent to the curve given by $y = x^2 + 5$ is parallel to the line $y = 6x - 4$.

4 Find the equation of the tangent to the curve $f(x) = \frac{1}{x}$ at the point $(4, \frac{1}{4})$.

5 The curve whose equation is $y = x^2 - 4x + 15$ has gradient 6 when $x = a$. Find the value of a.

6 Find the coordinates of the two points on the curve whose equation is $y = \frac{x^3}{3} - \frac{x^2}{2} - x + 5$ where the gradient is equal to **a** 1 and **b** 2.

7 The tangent to the parabola given by $y = x^2 + 3x - 8$ has gradient 7 at the point P. Find the coordinates of P.

8 Find the x-values of the points where the gradients of the curves $y = x^3$ and $y = 6x - 1.5x^2$ are the same.

9 A curve has equation $y = x^3 + \frac{9x^2}{2} - 9x + 5$. Find the x-coordinates of the points on the curve where the gradient is 1.

10 Find the equation of the tangent to the curve $y = 2x^4 + 3x^3 + 6x^2 - x - 4$ at the point whose x-value is 1. Check with your GDC.

7.4 Increasing and decreasing functions

If x lies in the interval (a, b) and $f(x)$ is increasing as x increases then we say $f(x)$ is an **increasing function** in the interval (a, b). If x lies in the interval (p, q) and $f(x)$ is decreasing as x increases then we say $f(x)$ is a **decreasing function** in the interval (p, q).

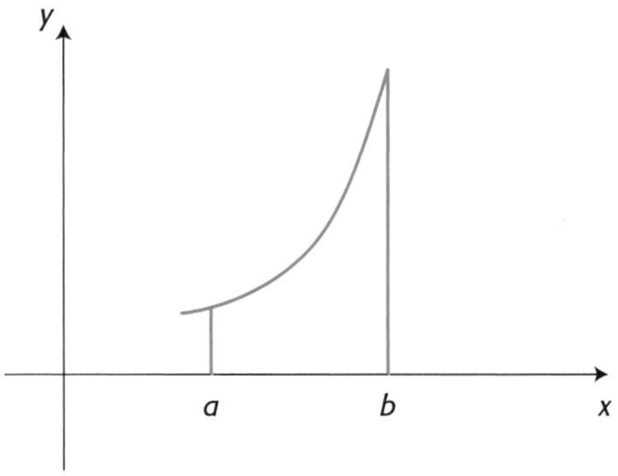

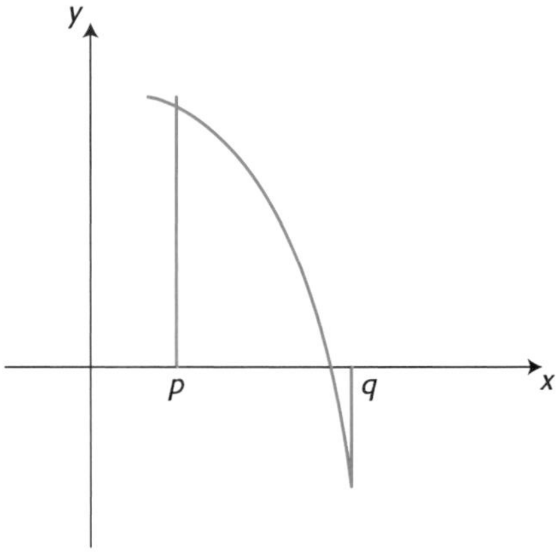

You can see that a function with positive gradient is an increasing function and a function with negative gradient is a decreasing one.

Consider the function $f(x) = x^3 + 1$.

The gradient $f'(x) = 3x^2$ and since $x^2 > 0$ for all real x apart from $x = 0$, the gradient is positive and so the function is increasing. The point $(0, 1)$, where the gradient is zero, is called a point of inflexion (see Section 7.5.1). This is not in the syllabus.

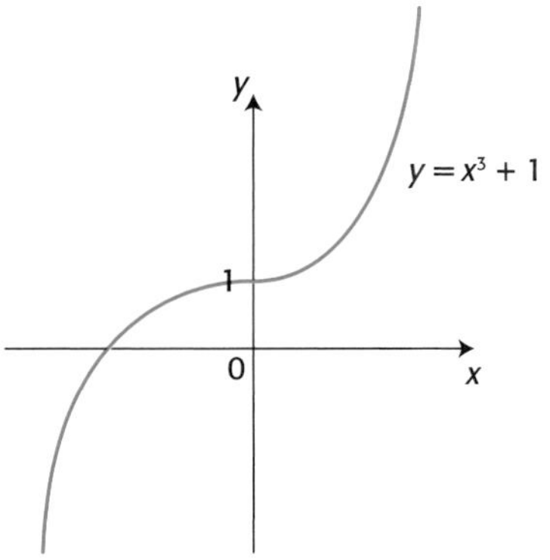

Example 7.4a

Sketch a graph of the function $s = f(t) = 4 - t^2$ for $0 \leqslant t \leqslant 4$.

Explain what is happening if t represents time in seconds and s metres represents the distance moved by a toy car from a fixed point O.

Solution
Use your GDC to sketch the curve. In **WINDOW**, set x- (or t-) values from 0 to 4 as instructed in the question.

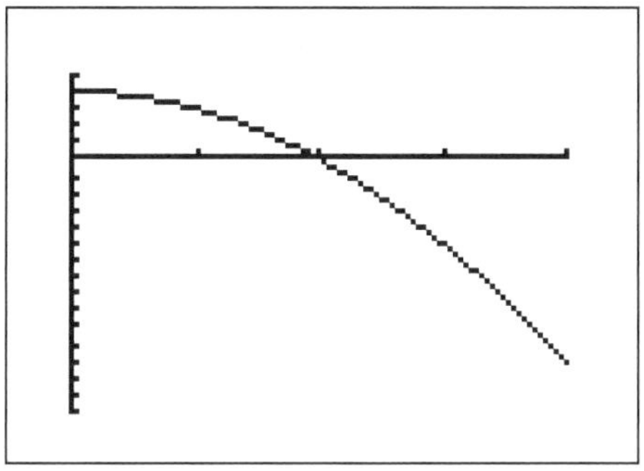

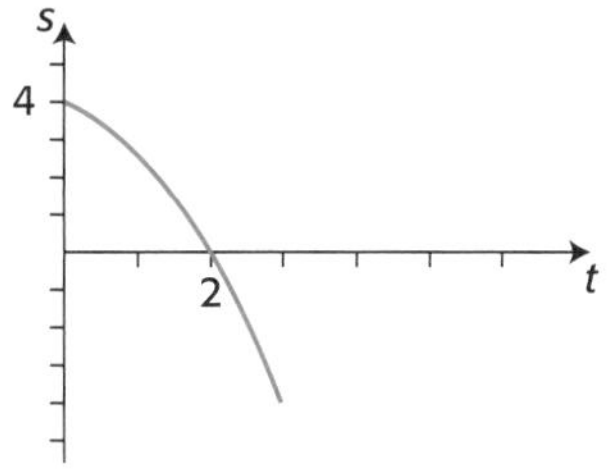

The graph is decreasing; the gradient $\frac{ds}{dt}$ is negative.
Since the gradient of a distance-time graph represents the velocity, we can say the velocity is negative. This means the toy car is going backwards.

The toy car starts 4 m away from O. It goes backwards until it reaches O after 2 sec, then it continues backwards, getting faster and faster (because the gradient gets steeper).

Example 7.4b
Sketch the graph of the function $y = x^3 + 1.5x^2 - 6x - 3$ for values of x from -3 to $+3$.

With domain $-3 < x < +3$ state the values of x for which y is an increasing function and the values of x for which y is a decreasing function.

Solution
Use your GDC to plot the graph. You must show a scale. You should also mark in the intercepts either with coordinates or make sure they are readable from your scale.

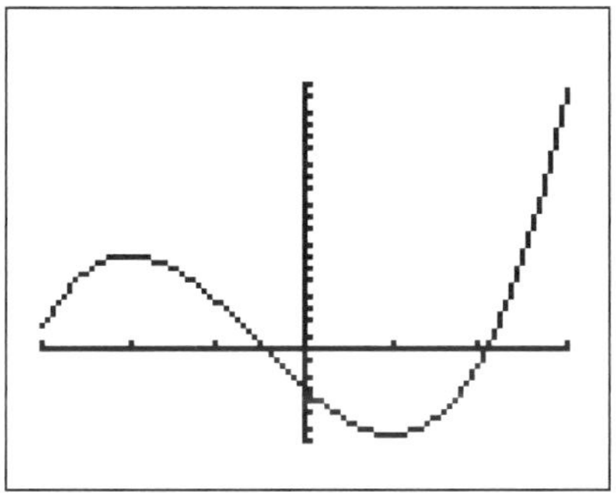

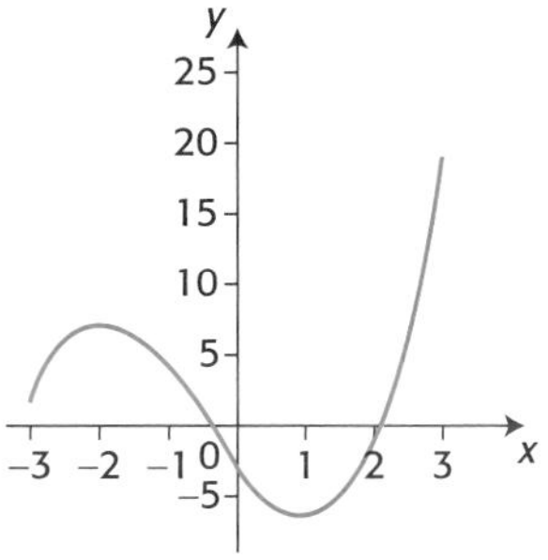

It is clear that the graph is increasing for $-3 < x < -2$, and also for $1 < x < 3$. The graph is decreasing for $-2 < x < 1$.

Example 7.4c

Sketch the graph of a function $f(x)$ which is increasing for $-2 < x < 3$ and decreasing for $3 < x < 6$.

Solution

Many different graphs are possible. One such is shown below.

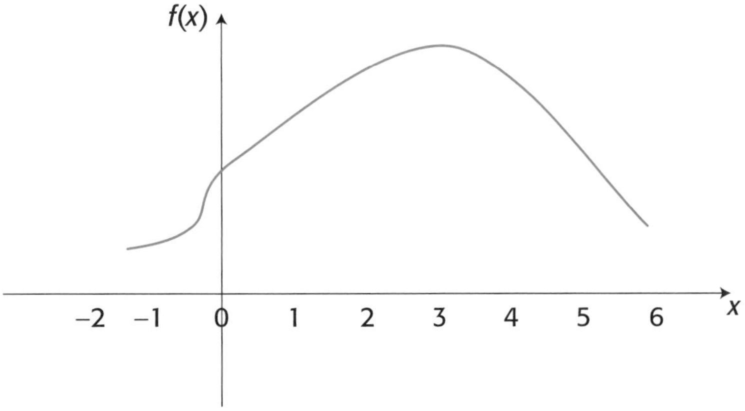

You can see that the graph does not have to be a regular shape, so long as it is increasing or decreasing all the time in the required interval.

You may have wondered what happens to the above graph at the point where $x = 3$. The graph is increasing for $x < 3$ and decreasing for $x > 3$. At $x = 3$ the gradient of the tangent is neither positive nor negative. It is zero. That means that $\frac{dy}{dx} = 0$ at this point. The point is called a **turning point** or **stationary point**.

Exercise 7.4

1 Sketch the graph of the function $f(x) = (x - 3)(x + 5)$ and find the set of values of x for which the function is
  a increasing   b decreasing, $x \in \mathbb{R}$.

2 Sketch the graph of the function $f(x) = 0.5x^4 + 3x - 6$ for $-3 \leqslant x < 3$. Find $f'(x)$ and give the set of values of x for which the function is
  a increasing   b decreasing.

3 The function $y = x^2 - ax + 5$ is decreasing for $x < 3$ and increasing for $x > 3$. Find the equation of the axis of symmetry and the value of a.

4 Find the set of values of x for which the function $y = 9 - x^2$ is increasing, $x \in \mathbb{R}$.

5 Find the set of values of x for which the function $y = (x - 3)^2$ is decreasing, $x \in \mathbb{R}$.

7.5 Zero gradient, local maxima and minima

7.5.1 Values of x where the gradient of a curve is zero: solution of $f'(x) = 0$

Consider this graph. Stationary or turning points occur where the gradient of the curve is zero, that is, where

$$\frac{dy}{dx} = 0.$$

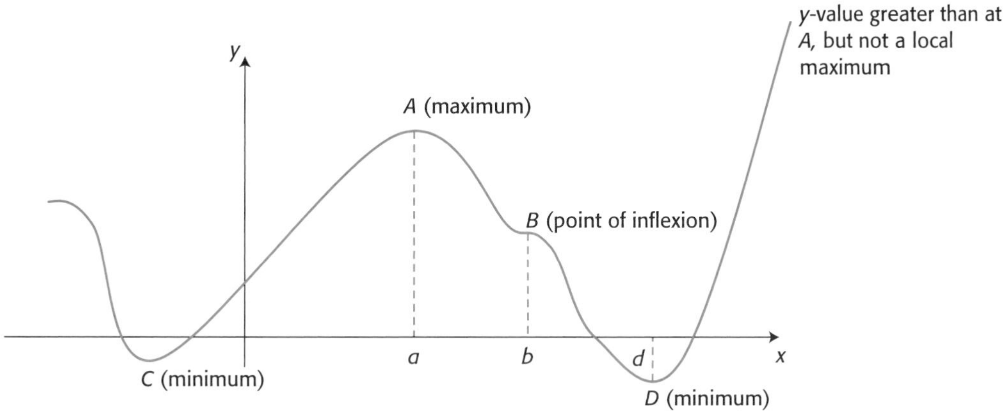

At the point A, x-coordinate a, the curve reaches a (local) maximum value. There are points on the curve which have a greater y-value, but the local maximum is where the gradient is zero and the gradient of the curve is positive immediately before and negative immediately after the point where $x = a$.

Similarly at the point D, x-coordinate d, the curve reaches a local minimum and is where the gradient is zero and the gradient of the curve is negative immediately before and positive immediately after the point where $x = d$.

At the points A, C and D, $\frac{dy}{dx} = 0$.

For **stationary** or **turning points**,

$$\frac{dy}{dx} = 0 \text{ or } f'(x) = 0$$

We often omit the word 'local' but we mean the local minimum or maximum.

The point B is interesting. Here $\frac{dy}{dx} = 0$, but the gradient is negative for $x < b$, and also negative for $x > b$. In other words, it is neither a maximum nor a minimum point. We say it is a **point of inflexion**.

It is often useful to find where the maximum or minimum points of a graph are. Some examples are given below, and these ideas can be used in your project as an alternative to a statistics-based project.

First we will do some examples.

Example 7.5.1

Find the coordinates of the turning point on the curve whose equation is given by $y = (2x - 3)(x + 2)$. Write down the equation of the tangent at that point.

Solution
Multiply out the brackets.

$$y = 2x^2 + 4x - 3x - 6 = 2x^2 + x - 6$$

$$\frac{dy}{dx} = 4x^1 + 1 = 4x + 1$$

For turning points $\frac{dy}{dx} = 0$, that is, $4x + 1 = 0 \Rightarrow x = -\frac{1}{4}$

Substitute back in the original equation to find the y-value.

$$y = \left(-\frac{1}{2} - 3\right)\left(-\frac{1}{4} + 2\right)$$

$$= \left(-\frac{1}{2} - 3\right)\left(1\frac{3}{4}\right) = -6.125$$

The coordinates of the turning point are $(-0.25, -6.125)$.

The question asks you to write down the equation of the tangent at the turning point. A sketch of the curve will help you see that the *tangent at a turning point is a horizontal line*. Its equation will be $y = -6.125$

Using your GDC

The sketch of the curve $y = 2x^2 + x - 6$ looks like this:

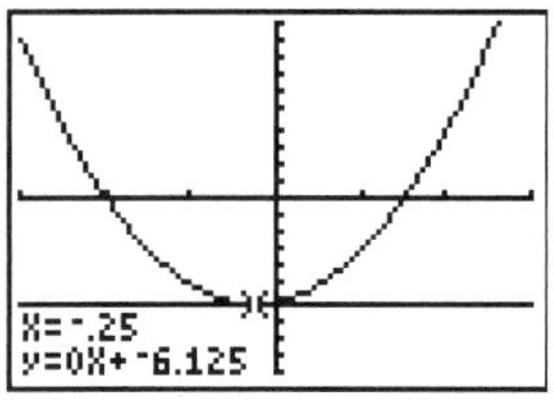

7.5.2 Local maximum and minimum points

So far we have found turning points, where the gradient of the tangent is zero. But how can we tell whether the turning point we have found is a maximum or a minimum? Or indeed, a point of inflexion?

Having found the coordinates of the turning point, one method is to sketch the graph on the GDC and look at it. Another method is to find the gradients at points on the curve which have an x-value slightly smaller and slightly larger than the turning point, and draw a small diagram with three straight lines representing tangents to the curve at the turning point and just to its left and right, taking particular care with the sign of the gradient. We don't need to know the exact values, just whether they are positive or negative. Be careful not to go too far to the right or left because other turning points can interfere with the decision you are trying to make.

Example 7.5.2a

The previous example has a turning point at (–0.25, –6.125). Determine whether it is a maximum or a minimum.

Solution

Take a value of x just smaller than –0.25 which is easy to work with, say $x = -1$.

Gradient $\frac{dy}{dx} = 4x + 1 = 4(-1) + 1 = -3$, that is, negative.

Take a value of x just greater than –0.25 which is easy to work with, say $x = 0$.

Gradient $\frac{dy}{dx} = 4x + 1 = 4\,(0) + 1 = 1$, that is, positive.

Fill in the table:

	$x < -0.25$	$x = -0.25$	$x > -0.25$
Sign of gradient	**negative**	**zero**	**positive**
	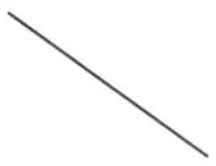	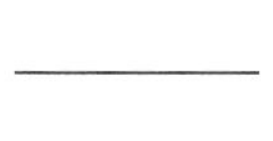	

You can see by looking at the table that the turning point is a (local) minimum. Alternatively you can see by looking at the graph on your GDC and observing that for $x < -0.25$ the function is decreasing and for $x > -0.25$ the function is increasing.

Example 7.5.2b

A sheet of thin card 50 cm by 100 cm has a square of side x cm cut away from each corner and the sides folded up to make a rectangular open box.

a Find the volume, V, of the box in terms of x.

b Using calculus, find the value of x which gives a maximum volume of the box.

Solution

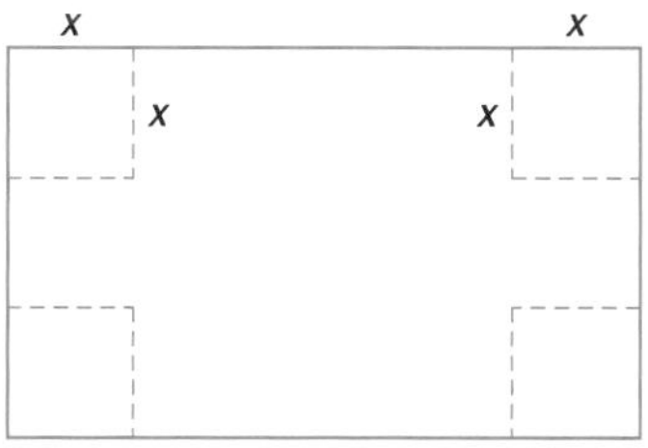

a Volume V = (length) × (width) × (height)
$= (100 - 2x)(50 - 2x)x\,\text{cm}^3$

b The word "maximum" reminds us that we should differentiate. Multiply out the brackets by multiplying the first two brackets first to give:

$$V = (5000 - 200x - 100x + 4x^2)x$$
$$= (5000 - 300x + 4x^2)x$$
$$= 5000x - 300x^2 + 4x^3$$

For turning points

$$\frac{dV}{dx} = 0 \Rightarrow 5000 - 600x + 12x^2 = 0 \Rightarrow 1250 - 150x + 3x^2 = 0$$

Using your GDC we find that $x = 10.57$ or 39.43.

x cannot be 39.43 because $50 - 2 \times 39.43$ is negative.

When $x < 10.57$, say $x = 10$, $\frac{dV}{dx} = 1250 - 1500 + 3(10)^2 = 50$, that is, positive.

When $x > 10.57$, say $x = 11$, $\frac{dV}{dx} = 1250 - 150(11) + 3(11)^2 = -37$, that is, negative.

	$x < 10.57$ (say $x = 10$)	**$x = 10.57$**	**$x > 10.57$ (say $x = 11$)**
Sign of gradient	**positive**	**zero**	**negative**
			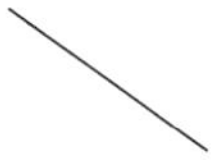

Thus we have a maximum volume when $x = 10.6$ cm.

Using your GDC

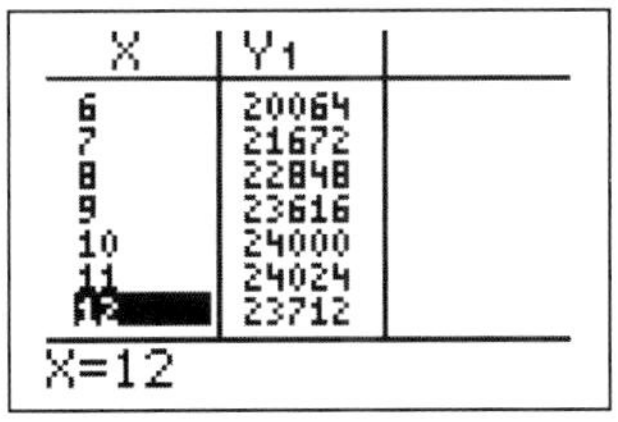

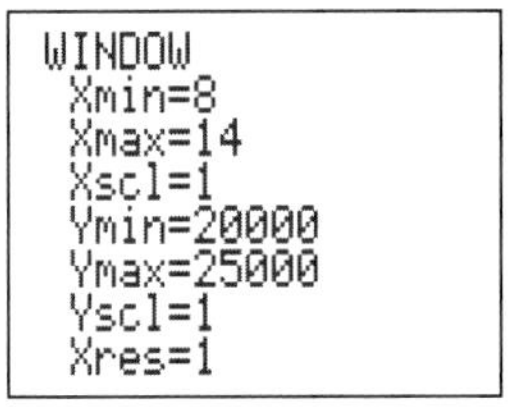

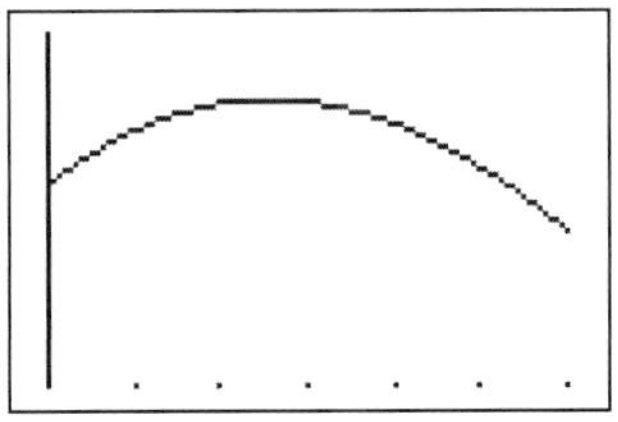

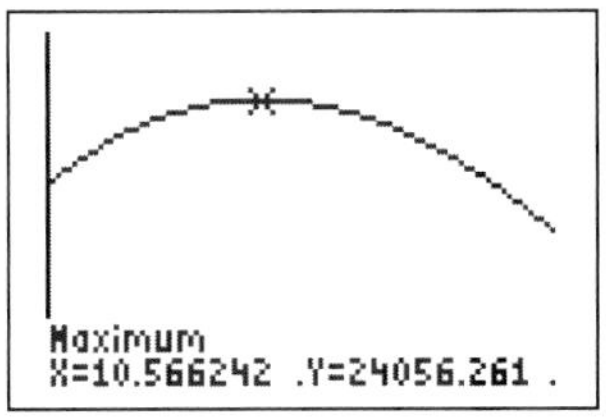

Thus we have a maximum volume when $x = 10.6$ cm, as before.

In Paper 2, working needs to be shown to gain a maximum score and so you should always show your method and check by GDC.

Example 7.5.2c

Ravi knows that the product of two positive numbers, x and y, is 200.

a Find a formula for their sum, S, in terms of x only.

b Use calculus to find the minimum value of S.

Solution

a $x \times y = 200 \Rightarrow y = \dfrac{200}{x}$

Sum $S = x + y = x + \dfrac{200}{x}$

b The word "minimum" suggests we must differentiate.

Write $S = x + 200x^{-1}$

For turning points $\dfrac{dS}{dx} = 0 \Rightarrow 1 + (-1)(200)x^{-2} = 1 - \dfrac{200}{x^2} = 0$

This equation can be solved by writing it as

$1 = \dfrac{200}{x^2} \Rightarrow x^2 = 200 \Rightarrow x = \sqrt{200} = 14.14$

$S = 14.14 + 200/14.14 = 28.3$

Now we must prove this is a minimum value and not a maximum value. Take values of x either side of 14.1, say 14 and 15 and substitute into the gradient $\dfrac{dS}{dx}$.

When $x < 14.1$, say $x = 14$, $\dfrac{dS}{dx} = 1 - \dfrac{200}{14^2} = -0.020$, that is, negative.

When $x > 14.1$, say $x = 15$, $\dfrac{dS}{dx} = 1 - \dfrac{200}{15^2} = 0.111$, that is, positive.

	$x < 14.1$ (say $x = 14$)	**$x = 14.1$**	**$x > 14.1$ (say $x = 15$)**
Sign of gradient	**negative**	**zero**	**positive**
	╲	—	╱

Therefore this is a minimum and the minimum value of S is 28.3.

You can check this using the GDC.

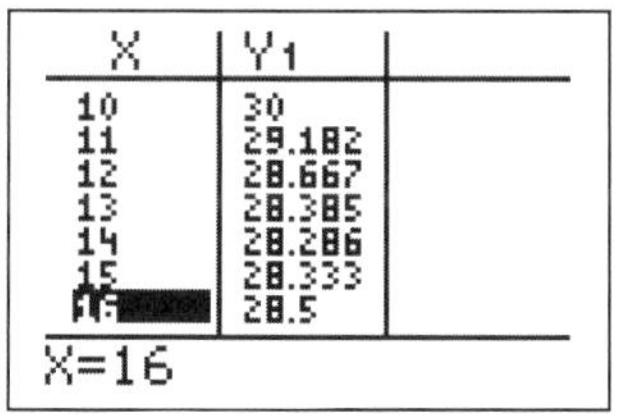

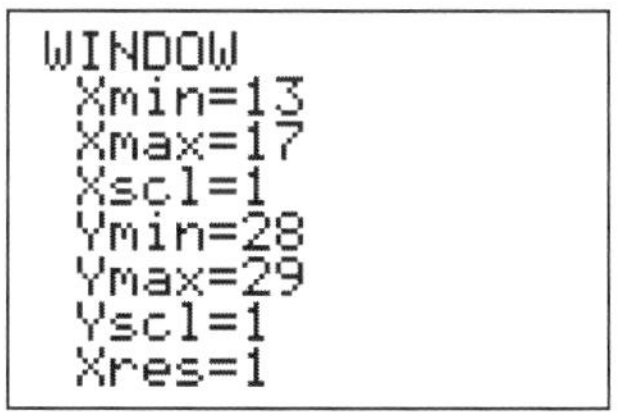

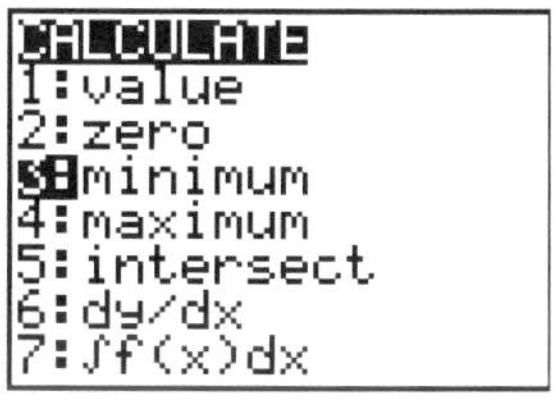

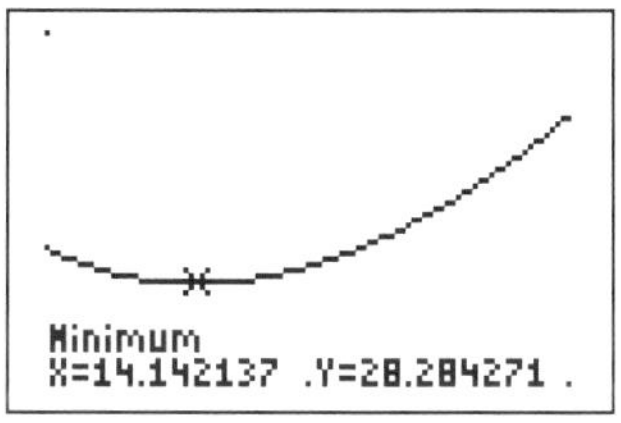

The minimum value of S is when $x = 14.1$ and is 28.3 to 3 s.f.

Exercise 7.5

1 Find the coordinates of the stationary point on the curve whose equation is $y = 3x^2 - 12x + 4$.

2 For the function with equation $f(x) = x^2 + \frac{2}{x}$, $x > 0$, find the coordinates of the turning point and determine whether it is a (local) maximum or minimum.

3 Find the coordinates of the points on the curve $y = x^3 - 6x^2 + 9x + 2$ at which y has a stationary value. Determine for each value you find, whether it is a maximum or a minimum.

4 Find the coordinates of the stationary points on each of these curves and determine their nature.

a $y = x^2 - 5x + 7$

b $y = 10 - 12x + 3x^2$

5 A rectangle has width x cm and length y cm. It has a constant area 20 cm^2.

a Write down an equation involving x, y and 20.

b Express the perimeter, P, in terms of x only.

c Find the value of x which makes the perimeter a minimum and find this minimum perimeter.

6 A closed box made of wood is $4x$ cm long, x cm wide and h cm high. The volume of the box is 320 cm^3.

a Write down an equation involving x, h and 320 and show that $h = \frac{80}{x^2}$.

b Write down an expression for the surface area, A, of the closed box in terms of x and h.

c Use the answer to part **a** to write down an expression for A in terms of x only.

d Use calculus to find the value of x which makes the surface area a minimum. Check your answer by GDC.

7 A closed cylindrical can of height h cm and radius x cm contains 350 ml of soft drink.

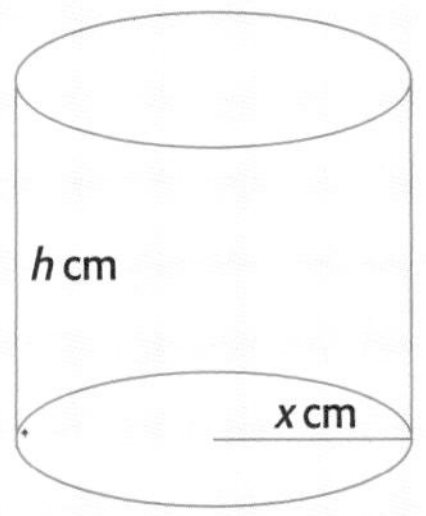

a Write down an equation involving 350, x and h.

b Express the surface area, S, in terms of x and h.

c Use part **a** to show that the surface area $S = 2\pi x^2 + \frac{700}{x}$.

d Find the value of x and the corresponding value of h that makes h cm the surface area a minimum.

e Find the minimum surface area giving your answer to the nearest integer.

8 At time t seconds after a ball is thrown vertically upwards, its height, h metres, above the point of projection is given by the formula $h = 18t - 5t^2$.

a Find the time when the ball is at its maximum height.

b Find the maximum height.

9 An open rectangular box is made from thin cardboard. The base is $2x$ cm long and x cm wide and the volume is 50 cm^3. Let the height be h cm.

a Write down an equation involving 50, x and h.

b Show that the area, y cm^2, of cardboard used is given by $y = 2x^2 + 150x^{-1}$.

c Find the value of x that makes the area a minimum and find the minimum area of cardboard used.

10 Across

1 Gradient of line joining (2, −6) and (5, 30)

2 $f(4)$ where $f(x) = 3x^2 + 0.5x - 3$

4 $f''(3)$ where $f(x) = 5x^4 - 3x^3 - 2x^2 + x$

5 Gradient of tangent to curve given by $y = x^4 - 2x^3 - 15x$ at the point (5, 300)

6 $11^2 + 10^2$

8 $f'(0.25)$ when $f(x) = \dfrac{-3}{x}$

9 $\dfrac{ds}{dt}$ when $t = 5$, $s = 7t + 4.9t^2$

Down

1 One of the intercepts of $y = x^2 - 13x - 30$ with the x-axis

3 Gradient of $y = \dfrac{3x^2}{2} + \dfrac{4}{7}$ at the point where $x = 24$

6 $f''(7)$ where $f(x) = \dfrac{1}{3}x^3 + 5x^2 + 2x - 4$

7 y-coordinate of minimum point on curve given by $y = x^2 + 2x + 17$

Theory of knowledge

Confessions of a Prep School Mathematics Teacher – The Mathematical Experience –

P. Davis & R. Hersh

In this book we have tried to give you a feel for the historical development of mathematics and the motivation and philosophy behind its construction through the ages.

The dispute between Newton and Leibniz touched on at the start of this topic has become a classic story in the history of mathematics. Do you think it is important to learn about this sort of material? Read the following (shortened) article and discuss.

Ted Williams (pseudonym) is the Chairman of the Mathematics Department in a fine private school in New England. Williams, who is in his early forties, teaches mathematics, physics and general science. He also coaches the boys' baseball team. He says that he prefers to teach mathematics rather than physics because it is hard to keep up with the new developments in physics. Williams has a master's degree in mathematics from an Ivy League school and has had an introductory college course in philosophy. He has read a bit of the history of mathematics. He says that his school expects him to do too much and he has very little time for reading.

Williams says that the history and philosophy of mathematics just don't come up in his classroom.

In answer to the question of whether mathematics is discovered or invented, he answered with a snap, "There's not much difference between the two. Why waste time trying to figure it out? The thing that is important is that doing mathematics is fun. That's what I try to put across to the kids."

When pressed harder on this question, he said, "Well, I think it's discovered."

In answer to the question as to whether there is a difference between pure and applied mathematics, Williams answered, "Pure mathematics is a game. It's fun to play. We play it for our own sake. It's more fun than applying it. Most of the mathematics I teach is never used by anyone. Ever. There's no mathematics in fine arts. There's no mathematics in English. There's no mathematics to speak of in banking. But I like pure mathematics. The world of mathematics is nice and clean. Its beautiful clarity is striking. There are no ambiguities."

"But there are applications of mathematics?"

"Of course."

"Why is mathematics applicable?"

"Because nature observes beautiful laws. Physicists didn't get far before mathematics was around."

"Is there beauty in mathematics?"

"Oh yes. For example, if you start with a few axioms for a field, you get a whole powerful theory. It's nice to watch a theory grow out of nothing."

"Is there such a thing as mathematical intuition?"

"Oh, yes. You see it in students. Some are faster than others. Some have more. Some have less. It can be developed but that takes sweat. Mathematics is patterns. If a person has no visual sense, he is handicapped. If one is 'tuned in', then one learns fast and the subject is fascinating. Otherwise it is boring. There are many parts of mathematics that bore me. Of course, I don't understand them."

"How would you sum it all up?"

"As a teacher I am constantly confronted by problem after problem that has nothing to do with mathematics. What I try to do is to sell mathematics to the kids on the basis that it's fun. In this way I get through the week."

Questions

1 Mr Williams has several very outspoken views on mathematics. Do you agree with all of his views, some of his views or none of his views?

2 Do you think that mathematics is invented or discovered? Give good reasons to back up your view.

3 How does mathematical knowledge differ from knowledge in the other areas?

Past examination questions for topic 7

Most of these are parts of questions because calculus was an option in the previous syllabus (up to November 2005) and only appeared on Paper 2 then. Parts of questions which are not on the new syllabus have been omitted. May 2006 questions are written for the new syllabus.

1 A function is defined by $f(x) = 3x^4 + \frac{2}{x} - \frac{x}{4} + 1, \quad (x \neq 0)$.

a Calculate the second derivative, $f''(x)$.

b Find the value of $f''(x)$ at the point $\left(1, \frac{23}{4}\right)$. *M06q14*

2 The function $g(x)$ is defined by

$g(x) = \frac{1}{8}x^4 + \frac{9}{4}x^2 - 5x + 7, \quad x \geq 0$.

a Find $g(2)$.

b Calculate $g'(x)$.

The graph of the function $y = g(x)$ has a tangent T_1 at the point where $x = 2$.

c i Show that the gradient of T_1 is 8.

ii Find the equation of T_1. Write the equation in the form $y = mx + c$.

d The graph has another tangent T_2 at the point $\left(1, \frac{35}{8}\right)$.

T_2 has zero gradient. Write down the equation of T_2.

e i Sketch the graph of $y = g(x)$ in the region $0 \leqslant x \leqslant 3$, $0 \leqslant y \leqslant 22$.

ii Add the two tangents T_1 and T_2 to your sketch, in the correct positions. *M06q3*

3 Consider the function $f(x) = x^3 + 7x^2 - 5x + 4$.

a Differentiate $f(x)$ with respect to x.

b Calculate $f'(x)$ when $x = 1$.

c Calculate the values of x when $f'(x) = 0$.

d Calculate the coordinates of the local maximum and the local minimum points.

e On graph paper, taking axes $-6 \leqslant x \leqslant 3$ and $0 \leqslant y < 80$, draw the graph of $f(x)$ indicating clearly the local maximum, local minimum and y-intercept. *N05q8(part)*

4 The cost of producing a mathematics textbook is \$15 (US dollars) and it is then sold for \$$x$.

a Find an expression for the profit made on each book sold.

A total of $(100\,000 - 4000x)$ books is sold.

b Show that the profit made on all the books sold is

$P = 160\,000x - 4000x^2 - 1500\,000$

c i Find $\frac{dP}{dx}$.

ii Hence calculate the value of x to make a maximum profit.

d Calculate the number of books sold to make this maximum profit. *N04q8(part)*

5 A function is given as $y = ax^2 + bx + 6$.

a Find $\frac{dy}{dx}$.

b If the gradient of this function is 2 when x is 6 write an equation in terms of a and b.

c If the point (3, −15) lies on the graph of the function find a second equation in terms of a and b. *N04q8(part)*

6 Consider the function $g(x) = x^4 + 3x^3 + 2x^2 + x + 4$.

Find

a $g'(x)$

b $g'(1)$. *N04q8(part)*

7 The height (cm) of a daffodil above the ground is given by the function $h(w) = 24w - 2.4w^2$, where w is the time in weeks after the plant has broken through the surface ($w \geq 0$).

a Calculate the height of the daffodil after two weeks.

b i Find the rate of growth, $\frac{dh}{dw}$.

ii The rate of growth when $w = k$ is 7.2 cm per week. Find k.

iii When will the daffodil reach its maximum height? What height will it reach?

c Once the daffodil has reached its maximum height, it begins to fall back towards the ground. Show that it will touch the ground after 70 days. *M04q8(part)*

8 Consider the function $f(x) = x^3 - 4x^2 - 3x + 18$.

a Find $f'(x)$.

b Find the coordinates of the maximum and minimum points of the function.

c Find the values of $f(x)$ for a and b in the table.

x	−3	−2	−1	0	1	2	3	4	5
f(x)	−36	*a*	16	*b*	12	4	0	6	28

d Using a scale of 1 cm for each unit on the x-axis and 1 cm for each 5 units on the y-axis, draw the graph of $f(x)$ for $-3 \leq x \leq 5$. Label clearly.

e The gradient of the curve at any particular point varies. Within the interval $-3 \leq x \leq 5$, state all the intervals where the gradient of the curve at any particular point is
i negative ii positive. *M03q8(part)*

9 Consider the function $f(x) = 2x^3 - 3x^2 - 12x + 5$.

a Find $f'(x)$.
b Find the gradient of the curve $f(x)$ when $x = 3$.
c Find the x-coordinates of the points on the curve where the gradient is equal to -12.
d Calculate the x-coordinates of the local maximum and minimum points.
e For what values of x is the value of $f(x)$ increasing? *N02q8(part)*

10 The length of the base of a cuboid is twice the width x, and its height is h centimetres, as shown in the diagram. Its total surface area is A cm² and its volume is V cm³.

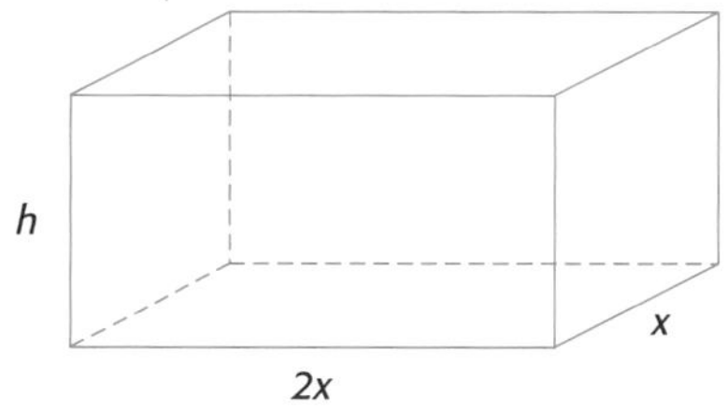

a Show that $A = 4x^2 + 6xh$.
b If $A = 300$, find an expression for h, in terms of x.
c If $A = 300$, show that the volume $V(x) = 100x - \dfrac{4x^3}{3}$
d Find $V'(x)$.
e Solve $V'(x) = 0$.
f Find the maximum volume.
g What is the height of the cuboid when the volume is maximized? *M02q8(part)*

TOPIC 8

Financial mathematics

In this topic standard three letter abbreviations will be used.
Some common ones are:
AUD----Australian dollar
EUR----euro
GBP----British pound
JPY----Japanese yen
USD----U S dollar
In addition, symbols can be used for the dollar \$, the euro €, and the pound £.
Other currencies may be used and these will be named individually as necessary and addressed with their three-letter symbols.
Most currencies have their basic amount split into hundredth units (for example pounds-pence or dollars-cents), but there are a few currencies which do not have hundredth units or that use thousandths instead.
Usually answers should be given correct to the nearest currency unit (2 decimal places).

Symbol for the Japanese yen

8.1 Currency conversions

Countries have many different names for their currencies. Because national economies also differ greatly, a unit of currency in one country almost always has a different value to a unit of currency in another country. This means that we have to apply conversion rates when we exchange currencies.

Exchange rates vary over time, even daily. They respond to world events such as war, peace and political upheaval as well as to local economic conditions, resource availability and speculation. The rates used here are those prevailing during mid-2006.

Currency conversion is very easy. It's just a matter of multiplication or division.

Money is like any other tradeable commodity.

If 2 cakes cost 2.50 USD then 1 cake costs $\frac{2.50}{2} = 1.25$ USD.

Similarly, if 2 euros cost 2.50 USD then 1 euro costs 1.25 USD.

The only slight complication arises because the exchange shops or banks who do the transaction for you want to make some money out of it. They can do this in one of two ways. Either they charge a fee called **commission** for the service, usually at a percentage rate of the total, or alternatively they charge different rates for the selling or buying of a currency, chosen to be in their favour.

Did you know that...?

In Papua New Guinea the currency is a Kina.
One Kina = 100 Toea.
The name Kina originates from the name of a rare kind of shell that was highly valued and was used for bartering, bride-price etc. before contact with other cultures.

Ruddy coloured Kina Shell of the Foi tribe, with point and scratch decoration at the rims.
Dimensions of shell (B × H): 205 mm × 175 mm

Example 8.1a

In the year 2000 Elke was planning to go on holiday from Germany to Greece. She changed 2000 euros to Greek drachma (GRD) at a rate of 1 EUR = 340.75 GRD.

a Calculate how many drachma she received.

Elke became ill. She had to cancel her holiday. She sold the drachma back to the bank at a rate of 1 EUR = 347.82 GRD.

b i How many euros did she receive?

ii How much did the bank gain from the two transactions?

c Calculate Elke's loss as a percentage of her original 2000 EUR.

> *Theory of knowledge*
> Why do we have different currencies?
> Would it not be easier to have one universal currency?
> What would be the advantages and disadvantages?

Solution

a $2000 \times 340.75 = 681\,500$ GDR

b i This time we have to divide as the transaction is in the other direction.

$\frac{681500}{347.82} = 1959.35$ EUR

ii The bank made $2000 - 1959.35 = 40.65$ EUR

iii Loss is $\frac{40.65}{2000} \times 100\% = 2.03\%$ (3 s.f.)

Example 8.1b

Jin is travelling on business from Korea to Japan.

The bank sells Japanese yen (JPY) at a rate of 1 Korean won (KRW) = 0.1219 JPY.

A commission is charged at a rate of 2% in KRW.

Jin pays 1 million KRW. How much does she receive in JPY (to the nearest JPY)?

Solution

The bank will keep 2% of 1 million KRW = 20 000 KRW.

So Jin changes 980 000 KRW to yen.

$980\,000 \times 0.1219 = 119\,462$ JPY

Example 8.1c

Zsolt travels from Budapest to London. He exchanges 300 000 forints (HUF) for British pounds before he leaves. The bank is selling GBP at a rate of 1 GBP = 411.78 HUF.

a Calculate how many GBP Zsolt receives.

b Zsolt spends 622.54 GBP in London. He has a choice for the remaining pounds.

He can

i change all the money back to HUF in London. The rate is 1 GBP = 405.55 HUF but the bank charges commission of 1.6% or

ii wait and do the exchange at home at a rate of 1 GBP = 397.85HUF

Calculate which option is better for Zsolt.

Solution

a $\dfrac{300\,000}{411.78} = 728.54\,\text{GBP}$

b After spending 622.54 GBP there is 728.54 − 620.54 = 108.00 GBP left.

i In London the bank takes 1.6% of the 108 GBP so he exchanges $0.984 \times 108 = 106.27$ GBP

This amount buys $106.272 \times 405.55 = 43\,097.80$ HUF

ii At home he receives $108 \times 397.85 = 42\,967.80$ HUF

Option **i** is better.

Exercise 8.1

Unless instructed otherwise, give all answers correct to 2 decimal places.

1 A bank in Cape Town advertises these rates for USD and South African Rand (ZAR):

We sell 1 USD = 7.1842 ZAR

We buy 1 USD = 6.9031 ZAR

a How many USD will you receive if you exchange 2500 ZAR?

b Suppose you spend 220 USD then change the remainder back to ZAR. How many ZAR will you receive?

2 Pramana will travel on business from Singapore to Indonesia.

The bank sells Indonesian rupiah (IDR) at a rate of 1 Singapore dollar (SGD) = 5778.6008 IDR.

A commission is charged at a rate of 1.8% in SGD.

Pramana pays 1400 SGD. Calculate how much he receives in IDR (correct to the nearest IDR).

3 In 2001 Iago travelled from Santiago, Chile to Madrid. He exchanged 400 000 Chilean peso (CLP) for Spanish pesetas (ESP) before he left. The bank was selling ESP at a rate of 1 ESP = 4.11918 CLP.

a Calculate how many ESP Iago received.

Iago spent 85 000 ESP in Madrid. He had a choice for the remaining pesetas. He could

b change all the money back to CLP in Madrid. The rate was 1 ESP = 4.0623 CLP and the bank was charging commission of 2.5% or

c wait and do the exchange at home at a rate of 1ESP = 3.9912 CLP

Calculate which option was better for Iago.

d Does Spain still use pesetas?

4 Annie lives in Manchester. She is going to visit friends in France. She goes to the bank and changes 400 GBP to euros at a rate of 1GBP = 1.4798 EUR

a Calculate how many euros she receives.

Annie gets a call from work. She has to cancel her trip. She sells the euros back to the bank at a rate of 1 GBP = 1.5160 EUR.

b i How many GBP does she receive?

ii How much has the bank gained from the two transactions?

c Calculate Annie's loss as a percentage of her original 400 GBP.

5 On 16 December 2004, the World Bank approved finance of \$60 million to assist with locust control in West Africa. Find an internet site which converts currencies. (There are several.) State the names of the currencies used in Gambia and Nigeria. Give the three letter abbreviations also. Convert 60 million USD to those currencies at the current rate.

8.2 Simple interest

Money invested in a bank or savings institution is called **capital**. When you save money in this way, the bank will use it to invest in various ways or lend it to businesses or individuals and charge them for the loan. Financial institutions make a lot of profit in this way, so to encourage you to invest with them, they offer a reward, called **interest**. Usually the amount of interest given will depend on the capital and on the length of time the capital has been invested.

The simplest relation between interest and capital is called **simple interest**.

To calculate the amount of simple interest received, multiply the capital by the percentage rate for a time period and by the number of time periods. Then divide by 100.

$$I = \frac{Crn}{100}$$

where C = capital
r = % rate
n = number of time periods
I = interest

The time period is often per year and you will see this written (from Latin) as per annum or just p.a.

Example 8.2a

Calculate the total interest received if an amount of 8000 GBP is invested at a simple interest rate of 4% per annum for a period of 10 years.

If the interest is added to the account how much capital is there after the 10 years?

Solution

Apply the formula with $C = 8000$, $r = 4$, $n = 10$

$$I = \frac{8000 \times 4 \times 10}{100} = 3200 \text{ GBP}$$

Add this amount to the original 8000 GBP.

The total in the account after 10 years is 11 200 GBP.

Example 8.2b

Calculate the yearly rate of simple interest needed for an amount of 600 USD to earn total interest of 120 USD if invested for 3 years.

Solution

Apply the formula with $C = 600$, $n = 3$, $I = 120$.

$$120 = \frac{600 \times r \times 3}{100} = 18r$$

Hence $r = \frac{120}{18} = \frac{20}{3} = 6.67$ (3 s.f.)

A rate of 6.67% is needed.

Example 8.2c

Calculate the time needed (in whole years) for a capital amount of 4000 EUR to double, if invested at a simple interest rate of

a 5% p.a. **b** 7% p.a.

Solution

a $C = 4000$, and this has to become 8000.

Hence the interest $I = 8000 - 4000 = 4000$.

$$4000 = \frac{4000 \times 5 \times n}{100} = 200n$$

Hence $n = \frac{4000}{200} = 20$ years

b The same process but with 7 replacing 5

$$4000 = \frac{4000 \times 7 \times n}{100} = 280n$$

Hence $n = \frac{4000}{280} = 14.3$ years (3 s.f.)

Now remember that the interest is only applied after whole years and after 14 years, the total will not quite have doubled. We have to round *up*, even though the result was less than 14.5. The correct answer is 15 years.

Connection with arithmetic sequences

The total amount, A, in an account after adding simple interest, I, to capital, C, for n time periods is

$$A = C + I = C + n\frac{Cr}{100}.$$

This should remind you of the formula for the nth term in an arithmetic sequence. (See Topic 2, Section 5, page 86)

$$u_n = u_1 + (n-1)d.$$

In fact the amounts in the bank account after each year do form an arithmetic sequence with first term $u_1 = C$, and common difference $d = \frac{Cr}{100}$.

Notice though that the way the n's appear does not correspond exactly. This is just because we think of the first term in the arithmetic sequence as occurring when $n = 1$ while for the simple interest calculation the case $n = 1$ is already the second term in the sequence and C is the amount when $n = 0$ before any interest was added.

The sum of the terms has no special meaning in the context of bank accounts, however, it takes on significance in examples like the next one.

Example 8.2d

John must repay a loan of 600 euros from his friend over a period of twelve months. In each month he will pay 50 euros plus an amount of interest equal to 6 euros in the first month, 5.50 in the second reducing by half a euro each month until it reaches 0.5 euros in the twelfth month.

Find how much John pays in total.

Solution

The payments form an arithmetic sequence with first term $u_1 = 50 + 6 = 56$ euros and with common difference $d = -0.5$.

The sum is $S_{12} = \frac{12}{2}(2 \times 56 + (12 - 1) \times (-0.5)) = 639$ euros.

Exercise 8.2

1 Calculate the total interest received if an amount of 2000 GBP is invested at a simple interest rate of 3.5% per annum for a period of 15 years. If the interest is added to the account how much capital is there after the 15 years?

2 a Calculate the time needed (in whole years) for a capital amount of 350 USD to reach 500 USD, if it is invested at a simple interest rate of 6% p.a.

b Suppose interest is applied instead at a rate of 3% every 6 months. Does the total reach 500 USD any quicker?

3 Calculate the yearly rate of simple interest needed for an amount of 1000 JPY to earn total interest of 200 JPY if invested for 4 years.

8.3 Compound interest, depreciation

8.3.1 Use of the compound interest formula

Adding simple interest to an investment means that the capital increases gradually. However, it is the initial capital value that is always used to calculate later interest payments. This seems unfair to the investor, so very often, the bank will offer compound interest payments instead. Now the interest is calculated as a percentage of the *new* capital amount through each period.

Compound interest is the interest calculated for a number of time periods in which the interest is added to the capital each time and the new capital total is used to calculate the next interest payment.

Take the letters to have the same meanings as in Section 8.2.

After *one* time period, the new capital is $C + \dfrac{Cr}{100} = C\left(1 + \dfrac{r}{100}\right)$.

The effect is to multiply the starting capital for the period by the quantity $\left(1 + \dfrac{r}{100}\right)$.

If we just want to know the interest for that first time period, we simply subtract C again to get $I = C\left(1 + \dfrac{r}{100}\right) - C = \dfrac{Cr}{100}$.

For the second time period we want to apply compound interest so we use the new capital to calculate the next interest payment. Using previous experience, we multiply the new capital by the quantity $\left(1 + \dfrac{r}{100}\right)$ and get

$$C\left(1 + \frac{r}{100}\right)\left(1 + \frac{r}{100}\right) = C\left(1 + \frac{r}{100}\right)^2$$

The total interest after 2 periods is $I = C\left(1 + \dfrac{r}{100}\right)^2 - C$.

Comparing this with the simple interest for 2 periods shows that the compounded amount is a little larger.

Each time we progress to the next time period we multiply the capital by a further factor $\left(1 + \dfrac{r}{100}\right)$. If we invest for n such time periods, there will be n such factors.

After n time periods of compounding the capital has become $C\left(1 + \dfrac{r}{100}\right)^n$ and the total interest is $I = C\left(1 + \dfrac{r}{100}\right)^n - C$.

Example 8.3.1a

Lars invests 10 000 Swedish krona (SEK) at a rate of 5.1% compounding yearly. Calculate the amount in Lars's account after 4 years and find how much of that amount is interest. Give answers correct to 2 d.p.

Solution

Total after 4 years is $T = 10\,000\left(1 + \dfrac{5.1}{100}\right)^4 = 12\,201.4$ SEK.

The interest is 12 201.43 – 10 000 = 2201.43 SEK.

Example 8.3.1b

Giovanni's bank manager told him that if he invests 3000 EUR now, compounding yearly, it will be worth 4600 EUR in 5 years' time.

What is the rate of interest p.a?

Solution

$$4600 = 3000\left(1 + \frac{r}{100}\right)^5$$

hence $\left(1 + \frac{r}{100}\right)^5 = \frac{4600}{3000}$

so $\left(1 + \frac{r}{100}\right) = \sqrt[5]{\frac{4600}{3000}}$

This results in $r = 100\left(\sqrt[5]{\frac{4600}{3000}} - 1\right) = 8.92$ (3 s.f.).

The rate is 8.92%.

Example 8.3.1c

Marina is saving to buy a small boat that costs 35 000 USD.
She has 28 000 USD in an account that pays 5.34% interest compounding yearly.
How long must Marina wait before she can buy the boat?

Solution

$$35\,000 = 28\,000\left(1 + \frac{5.34}{100}\right)^n \text{ hence } \left(1 + \frac{5.34}{100}\right)^n = \frac{35\,000}{28\,000} = \frac{35}{28} = \frac{5}{4}$$

This is a little different to previous calculations because the unknown variable n is in a power and we do not have available a method to extract it algebraically. When n is the variable, the compound interest formula is an example of an exponential function, (see Topic 4.4, page 194).

At this stage there is a choice. You can apply the solving facility of the GDC to this equation or you can use a trial and error method to guess a solution, then improve it.

If you know how to use the log function on your GDC then that is permitted, but that method is not included in the mathematical studies syllabus.

In both cases, you have to estimate a first guess. The best way to do that is to pretend that the interest is simple instead of compound.

In that case you would be solving $\frac{n \times 5.34}{100} = \frac{35\,000 - 28\,000}{28\,000} = \frac{1}{4}$

Only a rough guess is needed so $n = 5$ will do.

Use $n = 5$ as the guess in the solver to get a result of $n = 4.29$ (3s.f.).

Alternatively try out 5 in the compound formula.

This gives $\left(1 + \dfrac{5.34}{100}\right)^5 = 1.30$ (3 s.f.). This number is bigger than $\dfrac{5}{4}$ (which is 1.25). Hence 5 years is too long.

On the other hand trying 4 years in the formula gives

$\left(1 + \dfrac{5.34}{100}\right)^5 = 1.23$ this is less than 1.25. Hence the exact answer for n is between 4 and 5.

Whichever way you did this, you can now conclude that Marina needs to receive interest for 5 years before she has enough to buy the boat.

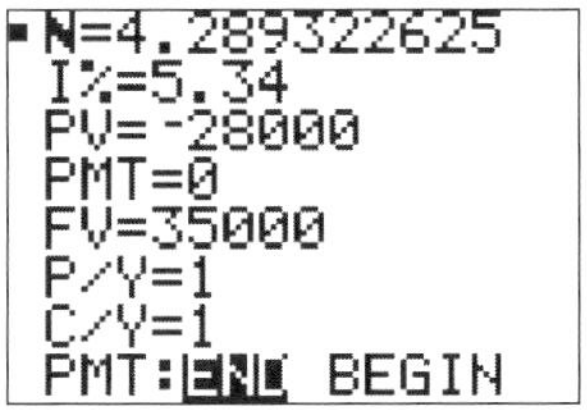

In practice, examples like this are most easily done using the financial application on your GDC. (See Topic 1.8, page 50). However, it is essential that you present explanations for the GDC result and not just an answer.

Subdivided time periods and nominal interest rate

There's a bit more to compound interest. Sometimes a bank will offer to add the interest more often than once a year. Typical periods are 6 months (*half-yearly*), 3 months (*quarterly*), *monthly* or *daily*. Beware though. The interest rate advertised will still be a yearly one, and even more confusing is that the rate is not the actual rate you would end up with if you invested for a year. A rate of this kind is called the **nominal rate**. To use a nominal yearly rate you decide how many of the compounding periods there are in a year and divide the nominal rate by that number. This will be the actual rate that you use for a single time period.

Example 8.3.1d

a The bank in Grabiton is advertising a nominal yearly rate of 5% with compounding applied quarterly. State the number of compounding periods for a 3-year investment and find the actual interest rate applied after each time period.

b Fleur invests 500 GBP in this bank for three years.

Calculate the total capital in her account after this time.

c Suppose Fleur invests this money at a rate of 5% p.a. compounding *only once* a year. How much less interest would she receive?

Solution

a There are 4 quarters in one year so in 3 years there are 12 compounding periods. The interest rate to apply at each quarter is

$$\frac{\text{nominal rate}}{\text{number of periods in one year}} = \frac{5\%}{4} = 1.25\%$$

b Use the compound interest formula for the total, with interest rate 1.25% and $n = 12$.

$$T = 500\left(1 + \frac{1.25}{100}\right)^{12} = 580.38 \text{ GBP}$$

c $$T = 500\left(1 + \frac{5}{100}\right)^{3} = 578.81 \text{ GBP}$$

Not a big difference. It is 1.57 GBP less.

Example 8.3.1e

Here is a real-life example. A bank in Australia is offering a "term-deposit" account with a choice of interest rates. As long as you leave your money with them for 2 years, you can get a nominal rate of

a 5.05% compounded monthly or

b 5.10% compounded quarterly or

c 5.15% compounded half-yearly or

d an actual yearly rate of 5.2%.

Abigail has 7000 AUD to invest. Which is her best option?

Solution

a 12 periods per year. 24 in two years. Monthly rate is $\frac{5.05\%}{12}$.

$$T = 7000\left(1 + \frac{5.05}{12 \times 100}\right)^{24} = 7742.30 \text{ AUD}$$

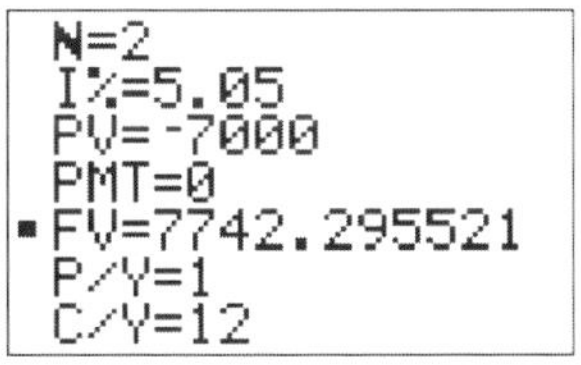

b 4 periods per year. 8 in two years. Quarterly rate is $\frac{5.10\%}{4}$.

$$T = 7000\left(1 + \frac{5.10}{4 \times 100}\right)^{8} = 7746.69 \text{ AUD}$$

c 2 periods per year, 4 in two years. Half-yearly rate is $\frac{5.15\%}{2}$.

$$T = 7000\left(1 + \frac{5.15}{2 \times 100}\right)^{4} = 7749.33 \text{ AUD}$$

d 2 periods in total. Use rate 5.20%.

$$T = 7000\left(1 + \frac{5.20}{100}\right)^{2} = 7746.93 \text{ GBP}$$

Option **c** is best.

Maybe you can see a pattern happening in these calculations.

The structure of the formula is the same as before but we use the total number of time periods for the power and the actual interest rate for one period in the bracket.

If we make the appropriate changes in the earlier compound interest formula we get:

$$\text{Total } T = C\left(1 + \frac{r}{100k}\right)^{kn} \text{ and interest } I = C\left(1 + \frac{r}{100k}\right)^{kn} - C$$

where as usual C is the capital, I is the interest and n is the number of years of investment but now r is the **nominal** interest rate and k is the number of compounding periods in 1 year.

An error seen often in examinations is to forget to apply one or other of these changes. If you then also only give an answer (for example using a GDC) with no working you will receive no marks at all in a Paper 1 question and very few in Paper 2. Working is essential. Write down the correct interest rate for one period and the number of time periods needed. Then show the compound interest formula with the right substitutions and you will get some marks.

Notice that setting $k = 1$ recovers the original compound formula.

Calculations with nominal interest rates can be performed with the finance application on your GDC. Finding the answer this way is acceptable as long as you show some working. A copy of the GDC screen is not really enough. Show the formula with substitutions.

Beware! On both types of GDC the value of N (or n) is the number of years not the number of time periods. The latter appears as C/Y. (See Topic 1.8, page 49).

Example 8.3.1f

Annoushka has 2736.74 EUR in her bank account.

She has left her money there for exactly 3 years at a nominal rate of 4.1% p.a. compounding daily. Calculate, correct to the nearest EUR, how much Annoushka put in the account when she opened it. (Assume there were no leap years in that time.)

Solution

$$2736.74 = C\left(1 + \frac{4.1}{36\,500}\right)^{3\times365}$$

$$C = \frac{2736.74}{\left(1 + \frac{4.1}{36\,500}\right)^{3\times365}} = 2420.02 \text{ EUR}.$$

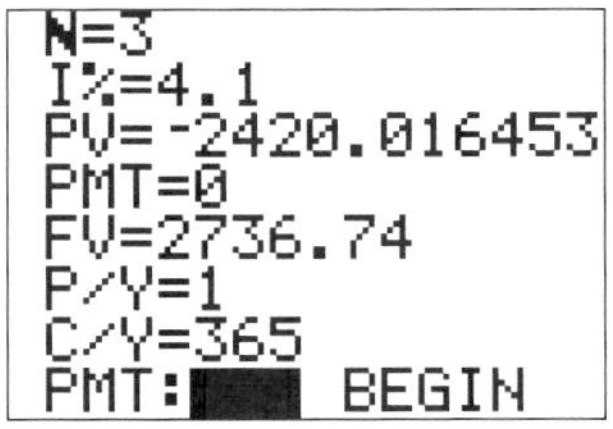

She must have deposited 2420 EUR.

Connection with geometric sequences

When we are **adding** regular equal amounts to a capital figure, as we did with simple interest, we form an arithmetic sequence.

For compound interest we are **multiplying** by equal amounts after each period. This forms a *geometric sequence*. (See Topic 2, Section 6, page 93)

In the formula $T = C\left(1 + \frac{r}{100}\right)^n$ we are multiplying by a factor $\left(1 + \frac{r}{100}\right)$ at each step.

Hence this factor is the common ratio for the geometric sequence. C is the first term (when $n = 0$) and the value of T is the $(n+1)$th term.

Example 8.3.1g

A geometric sequence has third term 212.24 and fifth term 220.82.

If these are the capital amounts in an account offering r% interest at each time period, after 3 and 5 time periods respectively, find r and the initial amount in the account C, correct to the nearest whole number.

Solution

Find r first as follows:

$$\frac{C\left(1+\frac{r}{100}\right)^5}{C\left(1+\frac{r}{100}\right)^3}=\frac{220.82}{212.24}$$

Hence $\left(1+\frac{r}{100}\right)^2=\frac{220.82}{212.24}$

and $r=100\left(\sqrt{\frac{220.82}{212.24}}-1\right)=2.00$ (2 s.f.)

so r is 2 to the nearest whole and the rate must be 2%.

With $r = 2$ we now have $C(1.02)^3 = 212.24$

hence $C=\frac{212.24}{1.02^3}=200$ (3 s.f.)

so $C = 200$ to the nearest whole number

The sum of the terms has no special meaning in the context of bank accounts, however, it takes on significance in examples like the one that follows.

Example 8.3.1h

Amanda has paid school fees for her son for seven years. In the first year, the fees were \$2000. They have increased by 5% p.a. every year.

Find how much Amanda has paid in total.

Solution

The fees form a geometric sequence with common ratio $r = 1.05$ and first term $u_1 = 2000$. We need the sum of the first seven terms. Use the sum formula for geometric sequences.

$$S_7=\frac{1.05^7-1}{1.05-1}=\$16\,284$$

Example 8.3.1i

Joe is saving for retirement. He pays 10 000 GBP into a bank account at the *start* of each year for n years. The account pays 7.1% interest compounding annually. Show that, after n years, the amount, A, that Joe has in the account is

$$A=10\,710\left(\frac{1.071^n-1}{0.071}\right)\text{GBP}$$

If n is 12, calculate A.

Solution

The first 10 000 GBP is invested for n years so it earns interest and accumulates to $10\,000(1.071)^n$.

The second 10 000 is only invested for $(n - 1)$ years so it accumulates to $10\,000(1.071)^{n-1}$.

This continues to the last investment of 10 000 GBP, which stays in the bank for one year and becomes 10 000 × 1.071.

Adding all these amounts together gives the total

$A = 10\,000\ (1.071 + 1.071^2 + 1.071^3 + \dots\dots\dots + 1.071^n)$

$= 10\,000 \times 1.071\ (1 + 1.071 + 1.071^2 + 1.071^3 + \dots\dots\dots + 1.071^{n-1})$

which is the sum of a geometric sequence and can be summed using the formula to give:

$$A = 10\,710\left(\frac{1.071^n - 1}{0.071}\right)$$

When we put $n = 12$, we get $A = 179\,940.85$ GBP.

8.3.2 Depreciation

While money invested in a safe bank account usually gains in amount, we sometimes need to calculate results for items which are *losing* value.

Loss of value is called **depreciation**.

Depreciation at a given rate over a known time period can be calculated with the same formula that we use for compound interest. The important difference to remember is that the rate r must now be *negative*.

Example 8.3.2a

Vijay has paid 300 000 Indian rupees (INR) for a car. The car depreciates at a rate of 9% p.a. Calculate the value, V, of the car in 4 years' time, giving your answer correct to the nearest INR.

Find the percentage loss over the 4-year period.

Solution

Use the compound interest formula with $r = -9$.

$$V = 300\,000\left(1 - \frac{9}{100}\right)^4 = 205\,724.88 \text{ INR}$$

Value V is 205 725 INR correct to nearest whole rupee.

The percentage loss is $\dfrac{300\,000 - 205\,725}{300\,000} = 31.4\%$

Example 8.3.2b

Mary has some shares in a telecommunications company. She paid 7.50 AUD per share 3 years ago but now each share is only worth 3.90 AUD. Calculate the depreciation rate over the 3 years.

Solution.

Here r is the unknown.

$$3.90 = 7.50\left(1 - \frac{r}{100}\right)^3$$

Rearranging:

$$1 - \frac{r}{100} = \sqrt[3]{\frac{3.9}{7.5}} \text{ which results in } r = 100\left(1 - \sqrt[3]{\frac{3.9}{7.5}}\right) = 19.6 \ (3\,\text{s.f.})$$

The depreciation rate is 19.6% p.a.

Example 8.3.2c

Anthony bought a house for 380 000 USD 7 years ago.

Since then, in his area, houses have increased in value by an average of 10% p.a. for the first 5 years but then lost value at a rate of 4% p.a. for the last 2 years.

What is the value, V, of the house now? Give the answer to the nearest 1000 USD.

Solution

This can be solved in two stages or as follows in a single step.

$$V = 380\,000\left(1 + \frac{10}{100}\right)^5\left(1 - \frac{4}{100}\right)^2 = 564\,000 \text{ USD (nearest 1000)}$$

Exercise 8.3

1 Monique has a stamp collection worth 2500 EUR. Her local stamp dealer estimates that it will grow in value by 1.8% a year over the next 5 years. Find how much the collection will be worth after 5 years.

2 Andy owns some land worth 30 000 AUD. The value of land is increasing at a rate of 2% every 3 months. Andy wants to sell for a minimum of 32 000 AUD. How long must he wait to sell?

3 Alla invested 5000 Russian roubles (RUB) at a rate of 6% p.a. After 3 years the rate changed. Alla left her money in the bank for a further two years at the new rate.
At the end of this time she had 6628.15 RUB.
Calculate **a** the amount in the account after the first 3 years
b the new interest rate.

4 Pablo has a rare painting. He paid 15 000 USD for it 5 years ago. The market for this kind of painting has not done well recently and the value has depreciated by 3.22% per year since he bought it. Calculate how much the painting is worth now, correct to the nearest hundred dollars.

5 Use your graphic display calculator to check the answers found in Examples 8.3.1e and 8.3.1f.

8.4 Loan and investment tables, inflation

8.4.1 Construction and use of tables: loan and repayment schemes

Financial institutions offer loans to individuals and to businesses. Typical reasons for taking a personal loan include purchase of a home, car, holiday, furnishings, shares etc. Just as when you deposit money in a bank they will offer you some interest, when you take out a loan, the bank will charge you interest. Usually the rate charged for a loan is higher than that given for a deposit. This is one of the ways in which banks make their profits. If you have a credit card, there will be rules about how interest is applied to the outstanding balance. The rates are often quite draconian but with wise use you can avoid interest payments altogether.

When you apply for a loan, you will usually have some idea how much you need. The bank will inform you of the interest rate, then you might negotiate about how long the loan will last and agree on the repayments. Typically for a home loan, you might make payments once a month (though sometimes more often). Interest might be calculated and added once a month also, though sometimes it accumulates daily.

The repayments are calculated in such a way that each one is enough to pay off the interest added for the most recent period and at the same time reduce the outstanding amount owing.

In this way the loan can be reduced to zero in exactly the time period agreed.

Banks understand that many people will not have the mathematical prowess to calculate their own repayments so they present these in tables.

Example 8.4.1a

Below is a table of monthly repayments for a home loan.

The interest rates given are nominal per annum rates but are to compound monthly.

The repayment is calculated on a loan of $10 000.

Explain how the table works and write down the monthly repayment for a home loan of $100 000 lasting for 20 years, at an interest rate of 7.5%.

How much money do you pay for this loan in total?

Time in years	6.0%	6.5%	7.0%	7.5%	8.0%	8.5%	9.0%
5	193.3280	195.6615	198.0120	200.3795	202.7639	205.1653	207.5836
10	111.0205	113.5480	116.1085	118.7018	121.3276	123.9857	126.6758
15	84.3857	87.1107	89.8828	92.7012	95.5652	98.4740	101.4267
20	71.6431	74.5573	77.5299	80.5593	83.6440	86.7823	89.9726
25	64.4301	67.5207	70.6779	73.8991	77.1816	80.5227	83.9196

Solution

The good news is that there is very little calculation to do.

We look in the row for 20 years and the column for 7.5%.

The figure there is \$80.5593.

This is the monthly payment for a loan of \$10 000. To scale this up to the repayment for a loan of \$100 000 just multiply by 10 $\left(\text{because } \frac{100\,000}{10\,000} = 10\right)$.

The answer is \$805.59 per month.

The total number of these repayments over 20 years will be $12 \times 20 = 240$.

Hence the total paid is $240 \times 805.59 = \$193\,341.60$.

(Notice this is almost double the amount of the original loan, a feature typical of most long-term loans.)

A word on accuracy. The final answer is rounded to 2 d.p. but the amounts in the table are given to higher accuracy. This is because it is anticipated that these amounts will be multiplied by some factor (in our example, by 10). The multiplication will move figures into higher positions to the left and so the third and fourth d.p. might affect the final result in the first and second places.

Example 8.4.1b

Using the table in Example 8.4.1a, compare the total repayments for a loan of 50 000 GBP at 8.0% nominal rate, compounding monthly, with time periods of

a 10 years **b** 25 years.

Solution

Note that the table is valid for any currency.

a First look in the row for 10 years and the column for 8.0%.

The figure there is 121.3276 GBP.

This is the monthly payment for a loan of 10 000 GBP. To scale this up to the repayment for a loan of 50 000 GBP multiply by 5.

The answer is 606.64 GBP per month.

b Now look in the row for 25 years and the column for 8.0%.

The figure there is 77.1816 GBP.

Scale this up to the repayment for a loan of 50 000 GBP.

The answer is 385.91 GBP per month.

This seems less than the monthly repayment in part **a** but:

- in **a** the total number of repayments over 10 years will be $12 \times 10 = 120$, hence the total paid is $120 \times 606.64 = 72\,796.80$ GBP
- in **b** the total number of repayments over 25 years will be $12 \times 25 = 300$, hence the total paid is $300 \times 305.91 = 115\,772.40$ GBP.

The extra time has increased the total payment by almost 43 000 GBP.

You might be wondering exactly how the amounts in the table were calculated. It is possible to work out a formula to generate the payments. The formula is not part of the mathematical studies syllabus (however, see the Challenge box on p. 398).

Maybe it is no surprise that your GDC can tell you the repayment amounts. This is discussed further in Topic 1.8.4 (page 53).

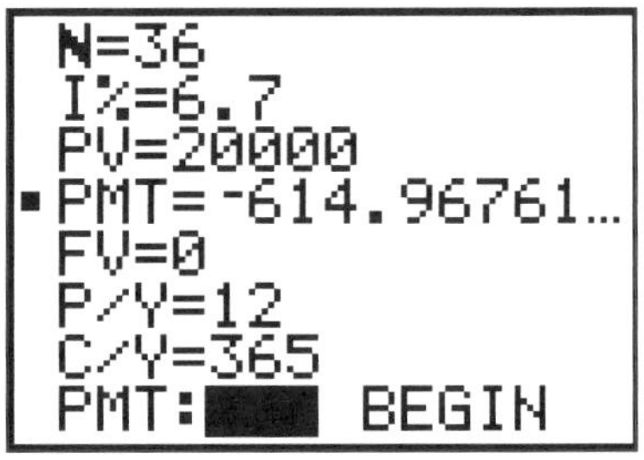

Here we show one screen, for finding the repayment on an amount of 20 000 (any currency), at a nominal yearly rate of 6.7% , compounding daily but with repayments monthly. The loan is to be paid off over 3 years. Notice how the C/Y variable reflects the daily compounding but P/Y shows the number of monthly repayments per year.

N is the total number of repayment periods.

The monthly repayment would be 614.97 and the total payment comes to $36 \times 614.97 = 22\,138.83$.

Example 8.4.1c

Use your graphic display calculator to fill in the missing values x, y, z, w and v in this loan table.

Time in years	4%	6%	y%
1	19.6253	z	20.2268
2	10.0088	10.2094	v
x	4.2451	w	4.8911

The interest rates given are nominal per annum rates but are to compound weekly The repayment is calculated on a loan of $1000 and is to be made on a weekly basis. The repayment time in the left-hand column is in years. The values for x and y should be whole numbers.

Solution

Use the finance application.

x must be found before w and y before v.

The screen for x is given in Topic 1.8.4. on page 53.

The other calculations can all be done by changing values on the same screen and recalculating with the changed values.

We find $x = 5$, $y = 10$, $z = 19.8246$, $w = 4.4541$, $v = 10.3106$.

Challenge (Not for the faint-hearted!)

The formula below gives the repayments, p, for a loan amount, C, at nominal rate, r, and with k compounding periods. It is assumed here that the repayment and compounding periods are the same. See if you can derive the formula.

$$p = C\left(\frac{r}{100k}\right)\frac{\left(1+\frac{r}{100k}\right)^{kn}}{\left(1+\frac{r}{100k}\right)^{kn}-1}$$ for the loan repayment, p, from scratch.

There is a hint with the answers for this section. This formula is not needed in examinations.

Theory of knowledge

Nobody ever plans to get too far into debt, of course. But it usually takes planning – and sometimes years of struggle – to get out of debt and establish new positive personal and financial habits.

While debt used to come with large purchases like houses and cars, today it's common for people to take out loans for every aspect of their lives – from second mortgages to fund other purchases, to loans to fund private school education.

From the use of credit cards for groceries and other out-of-pocket daily expenses to luxury items like resort vacations, consumer debt continues to grow while savings accounts dwindle nationwide, according to recent studies.

When you go to university you will probably have to take out a student loan.

Try to find out more about student loans. What is the average debt at the end of study?

How long does it take to repay the debt on average?

Do you think it is fair that people who have studied have to start their working career with a large debt to be repaid?

Take for example Zoran, who graduates at 23 years of age with a master's degree. During his study he has accumulated a debt of 15 000 euros which he has agreed to pay off over ten years. He finds a position that pays an annual salary of 30 000 euros a year with an annual increase of 3%.

Martin, however, left school at 16 and works at the local supermarket. His annual salary is 17 500 euros with 3% increase each year.

If both men retire when they are 60 years old, find their total incomes.

So, does studying hard really pay off?

8.4.2 Construction and use of tables: investment and saving schemes

Tables are used for displaying many kinds of financial information. From the point of view of the average bank customer, the loan tables already discussed are probably the most important. The reader is invited to perform an internet search for "investment tables" to discover other uses. Some of the information portrayed can require advanced technical expertise in finance, and is purely for experts. In general, mathematical studies candidates are expected to be able to extract information from simple tables, given some instruction about the meaning of the rows and column. Here we give just one simple example, which is really nothing new.

Example 8.4.2

This table shows future values for a £1000 fixed investment using various compound interest rates and time periods. All calculations involved are from Section 8.3.

The values are given only to 2 d.p. to save space.

Time in years	Annual nominal interest rate								
	5%			6 %			7%		
	Compounding period								
	Yearly	**Quarterly**	**Monthly**	**Yearly**	**Quarterly**	**Monthly**	**Yearly**	**Quarterly**	**Monthly**
1	1050.00	1050.95	1051.16	1060.00	1061.36	1061.68	1070.00	1071.86	1072.29
2	1102.50	1104.49	1104.94	1123.60	1126.49	1127.16	1144.90	1148.88	1149.81
5	1276.28	1282.04	1283.36	1338.23	1346.86	1348.85	1402.55	1414.78	1417.63
10	1628.89	1643.62	1647.01	1790.85	1814.02	1819.40	1967.15	2001.60	2009.66
15	2078.93	2107.18	2113.70	2396.56	2443.22	2454.09	2759.03	2831.82	2848.95
20	2653.30	2701.48	2712.64	3207.14	3290.66	3310.20	3869.68	4006.39	4038.74
25	3386.35	3463.40	3481.29	4291.87	4432.05	4464.97	5427.43	5668.16	5725.42

Write down the value of the investment after

a 2 years

b 20 years

if an amount of £2000 is invested at interest rates of

i 5% compounding yearly

ii 7% compounding monthly.

Solution

There are four cases asked for. Just read the values from the table for £1000.

The amounts must be doubled to apply to £2000.

a i 2 years at 5% yearly gives $1102.50 \times 2 = £2205.00$.

The other answers are worked out in the same way.

a ii £2299.62 b i £5306.60 b ii £8077.48

Did you know that...?

In order to get a loan your financial history and credit worthiness are reviewed? Those delinquent bills, court judgments and past or current financial problems are no secret to those deciding on your credit worthiness.

8.4.3 Inflation

Prices of goods and commodities change, usually upwards. The money in your pocket is likely to have less buying power in two years' time than it does now. This effect is called **inflation**. Governments tend to keep measures of inflation because the level affects peoples' happiness and hence their choice of government. The measure is commonly called the consumer price index, abbreviated as CPI. The CPI is a percentage increase for prices over a given time period. As such, the compound interest formula we already know can be used in inflation calculations.

Example 8.4.3a

From June 2005 to June 2006 the Australian Bureau of Statistics recorded a CPI of 4.4%.

a If a half-litre of milk cost 2.05 AUD in June 2005, how much did it cost in June 2006?

b Based on the same CPI, how much will the milk cost in June 2007?

c How much will it cost in June 2010?

Solution

a $2.05 \times \left(1 + \dfrac{4.4}{100}\right) = 2.14$ AUD

b $2.14 \times \left(1 + \dfrac{4.4}{100}\right) = 2.23$ AUD

c $2.14 \times \left(1 + \dfrac{4.4}{100}\right)^4 = 2.54$ AUD

www.CartoonStock.com

Example 8.4.3b

Based on the same 4.4% annual inflation rate, calculate what the cost (to the nearest dollar) of a new car would have been in June 2005, if the price in June 2007 is 22 000 AUD.

Solution

Let the cost in 2005 be C.

Then $C \times \left(1 + \dfrac{4.4}{100}\right)^2 = 22\,000$ AUD. Here we have to divide.

Hence $C = \dfrac{22\,000}{\left(1 + \dfrac{4.4}{100}\right)^2} = 20\,185$ AUD (nearest dollar).

The **real return** is found by subtracting the inflated value of an investment from the total after applying interest earned.
The **real average rate of return** is the real return expressed as a percentage of the original investment amount and averaged over the time of the investment.

When you invest money and calculate the future value after earning interest, you really should try to take into account the effect of inflation. If you invest $1000 dollars now and expect to have $1400 in five years' time, your investment will not really be worth $1400 because inflation will tend to erode the increase.

Example 8.4.3c

Prasad saves €2000 in an account offering 5.2% annual interest. Inflation is running at 3.6% p.a. Find the real return and the real rate of return after two years.

Solution

After two years, the total in the account is
$2000 \times 1.052^2 =$ €2213.41.

Applying the inflation rate to the €2000 gives
$2000 \times 1.036^2 =$ €2146.59.

The real return is $2213.41 - 2146.59 =$ €66.82 and the real average rate of return is $\dfrac{2213.41 - 2146.59}{2000 \times 2} \times 100\% = 1.67\%$ per year.

Exercise 8.4

1 The loan table given here is for a loan of ¥100 000, with the nominal interest rates given compounding quarterly. Calculate (correct to the nearest whole yen) the quarterly repayments for a loan of

Time (years)	3%	6%	9%
5	5403.063	5824.574	6264.207
10	2903.016	3342.710	3817.738
15	2075.836	2539.343	3053.533

a ¥1 000 000 over a period of 5 years at a rate of 3%
b ¥500 000 over a period of 10 years at a rate of 9%
c If the repayment on a ¥200 000 loan is ¥5079 per quarter and the rate is 6%, find the time taken to pay off the loan and calculate the total paid.

2 Use your GDC to fill in this repayment table with the given nominal rates per annum and fortnightly repayments compounding monthly. The loan is for 1000 currency units.

Time in years	2.5%	5.0%	7.5%	%	10.0%
1	38.9625				
2				20.8368	
					9.7844
8					

3 Use **a** the compound interest formula
b your graphic display calculator
to check some of the values in the table on page 400 in Section 8.4.2.

4 Using the table in Section 8.4.2 write down the amount in the account

a after 25 years if $100 is invested at 6% nominal rate compounding quarterly
b after 5 years if 3400 EUR is invested at 7% compounding yearly (correct to the nearest euro).

5 In a certain city in New Zealand a pair of men's shoes costs 149.95 New Zealand dollars (NZD) and a lounge suite costs 4200 NZD. Suppose the inflation rate has been stable at 4.1% p.a. in New Zealand for the last 2 years and is expected to remain the same for another year.

a What will the shoes cost in a year's time?
b Find how much the suite cost 2 years ago, giving your answer to the nearest 10 NZD.

6 Halima saves 6000 Egyptian pounds (EGP) in a bank offering 3.95% nominal interest rate p.a. compounding monthly. She leaves her money in the bank for five years. During that time the average rate of inflation is 1.9% p.a.
Calculate the real return and the real average rate of return for Halima's money.

Past examination questions for topic 8

Paper 1

1 The exchange rate from US dollars (USD) to French francs (FFR) is given by 1USD:7.5 FFR. Give the answers to the following correct to two decimal places.

a Convert 115 US dollars to French francs.

b Roger receives 600 Australian dollars (AUD) for 2430 FFR. Calculate the value of the US dollar in Australian dollars. *M02q11*

2 A family in Malaysia received a gift of 4000 AUD from a cousin living in Australia.

The money was converted to Malaysian ringgit. One ringgit can be exchanged for 0.4504 AUD.

a Calculate the amount of ringgit received.

The money was invested for 2 years and 6 months at 5.2% p.a., compounding monthly.

b Calculate the amount of interest earned from this investment. Give your answer to the nearest ringgit. *M05q11*

3 The rate of inflation from the beginning of 1995 has been 4.5% per year.

a A loaf of bread cost \$1.70 on January 1, 1996. What did it cost on January 1, 1999?

b A car cost \$40 000 on January 1, 1999. What would it have cost on January 1, 1997?
(Give your answer correct to the nearest thousand dollars.) *M01q6*

4 Keisha had 10 000 USD to invest. She invested m USD in the *Midland Bank*, which gave her 8% annual interest. She invested f USD at the *First National Bank*, which gave 6% annual interest. She received a total of 640 USD in interest at the end of the year.

a Write two equations that represent this information.

b Find the amount of money Keisha invested at each bank. *N02q11*

5 Zog from the planet Mars wants to change some Martian dollars (MD) into US dollars (USD). The exchange rate is 1 MD = 0.412 USD. The bank charges 2% commission.

a How many US dollars will Zog receive if she pays 3500 MD?

Zog meets Zania from Venus where the currency is Venusian rupees (VR). They want to exchange money and avoid bank charges. The exchange rate is 1 MD = 1.63 VR.

b How many Martian dollars, correct to the nearest dollar, will Zania receive if she gives Zog 2100 VR? *M03q3*

Paper 2

1 i Alex invests 3600 euros in an account that pays a nominal rate of 5.4% interest per year, compounding monthly. The interest is added to the account at the end of each month.

a Calculate the number of **whole** months it will take for Alex's investment to double.

b i Calculate the value of Alex's investment after nine years.

ii Find the rate of **simple interest** per year that would give the same value for the investment after nine years.

ii Annie is starting her first job. She will earn a salary of \$26 000 in the first year and her salary will increase by 3% every year.

a Calculate how much Annie will earn in her fifth year of work.

Annie spends \$24 800 of her earnings in her first year of work. For the next few years, inflation will cause Annie's living expenses to rise by 5% per year.

b i Calculate the number of years it will be before Annie is spending more than she earns.

ii By how much will Annie's spending be greater than her earnings in that year? *M06q4*

2 Ali, Bob and Connie each have 3000 USD (US dollars) to invest.

Ali invests his 3000 USD in a firm that offers simple interest at 4.5% per annum. The interest is added at the end of each year.

Bob invests his 3000 USD in a bank that offers interest compounded annually at a rate of 4% per annum. The interest is added at the end of each year.

Connie invests her 3000 USD in another bank that offers interest compounded half-yearly at a rate of 3.8% per annum. The interest is added at the end of each half year.

a Calculate how much money Ali and Bob have at the **beginning** of year 7.

b Show that Connie has 3760.20 USD at the beginning of year 7.

c Calculate how many years it will take for Bob to have 6000 USD in the bank.

At the beginning of year 7, Connie moves to England. She transfers her money into a bank there at an exchange rate of 1 USD = £0.711 (British pounds).
The Bank charges 2% commission.

d i Calculate the commission that the bank charges.

ii Calculate the amount of money, in £, that Connie transfers to the bank in England. *N05q4*

3 Miranti deposits $1000 into an investment account that pays 5% interest per annum.

a What will be the value of the investment after 5 years if the interest is reinvested?

b How many years would it take Miranti's investment to double in value?

At the beginning of each year Brenda deposits $1000 into an investment account that pays 5% interest per annum. Interest is calculated annually and reinvested.

c How much would be in Brenda's account after 5 years? *M00q5*

4 On Vera's 18th birthday, she was given an allowance by her parents. She had these choices.

Choice A: $100 every month of the year.

Choice B: A fixed amount of $1100 at the beginning of the year, to be invested at an interest rate of 12% per annum, compounded monthly.

Choice C: $75 the first month and an increase of $5 every month thereafter.

Choice D: $80 the first month and an increase of 5% every month.

a Assuming that Vera does not spend any of her allowance during the year, calculate, for each of the choices, how much money she would have at the end of the year.

b Which of the choices do you think Vera should make? Give a reason for your answer.

c On her 19th birthday Vera invests $1200 in a bank that pays interest at r% per annum, compounded annually. She would like to buy a scooter costing $1452 on her 21st birthday. What rate will the bank have to offer her to enable her to buy the scooter? *N02q3*

TOPIC 9

Internal assessment

9.1 Internal assessment criteria

The project is internally assessed by your teacher and externally moderated by the IBO using assessment criteria that relate to the objectives for group 5 mathematics.

9.1.1 Form of the assessment criteria

Each project will be assessed against the following seven criteria:

Criterion A	Introduction
Criterion B	Information/measurement
Criterion C	Mathematical processes
Criterion D	Interpretation of results
Criterion E	Validity
Criterion F	Structure and communication
Criterion G	Commitment

9.1.2 Applying the assessment criteria

Your project will be marked on how well it fulfils the seven assessment criteria listed above. So, in theory, it is possible for all students to receive top grades, unlike in some methods of assessment where performance is judged in relation to the work of other students (the top 10% of candidates in any one year score the highest grade, the next 20% get the next grade and so on.)

For each of the assessment criteria, different levels of achievement are described that concentrate on positive achievement. The person who marks your project will find, for each criterion, the level description that best matches your work and will grade it at that level. The level description represents the minimum requirement for you to achieve that level.

9.2 Academic honesty

The International Baccalaureate Organization regards honesty and integrity in all aspects of your work as overwhelmingly important. Before you undertake your internal assessment project you must familiarize yourself with the IBO academic honesty statement. You can find this statement at the beginning of this book. Please make the effort to read it and think about it.

The vast majority of IB Diploma Programme students respect these ideas without needing to be reminded. If you are tempted to break the rules though, be aware that candidates are often identified when they do this and the consequences can be severe. Teachers and examiners are very experienced at spotting copying and unreferenced material.

With particular reference to your internal assessment, you must ensure that ALL your sources are identified and that people's personal information is not published without permission. It does sometimes happen that a candidate commits plagiarism unintentionally, because he or she does not fully understand what this means, and this can still lead to trouble, so if in doubt, ask your teachers for advice.

9.3 Choice of topic

If you were to listen in on a group of teachers or examiners discussing the internal assessment project, you might well hear one of them sigh and say *"There is too much statistics, I wish there was more variety."* Of course there is a good reason why statistics projects are popular. There are very many situations you can analyse with statistics, and there is also quite a lot of useful statistical mathematics which is covered by the mathematical studies course. Furthermore a well-chosen statistical project will naturally suggest its own data for collection, will allow application of both simple and more sophisticated techniques and allow analysis of the validity of the techniques. The teacher who sighed knows these things very well and will certainly not advise you against a statistical project.

Using material outside the syllabus is not expected for your project but if you are willing to consider this then there are many techniques you can learn about given your mathematical studies course as a foundation. Your course gives you a good introduction to financial topics which can be extended by further reading without much trouble. You can use calculus to study rates of change of quantities (for example, depth of fluid draining from a receptacle). You already learn something about maximum and minimum calculations in your course and this material can be extended to study optimum values for quantities such as size of a can or box, to minimize use of material or cost. If you are willing to learn a bit about anti-differentiation, you can study areas and volumes of unfamiliar and irregular shapes. Many statistical projects use the normal distribution, which is no longer in the course but might still be useful in a project and occasionally the student t-test or even other statistical tests are seen. If linear regression does not work, you can easily try quadratic or higher regression or even try fitting exponential, logarithmic or trigonometric curves using your GDC. (But explain clearly what you are doing!) The probability you learn in mathematical studies can be used to analyse chances of success in games like poker or roulette.

A word of warning though. Sometimes an idea seems very nice and exciting and you might have a vision of a beautifully presented project with many pictures and designs. Such a project is great to read but it will not score highly unless the level of mathematics and analysis is adequate. Projects which can fall into this category include the use of the golden mean in art, architecture, music and biology and the closely related topic of Fibonacci sequences. By all means try such a project if your mathematics is secure – but be aware that to present enough good mathematics in these areas will require you to look beyond the mathematical studies syllabus at topics such as matrices, convergence of sequences and continuous fractions.

Another slightly more risky area seen occasionally is coding theory. There is some terrific mathematics here but the problem is that the explanations involving prime numbers and modular arithmetic are quite sophisticated and you need a clear and concise command of mathematical language to make things clear to the reader.

Try to find a topic that you are interested in and one that will generate sufficient data/information for you to perform both simple and sophisticated mathematical processes. Sport related investigations are a rich source of ideas. They are appealing to many people and data are often easily accessible from the internet. What makes a champion world cup team? How are tennis rankings worked out? How do players' salaries relate to performance? Does specially designed clothing really make a difference to speed? These are just a few possibilities.

If you are finding it difficult to come up with a title for your project your teacher should be able to show you a long list of ideas available from IBO web sites or at teacher workshops.

You must have a thorough understanding of the assessment criteria. Encourage your teacher to give you at least one past or sample project to mark. This will increase your understanding very efficiently.

The assessment criteria appeared at the beginning of this topic. The following comments relate to each of the criteria individually.

Criterion A: Introduction

This is probably the most crucial part of your project. Get this part right and the rest should follow.

Remember to give your project a title- not just *Internal assessment project*!

Write out a clear statement of the task – explain clearly what you are going to do for your project. You can also inform the moderator why you have chosen this task (but that is not compulsory).

Then write down in detail how you are going to achieve your aim; what processes will be used to gather information or data; what mathematical processes you will use and why they are relevant.

If you have one or more hypotheses then state them here too.

If the emphasis of your project changes after you have started it, remember to change the description in the introduction.

Criterion B: Information/Measurement

Always try to generate enough information or collect sufficient data. It is very difficult to perform any useful operations on a few pieces of data.

Make sure that you include in your project all the data/information that you collect.

If you use data from the Internet then remember to quote the site(s) that you use.

If you use a survey or questionnaire then make sure that you include a copy of it in your project. You can also include one or two completed forms – but not all of them!

Put your data/information into tables.

Criterion C: Mathematical processes

Make sure that you include at least one simple mathematical process in your project. Show the moderator that you know how to perform such processes by yourself and then you can use your GDC or computer to perform similar tasks.

Also remember to say why these processes are relevant to your project. Be careful! Check your work. Do not make errors in the simple mathematical processes. Remember that histograms have no spaces between the bars.

As your project develops you can introduce more sophisticated mathematical processes. Ensure that these processes are relevant otherwise they will not earn marks.

Don't include mathematical representations just for the sake of it. *All* the mathematics present should be relevant and have something to contribute to the project.

Projects with large numbers of pie charts or Venn diagrams and little else will not attract high marks. Drawing graphs for quantities which are not naturally related or connecting up points just for the sake of drawing a graph are not constructive exercises.

If you are performing a χ^2 test then make sure that you are using *frequencies* that are in your table of observed values. Find the expected values and check that they are suitable. You should have no values in the expected table that are less than 1, and no more than 20% should be less than 5. You can combine groups (as long as it still makes sense) until you get frequencies of 5 or more.

If you are drawing a scatter graph and you can see that there is no correlation, you can calculate the correlation coefficient to show that this is the case, however, in this situation it makes little sense to find the line of regression and even less sense to use it for prediction purposes.

Check that your answers are sensible – for example, a correlation coefficient of more than 1 or less than −1 should alert you to the fact that you have made a mistake. Probabilities outside the range 0 to 1 are also always wrong. Once again, check your mathematics – try not to lose marks by having errors in your calculations.

It is perfectly acceptable to use mathematical processes that are not in the syllabus but remember to explain why the process is relevant to what you are doing.

Criterion D: Interpretation of results

Immediately after you have performed the mathematical processes you should discuss the results/conclusions that you have arrived at.
Are they significant results? If so, why?
Are they what you expected? If not, why not?

Also make sure that you have a results/conclusions section where you discuss all your results/conclusions in more detail.

Criterion E: Validity

There are several things that you can do to satisfy this criterion.

How random was your data collection? Are there enough data?

Was there any bias? If so, what could you have done to compensate for this or remove the bias?

Were your questionnaires filled out honestly?

How valid were the processes that you were using? (If you were using a χ^2 test with more than 20% of the expected frequencies less than 5 then the test is less reliable and you should state this.)

If all the mathematics you have presented is standard material with no question as to its validity in the context of your work, then you should state this and write a few words to justify the statement. For example, there is no doubt about area and volume formulae for geometric figures, but is the item you are measuring truly a sphere or is that only an approximation? Methods of calculating maxima and minima of functions are tried and tested (but make sure the domain you are using is appropriate and the stationary point is truly relevant). Probability formulae are always valid if used in the correct context, but are you sure the events really are independent, mutually exclusive etc?

Theory of knowledge

After your teacher has marked all the projects, he or she will send a sample of the marking to an IB moderator. The moderator will re-mark the projects received, often agreeing fairly well with the teacher, though there will sometimes be minor differences. The moderators then all send samples to a single senior or chief moderator for further marking.

It is not appropriate here to describe the precise details of how the IBO handles the different marks, however, typically any exam board would be using linear regression to compare them and this might be linear regression for the whole data set or possibly piecewise linear regression, in which low, middle and high marks each acquire their own regression lines, with different slopes, but joining up at the boundaries.

Why do you think this process is applied? Is it a fair way of doing things? Perhaps you could even write your own I/A project on this by getting some people to mark and re-mark projects. Which kind of regression is better?

Try to propose further extensions or improvements to your project.

If you recognize weaknesses too late, then be honest and discuss them and try to describe how you would overcome them given more time.

Criterion F: Structure and communication

Set your project out neatly and lay it out under the various criterion headings – that way it follows a logical progression.

Use a spell-checker, especially if you are not writing in your first language.

If you are not writing in your first language try to find an expert who can check the grammar and syntax. Candidates often use syntax from their first language which sounds stilted or even unintelligible in another language.

Before each mathematical process include a table of the data that you will be using. The moderator likes this as it saves him/her searching for the relevant data.

Take care that your graphs are neat and labelled.

Criterion G: Commitment

This mark is set by your teacher and not moderated. So make sure that you impress your teacher by keeping to all the deadlines and showing lots of enthusiasm!

TOPIC 10

Mathematical studies, the Diploma Programme model and examination advice

10.1 Other subjects in the IB Diploma Programme

Don't think that because you are doing mathematical studies (as opposed to SL or HL), the mathematics you learn is of no use in other subject areas. Yes, it is true that more advanced mathematics is commonly used in science and economics, but the mathematical studies syllabus contains some very usable mathematics too.

Rounding and percentage errors are used frequently in science and economics. Arithmetic and geometric sequences appear naturally in many subject areas.

Geometric concepts appear everywhere in life. For example, in biology the volume to surface area ratio of a living creature is important and can be related to its metabolic rate. Architects also rely heavily on geometry and trigonometry.

The amount of statistics in the mathematical studies syllabus is substantial. This kind of material is used all the time by companies, banks, marketing departments, governments, journalists, aid agencies and almost anyone with a point to make or a clientele to convince. And sometimes, dare we say, it is used fallaciously. It pays to understand statistics and how they are used to persuade people of truths and lies.

If your main area of interest is a language, consider learning about the etymology of the number words in different languages. This is a fascinating subject, and because counting is such an important concept, much can be learned about the historic development of a language and the psychology of its speakers by studying the history of these words.

The history of mathematics is intimately intertwined with the history of civilizations and their achievements and good mathematicians often make important contributions to their societies as a whole, and help to shape the thinking of the age.

If you are interested in music, you will find mathematics permeating the history of this subject too and much of it is understandable using the mathematics you are learning now. In Europe from Bach to Stockhausen, mathematical ideas have been used both instinctively and more deliberately to construct instruments, understand the sounds they produce and to compose music. Did you know there is a theory that Mozart and his associates sometimes composed pieces using the throw of dice? Prime numbers are important in creating rhythmic devices for the *tabla* (Indian drum). Try an Internet search of the form "Mathematics and *** music" where *** might be Indian,

Chinese, Islamic, Biblical, Classical, Gamelan or many other possibilities.

If you are studying economics then you will probably already have met the material in our finance topic and perhaps gone beyond it.

Much of the world's great art contains mathematical ideas. From the design of the ancient pyramids through the symmetry and designs evident in Islamic architecture to the use of the golden mean in paintings by the great masters, there is mathematics that would be accessible to you.

Mathematical studies contains an introduction to mathematical logic. This is a little unusual at this level, but it is a very valuable exercise. Mathematics is a very precise language and logic helps us recognize the precise relation between a statement and its consequences.

An understanding of logic will also help you to communicate better and can be very valuable for those who need a command of the complicated language used in the application of law. Truth tables are an essential tool in the study of complicated electrical circuits also. This latter connection could give rise to a manageable internal assessment project for those willing to do the research. To investigate this, try looking up material on Boolean algebra and switching circuits.

The calculus you are learning is only a beginning. Calculus is used extensively, especially by engineers, physicists and chemists, but increasingly by biologists and economists too. Even with the little material covered here though, you can start to understand the idea of a "rate of change" and use it in other subject areas.

Mathematics is even used in sport. It might be statistics to analyse the performance of a team and predict their chances, geometry to study the design of a racing car, geometry and calculus for the study of ball trajectories or even in the design of streamlined clothing. Can you think of other possibilities? Try an internet search on "mathematics of juggling".

10.2 Theory of knowledge

In this book, the interaction between mathematics and TOK has been developed extensively and there is no need to add more here. See if you can think of some mathematical TOK ideas that the authors have missed.

10.3 Creativity, action, service

Have you considered using mathematics in any of your CAS activities? It's not such a silly idea. Maybe you are already doing this without realizing or perhaps quite deliberately. Your expertise is potentially quite valuable to your school or your local community, especially if you live in a remote area of a developing country.

A word of warning – do not portray yourself as a fully qualified expert. If experts are associated with your project, it is wise to defer to their opinions, especially in situations where you could be legally liable. Experts usually have legal indemnity; you do not. If in doubt, consult your CAS coordinator, teachers and local officials.

Here are just a few places where mathematics might come in useful.

- Statistics of sports results for your school or local teams, predictions on performance of your team against others
- Calculating volume of paint required to cover a wall, a gate, a fence, or even a whole house
- Planning ticketing and seating for school concerts, sport events and plays
- Accounting and investment for fund-raising projects
- Assistance in designing structures
- Layout of gardens in service projects
- Analysis of public opinion surveys for charities or even your family business

10.4 Extended essay

The current Chief Moderator for Mathematics Extended Essays advises that it is usually unwise for mathematical studies candidates to attempt an extended essay in mathematics. An extended essay is intended to challenge a candidate to research and write up advanced material on the subject chosen. The marking structure for essays incorporates 12 marks (out of 36) awarded only for subject specific achievements. In mathematics many of those 12 marks will be extremely difficult to earn for a mathematical studies candidate. The nature of the essay necessarily requires that candidates go well beyond the mathematical studies syllabus. By choosing to do an extended essay on mathematics, most mathematical studies candidates will be restricting themselves to a maximum likely achievement of 20 to 24 marks, and then only if the other aspects of the essay are handled extremely well.

It is not against IB regulations for a mathematical studies candidate to attempt an extended essay in mathematics and it does occasionally happen that such a candidate produces a good mathematical extended essay but this is really quite rare. In general, mathematical studies candidates should think very carefully before committing themselves to such an essay.

10.5 Examination advice

10.5.1 Examination tips

General

- Use pens and pencils which do not smudge.
- Make sure that your GDC batteries are new or fully charged. Maybe take some spare batteries to the examination.
- You are allowed to use the IBO information booklet (sometimes also called "formula booklet") during your mathematical studies examinations. Make sure you are familiar with all aspects of this booklet and know what is there and where to find it quickly. Correct selection and

substitution of these formulae will always be worth at least one or two marks.

- Plan your path through the paper in advance. The earlier questions are supposed to be easier. Allow more time for the last few questions than for the first few.
- Complete the questions you can do easily at the start. This can result in a "feel-good" frame of mind that gives you confidence to tackle harder things.
- If a question has units then assign the correct units to the answer. *Example*: 90 km/h, not just 90.
- Questions should increase in difficulty through the paper. In Paper 2, there may be an increase in difficulty through each question. Plan to try the easier questions first. Don't panic if you can't finish the last part of a question. It's probably the hardest part. Move on to something you can do.
- Always consider whether your answer makes sense. Distances should never be negative. Angles are also usually positive and in a triangle they should be less than 180 degrees. The length of a sports field is unlikely to be 1000 km. Similarly the distance between two planets is not 5 cm. Probabilities can never be outside the range 0 to 1. You are rather unlikely to score half a goal in soccer. Exam diagrams are often not drawn exactly to scale, but on the other hand, an angle in a question will not look like 170 degrees if it is meant to be 20 degrees.
  If you make an early mistake in a question, follow-through marks are available for doing the right thing with earlier wrong answers, however, you will not get full follow-through if the final answer is ridiculous. If an answer seems silly, ask yourself where it could have gone wrong. Perhaps you used radians instead of degrees for instance.

Reading instructions

- Read the rubric on the paper and the details of the questions very thoroughly.
- Make sure you understand exactly what a question is asking. Read each question part again before you enter the final answer especially in Paper 1.
- Be alert for special instructions such as specified accuracy.

Accuracy

- Exact answers can be left that way. Examples: 0.25, $\frac{3}{16}$, $\sqrt{2}$, $\sqrt{79}$, π.
- All other answers must be given correct to 3 significant figures (s.f.) unless the question instructs differently.
  Example:
  evaluate $\sqrt{4.28}$, answer 2.07 (3 s.f.).
- Answers not given to 3 s.f. accuracy will receive an *accuracy penalty* (AP) the first time this occurs.

- If a question specifies the level of accuracy then there will be a mark assigned for that procedure, which will be lost if you do not obey the instruction.
- Questions involving money will usually specify whether to correct to whole units, 2 d.p. or some other level. If in any doubt with money, 3 s.f. or 2 d.p. are safe but look for the instructions in the question.
- Keep extra accuracy for values part of the way through a question. Rounding prematurely can result in an error in the final answer, which will not just use up the AP, but will be judged wrong irrespective of other accuracy errors.

Working

In the aims and objectives for the course, you will find the words "communicate clearly" in several places. You will also find the words "demonstrate an understanding", "select and use appropriate strategies" and "formulate a mathematical argument".

All of these features require that you write more than just answers to convince the examiner that you are competent. Showing working is how you can do this. For this reason, the absence of adequate working is penalized in Paper 2. There is no definite rule about how many marks you might lose if you do not show working, but it can easily be as many as 5 or 6 in the paper and might be more.

- The command terms "Write down" or "State" mean that a quick answer is expected. Here you can give an answer with no working.
- For all other questions you should always write some working.
- Working can be a substituted formula, an equation, a factorization, a sketch graph, a Venn diagram or a set of elements, a table and many other possibilities.
- If a GDC is used to solve a question, you should still describe the mathematics used. This can be a sketch of a graph, a table of values, a logic expression, a Venn diagram or just a formula with substituted values.
- Notation used only on a GDC, which is not standard mathematical notation, is not acceptable in working. For example, if using the finance application to find a future value, you should write down the appropriate interest formula with values substituted, and not just list F/V, I, N etc. Similarly 2.53 E-6 is not acceptable for 2.53×10^{-6}.
- Working for a χ^2 test can be a table showing expected and observed values. Degrees of freedom should be mentioned if not asked for elsewhere in the question. The formula for χ^2 should be written and an example substitution shown (perhaps just for 1 or 2 terms) even if the GDC is then used to complete the calculation.

- Working is not required when you calculate a standard deviation. It is wise to show the formula for the mean though, unless the question just says "write down".

GDC: Common errors

- Angle mode set in radians. Always use degrees. Check it before the examination.
- Poor use of brackets, for example, writing $\frac{1}{1+x^2}$ instead of $\frac{1}{(1+x)^2}$.
- Incorrect choice of standard deviation. Know your GDC and its notation. It might differ from IBO notation. Use the *smaller* of the two values, regardless of what your GDC calls it.
- Incorrect order entered for lists when finding weighted means.
- TI only: make sure that diagnostics are ON.
- Finding no graph or only part of a graph because the window is inadequate.

Tips for Paper 1

- Paper 1 questions are answered on the question book. Extra sheets can be attached.
- Working has a designated space. Try to separate working for parts **a**, **b** etc. very clearly.
- In Paper 1 a correct answer written in the answer space with no working *does* receive full marks. But this is a risky course of action. If wrong you will get nothing. This includes a correct answer rounded too far.
- *Example*: correct answer only of 3.01 (3 s.f.) might get 3 marks but 3, 3.0, 3.1 with no working get nothing. Similarly the correct working leading to a wrong answer such as $\sqrt{3^2+(-1)^2}=\sqrt{8}$ should receive one mark while the answer $\sqrt{8}$ would receive nothing, even though it is clear that the correct formula was used with a sign wrong.
- Always transfer your answer (*only*) to the answer lines, especially if there is a lot of working.
- Read the question several times. In particular, read it through just before you enter your answer in the answer lines. Make sure that you are supplying precisely the answer required and not extra or irrelevant information. Examiners have to be sure that you have understood what is being asked. Extra information in the answer lines can lead to loss of marks, even if the correct answer is also present or is present in the working. Example: question asks for factors of a quadratic but candidate writes roots of the quadratic in the answer space. This is wrong and will lose a mark even in the factors can be seen in the working. See the list below for common mishaps.

Here is a list of incorrect answers in Paper 1 which are often seen and which will lose marks, even *if a correct answer is seen* in the working or in the answer box along with the wrong answer.

Question requires	**Candidate writes in answer space**
Factors of an equation	Roots of the equation
Roots	Factors
Interest, *I*	Final capital, *C*
Capital	Interest
Specified accuracy	Some other accuracy
x-value or *y*-value alone	Coordinate pair
Coordinate pair	*x*-value or *y*-value alone or pair in wrong order or pair without coordinate brackets
Accurate interval or range, for example $(-1, 1]$	-1 to 1
Answers **a** and **b**	Candidate interchanges answers

Tips for Paper 2

- Paper 2 questions require a blank answer book. Answers are never written on the question paper.
- Start each question (not each part though), on a new page. Do not cramp writing together.
- Random scribblings are not easy to judge for coherent thinking. Try to write meaningful mathematical sentences with a verb symbol (for example, $=$, $>$, $<$, $\in$). This applies to Paper 1 also.

Graphs, sketches and diagrams

- Graphs should be drawn with a sharp pencil on graph paper and should be clearly visible.
- Axes must always be clearly scaled and labelled. A mark is lost if this is not done.
- Draw axes and all other straight lines using a ruler.
- Points must be plotted clearly.
- Curves must be drawn curved, not as a series of connected straight lines. A penalty applies if a curve is drawn straight.
- Check the required scale in the question. If your graph seems very small, there is a good chance that you are using the wrong scale.
- Sketches of curves do not require the same care as graphs but they should still give a clear idea of the shape, indicating features such as turning points, axis crossings, asymptotes, intersections etc.
- Axes for a sketch do not need careful scales but some indication should be given of scale. You might get away with omitting labels but it is easy and safer to add them.
- Venn diagrams must have a universal set marked as *U* enclosing them, even if you don't know exactly what is the best choice for *U*.

Answers to exercises and examination questions

Topic 1

The TI 84+ has been used in this exercise.
The Casio screens would be similar in content.

1 a 8.06 b 1.28%

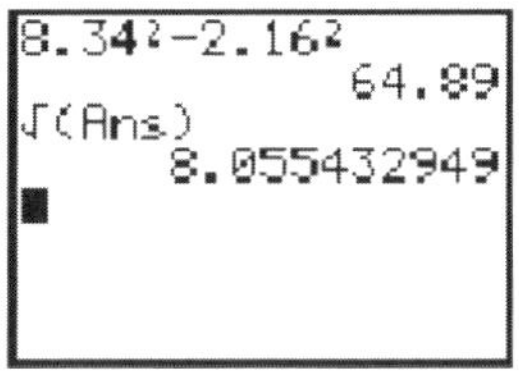

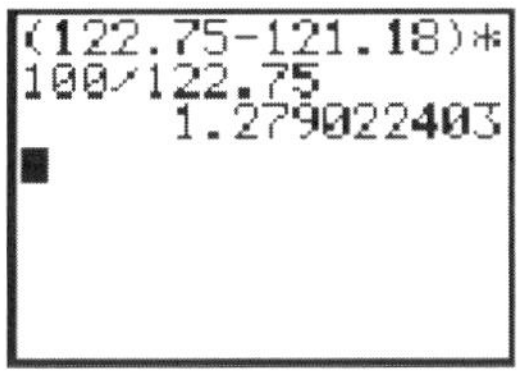

c 7.18 d 78.7

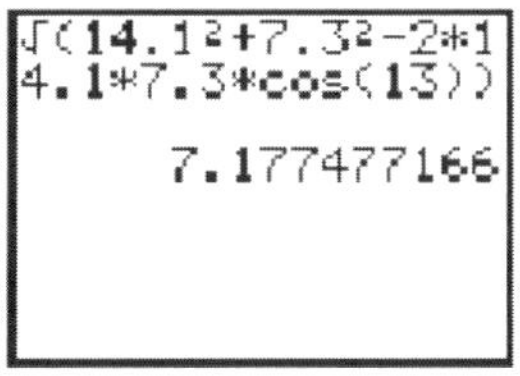

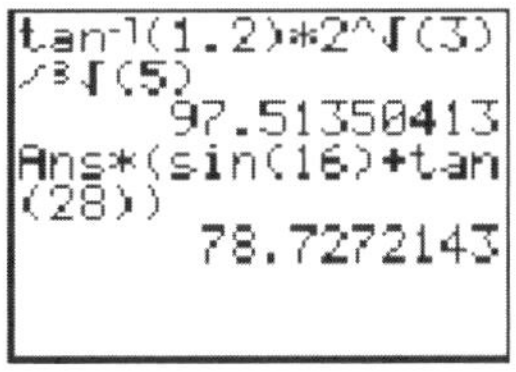

2 a i {23.8, 25.15, 26.50}
ii {−3.79, 3.60, −3.42} (3 s.f.)
b i 10 232.75 ii −4.402

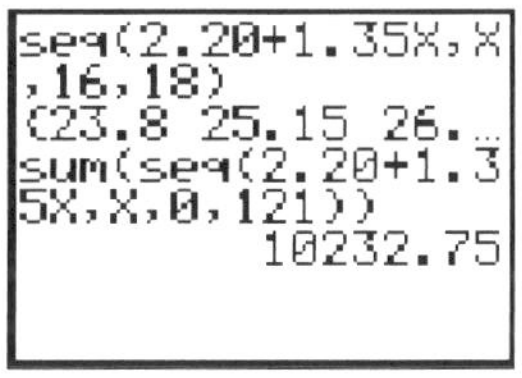

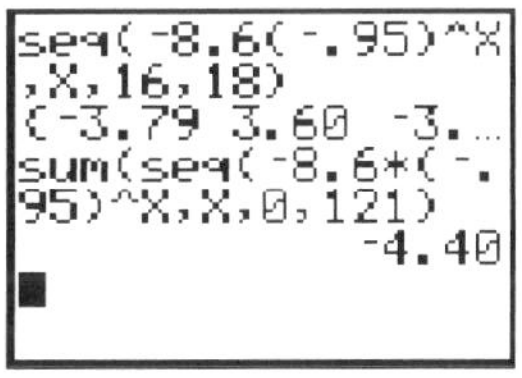

3 It is a contradiction. Can you see why? Only the last two screens are shown. Yours might be organised differently but the last one should be the same.

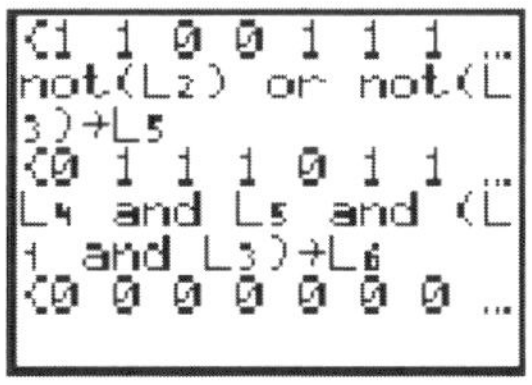

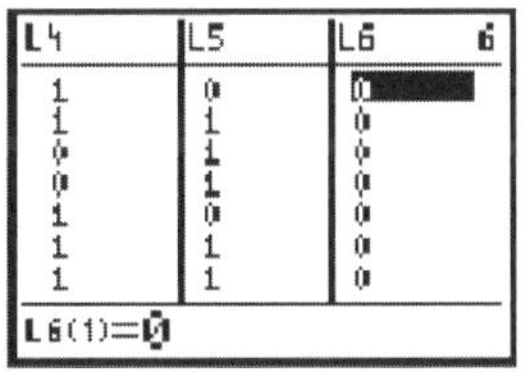

4 a The coordinates of the y-axis crossing are $(0, -7)$ and of the x-axis crossings are $(-1.42, 0)$ and $(2.23, 0)$.

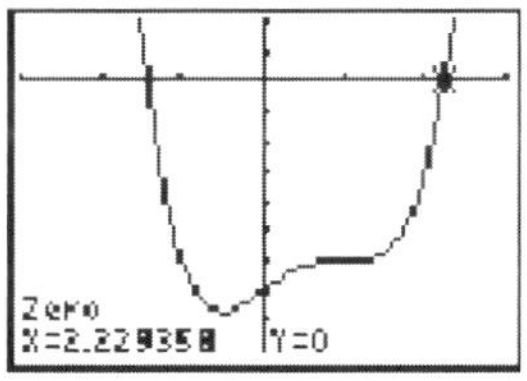

Min at $(-0.5, -7.7)$, point of inflexion at $(1, -6)$

b The positive x-value is at 58.5, shown graphically here.

The negative x-value is at −79.6, shown with the solver here.

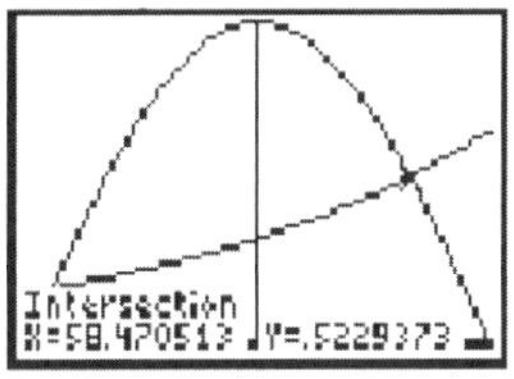

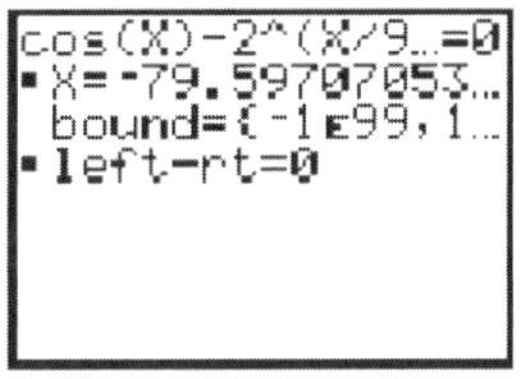

c Asymptotes: horizontal at $y = 4$, vertical at $x = -1$, zero at $x = \frac{1}{4}$, y-axis crossing $y = -1$.

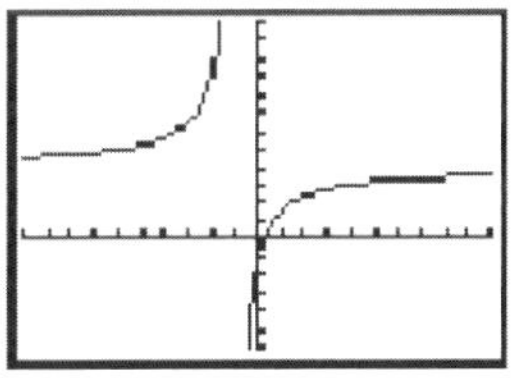

5 a 39.8° b 74.9° c 2.91 cm

Take care with brackets here.

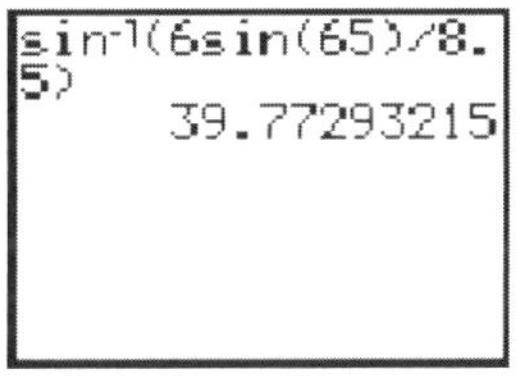

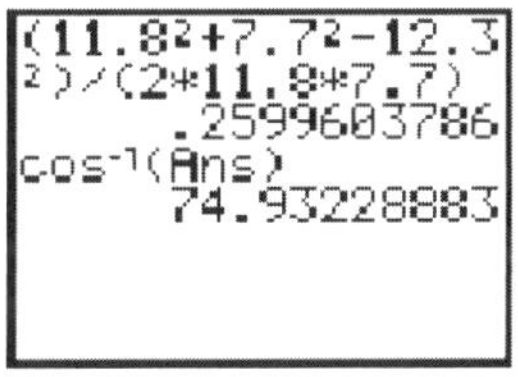

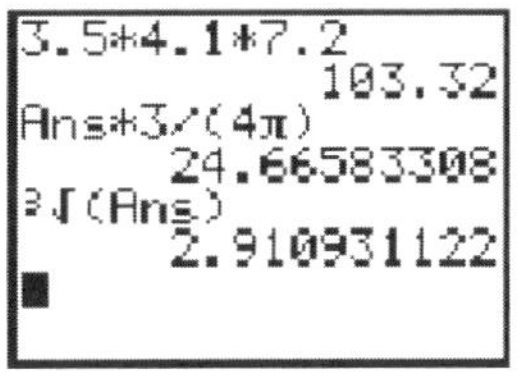

6 a

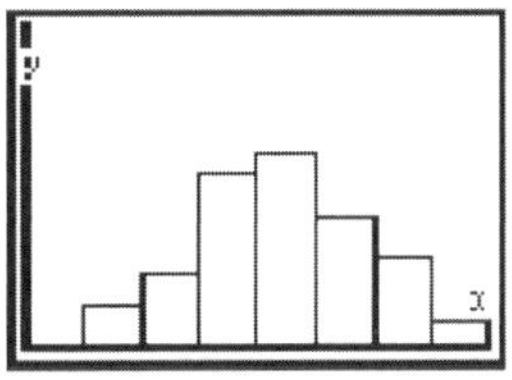

b

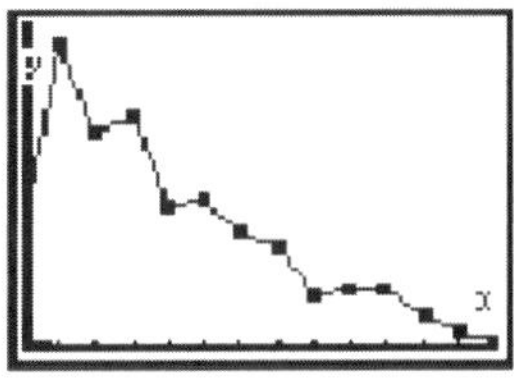

c Mean 14.45, median 12. Outliers at 48 and 52.

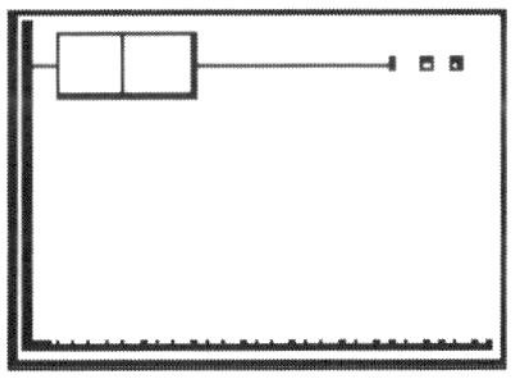

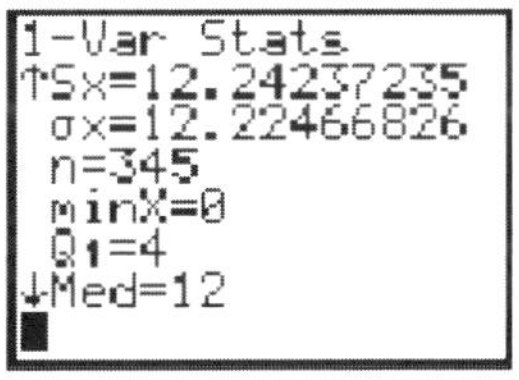

d $r = 0.960$. Strong correlation. Regression line $M = 1.09R - 1.28$.

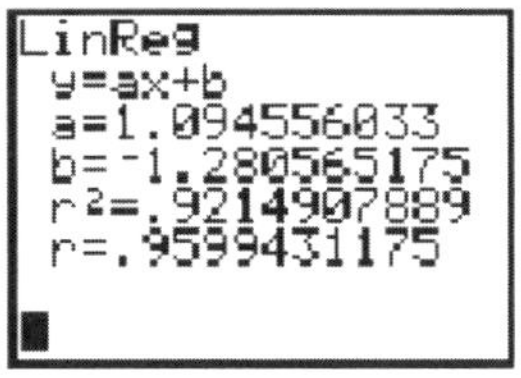

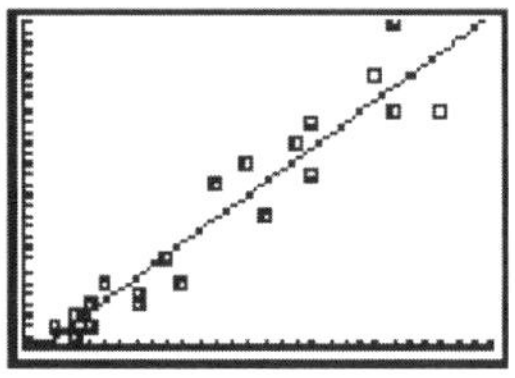

e i H_0: Nationality and favourite food are independent.

H_1: Nationality and favourite food are not independent.

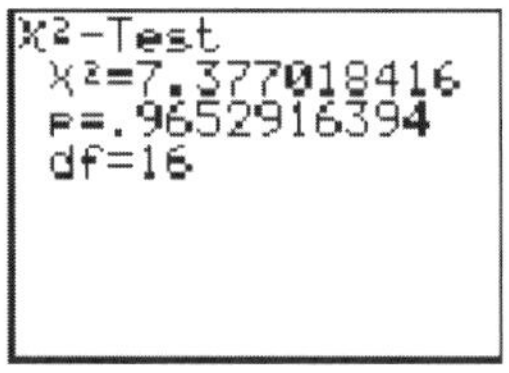

ii Degrees of freedom = 16

iii $p = 0.965 > 0.05$

$\chi^2 = 7.38 < 26.296$

iv We do not reject H_0. Food preference and nationality are independent at Jakob's school.

7 a −0.177 b Min is at (0.742, −1.18), Max is at (−0.628, 2.94).

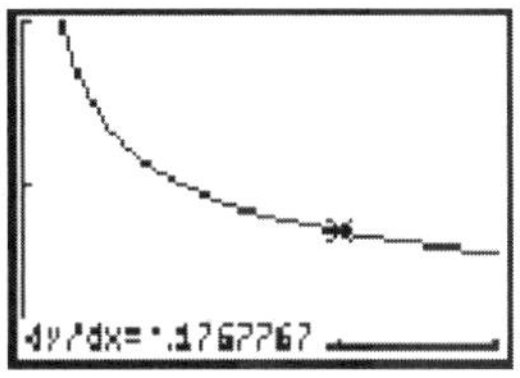

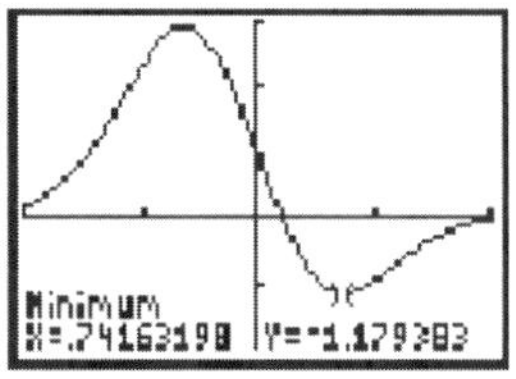

c A good y-window range is [−900, 900]. The first max is at (116, 104) and the first min is at (282, −276) (3s.f.).

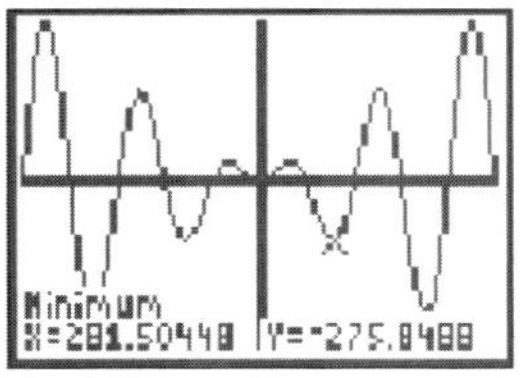

8 a 1467.06 b 78

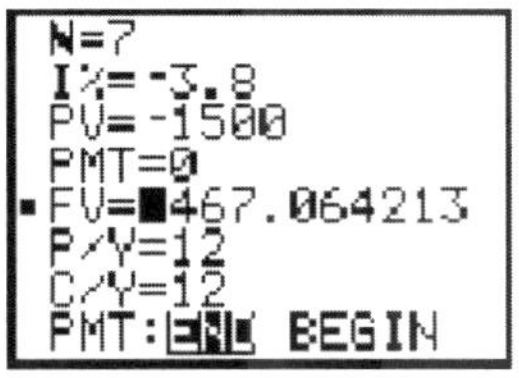

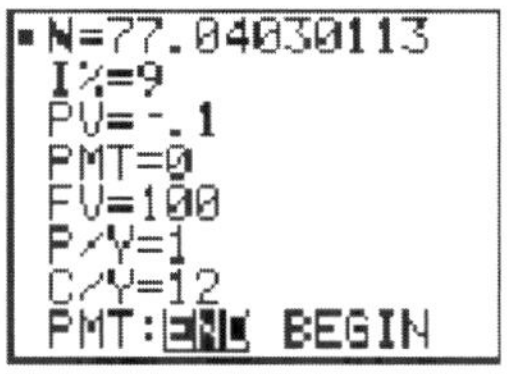

Topic 2

Exercise 2.1

1 a always b sometimes c never

2 For example $\frac{2}{3} \notin \mathbb{N}$, $2 \in \mathbb{N}$, $3 \in \mathbb{N}$

3 a

a	−3	4	$-\frac{1}{2}$	$-\sqrt{2}$	0.8	**−6**	x	***−t***
−a	**3**	**−4**	$\frac{1}{2}$	$\sqrt{2}$	***−0.8***	6	***−x***	t

b i <0 ii >0 iii >0 iv <0

4 a i, ii, iv

b i {−3} ii {2} iv {−10, 10}

5 a False b True

6 Division by zero is not defined.

7 For example a $\frac{1}{3}$ b $\frac{3}{4}$ c $\frac{1801}{900}$

8 a $-2.15 = -\frac{215}{100}$ b $4 = \frac{4}{1}$ c $1.\hat{8} = \frac{17}{9}$

9 a $\frac{1}{2}$ b $\frac{1}{8}$ c $\frac{3}{8}$ d $\frac{5}{4}$

10 a For example 0.917; 0.918; 0.920

b An infinite number

11 a $1+\sqrt{3}$ and $\frac{\pi}{2}$ are not rational numbers

b $\frac{16}{4}$; $1+\sqrt{3}$; $\frac{\pi}{2}$; 0; $\frac{2}{3}$

12 Table A

Number / Set	$1.03\hat{2}$	$\frac{30}{6}$	$-\frac{10}{5}$	$\frac{\sqrt{3}}{4}$
$\mathbb{N}$		✓		
$\mathbb{Z}$		✓	✓	
$\mathbb{Q}$	✓	✓	✓	
$\mathbb{R}$	✓	✓	✓	✓

Table B

Number / Set	−2.1	π	0	$-\frac{5}{10}$
$\mathbb{N}$			✓	
$\mathbb{Z}$			✓	
$\mathbb{Q}$	✓		✓	✓
$\mathbb{R}$	✓	✓	✓	✓

13 a rational b not rational c rational

14 a F b T c T d T

Exercise 2.2

1 a 240 b 1010 c 0 d 750

2 a 500 b 1300 c 18 200 d 400

3 a 2000 b 110 000 c 2000 d 20 000

4 a i 12.1 ii 12.05 b i 0.009 ii 0.0092

c i 5.991 ii 6.0

5 a i 42.3 ii 42.32 b i 0.176 ii 0.176 065

6 a i 110 ii 108.3 b i 40 000 ii 36 000
c i 0.0305 ii 0.03

7 a 5.15 b 7.98 c 104 d 2.51

8 a 12.2 b 269 000 c 0.02 d 900
e 0.6 f 0.2857

9 a 2.13 m b 80 000 c 1°C

10 a For example 3.45, 3.48 and 3.49
b For example 3.502, 3.51 and 3.54

11 a 0.12 b 5.04%

12 a 180 b 80 cm c 100 km/h

13 a 2629.96 b i 2630.0 ii 2600
c −4.94%

14 a 16.245 m^2 b 5 m and 3 m c 7.66%

15 a 60 b 60 cm c 100 km/h

16 a 1.76 m b 1.32%

Exercise 2.3

1 a $k = 3$ b $k = -3$ c $k = 8$ d $k = 2$

2 a 3.51×10^8 b 2.91×10^{-10} c 4.93×10^5
d 3.28×10^6 e 1.2×10^{-2}

3 a No product with power of ten shown
b as $11 \geqslant 10$ d as $0.32 < 1$ e as $2.3 \notin$

4 612×10^{-3}; 0.032×10^2; 8.94×10^2;
1.8×10^4

5 a 1 080 000 000 000 b 0.000 000 000 001
c 186 000 d 2 740 000 000

6 a 1.09×10^{12} b 6.46×10^{-4}
c 4.10×10^7 d -4.10×10^7
e 7.74×10^2

7 1.25×10^{23}

8 1.58×10^9 km

9 a 399.85 b 400 c 4×10^2

Exercise 2.4

1 a 0.304 25 km b 60 100 m c 0.2300 dam

2 a 2300 l b 3405 dl c 1.2 dal
d 2000 l

3 a 2 456 000 dg b 32 g c 10 t

4 a 2.591×10^9 m^2 b 1.89×10^8 m^2
c 6.255×10^{-3} km^2

5 a 5.634×10^5 s b 2.7×10^3 days
c 2.68×10^5 ms

6 a 120 b 12 cm

7 a 1370 m b i 70 000 m^2
ii 7 ha

8 a 2 h 18 min b 12:33 pm

9 a 4.17 m/s b 4320 km h^{-1}

10 a ½ h b 150 km/h c 141 km/h

11 b 233 °C

12

	Buenos Aires	Paris	Melbourne	Los Angeles	Casablanca	Ottawa	New Delhi
°F	**82**	**36**	**66**	55	**57**	32	**54**
°C	28	2	19	**13**	14	**0**	12

13 b 289 K

14 4×10^6 l

15 b i 3.08×10^{13} km ii 3.07×10^4 pc
iii 2.80×10^4 light-years.

Exercise 2.5

1 a −3, −6 b $\frac{5}{6}, \frac{6}{7}$ c 6, −6 d 37, 60

2 a ii and iii are arithmetic
b ii $\frac{3}{4}$ iii −2

3 a model 5: 11 dots; model 6: 13 dots
b i 2 ii 41

4 a $a_1 = 6$; $a_2 = 18$; $a_3 = 36$

5 a $d = -3$ b −2, −5 c −77
d Yes, it is the 19th term.

6 a $b_1 = 2$; $b_2 = 3.5$ b 1.5 c 32
d $b_{30} = 45.5$ e 6187.5

7 a $d = 2$; $u_1 = 8$ b $d = -2$; $u_1 = 6$

8 20 100

9 a 585 b −66.7 c 2.7

10 1860

11 a 2 b 22 c 660

12 a i $38 ii $46 b $118 c 18 hours

13 a $a = 6.5$ b 3.5 c $b = 13.5$ d 38

14 a $k = 3$ b 10, 15, 20

15 a i $u_1 = 5$ ii $u_2 = 7$ b $d = 2$ c $S_4 = 32$

16 a $P = 3 \times (a_1 + a_6)$ b 13 cm

17 a T b T c F d F e T

Exercise 2.6

1 b with common ratio 2
c with common ratio 0.2

2 a

sequence	arithmetic	geometric	Neither geometric nor arithmetic
10,9.7,9.4,9.1,…	✓		
$1, \frac{1}{2}, \frac{1}{3}, \frac{1}{4}, \ldots$			✓
12,1.2,0.12,0.012		✓	
1, 3, 4, 7, 11,…			✓

b −0.3 c 0.1

3 a $-\frac{1}{3}$ b 3 c −5 d π

4 a 10^{-7} b $-12\,288$ c 3.20 (3 s.f.)
 d ax^{10}

5 a 7 b 12 c 6

6 a 88 573 b 1.33

7 a 26.0 b 11.6

8 a -2 b 17 c 43 691

9 a $18 = 2r^2$ b i $r=3$ ii $x = 6$ iii 39 366

10 a 0.5 b 30 c the seventh term

11 a 5.5 cm b 7 times

12 a

n (number of bounces)	0	1	2	3	6
height after n bounces (in metres)	8	**7.2**	**6.48**	**5.83**	**4.25**

 b 14 times

13 a 162 b 242 c 2 pm

14 a $6000 b $3132.04 c 6 years

15 a, c and d

Exercise 2.7

1 a $x = 2, y = 3$ b $x = 7, y = 4$
 c $x = -14, y = -9$ d $x = 2\frac{1}{3}, y = 7\frac{1}{3}$

2 a $x = 7, y = 3$ b $x = 0, y = 1$
 c $x = 5, y = 3$ d $x = 49\frac{2}{7}, 98\frac{4}{7}$
 $y = 98.6$

3 a $x = 2, y = 2$ b $x = -\frac{2}{7}, y = 5\frac{4}{7}$
 c $x = 2, y = 9$ d $x = 0.235, y = -1.29$

4 Same answers as from 1 to 2.

5 a $(x+1)(x+4)$ b $(a+3)(a+6)$
 c $(x+5)(x+7)$ d $(x-1)(x-4)$
 e $(b-3)(b-6)$ f $(x-5)(x-7)$
 g $(x-1)(x+6)$ h $(c-2)(c+11)$
 i $(x-1)(x+13)$ j $(x+1)(x-6)$
 k $(d+1)(d-10)$ l $(x+1)(x-13)$

6 a $(2x+1)(x+2)$ b $(2a-1)(a+2)$
 c $2(x-1)(x-2)$ d $(3x+1)(x-2)$
 e $(3b-4)(b-2)$ f $(3y-10)(y+3)$
 g $(3-5x)(x+1)$ h $(c-2)(c+11)$
 i $(5x+4)(x+1)$

7 a $x = -1$ or $x = -4$ b $a = -3$ or $a = -6$
 c $x = -5$ or $x = -7$ d $x = 1$ or $x = -6$
 e $y = 2$ or $y = -11$ f $x = 1$ or $x = -13$
 g $x = 5$ or $x = -6$ h $p = 0.5$ or $p = -2$
 i $x = -1/3$ or $x = 2$

8 a $x = -4$ or $x = 4$ b $a = -3$ or $a = -6$
 c $x = -2.13$ or $x = -9.87$ d $x = -0.5$ or $x = 2/3$
 e no solution f $x = 2$ or $x = -5$
 g $x = -1.28$ or $x = 3.31$ h $p = 0.5$ or $p = -2$
 i $x = 1.28$ or $x = -2.93$

Answers to past examination questions topic 2

Paper 1

1 a 1.265×10^{-1} b 0.13 c 2.77%

2 a $6x + 3y = 163.17$
 $9x + 2y = 200.53$
 b $17.69 c $14.85

3 a 5.5 b $\frac{2}{3}$

4 a 2 ml b 5 460 410 000 joules c kg m s^{-1}

5 a $A = x^2 + x$ b $x^2 + x = 30$ c 5 and -6
 d 5 as lengths cannot be negative.

6 a $w = 1.3 \times 10^{-3}$
 b Statements **ii** and **iv** are incorrect

7 a £1273 b £6370.96

8 a i $q = 1, r = 2, s = 7$ ii 3
 b Two decimal places because 2 decimal places are given in the data.

9 a $500 b $5100

Paper 2

1 i a $u_1 = 59$ $u_2 = 55$ b $n = 19$ c $k = 11$
 ii a 0.957 m b 16.74 m

2 a i 2 minutes 12 seconds
 b 25 minutes 12 seconds

3 a i $1000 ii $5120 b $11500 c Option two

Topic 3

Exercise 3.1

1 a All of the groups of items given can be regarded as sets.
 b Finite **i–v** inclusive, **vii**, **x**, **xii**, **xiii**, Infinite **vi**, **viii**, **ix**, **xi**
 c i $[-80 \text{ to } 150]$ ii $(-\infty, 16.3)$
 iii $[16.3, \infty)$ iv $(0, 4\pi]$
 d i ii

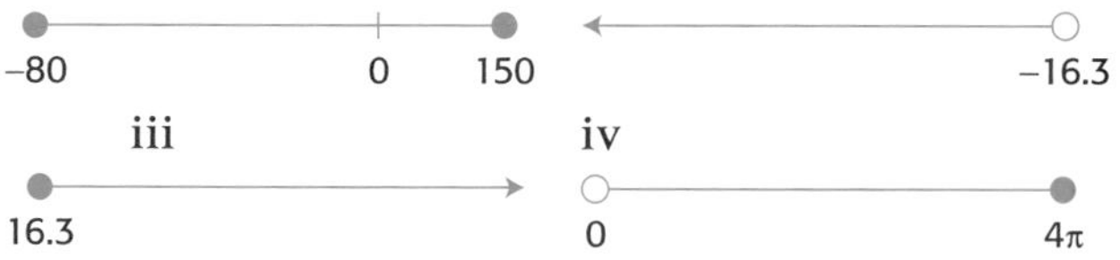

2 a i $\{2n-1 \mid n \in \mathbb{N}\}$ or $\left\{n \mid \frac{n+1}{2} \in \mathbb{N}\right\}$

Warning: using $2n + 1$ in the first of these or $\frac{n-1}{2}$ in the second, omits the number 1 from the set because 0 is not a natural number.

 ii $\left\{n \mid \frac{n}{5} \in \mathbb{N}\right\}$ or $(5n \mid n \in \mathbb{N})$

iii $\{2^z \mid z \in \mathbb{Z}\}$

iv $\{2^x \mid x \in \mathbb{R}\}$

v $\left\{x \mid \sqrt{x} \in \mathbb{R}\right\}$ or $\{x^2 \mid x \in \mathbb{R}\}$ or $[0, \infty)$ etc.

b i The vegetables listed in the left-hand set do all grow under the ground but the right-hand set contains many other vegetables not listed on the left, e.g. swedes, radishes, beetroot, parsnip, cassava, taro, chayote, chufa, yam etc. The sets cannot be equal.

ii $(-3)^2 = 9$ also, but -3 is missing from the left-hand set.

3 a There is an infinite choice. (1, 9) would do. So would [6, 8] etc.

b i Correct ii Incorrect iii Correct

iv Incorrect as 0 is a number, not a set.

v Incorrect. The left-hand set contains 1, the right-hand set does not.

c An internet search for 'root vegetables' finds many familiar and unfamiliar examples.

4 a The set is $J \cap P$ = {Sarah, Vicki, Jill, Abebe}.

b $P \cap J'$ = {Abdullah, Mei-Li, Chuck, Erasto}.

5 a i $[0, 3)$ ii $\mathbb{Z}$ iii $[-6, 4]$ iv $\{6, 5, 4\}$

v $[-1, 5]$ vi $\varnothing$ vii $\mathbb{R}$ viii $\mathbb{N}$

ix This is the union of all real numbers $\geqslant 1$ with all real numbers belonging to $(0,1)$. Hence the answer is $\mathbb{R}^+$.

x The subjects cover all of group 5 so this ought to be {All IB students at your school}.

b i $S \cap T = [0,1]$ ii $S \cap T' = [-1, 0)$

iii $T \cap S' = (1, 2]$ so

iv $(S \cap T') \cup (T \cap S') \cup (S \cap T)$ = the whole span of both intervals $= [-1, 2]$ which is also the union $S \cap T$. In general it must be the case that $(A \cap B') \cup (B \cap A') \cup (A \cap B) = A \cup B$.

c Both $S \cup S$ and $S \cap S = S$

d i $S \cap T = T$ was already proved in the hint.

ii $S \cup T = S$ as follows:

The LHS is the set $S \cap T$. Consider all the elements in this set. The collection consists of only members which are in both S and T. This means that we have at least everything in S because all members of S are also in T due to the subset relation. So once again we have everything in the set on the RHS. Is there anything extra coming from T? No because elements which are in T but not S are not included in the intersection. Hence the LHS and the RHS are the same and so $S \cap T = S$.

e Try for example

A = {Joseph, Nguyen, Anna}

B = {Joseph, Andy, Henry, Nguyen, Chloe, Anna, Ilya, Aiko, Yoshi}

Then $A \cup B$ = {Joseph, Andy, Henry, Nguyen, Chloe, Anna, Ilya, Aiko, Yoshi}= B.

while $A \cap B$ = {Joseph, Nguyen, Anna} = A.

6 $F \cap G'$ = {ä, é, ñ, å, µ, ß, Œ}.

Notice here that it is a bit tricky to list the elements in G'. We'd have to decide on what is the universal set and the list is likely to be long. In this exercise you simply have to say to yourself that this is the elements in F but not G. This is another reason why it is useful to have a special notation for relative complement. It avoids thinking about the actual details of U and making lists of irrelevant elements.

Exercise 3.2

1 a

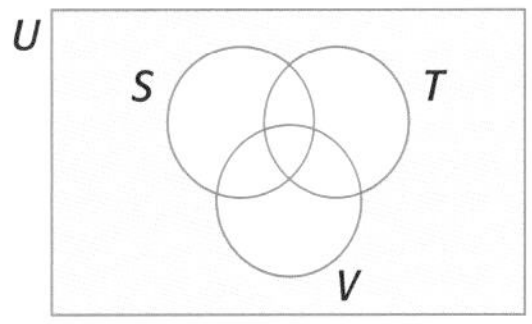

b

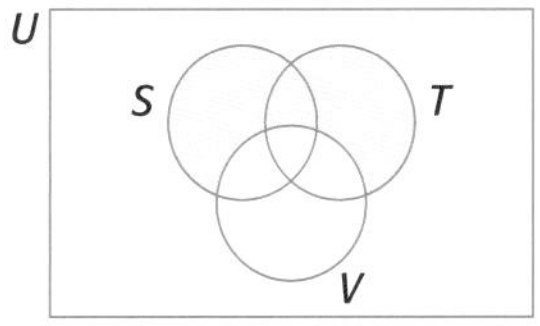

c

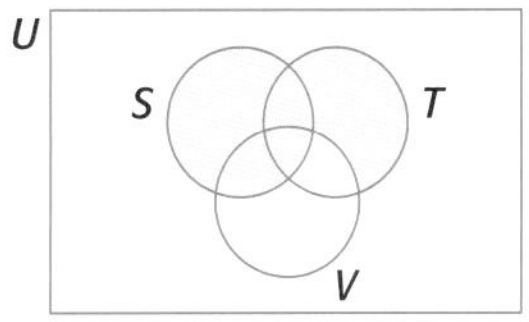

d This set is empty. Note: $T \cap (S \cap V')'$ shown below has no intersection with V'.

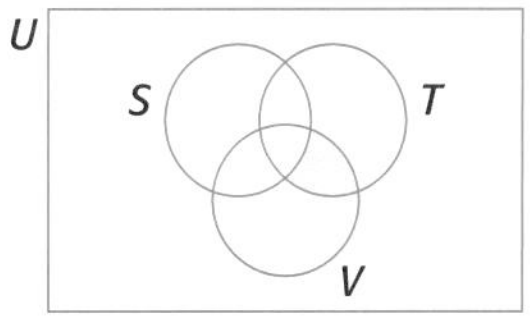

e

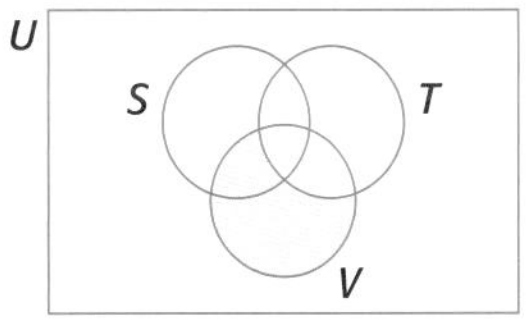

f

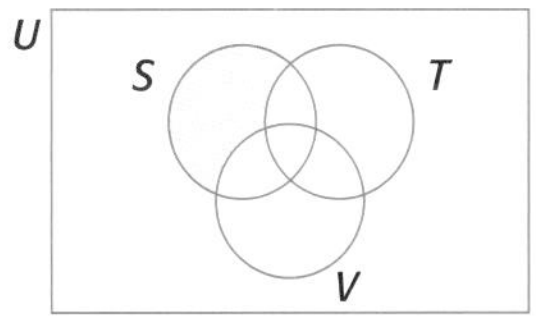

g $(S \cap T \cap V)$

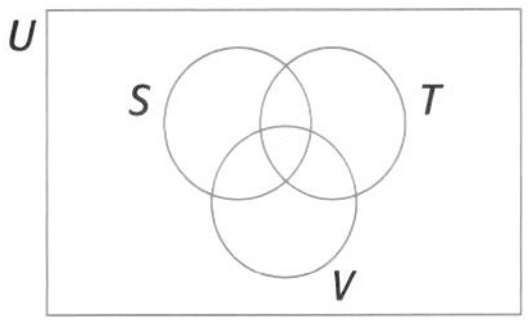

h $\varnothing' = U$ so this set is just T.

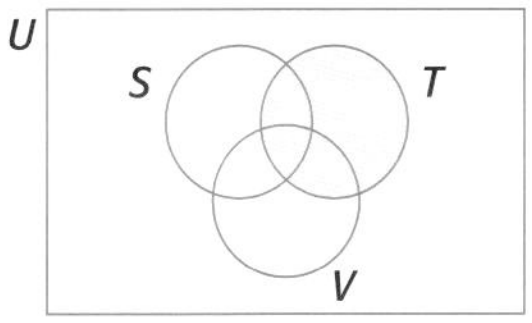

2 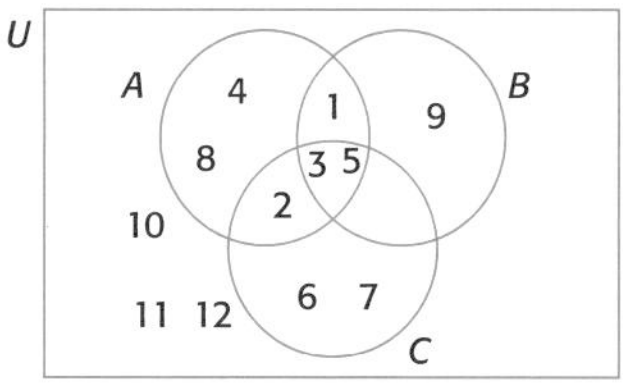

3 $\mathbb{R}$ is taken to be the universal set. The names have been placed *inside* their sets.

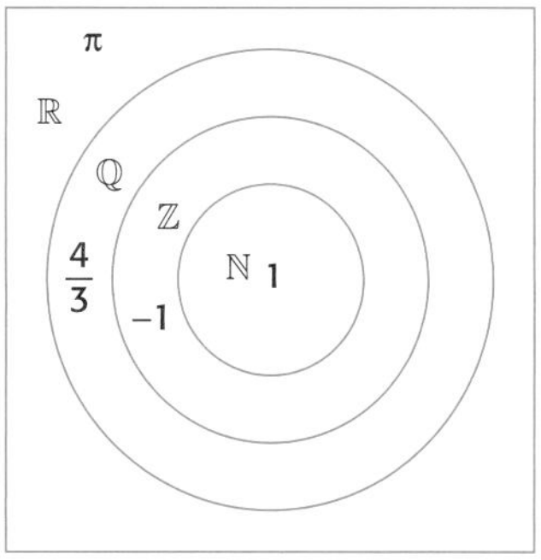

4 There is more than one way to describe this set. It's easiest to see it as a union of S and T with $S \cap V'$ removed. Hence
$(S \cup T) \cap (S \cap V)'$

5 Again combine pieces in stages: First the shaded area outside all three sets is $(S \cup T \cup V)'$. Add in the shaded part of T by making a union with $T \cap S'$ then the part shared by S and V (but not T) by making a union with $(S \cap V) \cap T'$.
The result is $(S \cup T \cup V)' \cup (T \cap S') \cup (S \cap V) \cap T'$.

6 Note: A and B have no intersection since a quadratic has either a maximum or a minimum but not both. Furthermore, the central part of C is empty because all quadratics have a maximum or a minimum.

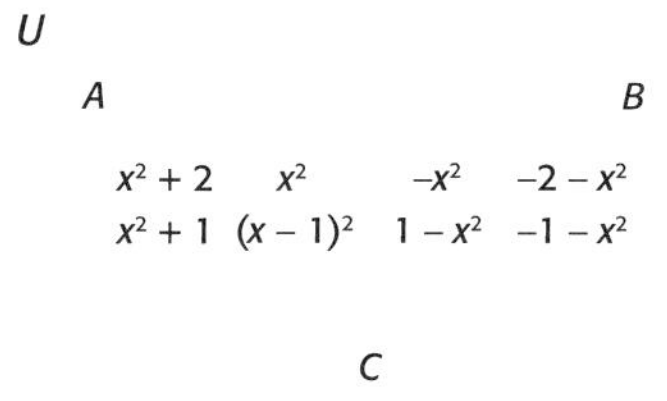

Exercise 3.3

1 With one dart you can score any number from 1 to 20, but also 25 or 50 in the central rings or double or treble any number from 1 to 20. (You might also miss the board!) So the sample space is $S = \{0, 1, 2, 3, \ldots\ldots 20, 25, 50, 22, 24, 26, 28, 30, 32, 34, 36, 38, 40, 21, 27, 33, 39, 42, 45, 48, 51, 54, 57, 60\}$.
A good universal set here is
$U = \{n \in \mathbb{Z} \mid 0 \leqslant n \leqslant 60\}$.
If we use this U then the numbers you cannot score are in $S' = \{23, 29, 31, 35, 37, 41, 43, 44, 46, 47, 49, 52, 53, 55, 56, 58, 59\}$.

2 The sample space of score combinations for the two players is $S = \{(15,0), (15,15), (0,15), (30,0), (30,15), (30,30), (15,30), (0,30), (40,0), (40,15), (40,30), (40,40), (30,40), (15,40), (0,40), (40,a), (a,40)\}$
$n(S) = 17$

3 **a** $G = \{1,2,3,4,5,6\}$ **b** $W = \{1,3,5\}$
c $L = W'$ (or W^c) $= \{2,4,6\}$
d 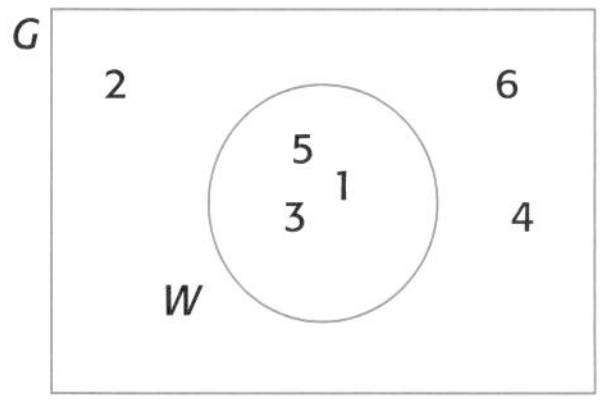

4 **a** $T = \{(1,1),(1,2),(1,3),(1,4),(1,5),(1,6),(2,2),(2,3),(2,4),(2,5),(2,6),(3,3),(3,4),(3,5),(3,6),(4,4),(4,5),(4,6),(5,5),(5,6),(6,6)\}$
$n(T) = 21$
b $S = \{(2,6), (3,5),(4,4)\}$
$n(S) = 3$

Exercise 3.4

1 **a** Yes this is a proposition. Most people believe it is false.
b Yes, a proposition. It is true.
c This is a question, not a proposition.
d **i** Yes a proposition, but it is indeterminate until we know x,
ii A proposition which is true,
iii Yes another true proposition.
e This is an exclamation and not a proposition.
f A proposition, but indeterminate.

Exercise 3.5

1 **a** **i** The maths teacher is not very strict.
ii The maths teacher is very strict and I work hard in class and I will do well in my maths test.
iii Neither is the maths teacher very strict nor do I work hard in class. (This is a bit easier to understand if we say instead: The maths teacher is **not** very strict **and** I do **not** work hard in class).
iv If the maths teacher is very strict, then I work hard in the maths class and if I work hard in the maths class, then I will do well in my maths test.
b **i** $p \Rightarrow \neg r$ **ii** $\neg p \veebar q$ **iii** $p \wedge q \Rightarrow r$

2 Take for example: p: The wind is strong,
q: *I will lose my hat*
r: *I will leave my hat at home*
Then $S = p \Rightarrow (q \veebar r)$.

3 **a** $S = P \cap T = \{0\}$

b **i** $(P \cup T)' = \mathbb{Z}$

ii $S' = \mathbb{R} \cap \{0\}'$, also often written as $\mathbb{R}/\{0\}$ or $\mathbb{R} - \{0\}$

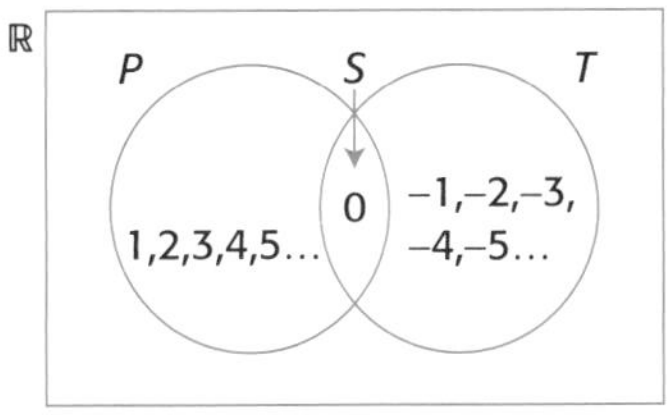

4 **a** {1, 2, 4, 6, 7, 8, 9, 12} **b** {1, 2, 4, 7, 8, 9}

c {6, 12} **d** {1, 2, 4, 6, 7, 8, 9, 12}

e {6, 10, 12, 13, 14, 15, 16}

f {6, 10, 12, 13, 14, 15, 16}

g {6, 10, 12, 13, 14, 15, 16}

a and **d** are the same.

e, **f** and **g** are the same.

Exercise 3.6

1 $\neg[p \wedge (\neg p \vee q)] \vee q$

p	q	$\neg p$	$\neg q$	$\neg p \vee q$	$p \wedge (\neg p \vee q)$	$\neg[p \wedge (\neg p \vee q)]$	$\neg[p \wedge (\neg p \vee q)] \vee q$
T	T	F	F	T	T	F	T
T	F	F	T	F	F	T	T
F	T	T	F	T	F	T	T
F	F	T	T	T	F	T	T

The last column is a tautology.

2 **a** The last column is a tautology.

p	q	$\neg p$	$\neg q$	$q \wedge \neg q$	$p \vee (q \wedge \neg q)$	$\neg p \vee [p \vee (q \wedge \neg q)]$
T	T	F	F	F	T	T
T	F	F	T	F	T	T
F	T	T	F	F	F	T
F	F	T	T	F	F	T

3 $\neg q \wedge [p \wedge (q \wedge \neg p)]$

p	q	$\neg p$	$\neg q$	$q \vee \neg p$	$p \wedge (q \wedge \neg p)$	$\neg q \wedge [p \wedge (q \vee \neg p)]$
T	T	F	F	T	T	F
T	F	F	T	F	F	F
F	T	T	F	T	F	F
F	F	T	T	T	F	F

The last column is a contradiction.

Exercise 3.7

1 The converse: *If I am not studying enough, then I am spending ages on the phone.*
The inverse: *If I an not spending ages on the phone, then I am studying enough.*
The contrapositive: *If I am studying enough, then I am not spending ages on the phone.*

2

p	q	r	$p \Rightarrow q$	$q \Rightarrow r$	$(p \Rightarrow q) \wedge (q \Rightarrow r)$ (Call this s)	$p \Rightarrow r$	$s \Rightarrow (p \Rightarrow r)$
T	T	T	T	T	T	T	T
T	T	F	T	F	F	F	T
T	F	T	F	T	F	T	T
T	F	F	F	T	F	F	T
F	T	T	T	T	T	T	T
F	T	F	T	F	F	T	T
F	F	T	T	T	T	T	T
F	F	F	T	T	T	T	T

The final column is a tautology. It is always true that if p implies q and q implies r, then p implies r.
For example: *p: I am 5 years old, q: I am a child, r: I like toys.*
Then the two compound propositions $p \Rightarrow q$: *If I am 5 years old then I am a child* and $q \Rightarrow r$ *If I am a child then I like toys,* when combined **always** have the result that $p \Rightarrow r$: *If I am 5 years old then I like toys.*

3

p	q	$p \Rightarrow q$	$q \Rightarrow p$	$(p \Rightarrow q) \vee (q \Rightarrow p)$
T	T	T	T	F
T	F	F	T	T
F	T	T	F	T
F	F	T	T	F

The shaded area can be described as: $(P \cap Q') \cup (Q \cap P')$

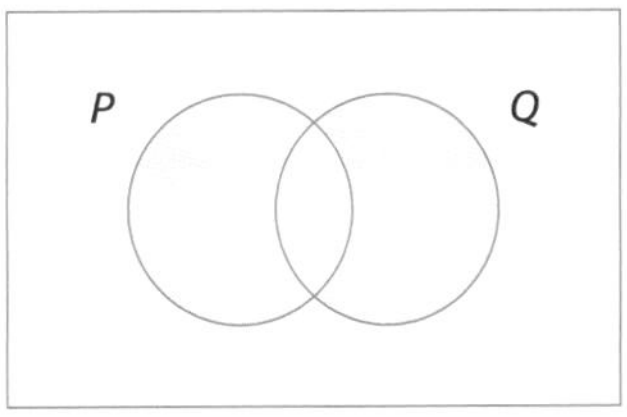

4

p	q	r	$\neg r$	$p \Rightarrow q$	$(p \Rightarrow q) \Rightarrow \neg r$
T	T	T	F	T	F
T	T	F	T	T	T
T	F	T	F	F	T
T	F	F	T	F	T
F	T	T	F	T	F
F	T	F	T	T	T
F	F	T	F	T	F
F	F	F	T	T	T

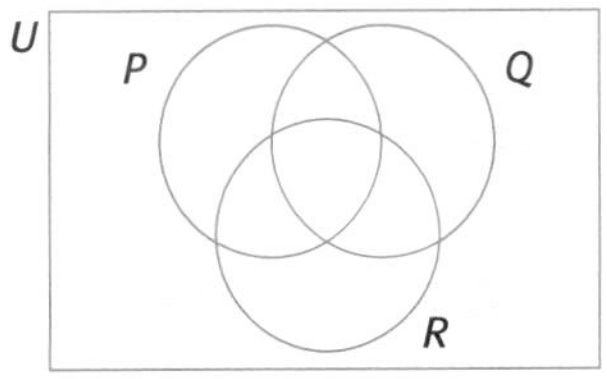

Remember that $p \Rightarrow q$ can also be written as $\neg p \vee q$ so we can rewrite the entire proposition as $\neg(\neg p \vee q) \vee \neg r$ which is also the same as $(p \wedge \neg q) \vee \neg r$. So the shaded area must be $(P \cap Q') \cup R'$.

5 The analogous logic expression is $(s \vee \neg t) \wedge [(t \wedge \neg r) \vee (s \wedge r)]$.

r	s	t	$\neg t$	$\neg r$	$s \wedge \neg t$	$t \wedge \neg r$	$s \wedge r$	$(t \wedge \neg r) \vee (s \wedge r)$	$(s \vee \neg t) \wedge [(t \wedge \neg r) \vee (s \wedge r)]$
T	T	T	F	F	T	F	T	T	T
T	T	F	T	F	T	F	T	T	T
T	F	T	F	F	F	F	F	F	F
T	F	F	T	F	T	F	F	F	F
F	T	T	F	T	T	T	F	T	T
F	T	F	T	T	T	F	F	F	F
F	F	T	F	T	F	T	F	T	F
F	F	F	T	T	T	F	F	F	F

and the Venn diagram is

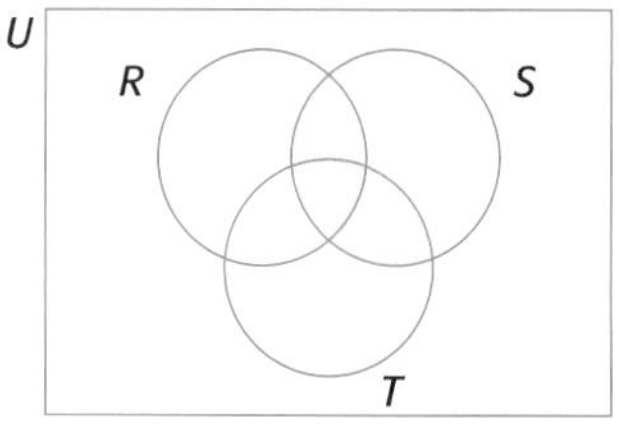

Hence by observation
$(S \cup T') \cap [(T \cap R') \cup (S \cap R)] = (S \cap R) \cup (S \cap T)$ and now the logic proposition can also be simplified by analogy to become $(s \wedge r) \vee (s \wedge t)$.

6

p	$\neg q$	$p \Rightarrow \neg p$	$p \Rightarrow p$
T	F	F	T
F	T	T	T

The last column is a tautology but the one before it is *not* a contradiction. The third column does not need to be a contradiction because p can be false, in which case there *is no* contradiction.

7

p	$\neg p$	$p \Leftrightarrow \neg p$	$p \Leftrightarrow p$
T	F	F	T
F	T	F	T

The last column is a tautology but now the third column *is* a contradiction.

Exercise 3.8

1 a $\frac{1}{52}$ b $\frac{4}{52} = \frac{1}{13}$ c $\frac{39}{52} = \frac{3}{4}$

d $\frac{26}{52} = \frac{1}{2}$ e $\frac{11}{52}$

2 a $\frac{4}{52} \times \frac{4}{51} = \frac{4}{663}$ b $\frac{4}{52} \times \frac{1}{51} = \frac{1}{663}$

c $1 \times \frac{12}{51} \times \frac{11}{50} \times \frac{10}{49} \times \frac{9}{48} = \frac{33}{16660} = 0.00198 (3\,\text{s.f.})$.

3 a $\frac{1}{4}$ b $\frac{1}{2}$ c $\frac{3}{4}$

4 The probability is just the ratio of the areas to that of the whole tray.

a 0.32 b $1 - 0.2 = 0.8$ c 0.52

Exercise 3.9

1 a 0.28

b

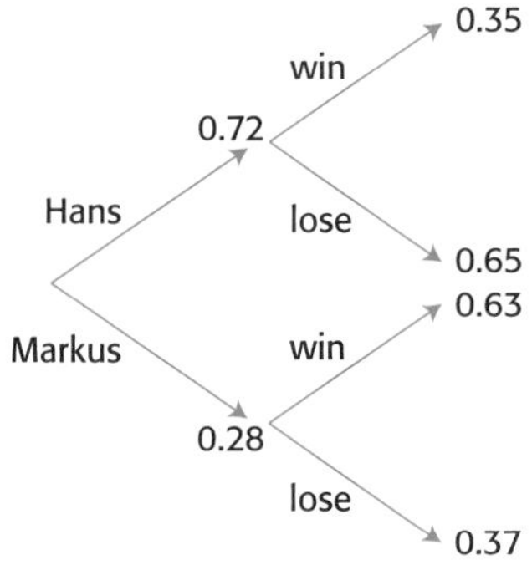

c 0.428 (3 s.f.)

2 a i 18 ii 12

b

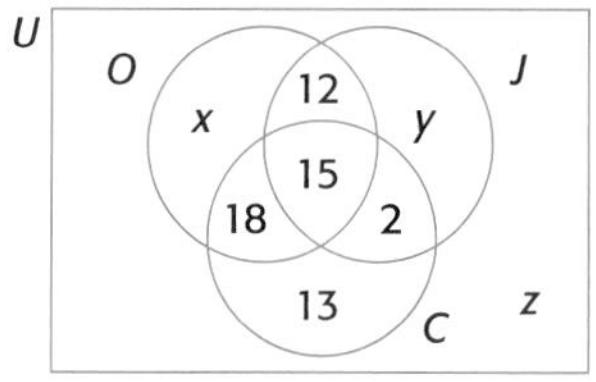

c $x + y = 13$, $x + z = 5$, $z + 73 = 78$ (or just $z = 5$).

d $z = 5$ so $x = 0$ and $y = 13$. 26 students do exactly one activity.

e i $\frac{15}{78} = 0.192 (3\,\text{s.f.})$ ii $\frac{60}{78} = 0.769 (3\,\text{s.f.})$

f $\frac{45}{73} = 0.616 (3\,\text{s.f.})$

3 a

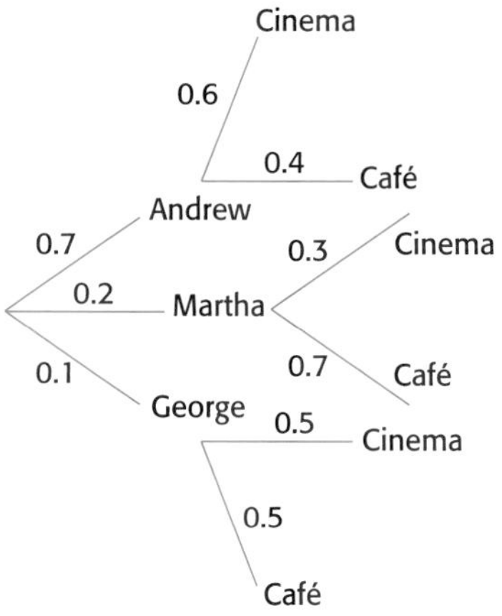

b i 0.05 ii 0.47 iii 0.19

4 a 0.36 **b** 0.6268 (0.627 3 s.f.)
c 0.1788 (0.179 3 s.f.)

Exercise 3.10

1 0.85

2 0.32

3 a $P(A) = 18/60$, $P(B) = 18/60$, $P(C) = 26/60$
b A and B since the intersection is zero. C and B since the intersection is zero.

4 0.0522

5 a 0.5 **b** 0.15 **c** 0.3 **d** 0.4
e neither **f** mutually exclusive.

6 a 3/11 **b** 5/11 **c** 3/5

7 a 0.6084 **b** 0.6854

8 a 0.396 **b** 0.6525

9 a 13/80 **b** 38/80 **c** $\frac{12}{38}$

10 a $P(A) = 1/13$, $P(B) = 0.5$, $P(C) = 0.25$, $P(D) = 0.25$
b 1/13 **c** 0.5
d A and D, A and C as $P(A) \times P(D) = P(A \cap D)$, $P(A) \times P(C) = P(A \cap C)$ and $P(A) \times P(B) = P(A \cap B)$
e C and D, B and D as $P(C \cap D) = 0$ and $P(B \cap D) = 0$

Answers to past examination questions topic 3

Paper 1

1

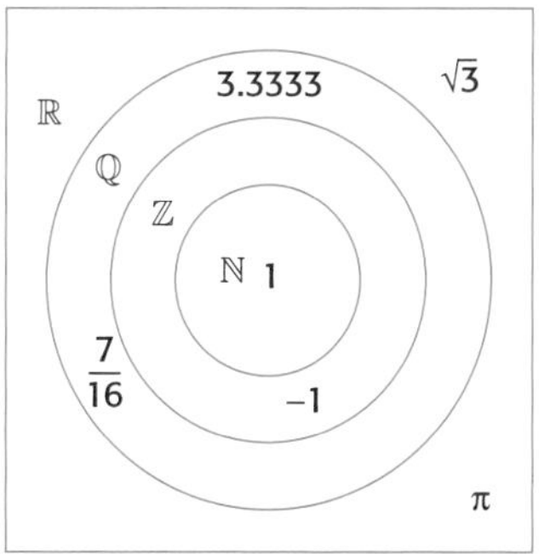

2 a If I am wearing my hat then the sun is not shining.

b

p	q	$\neg q$	$q \Rightarrow \neg p$
T	T	F	F
T	F	F	T
F	F	T	T
F	F	T	T

c $\neg p \Rightarrow q$

3 a $\frac{4}{15}$ (= 0.267 3 s.f.)
b $\frac{3}{35}$ (= 0.0857 3 s.f.) **c** 0

4 a 2/3 **b** 1/9 **c** 0.5 **d** 5/18

5 a 0.2 **b** $0.4 \times 0.65 \neq 0.2 \Rightarrow$ not independent
c $P(A \cap B) \neq 0 \Rightarrow$ not mutually exclusive

6 a i *If the number ends in zero then it is divisible by 5*
ii *If the number is divisible by 5 then it ends in zero,* (converse)
b i *If the number does not end in zero then it is not divisible by 5,* (inverse)
ii *If the number is not divisible by 5 then it does not end in zero,* (contrapositive)

Paper 2

1 a $A = \{1, 3, 7, 21\}$
b i $A \cup B = \{1, 3, 7, 14, 21\}$
ii C' is the set of all even numbers in U
$C' \cap B = \{14\}$
c $A \cap B \cap C = \{7, 21\}$
P(member of A and $A \cap B \cap C$) = $\frac{1}{2}$ (0.5)

2 a

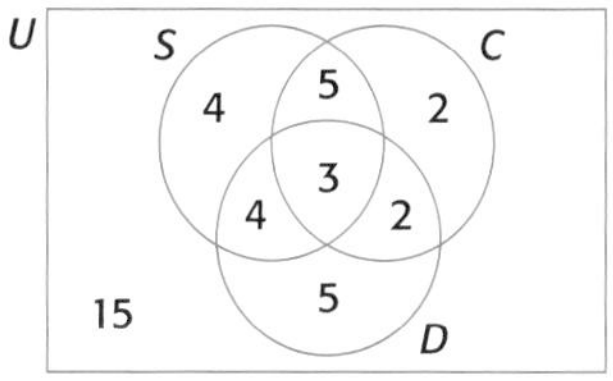

b $25 - (5 + 3 + 4 + 4 + 2 + 2) = 5$ **c** $\frac{1}{10}$ (= 0.1)

d $\frac{21}{40}$ (= 0.525), **e** $\frac{1}{2}$ (= 0.5)

3 a

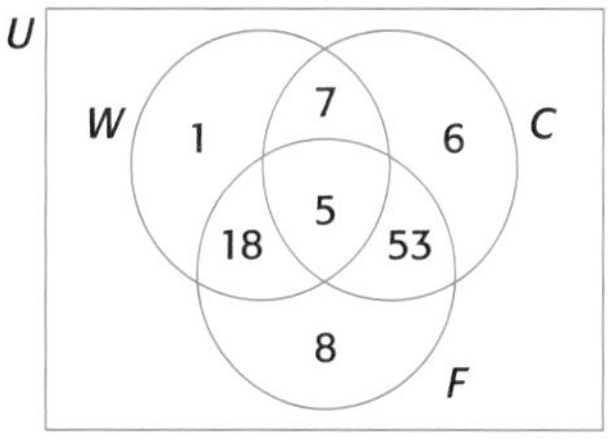

b 2 **c i** 0.71 **ii** 0.27 **iii** $\frac{8}{31} = 0.258$ (3 s.f.)

d $\frac{1}{495} = 0.002\,02$ (3 s.f.)

4 a 0.25 **b** 0.5 **c** 1/3

Topic 4

Exercise 4.1

1 a Domain {−2,−1, 0, 1, 2, 3} Codomain {25, 50, 75} Range {25, 50, 75}
b Domain {10, 20, 30, 40} Codomain {50, 90, 130, 170} Range {50, 90, 130, 170}
c Domain {1, 2, 3} Codomain {5, 7, 9, 11, 13} Range {5, 9, 13}, **a** and **b** are functions.

2

a

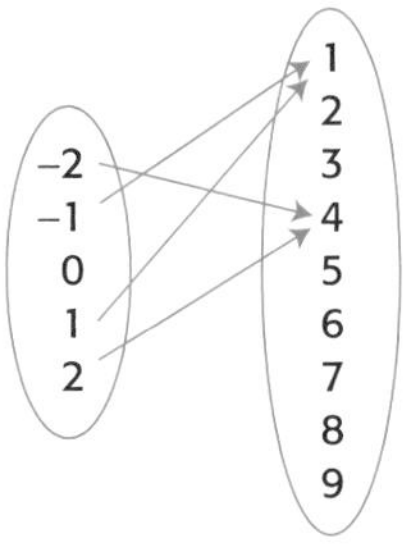

b

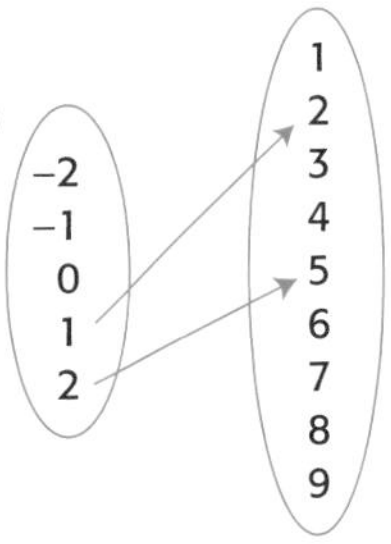

c

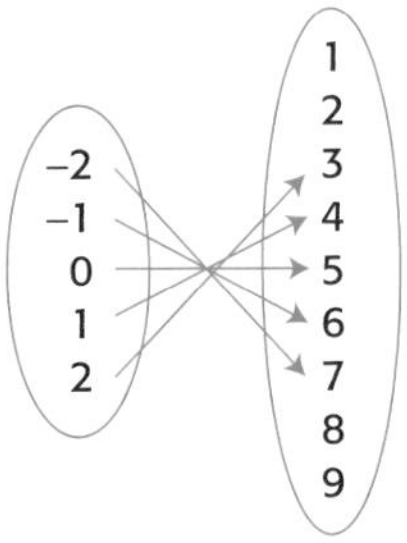

3 a $-3 \to -17$
$0 \to -2$
$3 \to 13$

b $-3 \to 25$
$0 \to -2$
$3 \to 25$

c $-3 \to 9$
$0 \to 3$
$3 \to -3$

4 a $f(x) = 2x - 2,\ x \in \mathbb{R}$ b $w(x) = x^2 + 3,\ x \in \mathbb{R}$
c $g(x) = 3x^3,\ x \in \mathbb{R}$

5 a $-19, -5, 9$ b $-3, 1, -3$ c $3, 1, 1$
d $-\frac{1}{2}$, undefined, $\frac{1}{2}$ e $-12, -10, 0$
f $8, 10, -4$

6 a Domain $\{0, 1, 2, 3, 4, 5\}$
Range $\{0, 10, 20, 30, 40, 50\}$
b Domain: $x \in \mathbb{R}$ Range: $y \geq 0,\ y \in \mathbb{R}$
c Domain: $x \leq 3,\ x \in \mathbb{R}$ Range: $y \leq 3,\ y \in \mathbb{R}$
d Domain: $-3 \leq x \leq 3$ Range: $-19 \leq y \leq -1$
e Domain: $x \geq 0,\ x \in \mathbb{R}$ Range: $y \geq 0,\ y \in \mathbb{R}$

7 a $2x - 3$

x	−2	−1	0	1	2
f(x)	−7	−5	−3	−1	1

b $2x^2 + 1$

x	−2	−1	0	1	2
g(x)	9	3	1	3	9

c $x(x - 1)$

x	−2	−1	0	1	2
S(x)	6	2	0	0	2

8 a

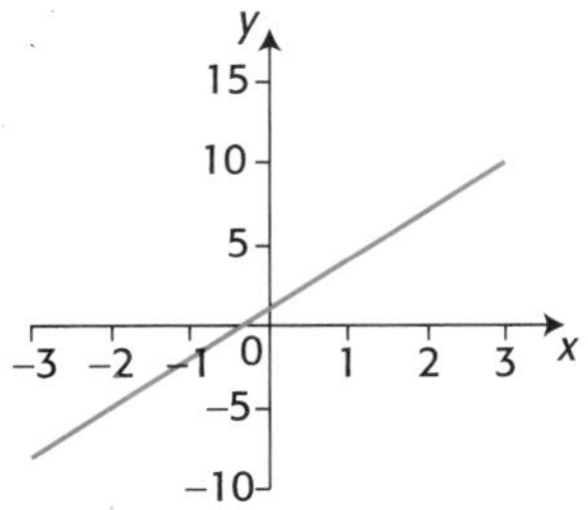

b

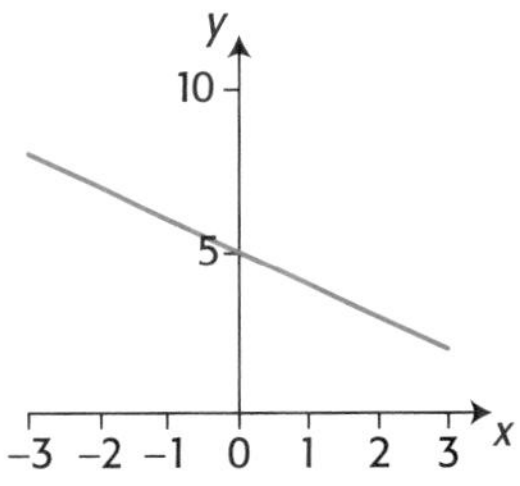

c

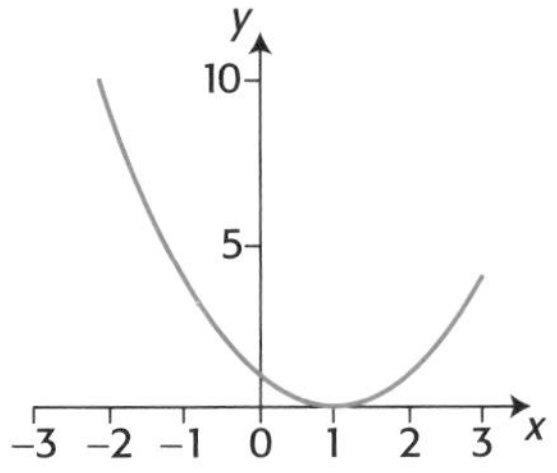

Exercise 4.2

1 a Domain: $-3 \leq x \leq 3$, Range: $-7 \leq y \leq 5$
b Domain: $x > 0$, Range: $y > 10$
c Domain: $-5 \leq x < 10$, Range: $-20 < y \leq 50$

2 a

x	−2	0	2
f(x)	−3	1	5

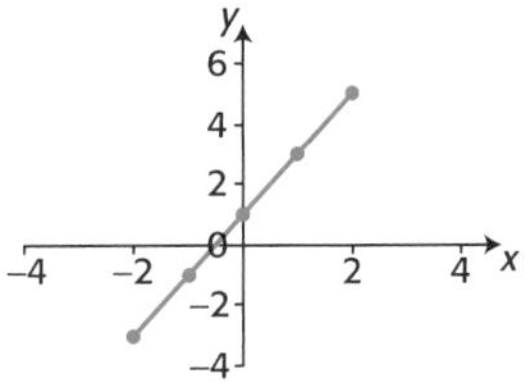

b

x	−10	0	10
g(x)	50	20	−10

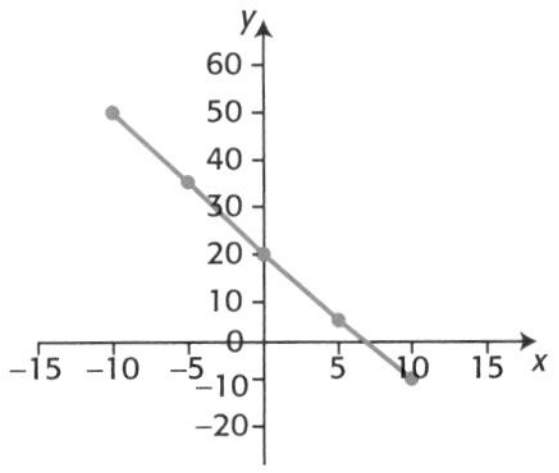

c

x	−3	0	3
g(x)	0.5	2	3.5

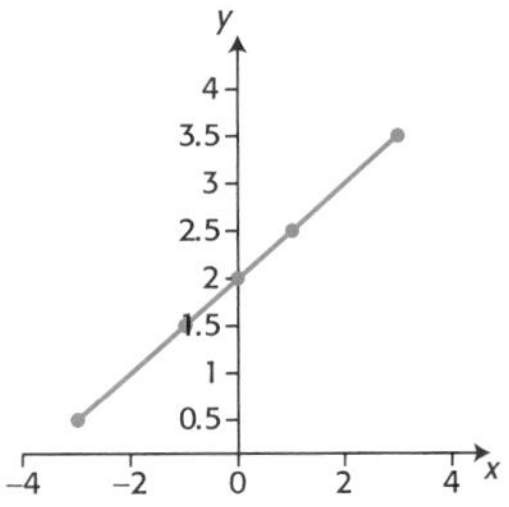

d

x	−100	0	100
g(x)	−4	16	36

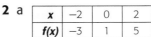

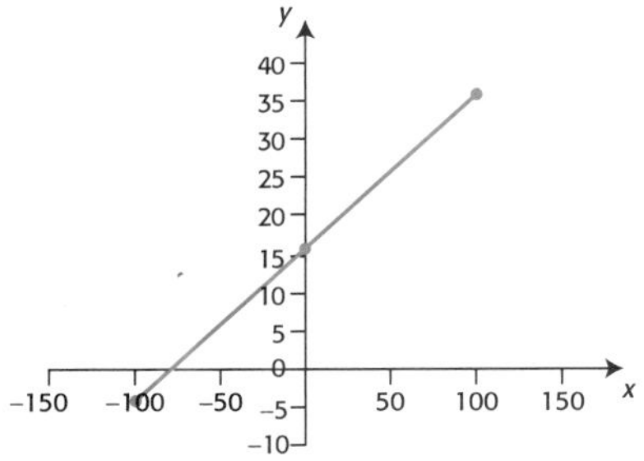

3

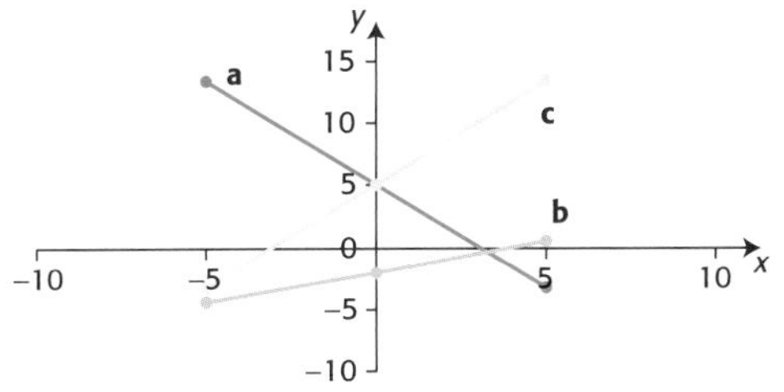

4 a gradient, +7 y-int, +3 b gradient, $+\frac{2}{3}$ y-int. – 3

c gradient, +2.5 y-int, +2 d gradient, –4 y-int. +1

e gradient, 0 y-int, +3 f gradient, –22 y-int. 66

5 a gradient, 5 y-int, 1 $y = 5x + 1$
$-5x + y = 1$

b gradient, –2 y-int, 3 $y = -2x + 3$
$2x + y = 3$

c gradient, 1.5 y-int, 2 $y = 1.5x + 2$
$-3x + 2y = 4$

d gradient, –2.5 y-int, 10 $y = -2.5x + 10$
$5x + 2y = 20$

6 a (1, 3) b (–1, 5) c (–2, –7)

7 a (3, 9)

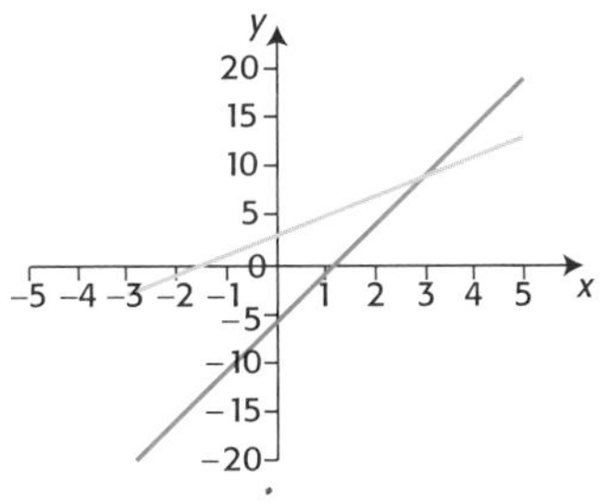

b (2.5, 4)

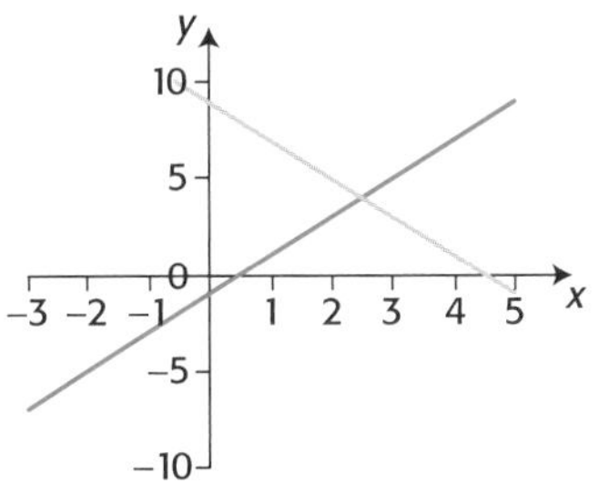

c (5, –2)

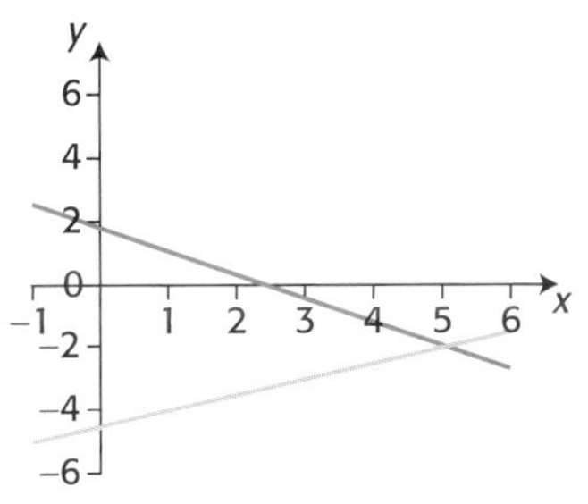

d (0, 2.5)

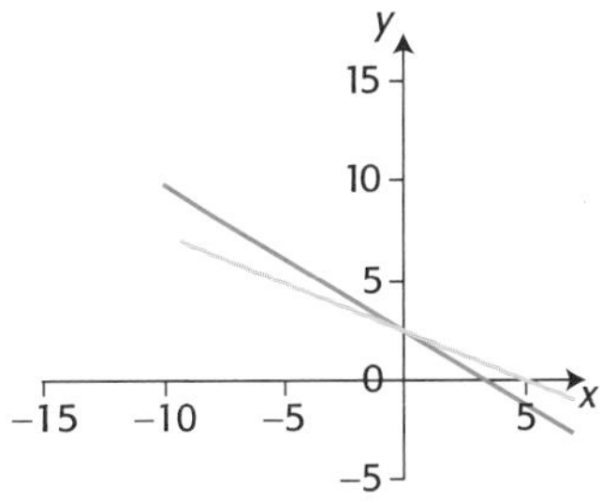

8 a i 51 500 rupiah 271 500 rupiah

ii

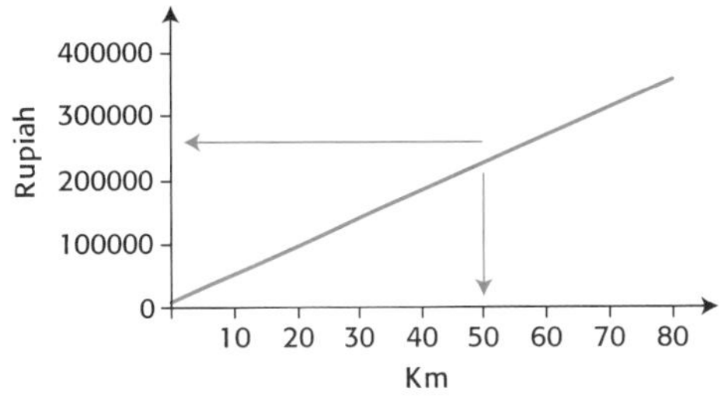

iii Cost of 50 km trip is approx. 230 000 rupiah.

b i Profit is £400

ii Break even when 140 items made and sold.

c i

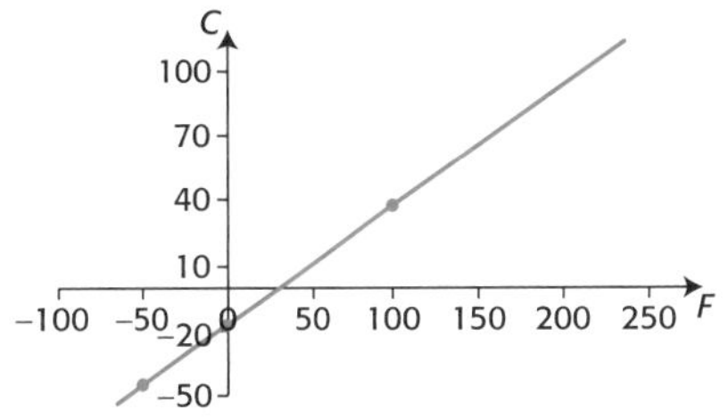

ii 38 °C –18 °C –35 °C

d i $R = 18.5x - 46\,000$

ii

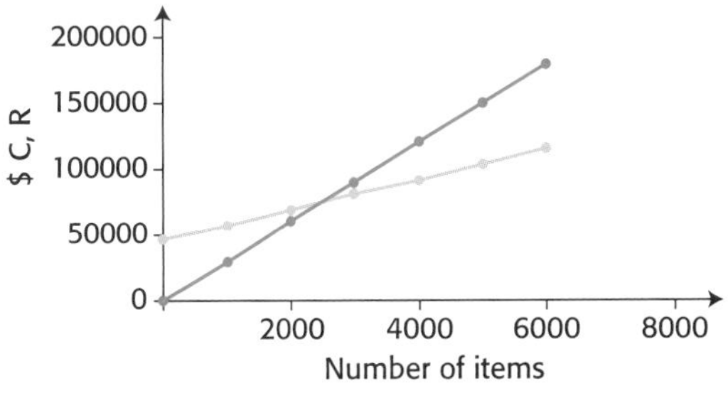

iii Profit is –$27 500 (loss)
Profit is =$46 500

iv Break even when 2486 items are produced and sold.

e i The equilibrium price is €62.50 and quantity is 3500 units

ii Excess demand is 1800 units

iii Excess supply is 1400 units

f i $V(t) = -150t + 12\,500$

ii 10 250 litres remain after 15 minutes.

iii 2000 litres remain after 70 minutes.

Exercise 4.3

1 a upwards, $x = -1$ b upwards, $x = -6$
c upwards, $t = 2$ d downwards, $x = -2.5$
e downwards, $x = 1.5$ f upwards, $n = 0.25$
g upwards, $x = 3.5$ h upwards, $t = 2$
i upwards, $x = 3$

2 a vertex $(-1, -9)$, min b $(2, -2)$, min
c $(1.25, -6.125)$, min d $(0.5, 1.25)$, max
e $(0.333, -4.33)$, min f $(5, 0)$, min
g $(0, -9)$, min h $(-5, -3)$, min
i $(5, 0)$, max

3 a x-intercepts 1, 6
y-intercept 6
vertex $(3.5, -6.25)$

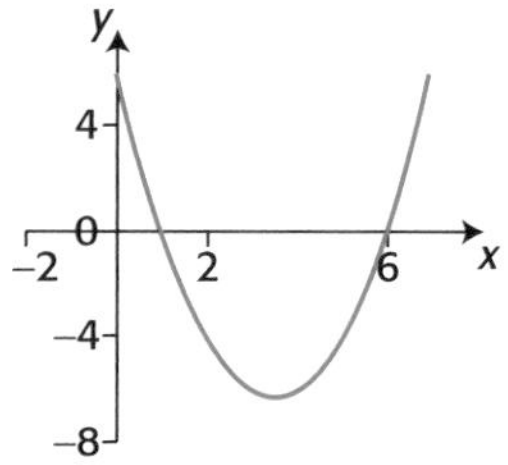

b x-intercepts -5, 2
y-intercept -10
vertex $(-1.5, -12.25)$

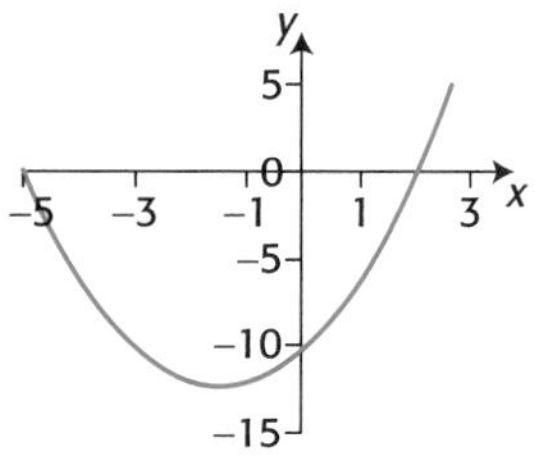

c x-intercepts -2, 5
y-intercept -40
vertex $(1.5, -49)$

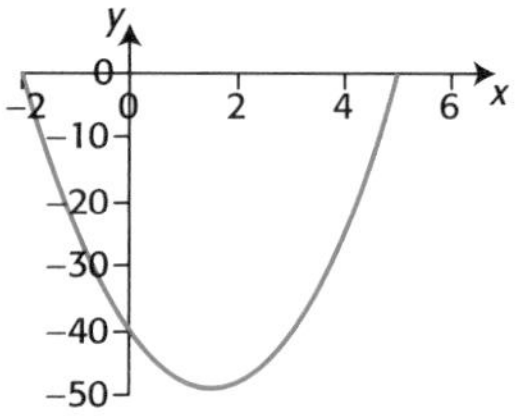

d x-intercepts -2, 3
y-intercept 6
vertex $(0.5, 6.25)$

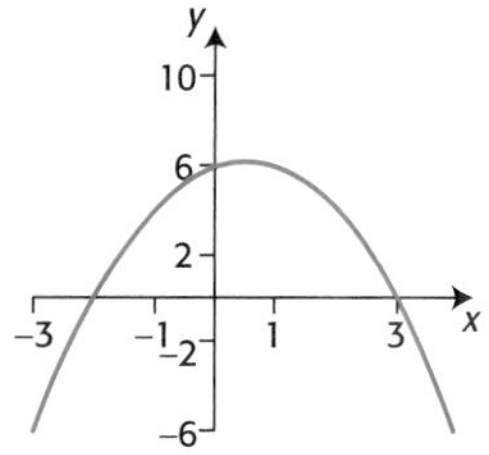

e x-intercepts -2, 4
y-intercept 4
vertex $(1, 4.5)$

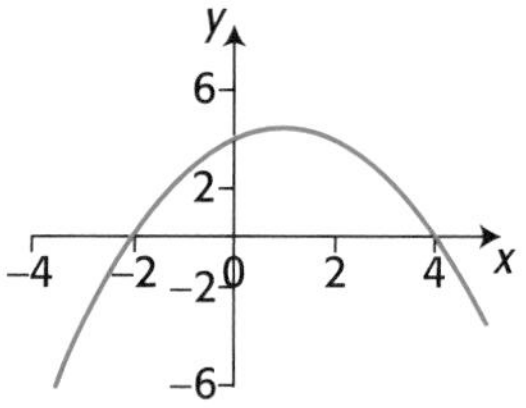

f x-intercepts -5, 2
y-intercept 10
vertex $(-1.5, 12.25)$

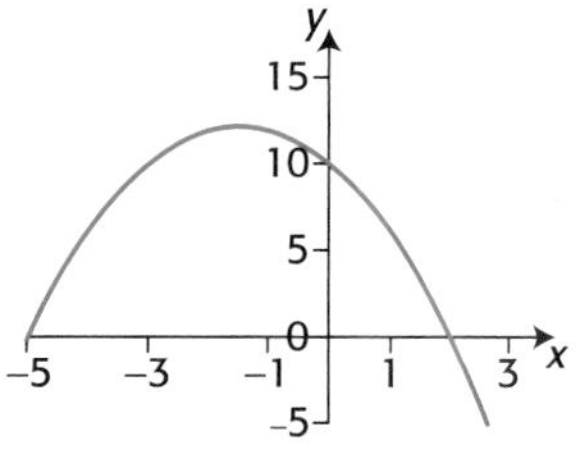

g x-intercept 3
y-intercept 9
vertex $(3, 0)$

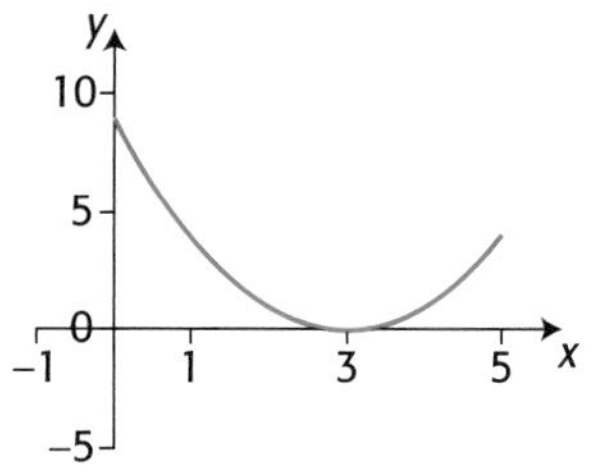

h x-intercepts 1.5
y-intercept -9
vertex $(1.5, 0)$

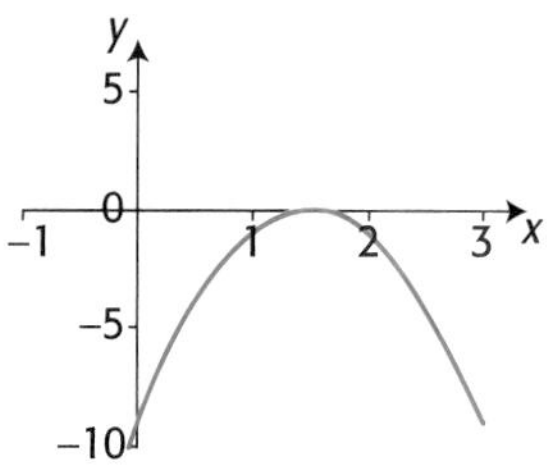

i x-intercepts -1.5, 5
y-intercept -15
vertex $(1.75, -21.125)$

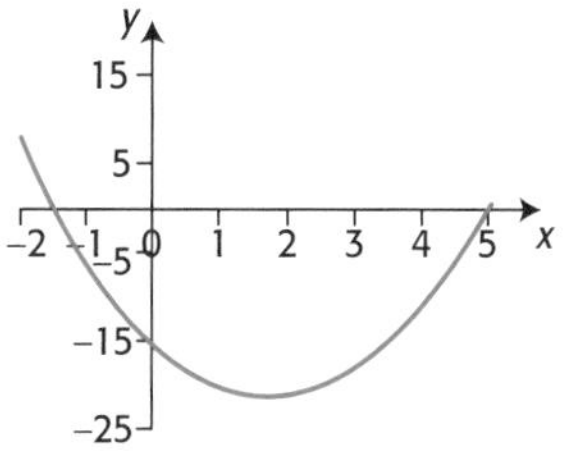

4 a $(x + 3)(x + 2)$ b $(x + 5)(x + 4)$
c $2(x + 6)(x + 1)$ d $(3 - t)(2 + t)$
e $2(x + 4)(x + 3)$ f $(x - 8)(x + 5)$
g $(7 - x)(2 + x)$ h $2(3 + x)(3 - x)$

5 a $y = x^2 - x - 2$ b $y = x^2 + 2x - 4$
c $y = -2x^2 + 4x + 6$ d $y = 2x^2 + 4x + 1$

6 a i $y = -2(x + 1)(x - 5)$ ii maximum is 18
b i $y = 3(x + 2)^2 + 3$ ii minimum is 3

7 a i

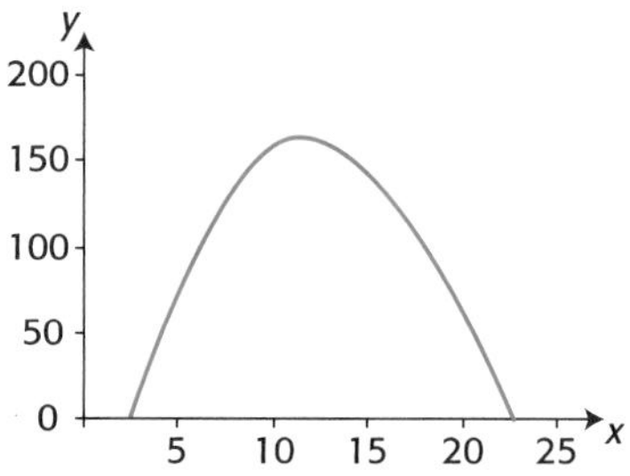

ii \$40, −\$60
iii \$40 max when 10 sold

b

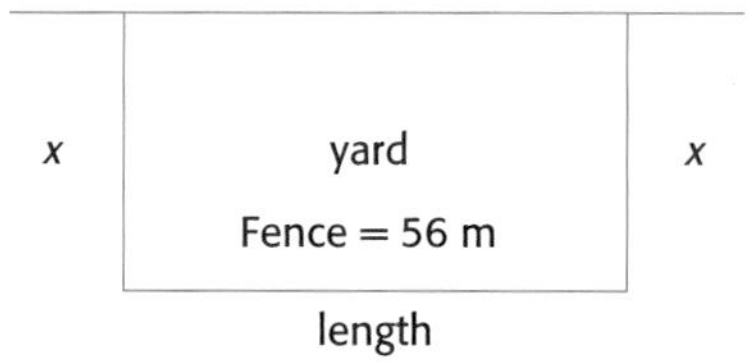

i Let the two equal sides be the width, x m. The length will be $(56 - 2x)$ m
ii $A = x(56 - 2x)$. Maximum area is 392 m^2 when dimensions are 14 m × 28 m

c i $H(10) = 400$ m ii $H(40) = 400$ m

ii

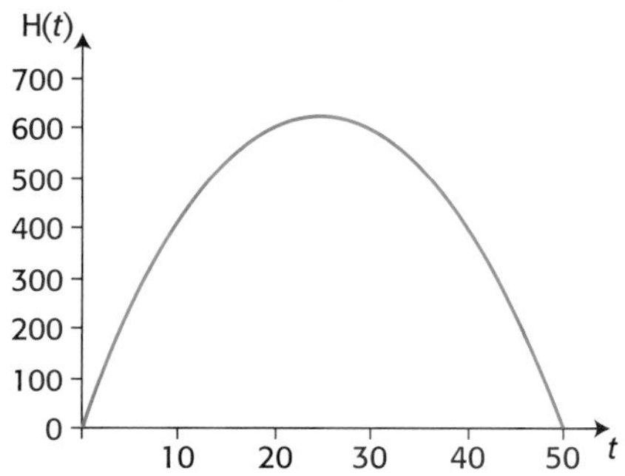

iii maximum height is 625 m iv 50 seconds

d i $\frac{n}{2}(2\times17+(n-1)\times13)=5390$

$n(34+13n-13)=10780$

$13n^2+21n-10780=0$

ii $n=-21$ or 28 so 28 terms sum to 5390.

e i $t=0$, $N=150$ bacteria

ii $t=5$, $N=570$ bacteria

iii $N=2000$, $t=15.9$ minutes

iv

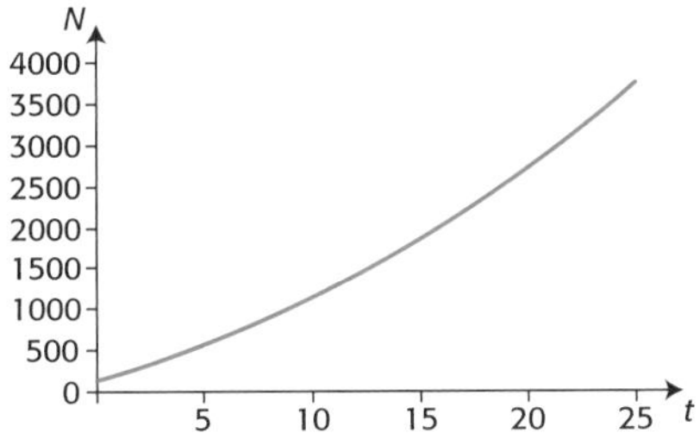

f i $0<x<18$ ii length is $18-x$

iii Area = length × width = $x(18-x)$

iv

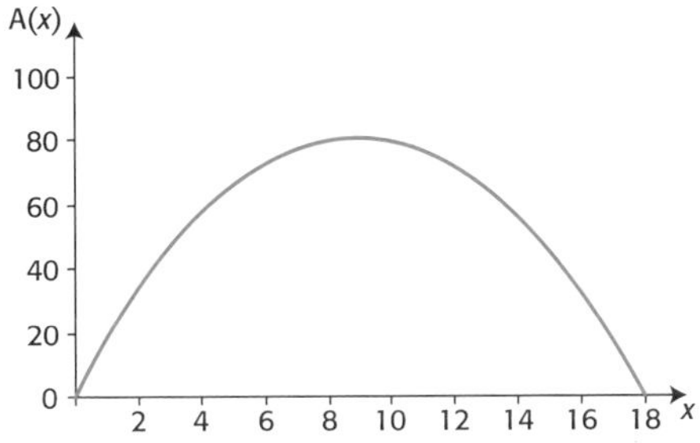

v maximum area when dimensions are 9 cm × 9 cm

Exercise 4.4

1 a 16 b 1 c $0.\overline{3}$
d 0.299 e ±8 f 0.06

2 a 0.008, 1, 125 b −0.008, −1, −125
c 125, 1, 0.008 d 0.381, 1, 2.63
e 2.63, 1, 0.381 f −125, −1, −0.008

3 a 32, 1, 0.313 b 10.125, $1.\overline{3}$, 0.176
c 57.7, 1, 0.017 d 1.82×10^{-12}, 0.00195, 2097152

4 a $k=12$ b $a=2.5$

5 a

x	−2	−1	0	1	2
y	−0.5	0	1	3	7

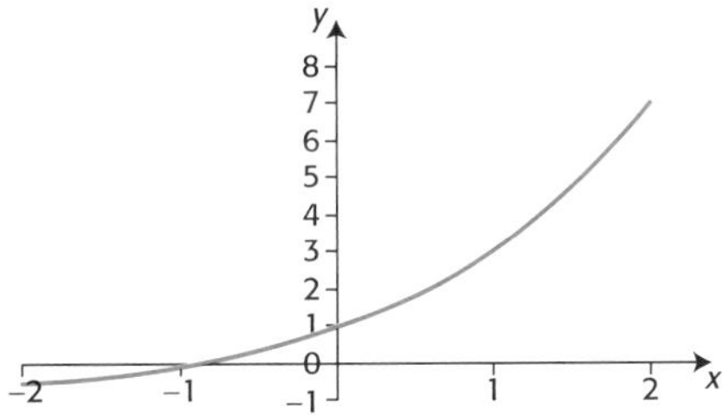

b

x	−2	−1	0	1	2
y	32	8	2	0.5	0.125

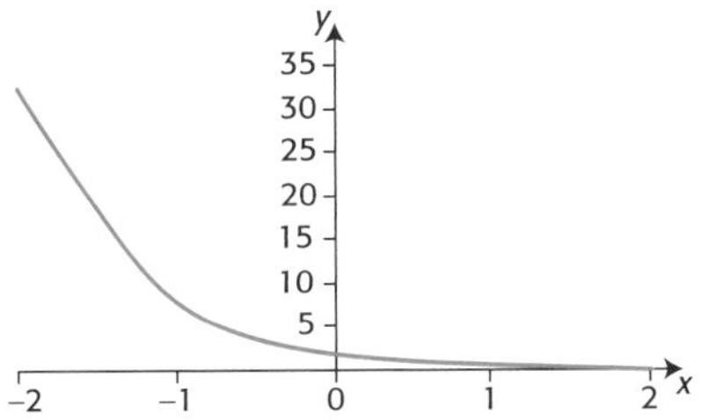

6

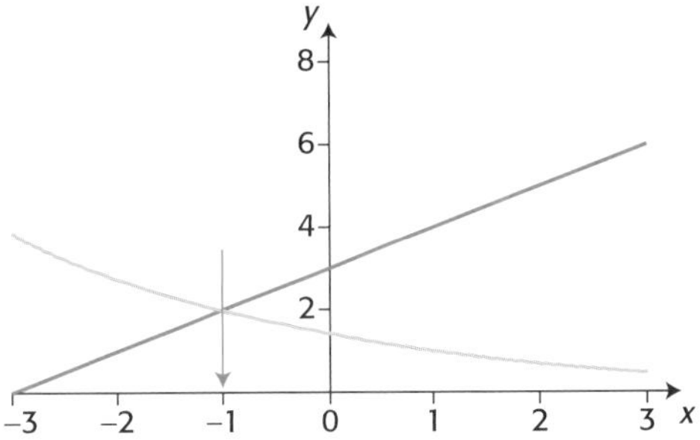

graphs intersect at (−1, 2) GDC (−0.930, 2.07)

7 a Diameter doubles in 6.3 years
b $t=15$, diameter is 15.6 cm

8 a i $P(0)=200$ ii 1.50 hours
iii 18.5 hours
iv $P(24)=12\ 800\ 000$ bacteria

b i $1.5(0.965)^{10}=1.05$ million hectares
ii $L(t)=1.5\times10^6\times0.965^t$
iii

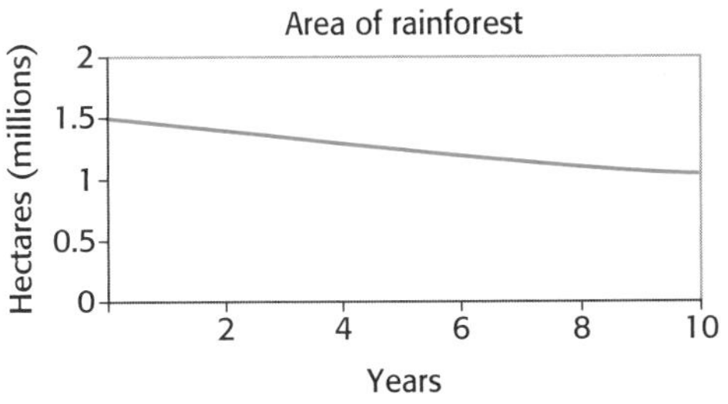

iv Area of rainforest < 0.5 million hectares after 30.8 (31) years.

c i a \$1082.43 b \$1485.95
ii Doubles in 8.75 years iii 9.04%

d i Original value is £14 559
ii 7.95 (8) years
iii Value is £2866.29. Profit on book value is £633.71

Exercise 4.5

1 a amplitude 1, period 360° b 2, 360°
c 2, 360° d 0.5, 360° e 0.2, 360°
f 1, 120° g 2, 720° h 1, 180°
i 0.5, 180°

2 a

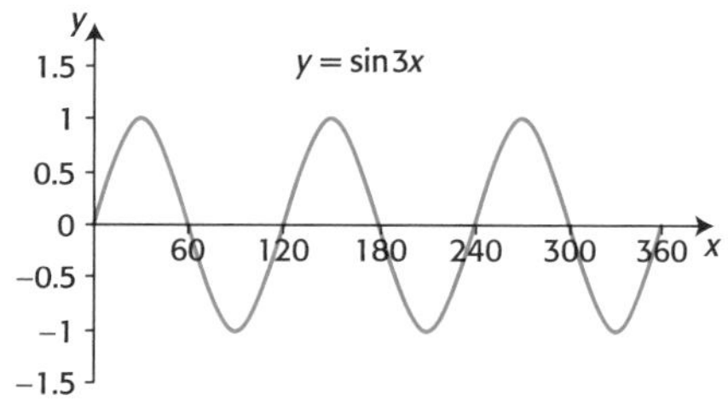

b

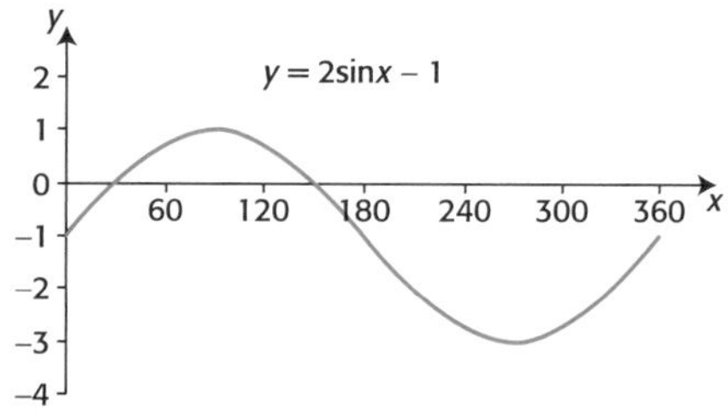

c

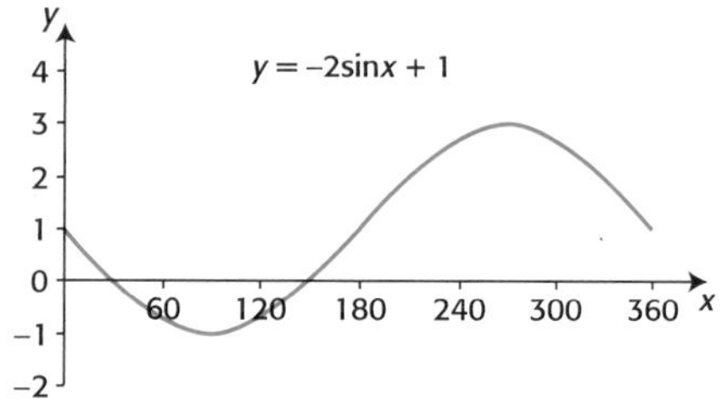

d

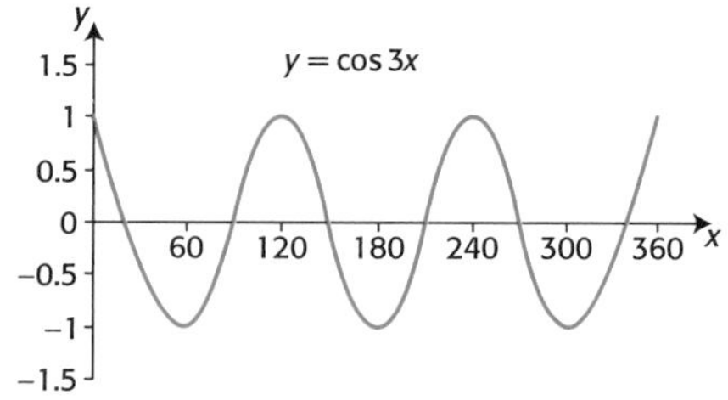

e

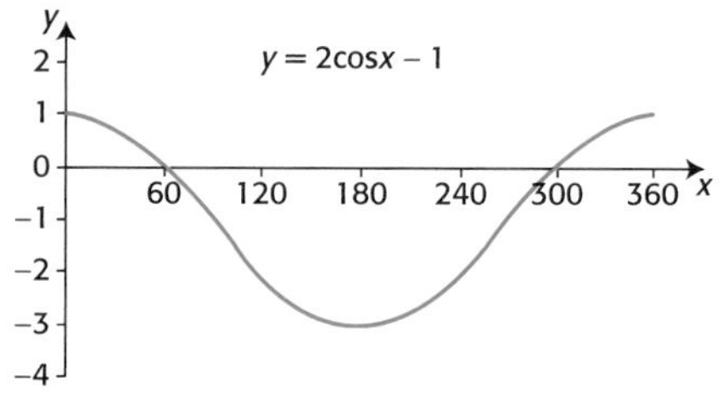

f

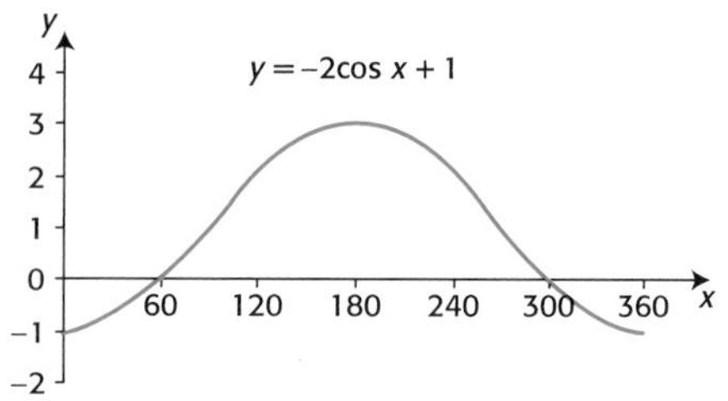

3 a

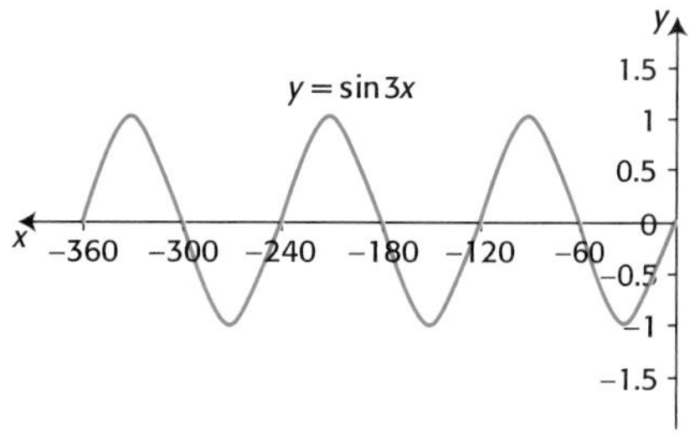

b

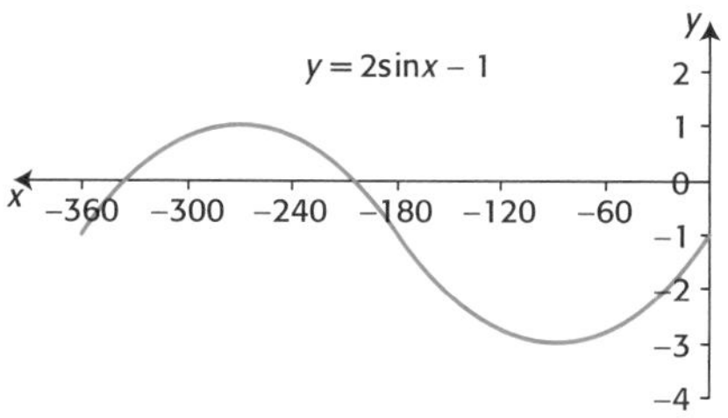

c

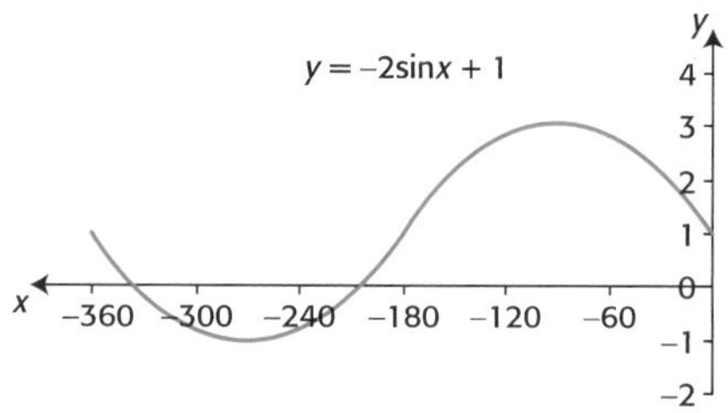

d

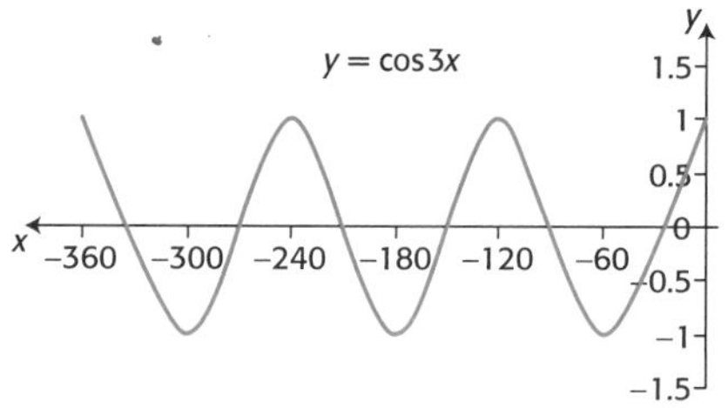

e

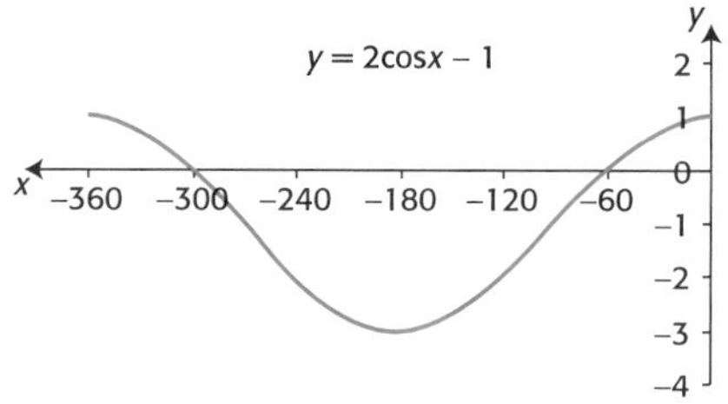

f

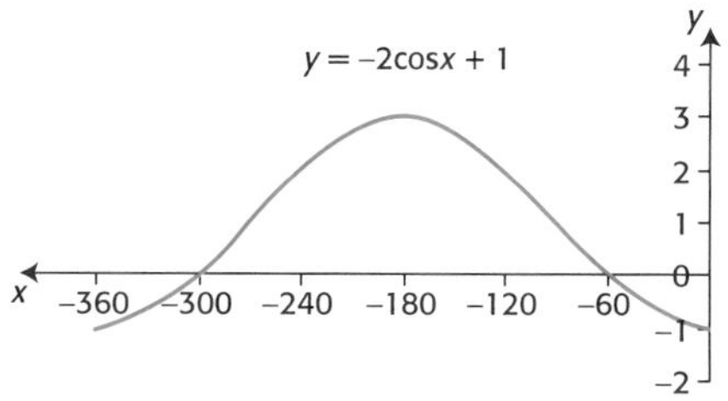

4 a $y = 2\sin x$ b $y = 1.5\cos x + 1$
c $y = -3\sin x + 2$ d $y = -\cos 2x - 1$

5 a -2 to $+2 = 4$ b -0.5 to $+2.5 = 3$
c -1 to $+5 = 6$ d -2 to $0 = 2$

6 a $-29.0°$ b $70.5°, 289.5°, 430.5°$
c $-56.4°, -123.6°$ d $\pm150°, \pm30°$
e $-63.4°, -26.6°, 116.6°, 153.4°$
f $-150°, -90°, -30°$

7 a i max 5.2 m after 1.8 seconds. min 4.8 m after 5.4 seconds.
ii 3.6 seconds

b i amplitude is 3 ii period 180°
iii minimum -1.8 iv 213.2°, 326.8°

c i max 4 after 180 seconds, min 2 after 60 seconds
ii amplitude is 1

iii

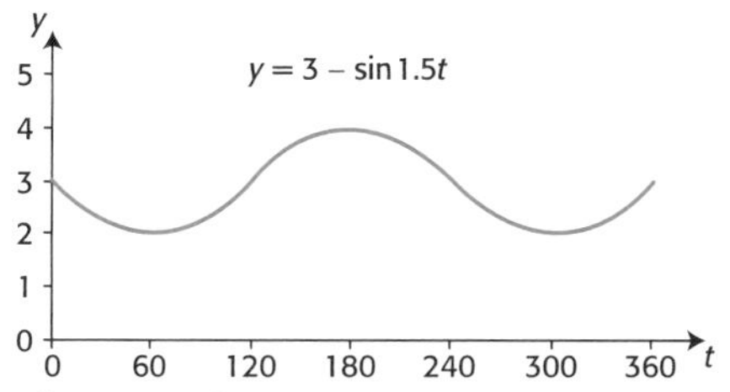

The period is 240°

d i 1000 caribou
ii 10 weeks 675, 50 weeks 675
iii min 350 after 6 weeks
iv after 12 weeks
v weeks 3 to 9 and 27 to 33, total of 12 weeks.

e i 11 °C ii 14 °C iii max 17 °C at 6 pm
iv from 6 am to 9:13 am and again from 2:47 am to 6 am the next day.

Exercise 4.7

1 x-intercepts –0.625, 0.728, 4.40. y-intercept 2 local max at (0,2) local min at (3, –11.5).

2 x-intercepts –2, 0.5, 2. y-intercept 2, local max at (–1, 4.5) local min at (1.33, –1.85).

3 touches x-axis at (1, 0) and (2, 0). y-intercept 4, local max at (1.5, 0.0625).

4 y-intercept –1. vertical asymptote $x = \frac{1}{3}$, horizontal asymptote $y = 0$.

as $x \to +\infty$, $y \to 0$ from above
as $x \to -\infty$, $y \to 0$ from below
as $x \to \frac{1}{3}$ from the right, $y \to +\infty$
as $x \to \frac{1}{3}$ from the left, $y \to -\infty$

5 y-intercept $\frac{1}{2}$, vertical asymptotes $x = 1$ and $x = 2$, horizontal asymptote $y = 0$

as $x \to +\infty$, $y \to 0$ from above
as $x \to -\infty$, $y \to 0$ from above
as $x \to 2$ from the right, $y \to +\infty$
as $x \to 2$ from the left, $y \to -\infty$
as $x \to 1$ from the right, $y \to -\infty$
as $x \to 1$ from the left, $y \to +\infty$

6 vertical asymptotes $x = 0$, horizontal asymptote $y = 0$

as $x \to +\infty$, $y \to +\infty$
as $x \to -\infty$, $y \to 0$ from below
as $x \to 0$ from the right, $y \to +\infty$
as $x \to 0$ from the left, $y \to -\infty$

7 x-intercepts 0, 2, y-intercept 0, min at (0.667, –1.089)

8 x-intercept 0, y-intercept 0, min at (2.72, 6.26) vertical asymptote $x = 1$

as $x \to 1$ from the right, $y \to +\infty$
as $x \to 1$ from the left, $y \to -\infty$

9 x-intercepts –180°, 0°, 180°, y-intercept 0, max 1 when x = –90° and 90°
min 0 when x = –180°, 0°, 180°

Exercise 4.8

a –0.404, 0.444, 4.46
b 4.55 **c** –0.434, 0.768
d 0.634, 2.37 **e** –1.64, 1.37 **f** 0.0520, 6.30
g 0.558, 4.77 **h** 2.55 **i** –21.2°, 21.2°

Answers to past examination questions topic 4

Paper 1

1 **a** $Q(0) = 25$ **b** $Q(20) = 13.7$
c average is 0.535 units per minute
d energy runs out after 54.0 minutes

2 **a**

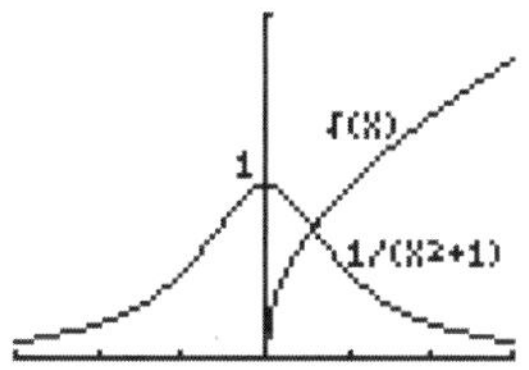

b One solution
c $x = 0.570$

3 **a** i period is 120° ii amplitude is 4
c $x = 10°$ or $x = 50°$

4 **a** original weight is 90 grams
b v = 5.625 grams **c** 20 years

5 **a** $c = 0$, $k = 5$ **b** maximum is at $y = 6.25$

6 **a** C, B, A **b** 180°
c range is $2 - a \le y \le 2 + a$

7 **a** $5x(6 - x)$ **b** A (6, 0) **c** $x = 3$

8 **a** period is 360° **b** minimum is –3
c $x = 90°$ or 450°

9 **a** 10 °C **b** –459.4 °F

10 **a** P is 2 units above the ground
b $2 \le y \le 4.25$

Paper 2

1 **a** i $a = 60$ $b = 48$

ii

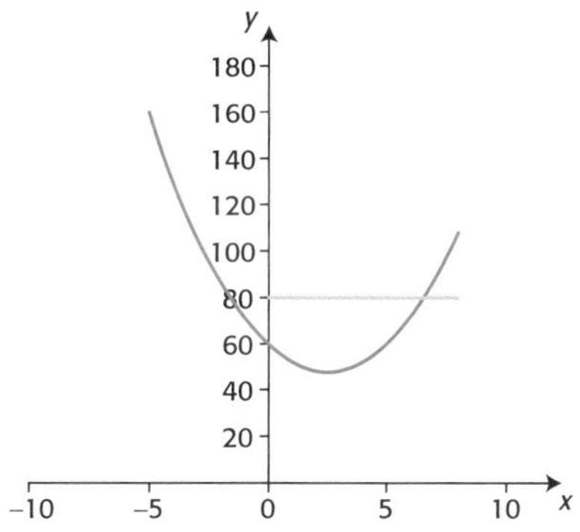

iii $x = \dfrac{-b}{2a} = \dfrac{-(-10)}{2(2)} = 2.5$

$f(2.5) = 2(2.5)^2 - 10(2.5) + 60 = 47.5$

iv $2.5 < x$ 8

b i Horizontal straight line in given domain, intercept at 80

ii (6.53, 80) iii $80 - 47.5 = 32.5$

2 a $a = 2.5$, $b = 13$

b

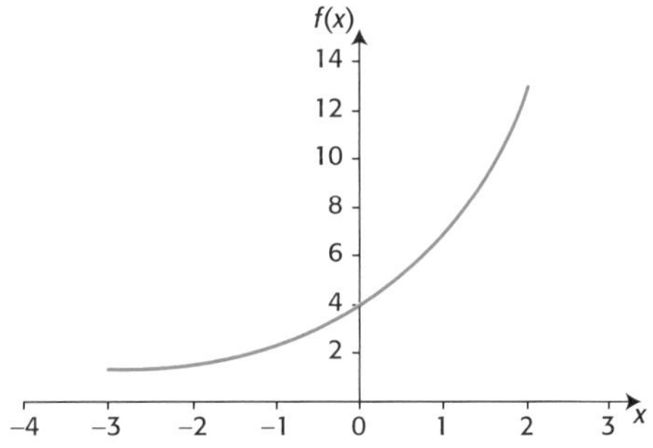

c range is $y > 1$

d $x = 1.58$

3 a $P = I - C$,

$150x - 0.6x^2 - (2600 + 0.4x^2)$

$= 150x - 0.6x^2 - 2600 - 0.4x^2$

$= -x^2 + 150x - 2600$

b maximum profit when 75 machines are made and sold.

c income is $150(75) - 0.6(75)^2 = \$7875$

Selling price $= \dfrac{7875}{75} = \$105$ each

d 21 machines needed to be made and sold for a positive profit.

4 a i height at 03:15 is 1.75 m

ii t = 1:40 and 6:15

b $2 < t < 6$ c $a = 1.5$, $b = 45$

d height at 13:00 ($t = 13$) is 4.06 m

e 12 noon

Topic 5

Exercise 5.1

1 $A(-2, 5)$, $B(-1, -2)$, $C(0, 4)$, $D(1, 3.5)$, $E(2, 0)$

2 a

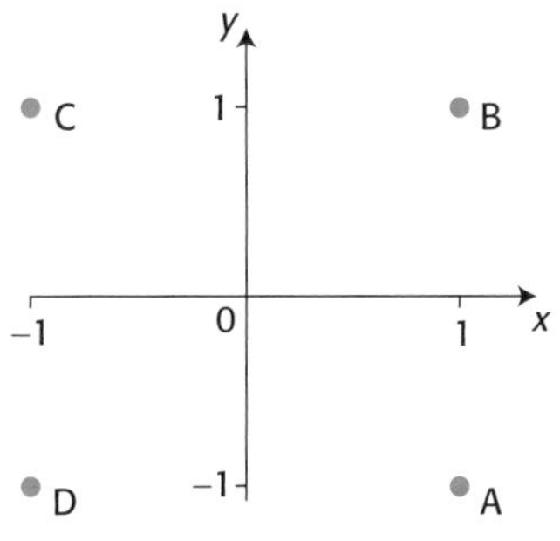

b Square c i 4 ii 8

3 a 7 b 6 c $\sqrt{13} = 3.61$ (3s.f.)

4 a

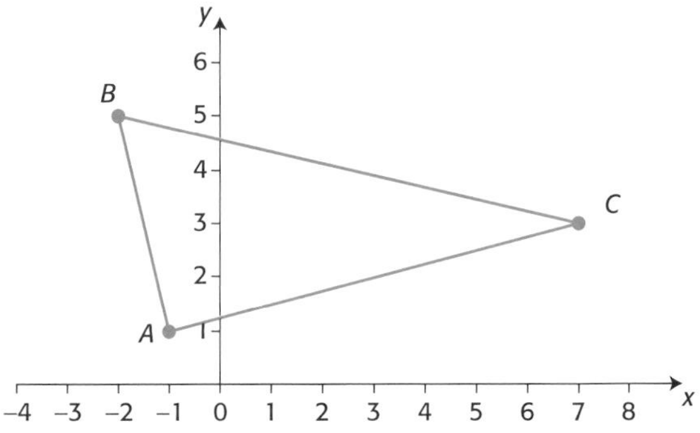

b $AB = \sqrt{17}$; $AC = \sqrt{68}$

5 a $M(1,2.5)$ b $B(5,1)$

6 a $y = -3$ or $y = 13$ b $x = -2$ or $x = 10$

7 a

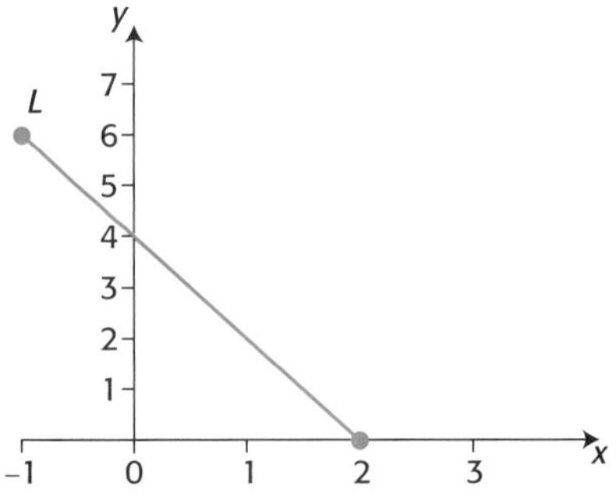

b $y = 2$ c 4.472

Exercise 5.2

1 a $m = -\dfrac{2}{3}$; $c = \dfrac{8}{3}$; b $m = \dfrac{1}{2}$; $c = -1$;

c $m = 0$; $c = 2.5$ d $m = \dfrac{2}{9}$; $c = -\dfrac{4}{3}$

2 a

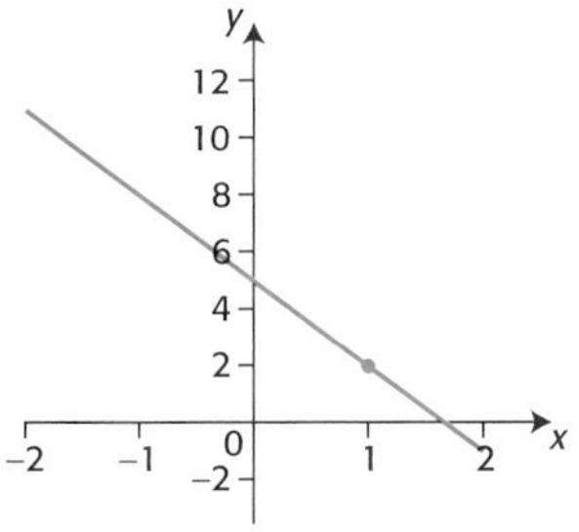

b

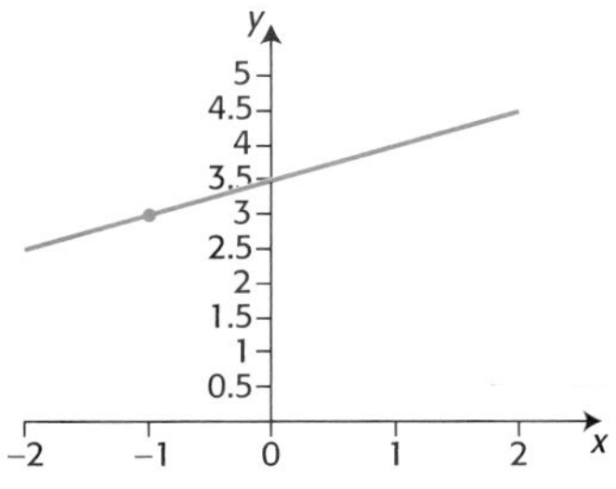

c

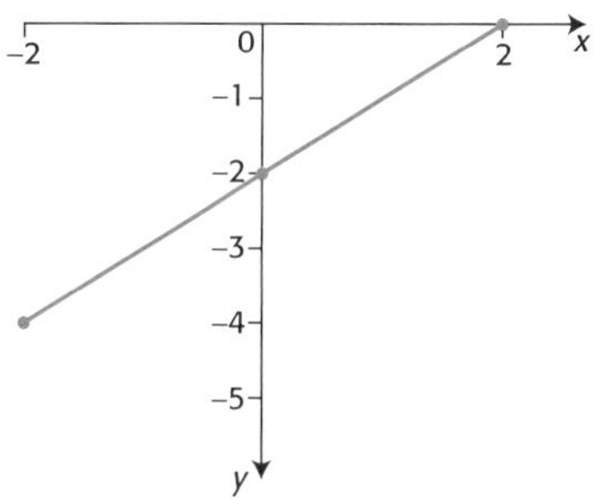

3 a $y=-3x+5$; **b** $y=0.5x+3.5$; **c** $y=x-2$

4 a $r=\dfrac{10}{3}$; $s=15$ **b** 3 **c** $x+3y-15=0$

5 Not collinear. Gradient of AB is not the same as gradient of AC.

6 a $y=5x+20$; **b** $y=-\dfrac{1}{4}x+3$;
c $y=-x+3$; **d** $y=-3$; **e** $x=2$

7 a

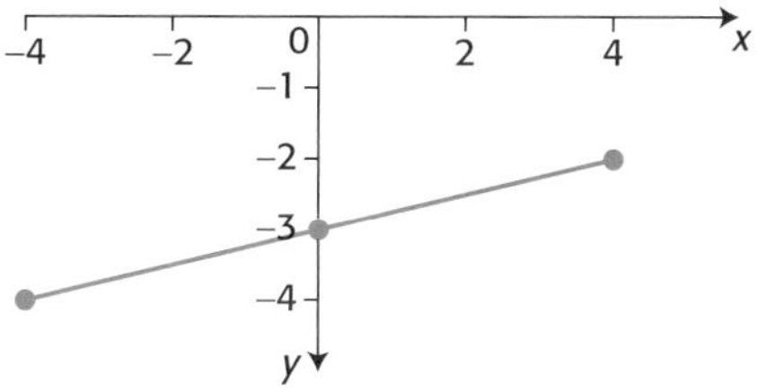

b

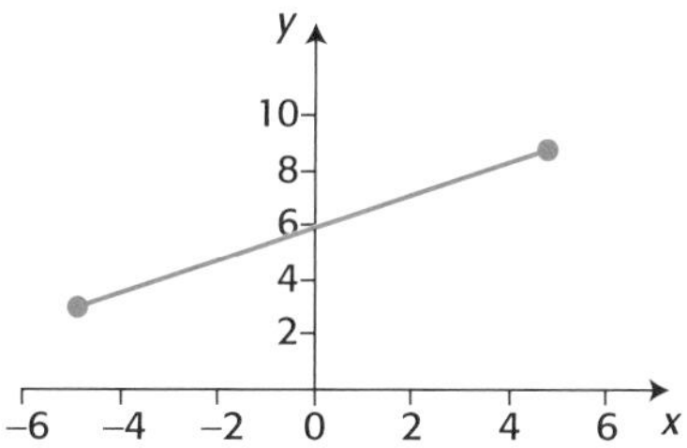

c

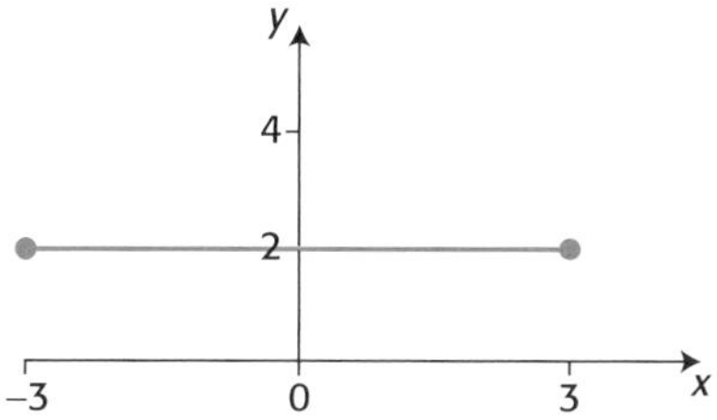

d

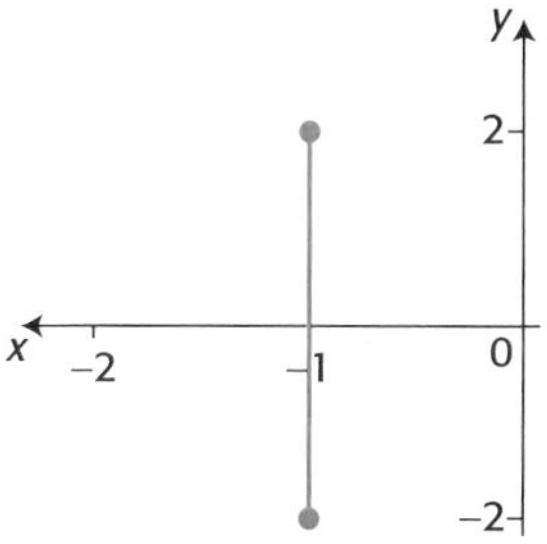

8 a $M(0, 4)$; **b** $y=-2x+4$; **c** $D\ (1, 2)$

9 $(-2, 1)$

10 a iii **b** i and iv **c** ii

11 a $O(4, -1)$ **b** 12.7 km
c gradients are negative reciprocals
d $S(1, -4)$ **e** 18 km²

12 a $L_2 : y=\dfrac{3}{5}x$ **b** $A(0, \dfrac{4}{5})$ **c** $y=\dfrac{4}{5}$

Exercise 5.3

1 a $\phi=54.5°$; **b** $\phi=9.6°$; **c** $\phi=78.0°$

2 a $AC=4.25$ cm $CB=9.06$ cm $\hat{B}+28–$
b $PQ=5.55$ cm $Q=59.73°$; $\hat{R}=30.27°$
c $CB=2.61$ cm $\hat{C}+\hat{B}+82.50–$

3

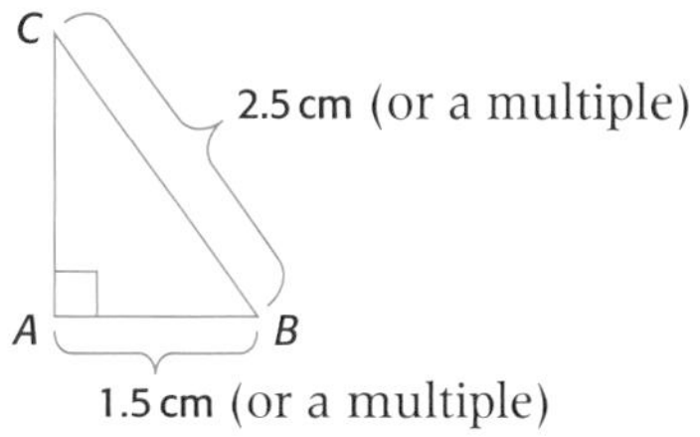

4 a

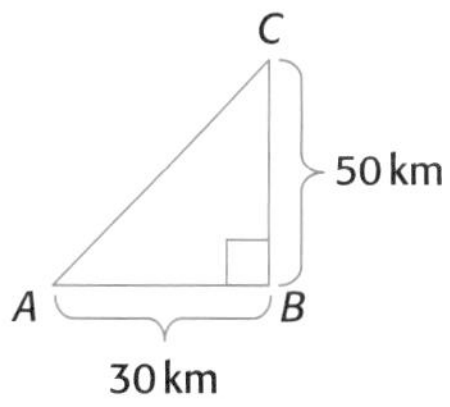

b 58.3 km **c** 59.0°

5 8 m

6 a and b

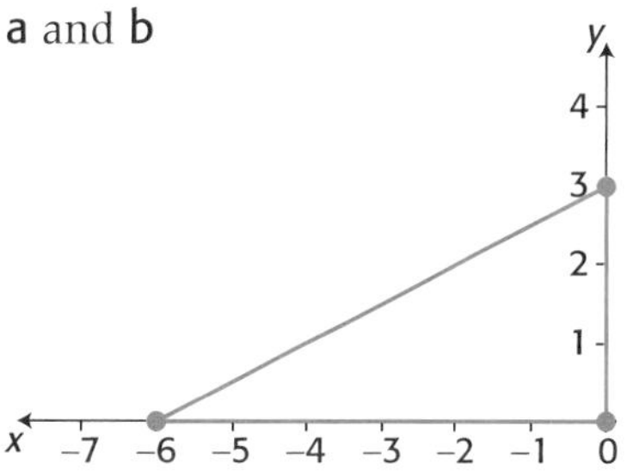

c $\frac{1}{2}$ **d** $\frac{1}{2}$

e Given that the scales used in both axes are equal, both the tangent of the angle and the gradient are also equal.

7 a 135° **b** 16.5 cm

8 39°

Exercise 5.4

1 a $a=11.1$ cm; $b=17.1$ cm **b** $r=7.46$ cm

2 a i $\beta=29.0°$ ii $A=14.5$ cm²
b i $\beta=32.5°$ ii $A=32.2$ cm²

3 $\hat{A}+30–$

4 a $O\hat{A}B = 57.2°$ b $16.4\,\text{cm}^2$ c $96.7\,\text{cm}^2$

5 a

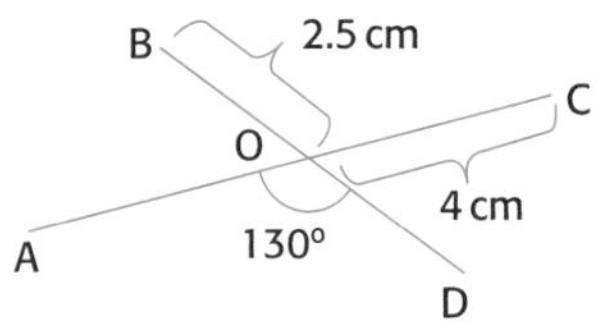

b parallelogram c $15.3\,\text{cm}^2$

6 a, b

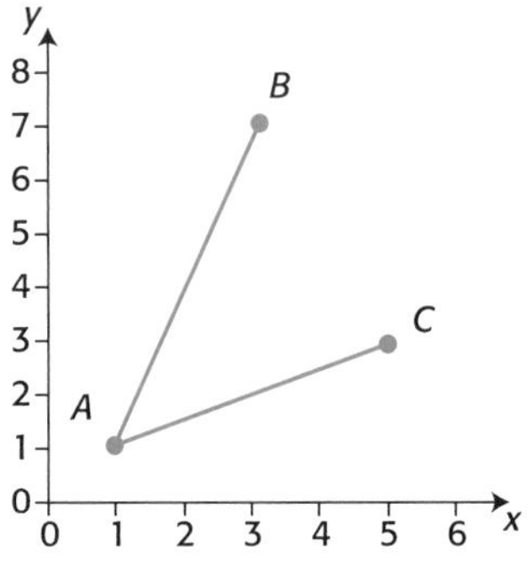

c $AB = \sqrt{40}$, $AC = \sqrt{20}$ $BC = \sqrt{20}$ d 45°

Exercise 5.5

1

6	6	0	✓
2	1	1	
8	8	0	✓
3	2	1	✓
5	5	0	✓
2	1	1	

2

Solid	Surface area	Volume
a	$316.6\,\text{cm}^2$	$312.8\,\text{cm}^3$
b	$14.5\,\text{dm}^2$	$3\,\text{dm}^3$
c	$417\,\text{cm}^2$	$633\,\text{cm}^3$
d	$2.76\,\text{m}^2$	$0.385\,\text{m}^3$

3 6 cm

4 a $V = x(35 - 2x)(28 - 2x)$
b i 3 cm ii $944\,\text{cm}^2$

5 a

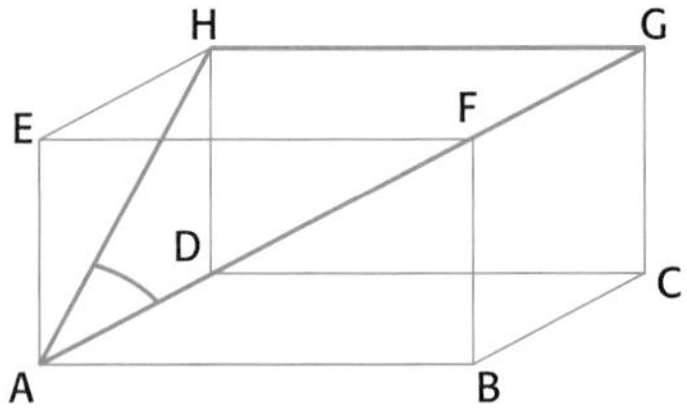

b

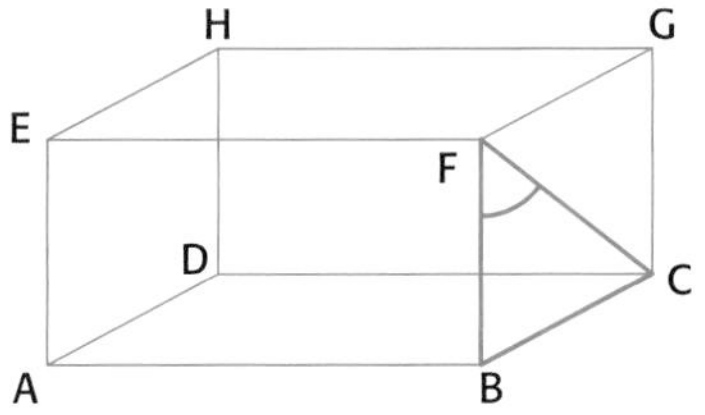

c

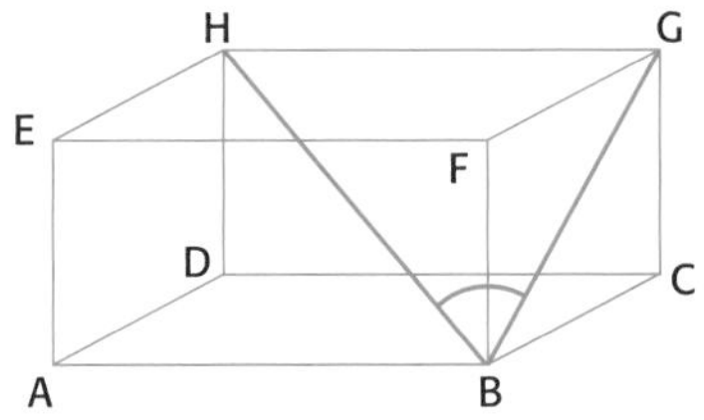

d

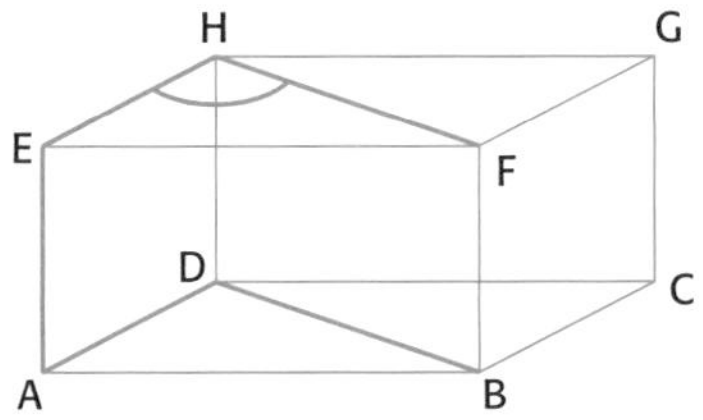

e

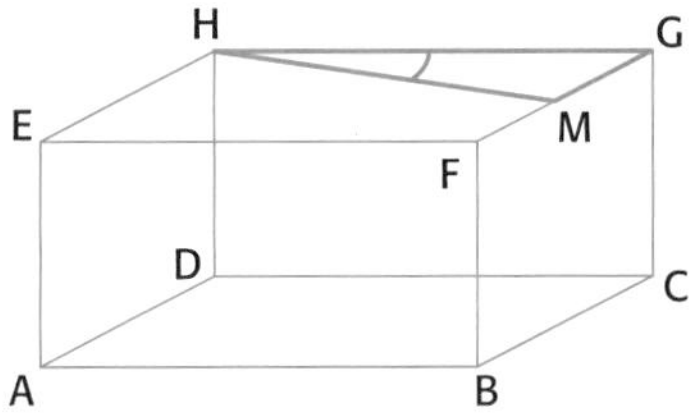

6 a $424\,\text{cm}^3$ b $311\,\text{cm}^2$

7 a

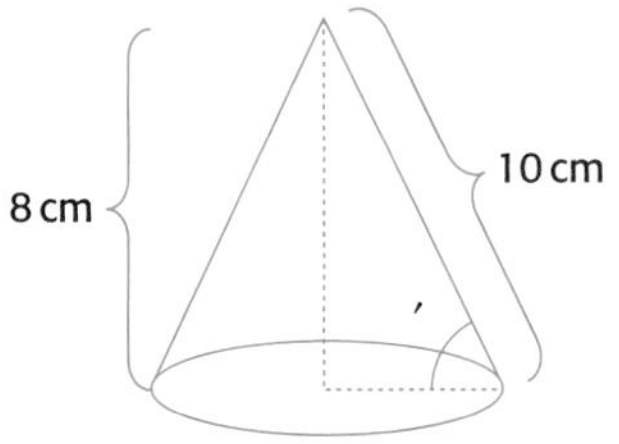

b 6 cm c 53°

8 a 1.12 m b $3.24\,\text{m}^2$ c 63.4°
d 54.7° e $\frac{1}{3}\,\text{m}^3$

Answers to past examination questions topic 5

Paper 1

1 a 6 cm b 9.83 cm (3 s.f.)
c $\sqrt{36\pi}$ or 10.6 cm (3 s.f.)

2 a $XM = 2\,\text{cm}$ b $DM = \sqrt{13}\,\text{cm}$ (=3.61 s.f.)
c $DMX = 56.3°$

3 a $B\hat{A}C = 85°$ b $BC = 9.57\,\text{m}$ (3 s.f.)
c $CBA = 38.6°$

4 a AC = 70 m b BC = 44.4°

5 a 3 b $m = -\frac{1}{3}$ c $b = 6$

6 a $y = -x + 2$ b $y = x$

Paper 2

1 a $\sqrt{128}$ cm = 11.3 (3 s.f.) b $ACH = 68.9°$
d $192\,\text{cm}^3$

2 a $-\dfrac{2}{11}$ b No. The gradients are not equal.

c $x-6y=-10$ d $T(2, 2)$

3 i a 385 m (3 s.f.) b 19 200 m^2 (3 s.f.)

ii 117 cm^3 (3 s.f.)

4 a $A(8, 0)$; $B(0, 6)$ b $M(4, 3)$ c $y=\dfrac{5}{4}x-2$

d i MC = 6.40 (3 s.f.) ii AC = 8.25 (3 s.f.)

e i CMA =91.8° (3 s.f.) ii 16.0 (3 s.f.)

Topic 6

Exercise 6.1

a continuous b discrete c discrete
d continuous e continuous f discrete
g continuous h continuous i discrete
j continuous k discrete l continuous

Exercise 6.3

1

Goals	*f*
0	8
1	8
2	7
3	4
4	1
5	1
6	1

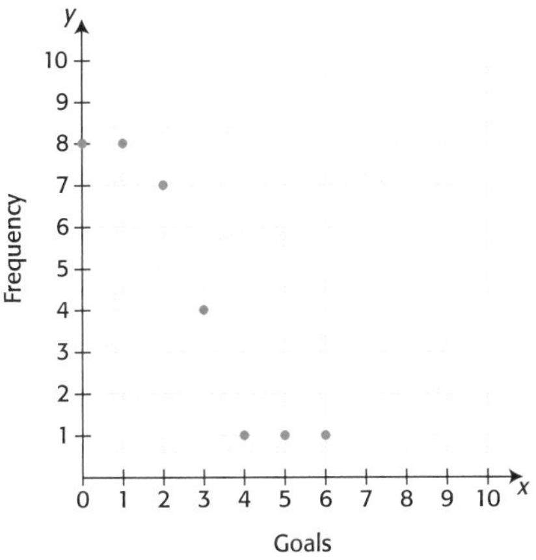

2

Sixes	*f*
0	2
1	2
2	2
3	2
4	4
5	10
6	10
7	8
8	3
9	3
10	2
11	1
12	1

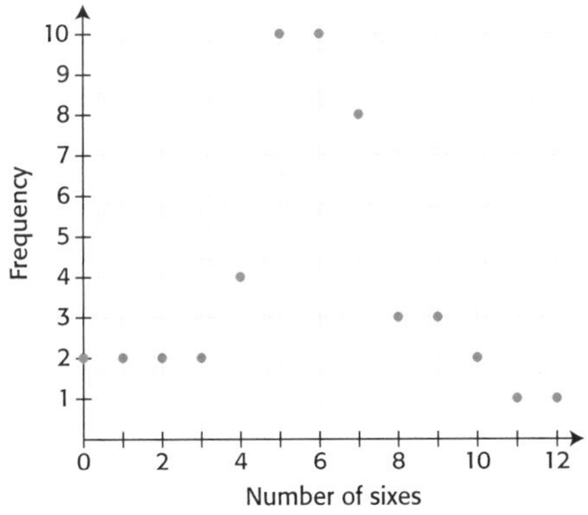

3

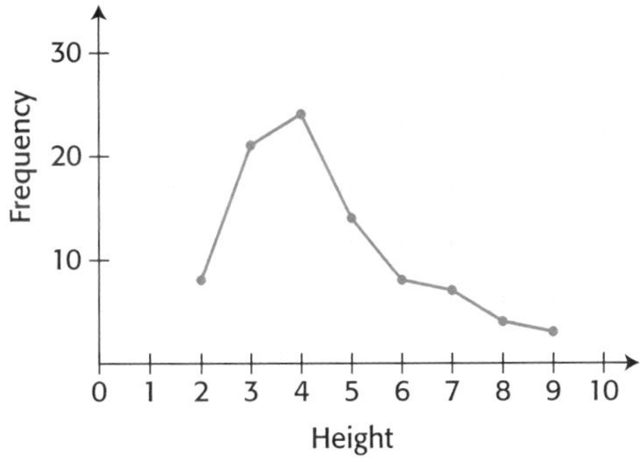

4 a one possible solution – other group sizes are possible.

People	*f*
1–10	13
11–20	12
21–30	8
31–40	8
41–50	9
51–60	2
61–70	4
71–80	4

b

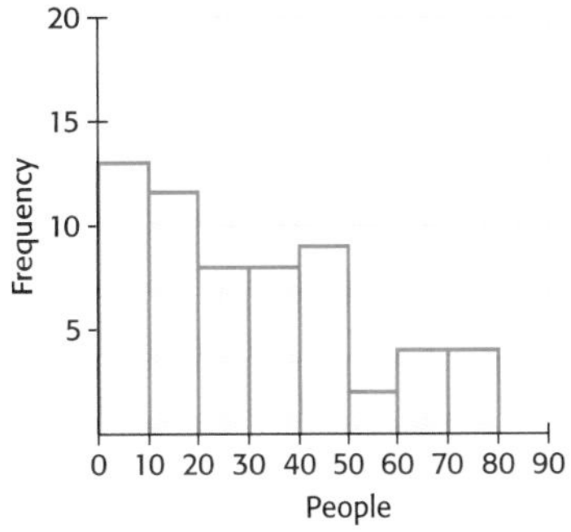

5

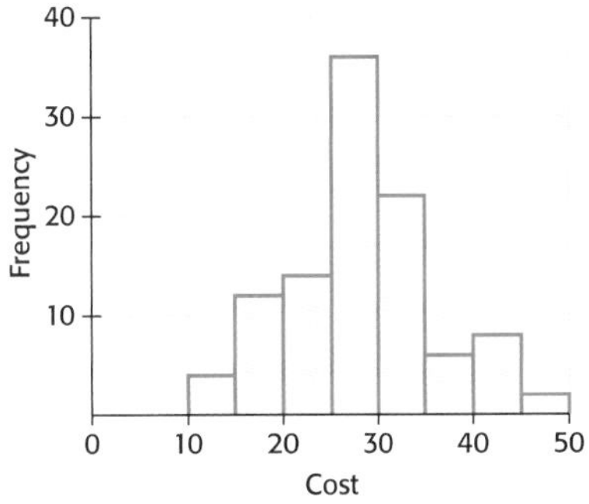

6

stem	leaf
0	6 8 9
1	0 0 3 4 5 6 8 8 9 9 9 9
2	0 1 5 5 6 7 7 7 8 8 9 9
3	1 2 2 2 3 3 6 8 9 9
4	0 1 1 1 1 2 3 5 6 8 9
5	1 1 2 3 5
6	1 1 3 3 4 5
7	0 2 3 3 3 8
8	0 1 2 3 5 8
9	2 4 5 7 8
10	0 1 2 3

7 a

Messages	*f*
0–9	7
10–19	9
20–29	14
30–39	12
40–49	15
50–59	8
60–69	8
70–79	4
80–89	3

b

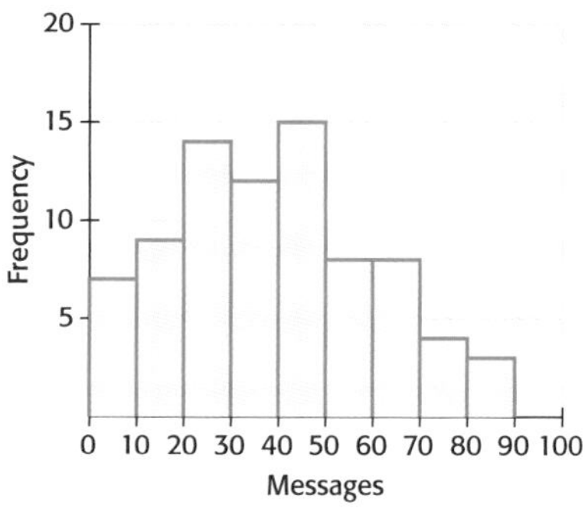

8 a

Grade	*f* (English)	*f* (Economics)
20–29	0	2
30–39	2	3
40–49	4	3
50–59	5	9
60–69	5	10
70–79	7	4
80–89	8	6
90–99	9	3

b

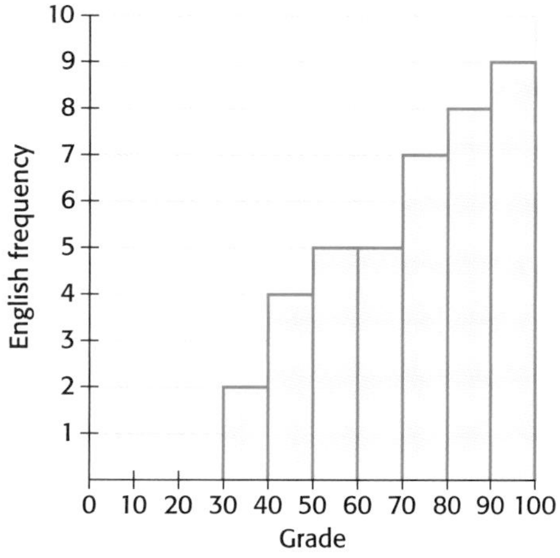

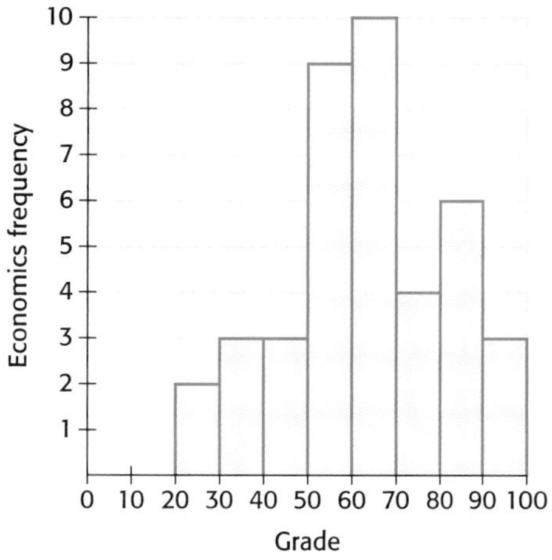

c The histogram for the English scores skews to the right, that is English marks tend to the higher scores, and the histogram for the Economics is more normally distributed.

Exercise 6.4

1 a ≈25 min b ≈14 min c ≈ 40%

2 a ≈144 cm b ≈16 cm c ≈125 cm d ≈7%

3 a 3 b 8 c 5 d 15

4 a 8, 9, 10

b

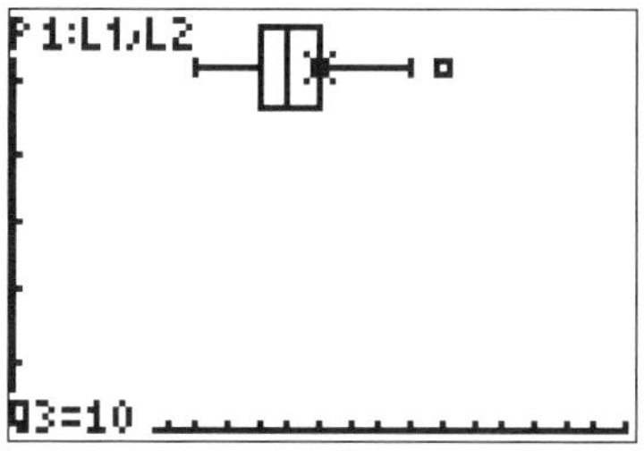

c $8 - 1.5 \times 2 = 5$

$10 + 1.5 \times 2 = 13$

So, 14 is an outlier

5 a

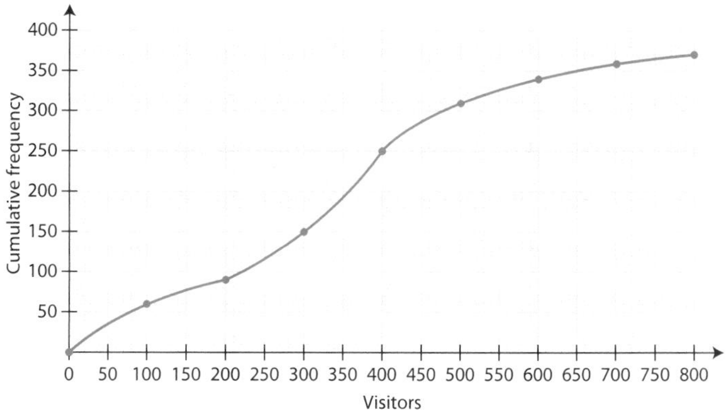

b median ≈335, IQR ≈250 c ≈14%

6 a

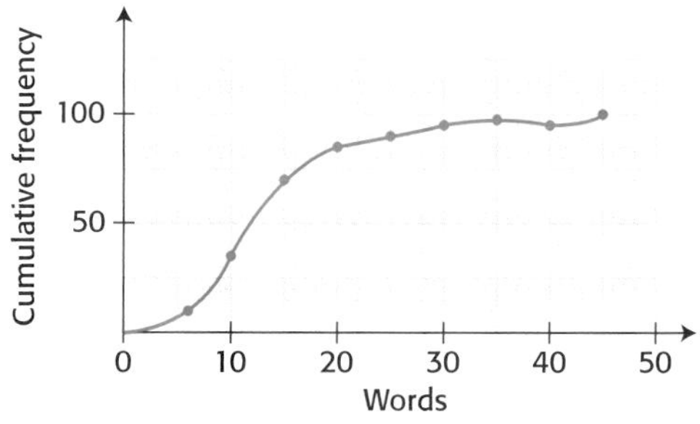

b LQ ≈9, median ≈12, UQ ≈16

c

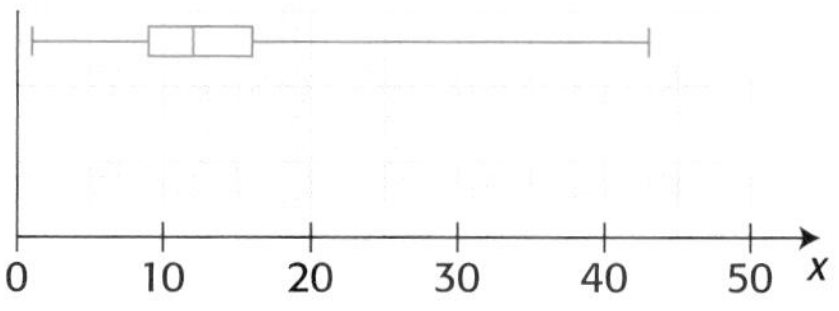

7 a

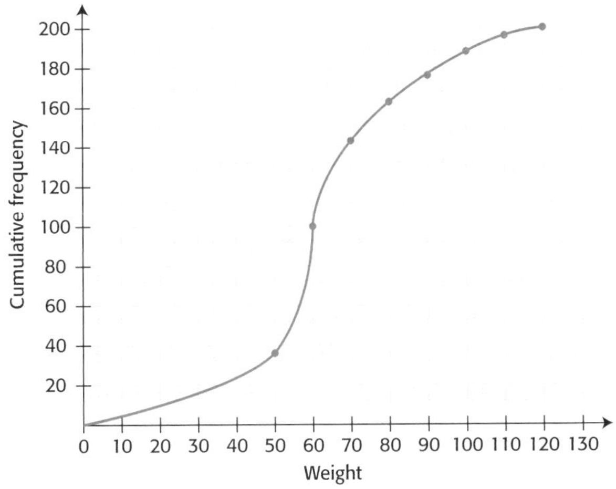

b i ≈60 ii ≈22 c ≈75% d ≈95 kg

e

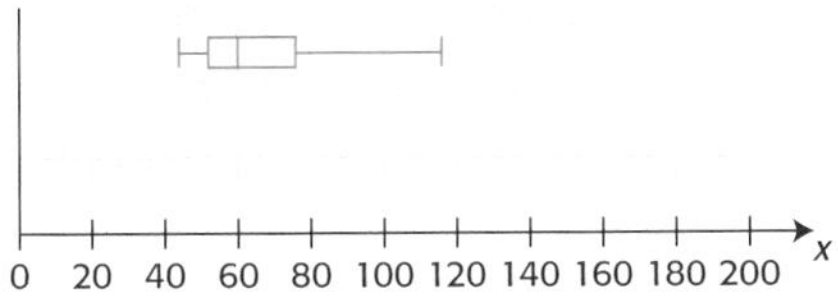

8 a

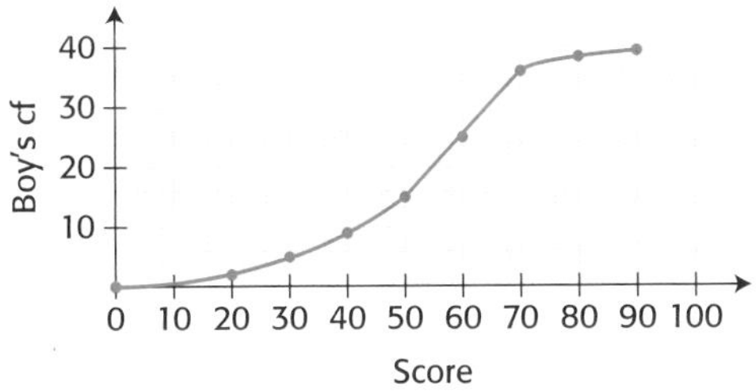

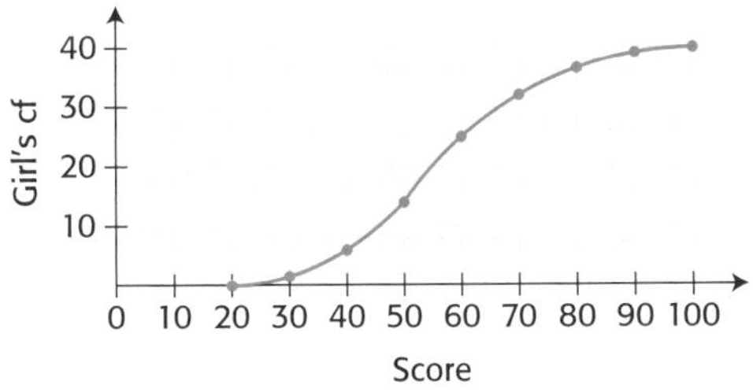

b Boy's median ≈ 55 and IQR ≈ 64 − 42 = 22
Girl's median ≈ 55 and IQR ≈ 67 − 45 = 22

c

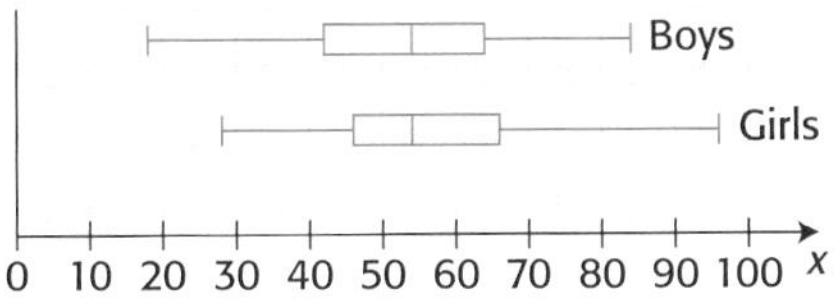

Even although both sets of marks have the same median and IQR, the girls marks are higher overall.

Exercise 6.5

1 a mode = 4, median = 4, mean = 4.18
b mode = 12, median = 11, mean = 11.3

2 a modal group is 60–65
b median ≈ 62.5 mean ≈ 64

3 $s = 5$, $t = 11$

4 a modal class is $3 \le s < 4$
b median ≈ 2.9 mean ≈ 2.94

5 65.4 kg **6** 0.541 kg **7** 8.77 kg

Exercise 6.6

1 a i range = 8 ii IQR = 2 iii sd = 2.41
b i range = 7 ii IQR = 2 iii sd = 1.67

2 a

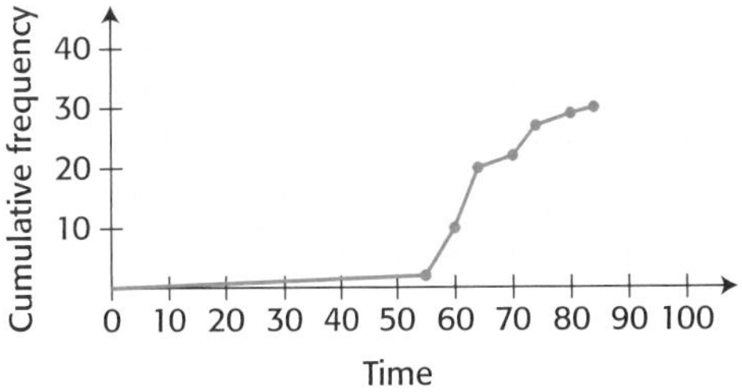

IQR ≈ 11

b sd ≈ 7.43

3 a

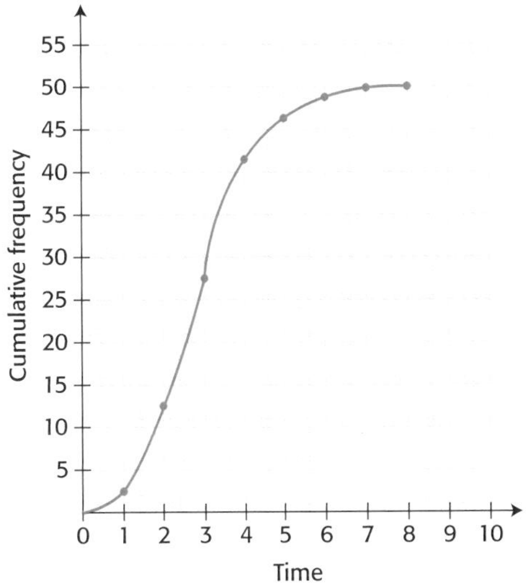

IQR ≈ 1.7

b sd ≈ 1.43

4 mean = 8.8, sd = 5.44

5 a mean = 12.7 b sd = 4.28

6 a $x = 45$ b sd = 18.6

7 a mean ≈ 5 b sd ≈ 1.84

8 mean ≈ 19.7 and sd ≈ 8.23

Sudoku solution

1	6	4	7	9	5	3	8	2
2	8	7	4	6	3	9	1	5
9	3	5	2	8	1	4	6	7
3	9	1	8	7	6	5	2	4
5	4	6	1	3	2	7	9	8
7	2	8	9	5	4	1	3	6
8	1	9	6	4	7	2	5	3
6	7	3	5	2	9	8	4	1
4	5	2	3	1	8	6	7	9

9 a Girls' mean ≈ 56.5 and sd ≈ 11.9
Boys' mean ≈ 51.8 and sd ≈ 24.4

b The mean for the girls is higher than that for the boys and their sd is smaller so the marks for the girls are less spread out than those for the boys.

Exercise 6.7A

1 a i

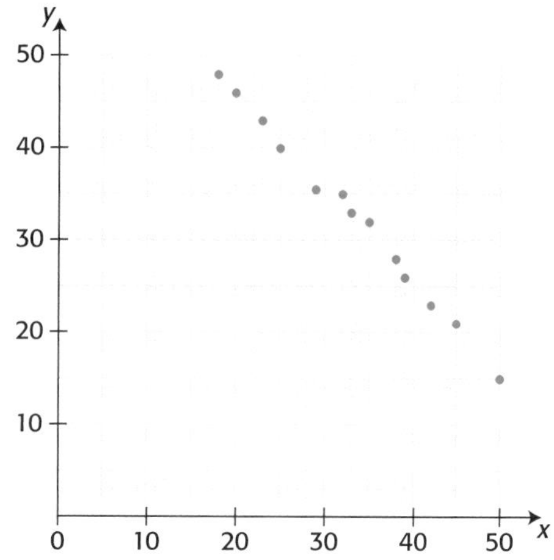

strong negative correlation

ii mean of $x = 33.1$, mean of $y = 32.6$

iii

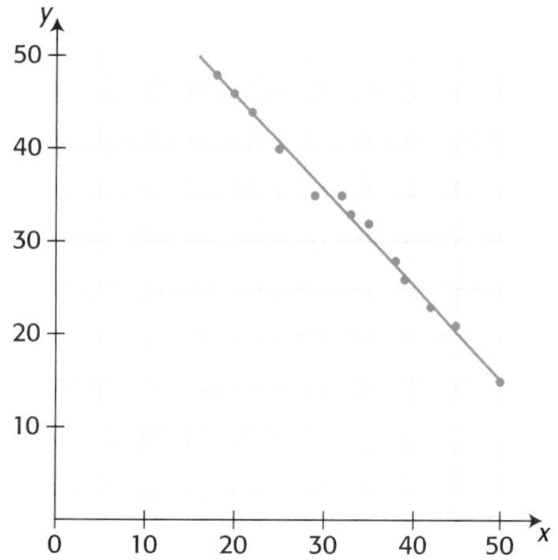

b i

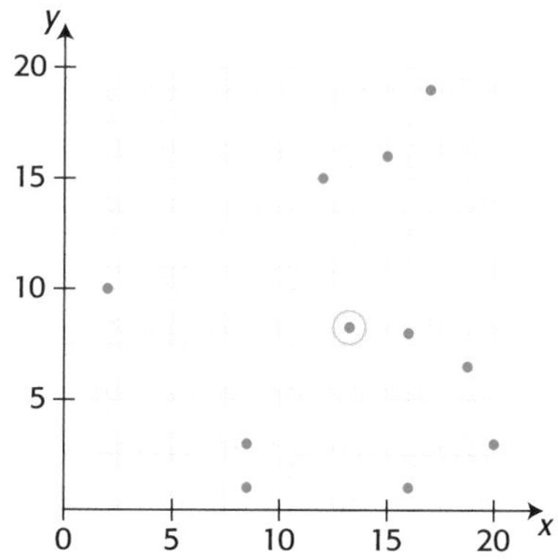

no correlation

ii mean of $x = 13.3$, mean of $y = 8.3$

iii no line of best fit

c i

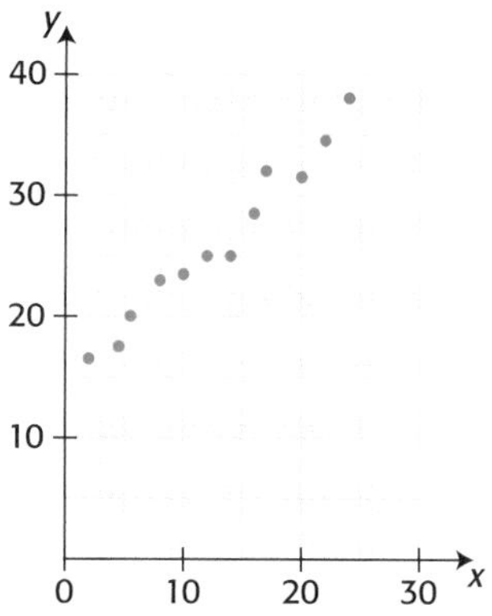

strong positive correlation

ii mean of $x = 13$, mean of $y = 26.5$

iii

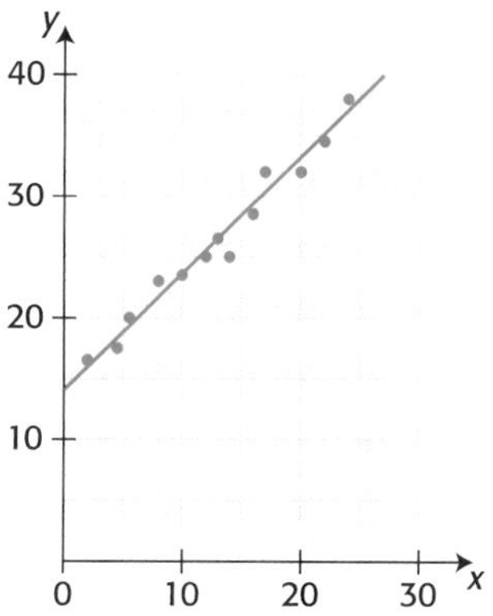

d i

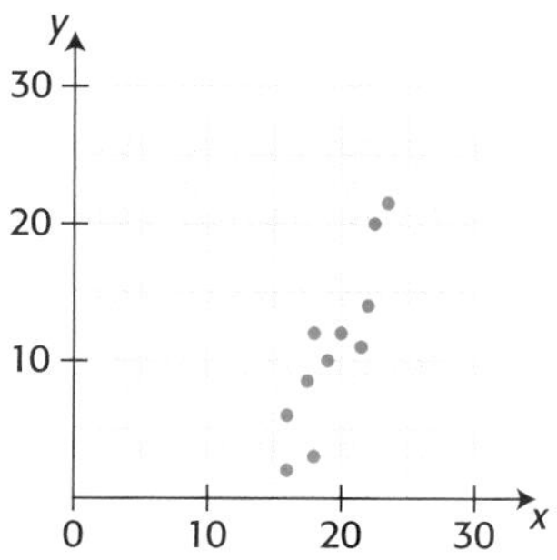

strong positive correlation

ii mean of $x = 19.5$, mean of $y = 10.6$

iii

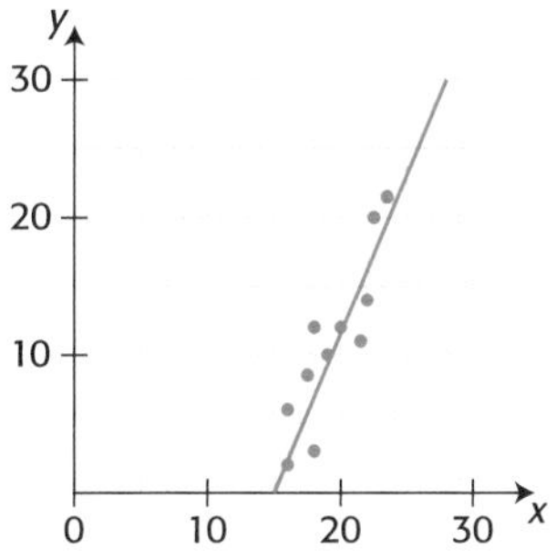

e i

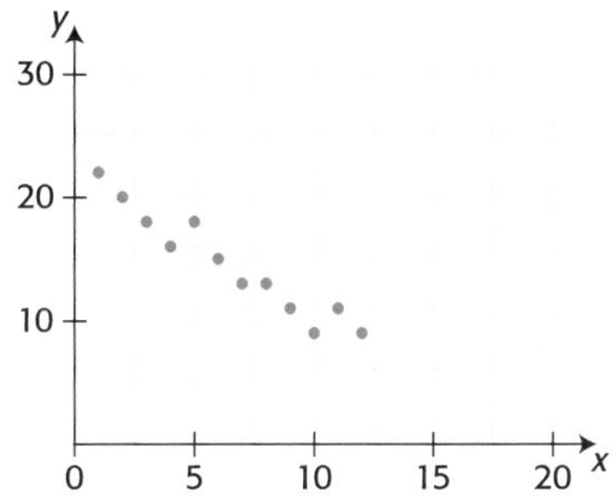

strong negative correlation

ii mean of $x = 6.5$, mean of $y = 14.3$

iii

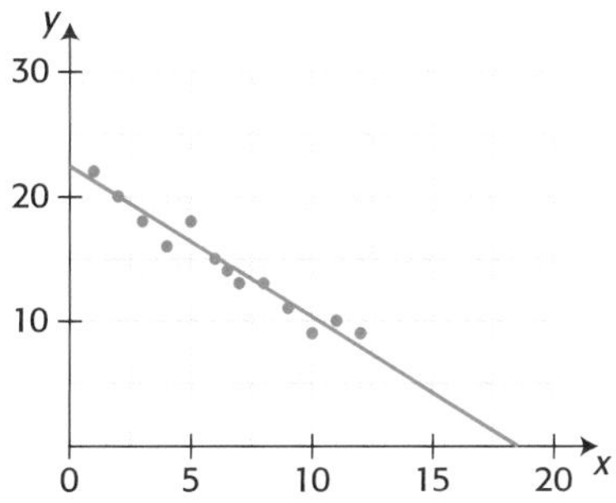

2 a

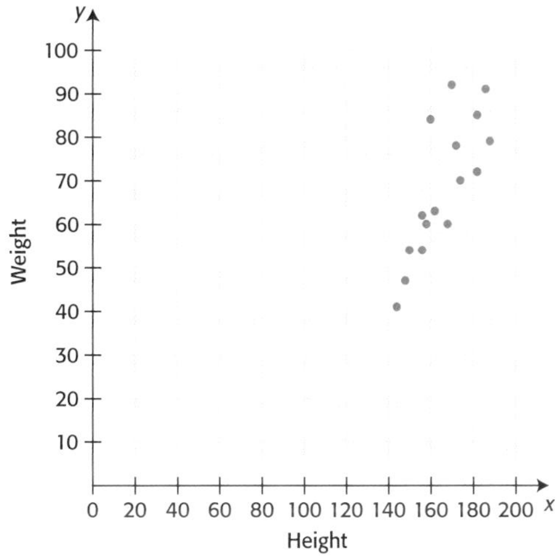

There is a strong positive correlation.

b mean height = 167 cm, mean weight = 68.3 cm

c

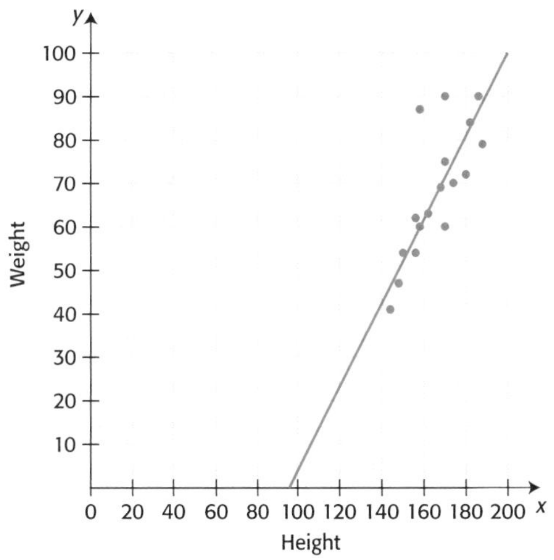

3 a

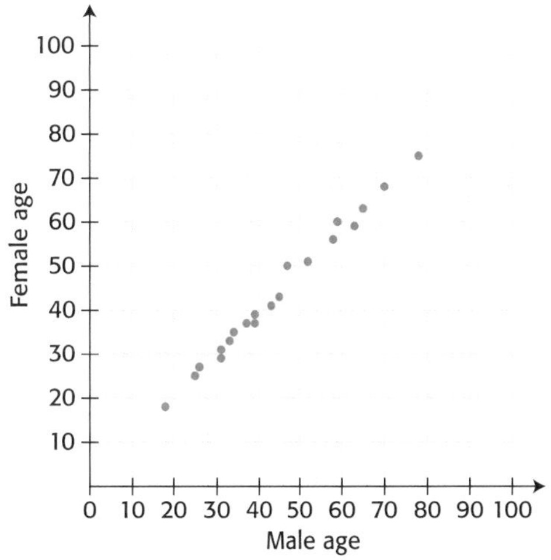

There is a strong, positive correlation

b mean male age = 44.7, mean female age = 43.6

c

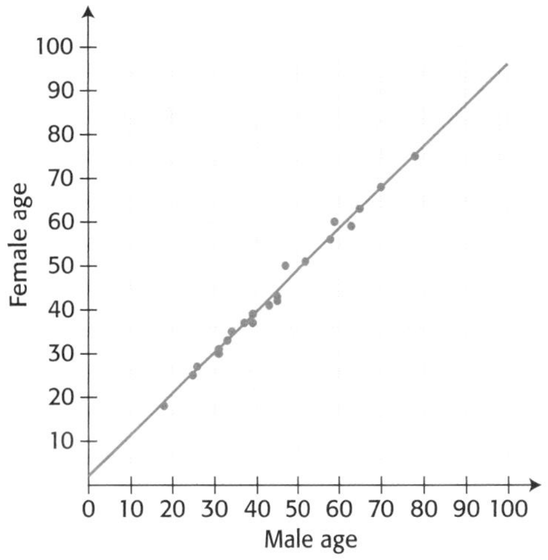

4 a

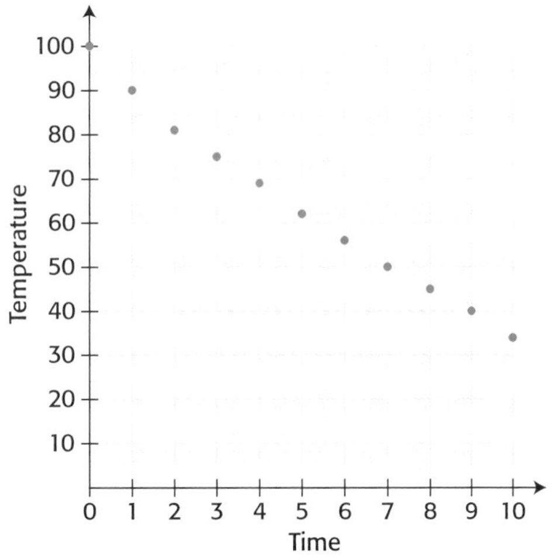

There is a very strong, negative correlation.

b mean time = 5, mean temperature = 63.4°c

c

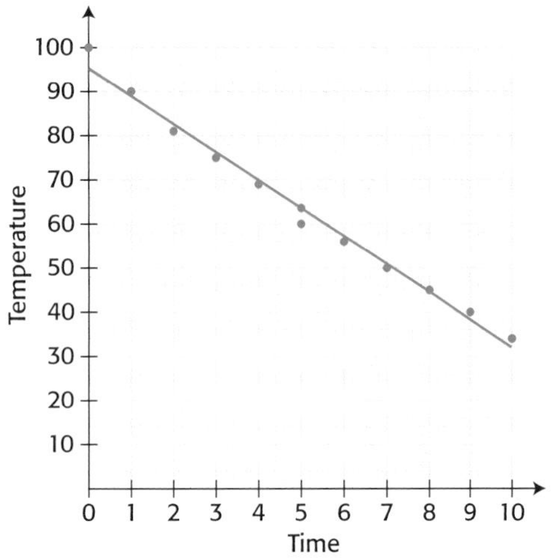

5 a

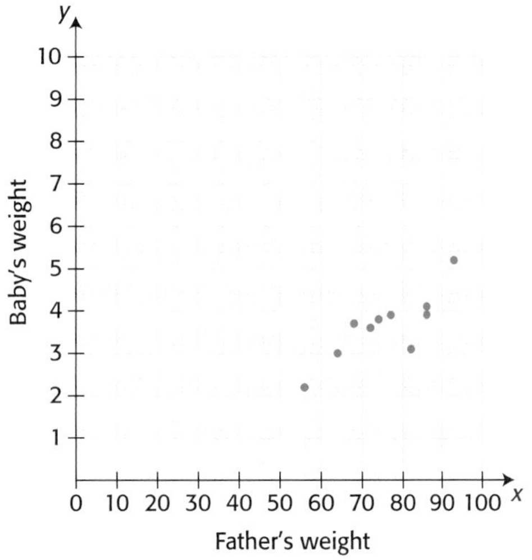

There is a fairly strong, positive correlation.

b mean weight of father = 75.8, mean weight of baby = 3.67

c

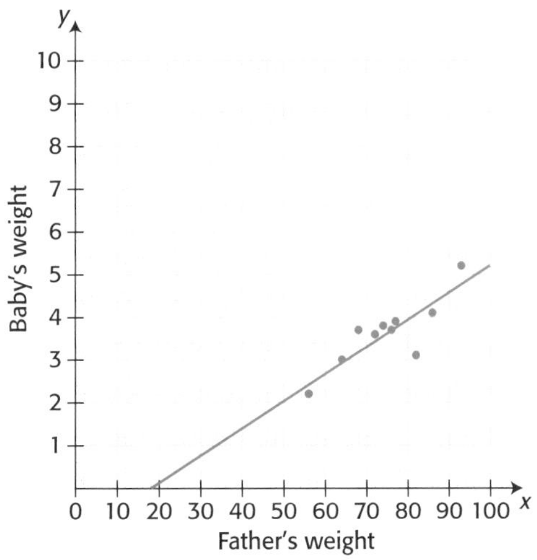

Exercise 6.7B

1 a strong, positive linear
b no linear correlation
c weak, negative linear
d no correlation
e strong, positive linear
f strong, negative linear
g no linear correlation

2 $r = -0.820$, so there is a fairly strong negative linear correlation.

3 $r = 0.982$, so there is a very strong positive linear correlation.

4 $r = 0.736$, so there is a moderate positive linear correlation.

5 $r = 0.976$, so there is a strong positive linear correlation.

6 $r = 0.914$, so there is a strong positive linear correlation.

7 $r = 0.986$, so there is a very strong positive linear correlation.

8 $r = -0.988$, so there is a very strong negative linear correlation.

9 $r = 0.267$, so there is a very weak positive linear correlation.

10 sd age = 14.8; sd time = 7.47; $r = 0.769$
From GDC r = 0.768 , so there is a fairly strong positive linear correlation.

Exercise 6.8

1 a $r = 0.990$ b $y = 4.63x + 30.9$ c 77.2 ml
d No, as it is outside the range of the data.

2 a $r = 0.993$ b $y = 8.73x + 132$ c €481.20

3 a $r = -0.767$ b $y = -0.147x + 17.0$
c 14.4 seconds

4 a $r = 0.993$ b $y = 1.27x + 12.0$ c 39.9

5 a $r = 0.940$ b $y = 0.002\,77x + 0.789$ c 2.17 m

6 a $y = 0.104x + 5.25$ b 15.65 AU$

c No as this is too far outside the data range.

7 a $y = 0.000\,586x + 59.2$ b 70.9

8 a $y = 0.922x + 7.00$ b 62

9 a $y = 0.524x - 0.501$ b 26.7

c No as this is too far outside the data range.

10 a $y = 0.0625x - 1.06$ b 3.94 kg

Exercise 6.9

1 a H_0 : hair colour and nationality are independent

H_1 : hair colour and nationality are not independent

b

	Black	Brown	Blonde	Ginger
American	20.2	17.6	15.1	5.1
Asian	25.4	22.2	19.0	6.4
European	41.4	36.2	30.9	10.5

c 6 d 82.0

e reject the null hypothesis as $82.0 > 12.592$. So, hair colour and nationality are related.

(or $p = 0.000\,000\,000\,000\,001\,37 < 0.05$ – so reject the null hypothesis)

2 a H_0 : hours spent watching TV and GPA are independent

H_1 : hours spent watching TV and GPA are not independent

b

	Low GPA	Average GPA	High GPA
0–9 hours	16.1	38.1	34
10–19 hours	18.9	44.8	40
> 20 hours	16.1	38.1	34

$85/270 \times 108/270 \times 270 = 34$

c $(3-1)(3-1) = 2 \times 2 = 4$ d 45.8

e $45.15 > 9.488$ so reject the null hypothesis, and conclude that the number of hours watching TV is related to the GPA.

(or $p = 0.000\,000\,002\,77 < 0.05$ so reject null hypothesis)

3 a H_0 : number of fish caught and time are independent

H_1 : number of fish caught and time are not independent

b

	Morning	Afternoon	Evening
0–3 fish	8.3	8.3	8.3
4–6 fish	10.7	10.7	10.7
> 7 fish	11	11	11

c 4 d 9.24

e $9.24 < 9.488$ so accept the null hypothesis that time of day and number of fish caught are independent.

(or $p = 0.0554 > 0.05$ so accept the null hypothesis)

4 a H_0 : the temperature and number of ice creams sold are independent

H_1 : the temperature and number of ice creams sold are not independent

b

	< 21	21–30	> 30
< 500	10.5	26.1	12.4
500–750	12.4	30.9	14.7
> 750	9.2	22.9	10.9

c 4 d 3.30

e accept the null hypothesis as $3.30 < 9.488$, so the number of ice creams sold is independent of temperature.

(or $p = 0.509 > 0.05$ – so accept the null hypothesis)

5 a H_0 : the cost and number of hours are independent

$H1$: the cost and number of hours are not independent

b

	<50	50–75	75–100	>100
<$2	9.9	14.9	21.8	19.5
$2 – $4	10.2	15.3	22.4	20.1
>$4	9.9	14.9	21.8	19.5

$66/200 \times 30/200 \times 200 = 9.9$

c 6 d 4.16

e accept the null hypothesis as $4.16 < 12.592$ – so the number of hours is independent of the cost of the battery

(or $p = 0.655 > 0.05$ so accept the null hypothesis)

6 a H_0 : the type of art one likes is independent of age

H_1 : the type of art one likes is not independent of age

b

	Modern	Landscape	Portrait	Still life
< 20 years	9.6	11.3	6.3	5.8
20–40	12.5	14.7	8.2	7.5
> 40	12.8	15.0	8.4	7.7

c 6 d 9.61

e $9.67 < 12.592$ so we accept the null hypothesis. The type of art one likes is independent of age.

(or $p = 0.142 > 0.05$ therefore accept the null hypothesis)

7 a H_0 : height and nationality are independent
H_1 : height and nationality are not independent

b

	European	African	Asian	American
< 1.70	15.6	9.4	10.9	14.1
1.70–1.80	24.1	14.4	16.8	21.7
>1.80	10.3	6.2	7.2	9.3

c 6 **d** 35.0

e 35.0 > 12.592 so we reject the null hypothesis. So, height and nationality are not independent.
(or $p = 0.00\,000\,429 < 0.05$ so reject the null hypothesis)

8 a $116/200 \times 114/200 \times 200 = 66.1$
$b = 49.9$, $c = 36.1$, $d = 47.9$

b 1 **c** 13.9

d 13.9 > 3.841 so we reject the null hypothesis. So, blood pressure is not independent of smoking.
(or $p = 0.000\,194 < 0.05$ so reject the null hypothesis)

9 a H_0 : walking age and gender are independent
H_1 : walking age and gender are not independent

b $28/140 \times 70/140 \times 140 = 14$ **c** 2 **d** 1.58

e 1.58 < 5.991 so we accept the null hypothesis. So walking age is independent of gender.
(or $p = 0.433 > 0.05$ accept the null hypothesis)

10 a $a = 50$, b = −17, $c = 289$, $d = 5.78$

b 24.1 **c** 1

d 24.1 > 3.841 so we reject the null hypothesis. Fox hunting is not independent of where a person lives.
(or $p = 0.000\,000\,901 < 0.05$ so reject the null hypothesis)

Answers to past examination questions topic 6

Paper 1

1 a 41.5

b

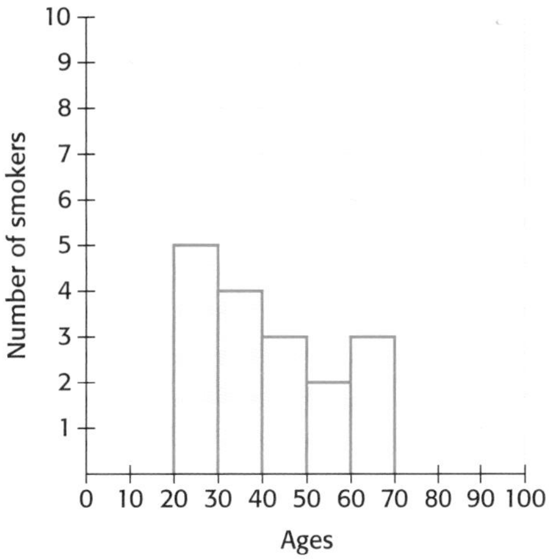

2 a i 145 **ii** 157 **iii** 167

b

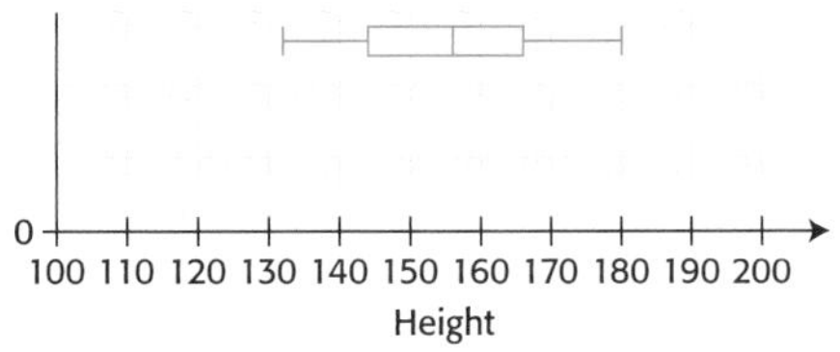

3

Stem	Leaf
16	1 1 5 6 7 9
17	3 3 5 7 7
18	0 3 4
19	2 5 5 7

Key 16|1 = 161 cm

4 a 50 **b** $m = 19$, $p = 9$, $q = 43$

5 a $a = 5$ **b** median = 5

6 a mode = 63 **b i** mean = 70.5 **ii** sd = 14.6

c 76 kg

7 a 20 **b** £15 000 **c** £23 000

8 a mean point plotted on diagram

b line of best fit drawn

c As leaf length increases then leaf width also increases.

9 a H_0: The time taken to prepare for a penalty shot is independent of the outcome.

b 2

c 0.073 < 0.10 so we reject the null hypothesis. The time taken to prepare is related to the outcome.

Paper 2

1 a

Time	C f
< 15.5	7
< 20.5	20
< 25.5	45
< 30.5	73
< 35.5	93
< 40.	100

b

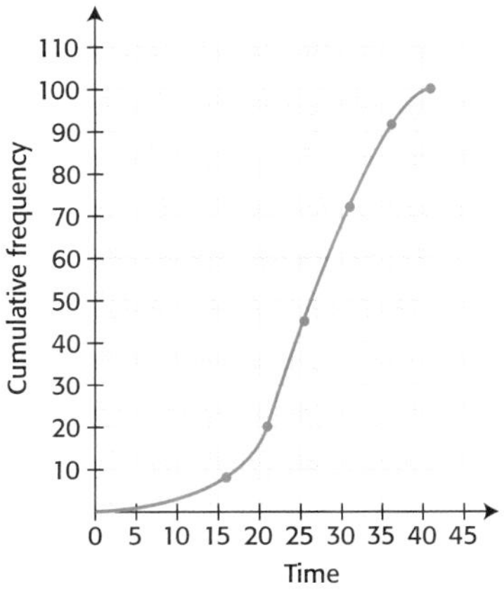

c i 12 ± 1 **ii** 31 ± 0.5

2 a 170–180 **b** mean = 171.15, sd = 11.1

c 170.9 cm

d Using the curve, $Q_3 - Q_1 = 12.8$ (±1 allowed). Using a GDC the answer is 10. Use of formula gives 14.6.

e 25%

3 b 1.06

c The second group has on average 1 child more than the first group but there is much more variation in the number of children as the standard deviation is larger for the second group than the first group.

d 0.16

e i 0.02 ii 0.28 iii 0.05

4 a

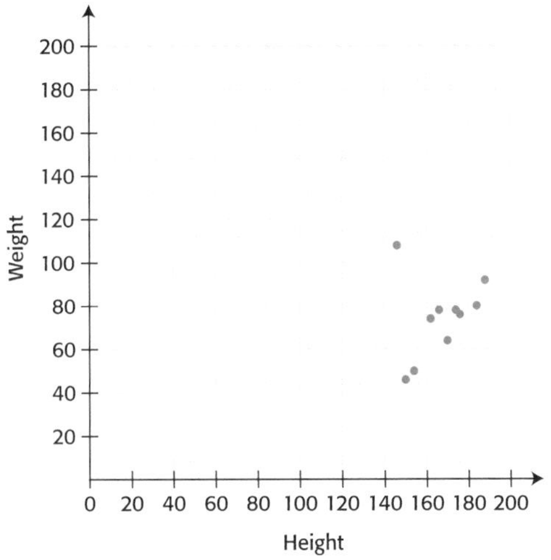

b 166.1 c 74.9

d ii $y = 0.276x + 29.1$ iii line drawn on graph

e i 81.5 kg ii 155.6 cm

f It will be a better fit.

5 a i 1

ii 3

b i sketch 1 – 0.04

sketch 2 – −0.20

sketch 3 – −0.85

ii reject 0.9 as there is no strong positive correlation

reject 1.60 as this is impossible.

6 a $y = 0.0705x - 3.22$ b 8 c 0.681

d moderate positive linear correlation.

7 a $p = 25.2$, $q = 16.8$, $r = 12.4$

b i Subject taken is independent of gender.

ii $(2 - 1)(3 - 1) = 2$ iii 5.99

c Accept the null hypothesis as $1.78 < 5.99$

8 a i H_1: The choice of candidate is dependent on where the voter lives

ii $a = 188$, $b = -22$

b i 8.14 (8.13) ii 1 iii 3.84

c i ii reject the null hypothesis as $8.14 > 3.84$. So where a voter lives does affect their vote.

Topic 7

Exercise 7.1

1 $\frac{5}{3}$ **2** 7.21 **3** −2

4

x-value	y-value	gradient
2	16.5	
2.1	19.1934	26.93
2.01	16.7587	25.87
2.001	16.5258	25.76

5 1 **6** 2

7

x-value	y-value	gradient
3	8	
3.1	8.41	4.1
3.01	8.0401	4.01
3.001	8.004001	4.001

8 $(0.0551 \div 0.005) = 11.02$

9 $(0.002\ 8126 \div 0.003) = 0.9375$

10 2.006, 2.0006, 2

Exercise 7.2

1 a $\frac{2}{5}x^{-1}$ b $\frac{1}{2}x^{-3}$ c $\frac{3}{4}x^{-7}$

d $\frac{1}{3}x^{-1}$ e $\frac{3}{5}x^{-2}$

2 a $\frac{3}{x^2}$ b $\frac{2}{3x}$ c x^2

d $\frac{1}{4x^2}$ e $\frac{4x}{3}$ f $\frac{4}{3x}$

3 a $-2x^{-2}$ b $4x^3$ c $27x^2$ d $-\frac{27}{x^4}$

4 a $5x^4 - 9x^2$ b $4x + 3$ c $-\frac{8}{x^3}$

d $6x^5 + 20x^3$

5 a $-\frac{8}{t^3}$ b $12t^3 - 10t$ c $2t - 3$ d 0

6 a $-12t^{-4} + 2$ b $-\frac{12}{t^4} - 8t - 2$

c $-\frac{7}{t^2} + 2$ d 1

7 a $-2x^{-2} - 8x^{-3}$ b $\frac{2}{5}$ c $\frac{3x}{2}$

d $-6x^{-3}$

8 $6x - 4$ **9** $12t^2 + 2 - \frac{3}{t^2}$ **10** 6

11 $-\frac{3}{7t^2} + 4t$ **12** a 35 cm b 7 cm c 2 cm/s

Exercise 7.3

1 Worked solution: $\frac{dy}{dx} = 10x - 8x^4 = 10 - 8 = 2$ when $x = 1$. Thus the gradient is 2.

Equation of tangent is $y = 2x + c$.

When $x = 1$, $y = 5 - 2 + 3 = 6$

Substitute into tangent equation: $6 = 2 + c$. $c = 4$

Equation of tangent is $y = 2x + 4$

2 $y = 20x - 32$ **3** 3

4 $y = -0.0625x + 0.5$ **5** 5

6 a $(-1, 5.17)$ $(2, 3.67)$ b $(-1.30, 4.72)$ $(2.30, 4.11)$

7 (2, 2) **8** $x = -2$, $x = 1$

9 $x = -3.86$, $x = 0.863$ **10** $y = 28x - 22$

Exercise 7.4

1 a $x > -1$ b $x < -1$

2 a $x > -1.14$ b $x < -1.14$

3 $x = 3$, $a = 6$ **4** $x < 0$ **5** $x < 3$

Exercise 7.5

1 (2, −8) **2** (1, 3) minimum

3 (1, 6) max (3, 2) min

4 a (2.5, 0.75) min b (2, −2) min

5 a $xy = 20$ b $P = 2x + \frac{40}{x}$ c 4.47, 17.9

6 a $4x^2h = 320$ b $A = 10xh + 8x^2$

c $A = \frac{800}{x} + 8x^2$ d 3.684

7 a $350 = \pi x^2 h$ b $S = 2\pi x^2 + 2\pi xh$

c $S = 2\pi x^2 + \frac{700}{x}$ d 3.82, 7.64 e 275

8 a 1.8 s b 16.2 m

9 a $50 = 2x^2h$ b 3.35 cm c 67.2 cm²

10

1	2		4	7
5		4	8	2
	3	3	5	
2	2	1		1
4	8		5	6

Answers to past examination questions topic 7

1 a $36x^2 + \frac{4}{x^3}$ b 40

2 a 8 b $\frac{1}{2}x^3 + \frac{9}{2}x - 5$ c ii $y = 8x - 8$

d $y = \frac{35}{8}$

e i, ii

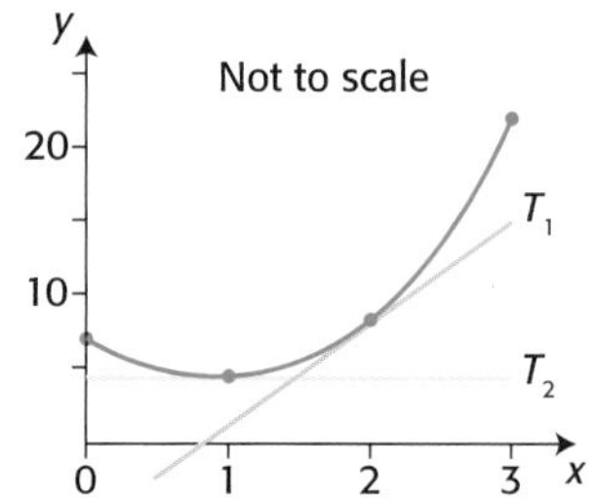

3 a $3x^2 + 14x - 5$ b 12 c $\frac{1}{3}$ or −5

d $\left(\frac{1}{3}, 3.15\right)$; (−5, 79)

e

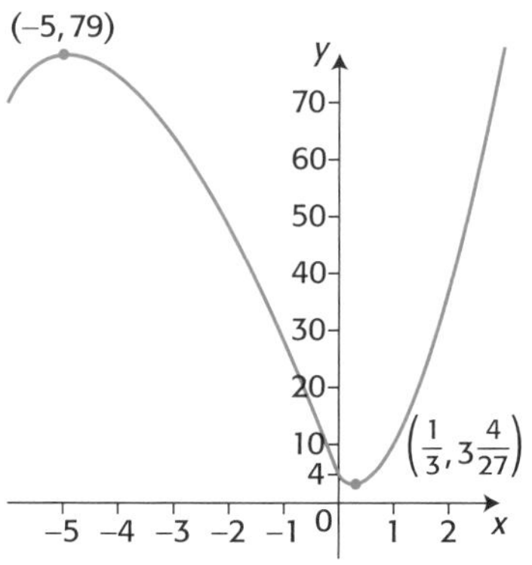

4 a $x - 15$ c i $160\,000 - 8000x$
ii 20 d 20 000

5 a $2ax + b$ b $12a + b = 2$ c $9a + 3b = -21$

6 a $4x^3 + 9x^2 + 4x + 1$ b 18

7 a 38.4 cm
b i $24 - 4.8w$ ii 3.5 iii 5 weeks, 60 cm

8 a $3x^2 - 8x - 3$ b $(-\frac{1}{3}, 18.5)$, (3, 0)

c 0, 18

d

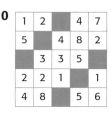

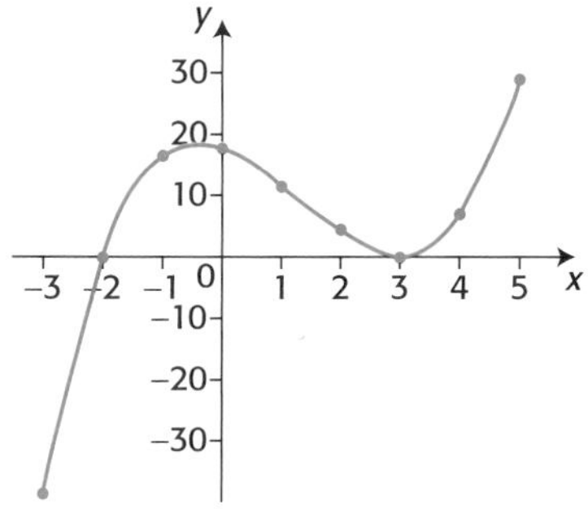

e i $-\frac{1}{3} < x < 3$

ii $x < -\frac{1}{3}$ and $x > 3$

9 a $6x^2 - 6x - 12$ b 24 c $x = 0$ and $x = 1$
d $x = -1$ max and $x = 2$ min e $x < -1$, $x > 2$

10 b $h = \frac{150 - 2x^2}{3x}$ d $100 - 4x^2$ e $x = 5$
f 333.3 cm³ g 6.67 cm

Topic 8

Exercise 8.1

1 a 347.99 USD b 883.50 ZAR

2 7 944 420 IDR

3 a 97 106.71 ESP b 47 951 54 CLP or
c 48 320.28 CLP so option **c** was better. d No

4 a 591.92 EUR b i 390.45 GBP ii 9.55 GBP
c 2.39%

5 Dalasi (GMD), Naira (NGN)

Exercise 8.2

1 Total interest is 1050 GBP. Total in account is then 3050 GBP.

2 a The calculation gives 7.14 years (3 s.f.) but we have to round up to 8 years.
b Changing to 6 monthly intervals at 3% gives 14.3 time periods (3 s.f.).

Rounding up to 15 but then substituting half years for the time period gives $7\frac{1}{2}$ years.

3 Exactly 5%.

Exercise 8.3

1 2722.25 EUR

2 0.8148 years or 9.78 months, but this is >9 months so must round up to 1 year.

3 a 5955.08 RUB b 5.50%

4 12 700 USD

5 The answers can be found in the text of the examples.

Exercise 8.4

1 a ¥54 031 b ¥19 089
c 15 years, total paid is ¥304 740.

2

Time in years	2.5%	5.0%	7.5%	8.0%	10.0%
1	38.9625	39.4669	39.9747	40.0767	40.4859
2	19.7245	20.2257	20.7342	20.8368	21.2501
5	8.1865	8.7000	9.2328	9.3416	9.7844
8	5.3064	5.8365	6.3972	6.5130	6.9878

3 The answers are all in the table in Section 8.4.2.

4 a $443.21 b 4769 EUR

5 a 156.10 NZD b 3880 NZD

6 Real return is 715.67 EGP. Real average rate of return is 2.39%.

Hint for the Challenge question on page 429. You need to sum a geometric sequence.

After 1 time period and repayment the amount owing is $C(1+\frac{r}{100k})-p$.

After 2 time periods and repayments it is

$$C\left(1+\frac{r}{100k}\right)^2-p\left(1+\frac{r}{100k}\right)-p \text{ etc.}$$

Work out the right expression for kn time periods. The coefficient of p is a geometric sequence. Sum the sequence then set the whole expression to 0 and solve for p.

Answers to past examination questions topic 8

Paper 1

1 a 862.50 FFR b 1.85 AUD

2 a 8880.99 (or 8881) b 1230 ringgit

3 a $1.94 b $37 000

4 a $m + f = 10000$, $0.08m + 0.06f = 640$
b $m = 2000$, $f = 8000$

5 a 1413.16 USD b 1288 MD

Paper 2

1 i a 155 months b i €5846.50 ii 6.93%
ii a $29 263.23 b i 3.46 years using the solver facility of your GDC
ii $298.20

2 a Ali $3810, Bob $3795.96
b Connie: $3000\left(1+\frac{3.8}{200}\right)^{12} = \$3760.20$ (2 dp)
c 18 years d i $75.20 ii $2620.03

3 a $1276.28 b 15 c $5801.91

4 a A: $1200, B: $1239.51, C: $1230, D: $1136.54
b Option B is best. It yields the largest amount after 1 year
c 10%

Index